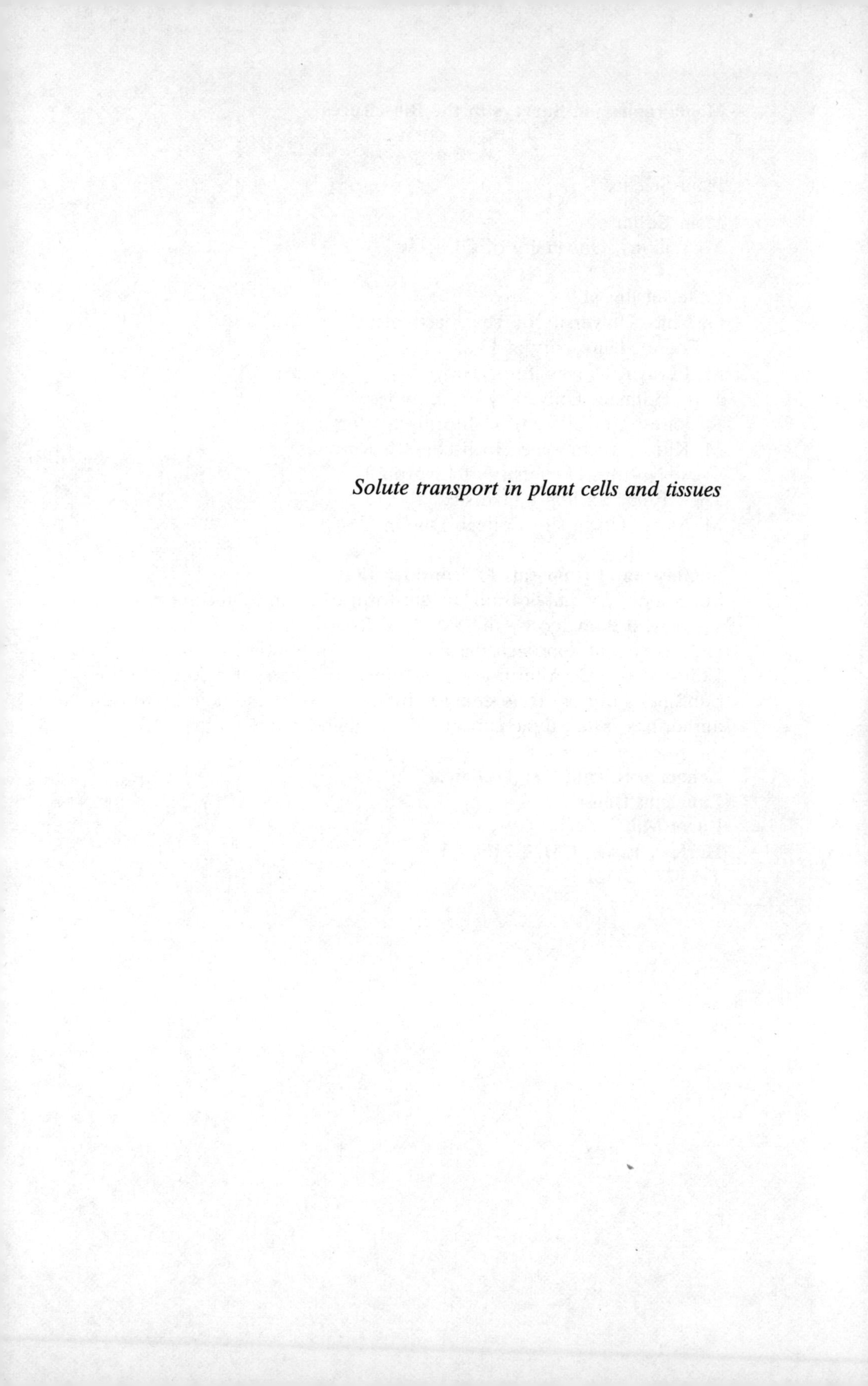

Solute transport in plant cells and tissues

Monographs and Surveys in the Biosciences

Plant Science

Submission of proposals for consideration
Suggestions for publication, in the form of outlines and representative samples, are invited by the Editorial Board for assessment. Intending authors should approach the main editor or another member of the Editorial Board. Alternatively, outlines may be sent directly to the publisher's offices. Refereeing is by members of the board and other authorities in the topic concerned, throughout the world.

Longman Scientific & Technical
Longman House
Burnt Mill
Harlow, Essex, CM20 2JE, UK
Tel 0279 26721

Solute transport in plant cells and tissues

Editors

D. A. Baker
Wye College, University of London
Ashford, Kent, TN25 5AH
United Kingdom

and

J. L. Hall
The University,
Southampton S09 5NH
United Kingdom

Copublished in the United States with
John Wiley & Sons, Inc., New York

Longman Scientific & Technical,
Longman Group UK Limited,
Longman House, Burnt Mill, Harlow,
Essex CM20 2JE, England
and Associated Companies throughout the world.

Copublished in the United States with
John Wiley & Sons, Inc., 605 Third Avenue, New York, NY 10158

First published 1988

British Library Cataloguing in Publication Data
Solute transport in plant cells and tissues.
— (Monographs and surveys in the biosciences).
1. Plant translocation
I. Baker, D. A. II. Hall, J. L. III. Series
581.19′12 QK871
ISBN 0-582-00580-9

Library of Congress Cataloguing in Publication Data
Solute transport in plant cells and tissues.

(Monographs and surveys in the biosciences)
Includes bibliographies and indexes.
1. Plant translocation. 2. Plant cells and tissues.
I. Baker, D. A. II. Hall, J. L. (John Lloyd)
III. Series.
QK871.S63 1988 581.1 87-16741
ISBN 0-470-20864-3 (USA only)

Set in Linotron 202 10/12 pt Times
Printed and bound in Great Britain
at the Bath Press, Avon.

Contents

Preface

It is now about twelve years since the publication of *Ion Transport in Plant Cells and Tissues* (Baker & Hall, 1975) which was intended to cover the then current concepts and developments in the various areas of ion transport studies. During the intervening years many major developments have taken place which have made much of the material in the earlier book redundant or incomplete. This book is, however, more than a mere update in that we have presented the changes in emphasis within transport physiology which are currently receiving the attention of researchers in this field. To this end we have again invited authors with specialist knowledge to contribute individual chapters within their particular areas of study. We were encouraged by the positive responses from our fellow researchers which have enabled us to present such a broad and comprehensive study of solute transport within plants. As with the earlier book, this text is intended to provide a digest of specialist knowledge for use by advanced undergraduate and postgraduate students following courses in the plant sciences, graduate students during their research programmes, and the teachers of advanced students preparing courses on solute transport.

The introductory chapter is presented to provide a coverage of the various biophysical relationships which are basic to many solute transport studies and some of the fundamental membrane phenomena involved. Mitochondria and chloroplasts, which have provided so much of our detailed knowledge on membrane transport processes, are considered together in Ch. 2, where the major solute exchanges which occur between these organelles and the cytosol are exemplified. Related events taking place across the cell boundary membranes, the plasma membrane and tonoplast, are the subject of Ch. 3 where recent studies on purified membrane vesicles are allowing plant cell membranes to be characterized in terms of their distinct transport properties. The algae, which have always provided a background basis for solute transport studies, have now emerged as a group studied in their own right as revealed by the detailed presentation of their transport mechanisms discussed in Ch. 4. Studies on the solute-transporting properties of fungal systems, particularly *Neurospora*, have contributed a great deal to current awareness of membrane behaviour, in particular the occurrence of ion-specific channels within the membrane and are presented in Ch. 5. The problems of interpreting membrane events in isolated cells and relating such findings to the behaviour of the membrane when the cells form an integrated tissue are the topic of Ch. 6, where the implications arising from studies employing the pressure probe are presenting a challenge to some of our preconceived ideas on cell turgor regulation. The pathway taken by solutes and the driving forces for their movement across the plant root has been the subject of considerable debate over recent years. The current interpretation of this complex system is the subject of Ch. 7. Chapter 8 considers the long-distance transport of photoassimilates through the phloem which has received considerable attention in recent years with a focusing of interest on the mechanisms of loading and unloading. At the whole plant level the integration of the various uptake and transport processes implies a control mechanism to regulate the distribution and circulation of the range of solutes involved (Ch. 9). Salinity presents one of the major problems in world agricultural production. In recent years considerable progress has been made in elucidating the mechanism whereby halophytes can survive such salinities, as revealed in Ch. 10. An interesting system is provided by the CAM plants, which are the topic of Ch. 11. These unique plants are able to survive adverse conditions such as salinity (cf. Ch. 10) by dark fixation of CO_2, which has important implications for their solute transport mechanisms. The vast majority of plants fix CO_2 in the light, when open stomata permit gaseous exchange. The mechanisms which regulate the stomatal movements which control the aperture are the subject of Ch. 12. Some plants survive salinity (cf. Ch. 10) by expelling excess salts through salt glands. This salt-secreting process provides a valuable experimental system and the mechanism whereby these glands secrete relatively large quantities of salts is discussed in Ch. 13. Only a relatively few plants have salt glands, but nectaries (Ch. 14) occur extensively among higher plants, often associated with insect attraction. The mechanisms whereby large fluxes of mainly sugars occur, is of fundamental interest in solute transport.

All of these broad topics have been presented within the relatively narrow limits of this book and the reader is encouraged to follow the extensive literature which has been cited by the various contributors. As a subject such as this expands and publications proliferate, sometimes knowledge is obtained at the expense of understanding. It is our hope and belief that readers of this book will be introduced to an understanding of the processes involved in the transport of solutes by plant cells and tissues, as we currently perceive them, and will avail themselves of the many detailed reviews and research papers referred to in the individual chapters here.

We would like to thank our fellow researchers in transport processes who have contributed their specialist knowledge in this book. The topics were selected by us and the detail and precise content provided by them. Their preparation of the individual chapters has been of a high standard and has made our editorial task a stimulating and pleasurable one. We are particularly grateful to Mrs Sue Briant and Miss Carolyn Cole who provided patient and willing secretarial assistance during the production of the text.

Wye and Southampton, 1987 **D. A. B. and J. L. H.**

List of contributors

D. A. Baker,
Department of Biological Sciences,
Wye College, University of London,
Ashford,
Kent, TN25 5AH, England.

D. T. Clarkson,
Department of Agriculture and Horticulture,
University of Bristol,
Long Ashton Research Station,
Long Ashton, Bristol, BSl8 9AF, England.

C. D. Faraday,
Department of Botany and Plant Sciences,
University of California,
Riverside,
California 92521, USA.

N. Findlay,
School of Biological Sciences,
The Flinders University,
Bedford Park,
South Australia 5042, Australia.

T. J. Flowers,
School of Biological Sciences,
University of Sussex,
Falmer,
Brighton, BN1 9QG, England.

J. L. Hall,
Department of Biology,
Building 44,
The University,
Southampton, S09 5NH, England.

T. E. Humphreys,
Vegetable Crops Department,
1255 HS/PP Building,
University of Florida,
Gainesville, Florida 32611, USA.

U. Luttge,
Institut für Botanik,
Technische Hochschule,
Schnittspahn. str. 3–5,
D–6100 Darmstadt, Federal Republic of Germany.

E. A. C. MacRobbie,
Department of Botany,
Downing Street,
University of Cambridge,
Cambridge,CB2 3EA, England.

J. W. Oross,
Department of Botany and Plant Sciences,
University of California
Riverside, California 92521, USA.

M. G. Pitman,
CSIRO,
Institute of Biological Resources,
P.O.Box 225, Limestone Avenue,
Dickson, ACT 2602, Australia.

R. Poole,
Department of Biology,
McGill University,
1205 Avenue Docteur Penfield,
Montreal H3A 1B1, Canada.

J. N. Prebble,
Department of Biochemistry,
Royal Holloway and Bedford New College, University of London,
Egham Hill, Surrey TW20 OEX, England.

J. A. Raven,
Department of Biology,
University of Dundee,
Dundee, DEl 4HN, Scotland.

D. Sanders,
Department of Biology,
University of York,
Heslington, York, Y0l 5DD, England.

J. A. C. Smith,
Department of Botany,
University of Edinburgh,
The King's Buildings,
Mayfield Road,
Edinburgh, EH9 3JH, Scotland.

W. W. Thomson,
Department of Botany and Plant Sciences,
University of California,
Riverside,
California 92521, USA.

A. D. Tomos,
Department of Biochemistry and Soil Science,
University College of North Wales,
Bangor,
Gwynedd, LL57 2UW, Wales.

R. G. Wyn Jones,
Department of Biochemistry and Soil Science,
University College of North Wales,
Bangor,
Gwynedd, LL57 2UW, Wales.

A. R. Yeo,
School of Biological Sciences,
University of Sussex,
Falmer, Brighton, BN1 9QG, England.

Abbreviations, symbols and units

ABA	abscisic acid
ADP	adenosine-5′-diphosphate
AEC	adenylate energy charge
AFS	apparent free space
AMP	adenosine-5′-monophosphate
ATP	adenosine-5′-triphosphate
ATPase	adenosine triphosphatase
AZ	azetidine-2-carboxylic acid
BSA	bovine serum albumin
C_3	photosynthetic carbon reduction cycle
C_4	photosynthetic dicarboxylic acid pathway
CAM	Crassulacean acid metabolism
CCCP	carbonylcyanide-*m*-chlorophenyl hydrazone
CHM	cycloheximide
CN^-	cyanide
DCCD	*N*,*N*′-dicyclohexylcarbodiimide
DCMU	3′(3, 4 dichloropheny1)-1′,1-dimethyl urea
DES	diethylstilbestrol
DIDS	4,4′-diisothiocyano-2,2′-stilbenesulphonic acid
DMO	5,5-dimethyl-2,4-oxazolidine
DNA	deoxyribonucleic acid
DNP	2,4-dinitrophenol
EDTA	ethylenediaminetetracetic acid
EGTA	ethyleneglycol-bis (β-aminoethylether) *N*, *N*, *N*′, *N*′-tetraacetic acid
ER	endoplasmic reticulum
FC	fusicoccin
FCCP	carbonylcyanide *p*-trifluoromethoxyphenylhydrazone
Fe–S	iron-sulphur centres
FITC	fluorescein isothiocyanate

FMN	flavin mononucleotide
FPA	DL-*p*-fluorophenylalanine
I–V	current–voltage
Mes	2-(*N*-morppholino) ethanesulphonic acid
NAD^+	nicotinamide adenine dinucleotide
NADH	reduced form of above
$NADP^+$	nicotinamide adenine dinucleotide phosphate
NADPH	reduced form of above
NBD-Cl	7-chloro-4-nitrobenzo-2-oxa-1,3-diazole
NEM	*N*-ethylmaleimide
NR	nitrate reductase
OAA	oxaloacetate
p	proton motive force ($\triangle p$)
ppt	parts per thousand inorganic phosphate
PAGE	polyacrylamide gel electrophoresis
PAR	photosynthetically active radiation
PCMBS	*p*-chloromercuribenzene sulphonic acid
PCRC	photosynthetic carbon reduction cycle
PEP	phosphoenolpyruvate
PFK	phosphofructokinase
PGA	phosphoglyceric acid
PMF	proton motive force
PP_i	inorganic pyrophosphate
PSI	photosystem I
PSII	photosystem II
RNA	ribonucleic acid
RUBISCO	ribulose bisphosphate/carboxylase-oxygenase
RuBP	ribulose bisphosphate
SCN^-	thiocyanate
SDS	sodium dodecylsulphate
SER	smooth endoplasmic reticulum
TEA	tetraethylammonium
TNP-ATP	2′,3′-*O*-(2,4,6-trinitrophenyl)-adenosine 5′-triphosphate
$TPMP^+$	triphenylmethylphosphonium ion
Tris	tris (hydroxymethyl) aminomethane
UDPG	uridine 5′-diphosphate glucose
VAM	vesicular–arbuscular mycorrhiza

Symbols

Symbol	*Description*	*Unit*
A	area	m^2
a	activity	$mol\ m^{-3}$ (mM)
C	capacity	F
c	concentration	$mol\ m^{-3}$ (mM)

D	diffusion coefficient	$m^2\ s^{-1}$
G	Gibbs' free energy	J
g	membrane conductance	$S\ m^{-2}$
I	current	A
J	generalized flow	
J_j	net flux	
J_s	solute flow	$mol\ m^{-2}\ s^{-1}$
J_v	volume flow	$m^3\ m^{-2}\ s^{-1}$ ($m\ s^{-1}$)
K_i	inhibitor constant	$mol\ m^{-3}$
K_m	Michaelis–Menten constant	$mol\ m^{-3}$
k	rate constant	
l	length	m
ln	logarithm to base e	
log	logarithm to base 10	
n	number of moles	
P	permeability coefficient	$m\ s^{-1}$
P	turgor pressure	Pa
R	electrical resistance	ohm, Ω
T	temperature	K or °C
t	time	s
u	electrical mobility	$m^2\ V^{-1}\ s^{-1}$
V	volume	m^3
V_{max}	maximum reaction velocity	$mol\ s^{-1}$
v	velocity of reaction	$mol\ s^{-1}$
X	generalised force	
z	valency	
γ	activity coefficient	
ϵ	volumetric elastic nodules	Pa
μ	chemical potential	$J\ mol^{-1}$
$\bar{\mu}$	electrochemical potential	$J\ mol^{-1}$
π	osmotic pressure	Pa
ϕ	net flux	$mol\ m^{-2}\ s^{-1}$
ψ	water potential	Pa
Ψ_π	osmotic potential	Pa
Ψ	electrical potential	V
[ ]	denotes concentration	
$\triangle$	difference in	

Subscripts

c	cytoplasmic
i	inside

j	species
M	membrane
o	outside
t	tonoplast
v	vacuole
w	water
x	xylem

Chemical symbols (H, K, Cl) when used as subscripts specify the particular ion.

Constants

Constant	*Description*	*Value*
F	Faraday constant	96 490 coulombs mol^{-1}
R	gas constant	8.314 J mol^{-1} K^{-1}
RT/F		25.3 mV at 20°C

Units

The International System of Units (SI) is used wherever possible in this book. In some instance, where data are reproduced, other units are quoted with their SI equilvalent values.

Prefixes

M	mega	10^6
k	kilo	10^3
c	centi	10^{-2}
m	milli	10^{-3}
μ	micro	10^{-6}
n	nano	10^{-9}
p	pico	10^{-12}

Unit	*Description*
A	ampere, electrical current
bar	unit of pressure, 10^5 Pa
°C	degree Celsius, temperature
d	day
Da	dalton
eq	equivalent mol × valency
F	farad, electrical capacitance
g	gram

h	hour
J	joule
l	litre
K	kelvin (degrees)
m	metre
mho	unit of conductance
min	minute, time
mol	mass equal to molecular weight in grams
osmol	sum of mole contribution to osmotic pressure
Pa	pascal, pressure
S	siemen, electrical conductance
s	second, time
V	volt, electrical potential difference
y	year
Ω	unit of resistance

1 Introduction and general principles

D. A. Baker and J. L. Hall

1.1 Introduction

Plants have evolved not only as the primary synthesizers of organic compounds but also as selective accumulators of inorganic nutrients from the earth's crust. Such mining of the physical environment is restricted to green plants and some microorganisms, with other life forms being directly or indirectly dependent on this process for their source of mineral nutrients. This initial accumulation of ions by plants is usually spatially separated from the regions of photosynthesis, thus necessitating a transport to the photosynthetic parts of the various inorganic solutes acquired. Conversely the requirement of the accumulation process for a supply of energy-rich materials is met by their transport, usually in the opposite direction, from the photosynthetic areas.

These solute transport processes within plants have been studied at the subcellular, cellular, tissuc and whole organisms levels. The basic problems of resolving the nature of the driving forces and the form of the energy supply for the transport mechanisms remain the same for all systems, but the methodology applied, and results obtained, vary widely with the experimental materials employed. This reflects the variation in the solute-transporting properties which have selectively evolved in response to both internal and external environmental pressures. Whereas animals maintain their cell ionic composition in the relatively constant internal environment of the blood plasma, which has an ionic composition similar to dilute sea water, most higher plants live and grow in environments in which the ionic composition of the soil solution often differs markedly from that of sea water and therefore require complex selection mechanisms to achieve the necessary ionic balance. In these terms plant cells have become truly terrestrial.

Most of the fundamental relationships, which have been developed to describe the transport of solutes across membranes, are based on models of physical diffusion barriers and passive equilibrium conditions. Only rarely do biological membranes behave in the manner predicted by such models, and transport physiologists concentrate on describing the deviation of the living system from the inert physical model. Attempts to characterize the active transport properties of biological membranes using the criteria of irreversible thermodynamics have not so far been particularly successful, the complexities of the living systems giving rise to parameters and interactions which are not at present readily measured. The importance of the biophysical descriptions based on thermodynamics is that they provide a reference point, an equilibrium zero for descriptions of solute transport across plant cell membranes. Thus they are of major conceptual importance and provide us with a diagnostic tool for describing some of the phenomena of living membranes and their solute-transporting properties.

1.2 The forces moving solutes

The driving forces for solute transport at the membrane level in plants are the same as those in other biological membrane systems. However, the presence of a cell wall and a membrane-bound vacuole add considerably to the technical problems of measuring the various transport parameters in plant cells. Basic transport equations (see below) may be applied to the transport of solutes across plant cell membranes, but the application of such an approach to the cells and tissues of higher plants is often severely limited. The cells are frequently rather too small for accurate electrophysiological measurements to be taken and until relatively recently many of the basic investigations have been made on the transport properties of large algal coenocytes (see Ch. 5)

For uncharged solutes, the driving force for transport is the gradient of chemical potential only. An ion in aqueous solution is acted on by at least two physical forces – one arising from chemical potential gradients and the other from electrical potential differences. Chemical potential is related to the concentration of the ion and electrical potential is the result of the net positive or negative charge carried by the ion. When a salt is added to water it diffuses through the solution from regions of higher concentration to those of lower concentration until a uniform concentration is achieved. In general, one of the ion species will have a higher mobility than the other and will tend to diffuse more quickly than its oppositely charged partner and thus cause a slight separation of charges. This sets up an electrical potential gradient leading to a diffusion potential. As a result the faster moving ion of the pair is slowed down and the slower one speeded up until they both move at the same rate. Thus, a dissociated salt diffusing in a solution behaves as a single substance and has a characteristic diffusion coefficient. According to Fick's law, the rate of diffusion dv/dt is related to the concentration gradient dc/dx thus:

$$dv/dt = -DA\ dc/dx$$

where D is the diffusion coefficient and A the area across which diffusion occurs. The negative sign is a convention to indicate that diffusion occurs from a higher to a lower concentration.

The cell surface membrane (the plasmalemma or plasma membrane) is usually the main barrier for the diffusion of molecules into and out of cells. Diffusion of ions across such a membrane results in the development of a diffusion potential across it which is termed the 'membrane diffusion potential'.

The driving force for ion migration is the gradient of electrochemical potential. For an ion species j the electrochemical potential μ_j is given by the following component energies:

$$\bar{\mu}_j = \mu_j^* + RT \ln \gamma_j c_j + z_j F \psi + P\bar{V}_j \qquad (1.1)$$

where μ_j^* is the chemical potential of the ion j in its standard state, R the gas constant, T the temperature in degrees kelvin, γ_j the activity coefficient, c_j the chemical concentration, z_j the valency (with sign), F the Faraday constant, ψ the electric potential, P the pressure in excess of hydrostatic, $\bar{V}_j$ the partial molal volume. $\gamma_j c_j$ is equal to the chemical activity, a_j, the collective term $RT \ln a_j$ sometimes being used. If the ion is in an ideal solution only the concentration c_j need be employed and the collective term is then $RT \ln c_j$. In practice $P\bar{V}_j$ is rarely used, having a negligible effect except at pressures of hundreds of bars and will be omitted from further consideration here.

The force X_j on an ion species j is the negative of the electrochemical potential gradient, $-d\bar{\mu}_j/dx$ obtained by differentiation of Eq. (1.1), i.e.

$$X_j = \frac{-d\bar{\mu}_j}{dx} = \frac{-RT}{a_j}\frac{da_j}{dx} - z_j F \frac{d\psi}{dx} \qquad (1.2)$$

where x is the distance across which movement is considered.

Thus the driving force consists of two terms, one depending on the concentration gradient and one on the electrical potential gradient. In the absence of electrical potential gradients ($d\psi/dx = 0$), and for neutral solutes ($z_j = 0$), Eq. (1.2) reduces to a simple diffusional driving force described by Fick's law.

1.2.1 Nernst equation

An ion will be in equilibrium across a membrane when its electrochemical potential is the same on the two sides of the membrane and thus no change in free energy occurs if the ions move from one side to another provided the net flux ϕ_j is zero. ϕ_j is velocity times concentration where velocity is given by the mobility, u_j, multiplied by the force; i.e.

$$\phi_j = (u_j X_j)c_j \tag{1.3}$$

Thus at equilibrium

$$\phi_j = 0 = -u_j c_j \left(\frac{RT}{a_j}\frac{da_j}{dx} + z_j F \frac{d\psi}{dx}\right) \tag{1.4}$$

and therefore

$$z_j F \frac{d\psi}{dx} = -\frac{RT}{a_j}\frac{da_j}{dx} \tag{1.5}$$

Integration from outside (o) to inside (i) across the membrane and rearranging, Eq. (1.5) becomes

$$\psi^i - \psi^o = \frac{RT}{z_j F} \ln \frac{a_j^o}{a_j^i} \tag{1.6}$$

Putting $\psi^i - \psi^o = \psi_{N_j}$, inserting actual numerical values for R, F and replacing the natural logarithm by 2.303 log, where log is the common logarithm to the base 10, Eq. (1.6) becomes

$$\psi_{N_j} = \frac{59.2}{z_j} \log \left(\frac{a_j^o}{a_j^i}\right) \text{ mV at 25 °C} \tag{1.7}$$

Equation (1.7) is Nernst's equation and the potential ψ_{N_j} is the Nernst potential for ion species j. For some calculations the ratio of the activity coefficients γ_j^o/γ_j^i may be assumed to equal 1 and the ratio a_j^o/a_j^i becomes c_j^o/c_j^i, the ratio of concentrations. This assumption is justified when the ionic strengths on the two sides of the membrane are approximately equal but leads to errors when the outside solution is more dilute and then the internal concentrations must be corrected for activity.

The Nernst potential for an individual ionic species j may be calculated from Eq. (1.7) using the ratio of activities (a_j^i/a_j^o) or concentrations (c_j^o/c_j^i) and can be compared with the measured potential difference across the membrane ψ_M. If $\psi_M = \psi_{N_j}$, a passive equilibrium situation exists, while if $\psi_M \neq \psi_{N_j}$ then energy must be expended to maintain an equilibrium. The difference between ψ_M and ψ_{N_j}, $\Delta\psi_j$ is the minimum energy required in millivolts (mV), (1 mV equals 975 J mol^{-}). The sign of $\Delta\psi_j$ indicates the direction of the net passive driving force which must be opposed by a metabolically driven active transport if equilibrium is to be maintained.

	$\Delta\psi_j$ value	**Direction of passive driving force**
Cations	+	i → o
	−	o → i

Anions	+	o → i
	−	i → o

1.2.2 Passive ion fluxes across membranes

Equation (1.4) is the equilibrium condition of the general expression for the net flux of an ion species j moving passively. Neglecting γ this can be written:

$$\phi_j = -u_j RT \frac{dc_j}{dx} - u_j c_j z_j F \frac{d\psi}{dx} \quad (1.8)$$

In order to calculate passive ion fluxes, certain simplifying assumptions must be made. The electrical potential is assumed to be a linear function of the distance inside the membrane. This means that $d\psi/dx$ is a constant equal to ψ_M/δ, where ψ_M is the electrical potential difference and δ is the membrane thickness, and the electrical potential steps at the two boundaries of the membranes with the bathing solutions can be ignored or are equal and opposite. Making these rather arbitrary assumptions, Eq. (1.8) can be integrated from one side of the membrane to the other to give:

$$\frac{RT}{z_j F\psi_M} = \ln \frac{\phi_j + u_j c_j^o z_j F\psi_M/\delta}{\phi_j + u_j c_j^i z_j F\psi_M/\delta} \quad (1.9)$$

Taking exponentials of both sides of Eq. (1.9) and rearranging:

$$\phi_j = -\left(\frac{u_j z_j F\psi_M}{\delta}\right)\left(\frac{c_j^o - c_j^i \exp (z_j F\psi_M/RT)}{1 - \exp (z_j F\psi_M/RT)}\right) \quad (1.10)$$

Application of Eq. (1.10) is difficult in biological systems. In particular, mobilities, solubilities and δ, the membrane thickness, are usually uncertain and in practice u_j and δ are combined to give a permeability coefficient, P_j, which is given by:

$$P_j = \frac{u_j K_j RT}{\delta} \quad (1.11)$$

where u_j is the mobility, K_j the water : membrane partition coefficient, R, T and δ as before. P_j is relatively easy to estimate from tracer experiments as the sum of the permeabilities of a number of pathways which may be followed by an ion crossing a membrane, and is therefore subject to some environmental modifications. Incorporating P_j, Eq. (1.10) then becomes:

$$\phi_j = -P_j \frac{z_j F\psi_M}{RT} \left(\frac{c_j^o - c_j^i \exp (z_j F\psi_M/RT)}{1 - \exp (z_j F\psi_M/RT)}\right) \quad (1.12)$$

Although the derivation of Eq. (1.12) involves some cumbersome mathematical manipulations, it is of extreme importance for the description of both passive ion fluxes and membrane diffusion potentials.

1.2.3 Ussing–Teorell equation

The Nernst equation is applicable only when the ions are in flux equilibrium conditions across the membrane. For non-equilibrium conditions, active transport may be identified by the flux ratio for an ionic species across a membrane, measured using suitable radioisotopes and following the initial influx and efflux to give the resultant net flux:

$$\text{net flux}(\phi_j) = \text{influx}(\phi_j^{oi}) - \text{efflux}(\phi_j^{io}) \quad (1.13)$$

It is conventional to identify the unidirectional influx and efflux with the two terms of Eq. (1.13). The rationale behind this is as follows: influx is measured by labelling the outside solution with an appropriate isotope and the internal concentration c_j^i for this isotope equals zero at the beginning of the experiment. The initial net flux (ϕ_j) of the isotope will thus reflect ϕ_j^{oi} only. Hence

$$\phi_j^{oi} = -P_j \frac{z_j F \psi_M / RT}{1 - \exp(z_j F \psi_M / RT)} \quad (1.14)$$

By a similar argument, if material is labelled with radioisotope and then transferred to an unlabelled solution of the same composition, the initial net flux (ϕ_j) of the isotope will reflect ϕ_j^{io} only, and therefore

$$(\phi_j^{io}) = -P_j \frac{z_j F \psi_M / RT}{1 - \exp(z_j F \psi_M / RT)} c_j^i \exp(z_j F \psi_M / RT) \quad (1.15)$$

The ratio of the two fluxes is obtained by dividing Eq. (1.14) by Eq. (1.15) which takes on the relatively simple form:

$$\frac{(\phi_j^{oi})}{(\phi_j^{io})} = \frac{c_j^o}{c_j^i \exp(z_j F \psi_M / RT)} \quad (1.16)$$

Equation (1.16) was derived independently by Ussing (1949) and Teorell (1949) and is known as the Ussing–Teorell equation. It is the basis of an important test for the independent, passive movement of ions across a membrane. By taking logarithms of the two sides, Eq. (1.16) can be expressed:

$$RT \ln\left(\frac{\phi_j^{oi}}{\phi_j^{io}}\right) = (\bar{\mu}_j^o) - (\bar{\mu}_j^i) \quad (1.17)$$

When μ_j^o equals μ_j^i, $\phi_j = 0$ and Eq. (1.16) reduces to Nernst's equation (1.7). Thus, when ψ_M equals ψ_{Nj}, ϕ_j^{oi} will equal ϕ_j^{io} and no net passive flux of ion j is expected and no energy need be expended in moving the ion from one side of the membrane to the other. When Eq. (1.17) is not satisfied, such ions are not moving passively or, in some cases, not moving independently from other fluxes. Active transport, using energy derived from metabolism, is usually the way in which solutes are moved to regions of higher chemical potential, by doing work against the passive gradient.

1.2.4 Goldman equation

In some plant cells, particularly algae, the total ionic flux is mainly due to movements of K^+, Na^+ and Cl^-, although H^+ and OH^- fluxes can also be considerable. The fluxes of these ions across the membrane are caused by gradients in the chemical potentials which create electrical potential differences across the membrane. This electrical potential difference is termed a 'diffusion potential' and arises as a result of the differential mobilities of the ions involved and the condition of electrical neutrality which is maintained. The faster moving ion is slowed down and the slower moving ion of opposite charge is speeded up until they are both moving at the same rate and therefore no electric charge is carried across the membrane.

The potential difference in terms of the concentrations and permeabilities may be expressed when there is no net electric current. For the three ion species K^+, Na^+ and Cl^-, substitution of the net flux of each species into Eq. (1.12) so that $\phi_K + \phi_{Na} - \phi_{Cl} = 0$ is given by

$$P_K\left(\frac{c_K^o - c_K^i \exp(F\psi_M/RT)}{1 - \exp(F\psi_M/RT)}\right) + P_{Na}\left(\frac{c_{Na}^o - c_{Na}^i \exp(F\psi_M/RT)}{1 - \exp(F\psi_M/RT)}\right) + P_{Cl}\left(\frac{c_{Cl}^o - c_{Cl}^i \exp(-F\psi_M/RT)}{1 - \exp(-F\psi_M/RT)}\right) = 0 \quad (1.18)$$

In the above equation valencies have been given actual numerical values and $F\psi_M/RT$ has been cancelled from each term. Further cancellation of $1/(1 - \exp(F\psi_M/RT))$ from each term of Eq. (1.18) gives

$$P_K c_K^o - P_K c_K^i \exp(F\psi_M/RT) + P_{Na} c_{Na}^o - P_{Na} c_{Na}^i \exp(F\psi_M/RT) - P_{Cl} c_{Cl}^o \exp(F\psi_M/RT) + P_{Cl} c_{Cl}^i = 0 \quad (1.19)$$

Solving Eq. (1.19) for $\exp(F\psi_M/RT)$ and taking logarithms, the following constant-field equation is obtained for the membrane potential:

$$\psi_M = \frac{RT}{F} \ln \left(\frac{P_K c_K^o + P_{Na} c_{Na}^o + P_{Cl} c_{Cl}^i}{P_K c_K^i + P_{Na} c_{Na}^i + P_{Cl} c_{Cl}^o}\right)$$

or

$$= \frac{RT}{F} \ln \left(\frac{c_K^o + (P_{Na}/P_K) c_{Na}^o + (P_{Cl}/P_K) c_{Cl}^i}{c_K^i + (P_{Na}/P_K) c_{Na}^i + (P_{Cl}/P_K) c_{Cl}^o}\right) \quad (1.20)$$

Equation (1.20) is the Goldman or Hodgkin–Katz equation which is widely used in interpreting membrane potential differences. The equation predicts that the membrane potential is determined by the passively moving ions, but undoubtedly the active transport of ions will contribute to this potential. If the active transport involves an ion–carrier complex which is charged on one or both of its journeys across the membrane, leading to a net transfer of charge across that membrane, then this process of ion-pumping will contribute to the membrane potential. This may be achieved by the independent, unidirectional transport of an ion (uniport), or by an exchange pump with a stoichiometry other than one-to-one. Such a pump is termed 'electrogenic' (as opposed to neutral pumps where there is no net charge transfer across the membrane) and an additional term may be added to the membrane potential ψ_M. Thus, in the presence of an electrogenic pump,

$$\psi_M = \psi_{eq} + \psi_X \tag{1.21}$$

where ψ_{eq} is a diffusion equilibrium potential and ψ_x an additive electrogenic mechanism. $\psi_x = F(\phi_x/g_m)$ where ϕ_x is the electrogenic ion flux and g_m the conductance of the membrane. Thus Eq. (1.20) becomes

$$\psi_M = \frac{RT}{F} \ln \left(\frac{P_K c_K^0 + P_{Na} c_{Na}^0 + P_{Cl} c_{Cl}^i}{P_K c_K^i + P_{Na} c_{Na}^i + P_{Cl} c_{Cl}^0} \right) + F \frac{\phi_x}{g_m} \tag{1.22}$$

In most biological membranes $P_K \approx P_{Cl} > P_{Na}$, and the Goldman equation often gives a similar result to the Nernst equation (1.7) for K^+ and Cl^-. A difference between calculated and predicted membrane potentials does not always indicate an electrogenic pump and additional evidence is required before electrogenicity can be unequivocally claimed. Evidence of electrogenicity is provided by the very rapid depolarization of a membrane with metabolic inhibitors, which often reduce the value of ψ_M close to that of the predicted Goldman value. When the inhibitor is removed the electrogenic pump commences once again and the cell returns to its resting potential. The demonstration and characteristics of electrogenic pumps are discussed fully by Spanswick (1981).

1.2.5 Membrane conductance

When small currents flow through the membrane, the membrane potential is not shifted far from its resting value as given by the Goldman equation (1.20). The conductance of the membrane in terms of permeabilities and activities is given by

$$g_m = \frac{dJ}{d\psi_M} = -\frac{F^2}{RT} \frac{\ln (C^o/C^i)\, C^o}{1 - C^o/C^i} \tag{1.23}$$

where

$$C^{o,i} = P_K c_K^{o,i} + P_{Na} c_{Na}^{o,i} + P_{Cl} c_{Cl}^{o,i}$$

if we are concerned only with current carried by these ions. Under these conditions g_m is the slope conductance.

When larger currents are passed, the expression for conductance is more complicated and the chord conductance must be calculated. This conductance should increase when current is passed in one direction and should decrease when passed in the opposite direction, a phenomenon equivalent to rectification. If then the membrane is permeable to monovalent cations only, the chord conductance g'_m is given by the following expression:

$$g'_m = \frac{F^2\psi_M C^o_+ [1 - \exp(F\Delta\psi_M/RT)]}{RT\Delta\psi_M[1 - \exp(F\psi_M/RT)]} \tag{1.24}$$

When the Nernst condition for electrochemical equilibrium is satisfied the partial conductance expected from an observed unidirectional flux is given by:

$$g_j = (z_j^2 F^2/RT)\phi_j \tag{1.25}$$

This assumes that the influx and efflux of ion species j are independent of each other and of other fluxes.

A ϕ_K of 10 nmol m^{-2} s^{-1} will contribute 38 mmho m^{-2} partial conductance at 20 °C, and its reciprocal, the resistance, R_K would be 260×10^4 kΩ m^2. Estimates of ϕ_j or R_j from Eq. (1.25) may be compared with direct electrical measurements obtained by passing a known current across the membrane. In giant algal cells g_K values of 40 m mho m^{-2} have been calculated for the plasma membrane, whereas measured values give 500–1000 mmho m^{-2} (Hope, 1971). This observation has two possible explanations. Either the K^+ does not carry all the current, or the ion movement is not independent of the transport of other ions. Considering the first possibility, ϕ_{Na}, ϕ_{Cl} and ϕ_{Ca} would not contribute more than a few mmho m^{-2}.However, it is probable that ϕ_H contributes a large part of the conductance in algal cells and probable that ϕ_K is not totally independent of other fluxes.

Current passing through a membrane will, according to its direction of flow, either depolarize or hyperpolarize the resting potential of the membrane. Hyperpolarization cannot proceed beyond ψ_M values of about −300 mV due to the occurrence of the 'punch-through' phenomenon (Fig. 1.1). The membrane behaves in a manner which suggests a semiconductor composed of alternating layers of opposite fixed charge between which there is a depletion layer. As ψ_M increases, the depletion layer widens until it extends to the outer limits of the membrane at which point 'punch-through' occurs and a sudden increase in current is observed.

1.2.6 Donnan potential

Another type of electrical potential difference encountered in biological systems is that associated with immobile or fixed charges in a solid phase adjacent to an aqueous phase and this is termed the 'Donnan potential'. Such a phase boundary occurs in the cell walls of plant cells. The region containing the immobile charged particles is generally referred to as the 'Donnan phase'. In cell walls the presence of a large number of immobile carboxyl groups ($RCOO^-$) associated with pectin and other compounds

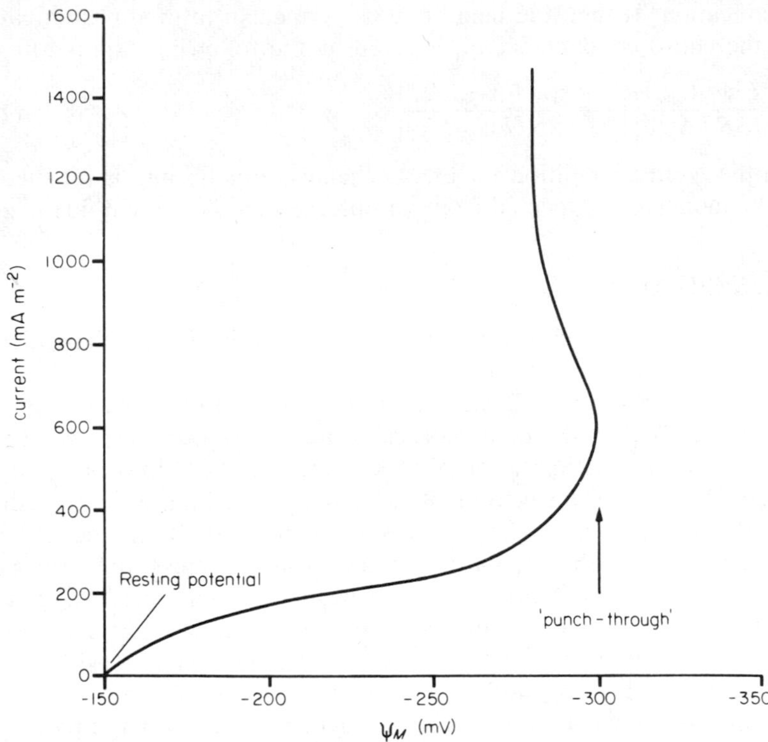

Fig. 1.1 The effect of an increased current on the membrane potential ψ_M. The resting potential was -150 mV. 'Punch-through' takes place when $\psi_M = -300$ mV.

provides a cation-exchange system. The anion-exchange capacity, which is usually less than that for cations, is due to the presence of fixed organic cations, such as amines, within the cell wall matrix. Donnan phases also occur in the cytoplasm where the immobile charges are mainly due to proteins. These proteins are fixed in the sense that they are unable to diffuse across either the plasmalemma or the tonoplast.

They carry a net negative charge at the prevailing cytoplasmic pH (about 7.2) and thus provide an asymmetry of mobile ions across the membrane to give an equilibrium, termed the 'Donnan equilibrium' after the physical chemist F. G. Donnan who described the phenomenon in 1911, in which the distribution of mobile anions and cations is unequal such that for potassion and chloride:

$$\frac{c_{\mathrm{K}}^{\mathrm{i}}}{c_{\mathrm{K}}^{0}} = \frac{c_{\mathrm{Cl}}^{0}}{c_{\mathrm{Cl}}^{\mathrm{i}}} \tag{1.26}$$

This immobility of the protein anion will cause an increase in $c_{\mathrm{K}}^{\mathrm{i}}$ and a decrease in $c_{\mathrm{Cl}}^{\mathrm{i}}$ relative to c_{K}^{0} and c_{Cl}^{0} which will give rise to an electrical potential of negative sign in the same manner as described above for the

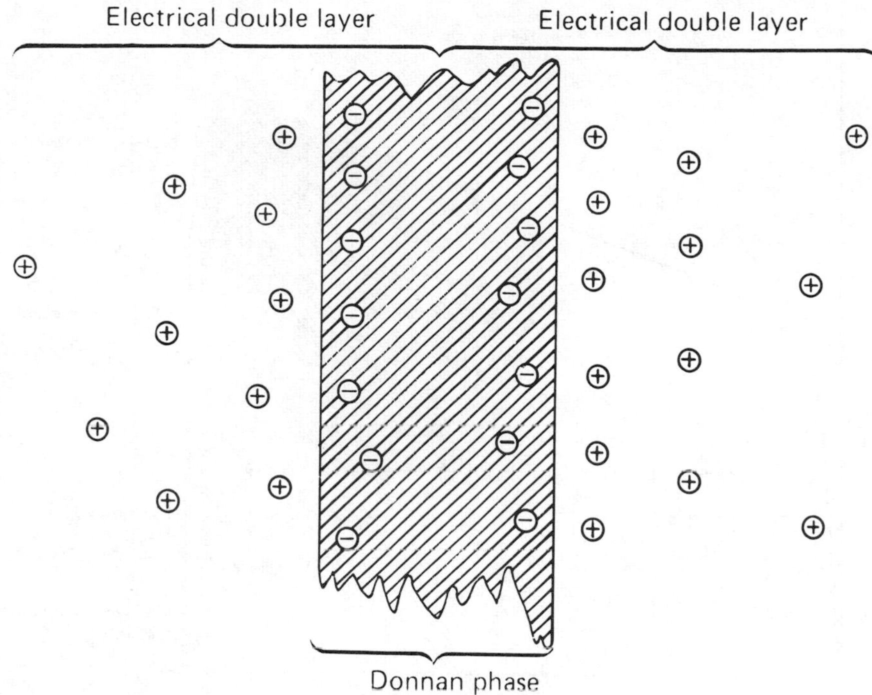

Fig. 1.2 The distribution of positively charged ions occurring on either side of a Donnan phase containing immobile negative charges.

diffusion potential. The principles of the Donnan equilibrium and the resultant electrical potential, the Donnan potential ψ_D, are presented in Fig. 1.2. The relationship may be expressed:

$$\psi_D = \frac{RT}{F} \ln \left(\frac{c_K^0}{c_K^i}\right) \qquad (1.27)$$

an expression identical with the Nernst equation (1.7).

When the immobile or fixed ions are arranged in a layer the mobile ions of opposite sign occur in an adjacent layer, the two together being referred to as an 'electrical double layer', such as is shown in Fig. 1.2. From a knowledge of the ratio of the concentrations of any mobile ion across the electrical double layer the electrical potential difference, or Donnan potential can be calculated, again using the Nernst equation.

1.2.7 Free space uptake

If a plant tissue is first washed in water and then immersed in a solution of a salt, there is a rapid initial uptake which is usually completed in 10–20 min followed by a less rapid steady uptake which may continue for

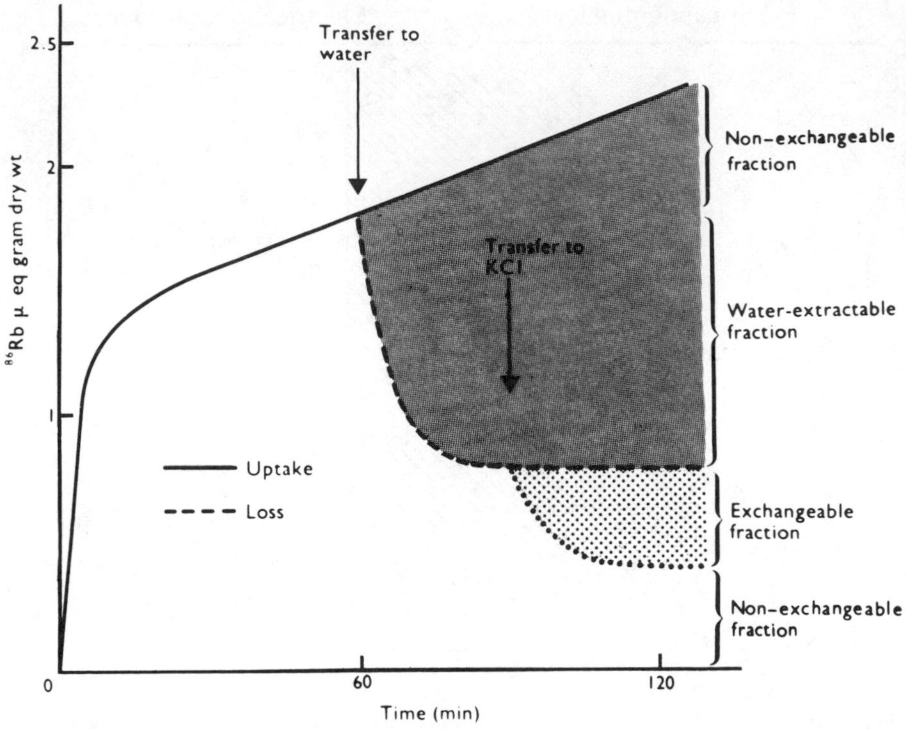

Fig. 1.3 The uptake and loss of potassium by maize roots. The initial rapid uptake may be reversed by washing out the water-extractable and exchangeable fractions with water and unlabelled KCl, respectively. The slow steady uptake process evident after 15 min is non-exchangeable. (From Sutcliffe and Baker, 1974.)

several hours or even days (Fig. 1.3). The initial rapid uptake is reversible, non-selective and independent of metabolism. If the tissue is transferred back to water a large proportion of the initial uptake may be washed out and this is referred to as the 'water-extractable fraction'. A further fraction will be extracted if the tissue is washed in a salt solution with which adsorbed ions can exchange. This latter fraction, the 'exchangeable fraction', may be measured more conveniently if the original solution is labelled with a radioisotope (Fig. 1.3).

It is now generally agreed that the rapid initial uptake represents ion movement into water-filled spaces in the cell walls of the tissue and it is usually referred to as 'uptake into the free space', that is into the freely accessible part of the cells. The water-extractable fraction consists of mobile ions in the aqueous phase, or water-free space, in the cell wall, while the exchangeable fraction comprises those ions which become adsorbed in the electrical double layer, or Donnan free space, which is also in the cell wall.

There have been attempts in the past to estimate the volume of tissue occupied by the free space. This can be done by assuming that the concentration of solutes in the free space is the same as that in the bathing medium and converting a quantity of ions taken up per unit volume of tissues to a volume of solution it would occupy. As such an assumption is incorrect, owing to the high concentration of ions associated with the electrical double layer, the volumes calculated for the free space in this way are greatly in excess of the actual free space volumes. To avoid this error the concept of apparent free space (AFS) was introduced. This may be defined as the calculated volume of a cell or tissue which would be occupied if the ion concentration in that volume were the same as that of the bathing medium. This somewhat cumbersome concept is now generally little used and water-extractable and ion-exchangeable fractions of the free space expressed in quantities, not volumes, are the terms used by modern researchers in this field.

1.3 Ion fluxes and compartmentation

The influx and efflux of ions across plant cell membranes may be measured and compared with the behaviour of model systems. The simplest model is a one-compartment system in which a cell of volume V, surface area A, containing a solution concentration c_i of solute j, is bounded by a membrane. It is assumed that rapid diffusion occurs on either side of the surface boundary and thus the gradient of solute j is across the membrane only. The fluxes of j, or a radioactive version of j*, across the boundary membrane are depicted in Fig. 1.4.

Influx is determined by measuring the initial rate of increase of internal radioactivity when unlabelled cell is placed in a radioactive solution. Thus at time zero, $S_j = 0$ and S_o = constant. If the fluxes of solute j across the boundary are constant and unidirectional then $\phi_{oi} = \phi_{io} = 0$. Putting $Q_i = c_i V_i$, the total quantity of j in the inside compartment. With the above conditions, for uptake

$$\frac{dS_i}{dt} = A(\phi^*_{io} - \phi_{io})/Q_i = A(S_o\phi_{oi} - S_i\phi_{io})/Q_i = A(S_o - S_i)\phi/Q_i \tag{1.28}$$

from which

$$S_i = S_o[1 - e^{-\phi At/Q_i}] \tag{1.29}$$

and thus the internal specific activity will increase exponentially until $S_i = S_o$ at $t = \infty$. The initial uptake is linear and thus

$$\phi = \frac{\Delta(S_i Q_i)}{AS_o\Delta t} \tag{1.30}$$

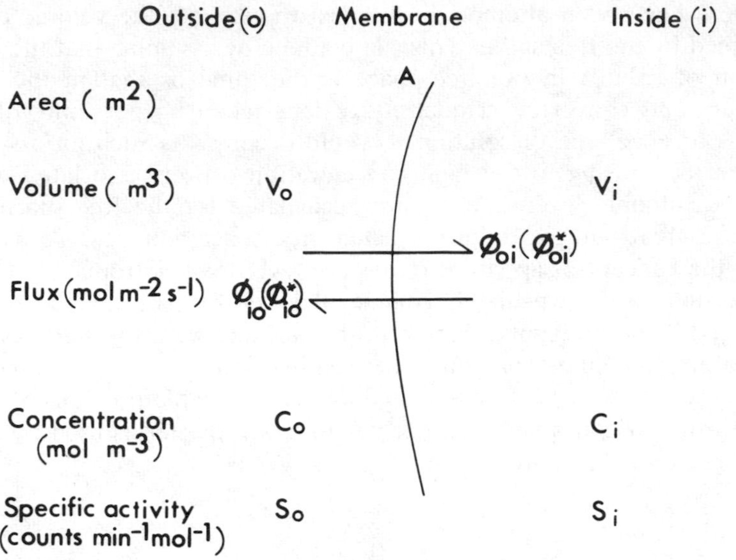

Fig. 1.4 The quantities considered in relating specific activity to time during the course of labelling or eluting a one-compartment system.

When $S_i \neq 0$ and tends towards S_o.

$$\phi_j = \frac{-2.203 Q_i \log(1 - S_i/S_o)}{t'A} \tag{1.31}$$

where t' is the duration of the uptake.

Efflux is determined by allowing the cell to accumulate the radioisotope from a bathing solution until the internal specific activity S_i has reached some convenient value ${}^oS_i$ at $t = 0$. The cell is then transferred to an identical non-radioactive bathing solution which is replaced frequently so that $S_o = 0$.

Under these conditions

$$\frac{dS_i}{dt} = -A\phi^*_{io}/Q_i = -A\phi S_i/Q_i \tag{1.32}$$

Integration gives

$$S_i = {}^oS_i e^{\phi At/Qi} \tag{1.33}$$

The initial appearance of radioactivity in the external solution is linear and can be estimated by the following relationship:

$$\phi = \frac{\Delta(S_o Q_o)}{AS_i \Delta t} \tag{1.34}$$

When the log of S_i is plotted against time, a straight line is obtained of slope $-\phi A/Q_i$. This is related to the 'half-time' for the exchange by the following:

$$t_{\frac{1}{2}} = 0.693\ Q_i/\phi A \tag{1.35}$$

where $t_{\frac{1}{2}}$ is the time for S_i to reach $S_o/2$ for uptake or for S_i to reach ${}^{o}S_i$ for efflux, $\phi A/Q_i$ is termed the 'rate constant' k, with dimension time^{-1}.

Substituting k in Eq. (1.35) gives

$$t_{\frac{1}{2}} = 0.693/k \tag{1.36}$$

The simple situation does not pertain, however, even in a single plant cell. There are at least three compartments – cell wall, cytoplasm and vacuole. The non-membrane-bound cell wall equilibrates rapidly with the external solution and can be corrected for leaving two membrane-bound compartments in series, the cytoplasm and the vacuole (see Cram, 1975).

The method for compartmental analysis is dependent on the relative permeabilities of the plasmalemma and tonoplast. If the plasmalemma is more permeable than the tonoplast, then during an influx period the cytoplasm will fill up with isotope first, followed by the vacuole, thus enabling the separate influxes to be measured. If the cell is loaded with radioactive tracer and the time course of efflux followed, three phases can be observed in the loss of radioactivity. The logarithm of the amount of isotope remaining in the tissue plotted against time yields a straight line after 2–4 h which is believed to represent the steady rate of efflux from the vacuole (Fig. 1.5). Extrapolation to $t = 0$ gives the initial amount of isotope present in the vacuole, and the slope of the line gives k_v, the rate constant of efflux from the vacuole, assuming that there is a steady-state transfer of isotope through the cytoplasm. By subtraction of the amount of isotope present in the vacuole from the total amount in the cell, the time course of the efflux from the cell wall and cytoplasm is obtained, the straight line obtained after that 15 min representing the cytoplasm. Extrapolation to $t = 0$ yields the amount of isotope initially present in the cytoplasm and the slope of the line gives k_c, the rate constant of the efflux from the cytoplasm.

The situation becomes more complex to analyse if the tonoplast is much more permeable than the plasmalemma. Under these conditions the plasmalemma becomes the rate-limiting membrane for both influx and efflux determinations. To obtain the fluxes across the tonoplast the specific activities of the cytoplasm and the vacuole must be estimated separately. MacRobbie (1966) has shown that at time t' when the specific activities of cytoplasm and vacuole are rising together as a result of bathing the cell in a radioactive solution

$$\phi_t = Q_v S_v / t'(S_c - S_v) \tag{1.37}$$

where subscripts c, t and v refer to cytoplasm, tonoplast and vacuole, respectively.

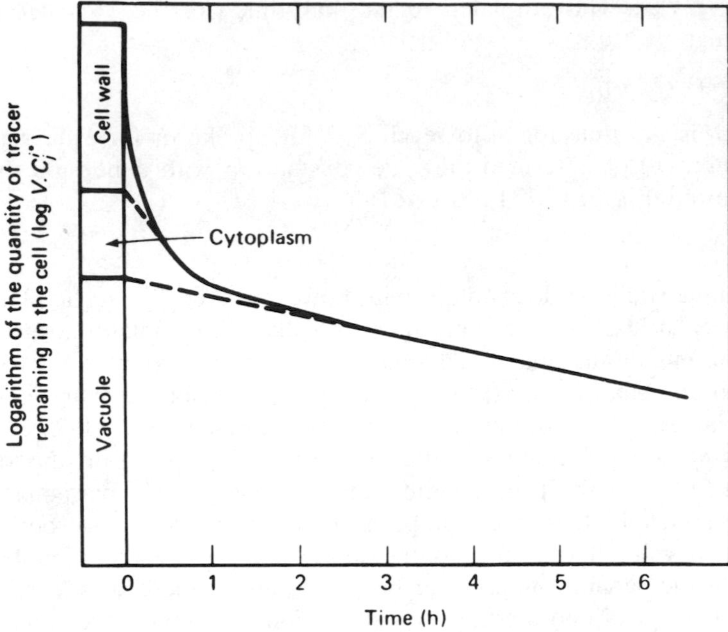

Fig. 1.5 Efflux curves for a plant cell. The extrapolation of the linear portions to zero time gives the quantity of tracer in each phase at the start of the elution. (From Clarkson, 1974.)

1.4 Kinetics of ion transport

The permeability of biological membranes to solute molecules, P_j, is known to decrease with increasing polarity of the permeant. This is due primarily to the low partition coefficient, K_j, of polar solutes which is a result of the high lipid content of cell membranes. Many solutes, particularly ions, must therefore have their movement ascross the plasmalemma facilitated in some way if they are to accumulate within the cell at observed concentrations. Some form of carrier molecule is envisaged which is soluble in the lipid membrane, combines with the permeating solute and transports it across the membrane. When such movement is down the gradient of chemical or electrochemical potential, the carrier process may be regarded as a facilitated diffusion and the carrier need not be energized, the transport is 'passive'. However, when solute movement is against the prevailing potential gradient the carrier must be energized and an 'active' process is invoked, work being done by the carrier to move the solute in a thermodynamically uphill direction.

It has been observed that the kinetics of ion of transport across membranes in a wide range of cells are similar to those of enzymic catalysis and that reversible binding to a carrier mediates the accumulation process.

In both cases, ion transport and enzymic catalysis, the mechanism is believed to involve the attachment of a substrate (ion or enzyme substrate) to an active agent (ion carrier or enzyme substrate) to an active agent (ion carrier or enzyme) and following transport or catalysis the active agent is released to recombine with the substrate once again.

The rate of carrier-mediated ion transport can be characterized by two factors, one the maximum rate of transport which can be achieved when all available carrier sites are loaded, V_{max}, and the other the fraction of the carrier actually loaded at a given substrate concentration, θ. V_{max} can be calculated from the asymptote reached when rate of absorption is measured over a range of concentrations (Fig. 1.6) and θ, the fraction of sites occupied at a given ion concentration [S] is obtained from the Langmuir adsorption equation

$$\theta = \frac{[S]}{K_m + [S]} \tag{1.38}$$

where K_m is the dissociation constant of the carrier–ion complex, characteristic of particular ion crossing a specific membrane and is expressed in units of concentration mol m^{-3}. The rate of absorption, v, is given by the product of the two factors V_{max} and θ to give

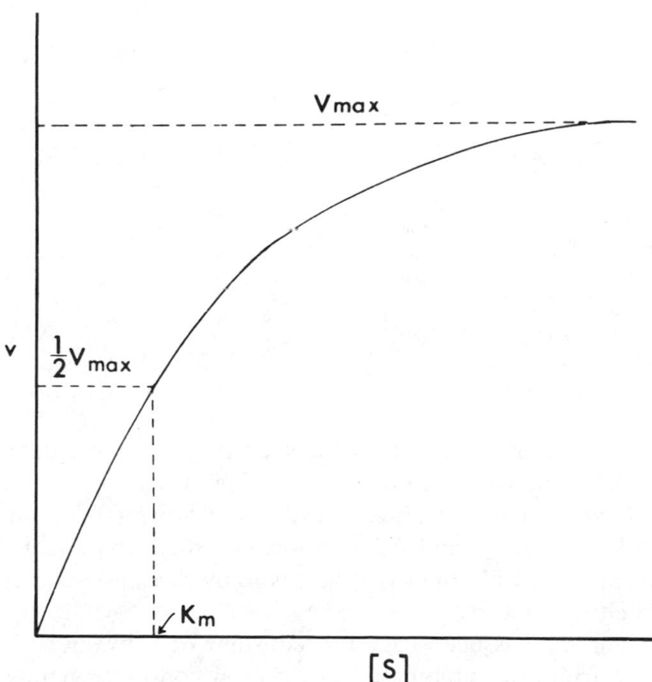

Fig. 1.6 The relationship between the external solute concentration [S] and the rate of absorption, v.

$$v = \frac{V_{max}\,[S]}{K_m + [S]} \tag{1.39}$$

This relationship is based on the assumption that the model for uptake is as follows

$$S + C \underset{k_{-1}}{\overset{k_1}{\rightleftharpoons}} SC \underset{k_{-2}}{\overset{k_2}{\rightleftharpoons}} C + S \tag{1.40}$$

where S represents the ion, C the carrier and k_1, k_{-1}, k_2 and k_{-2} the rate constants of the reaction. Equation (1.40) is dependent on there being no counterflow in the opposite direction, k_{-2} being negligible.

By analogy with enzyme kinetics, K_m is equal to the concentration of the ion [S] at which v reaches half the theoretical maximal rate. Substituting K_m for [S] in Eq. (1.39) we have:

$$v = \frac{K_m V_{max}}{2K_m} = \frac{V_{max}}{2} \tag{1.41}$$

Equation (1.39) is the Michaelis–Menten equation which can be rearranged to yield straight-line plots. These are the double reciprocal form due to Lineweaver & Burk (1934):

$$\frac{1}{v} = \frac{K_m}{V_{max}[S]} + \frac{1}{V_{max}} \tag{1.42}$$

and the single reciprocal form of Eadie and of Hofstee (1952):

$$\frac{v}{[S]} = \frac{V_{max}}{K_m} - \frac{v}{K_m} \quad \text{or } v = V_{max} - \frac{K_m}{v[S]}$$

also

$$\frac{[S]}{v} = \frac{K_m}{V_{max}} + \frac{[S]}{V_{max}} \tag{1.43}$$

The way in which these equations may be employed for the determination of V_{max} and K_m is indicated in Fig. 1.7. Plotting of data by Eq. (1.42) gives $1/v$ versus 1/[S], which appears to have some advantages, but determination of values for V_{max} and K_m requires statistical treatment. Data can be fitted to Eq. (1.43) plotting v/[S] versus v, by the least squares method, using v^4 weighting factors.

The constant K_m has been called a 'constant of convenience' in that it may not be a true rate, affinity or dissociaiton constant, merely an operational term. However, it enables investigators to characterize the affinity

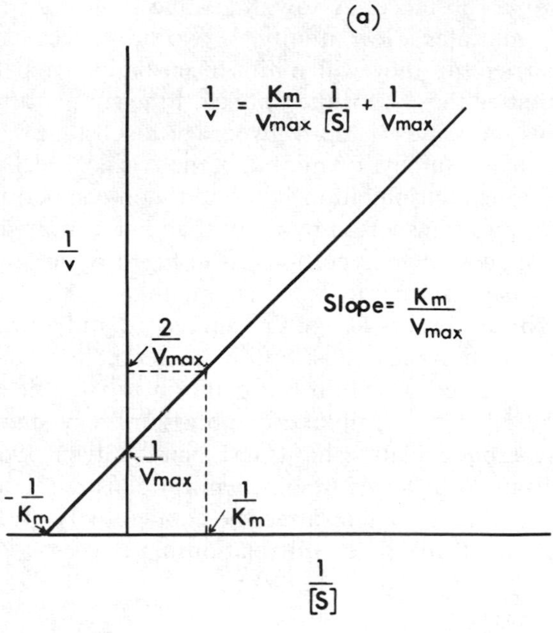

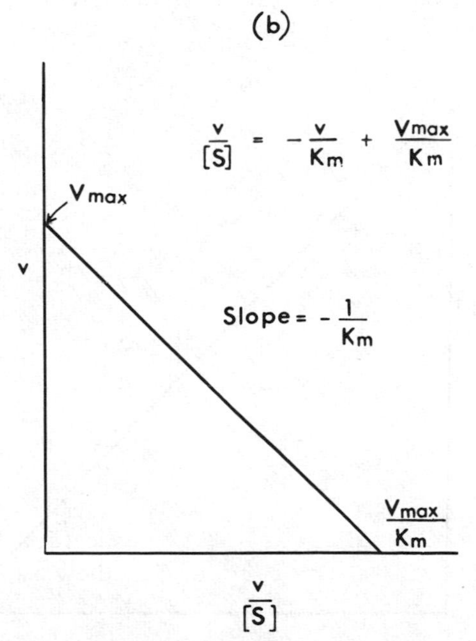

Fig. 1.7 Linear transformations of the Michaelis–Menten equation. [(a) Lineweaver and Burk (1934): (b) Eadie, Hofstee (1952).]

of a 'carrier' for a substance. A low K_m value indicates a high affinity, while a high K_m indicates a low affinity. If two substances have an affinity for the same carrier site they will inhibit each other competitively. With this type of inhibition the K_m for the uptake of the substance is altered and the V_{max} remains the same. If the transport of a substance is inhibited by a compound, such as a sulphydryl inhibitor, there is no effect on the K_m but V_{max} is reduced. Such an inhibition is termed 'non-competitive.'

If a substance were transported by more than one carrier site the uptake isotherm may not be a single hyperbole, as in Fig. 1.6, but might show two or more hyperbolae resulting in a similar number of K_m and V_{max} values. Such multiple-carrier processes, often limited to only two, have been indicated in plant and animal cells, and in bacteria. In a number of cases data have been obtained which do not conform to the Michaelis–Menten equation. In the erythrocyte, uptake of potassium may show a sigmoidal rather than a hyperbolic relationship (Garrahan & Glynn, 1967). Such data have been interpreted in terms of biochemical allosteric interactions and Hill plots have been used to determine the cooperativity of the interaction (Fig. 1.8; Koshland, 1970). The Hill relationship is

$$v = V_{max} \Big/ \left(1 + \frac{K}{[S]^n}\right) \tag{1.44}$$

where [S] is the activator concentration, n the interaction coefficient, K

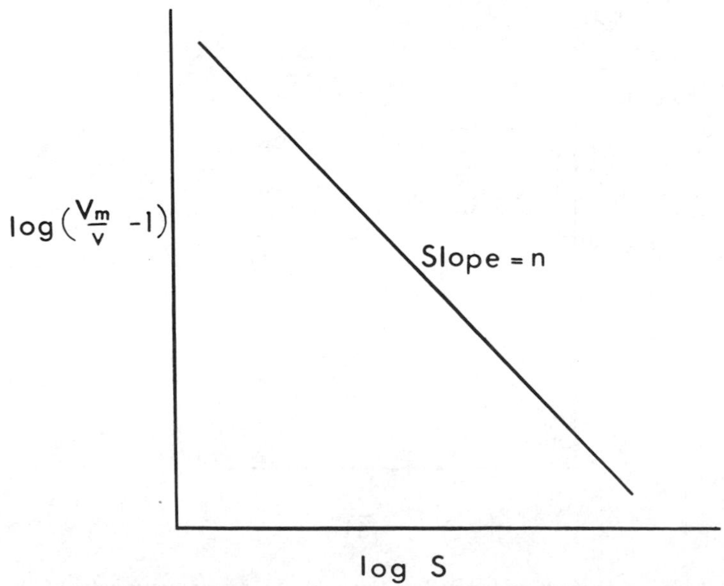

Fig. 1.8 Hill plot for the determination of the cooperativity of the interaction between substrate and enzyme.

the apparent dissociation equilibrium constant, and v and V_{max} are as previously defined. When $n > 1$ positive cooperativity is indicated, while $n < 1$ indicates negative cooperativity and $n = 1$ is equivalent to the Michaelis–Menten equation. Such comparisions indicate that ion absorption isotherms which are the result of active processes need not follow the Michaelis–Menten formalism, nor can too much significance be attached to the observation that a particular isotherm does so. Passive fluxes, particularly if carrier mediated, may under certain conditions manifest saturation kinetics and therefore produce a rectangular hyperbola and apparent Michaelis–Menten kinetics.

An interesting alternative approach to ion-uptake kinetics has been developed by Thellier (1970) who has analysed absorption kinetics, including the dual mechanisms outlined above, without invoking the concept of carrier sites. The overall uptake is envisages as

$$S_o \overset{\text{cell}}{\rightleftharpoons} S_i$$

where S is the substrate (ion) outside and inside the cell. The speed of the process v is formally equivalent to an electric intensity I and the magnitude is given by

$$\Delta\psi = 2.3A \log B \frac{[S_o]}{[S_i]} \qquad (1.45)$$

where ψ is the formal equivalent of the electrical potential difference, A equals RT/zF when electric charges are transferred and B is a constant characteristic of the thermodynamic state of the cell.

With a process which obeys Ohm's law

$$I = \Delta\psi/r \qquad (1.46$$

where r is the resistance, and then

$$v = 2.3\frac{A}{r} \log \frac{B[S_o]}{[S_i]} \qquad (1.47)$$

When the process is non-ohmic as with varistant semiconductors or with tissue when $[S_o]$ is high

$$I = \Delta\psi/r + (\lambda\Delta\psi)^{m,} \text{ where } m > 1 \qquad (1.48)$$

and

$$v = 2.3\frac{A}{r} \log B \frac{[S_o]}{[S_i]} + \left(2.3\lambda A \log B \frac{[S_o]^m}{[S_i]}\right) \qquad (1.49)$$

where λ and m are parameters characteristic of cell structures catalysing the processes.

When the above equations are applied to data on ion absorption showing the dual isotherms, an ohmic process is indicated when $[S_o]$ is low and a non-ohmic process when $[S_o]$ is high. This result implies that dual isotherms are the result of structural changes within the membrane which is seen to behave as a semiconductor, thus making it unnecessary to invoke the presence of two or more carrier systems.

1.5 Active transport – definition and description

Early workers in the field of ion transport believed that the finding of a higher concentration of ions on the inside of a biological membrane than outside in the medium was evidence of an active accumulation of those ions. However, the inadequacy of this definition and its lack of specificity eventually became apparent and it was replaced by the more rigourous definition of Ussing (1949) who defined active transport as the process by which an ion is moved against a gradient of electrochemical potential. Movement of the ion is therefore dependent on the decrease in free energy of a metabolic process. When the membrane conditions are known in sufficient detail the Ussing–Teorell flux-ratio equation (1.16) can be used to predict whether the net flux of an ion is passive or active. As the magnitude, but not the sign, of the logarithm of the flux ratio may be changed by non-independence and interaction of ions in the membrane, active transport is identified by a flux ratio whose logarithm has the wrong sign provided there is no obvious flow component to which the transport could be coupled.

Kedem (1961) has defined active transport in terms of irreversible thermodynamics as an entrainment between a transport flux and a metabolic reaction. This requires the inclusion of a reaction flux, ϕ_r, driven by its affinity, A_r, in the phenomenonological equations for the system. Active transport is then characterized by a non-zero coupling coefficient, R_{jr}, between the flow of substance j and the reaction. MacRobbie (1970) has discussed the usefulness of this definition of active transport, concluding that the difficulties of applying the thermodynamic conditions correctly imposes limits upon this approach and that the usual practical test for active transport remains that of the older Ussing definition.

A wide variety of studies has revealed that the distribution of ions across plant cell membranes is the result of both passive and active ion transport. A large proportion of these ions are accumulated in the central vacuole, which generally occupies 90 % or more of the cell volume. The vacuolar contents vary from the complex mixture of inorganic and organic solutes found in higher plants to the simple salt solution found in many algae (see Hall, 1983; Leigh, 1983).

A feature of many plant cells is the low level of Na^+ in comparison with K^+ and an active Na^+ extrusion pump has been found in those cells which have so far been studied in sufficient detail. An active K^+ influx often accompanies the Na^+ efflux, although in many cases the vacuolar K^+ is found to be near or at passive equilibrium with the level in the bathing medium. K^+ has an important function in maintaining the correct ionic environment for metabolism, and the plant appears to require a cytoplasmic concentration in the range of from 100–200 mol m^{-3} (Leigh & Wyn Jones, 1984). The situation in halophytes is not always in complete agreement with the above general picture and high Na^+/K^+ ratios are often found (see Ch. 10). Other ion species are found to be dependent upon active transport for their accumulation, an influx pump for anions and in particular Cl^- being a feature of all plant cells so far studied.

We can distinguish two types of free energy yielding processes in active transport. In 'primary active transport', energy is supplied directly by a chemical reaction involving the transport system. 'Secondary active transport', in contrast, is driven by energy stored in ionic gradients ultimately created by primary active transport; the uphill movement of one solute is coupled to and driven by the flow of a second solute down its electrochemical gradient. This process is termed 'co-transport' (or 'symport') or 'counter-transport' (or 'antiport') depending on whether the solutes move in the same direction or opposite directions. Na^+ is usually the driving ion in animal cells, whereas bacteria and plants use proton gradients.

The primary active transport process in plants is now widely believed to be the electrogenic transport of protons (see Sze, 1985; and Chs. 2 and 3). The electrochemical proton gradient or proton motive force provides the driving force for the transport of a variety of inorganic nutrients and organic assimilates, as well as its role in a range of physiological processes (Raven & Smith, 1979; Sze, 1985). Two driving mechanisms may operate to produce this proton gradient in plant cells. One depends on ATP hydrolysis catalysed by membrane-boune ATPases, and the other on the oxidation of reduced substrates by redox reactions. Although the former has been the most thoroughly investigated so far (see Ch. 3), Lin (1985) envisages that the two processes may operate in parallel to generate the necessary proton gradient across the plasmalemma. This will drive the P_i—OH^- and Cl^-/SO_4^{2-}—OH^- antiporters or Cl^-/SO_4^{2-}—H^+symporters for for anion flux, and electrostatic binding and monovalent cation transport will result from the extra negative charge generated by proton pumps (see Lin, 1985; and Ch. 3). Organic solute transport is coupled to the downhill and inward movement of protons in a secondary active transport system (e.g. see Ch. 8). However, the alternatives to this proton–symport theory (e.g. the transport of solutes fueled directly by ATP hydrolysis) should perhaps not be overlooked (Reinhold & Kaplan, 1984).

1.6 Future prospects

Over the last 15 years or so major advances have been made in our understanding of the mechanisms of solute transport across plant membranes. Much of this can be attributed to the development of techniques for the fractionation of plant cells into their subcellular components. Techniques are now available for the isolation and purification of all the major membranes and organelles of higher plant cells (see Hall & Moore, 1983). In particular, transport physiologists have been able to analyse purified vesicle preparations of plasmalemma and tonoplast for ATPase and proton-pumping activities (see Ch. 3), and to study the transport properties of isolated vacuoles (Leigh, 1983) and protoplasts (Lin, 1985). The preparation of monoclonal antibodies to plant plasma membrane antigens (Norman *et al.*, 1986) opens up a new approach to the investigation of the molecular properties of this membrane. Nevertheless, although the use of these subcellular fractions offers distinct advantages over intact tissue systems in terms of access and simplicity, caution is needed since there is evidence that their behaviour can differ somewhat from that of whole cells and tissues (e.g. West, 1980; Lefebvre & Clarkson, 1984). However, these techniques are still in their infancy and represent enormous potential for the future. In particular, the isolation and identification of transport proteins from specific membranes and their reconstitution into artificial membranes will allow their specific functions to be investigated. The production of transport mutants may also provide valuable information towards the biochemical characterization of transport mechanisms (Glass, 1983).

In relation to plant cell growth and development, reports of the occurrence of proton and Ca^{2+} pumps on other cellular membranes such as the Golgi system (e.g. Chanson & Taiz, 1985) and endoplasmic reticulum (e.g. Bush & Sze, 1986) are of great interest, as are reports of the regulation of proton-pumping ATPases by auxin (e.g. Gabathuler & Cleland, 1985) and phytochrome (e.g. Roth-Bejerano & Hall, 1986). Thus there is enormous scope for studies on the regulation of proton pumps.

Such information obtained from subcellular systems needs to be matched to a more detailed knowledge of the concentration of solutes and electric potentials in the various phases of plant cells, and the direction of movement of solutes across the membrane boundaries between these phases. There are still considerable problems in the quantification of X-ray microanalysis data of plant cells since the technical problems are very considerable (see Robards, 1985); low-temperature scanning electron microscopy is the most convenient technique although the information is, at best, semiquantitative. Yet there is a pressing need to know much more about the quantitative localization of ions in plant cells (Clarkson & Hanson, 1980).

However, a very different approach, that of patch–clamping, which involves the electrical analysis of a very small patch of membrane with only a few ion channels, promises to provide detailed information on the nature and regulation of ion channels in specific membranes (see Schroeder *et al.*, 1984; and Ch. 12). Nevertheless, the relationship between the protein carriers described by biochemists and the ion channels of the electrophysiologists needs to be resolved in the near future.

At the tissue and whole-plant level there is even more scope for innovative approaches. There is a growing awareness of the structural and functional diversity, not only between species, but between different tissues and cells in the same plant. A combined structural and physiological approach to a particular problem is perhaps still much less common than is desirable. This, together with innovative techniques, can lead to rapid advances in our understanding of a particular system, e.g. the empty seed coat technique for the study of phloem unloading (Thorne, 1985). There is much to learn of the regulation of solute transport in the whole plant. For example, Glass (1983) has questioned the ability of the proton gradient hypothesis as the driving force for cell transport to account for the regulation of the uptake of a range of diverse ions. The factors controlling transport across the root to the xylem and the movement of water and solutes between root and shoot and vice versa are still poorly understood (see Chs. 7 and 9). The role of hormones in these processes is still very speculative and awaits the development of techniques for the analysis of pure xylem and phloem saps from all parts of the plant. There is also an increasing interest in the transport of xenobiotic substances, since a greater understanding of this process could lead to a greater efficiency of crop protection chemicals (see Cronshaw *et al.*, 1986).

References

Bush, D. R. & Sze, H. (1986). Calcium transport in tonoplast and endoplasmic reticulum vesicles isolated from cultured carrot cells. *Plant Physiology* **80**, 549–55.

Chanson, A. & Taiz, L. (1985). Evidence for an ATP-dependent proton pump on the Golgi of corn coleoptiles, *Plant Physiology* **78**, 232–40.

Clarkson, D. T. (1974). *Ion Transport and Cell Structure in Plants*. London, McGrow-Hill.

Clarkson, D. T. & Hanson, J. B. (1980). The mineral nutrition of higher plants. *Annual Review of Plant Physiology* **31**, 239–98.

Cram, W. J. (1975). Storage tissues. In Baker, D. A. & Hall, J. L. (eds.). *Ion Transport in Plant Cells and Tissues*, pp. 161–91. Amsterdam, North Holland.

Cronshaw, J., Lucas, W. J. & Giaquinta, R. T. (eds.) (1986). *Phloem Transport*. New York, Alan R. Liss.

Eadie, G. S. (1952). On the evaluation of the constants V_m and K_m in enzyme kinetics. *Science* **116**, 688.

Gabathuler, R. & Cleland, R. E. (1985). Auxin regulation of a proton translocating ATPase in pea root plasma membrane vesicles. *Plant Physiology* **79**, 1080–5.

Garrahan, P. J. & Glynn, I. M. (1967). The sensitivity of the sodium pump to external sodium. *Journal of Physiology* **192**, 175–88.

Glass, A. D. M (1983). Regulation of ion transport. *Annual Review of Plant Physiology* **34**, 311–26.

Hall, J. L. (1983) Cells and their organization: current concepts. In Steward, F. C. (ed.), *Plant Physiology*, vol. 7, 3–156. New York, Academic Press.

Hall, J. L. & Moore A. L. (eds.) (1983). *Isolation of Membranes and Organelles from Plant Cells*. London, Academic Press.

Hope, A. B. (1971). *Ion Transport and Membranes – a Biological Outline*. London, Butterworths.

Kedem, O, (1961). In Kleinzeller, A & Kotyk, A. (eds.). *Membrane Transport and Metabolism, pp. 87–90. London, Academic Press.*

Koshland, D. E. (1970). The molecular basis of enzyme regulation. In Bayer, P. D. (ed.). *The Enzymes*, vol. 1, pp. 341–96. London, Academic Press.

Lefebvre, D. D. & Clarkson, D. T. (1984). Characterization of orthophosphate absorptiona by pea root protoplasts. *Journal of Experimental Botany* **35**, 1265–76.

Leigh, R. A. (1983). Methods, progress and potential for the use of isolated vacuoles in studies of solute transport in higher plant cells. *Physiologia Plantarum* **57**, 390–6.

Leigh, R. A. & Wyn Jones, R. G. (1984). A hypothesis relating critical potassium concentrations for growth to the distribution and functions of this ion in the plant cell. *New Phytologist* **97**, 1–13.

Lin, W. (1985). Energetics of membrane transport in protoplasts. *Physiologia Plantarum* **65**, 102–8.

Lineweaver, H. & Burk, D. (1934). The determination of enzyme dissociation constants. *Journal of the American Chemical Society* **56**, 658–66.

MacRobbie, E. A. C. (1966). Metabolic effects on ion fluxes in *Nitella translucens* I. Active influxes. *Australian Journal of Biological Sciences* **19**, 363–70.

MacRobbie, E. A. C. (1970). The active transport of ions in plant cells. *Quarterly Review of Biophysics* **3**, 251–94.

Norman, P. M., Wingate, V. P. M., Fitter, M. S. & Lamb, C. J. (1986). Monoclonal antibodies to plant plasma membrane antigens. *Planta* **167**, 452–9.

Raven, J. A. & Smith, F. A. (1979), Intracellular pH and its regulation. *Annual Review of Plant Physiology* **30**, 289–311.

Reinhold, L. & Kaplan, A. (1984). Membrane transport of sugars and amino acids. *Annual Review of Plant Physiology* **35**, 45–83.

Robards, A. W. (1985). The use of low temperature methods of structural and analytical studies of plant transport processes. In Robards, A. W. (ed.). *Botanical Microscopy*. Oxford University Press.

Roth-Bejerano, N. & Hall, J. L. (1986). Photoregulation of proton extrusion by cucumber hypocotyls and the role of ATPase activity. *Journal of Plant Physiology* **122**, 329–36.

Schroeder, J. I. Hedrick, R. & Fernandez, J. M. (1984). Potassium-selective single channels in guard cell protoplasts of *Vicia faba*. *Nature* **312**, 361–2.

Spanswick, R. M. (1981). Electrogenic ion pumps. *Annual Review of Plant Physiology* **32**. 267–89.

Sutcliffe, J. R. & Baker, D. A. (1974). *Plants and Mineral Salts*. London, Arnold.

Sze, H. (1985). H^+-translocating ATPases: advances using membrane vesicles. *Annual Review of Plant Physiology* **36**, 175–208.

Teorell, T. (1949). Membrane electrophoresis in relation to bioelectrical polarization effects. *Archives des Sciences Physiologiques* **3**, 205–19.

Thellier, M. (1970). An electrokinetic interpretation of the functioning of biological systems and its application to the study of mineral salts absorption. *Annals of Botany* **34**, 983–1009.

Thorne, J. H. (1985). Phloem unloading of C and N assimilates in developing seeds. *Annual Review of Plant Physiology* **36**, 317–43.

Ussing, H. H. (1949). The distinction by means of tracers between active transport and diffusion. *Acta Physiologica Scandinavica* **19**, 43–56.

West, I. C. (1980). Energy coupling in secondary active transport. *Biochimica et Biophysica Acta* **604**, 91–126.

2 Mitochondria and chloroplasts

J. N. Prebble

2.1 Introduction

Both mitochondria and chloroplasts are widely thought to have evolved from prokaryotes which at an early point in geological time invaded other living cells setting up a symbiotic relationship with their hosts. That symbiotic relationship expresses itself in the metabolic interdependence of the cytosol and organelle (John & Whatley, 1975; Whatley *et al.*, 1979; Whatley & Whatley, 1981). Both organelles show two types of interrelated transport processes. First there are the bioenergetic systems involving proton translocation. Secondly there are systems which relate the internal metabolism of the organelle with that in the cytosol through the specific transport of metabolites across external membranes, the inner mitochondrial membrane and the inner membrane of the chloroplast envelope. In addition and as a result perhaps of the evolution of the particle, while some of the organelle's proteins are synthesized within, the majority are synthesized in the cytosol and must be imported across membranes.

2.2 Organelle structure

2.2.1 Mitochondrial structure

The mitochondrion consists of two membranes, the inner and outer which together define two spaces, the intermembrane space and the matrix space (Fig. 2.1a). The cristae, so obvious in electron micrographs, are generally regarded as part of the inner membrane formed by invagination and the cristal space therefore becomes part of the intermembrane space. A series of experiments in the 1960s measuring the extent to which sucrose permeates the organelle led to the conclusion that the mitochondrion

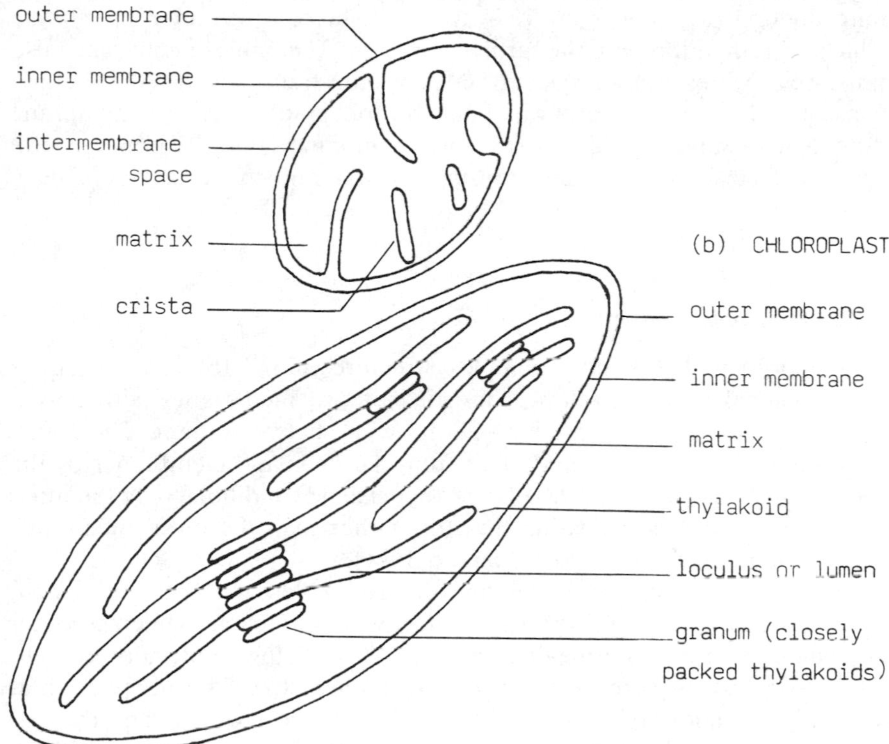

Fig. 2.1 (a) Structure of the mitochondrion. (b) Structure of the chloroplast.

possessed sucrose-permeable space, equivalent to the intermembrane space and sucrose impermeable space equivalent to the matrix (Klingenberg & Pfaff, 1966). Further work on permeation showed that the outer membrane was permeable up to a molecular weight limit of 5000 to 10 000 whereas the inner membrane behaved as a classical selectively permeable membrane. Proteins responsible for this non-specific permeability have been isolated from mitochondrial outer membranes (e.g. rat liver and *Neurospora*). When purified and inserted into phospholipid vesicles they render the vesicle non-specifically permeable to small molecules. The proteins (M_r about 30 000) are similar to those found in bacterial outer membranes and have been termed 'porins' (Zalman *et al.*, 1980; Freitag *et al.*, 1982a). A recently identified porin (M_r 35 000) from pig heart binds *N*, *N*-dicyclohexylcarbodiimide (DCCD), an inhibitor of proton-translocating ATPases (De Pinto *et al.*, 1985).

Thus the intermembrane space is freely accessible to substrates in the cytosol (but not proteins) whereas in general the matrix is only readily accessible to those hydrophilic cytosolic components for which there is an appropriate transport system, or to hydrophobic substances. So far as mitochondria are concerned it is transport across the inner membrane which will therefore be the prime concern. The inner membrane also possesses the respiratory system and a proton-translocating ATPase and consequently *in vivo* there will be a proton gradient and a membrane potential, positive outside, across the membrane. For a recent general review of structure and metabolism in plant mitochondria, see Douce (1985).

2.2.2 Chloroplast structure

The chloroplast has a more complex structure (Fig. 2.1b). The organelle is surrounded by an envelope consisting of two membranes with only a narrow space, of the order of 10–20 nm wide, between them. The envelope may be isolated apparently free from the thylakoid membranes by the method of Douce *et al.* (1973). Under hypertonic conditions only the inner membrane contracts indicating that this rather than the outer membrane is the osmotic barrier (Heldt & Sauer, 1971).

In general, permeation of the envelope resembles that of the mitochondrion since the inner membrane is selectively permeable while the outer is apparently freely permeable to substrates of low molecular weight. These conclusions were based in part on studies of sucrose permeation as with mitochondria (Heldt & Rapley, 1970a). More recently the envelope membranes have been separated and characterized (Cline *et al.*, 1981). There is also a pH gradient across the envelope, the exterior being at a lower pH than the interior although the mechanism for this is quite different to that of the mitochondrion.

Plant chloroplasts possess within the envelope a series of flattened sacs, thylakoids, which are themselves osmotically sensitive. The thylakoid membranes possess the photochemical pigments, the associated electron transport chain and the ATPase. In the light,the membrane pumps protons into the interior lumen (loculus) of the thylakoid from the matrix space (or stroma) of the chloroplast creating a potential and pH gradient across the thylakoid membrane. For a review of chloroplast structure, see Coombs & Greenwood (1976).

2.3 Development of ideas on permeation

Little was known of mitochondrial metabolite transport systems before the late 1960s and in chloroplasts our knowledge dates from the early 1970s. In mitochondria, early studies were carried out almost exclusively with animal mitochondria and the findings in this field provided the basis for work on plant mitochondria and stimulated an examination of such processes in chloroplasts. The preparation of relatively pure plant mitochondria has proved more difficult than in animal tissues and this applied particularly to leaves. However, in recent years several methods which give intact preparations have been described (see, for example, Hrubec *et al.*, 1985).

In discussing the transport of substances across membranes several factors need to be borne in mind. Experimentally it is important to distinguish between transport across the membrane and adsorption on the membrane surface. The latter may be influenced by conformational changes in membrane proteins which may mask or expose reactive groups. This effect has featured strongly in the discussion of the mechanism of oxidative phosphorylation (for example, see Boyer, 1975). Where a specific ion is under consideration, it is necessary to determine whether it crosses the membrane in the charged or uncharged state. For example ammonia, present as the ammonium ion, NH_4^+ in solution will migrate across the membrane as NH_3 thus giving rise to a proton gradient (Fig. 2.2). Similar considerations apply to carboxylic acids.

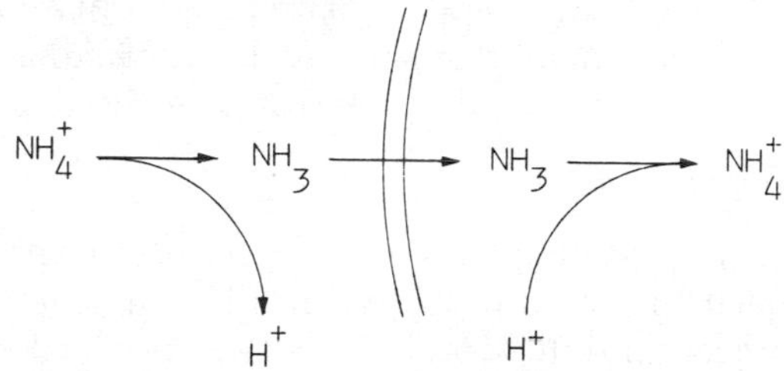

Fig. 2.2 Creation of a proton gradient by ammonia.

The movement of a charged ion across a membrane will create a potential across that membrane. For example, movement of a positively charged ion into an organelle will create a potential positive on the inside, negative on the outside. This creates a condition promoting the uptake of anions or the loss of cations, such ion movements tending to dissipate the potential. Where there is an electron-transport system, such as the respiratory chain in the mitochondrial inner membrane or the electron-transport chain in the thylakoid membrane of the chloroplast, there will be a potential and proton gradient across the membrane. The ATPase (ATP synthase) in such membranes will affect the magnitude of the proton gradient and membrane potential through its ability to dissipate the gradient and potential by synthesizing ATP or to increase it by hydrolysing ATP. However, in the intact system where electron transport normally contributes to the potential and it is dissipated by driving ATP synthesis, there will be a permanent membrane potential and proton gradient influenced by the ATP/ADP ratio and the phosphate concentration. This will therefore promote transport processes associated with a transmembrane flux of protons or other cations and, in the opposite direction of OH^- or other anions. The energy associated with such a gradient, variously referred to as the 'proton motive force', Δp, is given by $\Delta p = \Delta\psi - 2.3\ RT/F.\ \Delta pH$, where $\Delta\psi$ is the membrane potential. This energy is available for ATP synthesis according to the chemiosmotic theory (which in this chapter is assumed to be the mechanism of phosphorylation), 'reverse electron transport' and for driving the transport of metabolites and ions across membranes. The presence of such a gradient assumes that the membrane is impermeable to protons, an assumption questioned earlier for plant mitochondria though more recently verified (Moore & Wilson, 1978).

2.4 Proton-pumping ATPases (ATP synthases)

2.4.1 General properties of ATPases

Proton-pumping ATPases are found in the bacterial plasma membrane, in the mitochondrial inner membrane and in the thylakoid membrane and normally require Mg^{2+} for activity. They catalyse the reaction:

$$ADP + P_i = ATP + H_2O$$

This is, however, coupled obligatorily to the transfer of protons across the membrane in which the enzyme is located. The stoichiometry of proton translocation is uncertain but is probably either two or three protons per ATP synthesized or hydrolysed.

The first enzyme to be studied was that in mammalian mitochondria. The enzyme could be seen as a stalked sphere attached to the inside of the inner mitochondrial membrane when preparations were treated with ammonium molybdate for negative staining. The enzyme can be divided into an integral hydrophobic protein complex tightly bound into the structure of the membrane, F_0, and a peripheral hydrophilic protein, F_1, bound to F_0 through the 'stalk' sector. F_1, which can readily be removed by washing at high ionic concentration, has ATPase activity. The tripartite structure can be isolated using detergent (Soper *et al.*, 1979). F_1 is composed of several polypeptides, the major components being α, β, γ, δ and ε. The composition is probably α_3, β_3, γ, δ, ε. The catalytic site is on the β-subunit but may also involve the α-subunit. The γ-subunit has been suggested to have a role in proton translocation (Moroney & McCarty, 1982). F_0 is composed of at least three polypeptides (a, b and c) but the composition may not be the same for all ATPases. The F_1

portion of the enzyme is able to bind up to six molecules of adenine nucleotide of which three are firmly bound.

A potent inhibitor of membrane-bound ATPases, DCCD, binds covalently to subunit c which is a proteolipid. DCCD appears to inhibit the ATPase by blocking proton transport through F_0. Thus F_0 can be shown to render membranes permeable to protons. The addition of either F_1 or DCCD will substantially reduce proton conductance of the membrane.

The genes coding for the ATPase in *E. coli* have been identified and mapped. They form an operon mapping at 83 min as shown in Fig. 2.3. The amino acid sequences have been determined (see Futai & Kanazawa, 1983; Gibson, 1983; Senior & Wise, 1983, Hoppe & Sebald 1984, for reviews).

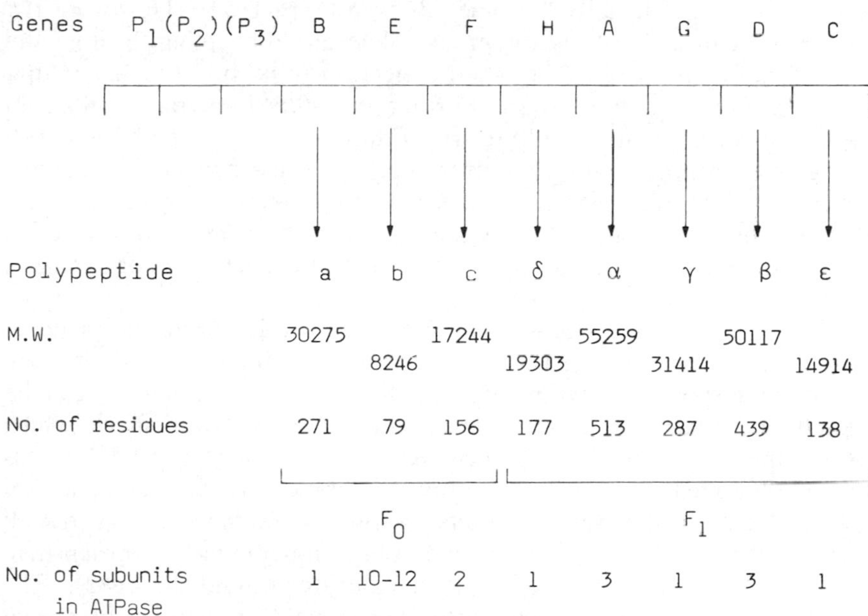

Fig. 2.3 The *unc* (*atp*) operon in *Escherichia coli*.

The chemiosmotic theory (Mitchell, 1961) explains the synthesis of ATP in association with electron transport in terms of Δp; that is, electron transport pumps protons across a membrane creating Δp and the protons return through the ATPase dissipating Δp. This relationship between electron transport and the ATPase (ATP synthase) is reversible. The apparent transfer of protons across mitochondrial and chloroplast membranes can be readily observed and the chemiosmotic theory, for which Peter Mitchell received a Nobel prize in 1978, is widely accepted. However, the scientific community has never regarded the award of a Nobel prize as conferring infallibility and there remains a scepticism about the true mechanism of oxidative phosphorylation. This is currently expressed in terms of localized or delocalized coupling between electron transport and ATP synthesis (for a review, see Ferguson, 1985). The transfer of protons to create a Δp which drives ATP synthesis is clearly delocalized since there is no requirement for the ATPase to be structurally linked with the electron-transport-driven proton pump. Other views of the system see a close association between the ATPase and the energy-conserving oxidations as an essential part of the mechanism. An ingenious hypothesis which relates the proton gradient to both the ATPase and to the oxidation–reduction reactions but in which energy is transferred directly from the oxidation–reduction system to the ATPase (without involving Δp) has been put forward by Slater *et al.* (1985). As the authors point out it bears some relationship to the chemical theory originally formulated by Slater (1953). However, the hypothesis involves localized rather than delocalized coupling. Neither the mechanism of proton pumping by the electron transport system nor that of the ATPase are understood. Further the stoichiometric relationship between protons pumped and electron transport or ATP synthesis is still not agreed. Such a situation therefore leaves the mechanism of oxidative phosphorylation still open to debate. In this chapter, however, we will concern ourselves solely with systems which transfer protons across membranes leaving aside the wider question of the mechanism of oxidative phosphorylation.

There is substantial evidence that the ATPase will translocate protons. Thus, F_0 from *E. coli*, when incorporated into liposomes, will translocate protons in response to a membrane potential. Such a potential can be created by the use of a K^+ gradient and the ionophore valinomycin which renders the membrane permeable to K^+; the migration of the K^+ across the liposome membrane will set up the membrane potential. The addition of F_1 to the liposome system inhibits the proton conductance but if ATP is present, the ATPase appears to act as a proton-pumping system sensitive to both DCCD and uncouplers (Schneider & Altendorf, 1982). The ATPase complex, if incorporated into phospholipid vesicles can be shown to translocate protons; moreover it will synthesize ATP in response to an applied Δp. The synthesis of ATP by the ATPase complex from thermophilic bacteria when reconstituted into phosphoplipid vesicles has been demonstrated when Δp was applied using K^+/valinomycin and an acid wash. This system would also generate Δp when hydrolysing ATP (Sone *et al.*, 1977). The rate of ATP synthesis is, however, rather low (see Casey, 1984 for a discussion). Similar criticisms apply to reconstituted systems involving bacteriorhodopsin (which creates Δp in the light) and the ATPase from chloroplasts (Winget *et al.*, 1977) or yeast mitochondria (Ryrie *et al.*, 1979).

Two basic approaches to the question of mechanism have been taken.

Experiments of Boyer *et al.* (1973) with submitochondrial particles suggested that ^{18}O in H_2O would exchange with that in p_i but that a substantial amount of this exchange although sensitive to oligomycin, a specific inhibitor of the ATPase, was not sensitive to uncouplers (unlike the P_i/ATP exchange for example). Boyer concluded that the formation of enzyme-bound ATP from P_i and ADP was not energy dependent but that the energy-requiring step for oxidative phosphorylation was the release of ATP (see Gresser *et al.*, 1982). In support of this view the isolated and membrane-bound F_1 from chloroplasts can be shown to synthesize bound ATP (Feldman & Sigman, 1982, 1983). The ATPase can be demonstrated to undergo a conformational change as shown for example by the change in binding and fluorescence of aurovertin on treatment with ATP (Issartel *et al.*, 1983). This has led to a cooperative view of the enzymc in which three equivalent catalytic sites undergo a sequential binding change mechanism. Thus while one site converts tightly bound P_i and ADP to ATP, the remaining sites possess loosely bound reactants. The binding of ATP at one site promotes the release of ADP + P_i from another (O'Neal & Boyer, 1984).

A second basic approach stems from the observations on proton transport noted above. Mitchell (1974) suggested that the function of the protons would be either the protonation of the phosphorus centre on the low pH side or protonation of an enzyme-bound oxide ion, e.g.

$$2H^+_{(out)} + \left[P(O^-)_3(=O)\right]_{(in)} + {}^-OADP_{(in)} \rightleftharpoons H^+ + H{-}O{-}P(O^-)_2(=O) + {}^-OADP$$

$$\rightleftharpoons H_2O_{(out)} + \left[(O^-)_2P(=O){-}OADP\right]_{(in)}$$

Thus the synthesis of ATP would be directly driven by the proton gradient. An alternative view of the role of proton movements is to see them as inducing a conformational change in the ATPase rather than involved directly in catalysis (Boyer, 1975; Rosen *et al.*, 1979). Recently Mitchell (1985) has argued for a model (the rolling well and turnstile hypothesis) in which the protons function both in the direct catalysis and in conveying the reactants to and from the active site. The protons also induce rotations of the subunits of the ATPase as a part of the overall catalytic process. Thus the rotation of the α- and β-subunits results in binding and release of reactants in sequence.

2.4.2 The mitochondrial ATPase

The mammalian and yeast enzymes possess in addition to the five subunits found in the bacterial enzyme an inhibitor protein IF_1 (M_r 9578 in bovine heart) which is very sensitive to attack by trypsin (Frangione *et al.*, 1981). This protein, present at one IF_1 per F_1 appears to bind to β-subunits and inhibit ATP hydrolysis and also the initial rate of ATP synthesis although not steady-state synthesis (Schwerzmann & Pedersen, 1981). Two other polypeptides found in the mitochondrial ATPase and thought to constitute the 'stalk' are the oligomycin sensitivity-conferring protein (OSCP) and coupling factor F_6 (M_r 20 967 and approximately 8000, respectively, from beef). A further protein, F_B, is probably also involved (see Hatefi, 1985, for a summary). The composition of F_0 is uncertain.

Plant mitochodrial ATPase (M_r 380 000) has been isolated from several sources including pea (Whisson & Spencer, 1983) and maize. The maize enzyme possesses five subunits of molecular weights 58 000 (α), 56 000 (β) 35 000 (γ), 22 000 (δ) and 8000–12 000 (ε) (Hack & Leaver, 1983; Spitsberg *et al.*, 1985) while six have been observed in pea (Horak & Packer, 1985). The existence of a trypsin-sensitive inhibitory protein has been demonstrated in potato where this protein seems firmly bound (Jung & Laties, 1976). While little is known about F_0 in plant mitochondria, Hack & Leaver (1984) have identified the DCCD-binding proteolipid (M_r 6000). So far, studies of plant mitochondrial ATPases have contributed little to our knowledge of this enzyme complex. The enzyme is, however, structurally distinct from that in chloroplasts although in peas, the mitochondrial F_1 -ATPase will function on CF_1 depleted chloroplast membranes. A particularly interesting characteristic is the effect of Ca^{2+} which depresses ATPase activity. The phytochrome system appears to regulate Ca^{2+} release from mitochondria (see Sect. 2.6.3) and the activity of the enzyme has been shown to be affected by red and far-red light (Serlin *et al.*, 1984; Horak & Packer, 1985; Spitsberg *et al.*, 1985)

2.4.3 The chloroplast ATPase

This enzyme has a similar structure to those already discussed (α_3, β_3, γ, δ, ε). The CF_0-CF_1 complex is bound in the thylakoid membrane with CF_1 on the matrix side. The H^+/ATP ratio has been estimated at 1.9 (see Lemaire *et al.*, 1985). The polypeptides of CF_1 which has a molecular weight of about 400 000 have been separated and have molecular weights of 59 000 (α), 56 000 (β), 37 000 (γ), 17 500 (δ) and 13 000 (ε) (Moroney *et al.*, 1983). The genes for α-, β-and ε-subunits are located on the chloroplast genome while γ and β-subunits are encoded in nuclear genes (Nelson *et al.*, 1980; Nechushtai *et al.*, 1981). CF_0 is composed of at least three subunits (M_r 15 000, 12 500 and 8000) but the composition is uncertain. However, a DCCD-binding proteolipid (M_r 8000) has been identified.

This proteolipid if reconstituted into liposomes renders them permeable to protons (Sigrist-Nelson & Azzi, 1980).

The chloroplast ATPase is a latent enzyme but is activated in the light. The inhibitor of the ATPase appears to be the ε-subunit (Richter *et al.*, 1984, 1985). Activation of ATPase activity in the isolated CF_1 may also be achieved by treatment with trypsin, heat, dithiothreitol, detergent or alcohols. Activation by light appears to be brought about by energizing the membrane (the formation of $\triangle$pH) which in turn results in loss of bound nucleotides and a change in conformation (Schlodder & Witt, 1981). Reduction of a thiol group possibly by thioredoxin is also involved (Shahak, 1985).

The envelope ATPase has been isolated. Its activity is stimulated by Ca^{2+}, Mg^{2+} and calmodulin. It has a molecular weight of 260 000 and is probably composed of four identical subunits each of 65 000 molecular weight. Whether it is involved in proton transport is not clear (see Douce *et al.*, 1973; Nguyen & Siegenthaler 1985).

2.5 Proton transport by electron-transport systems

2.5.1 Mitochondrial respiratory chain

The respiratory chain, located in the inner mitochondrial membrane, may be considered as a group of lipoprotein complexes. These complexes have been isolated in a functional state from animal mitochondria, particularly beef heart and their properties have been described in detail (for a review of the mammalian respiratory chain, see Hatefi, 1985). However, in each case there is still considerable uncertainty about the mechanisms of oxidation–reduction involved. Of the four principal complexes in mammalian mitochondria, it is now agreed that three of them pump protons although the stoichiometry of the process is a matter for debate. The H^+/e ratio for each complex lies between 1 and 2. The succinate dehydrogenase complex (complex II) is inactive in this respect. The mechanism for proton transport is also unclear but a simple approach to this problem is shown in Fig. 2.4 in which the electron-transport chain is folded in a loop with oxidation-reduction involving both electrons and protons in one direction and electrons only in the other. While such a system might in principle operate in complex I (NADH dehydrogenase) it is certainly too simple a view for complexes III and IV.

The outline of the respiratory chain found in plant mitochondria is shown in Fig. 2.5. It should be noted that the succinate dehyrogenase, the external NADH dehydrogenase (with a catalytic site for NADH located on the outside of the inner membrane) and the cyanide-insensitive oxidase complex do not translocate protons and will therefore not be considered further.

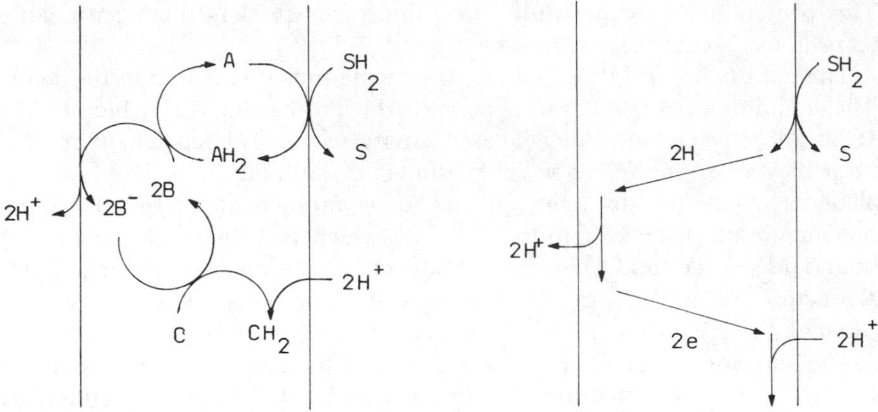

Fig. 2.4 Proton translocation brought about by a sequence of transmembrane oxidation-reduction reactions. The substrate (SH_2) is oxidized by hydrogen carriers (A) which are themselves oxidized on the opposite side of the membrane by electron carriers (B). This type of 'loop' mechanism is schematically summarized in the right-hand diagram.

Complex I, the NADH–ubiquinone oxidoreductase, in mammals is composed of some twenty-six different polypeptides as well as phospholipids (Heron *et al.*, 1979).It possesses FMN as its prosthetic group together with several (between four and six) iron–sulphur (Fe–S) centres. Estimates of proton translocation by the complex vary but may be up to a value of $H^+/e = 2.0$ (Wikstrom, 1984). The isolated complex can be incorporated into phospholipid vesicles and will translocate protons in association with the oxidation of NADH and reduction of ubiquinone (Ragan & Hinkle, 1975). The mechanism for translocation may be a simple redox loop as proposed by Mitchell or a mechanism based on protonation and deprotonation of proteins during this oxidation–reduction cycle, such a process being linked to a change in orientation of the H^+-binding site (see Wikstrom & Krab, 1979; Malmström, 1985).

The plant enzyme has been little studied except for the Fe–S centres which appear to be similar to those in the mammalian dehydrogenase (Cammack & Palmer, 1973).

The ubiquinone–cytochrome *c* oxidoreductase (complex III) of mammals is composed of phospholipid, and from eight to ten different polypeptides binding a number of prosthetic groups including two *b* cytochromes and cytochrome c_1. The two cytochromes, b_{562}, b_{566}, appear to be chemically identical although functionally different. There is also a single Fe–S centre. The complex does not behave as a linear oxidation–reduction system since under some conditions it is possible to demonstrate the oxidation of cytochrome c_1 coupled to the reduction of one of the two *b* cytochromes. Two similar mechanisms to accommodate this observation

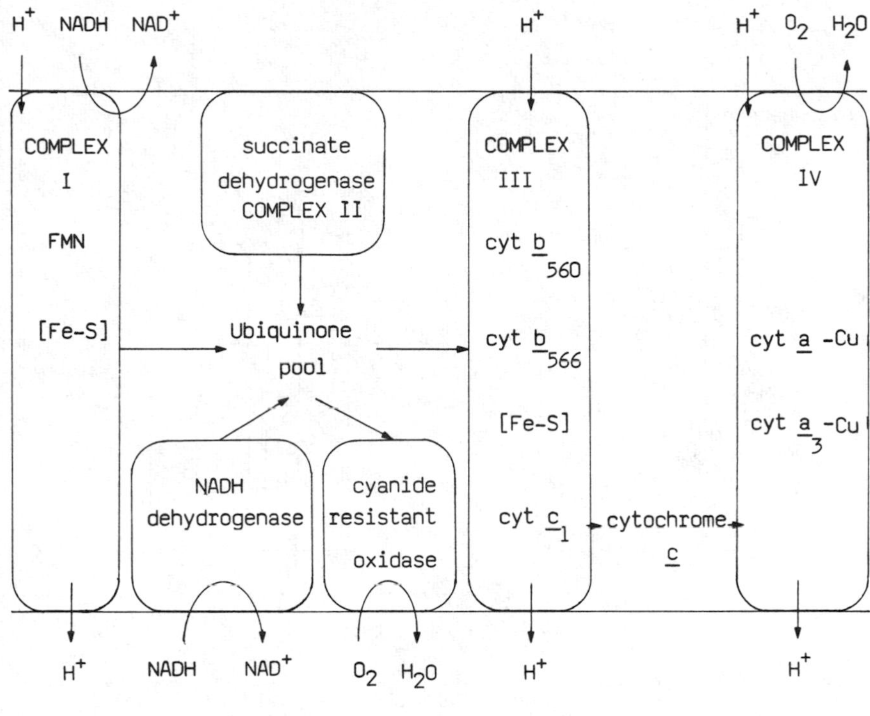

Fig. 2.5 Mitochondrial respiration chain : proton translocation. Protons are pumped by complexes I, III and IV. The number of protons translocated are not shown. Although the externally orientated NADH dehydrogenase is shown here reducing the ubiquinone pool it may reduce complex III directly.

and to account for the ability of the complex to translocate protons have been proposed (Fig. 2.6). In essence the reduced ubiquinone donates one electron to the Fe–S centre and one to the *b* cytochrome complex. The arrangement of the carriers across the membrane results in the transport of hydrogens in one direction and electrons in the other thus providing for a proton pump. The observed H^+/e ratio is 2, consistent with the scheme in Fig. 2.6.

In plants the complex has been little investigated. An Fe–S centre has been detected by Bonner & Prince (1984). The *b* cytochromes are less fully understood but cytochromes b_{560} and b_{566} appear to be the key cytochromes of the complex together with cytochrome c_1 (see Douce, 1985, for a discussion of this problem). A recent partial purification of the complex from potato suggests that many of its properties may be similar to those of the mammalian enzyme (Esposti *et al.*, 1985).

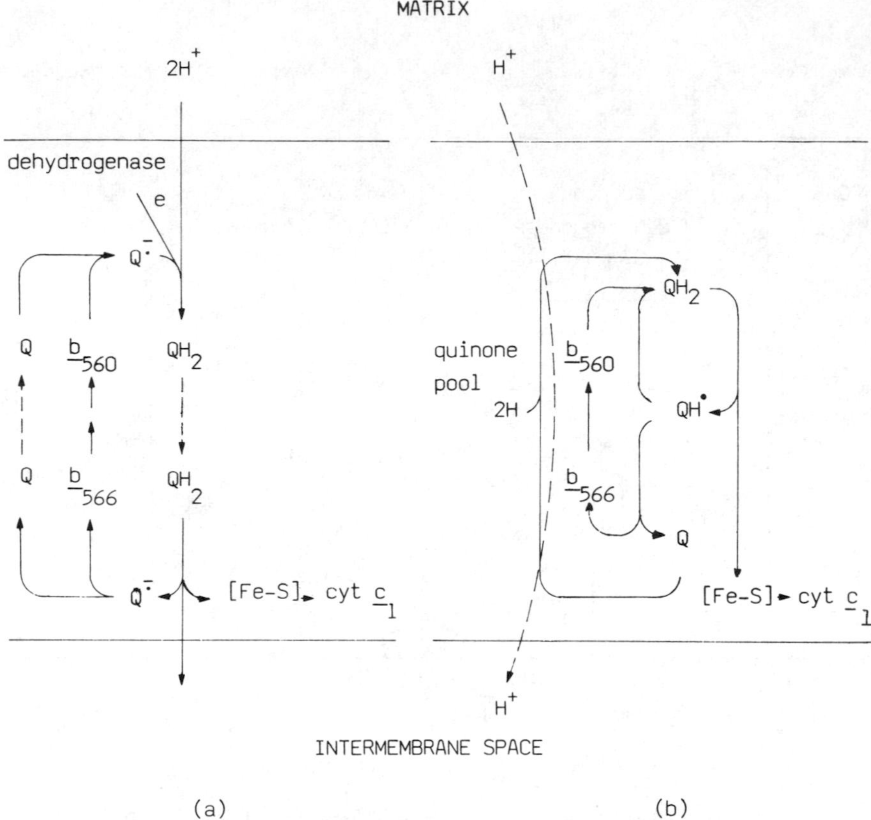

Fig. 2.6 Two suggested cycles for the operation of complex III: (a) The Q-cycle. (b) The *b*-cycle. Q represents the ubiquinone pool in the Q-cycle and protein-bound quinone in the *b*-cycle. $Q^{\dot{-}}$ represents the ubisemiquinone anion and QH˙ ubisemiquinone. Antimycin in either scheme blocks the oxidation of cytochrome b_{560}. Note that in the Q-cycle the quinone migrates across the membrane whereas in the *b*-cycle it is protein-bound. In the Q-cycle proton translocation is directly associated with the reduction and oxidation of the ubiquinone. In the *b*-cycle it is assumed that proton translocation is coupled to the redox reactions of the *b*-cytochromes.

Cytochrome oxidase (complex IV) in beef heart is composed of at least eight different polypeptides and may be present as a dimer. The complex possesses two cytochrome *a*–Cu pairs which are functionally different. Incorporation of the isolated complex into phospholipid vesicles results in proton pumping coupled to the oxidation of cytochrome *c*. In mitochondria it can be shown that $H^+/e = 1$ for cytochrome *c* oxidation by oxygen (Wikstrom, 1977). The mechanism for such pumping is unclear. At first sight there are no hydrogen carriers which might act in a redox loop system. However, Mitchell *et al.* (1985) have suggested that bound reduced oxygen could act as a proton carrier either in a redox loop or redox cycle mechanism. The redox loop shown in Fig. 2.7 gives two separate roles to oxygen; firstly a proton carrier associated with a cytochrome *a*–Cu pair and secondly as a substrate for reduction to water. Alternatively a form of proton pump like that already referred to in Complex I may be involved. Interestingly, DCCD, an inhibitor of proton translocation in the ATPase–F_0 complex, also blocks the proton pump of the oxidase. Again there have been relatively few studies of the cytochrome oxidase from higher plants but the complex has been purified and possesses at least five different polypeptides (Maeshima & Asahi, 1978; Matsuoka, *et al.*, 1981). Such results would suggest that the plant enzyme has a simpler structure than the animal one but studies of these complexes have not yet reached an advanced state.

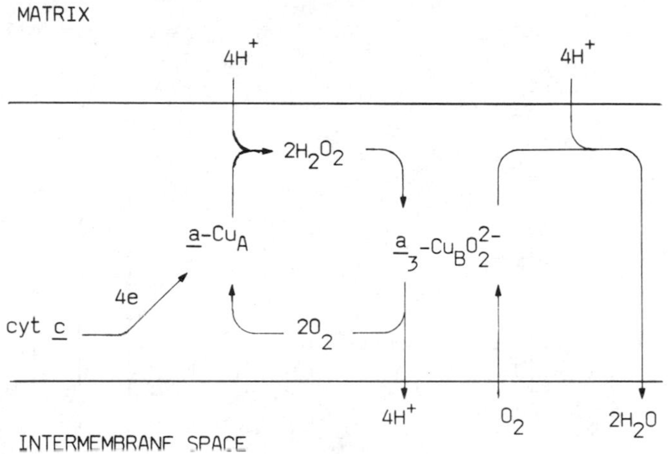

Fig. 2.7 A hypothetical scheme for the translocation of protons in complex IV. The cytochrome *a*–Cu_A pair is reduced by electrons from cytochrome *c*. The transfer of reducing equivalents from *a*–Cu_A to a_3–Cu_B involves hydrogens associated with oxygen (H_2O_2) and results in proton translocation. (For details of the scheme, see Mitchell *et al.*, 1985.)

Attempts to measure the H^+/O ratio in plants have not produced completely satisfactory results but Moreau & Davy de Virville (1985) have produced data which suggest a value of 6 for the intramitochondrial oxidation of malate. This value is at the lower end of the range of values obtained for animal mitochondria where an NAD-reducing substrate is used. There is therefore at present no reason to suppose that proton pumping by plant mitochondria will prove to have a different stoichiometry to that found in animals.

Estimates of Δp in plant mitochondria have varied but overall have given values of up to about 260 mV. A substantial proportion of this is provided by $\Delta\psi$ which has been estimated at about 220–50 mV as compared to estimates of $\Delta\psi$ in animal mitochondria of 150–80 mV (Ducet *et al.*, 1983; Mandolino *et al.*, 1983; Diolez & Moreau, 1985). Hence the value of ΔpH must be small and in the studies of Ducet and coworkers was not greater than 0.2.

2.5.2 Chloroplast electron transport chain

A diagram of the electron transport chain is shown in Fig. 2.8. Essentially the chain consists of three complexes, photosystem II (PSII), cytochrome b_6–f complex and photosystem I (PSI). The properties of the complexes have been reviewed by Haehnel (1984) and by Crofts & Wraight (1983).

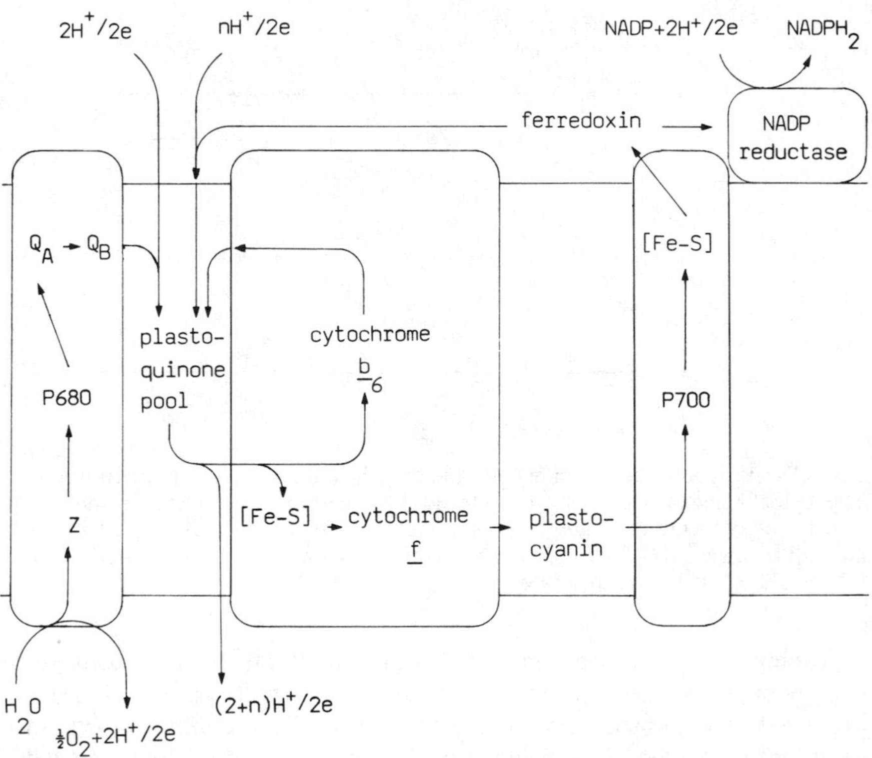

Fig. 2.8 Outline of the electron transport chain in the thylakoid. There is a linear electron transfer pathway between water and NADP+ which is coupled to proton translocation through the plastoquinone pool. There is the possibility of a cyclical pathway through the cytochrome *b*-cytochrome *f* complex similar to that shown in Fig. 2.6. Finally there is the cyclical pathway involving ferredoxin which is capable of reducing the plastoquinone pool with associated proton translocation.

The action of PSII is concerned with the oxidation of water to generate molecular oxygen, protons which are released into the thylakoid loculus and electrons which are involved in a photochemical reduction of plastoquinone. The plastoquinone reactions are complex involving a plastoquinone–Fe^{2+} complex (Q_A) and a secondary acceptor Q_B, also a plastoquinone. The reduction of Q_B involves the uptake of two protons from the matrix generating Q_BH_2 which is then released to become part of the plastoquinone pool. Thus the operation of PSII results in the uptake of protons from the matrix and release into the loculus with an expected $H^+/e = 1$.

The cytochrome b_6–f complex is non-photochemical but is of a type already seen in the mitochondrial electron transport chain (complex III) and also found in photosynthetic bacteria. The complex comprises a *c*-type cytochrome, cytochrome *f*, two cytochrome b_6 haems with different mid-point potentials and a high potential Fe–S centre together with plastoquinone. The oxidation of the plastoquinol by the complex results in the release of protons into the loculus. The complex may be capable of carrying out a Q-cycle similar to that seen in complex III above, and would therefore be capable of additional proton translocation.

The third complex, PSI, involves the oxidation of the b_6–f complex and the photochemical reduction of ferredoxin, itself an Fe–S protein. Ferredoxin may take part in both a cyclic electron flow by reducing the plastoquinone pool and in reducing $NADP^+$, a reaction involving withdrawal of protons from the matrix. It should be noted that the cyclic flow of electrons leads to further proton pumping in association with the b_6–f complex. Satisfactory measurements of proton translocation by the chloroplast electron transport chain have proved difficult. The value for H^+/e is probably 2. However, attempts to block the full functioning of the Q-cycle in the b_6–f complex do not produce consistent results (Crofts & Wraight, 1983). The proton motive force in thylakoids at steady state appears to be mainly due to ΔpH and the value of $\Delta\psi$ is low (Ferguson & Sorgato, 1982). Of particular significance is the failure to demonstrate a satisfactory relationship between the kinetics of those reactions which are proposed to generate a proton gradient and the disappearance of protons from the bulk phase. This suggests that there may be a complex relationship between the protolytic reaction and the binding of protons involving a pathway between the reaction site and the bulk phase (Crofts & Wraight, 1983). Indeed, the existence of a 'bulk phase' inside the native thylakoid has itself been questioned by Hong & Junge (1983).

Other ions known to cross the thylakoid membrane include Mg^{2+} and Cl^-. The former has been regarded as having a major role in regulating enzyme activity in the matrix. Cl^- is seen as having a role in the water-splitting reaction of PSII; its retention within the thylakoid appears to be linked to ΔpH (Theg & Homann, 1982).

2.6 Transport across the mitochondrial inner membrane

As noted earlier, the inner mitochondrial membrane is the permeability barrier. Plant mitochondria can be shown to oxidize a wide variety of substrates including pyruvate, citrate, α-ketoglutarate, succinate, malate, glutamate, proline and in C_3 plants, glycine. The major substrates for the citric acid cycle located in the mitochondrial matrix are pyruvate and malate (see Fig. 2.9). Plant mitochondria are distinguished from those in mammals by the ability rapidly to metabolize malate. This is mainly due to the possession of an NAD-dependent malate enzyme (forming pyruvate) in addition to the malate dehydrogenase which forms oxaloacetate and which is the main route in the mammalian system. The malate dehydrogenase reaction has an equilibrium strongly in favour of malate and hence metabolism by this route is dependent on the conversion of oxaloacetate (OAA) to citrate. Here again plants provide an alternative in that the mitochondrial inner membrane possesses a specific OAA translocator (not found in mammalian systems) which can export the OAA from the organelle.

A further distinctive feature of plant mitochondria is that unlike those in, say, liver cells, they are not the main site of fatty acid oxidation which in plant cells is located in peroxisomes or glyoxysomes. Thus in tissues such as castor bean endosperm which on germination metabolize substantial amounts of lipid reserves, the fatty acids are oxidized in glyoxysomes not mitochondria.

The mitochondrial inner membrane may be permeable to NAD^+ but it is not permeable to NADH. The transport of reducing equivalents across the membrane may be a function of substrate transport systems as it is in animal mitochondria where for example the malate–aspartate (Borst) shuttle transports reducing equivalents into the organelle by exchanging the more reduced glutamate for efflux of aspartate.

Thus the role of the inner mitochondrial membrane will include the uptake of substrates for oxidation such as pyruvate and malate, systems for the transfer of reducing equivalents and the exchange of mitochondrially synthesized ATP for ADP. There is also a role in association with chloroplast metabolism both in C_4 plants where malate decarboxylation may occur in the mitochondria or in C_3 plants where mitochondria play a part in photorespiration. The principal permeation systems proposed for plant mitochondria are summarized in Table 2.1.

2.6.1 Transport of substrates

Pyruvate is a major substrate for mitochondrial metabolism and particularly for the pyruvate dehydrogenase of the matrix. This requires its transport across the mitochondrial inner membrane and although early studies

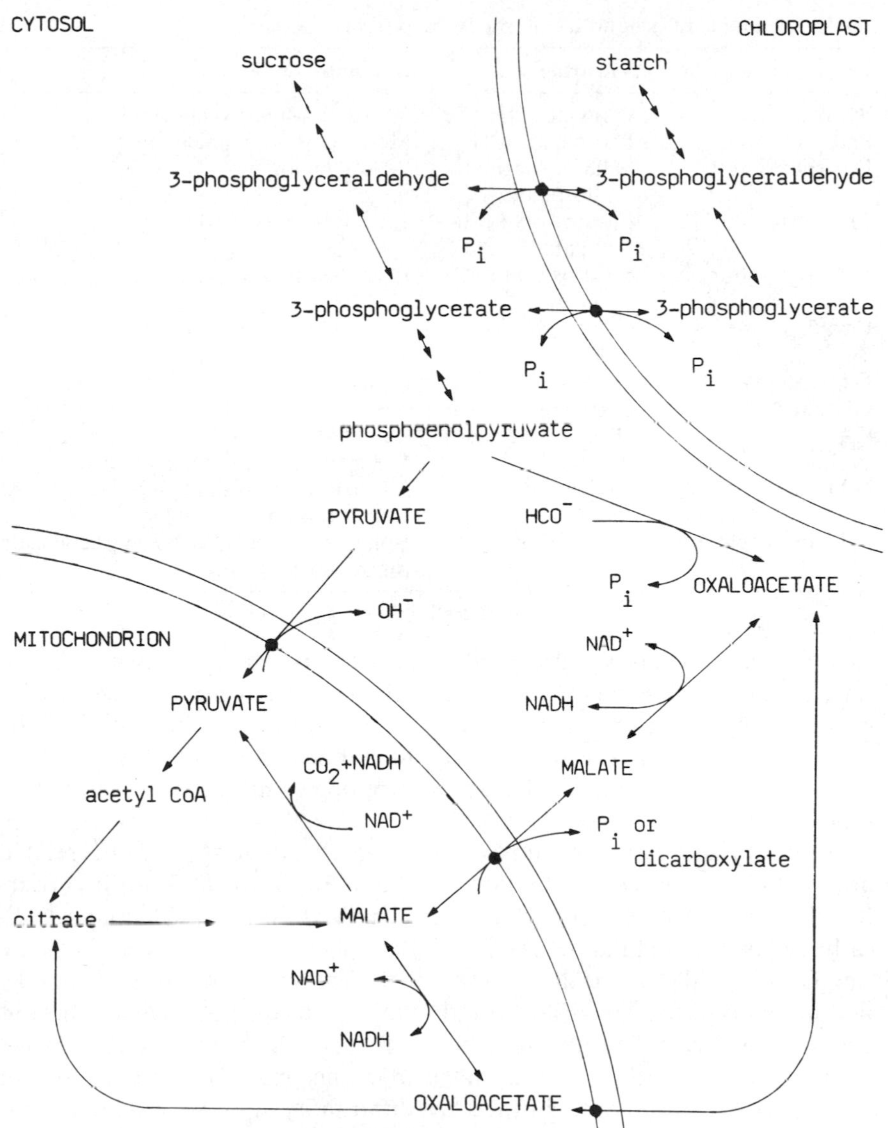

Fig. 2.9 Mitochondrial metabolism of pyruvate and malate showing the relationships of the pyruvate and dicarboxylate translocators to carbohydrate metabolism.

in mammalian mitochondria suggested a simple diffusion mechanism, several workers provided evidence for a carrier system sensitive to specific inhibitors such as α-cyano-4-hydroxycinnamate, phenylpyruvate and α-ketoisocaproate (Halestrap & Denton, 1974; Land & Clark, 1974; Mowbray 1974). Subsequent studies with mitochondria from etiolated corn

Table 2.1 Plant mitochondrial permeation systems

Translocator	Substrate	Inhibitors
Pyruvate	Pyruvate, OH^-	α-Cyano-4-hydroxycinnamate
Phosphate	Phosphate, OH^-	Mersalyl, N-ethylmaleimide
Dicarboxylate	Malate, succinate, phosphate	*n*-Butylmalonate
Tricarboxylate	Citrate, isocitrate, malate	1,2,3-Benzenetricarboxylate
α-Ketoglutarate	α-Ketoglutarate, malate	Phenylsuccinate
Citrate	Citrate, isocitrate, H^+	–
Oxaloacetate	Oxaloacetate	Phthalonate
Glutamate	Glutamate	–
Glycine	Glycine (? serine)	–
Proline	Proline	L-thiazolidine-4-carboxylic acid
NAD^+	NAD^+	*N*-4-azido-2-nitrophenyl-4-aminobutyryl-3′-NAD^+
Adenine nucleotide	ADP, ATP	Bongkrekic acid, Carboxyatractyloside (atractyloside)

shoots (Day & Hanson, 1977) and green and etiolated pea tissue (Proudlove & Moore, 1982) demonstrated a similar carrier system sensitive to α-cyano-4-hydroxycinnamate. Uptake shows saturation kinetics but the maximum rates of pyruvate transport have been somewhat low although Douce (1985) has claimed higher rates for mung bean hypocotyl and potato tuber mitochondria.

Studies in animal mitochondria led to the elucidation of an interrelated group of translocating systems summarized in Fig. 2.10. This shows carriers for P_i, phosphate-dicarboxylate, dicarboxylate-tricarboxylate and dicarboxylate-α-ketoglutarate. As Fig. 2.10 implies, uptake of a dicarboxylate can the place either in exchange for another dicarboxylate or for P_i. Similarly, net uptake of a tricarboxylate would involve exchange for malate and this could in turn be taken up in exchange for P_i. The animal tricarboxylate system is sensitive to 1,2,3-benzenetricarboxylic acid. The existence of two P_i carriers has been distinguished by differential sensitivity to inhibitors such as *n*-butylmalonate which blocks the phosphate-dicarboxylate exchange and *N*-ethylmaleimide which blocks the simple P_i uptake. All these exchange systems are electroneutral; the *N*-ethylmaleimide-sensitive P_i uptake takes place either in association with a proton or in exchange for a hydroxyl ion and there is at present, no way of distinguishing between these two latter possibilities. In bean mitochondria, a P_i carrier sensitive to mersalyl and *N*-ethylmaleimide and resembling the animal system has been demonstrated (De Santis *et al.*, 1975, 1976). These workers also demonstrated the existence of the dicarboxylate carrier sensitive to *n*-butylmalonate together with a 1,2,3-benzenetricarboxylate-sensitive citrate

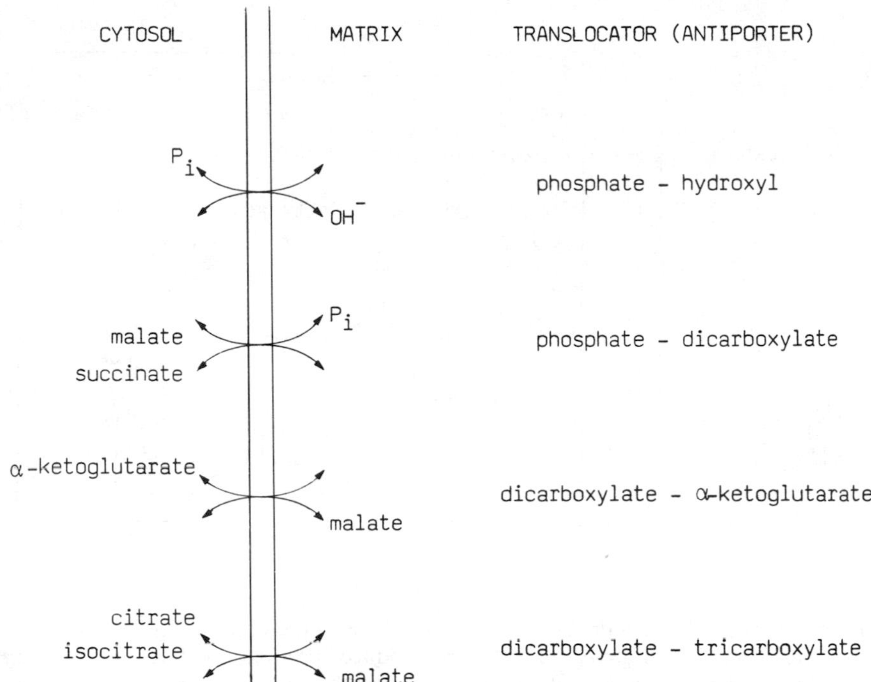

Fig. 2.10 Transport of phosphate, dicarboxylates and tricarboxylates through the mitochondrial inner membrane. The transport of dicarboxylates is dependent on phosphate. The transport of α-ketoglutarate, citrate and isocitrate are dependent on malate (and hence phosphate). The systems are reversible.

transporter and an α-ketoglutarate transporter which could be inhibited by phenylsuccinate. Thus this study implied a series of transport systems in plants similar to those found in mammalian mitochondria.

Subsequent studies have generally confirmed these findings in other tissues, for example in castor bean mitochondria (Chappell & Beevers, 1983) where the conversion of fat to carbohydrate involves dicarboxylate transport into mitochondria (see Fig. 2.11).

Apparently, citrate uptake may proceed by two mechanisms, one being the tricarboxylate carrier described above. The other mechanism was initially suggested in a study by Wiskich (1974) in which swelling of beet-root mitochondria in ammonium citrate did not depend on the presence of malate. He also found that mitochondrial citrate uptake in castor bean and wheat coleoptile mitochondria was independent of P_i and malate. Later Jung and Laties (1979) showed that citrate could be accumulated in potato mitochondria in the absence of P_i and malate. A study using both swelling studies and uptake of labelled substrates showed that in corn mitochondria inhibition of the P_i and dicarboxylate carriers did not elim-

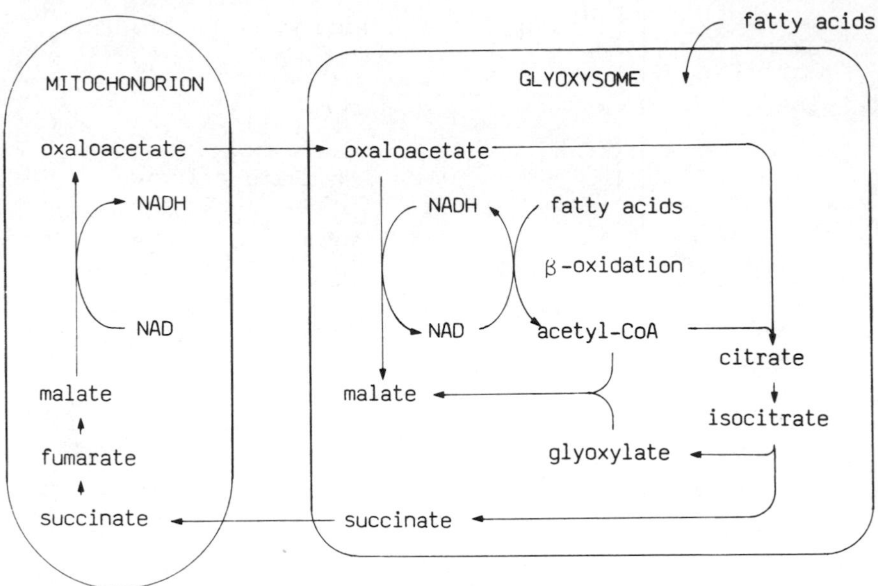

Fig. 2.11 The role of mitochondria in castor bean metabolism. The oxidation of fatty acids proceeds in glyoxysomes which produce succinate, malate and reducing equivalents. The succinate is metabolized in the mitochondrion. A malate-oxaloacetate shuttle is suggested for the transfer of reducing equivalents to the mitochondrion based on Mettler & Beevers (1980) and Ebbighausen *et al.* (1985).

inate citrate uptake; nevertheless, a low level of citrate–malate exchange was observable (Birnberg *et al.*, 1982). These workers concluded that, in addition to the tricarboxylate carrier, there was a citrate/H^+ co-transport and that the system also transported isocitrate (Birnberg & Hanson, 1983).

A second system which differs from that in animal mitochondria is a specific transporter for OAA. In most animal tissues there is at most only a low level of OAA transport mainly attributed to the α-ketoglutarate but also the dicarboxylate carrier systems, (Gimpel *et al.*, 1973; Passarella *et al.*, 1977). In many plant mitochondria there is rapid OAA transport as demonstrated in mitochondria from spinach leaf (Neuberger & Douce, 1980), pea leaf (Oliver & Walker, 1984) and in various pea tissues, spinach leaves and potato tuber (Ebbighausen *et al.*, 1985). This system is competitively sensitive to phthalonate (Day & Wiskich, 1981a) which incidentally does not appear to inhibit the α-ketoglutarate system, and insensitive to *n*-butylmalonate which inhibits the dicarboxylate carrier (Day & Wiskich, 1981b). The system does not appear to transport malate, succinate, α-ketoglutarate or P_i and there appears to be no requirement for a counter ion (Oliver & Walker, 1984).

A third system which shows significant differences from that found in animal mitochondria is that concerned with glutamate transport. In animals there are two glutamate carriers. The simplest exchanges glutamate for hydroxyl ions. This carrier has been isolated, incorporated into liposomes and shown to carry out a glutamate–hydroxyl ion exchange (Gautheron & Julliard, 1979). The second system exchanges glutamate for aspartate. Here transport of the glutamate is associated with a proton and hence in coupled mitochondria which have a proton gradient across the inner membrane with a higher concentration of protons outside, glutamate will normally enter and aspartate leave the mitochondria. Day & Wiskich (1977) found that glutamate transport is carried out in exchange for dicarboxylates and this has led to the view that there is a glutamate–dicarboxylate carrier. However, Proudlove and Moore (1982) found that several amino acids, aspartate, glutamate, serine and glycine, permeate the inner membrane but none of them show the substrate saturation kinetics expected of a carrier system. Thus the process of glutamate transport is in need of further investigation. Uncertainty also surrounds the question of an aspartate–glutamate exchange which it is suggested is not electrogenic and therefore will function to import or export aspartate (see Journet *et al.*, 1980, 1982 and also a discussion by Douce, 1985).

A fourth feature peculiar to plant mitochondria is the transport system or systems proposed by Cavalieri and Huang (1980) for neutral amino acids. Working with mung bean mitochondria they showed a swelling due to uptake of L-proline, L-serine, L-threonine (stereospecific for these L forms) and glycine which was sensitive to thiol-modifying reagents; alanine and methionine behaved similarly but not identically with the other amino acids. Later a specific carrier for glycine was proposed (Walker *et al.*, 1982) in a study which confirmed and extended the earlier findings. A transport system for glycine, serine and proline was proposed by Yu *et al.*, (1983) in spinach leaf mitochondria. However, at high extramitochondrial concentrations (greater than 0.5 mol m^{-3}) of the amino acid, diffusion was the major mechanism for amino acid entry. Nevertheless, Day & Wiskich (1980) and Proudlove & Moore (1982) have concluded that glycine enters by diffusion. Mitochondrial uptake of glycine is of significance in glycollate metabolism. Phosphoglycollate is formed as a by-product of the RuBP carboxylase reaction in chloroplasts. This reaction is the basis for photorespiration. The phosphoglycollate is converted to glycollate in the chloroplast and to glycine in peroxisomes. The glycine is then converted to serine in the mitochondria (see Fig. 2.12).

The transport of proline is, however, clearer. Plant mitochondria possess a proline dehydrogenase as do animal mitochondria (Johnson & Strecker, 1962; Boggess *et al.*, 1978). In corn mitochondria, the proline dehydrogenase is bound to the inside of the inner membrane (Elthon & Stewart, 1982). The proline uptake system is stereospecific and is inhibited by the proline analogue L-thiazolidine-4-carboxylic acid. Uptake appears to be driven by ΔpH (Elthon *et al.*, 1984). Of particular interest is the relationship of mitochondrial proline metabolism to water stress observed in many

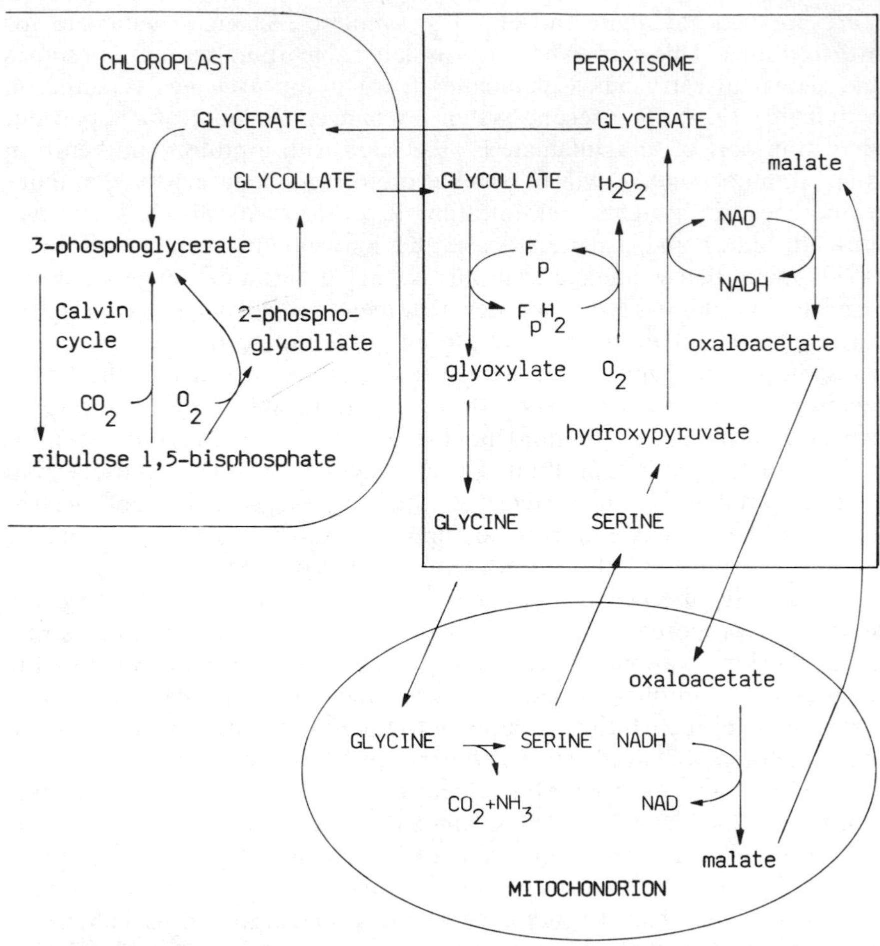

Fig. 2.12 Transport of substrates across the chloroplast and mitochondrial membranes in photorespiration. A reaction of the RuBP carboxylase enzyme with oxygen rather than CO_2 results in the formation of phosphoglycollate. Phosphoglycollate is converted through a sequence of reactions in the peroxisome, mitochondrion and chloroplast to 3-phosphoglycerate. The metabolic system predicts translocators for glycerate, glycollate, glycine and serine. The reducing equivalents required in the peroxisome are supplied from the mitochondrion by a malate-oxaloacetate shuttle as proposed by Ebbighausen *et al.* (1985).

plants. Proline accumulation occurs in many plant leaves during stress but is rapidly relieved by rewatering (See Chs. 6 and 10). Mitochondrial proline oxidation is substantially decreased during water stress (Stewart *et al.*, 1977) but there is also a possible effect on proline transport into mitochondria (Sells & Koeppe, 1981).

A dramatic difference between animal and plant mitochondria is found in the oxidation of NADH. Animal mitochondria do not oxidize extramitochondrial NADH directly since the catalytic site of the NADH dehydrogenase is on the inside of the inner membrane and the membrane is impermeable to NAD^+, NADH, $NADP^+$ and NADPH. Reducing equivalents are transferred across the membrane in the form of reduced substrates such as malate. In plants, in addition to the NADH dehydrogenase oxidizing matrix NADH, there is also an inner membrane enzyme oxidizing extramitochondrial NADH. This dehydrogenase is stimulated by Ca^{2+} (Moore & Åkerman, 1982) and may pass its electrons directly to complex III. There is also evidence for extramitochondrial NADPH oxidation by an inner membrane dehydrogenase independent of that just described (Møller & Palmer, 1981).

Recent evidence suggests (Neuberger *et al.*, 1985) that NAD^+ may be able to enter the plant mitochondrial matrix by a carrier mechanism. Both NAD^+ uptake into and loss from the mitochondrial NAD^+ pool was demonstrated in potato and both processes were inhibited by the analogue *N*-4-azido-2-nitrophenyl-4-aminobutyryl-3′-NAD^+. Uptake was inhibited by uncouplers and showed saturation kinetics.

Like animal mitochondria, plant mitochondria may possess a carrier for carnitine and acetyl carnitine although the inner membrane is impermeable to palmitoyl coenzyme A and coenzyme A (Wood *et al.*, 1984).

2.6.2 Adenine nucleotide transport

In early studies with animal mitochondria, it was shown that there was an exchange of adenine nucleotide (ADP and ATP but not AMP, Mg^{2+}–ATP or other nucleotides) across the inner mitochondrial membrane. In coupled (but not uncoupled) mitochondria, external ADP was preferentially exchanged for internal ATP, a process which is specifically inhibited by atractyloside (Pfaff *et al.*, 1965). Subsequent studies (La Noue *et al.*, 1978) showed that the exchange of adenine nucloetides was electrogenic rather than proton conducting in that $ATP^{(n+1-)}$ was exchanged for ADP^{n-}. It was suggested that the additional charge on ATP would result in preferential uptake of ADP in coupled mitochondria which have a membrane potential, positive on the outside (c side) and negative on the inside (m side). A second approach to the asymmetric behaviour of the adenine nucleotide carrier system derives from the fact that it shows different conformations on each side of the membrane. Thus at the outer surface, it has a greater affinity for ADP while at the inner surface it has a greater affinity for ATP. Atractyloside, carboxyatractyloside and bongkrekic acid specifically inhibit the translocation system. Atractyloside competes with ADP at the outer surface while bongkrekate inhibits on the matrix side non-competitively. Carboxyatractyloside also acts on the outside.

In plants there is disappointingly little information on this system.

Adenine nucleotides are rapidly exchanged across the inner membrane by a translocation system sensitive to inhibitors (Vignais *et al.*, 1976; Earnshaw, 1977). The plant translocator is highly sensitive to bongkrekic acid and carboxyatractyloside, but less sensitive to atractyloside. In plants but not in mammalian mitochondria, ADP will relieve carboxyatractyloside inhibition (Lima & Denslow, 1979). In addition to the exchange process a system for net uptake of ADP has been proposed by Abou-Khalil & Hanson (1977), although the importance of this is not clear.

In general, progress with the understanding of translocation systems has been much greater in the case of animal mitochondria than with plants and this is particularly evident in the case of the adenine nucleotide carrier (see Vignais, 1976; La Noue & Schoolwerth, 1979). Thus the carrier protein which is a major part of the inner membrane protein (9 % of the total protein and 12 % of the membrane protein in beef heart mitochondria) has been isolated by several methods including isolation of an atractyloside-binding protein (Brandolin *et al.*, 1974), a carboxyatractyloside-binding protein (Klingenberg *et al.*, 1978), a bongkrekate–protein complex (Aquila *et al.*, 1978) and by photoaffinity labelling with an ADP analogue (Lauquin *et al.*, 1978). The isolated protein is a dimer, each subunit having a molecular weight of 30 000 in mammals, 37 000 in yeast (Lauquin *et al.*, 1978). The amino acid composition shows the presence of a large proportion of basic amino acids (lysine and arginine) as well as substantial hydrophobicity. The carboxyatractyloside and bongkrekate complexes differ in their properties but the bongkrekate-binding form of the complex can be converted into the carboxyatractyloside-binding form in the presence of adenine nucleotides. The two forms are presumed to represent the conformations of the protein found on the c and m faces of the membrane. Immunological studies of the protein present in various mammalian organs suggest that there may be tissue-specific differences in addition to the m- and c-conformations (Schultheiss & Klingenberg, 1984). Adenine nucleotide transport across chemically defined membranes has been demonstrated by adding the purified translocator to lecithin liposomes. Such a system can be used to demonstrate that the protein can carry out a transmembrane ADP/ATP exchange (Kramer & Klingenberg, 1977, 1979). The liposome technique provides a means to demonstrate that the purified protein possesses the biological activity attributed to it as well as showing that the lipid of the membrane influences to a limited extent the properties of the protein. However, it also makes possible a variety of further studies which lead to a deeper understanding of such a membrane-translocating system.

For many years there have been two explanations of the observation that in coupled mitochondria the uptake of ATP is suppressed as compared to ADP while the efflux of ATP is favoured rather than ADP. One explanation is that the membrane potential induces a conformational change in the protein such that the c-form has higher affinity for ADP and the m-form a higher affinity for ATP. Alternatively, the preferential uptake can be accounted for by the fact that the rate constants for transport are different in the two directions (due to the difference in negative charge carried by the nucleotide) so that the substrate-binding sites are enriched on one side of the membrane.

A kinetic analysis using the liposome system to which a membrane potential can be applied has shown that the second mechanism does apply. About forty times more ATP-loaded binding sites were found to face outwards as compared to inwards but ADP-loaded sites were more evenly distributed. Thus it can be argued that the basic mechanism is an electrophoretic one, the carrier–ATP complex possessing an extra negative charge and therefore distributing itself towards the outside. Such a system would explain the apparent difference in affinity observed in earlier studies. Thus the primary function of the membrane potential in promoting ADP uptake and ATP efflux is to bring about an asymmetric arrangement of the carrier and hence of the c- and m-forms. The fundamental reason for carrier performance is electrophoretic (Klingenberg, 1980; Kramer & Klingenberg, 1982). Studies of this kind with pure systems move a step closer to understanding the mechanism of a protein-mediated translocation system as for example discussed by Klingenberg (1981), where the role and significance of the protein as a dimer is discussed. As pointed out by Klingenberg (1980), 'the energy-dependent ADP–ATP exchange on mitochondria appears to be an example where the energy transduction on transport through a biomembrane will be first understood.'

2.6.3 Calcium uptake

In mammalian mitochondria, Ca^{2+} is taken up in an energy-dependent manner but the uptake mechanism itself does not involve the simultaneous uptake of an anion. The uptake can be driven by respiration or by ATP hydrolysis and the migration across the membrane of the positively charged Ca^{2+} appears to be electrophoretic in response to the membrane potential. Any one of several proton-conducting anions may be used experimentally as the counter ion including P_i and acetate.

Some plant mitochondria also take up Ca^{2+} although there is considerable variation depending on the species, plant organ and age of the plant (Dieter & Marmé, 1980). Uptake appears to be obligatorily linked to P_i and other anions such as acetate will not replace P_i (Day *et al.*, 1978). It has been suggested that Ca^{2+} may be transported as a phosphate complex (Wilson & Minton, 1974) rather than by the simple electrophoretic carrier system proposed for mammalian mitochondria (Åkerman & Moore, 1983). The process is energy-dependent and is sensitive to uncouplers (protonophores) such as carbonylcyanide-*m*-chlorophenylhydrazone (CCCP) and the specific inhibitor of mammalian Ca^{2+} uptake, ruthenium red (Ferguson *et al.*, 1985) Sensitivity to ruthenium red has, however, been disputed

(Roux *et al.*, 1981; Akerman & Moore, 1983). Although Na^+ can bring about Ca^{2+} release in mammalian mitochondria, this does not occur in plants (Dieter & Marme, 1980).

Like the plasma membrane in plants (Hale & Roux, 1980), an additional factor in Ca^{2+} uptake in mitochondria is regulation by phytochrome as evidenced by the effects of red and far-red light on the system. Red light diminishes uptake while far-red restores it. In the presence of ruthenium red (which inhibited Ca^{2+} uptake in these experiments), red light induced an efflux of Ca^{2+} whereas far-red almost abolished the efflux (Roux *et al.*, 1981). It has therefore been suggested that mitochondria play a key role in controlling cytosolic Ca^{2+} concentrations and this in turn exerts a control on various enzyme activities.

Information on the transport of other cations into mitochondria is scant although among the divalent ions, Sr^{2+} is claimed to be taken up more strongly, Ba^{2+} less strongly than Ca^{2+}. Mg^{2+} uptake is weak (Wilson & Minton, 1974).

2.7 Transport across the chloroplast envelope

The metabolism of chloroplasts centres round their role in carbon fixation and the conversion of light energy to chemical energy. In C_3 plants (those that fix CO_2 by the classical Calvin cycle), CO_2 and P_i enter the chloroplast and the major product exported is 3-phosphoglyceric acid (PGA). In C_4 plants, the mesophyll chloroplasts fix CO_2 into an organic acid which is exported to the bundle sheath cells where it is decarboxylated. The CO_2 is then fixed by the Calvin cycle (see Fig. 2.14). Thus the role of the envelope may be expected to reflect these various forms of chloroplast metabolism (see Table 2.2 for a summary of the main permeation systems). The widely used method for investigating metabolite uptake by chloroplasts involves incubating the organelle suspension with substrate in the top layer of a three-layer system. The middle layer is composed of a silicone oil and the bottom layer of a denaturing agent such as perchloric acid. After an appropriate incubation period (say, 30 s), the system is rapidly centrifuged and hence the organelles are removed from the substrate. The organelle pellet can be assayed for various substrates (Heldt, 1980).

Unlike the mitochondrion where the inner membrane surrounds a single compartment, the chloroplast envelope surrounds two compartments, the matrix and, surrounded by the thylakoid membranes, the loculus or lumen. The inner membrane of the envelope is relatively impermeable to protons (Heber & Krause, 1971). The major effect of light on this system is to activate the thylakoid electron-transport system which translocates protons

Table 2.2 Chloroplast permeation systems (Weak substrates are shown in brackets)

Translocator	Substrate	Inhibitors
Phosphate (C_3 type)	Phosphate, 3-phosphoglyceraldehyde, 3-phosphoglyceric acid, dihydroxyacetone phosphate, (phosphoenolpyruvate)	Pyridoxal 5-phosphate trinitrobenzene sulphonate PCMBS
Phosphate (C_4 type)	Phosphate, phosphoenolpyruvate, 3-phosphoglyceric acid, 3-phosphoglyceraldehyde, dihydroxyacetone phosphate	pyridoxal 5-phosphate
Dicarboxylate	α-Ketoglutarate, malate, fumarate, glutamate, succinate (? aspartate)	
Oxaloacetate	Oxaloacetate (malate, α-ketoglutarate)	
Aspartate	Aspartate	PCMBS mersalyl diethylpyrocarbonate
Glycerate	Glycerate	PCMBS
Pyruvate	Pyruvate (? C_4 plants only)	3-PGA chloride, phosphate, phenylpyruvate, α-ketoisovalerate α-cyano-4-hydroxycinnamic acid
Glucose	D-Glucose, maltose (possibly other sugars)	
Adenine nucleotide	ADP, ATP, phosphoenol pyruvate, ? pyrophosphate	Bongkrekic acid (carboxyatractyloside)

into the thylakoid resulting in an increase of the stroma pH (Heldt *et al.*, 1973) and of the Mg^{2+} concentration. Both factors have a significant effect on stroma metabolism and in addition may influence transport.

2.7.1 Transport of inorganic substances

The chloroplast envelope like lipid membranes in general is readily permeable to CO_2 (Blank & Roughton, 1960; Heber & Heldt, 1981). The question as to whether the major permeant form of inorganic carbon is CO_2 or HCO_3^- has been disputed (Poincelot, 1974); however, HCO_3^- appears to permeate only slowly (Werdan *et al.*, 1972; Shiraiwa & Miyachi, 1978). The study of CO_2 uptake is complicated by its binding to chloroplast proteins, particularly RuBP carboxylase which accounts for more than half the total protein in the stroma. Thus the stimulating effect of P_i on inorganic carbon uptake has been attributed to the requirement of P_i for activation of the carboxylase and hence stimulation of CO_2 binding by the carboxylase (Sicher, 1984).

Little is known about permeation by cations; However, Na^+, Rb^+ and K^+ will permeate chloroplasts (Kaiser *et al.*, 1980). Huber and Maury (1980) found that Mg^{2+} would induce both acidification of the stroma (not dependent on photosynthesis) and the efflux of K^+. This led to the conclusion that the envelope possessed a K^+/H^+ exchange system. The Mg^{2+} effect could also be brought about by Ca^{2+} or Mn^{2+} but not Ba^{2+} (Maury *et al.*, 1981).

Although Gimmler *et al.* (1975) had claimed that the envelope was impermeable to Mg^{2+}, Deshaies *et al.* (1984) have obtained evidence that when suspended in a medium lacking free Mg^{2+}, chloroplasts lost up to 75 % of their stromal Mg^{2+} to the medium although the organelles remained intact. These depleted chloroplasts were able to take up Mg^{2+} very rapidly.

Evidence suggests that Cl^- permeates the envelope readily as do several other halogenides, NO_3^- and NO_2^- (see Heber & Heldt, 1981). There is, however, no evidence for specific uptake systems.

As noted earlier, in the light the pH of the matrix is significantly higher than the cytosol. The pH gradient will collapse to the value found in the dark if an uncoupler such as CCCP is added (Heber & Krause, 1971). The maintenance of the envelope gradient appears essential for photosynthesis (Enser & Heber, 1980) but the source of the light-activated proton pump is not clear. The envelope has been shown to possess a Mg^{2+}-ATPase activity although this is insensitive to DCCD (Douce *et al.*, 1973) and only partially sensitive to oligomycin (Maury *et al.*, 1981). There is no satisfactory evidence that the ATPase can act as a proton pump (see Sect. 2.4.3).

2.7.2 The phosphate translocators

Chloroplasts which possess a Calvin cycle although capable of starch synthesis export the organic products of CO_2 fixation mainly as 3-PGA. Heldt and Rapley (1970b) found a P_i translocator capable of exchanging 3-PGA, dihydroxyacetone phosphate or 3-phosphoglyceraldehyde for P_i. Such a system is located in the inner envelope membrane. The translocator which is specific and will not react with 2-PGA is inhibited by PCMBS, pyridoxal 5-phosphate and trinitrobenzene sulphonate. The exchange system involves the transport of a proton with 3-PGA but not P_i or the triose phosphates (dihydroxyacetone phosphate and 3-phosphoglyceraldehyde) (Fliege *et al.*, 1978). The protein responsible for transport has been isolated by ionic detergent treatment of the isolated envelope after labelling the translocator with 2,4,6-trinitrobenzene sulphonate which labelled only a single polypeptide with a molecular weight of 29 000. The binding studies suggest the presence of a lysine residue at the active site and there is also evidence for arginine. Presumably the positively charged residues are involved in binding the anion substrates (Flugge & Heldt, 1978). Evidence from studies on pH dependence suggest that P_i is transported with two negative charges. Recently the replacement of the ionic detergent by Triton X-100 has made possible the isolation of a functional protein which appears to have a molecular weight of 61 000 (after allowing for bound Triton) and is therefore assumed to be a dimer (Flugge, 1985). Incorporation of the protein into liposomes has shown a specific transport of P_i which can be inhibited with pyridoxal phosphate. The uptake of P_i into the liposome is dependent on exchange with internal P_i or 3-PGA but not 2-PGA (Flugge & Heldt, 1981). Transport across liposome membranes is influenced by a pH gradient. An inward directed gradient where the *external* pH is lower promotes 3-PGA but decreases P_i and triose phosphate uptake. Where the *internal* pH is lower than the external medium, P_i and triose phosphate uptake is enhanced. The effect of $\triangle$pH is to alter the apparent K_m of the translocator for P_i and triose phosphate but in the case of 3-PGA the apparent V_{max} is altered (Flugge *et al.*, 1983). The relevance of these findings resides in the fact the light causes an increase in matrix pH by movement of protons into the thylakoid loculus (Heldt *et al.*, 1973). Light also causes a decrease in 3-PGA export from chloroplasts so that triose phosphate is preferentially released even though the matrix concentration of the latter is significantly lower than that of PGA (Stitt *et al.*, 1980). Since under physiological conditions 3-PGA has one more negative charge than the other substrates, the transport of this substance would be expected to be associated with a proton.

The function of the translocator, which has been reviewed recently (Flugge & Heldt, 1984) is summarized in Fig. 2.13. The exchange of triose phosphate for P_i in the light will serve both to export the product of photosynthesis and to maintain the chloroplast P_i concentration. A triose phosphate/3-PGA shuttle will export reducing equivalents and also enable ATP synthesis to take place in the cytosol at the expense of energy trapped by photosynthesis.

While the description above provides a satisfactory view for phosphate uptake in C_3 plants, a rather different perspective arises for C_4 plants such as *Digitaria sanguinalis* and *Zea mays* where the principal export of mesophyll chloroplasts is phosphoenolpyruvate (PEP) which is formed from imported P_i and pyruvate (see Fig. 2.14). Isolated chloroplasts preloaded with labelled PGA or PEP will release these compounds when external P_i, 3-PGA, PEP or dihydroacetone phosphate are provided. The exchange of P_i for PEP is inhibited by pyridoxal phosphate and the uptake of 3-PGA is inhibited by P_i and PEP. The results are consistent with the P_i translocator as the basis for PEP/P_i exchange (Huber & Edwards, 1977b; Day

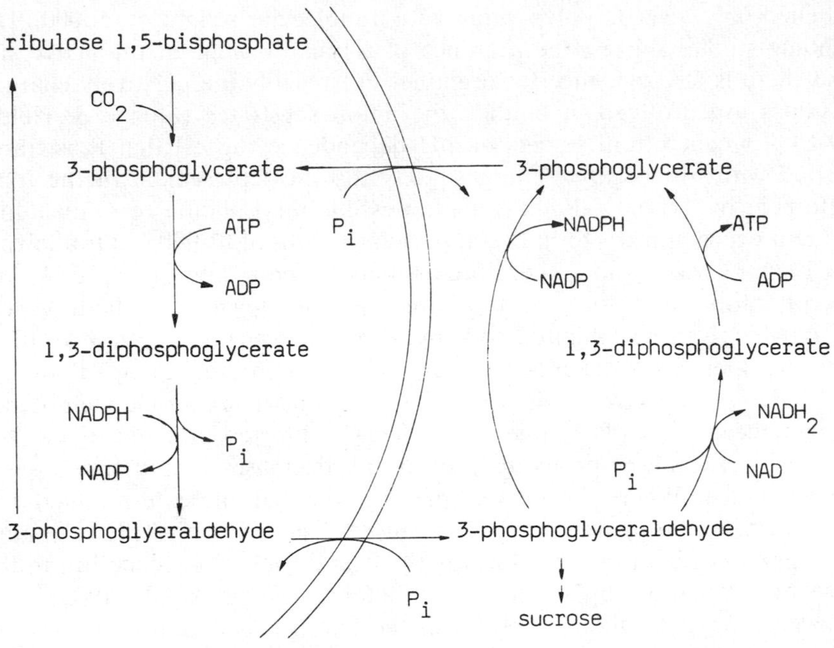

Fig. 2.13 The chloroplast phosphate translocator. The diagram shows the translocator as a means of exporting the products of CO_2 fixation in C_3 plants while maintaining the concentration of phosphate within the chloroplast and as a shuttle for the export of ATP and reducing equivalents. NADPH required for biosynthetic reactions in the cytosol is formed by means of an irreversible NADP-dependent phosphoglyceraldehyde dehydrogenase.

& Hatch, 1981b). However, the P_i translocator from spinach chloroplasts has a significantly lower affinity for PEP than that from C_4 plants and it would appear that the two P_i translocators are not identical.

2.7.3 Dicarboxylate and other chloroplast translocators

In addition to the P_i translocators, several other systems for metabolite transport have now been described. The C_4 pathway of photosynthesis (see Fig. 2.14) clearly implies that chloroplasts should be able to import and export dicarboxylates. However, both C_4 and C_3 plants are able to transport dicarboxylates across their chloroplast envelopes.

The dicarboxylate translocator was originally described by Heldt and Rapley (1970a). It was found to transport malate, succinate, fumarate, aspartate, α-ketoglutarate and glutamate by an exchange process. Since each of these substances acted as competitive inhibitors of dicarboxylate transport, it could be assumed that a single carrier was involved although kinetic evidence suggested that aspartate might require a separate carrier

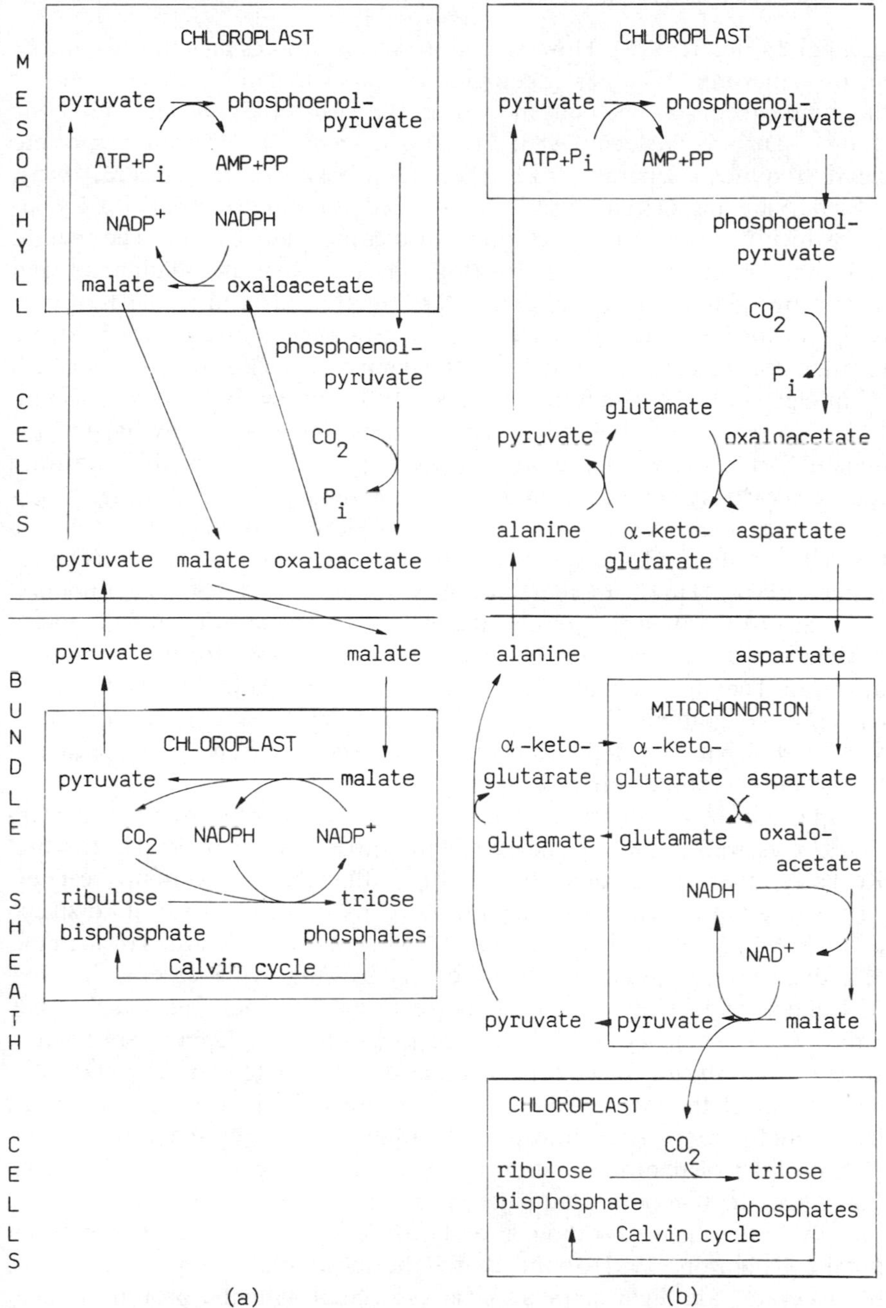

Fig. 2.14 Pathways for C_4 photosynthesis. (a) NADP-dependent malic enzyme type (*Digitaria sanguinalis, Zea mays*). (b) the NAD-dependent malic enzyme type. A third type, the phosphoenolpyruvate caboxykinase type, is not shown. The diagrams illustrate the need for transport of substrates across organelle membranes in this type of photosynthesis where the initial CO_2 fixation occurs in the mesophyll cells but CO_2 is released in the bundle sheath cells for fixation by the ribulose bisphosphate carboxylase.

(Lehner & Heldt, 1978). However, more recent studies have thrown doubt on this question. Whether α-ketoglutarate and malate are substrates for the same carrier has been questioned on the basis of kinetic and metabolic studies (Dry & Wiskich, 1983; Proudlove *et al.*, 1984) while aspartate clearly requires a separate carrier (Werner-Washburne & Keegstra, 1985).

Somerville and Ogren (1983) have studied a mutant of *Arabidopsis thaliana* which is deficient in the chloroplast dicarboxylate carrier. The mutant lacks the ability to take up α-ketoglutarate, aspartate, glutamate and malate but glutamine and P_i uptake are unaffected. Glutamine transport is significant here since it had earlier been argued that the dicarboxylate translocator mediated the uptake of this amine into chloroplasts of spinach (Gimmler *et al.*, 1974) and pea (Barber & Thurman, 1978). Oxaloacetate which is readily metabolized by chloroplasts and may be exchanged for malate (Anderson & House, 1979) was suggested as a possible substrate for the dicarboxylate translocator (Lehner & Heldt, 1978). In some chloroplasts (e.g. maize where the mesophyll chloroplasts need to exchange OAA for malate) OAA appears to be very actively transported (Day & Hatch, 1981a). However, more recently Hatch *et al.* (1984) have pointed out that OAA would be present at too low a concentration for the dicarboxylate carrier to transport OAA at the required rate for photosynthesis in maize. They have obtained evidence for a translocator with high specificity and strong affinity for OAA both in maize and also in spinach. Earlier Woo (1983) had also concluded that OAA was not transported primarily by the dicarboxylate carrier in spinach.

Aspartate in pea chloroplasts may also be transported by its own translocator as well as by the dicarboxylate carrier. Werner-Washburne and Keegstra (1985) have found both a high-affinity and low-affinity carrier. The high-affinity carrier is distinguished by its sensitivity to inhibitors such as PCMBS and diethyl pyrocarbonate whereas the low-affinity carrier is sensitive only to the latter. Thus there is growing evidence that the view of Lehner and Heldt is proving correct, namely that 'our results may indicate the existence of different carriers with overlapping specificity'. Indeed it is difficult to see how a single dicarboxylate carrier would satisfactorily meet the metabolic needs of the organelle if so many dicarboxylates could potentially be transported simultaneously, by the same system.

A product of photorespiration is glycerate formed in peroxisomes (see Fig. 2.12). It is imported into the chloroplast and converted to PGA by the glycerate kinase. Although early studies failed to show significant uptake Robinson (1982) found that light substantially stimulated uptake of glycerate. The light stimulation is associated with the proton gradient across the envelope since a proton ionophore, FCCP abolished the light effect. ATP is not involved in glycerate uptake and it may be concluded that the envelope possesses a glycerate transporter which is coupled to the proton gradient independent of the P_i transporter and inhibited by PCMBS and 3-PGA (Robinson, 1984).

In *Digitaria sanguinalis* mesophyll chloroplasts, a specific uptake of pyruvate has been demonstrated. The chloroplasts take up pyruvate as part of a C_4 pathway of photosynthesis. Uptake is electrogenic and requires the presence of valinomycin and K^+ in the medium in order to demonstrate the system independently in the dark, leading to the conclusion that the charged anion is transported. The carrier system is inhibited by Cl^-, P_i, phenylpyruvate, α-ketoisovalerate, α-cyano-4-hydroxycinnamic acid and α-cyanocinnamic acid. A comparable system could not be detected in chloroplasts from spinach which employs a C_3 pathway (Huber & Edwards, 1977a).

The transport of amino acids into chloroplasts is a somewhat confused field. Gimmler *et al.* (1974) found that neutral amino acids (alanine, glycine, serine, proline, threonine and valine) entered spinach chloroplasts by non-specific diffusion. In contrast McLaren and Barber (1977), using the silicone oil centrifugation technique, obtained some evidence for specific uptake of leucine and isoleucine in pea leaf chloroplasts. Aspartate as noted above may have its own translocator while glutamate appears to be transported by the dicarboxylate carrier. Glutamine is transported across chloroplast envelopes but the carrier remains to be identified.

Starch is deposited in chloroplasts and is mobilized mainly by conversion to triose phosphate and 3-PGA. There is, however, some evidence for limited conversion of starch to glucose and maltose; these sugars together with some other monosaccharides can permeate the envelope (Heldt *et al.*, 1977). A carrier-mediated diffusion process has been proposed (Schafer, *et al.*, 1977; Herold *et al.*, 1981). Ascorbate uptake by spinach chloroplasts has also been described (Beck *et al.*, 1983).

Heldt (1969) decribed a weak adenine nucleotide translocator in spinach chloroplast envelopes and this lead to the view that chloroplasts were unable to exchange substantial amounts of the nucleotide. However, a more active translocator has been found in the C_3 chloroplasts from pea and C_4 chloroplasts from *Digitaria sanguinalis* (Huber & Edwards, 1976; Robinson & Wiskich, 1977).

It was suggested that the main function of such a system was to provide ATP for chloroplasts in the dark when the thylakoid ATPase was not functional. However, recent studies have shown that in pea chloroplasts, not only is there substantial adenine nucleotide exchange but the system will also transport PEP and probably PP_i. The translocator is strongly inhibited by bongkrekate but only weakly by carboxyatractyloside (Woldegiorgis *et al.*, 1985). The metabolic role of this system still remains a matter for speculation.

2.8 Uptake of proteins into subcellular organelles

2.8.1 Transport of proteins across the chloroplast envelope

Chloroplasts are capable of synthesizing some of their own proteins based on information coded in the chloroplast genome. This system has many similarities with that found in prokaryotes and is closer in its properties to the latter than to protein synthesis in the cytosol. Thus the plastid genome codes for the ribosomal RNAs, the transfer RNAs and a number of additional proteins including the large subunit of RuBP carboxylase, the P700 apoprotein, several PSII proteins, cytochromes f and b_6, some subunits of the ATPase and some components of the protein-synthesizing machinery. The list of genes is almost certainly not yet fully known; based on its size it is estimated that the genome could code for up to one hundred proteins of molecular weight 30 000 (Rochaix, 1985). Nevertheless, it is clear that the majority of chloroplast proteins are synthesized in the cytosol and encoded in the nuclear DNA. These polypeptides will therefore need to cross the chloroplast envelope in order to reach their functional location. Further it is clear that the efficiency displayed by living organisms will only be achieved if the transport of polypeptides across the envelope membranes is essentially limited to those polypeptides which function in the chloroplast (Ellis, 1983). The mechanisms involved in uptake are largely unknown but four aspects can be identified:

1. A recognition event which enables only the appropriate proteins to enter the chloroplast.
2. Translocation across the membrane.
3. Modification of the transported protein.
4. Assembly at the functional site.

The ribulose 1,5-bisphosphate (RuBP) carboxylase is the major protein found in leaves (Fraction 1 protein) and is composed of large subunits synthesized within the chloroplast and small subunits synthesized in the cytosol. The small subunit of molecular weight 14 000 is synthesized as an 18 000–20 000 molecular weight precursor in the cytosol. This larger precursor synthesized on free ribosomes in the cytosol in taken up by pea or spinach chloroplasts and is processed to the mature carboxylase. Unlike the situation proposed in the signal hypothesis for protein secretion by cells, the uptake of the polypeptide does not require the presence of ribosomes bound to the envelope of the chloroplast (Cashmore *et al.*, 1978; Chua & Schmidt, 1978; Highfield & Ellis, 1978). Similar principles apply to other proteins such as plastocyanin with a precursor molecular weight 15 kDa larger than the mature protein and ferredoxin-$NADP^+$ oxido-reductase which is 8 kDa larger than the mature product (Grossman *et al.*, 1982). The precursor proteins appear to bind to proteins located on the envelope surface since binding and transport can be blocked by pretreatment of chloroplasts with protease (Cline *et al.*, 1985). Uptake appears to be energy dependent, and can be stimulated with either light or ATP (Grossman *et al.*, 1980). Precursor proteins will bind preferentially (as compared to other proteins) to isolated chloroplast envelopes; the binding does not require ATP but is sensitive to protease treatment (Pfisterer *et al.*, 1982). However, the envelope proteins determining binding have not so far been identified. The amino acid sequence of the precursor of the RuBP carboxylase synthesized *in vitro* has been determined for the alga *Chlamydomonas* and forty-four amino acid residues at the *N*-terminal end have been identified as the 'transit' sequence. An endopeptidase found in *Chlamydomonas* was shown to convert the precursor to the mature polypeptide by removal of the transit sequence (Schmidt *et al.*, 1979).

More recently Robinson and Ellis (1984a) have partially purified the protease involved in processing imported precursor polypeptides. The enzyme isolated from pea chloroplasts was found to process different precursors (RuBP carboxylase and plastocyanin from wheat and barley) but did not attack other non-precursor proteins. The isolation procedure suggests that the enzyme was derived from the stroma where processing is presumed to occur. With RuBP carboxylase, processing involved two steps, both catalysed by the same enzyme; the M_r = 20 000 to 18 000 step and M_r = 18 000 to 14 000 step (Robinson & Ellis, 1984b).

The chlorophyll *a*/*b* binding protein is composed of two related polypeptides which have also been shown to exist in precursor forms outside the chloroplast. The precursors are about 4–5 kDa larger than the mature proteins. The precursors are taken up by chloroplasts, processed into the mature polypeptides and inserted into the thylakoid membrane to form part of the functional light-harvesting complex (Schmidt *et al.*, 1981). The gene for one of the polypeptides has been sequenced and the transit sequence of the *N* terminal identified as thirty-seven or thirty-eight amino acids which is shorter than the corresponding sequences for RuBP carboxylases which vary between forty-four and fifty-seven amino acids (Cashmore, 1984).

Although there is considerable variation in the size and amino acid sequence of transit sequences (see Mishkind *et al.*, 1985 for a summary) the *Chlamydomonas* RuBP carboxylase small subunit precursor is efficiently transported into pea and spinach chloroplasts. However, removal of all or even part of the transit sequence blocked uptake. Comparison of sequences from the *Chlamydomonas* protein with those in various plants has enabled Mishkind *et al.* (1985) to suggest a nine amino acid sequence which is conserved in all transit sequences of small subunit precursors of the carboxylase although there is no recognizable homology seen with the transit sequence of the chlorophyll *a*/*b*-binding protein (Cashmore, 1984).

It has been demonstrated that the transit peptide is responsible for the specific uptake of proteins into chloroplasts by fusion of the transit sequence from pea RuBP carboxylase small subunit to bacterial neomycin phosphotransferase. Using the *Agrobacterium* tumour-inducing (Ti) plasmid as a vector for transforming tobacco cells it was demonstrated that whereas the neomycin phosphotransferase could not be introduced into chloroplasts *in vivo*, the enzyme linked to the transit sequence could. Further the transit sequence–phosphotransferase complex was processed to the mature phosphotransferase. Since there is no sequence homology between the small subunit of the RuBP carboxylase and the phosphotransferase it can be concluded that the information for uptake resides solely in the transit sequence itself (Broeck *et al.*, 1985).

2.8.2. Transport of proteins into mitochondria

As with the chloroplast, most mitochondrial proteins are encoded in the nucleus and are synthesized outside the organelle on cytoplasmic ribosomes (for a review see Hay *et al.*, 1984). Higher plant mitochondria possess a DNA molecule substantially larger than that found in higher animal mitochondria. Although the structure of this DNA is unclear, it has been shown to code for a number of proteins including constituents of the ATPase, ribosomes, cytochrome b–c_1 complex, cytochrome oxidase, etc. (see Douce, 1985, for a recent summary of mitochondrial biogenesis). Nevertheless, this intramitochondrial protein synthesis accounts for only about 10 % of the organelle's proteins, the majority being imported across the mitochondrial membrane from the cytosol. There is, however, little information on higher plant mitochondrial biogenesis. Evidence for protein synthesis on mitochondrially bound ribosomes coupled with direct injection of the newly synthesized polypeptide is lacking. Mitochondrial proteins can be shown to be synthesized on free ribosomes in wheat (Zitomer & Hall, 1976), although there has been considerable interest in mitochondrially bound ribosomes (see Chua & Schmidt, 1979, for a review of this argument).

Most mitochondrial proteins synthesized in the cytosol are initially larger than the mature intramitochondrial polypeptides. Thus in yeast the primary precursors of the α-and β-subunits of the ATPase are 6 kDa larger while the γ-subunit is 2 kDa larger. Cytochrome *c* oxidase subunits V and VI are 2.5 kDa or more larger (Lewin *et al.*, 1980). The transit sequence of the precursor is attached to the *N*-terminal of the polypeptide and appears to be removed after uptake by a protease. This enzyme has been isolated from the mitochondrial matrix in yeast and elsewhere. It has been shown to process several precursors including cytochrome oxidase subunit V and an ATPase subunit. The enzyme is a metalloprotein activated by Co^{2+} or another metal (McAda & Douglas, 1982; Bohni *et al.*, 1983; Cerletti *et al.*, 1983). Processing of precursors which migrate towards the intermembrane space such as cytochrome b_2 appears to involve two steps, the first due to the protease described above but the second catalysed by an enzyme with very different properties and about which little is known (Daum *et al.*, 1982).

However, neither apo-cytochrome *c* nor the adenine nucleotide translocator appear to be synthesized as a larger precursor (Zimmerman *et al.*, 1979). The uptake of apo-cytochrome *c* and its conversion to the mature holo-cytochrome *c* by addition of haem has been demonstrated using the radiolabelled apo-protein. The kinetics of the uptake suggest that there are a limited number of binding sites on the mitochondrial surface. The treatment of mitochondria with a proteolytic enzyme destroys the receptor. There may be more than one receptor since apo-cytochrome *c* does not inhibit the uptake of the precursors of the adenine nucleotide translocator (Zimmerman *et al.*, 1981; Teintze *et al.*, 1982) although unlabelled apo-cytochrome *c* will inhibit the entry of labelled cytochrome *c*. Uptake of cytochrome *c* but not the precursors of the adenine nucleotide translocator, the ATPase proteolipid or the subunits of the b–c_1 complex is independent of the membrane potential and unaffected by uncouplers. The energy requirement for uptake of ATPase or b–c_1 subunits has been demonstrated in various ways and it is clear the Δ_p is necessary for uptake. Vesicles prepared from outer membranes bind precursor proteins (but not intermediate or mature forms) much more strongly than those formed from inner membranes. Binding does not require an energized inner membrane. The binding protein can be removed from the outer membrane and incorporated into a phospholipid vesicle which will then bind precursor mitochondrial protein (Riezmen *et al.*, 1983).

Proteins found in the outer membrane do not appear to have larger precursors. Thus the 29 kDa protein which is believed to be a porin in yeast and the 31 kDa protein from *Neurospora* is incorporated into mitochondrial outer membranes without processing. Neither ATP nor a membrane potential are required (Freitag *et al.*, 1982b; Gasser & Schatz, 1983). Another protein from the yeast outer membrane which is synthesized on cytosolic ribosomes and is not processed has a molecular weight of 70 000. The gene responsible has been identified and a series of altered genes prepared. These have been introduced into yeast and the incorporation into outer mitochondrial membranes assessed. All the information for targeting this protein on to the outer membrane was shown to be contained within the *N*-terminal amino acids (the protein contains 617 amino acids). The deletion of this region left the protein in the cytosol. The authors of this study have distinguished between targeting the protein on the membrane and anchoring it within the membrane. Both functions are associated with the terminal forty-one amino acids but different deletions suggest that targeting and anchoring concern separate regions, anchoring being associated with residues 1–11 (Hase *et al.*, 1984). Thus although overall processes for uptake of proteins into organelles are becoming much clearer, little is understood about the detailed mechanisms for their transport across mitochondrial membranes.

2.9 Conclusions

Our understanding of substrate transport across mitochondrial and chloroplast membranes has developed rapidly in recent years. Nevertheless it has to be admitted that in no case do we understand the mechanism of transport of any substrate across the membrane and in only two cases, the adenine nucleotide carrier of mammalian mitochondria and the phosphate carrier of the plant chloroplast do we have any detailed knowledge of the protein involved. In respect of proteins themselves, even less is known of the mechanism for import. In the case of proton transport the situation is more advanced and a number of very plausible schemes particularly the Q-cycle are available. Nevertheless the precise mechanism of proton pumping is in no case clear.

In general, because of the ease of preparation of animal mitochondria as compared with those of plants, studies of the latter have developed only slowly. The assumption frequently made in the early 1970s that plant mitochondria behave like animal ones is now increasingly questioned. Indeed the whole question of transport in plant organelles now appears a much more rewarding area of investigation.

Acknowledgements

I would like to thank Dr John Bowyer for many helpful criticisms of the manuscript.

References

Abou-Khalil, S. & Hanson, J. B. (1977). Net adenosine diphosphate accumulation in mitochondria. *Archives of Biochemistry and Biophysics* **183**, 581–7.

Åkerman, K. E. O. & Moore, A. L. (1983). Phosphate dependent, ruthenium red insensitive Ca^{2+} uptake in mung bean mitochondria. *Biochemical and Biophysical Research Communications* **114**, 1176–81.

Anderson, J. W. & House, C. M. (1979). Polarographic study of oxaloacetate reduction by isolated pea chloroplasts. *Plant Physiology* **64**, 1058–63.

Aquila, H., Eiermann, W., Babel, W. & Klingenberg, M. (1978). Isolation of the ADP/ATP translocator from beef heart mitochondria as the bongkrekate protein complex. *European Journal of Biochemistry* **85**, 569–60.

Barber, D. J. & Thurman, D. A. (1978). Transport of glutamine into isolated pea chloroplasts. *Plant, Cell and Environment* **1**, 297–303.

Beck, E., Burkert, A. & Hofmann, M. (1983). Uptake of L-ascorbate by intact spinach chloroplasts. *Plant Physiology* **73**, 41–5.

Birnberg, P. R. & Hanson, J. B. (1983). Mechanisms of citrate transport and exchange in corn mitochondria. *Plant Physiology* **71**, 803–9.

Birnberg, P. R., Jayroe, D. L. & Hanson, J. B. (1982). Citrate transport in corn mitochondria. *Plant Physiology* **70**, 511–16.

Blank, M. & Roughton, F. J. W. (1960). The permeability of monolayers to carbon dioxide. *Transactions of the Faraday Society* **56**, 1832–41.

Boggess, S. F., Koeppe, D. E. & Stewart, C. R. (1978). Oxidation of proline by plant mitochondria. *Plant Physiology* **62**, 22–5.

Bohni, P. C., Daum, G. & Schatz, G. (1983). Import of proteins into mitochondria. Partial purification of a matrix-located protease involved in cleavage of mitochondrial precursor polypeptides. *Journal of Biological Chemistry* **258**, 4937–43.

Bonner, W. D. & Prince, R. C. (1984). The Rieske iron–sulphur cluster of plant mitochondria. *FEBS Letters* **177**, 47–50.

Boyer, P. D. (1975). A model for conformational coupling of membrane potential and proton translocation to ATP synthesis and to active transport. *FEBS Letters* **58**, 1–6.

Boyer, P. D., Cross, R. L., & Momsen, W. (1973). A new concept for energy coupling in oxidative phosphorylation based on a molecular explanation of the oxygen exchange reactions. *Proceedings of the National Academy of Sciences, USA* **70**, 2837–9.

Brandolin G., Meyer C., Defaye, G., Vignais, P. M. & Vignais, P. V. (1974). Partial purification of an atractyloside-binding protein from mitochondria. *FEBS Letters* **46**, 149–53.

Broeck, G., Timko, M. P., Kausch, A.P., Cashmore, A. R. Montagu, M. & Herrera- Estrella, L. (1985). Targeting of a foreign protein to chloroplasts by fusion to the transit peptide from the small subunit of ribulose 1,5-bisphosphate-carboxylase. *Nature* **313**, 358–63.

Cammack, R. & Palmer, J. M. (1973). EPR studies of iron–sulphur proteins of plant mitochondria. *Annals of the New York Academy of Sciences, USA* **222**, 816–23.

Casey, R. P. (1984). Membrane reconstitution of the energy conserving enzymes of oxidative phosphorylation. *Biochimica et Biophysica Acta* **768**, 319–47.

Cashmore, A. R. (1984). Structure and expression of a pea nuclear gene encoding a chlorophyll *a*/*b*-binding polypeptide. *Proceedings of the National Academy of Sciences, USA* **81**, 2960–4.

Cashmore, A. R., Broadhurst, M. K. & Gray, R. E. (1978). Cell-free synthesis of leaf protein: identification of an apparent precursor of the small subunit of ribulose-1,5-bisphosphate carboxylase. *Proceedings of the National Academy of Sciences, USA* **75**, 655–9.

Cavalieri, A. J. & Huang, A. H. C. (1980). Carrier protein-mediated transport of neutral amino acids into mung bean mitochondria. *Plant Physiology* **66**, 588–91.

Cerletti, N., Bohni, P. C. & Suda, K. (1983). Import of proteins into mitochondria. Isolated yeast mitochondria and a solubilized matrix protease correctly processed cytochrome *c* oxidase subunit V precursor at the NH_2 terminus. *Journal of Biological Chemistry* **258**, 4944–9.

Chappell, J. & Beevers, H. (1983). Transport of dicarboxylic acids in castor bean mitochondria. *Plant Physiology* **72**, 434–40.

Chua, N. H. & Schmidt, G. W. (1978). Post translational transport into intact chloroplasts of a precursor to the small subunit of ribulose-1,5-bisphosphate carboxylase. *Proceedings of the National Academy of Sciences*, USA **75**, 6110–14.

Chua, N. H. & Schmidt, G. W. (1979). Transport of proteins into mitochondria and chloroplasts. *Journal of Cell Biology* **81**, 461–83.

Cline, K., Andrews, J., Mersey, B., Newcomb, E. H. & Keegstra, K. (1981). Separation and characterisation of inner and outer envelope membranes of pea chloroplasts. *Proceedings of the National Academy of Sciences USA*, **78**, 3595–9.

Cline, K., Werner-Washburne, M., Lubben, T. H. and Keegstra, K. (1985). Precursors to two nuclear-encoded chloroplast proteins bind to the outer envelope membrane before being imported into chloroplasts. *Journal of Biological Chemistry* **260**, 3691–6.

Coombs, J. & Greenwood, A. D. (1976). Compartmentation of the photosynthetic apparatus. In Barber, J. (ed.). *The Intact Chloroplast*, pp. 1–51. Amsterdam, Elsevier.

Crofts, A. R. & Wraight, C. A. (1983). The electrochemical domain of photosynthesis. *Biochimica et Biophysica Acta* **726**, 149–85.

Daum, G., Gasser, S. M. & Schatz, G. (1982). Import of proteins into mitochondria. Energy dependent two-step processing of the intermembrane space enzyme cytochrome b_2 by isolated yeast mitochondria. *Journal of Biological Chemistry* **257**, 13075–80.

Day, D. A., Bertagnolli, B. L. & Hanson, J. B. (1978). The effect of calcium on the respiratory responses of corn mitochondria. *Biochimica et Biophysica Acta* **502**, 289–97.

Day, D. A. & Hanson, J. B. (1977). Pyruvate and malate transport and oxidation in corn mitochondria. *Plant Physiology* **59**, 630–5.

Day, D. A. & Hatch, M. D. (1981a). Dicarboxylate transport in maize mesophyll chloroplasts. *Archives of Biochemistry and Biophysics* **211**, 738–42.

Day, D. A. & Hatch, M. D. (1981b). Transport of 3-phosphoglyceric acid, phosphoenolpyruvate and inorganic phosphate in maize mesophyll chloroplasts and the effect of 3-phosphoglyceric acid on malate and PEP production. *Archives of Biochemistry and Biophysics* **211**, 743–9.

Day, D. A. & Wiskich, J. T. (1977). Glutamate transport by plant mitochondria. *Plant Science Letters* **9**, 33–6.

Day, D. A. & Wiskich, J. T. (1980). Glycine transport by pea leaf mitochondria. *FEBS Letters* **112**, 191–4.

Day, D. A. & Wiskich, J. T. (1981a). Effect of phthalonic acid on respiration and metabolite transport in higher plant mitochondria. *Archives of Biochemistry and Biophysics* **211**, 100–7.

Day, D. A. & Wiskich, J. T. (1981b). Glycine metabolism and oxaloacetate transport by pea leaf mitochondria. *Plant Physiology* **68**, 425–9.

De Pinto, Y., Tommasino, M., Benz, R. & Palmieri, F. (1985). The 35 kDa DCCD-binding protein from pig heart mitochondria is the mitochondrial porin. *Biochimica et Biophysica Acta* **813**, 230–42.

De Santis, A., Arrigoni, O. & Palmieri, F. (1976). Carrier-mediated transport of metabolites in purified bean mitochondria. *Plant and Cell Physiology* **17**, 1221–33.

De Santis, A., Borraccino, G., Arrigoni, O. & Plamieri, F. (1975). The mechanism of phosphate permeation in purified bean mitochondria. *Plant and Cell Physiology* **16**, 911–23.

Deshaies, R. J., Fish, L. E. and Jagendorf, A. T. (1984). Permeability of chloroplast envelopes to Mg^{2+}. Effects of protein synthesis. *Plant Physiology* **74**, 956–61.

Dieter, P. & Marmé, D. (1980). Ca^{2+} Transport in mitochondrial and microsomal fractions from higher plants. *Planta* **150**, 1–8.

Diolez, P. & Moreau, F. (1985). Correlation between ATP synthesis, membrane potential and oxidation rate in potato mitochondria. *Biochimica et Biophysica Acta* **806**, 56--63.

Douce, R. (1985). *Mitochondria in Higher Plants: Structure, Function and Bioenergetics*. Orlando, Fl., Academic Press.

Douce, R., Holtz, R. B. & Benson, A. A. (1973). Isolation and properties of the envelope of spinach chloroplasts. *Journal of Biological Chemistry* **248**, 7215–22.

Dry, I. B. & Wiskich, J. T. (1983). Characterisation of dicarboxylate stimulation of ammonia-, glutamine- and 2-oxoglutarate-dependent O_2 evolution in isolated pea chloroplasts. *Plant Physiology* **72**, 291–6.

Ducet, G., Gidrol, X. & Richaud, P. (1983). Membrane potential changes in coupled potato (*Solanum tuberosum*) mitochondria. *Physiologie Végétale* **21**, 385–94.

Earnshaw, M. J. (1977). Adenine nucleotide translocation in plant mitochondria. *Phytochemistry* **16**, 181–4.

Ebbighausen, H., Jia, C. & Heldt, H. W. (1985). Oxaloacetate translocator in plant mitochondria. *Biochimica et Biophysica Acta* **810**, 184–99.

Ellis, R. J. (1983). Chloroplast protein synthesis: principles and problems. *Subcellular Biochemistry* **9**, 237–61.

Elthon, T. E. & Stewart, C. R. (1982). Proline oxidation in corn mitochondria. Involvement of NAD, relationship to ornithine metabolism and sidedness on the inner membrane. *Plant Physiology* **70** 567–72.

Elthon, T. E., Stewart, C. R. & Bonner, W. D. (1984). Energetics of proline transport in corn mitochondria. *Plant Physiology* **75**, 951–5.

Enser, U. & Heber, U. (1980). Metabolite regulation by pH gradients. Inhibition of photosynthesis by direct proton transfer across the chloroplast envelope. *Biochimica et Biophysica Acta* **592**, 577–91.

Esposti, M. D., Flamini, E. & Zannoni, D. (1985). Functional characterisation and partial purification of the ubiquinol–cytochrome *c* oxidoreductase from higher plant mitochondria from *Helianthus tuberosus*. *Plant Physiology* **77**, 758–64.

Feldman, R. I. & Sigman, D. S. (1982). The synthesis of enzyme-bound ATP by soluble chloroplast coupling factor 1. *Journal of Biological Chemistry* **257**, 1676–83.

Feldman, R. I. & Sigman, D. S. (1983). The synthesis of ATP by the membrane-bound ATP synthase complex from $^{32}P_i$ under completely uncoupled conditions. *Journal of Biological Chemistry* **258**, 12178–83.

Ferguson, S. J. (1985). Fully delocalised chemiosmotic or localised proton flow pathways in energy coupling. *Biochimica et Biophysica Acta* **811**, 47–95.

Ferguson, I. B., Reid, M. S. & Romani, R. J. (1985). Effects of low temperature and respiratory inhibitors on calcium flux in plant mitochondria. *Plant Physiology* **77**, 877–80.

Ferguson, S. J. & Sorgato, M. C. (1982). Proton electro-chemical gradients and energy-transduction processes. *Annual Review of Biochemistry* **51**, 185–217.

Fliege, R., Flugge, U., Werdan, K. & Heldt, H. W. (1978). Specific transport of inorganic phosphate, 3-phosphoglycerate and triose phosphates across the inner membrane of the envelope in spinach chloroplasts. *Biochimica et Biophysica Acta* **502**, 232–47.

Flugge, U. I. (1985). Hydrodynamic properties of the Triton X-100 solubilised chloroplast translocator. *Biochimica et Biophysica Acta* **815**, 299–305.

Flugge, U. I., Gerber, J. & Heldt, H. W. (1983). Regulation of the reconstituted chloroplast phosphate translocator by an H^+ gradient. *Biochimica et Biophysica Acta* **725**, 229–37.

Flugge, U. I. & Heldt, H. W. (1978). Specific labelling of the active site of the phosphate translocator in spinach chloroplasts by 2,4,6-trinitrobenzene sulphonate. *Biochimica et Biophysica Acta* **84**, 37–44.

Flugge, U. I. & Heldt, H. W. (1981). The phosphate translocator of the chloroplast envelope. Isolation of the carrier protein and reconstitution of transport. *Biochimica et Biophysica Acta* **638**, 296–304.

Flugge, U. I. & Heldt, H. W. (1984). The phosphate–triose, phosphate–phosphoglycerate translocator of the chloroplast. *Trends in Biochemical Science* **9**, 530–3.

Fragione, B., Rosenwasser, E., Penefsky, H. S. & Pullman, M. E. (1981). Amino acid sequence of the protein inhibitor of mitochondrial adenosine triphosphatase. *Proceedings of the National Academy of Sciences USA* **78**, 7403–7.

Freitag, H., Janes, M. & Neupert, W. (1982a). Biosynthesis of mitochondrial porin and insertion into the outer mitochondrial membrane of *Neurospora crassa*. *European Journal of Biochemistry* **126**, 197–202.

Freitag, H., Neupert, W. & Benz, R. (1982b). Purification and characterisation of a pore protein of the outer membrane from *Neurospora crassa* mitochondria. *European Journal of Biochemistry* **123**, 629–36.

Futai, M. & Kanazawa, H. (1983). Structure and function of proton-translocating adenosine triphosphatase (F_0 F_1): Biochemical and molecular biological approaches. *Microbiological Reviews* **47**, 285–312.

Gasser, S. M. & Schatz, G. (1983). Import of proteins into mitochondria. *In vitro* studies on the biogenesis of the outer membrane. *Journal of Biological Chemistry* **258**, 3427–30.

Gautheron, D. C. & Julliard, J. H (1979). Isolation of a glutamate carrier system from pig heart mitochondria and incorporation into liposomes. *Methods in Enzymology* **56**, 419–43.

Gibson, F. (1983). Biochemical and genetic studies on the assembly and function of the F_1–F_0 adenosine triphosphatase of *Escherichia coli*. *Biochemical Society Transactions* **11**, 229–40.

Gimmler, H., Schafer, G. & Heber, U. (1975). Low permeability of the chloroplast envelope toward cations. In Avron M. (ed.) *Proceedings of the Third International Congress on Photosynthesis*, pp. 1381–91. New York, Elsevier.

Gimmler, H., Schafer, G., Kraminer, H. & Heber, U. (1974). Amino acid permeability of the chloroplast envelope as measured by light scattering volumetry and amino acid uptake. *Planta* **120**, 47–61.

Gimpel, J. A., De Haan, E. J. & Tager, J. M. (1973). Permeability of isolated mitochondria to oxaloacetate. *Biochimica et Biophysica Acta* **292**, 582–91.

Gresser, M. J., Myers, J. A. & Boyer, P. D. (1982). Catalytic site cooperativity of beef heart mitochondrial F_1 adenosine triphosphatase. Correlation of initial velocity, bound intermediate and oxygen exchange measurements with an alternating site model. *Journal of Biological Chemistry* **257**, 12030–8.

Grossman, A., Bartlett, S. & Chua, N. H. (1980). Energy- dependent uptake of cytoplasmically synthesised polypeptides by chloroplasts. *Nature* **285**, 625–8.

Grossman, A. R., Bartlett, S. G., Schmidt, G. W., Mullet, J. E. & Chua, N.-H. (1982). Optimal conditions for post translational uptake of proteins by isolated chloroplasts. *In vitro* synthesis and transport of plastocyanin, ferredoxin-NADP$^+$ oxidoreductase and fructose-1,6-bisphosphatase. *Journal of Biological Chemistry* **257**, 1558–63.

Hack, E. & Leaver, C. J. (1983). The α-subunit of the maize F_1-ATPase is synthesised in the mitochondrion. *EMBO Journal* **2**, 1783–9.

Hack, E. & Leaver, C. J. (1984). Synthesis of a dicyclocarbodiimide-binding proteolipid by cucumber (*Cucumis sativus* L.) mitochondria. *Current Genetics* **8**, 537–42.

Haehnel, W. (1984). Photosynthetic electron transport in higher plants. *Annual Review of Plant Physiology* **35**, 659–93.

Hale, G. C., II & Roux, S. J. (1980). Photoreversible calcium fluxes induced by phytochrome in oat coleoptile cells. *Plant Physiology* **65**, 658–62.

Halestrap, A. P. & Denton, R. M (1974). Specific inhibition of pyruvate transport in rat liver mitochondria and human erythrocytes by α-cyano-4-hydroxycinnamate. *Biochemical Journal* **138**, 313–16.

Hase, T., Muller, U., Riezman, H. & Schatz, G. (1984). A 70 kD protein of the yeast mitochondrial outer membrane is targeted and anchored via its extreme amino terminus. *EMBO Journal* **3**, 3157–64.

Hatch, M. D., Droscher, L., Flugge U. I. & Heldt, H. W. (1984). A specific translocator for oxaloacetate transport in chloroplasts. *FEBS Letters* **178**, 15–19.

Hatefi, Y. (1985). The mitochondrial electron transport and oxidative phosphorylation system. *Annual Review of Biochemistry* **54**, 1015–69.

Hay, R., Bohni, P. & Gasser, S. (1984). How mitochondria import proteins. *Biochimica et Biophysica Acta* **779**, 65–87.

Heber, U. & Heldt, H. W. (1981). The chloroplast envelope: structure, function and role in leaf metabolism. *Annual Review of Plant Physiology* **32**, 139–68.

Heber, U. & Krause, G. H. (1971). Transfer of carbon phosphate energy and reducing equivalents across the chloroplast envelope. In Hatch, M. D., Osmond, C. B. & Slatyer, R. O. (eds.). *Photosynthesis and Photorespiration*, pp. 218–23. New York, Wiley-Interscience.

Heldt, H. W. (1969). Adenine nucleotide translocation in spinach chloroplasts. *FEBS Letters* **5**, 11–14.

Heldt, H. W. (1980). Measurement of metabolite movement across the envelope and of the pH in the stroma and the thylakoid space in intact chloroplasts. *Methods in Enzymology* **69**, 604–13.

Heldt, H. W., Chon, C. J., Maronde, D., Herold, A. Stankovic, Z. S., Walker, D. A., Kraminer, A., Kirk, M. R. & Heber, U. (1977). Role of orthophosphate and other factors in the regulation of starch formation in leaves and isolated chloroplasts. *Plant Physiology* **59**, 1146–55.

Heldt, H. W. & Rapley, L. (1970a). Unspecific permeation and specific uptake of substances in spinach chloroplasts. *FEBS Letters* **7**, 139–42.

Heldt, H. W. & Rapley, L. (1970b). Specific transport of inorganic phosphate, 3-phosphoglycerate and dihydroxyacetone phosphate and of dicarboxylates across the inner membrane of spinach chloroplasts. *FEBS Letters* **10**, 143–8.

Heldt, H. W. & Sauer, F. (1971). The inner membrane of the chloroplast envelope as the site of specific metabolite transport. *Biochimica et Biophysica Acta* **234**, 83–91.

Heldt, H. W., Werdan, K., Milovancev, M. & Geller, G. (1973). Alkalization of the chloroplast stroma caused by light-dependent proton flux into the thylakoid space. *Biochimica et Biophysica Acta* **314**, 224–41.

Herold, A., Leegood, R. C., McNeil, P. H. & Robinson, S. P. (1981). Accumulation of maltose during photosynthesis in protoplasts isolated from spinach leaves treated with mannose. *Plant Physiology* **67**, 85–8.

Heron, C., Smith, S. & Ragan C. I. (1979). An analysis of the bovine heart mitochondrial NADH-ubiquinone oxidoreductase by two-dimensional polyacrylamide-gel electrophoresis. *Biochemical Journal* **181**, 435–43.

Highfield, P. E. & Ellis, R. J., (1978). Synthesis and transport of the small subunit of chloroplast ribulose bisphosphate carboxylase. *Nature* **271**, 420–4.

Hong, Y. Q. & Junge, W. (1983). Localised or delocalised protons in photophosporylation? On the accessibility of the thylakoid lumen for ions and buffers. *Biochimica et Biophysica Acta* **722**, 197–208.

Hoppe, J. & Sebald, W. (1984). The proton conducting F_0 part of bacterial ATP synthases. *Biochimica et Biophysica Acta* **768**, 1–27.

Horak, A. & Packer, M. (1985). Coupling factor activity of the purified pea mitochondrial F_1-ATPase. *Biochimica et Biophysica Acta* **810**, 310–18.

Hrubec, T. C., Robinson, J. M. & Donaldson, R. P. (1985). Isolation of mitochondria from soybean leaves on discontinuous Percoll gradients. *Plant Physiology* **77**, 1010–12.

Huber, S. C. & Edwards, G. E. (1976). A high activity ATP translocator in mesophyll chloroplasts of *Digitaria sanguinalis*, a plant having the

C-4 dicarboxylic acid pathway of photosynthesis. *Biochimica et Biophysica Acta* **440**, 675–87.

Huber, S. C. & Edwards, G. E. (1977a). Transport in C_4 mesophyll chloroplasts. Characterisation of the pyruvate carrier. *Biochimica et Biophysica Acta* **462**, 583–602.

Huber, S. C. & Edwards, G. E. (1977b). Transport in C_4 mesophyll chloroplasts. Evidence for an exchange of inorganic phosphate and phosphoenol pyruvate. *Biochimica et Biophysica Acta* **462**, 603–12.

Huber, S. C. & Maury, W. (1980). Effects of magnesium on intact chloroplasts. 1. Evidence for activation of (sodium) potassium/proton exchange across the chloroplast envelope. *Plant Physiology* **65**, 350–4.

Issartel, J. P., Klein, G., Satre, M. and Vignais, P. V. (1983). Aurovertin binding sites on beef heart mitochondrial F_1-ATPase. Study with [^{14}C] aurovertin D of the binding stoichiometry and of the interaction between aurovertin and the natural ATPase inhibitor for binding to F_1. *Biochemistry* **22**, 3492–7.

John, P. & Whatley, F. R. (1975). *Paracoccus denitrificans* and the evolutionary origin of the mitochondrion. *Nature* **254**, 495–8.

Johnson, A. B. & Strecker, H. J. (1962). The interconversion of glutamic acid and proline. IV. The oxidation of proline by rat liver mitochondria. *Journal of Biological Chemistry* **237**, 1876–82.

Journet, E., Bonner., W. D. & Douce, R. (1982). Glutamate metabolism triggered by oxaloacetate in intact plant mitochondria. *Archives of Biochemistry and Biophysics* **214**, 366–75.

Journet, E., Neuberger, M. & Douce, R. (1980). Role of glutamate-oxaloacetate transaminase and malate dehydrogenase in the regeneration of NAD^+ for glycine oxidation by spinach leaf mitochondria. *Plant Physiology* **67** 467–9.

Jung, D. W. & Laties, G. G. (1976). Trypsin-induced ATPase activity in potato mitochondria. *Plant Physiology* **57**, 583–8.

Jung, D. W. & Laties, G. G. (1979). Citrate and succinate uptake by potato mitochondria. *Plant Physiology* **63**, 591–7.

Kaiser, W. M., Urbach, W. & Gimmler, H. (1980). The role of monovalent cations for photosynthesis of isolated intact chloroplasts. *Planta* **149**, 170–5.

Klingenberg, M. (1980). The ADP–ATP translocation in mitochondria, a membrane potential controlled transport. *Journal of Membrane Biology* **56**, 97–105.

Klingenberg, M., (1981). Membrane protein oligomeric structure and transport function. *Nature* **290**, 449–54.

Klingenberg, M. & Pfaff E. (1966). Structural and functional compartmentation in mitochondria. In Taga, J. M., Papa, S., Quagliariello, E. & Slater, E. C. (eds.) *Regulation of Metabolic Processes in*

Mitochondria, pp. 180–201. Amsterdam, Elsevier.

Klingenberg, M., Riccio, P. & Aquila, H. (1978). Isolation of the ADP, ATP carrier as the carboxyatractylate–protein complex from mitochondria. *Biochimica et Biophysica Acta* ***503***, 193–210.

Kramer, R. & Klingenberg, M. (1977). Reconstitution of adenine nucleotide transport with purified ADP, ATP-carrier protein. *FEBS Letters* **82**, 363–7.

Kramer, R. & Klingenberg, M. (1979). Reconstitution of adenine nucleotide transport from beef heart. *Biochemistry* **18**, 4209–15.

Kramer, R. & Klingenberg, M. (1982). Electrophoretic control of reconstituted adenine nucleotide translocation. *Biochemistry* **21**, 1082–9.

Land, J. M. & Clark, J. B. (1974). Inhibition of pyruvate and β-hydroxybutyrate oxidation in rat brain mitochondria by phenylpyruvate and α-ketoisocaproate. *FEBS Letters* **44**, 348–51.

La Noue, K., Mizani, S. M. & Klingenberg, M. (1978). Electrical imbalance of adenine nucleotide transport across the mitochondrial membrane. *Journal of Biological Chemistry* **253**, 191–8.

La Noue, K. F. & Schoolwerth, A. C. (1979). Metabolite transport in mitochondria. *Annual Review of Biochemistry* **48**, 871–922.

Lauquin, G. J. M., Brandolin, G., Lunardi, G. & Vignais, P. V. (1978). Photoaffinity labelling of the adenine nucleotide carrier in heart and yeast mitochondria by an arylazido ADP analog. *Biochimica et Biophysica Acta* **501**, 10–19.

Lehner, K. & Heldt, H. W. (1978). Dicarboxylate transport across the inner membrane of the chloroplast envelope. *Biochimica et Biophysica Acta* **501**, 531–44.

Lemaire, C., Girault, G. & Galmiche, J. M. (1985). Flash induced ATP synthesis in pea chloroplasts in relation to proton flux. *Biochimica et Biophysica Acta* **807**, 285–92

Lewin, A. S., Gregor I., Mason, T. L., Nelson, N. & Schatz, G. (1980). Cytoplasmically made subunits of yeast mitochondrial F_1-ATPase and cytochrome *c* oxidase are synthesised as individual precursors, not as polyproteins. *Proceedings of the National Academy of Sciences USA* **77**, 3998–4002.

Lima, M. S, & Denslow, N. D. (1979). The effect of atractyloside and carboxyatractyloside on adenine nucleotide translocation in mitochondria of *Vigna sinensis* L. *Savi cv. Serido*. *Archives of Biochemistry and Biophysics* **193**, 368–72.

McAda, P. C. & Douglas, M. G. (1982). A neutral metallo-endoprotease involved in the processing of an F_1 ATPase subunit precursor in mitochondria. *Journal of Biological Chemistry* **257**, 3177–82.

McLaren J. S. & Barber, D. J (1977). Evidence for carrier-mediated transport of L-leucine into isolated Pea (*Pisum sativum* L.) chloroplasts. *Planta* **136** 147–51.

Maeshima, M. & Asahi, T. (1978). Purification and characterisation of sweet potato cytochrome *c* oxidase. *Archives of Biochemistry and Biophysics* **187**, 423–430.

Malmström, B. G. (1985). Cytochrome *c* as a proton pump. A transition-state mechanism. *Biochimica et Biophysica Acta* **811**, 1–12.

Mandolino, G., De Santis, A. & Melandri, B. A. (1983). Localized coupling in oxidative phosphorylation by mitochondria from Jerusalem artichoke (*Helianthus tuberosus*). *Biochimica et Biophysica Acta* **723**, 428–39.

Matsuoka, M., Maeshima, M. & Asahi, T. (1981). The subunit composition of pea cytochrome *c* oxidase. *Journal of Biochemistry* **90**, 649–55.

Maury, W. J., Huber S. C. & Moreland, D. E. (1981). Effects of magnesium on intact chloroplasts. II. Cation specificity and involvement of the envelope ATPase in (sodium) potassium/proton exchange across the envelope. *Plant Physiology* **68**, 1257–63.

Mettler, I. J. & Beevers, H. (1980). Oxidation of NADH in glyoxysomes by a malate–aspartate shuttle. *Plant Physiology* **66**, 555–60.

Mishkind, M. L., Wessler, S. R. & Schmidt, G. W. (1985). Functional determinants in transit sequences: import and partial maturation by vascular plant chloroplasts of the ribulose-1,5-bisphosphate carboxylase small subunit of *Chlamydomonas*. *Journal of Cell Biology* **100**, 226–34.

Mitchell, P. (1961). Coupling of phosphorylation to electron and hydrogen transfer by a chemi-osmotic type of mechanism. *Nature* **191**, 144–8.

Mitchell, P. (1974). A chemiosmotic molecular mechanism for proton translocating adenosine triphosphatases. *FEBS Letters* **43**, 189–94.

Mitchell, P. (1985). Molecular mechanisms of protonmotive F_0F_1 ATPases. Rolling well and turnstile hypothesis. *FEBS Letters* **182**, 1–7.

Mitchell, P., Mitchell, R., Moody, A. J., West, I. C., Baum, H. & Wrigglesworth, J. M. (1985). Chemiosmotic coupling in cytochrome oxidase, possible protonmotive O loop and O cycle mechanisms. *FEBS Letters* **188**, 1–7.

Møller, I. M. & Palmer, J. M. (1981). Properties of the oxidation of exogenous NADH and NADPH by plant mitochondria. Evidence against a phosphatase or a nicotinamide nucleotide transhydrogenase being responsible for NADPH oxidation. *Biochimica et Biophysica Acta* **638**, 225–33.

Moore, A. L. & Åkerman, K. E. O. (1982). Ca^{2+} stimulation of the external NADH dehydrogenase in Jerusalem artichoke (*Helianthus tuberosum*) mitochondria. *Biochemical and Biophysical Research Communications* **109**, 513–17.

Moore, A. L. & Wilson, S. B. (1978). An estimation of the proton conductance of the inner membrane of turnip (*Brassica napus* L.) mitochondria. *Planta* **141**, 297–302.

Moreau, F. & Davy de Virville, J. (1985). Stoichiometry of proton translocation coupled to substrate oxidation in plant mitochondria. *Plant Physiology* **77**, 118–23.

Moroney, J. V., Lopresti, L., McEwen, B. F., McCarty, R. E. & Hammes, G. (1983). The M_r-value of chloroplast coupling factor 1. *FEBS Letters* **158**, 58–62.

Moroney, J. V. & McCarty, R. E. (1982). Light-dependent cleavage of γ-subunit of coupling factor 1 by trypsin causes activation of Mg^{2+}-ATPase activity and uncoupling of phosphorylation in spinach chloroplasts. *Journal of Biological Chemistry* **257**, 5915–20.

Mowbray, J. (1974). Evidence for the role of a specific monocarboxylate transporter in the control of pyruvate oxidation by rat liver mitochondria. *FEBS Letters* **44**, 344–7.

Nechushtai, R., Nelson, N., Mattoo, A. K. & Edelman, M. (1981). Site of synthesis of subunits to photosystem I reaction center and the proton ATPase in *Spirodela*. *FEBS Letters* **125**, 115–19.

Nelson, N., Nelson, H. & Schatz, G. (1980). Biosynthesis and assembly of the proton-translocating adenosine triphosphatase complex from chloroplasts. *Proceedings of the National Academy of Sciences, USA* **77**, 1361–4.

Neuberger, M., Day, D. A. & Douce, R. (1985). Transport of NAD^+ in Percoll-purified potato tuber mitochondria. *Plant Physiology* **78**, 405–10.

Neuberger, M. & Douce, R. (1980). Effect of bicarbonate and oxaloacetate on malate oxidation by spinach leaf mitochondria. *Biochimica et Biophysica Acta* **589**, 176–89.

Nguyen, T. D. & Siegenthaler, P. (1985). Purification and some properties of an Mg^{2+}–Ca^{2+} and calmodulin stimulated ATPase from spinach chloroplast envelope membranes. *Biochimica et Biophysica Acta* **840**, 99–106.

Oliver, D. J. & Walker, G. H. (1984). Characterisation of the transport of oxaloacetate by pea leaf mitochondria. *Plant Physiology* **76**, 409–13.

O'Neal, C. C. & Boyer, P. D. (1984). Assessment of the rate of bound substrate interconversion and ATP acceleration of product release during catalysis by mitochondrial adenosine triphosphatase. *Journal of Biological Chemistry* **259**, 5761–7.

Passarella, S., Palmieri, F. & Quagliariello, E. (1977). The transport of oxaloacetate in isolated mitochondria. *Archives of Biochemistry and Biophysics* **180**, 160–8.

Pfaff, E., Klingenberg, M. & Heldt, H. W. (1965). Unspecific permeation and specific exchange of adenine nucleotides in liver mitochondria. *Biochimica et Biophysica Acta* **104**, 312–15.

Pfisterer, J., Lachmann, P. & Kloppstech, K. (1982). Transport of proteins into chloroplasts. Binding of nuclear-coded chloroplast proteins to the chloroplast envelope. *European Journal of Biochemistry* **126**, 143–8.

Poincelot, R. P. (1974). Uptake of bicarbonate ion in darkness by isolated chloroplast envelope membranes and intact chloroplasts of spinach. *Plant Physiology* **54**, 520–6.

Proudlove, M. O. & Moore, A. L. (1982). Movement of amino acids into isolated plant mitochondria. *FEBS Letters* **147**, 26–30.

Proudlove, M. O., Thurman, D. A. & Salisbury, J. (1984). Kinetic studies on the transport of 2-oxoglutarate and L-malate into isolated pea chloroplasts. *New Phytologist* 961–5.

Ragan, C. I. & Hinkle, P. C. (1975). Ion transport and respiratory control in vesicles formed from reduced NAD–CoQ reductase and phospholipids. *Journal of Biological Chemistry* **250**, 8472–6.

Richter, M. L., Patrie, W. J. & McCarty, R. E. (1984). Preparation of the ϵ-subunit and ϵ-subunit-deficient chloroplast coupling factor 1 in reconstitutively active forms. *Journal of Biological Chemistry* **259**, 7371–3.

Richter, M. L., Snyder, B., McCarty, R. E. & Hammes G. G. (1985). Binding stoichiometry and structural mapping of the ϵ-polypeptide of chloroplast coupling factor 1. *Biochemistry* **24**, 5755–3.

Riezman, H., Hay, R., Witte, C., Nelson, N. & Schatz, G. (1983). Yeast mitochondrial outer membrane specifically binds cytoplasmically synthesised precursors of mitochondrial proteins. *EMBO Journal* **2**, 1113–18.

Robinson, C. & Ellis, R. J. (1984a). Tranport of proteins into chloroplasts. Partial purification of a chloroplast protease involved in the processing of imported polypeptides. *European Journal of Biochemistry* **142**, 337–42.

Robinson, C. & Ellis, R. J. (1984b). Transport of proteins into chloroplasts. The precursor of small subunit of ribulose bisphosphate carboxylase is processed to the mature size in two steps. *European Journal of Biochemistry* **142**, 343–6.

Robinson, S. P. (1982). Transport of glycerate across the envelope membrane of isolated spinach chloroplasts. *Plant Physiology* **70** 1032–8.

Robinson, S. P. (1984). Lack of ATP requirement for light stimulation of glycerate transport into intact isolated chloroplasts. *Plant Physiology* **75** 425–30.

Robinson, S. P. (1985). The involvement of stromal ATP in maintaining the pH gradient across the chloroplast envelope in the light. *Biochimica et Biophysica Acta* **806**, 187–94.

Robinson, S. P. & Wiskich, J. T. (1977). Pyrophosphate inhibition of carbon dioxide fixation in isolated pea chloroplasts by uptake in exchange for endogenous adenine nucleotides. *Plant Physiology* **59**, 422–7.

Rochaix, J. D. (1985). Genetic organisation of the chloroplast. *International Review of Cytology* **93**, 57–91.

Rosen, G., Gresser, M., Vinkler, C. & Boyer P. D. (1979). Assessment of total catalytic sites and the nature of bound nucleotide participation in photophosphorylation. *Journal of Biological Chemistry* **254**, 10654–61.

Roux, S. J., McEntire, K., Slocum, R. D., Cedel, T. E. & Hale, C. C., II (1981). Phytochrome induces photoreversible calcium fluxes in a purified mitochondrial fraction from oats. *Proceedings of the National Academy of Sciences, USA* **78**, 283–7.

Ryrie, I. J., Critchley, C. & Tillberg, J. (1979). Structure and energy-linked activities in reconstituted bacteriorhodopsin-yeast ATPase proteoliposomes. *Archives of Biochemistry and Biophysics* **198**, 182–94.

Schafer, G., Heber, U. & Heldt, H. W. (1977). Glucose transport into spinach chloroplasts. *Plant Physiology* **60**, 286–9.

Schlodder, E. & Witt, H. T. (1981). Relation between the initial kinetics of ATP synthesis and of conformational changes in the chloroplast ATPase studied by external fluid pulses. *Biochimica et Biophysica Acta* **635**, 571–84.

Schmidt, G. W., Bartlett, S. G., Grossman, A. R., Cashmore, A. R. & Chua, N.-H. (1981). Biosynthetic pathways of two polypeptide subunits of the light-harvesting chlorophyll *a/b* protein complex. *Journal of Cell Biology* **91**, 468–78.

Schmidt, G. W., Devillers-Thiery, A., Desruisseaux, H., Blobel, G. & Chua, N.-H. (1979). NH_2-terminal amino acid sequences of precursor and mature forms of the ribulose-1,5-bisphosphate carboxylase small subunit from *Chlamydomonas reinhardtii*. *Journal of Cell Biology* **83**, 615–22.

Schneider, E. & Altendorf, K. (1982). ATP synthetase (F_1F_0) of *Escherichia coli* K-12. High yield preparation of functional F_0 by hydrophobic affinity chromatography. *European Journal of Biochemistry* **126**, 149–53.

Schultheiss, H. P. & Klingenberg, M. (1984). Immunochemical characterisation of the adenine nucleotide translocator. Organ specificity and conformation specificity. *European Journal of Biochemistry* **143**, 599–605.

Schwerzmann, K. & Pedersen P. L. (1981). Proton–adenosine triphosphatase complex of rat liver mitochondria: effect of energy state on its interaction with the adenosine triphosphatase inhibitory peptide. *Biochemistry* **20**, 6305–11.

Sells, G. D. & Koeppe, D. E. (1981). Oxidation of proline by mitochondria isolated from water stressed maize shoots. *Plant Physiology* **68**, 1058–63.

Senior, A. E. & Wise, J. G. (1983). The proton–ATPase of bacteria and mitochondria. *Journal of Membrane Biology* **73**, 105–124.

Serlin, B. S., Sopory, S. K. & Roux, S. J. (1984). Modulation of oat mitochondrial ATPase activity by Ca^{2+} and phytochrome. *Plant Physiology* **74**, 827–33.

Shahak, Y. (1985). Differential effect of thiol oxidants on the chloroplast H^+-ATPase in the light and in the dark. *Journal of Biological Chemistry* **260**, 1459–1464.

Shiraiwa, Y. & Miyachi, S. (1978). Form of inorganic carbon utilised for photosynthesis across the chloroplast membrane. *FEBS Letters* **95**, 207–10.

Sicher, R. C. (1984). Characteristics of light-dependent inorganic carbon uptake by isolated spinach chloroplasts. *Plant Physiology* **74**, 962–6.

Sigrist-Nelson, K. & Azzi, A. (1980). The proteolipid subunit of the chloroplast adenosine triphosphatase complex. Reconstitution and demonstration of proton-conductive properties. *Journal of Biological Chemistry* **255**, 10638–43.

Slater, E. C. (1953). Mechanism of phosphorylation in the respiratory chain. *Nature* **172**, 975–8.

Slater, E. C., Berden, J. A. & Herweijer, M. A. (1985). A hypothesis for the mechanism of respiratory-chain phosphorylation not involving the electrochemical gradient of protons as obligatory intermediate. *Biochimica et Biophysica Acta* **811**, 217–31.

Somerville, S. E. & Ogren, W. L. (1983). An *Arabidopsis thaliana* mutant defective in chloroplast dicarboxylate transport. *Proceedings of the National Academy of Sciences, USA* **80**, 1290–4.

Sone, N., Yoshida, M., Hirata, H. & Kagawa, Y. (1977). Adenosine triphosphate synthesis by electrochemical proton gradient in vesicles reconstituted from purified adenosine triphosphatase and phospholipids of a thermophilic bacterium. *Journal of Biological Chemistry* **252**, 2956–60.

Soper, J. W., Decker, G. L. & Pedersen, P. L. (1979). Mitochondrial ATPase complex. A dispersed cytochrome-deficient oligomycin-sensitive preparation from rat liver containing molecules with a tripartite structural arrangement. *Journal of Biological Chemistry* **254**, 11170–6.

Spitsberg, V. L., Pfeiffer, N. E., Partridge, B., Wylie, D. E. & Schuster, S. M. (1985). Isolation and antigenic characterisation of corn mitochondrial F_1-ATPase. *Plant Physiology* **77**, 339–45.

Stewart, C. R., Boggess, S. F., Aspinall, D. & Paleg, L. G. (1977). Inhibition of proline oxidation by water stress. *Plant Physiology* **59**, 930–2.

Stitt, M., Wirtz, W. & Heldt, H. W. (1980). Metabolite levels during induction in the chloroplast and extra chloroplast compartments of spinach protoplasts. *Biochimica et Biophysica Acta* **593**, 85–102.

Teintze, M., Slaughter, M., Weiss, H. & Neupert, W. (1982). Biogenesis of mitochondrial ubiquinol: cytochrome *c* reductase (Cytochrome bc_1 complex). Precursor proteins and their transfer into mitochondria. *Journal of Biological Chemistry* **257**, 10364–71.

Theg, S. M. & Homann, P. H. (1982). Light-, pH- and uncoupler-dependent association of chloride with chloroplast thylakoids. *Biochimica et Biophysica Acta* **679**, 221–34.

Vignais, P. V. (1976). Molecular and physiological aspects of adenine nucleotide transport in mitochondria. *Biochimica et Biophysica Acta* **456**, 1–38.

Vignais, P. V., Douce, R., Lauquin, G. L. M., & Vignais, P. M. (1976). Binding of radioactively labelled carboxyatractyloside, atractyloside and bongkrekic acid to the ADP translocator of potato mitochondria. *Biochimica et Biophysica Acta* **440**, 688–96.

Walker, R. G. H., Sarojini, G. & Oliver, D. J. (1982). Identification of a glycine transporter from pea leaf mitochondria. *Biochemical and Biophysical Research Communications* **107**, 856–61.

Werdan, K., Heldt, H. W. & Geller, G. (1972). Accumulation of bicarbonate in intact chloroplasts following a pH gradient. *Biochimica et Biophysica Acta* **283**, 430–41.

Werner-Washburne, M. & Keegstra, K. (1985). L-Aspartate transport into pea chloroplasts. *Plant Physiology* **78**, 221–7.

Whatley, J. M., John, P. & Whatley, F. R. (1979). From extracellular to intracellular: establishment of mitochondria and chloroplasts. *Proceedings of the Royal Society, London, B* **204**, 165–87.

Whatley, J. M. & Whatley, F. R. (1981). Chloroplast evolution. *New Phytologist* **87**, 233–247.

Whisson, M. B. & Spencer, M. S. (1983). Solubilisation and partial purification of *N*, *N′*-dicyclohexylcarbodiimide-sensitive ATPase of pea cotyledon mitochondria. *Plant Physiology* **71**, 707–11.

Wikstrom, M. K. F. (1977). Proton pump coupled to cytochrome *c* oxidase in mitochondria. *Nature* **266**, 271–3.

Wikstrom, M. (1984). Two protons are pumped from the mitochondrial matrix per electron transferred between NAD and ubiquinone. *FEBS Letters* **169**, 300–4.

Wikstrom, M. & Krab, K. (1979). Proton pumping cytochrome oxidase. *Biochimica et Biophysica Acta* **549**, 177–222.

Wilson, R. H. & Minton, G. A. (1974). The comparative uptake of Ba^{2+} and other alkaline earth metals by plant mitochondria. *Biochimica et Biophysica Acta* **333**, 22–7.

Winget, G. D., Kanner, N. & Racker, E. (1977). Formation of ATP by the adenosine triphosphatase complex from spinach chloroplasts reconstituted together with bacteriorhodopsin. *Biochimica et Biophysica Acta* **460**, 490–9.

Wiskich, J. T. (1974). Substrate transport into plant mitochondria: swelling studies. *Australian Journal of Plant Physiology* **1**, 177–81.

Woldegiorgis, G., Voss, S., Shrago, E., Werner-Washburne, M. & Keegstra, K. (1985). Adenine nucleotide translocase-dependent anion transport in pea chloroplasts. *Biochimica et Biophysica Acta* **810**, 340–5.

Woo, K. C. (1983). Effect of inhibitors on ammonia, 2-oxoglutarate and oxaloacetate dependent O_2 evolution in illuminated chloroplasts. *Plant Physiology* **71**, 112–17.

Wood, C., Jalil, N. H., McLaren, I., Yong, B. C. S, Ariffin, A., McNeil, P. H., Burgess, N. & Thomas, D. R. (1984). Carnitine long chain acyl transferase and oxidation of palmitate, palmitoyl coenzyme A and palmitoyl carnitine by pea mitochondria preparations. *Planta* **161**, 255–60.

Yu, C., Claybrook, D. L. & Huang, A. H. C. (1983). Transport of glycine, serine and proline into spinach leaf mitochondria. *Archives of Biochemistry and Biophysics* **227**, 180–7.

Zalman, L. S., Nikaido, H. & Kagawa, Y. (1980). Mitochondrial outer membrane contains protein producing non-specific diffusion channels. *Journal of Biological Chemistry* **255**, 1771–4.

Zimmerman, R., Hennig, B. & Neupert, W. (1981). Different transport pathways of individual precursor proteins in mitochondria. *European Journal of Biochemistry* **116**, 455–60.

Zimmerman, R., Paluch, U. & Neupert, W. (1979). Cell-free synthesis of cytochrome *c*. *FEBS Letters* **108**, 141–6.

Zitomer, R. S. & Hall, B. D. (1976). Yeast cytochrome *c* messenger RNA *Journal of Biological Chemistry* **251**, 6320–6.

3 Plasma membrane and tonoplast

R. J. Poole

3.1 Introduction: Functions of plasma membrane and tonoplast

The plasma membrane and tonoplast are the limiting membranes of the two major compartments of the plant cell, the cytoplasm and vacuole, and share in the functions of cytoplasmic homeostasis, nutrition and osmoregulation. These functions, which require a coordinated activity of the two membranes, may be illustrated by the following examples:

1. Both membranes pump Na^+ and Ca^{2+} out of the cytoplasm. However, the relative importance of the two membranes depends on the situation. For example, Na^+ efflux at the plasma membrane predominates in glycophytes and especially in cells with a high ion content, whereas Na^+ transport at the tonoplast is more important in halophytes (Flowers *et al.*, 1977) or in salt-depleted glycophytes (Pitman *et al.*, 1968; see also Ch. 10).
2. Nutrients taken up at the plasma membrane in excess of immediate needs are transported across the tonoplast. Thus the two membranes may act in series to accumulate solutes in the vacuoles. On the other hand, useful nutrients must later be retrieved from the vacuole. For example, Aslam *et al.* (1976) showed that NO_3 stored in leaf vacuoles is retrieved under the influence of light.
3. The large-scale transport of osmotic solutes across both membranes in the course of cell enlargement or turgor movements can change the volume of the vacuole several-fold without changing the volume of the cytoplasm (Campbell & Garber, 1980). How transport at the two membranes is coordinated so as to control cytoplasmic volume is not known.

Along with these shared functions of plasma membrane and tonoplast, each membrane performs additional specific roles. The plasma membrane is the main boundary, as far as solutes are concerned, between cell and environment, and must carry receptors for various exogenous signals such as hormones, toxins, and elicitors of the cellular responses to pathogens. Thus, sites at the plasma membrane for auxin transport (Goldsmith, 1982) and for fusicoccin (FC) binding (Dohrmann *et al.*, 1977) have been indicated. The plasma membrane is also the site of secretion of cell wall materials by exocytosis, and of some aspects of polysaccharide metabolism, e.g. glucan synthase II (Quail, 1979). The vacuole, on the other hand, is recognized as a lysosomal compartment (Matile, 1978) and the tonoplast shares some functions of lysosomal membranes. For example, proton transport at the tonoplast not only provides a driving force for other transport processes (see below) but also facilitates some post-translational processing of proteins (Stinissen *et al.*, 1985). The vacuole also serves to store proteins, particularly in legumes, and to store a variety of secondary metabolites which function as insecticides, fungicides, etc., to which the tonoplast must be resistant.

To understand the physiological roles of plasma membrane and tonoplast, we need to study the properties of each membrane in isolation, as well as the interaction of their activities in the cell. The coordination and physiological consequences of transport at these membranes will be discussed in other chapters. The present chapter will be concerned with the individual properties of each membrane, and especially with what has been learned from the study of plasma membrane and tonoplast isolated from the cell in the form of either membrane fragments or vesicles.

Notwithstanding the varied functions of the plasma membrane and tonoplast outlined above, at both of these membranes a dominant role is played by the proton-motive force, or electrochemical gradient of H^+ ($\Delta\bar{\mu}_{H^+}$) a concept already discussed in detail for chloroplasts and mitochondria (Ch. 2). A major theme in this chapter will therefore be the mechanisms involved in the generation and utilization of ΔpH and $\Delta\psi$, mechanisms which are in fact different for each membrane.

3.2 Membrane structure and turnover

This topic is discussed in detail in a recent monograph by Robinson (1985). The plasma membrane and tonoplast have the basic structure common to all cellular membranes, consisting of a continuous lipid bilayer into which globular proteins are inserted. Both membranes are extensively glycosylated on their exoplasmic surfaces. Both plasma membrane and tonoplast have a complex and variable complement of lipids, including phospholipids (especially phosphatidyl choline and phosphatidyl ethanolamine) along with large amounts of glycolipids, sterols and sterol glycosides. In each case, the lipid mixture permits a degree of fluidity, which can be measured by the sharpness of ESR signals from spin-labelled probes, or by fluorescence depolarization. Whereas the tonoplast appears to be particularly liquid and expandable, the plasma membrane is relatively rigid, with a lower fluidity than other cellular membranes. This is reflected in its lipid composition: the plasma membrane has a smaller proportion of unsaturated fatty acids in its phospholipids, and a higher proportion of sterols than other membranes. In the electron microscope, the plasma membrane appears to be the thickest membrane in the plant cell.

Since lipid bilayers are essentially impermeable to most metabolically important substances (except water), almost all transport properties of cellular membranes may be attributed to specific transport proteins inserted into the bilayer. A freeze–fracture electron microscope study of the red alga *Porphyridium* by Knoth and Wiencke (1984) showed an increase in the number of membrane particles in the tonoplast during a period of adaptation (some hours) after imposing a salt stress. The particles were thought to represent new transport proteins inserted into the tonoplast to cope with the increased salt load in the cytoplasm.

The assortment of polypeptides in a given membrane may be visualized by polyacrylamide gel electrophoresis in the presence of the detergent sodium dodecyl sulphate. Such studies show not only that the plasma membrane and tonoplast contain different mixtures of proteins, but also that the protein composition of the plasma membrane changes during normal development (Booz & Travis, 1980) and also changes in response to new environmental conditions, such as low temperature (Yoshida & Uemura, 1984).

Most membrane proteins in plants, as in animals, are synthesized on endoplasmic reticulum-bound polysomes, and processed and transported by the ER and Golgi apparatus. The final steps in the synthesis of several membrane lipids are also in the ER and Golgi. It is assumed therefore that the growth, adaptation and turnover of plasma membrane and tonoplast involve the flow of membrane vesicles from the Golgi and/or the ER , and membrane fusion at the cell surface or vacuole. It has been estimated that membrane in the form of secretory vesicles may be added to the plasma membrane at rates of up to 10 % of the plasma membrane per minute (Robinson, 1985). Since this is much faster than the rates of turnover of membrane proteins (from a few hours to a few days) it is assumed that exocytosis is accompanied by endocytosis and recycling of membrane vesicles. Some evidence has been obtained for the recycling of plasma membrane to the *trans*-face of the Golgi apparatus (Robinson, 1985). Little is known at present about how specific membrane proteins are targeted to the plasma membrane or tonoplast, or about the sorting of protein molecules which enables each organelle to retain its characteristic proteins in the face of a constant exchange of membrane between organelles.

3.3 Isolation of plasma membrane and tonoplast

Understanding cell activities always requires a combination of *in vivo* and *in vitro* studies. In the case of membrane activities, difficulties in isolation and handling of membrane fractions have greatly limited our knowledge of membrane composition, enzymology and transport properties. These technical problems are gradually being overcome, and an increasing amount of information is coming from studies on isolated membrane (see Hall & Moore, 1983).

3.3.1 Plasma membrane markers

The first requirement for isolation of a subcellular component is the identification of suitable markers (Quail, 1979; Robinson, 1985) to recognize the desired fraction and to detect contamination by other cell constituents. Traditional markers for the plasma membrane include phosphotungstate–chromate staining, high sterol content and glucan synthase II activity. Following the characterization of a K^+-stimulated ATPase at the plasma membrane (Hodges *et al.*, 1972), this ATPase has also been used as a marker. The plasma membrane ATPase is now more reliably recognized by its sensitivity to vanadate, especially in the presence of molybdate, which inhibits other vanadate-sensitive phosphatases (Gallagher & Leonard, 1982). Additional markers which have been used include FC or naphthylphthalamic acid binding, UDPG–sterol glucose transferase, and in the case of plasma membrane isolated from protoplasts, surface labelling with ^{125}I. The distributions of these markers amongst various isolated fractions do not always coincide. This might indicate that some of the markers are present in other cellular membranes such as Golgi vesicles, or that the plasma membrane itself is not homogeneous, and can be subfractionated into patches with different characteristics. It appears that auxin transport sites, for example, are restricted to one end of cells involved in the polar transport of auxin (Goldsmith, 1982).

3.3.2 Tonoplast markers

Isolation of tonoplast from homogenates was not possible until recently because of the lack of known markers. This has been remedied by studies on tonoplast enzyme activities following the isolation of intact vacuoles either from sliced tissue (Leigh & Branton, 1976) or from protoplasts (Wagner & Seigelman, 1975). A characteristic anion-sensitive ATPase and a cation-sensitive pyrophosphatase identified by Walker & Leigh (1981a, b) and others have been used as tonoplast markers. When tissue homogenates are fractionated by sucrose density gradient centrifugation (see below), these enzyme activities show the same density distribution as do fragments from isolated intact vacuoles (e.g. Bennett *et al.*, 1984; Rea & Poole, 1985).

3.3.3 Protectants during isolation

A second requirement for subcellular fractionation is the stabilization of membranes against degradation, especially by hydrolytic enzymes and toxic secondary metabolites released from broken vacuoles (e.g. Loomis, 1974). Typically, the plant tissue is homogenized at 0 °C in the presence of an osmoticum and pH buffer, and with the addition of various protective agents. These commonly include EDTA (or EGTA plus a low concentration of Mg^{2+}) to prevent membrane aggregation and to bind heavy metals, reducing agents to prevent harmful oxidative reactions, polyvinylpyrrolidone to bind phenolics released from the vacuoles, and bovine serum albumin to bind free fatty acids. Still more additives may be used to inhibit lipid breakdown (Scherer & Morré, 1978) or proteolysis. The need for particular chemical protectants varies with the tissue or species used.

3.3.4 Purification of plasma membrane and tonoplast fractions

Cellular membranes and organelles are most often separated by a series of centrifugation steps. Typically, the homogenate is centrifuged at low speed (13 000 × *g*) to remove large organelles and cell walls, then at higher speed (80 000 × *g*) to sediment the remaining membranes. This microsomal pellet is layered on a sucrose density gradient and centrifuged at 80 000 × *g* for 2 h. Plasma membrane markers are then found to be concentrated between 34 and 45 % sucrose, but this can vary even in the same material at different stages of development (Poole *et al.*, 1984), and tonoplast markers are found near the top of the gradient, at 10–25 % sucrose. The extent of separation of plasma membrane and tonoplast, and the degree of contamination by ER, Golgi, mitochondria or other membranes varies with the material, and often is not accurately known even after assaying for various markers. The purity of plasma membrane fractions as judged by specific staining may be about 75 % (Hodges *et al.*, 1972).

Various other methods have been reported for membrane fractionation. These include the use of dextran or other substances for density gradient centrifugation (Sze, 1985). An advantage of dextran over sucrose for density gradients is that the osmotic pressure can be kept approximately constant through the gradient, thus avoiding possible membrane damage by swelling and shrinking of vesicles. A possible alternative for plasma membrane isolation is a phase partition method using two immiscible polymers, dextran and polyethylene glycol, in aqueous solution. This method separates membranes by surface properties rather than density. It is difficult to achieve both good yield and purity, but the method is finding some application (Yoshida *et al.*, 1983; Larsson, 1985). Plasma membranes have also been isolated from protoplasts by binding to cationic silica microbeads (Polonenko & Maclachlan, 1984).

3.4 Membrane vesicles

When tissue is homogenized, broken pieces of membrane have a natural tendency to round up and seal to form hollow spheres or vesicles. This is a thermodynamically stable configuration for membrane lipids, as shown by the spontaneous formation of vesicles (liposomes) when a suspension of phospholipids is agitated by ultrasonic vibration. Membrane vesicles derived from the plasma membrane or tonoplast are very useful for studying transport across these membranes, primarily because one has much better experimental control of the solutes (transport substrates, ATP, pH, etc.) on each side of the membrane than in the intact tissue. In addition, the large surface/volume ratio of membrane vesicles allows changes in vesicle contents to be detected within seconds after activation of a transport system. It also allows the internal solution to be equilibrated with any desired medium by incubating the vesicles in the appropriate solution for, say, 30 min.

Accumulation of a radioactive substrate into membrane vesicles may be measured by scintillation counting after collecting the vesicles on a filter. Internal acidification by a proton pump may be detected by the accumulation of a labelled weak base, and the membrane potential may be determined from the distribution of a lipid-soluble ion such as thiocyanate. More conveniently, a number of dyes have been found to respond to $\triangle$pH or membrane potential $\triangle\psi$ in membrane vesicles by changes in their absorbance or fluorescence, thus giving a continuous non-destructive measure of these important parameters. Although in some cases the response of these dyes to $\triangle$pH or $\triangle\psi$ has been given a theoretical basis (Lee *et al.*, 1982), the change in absorbance or fluorescence is usually treated empirically. For a given membrane preparation, the dye response can be calibrated by measuring absorption or fluorescence, when a known ΔpH or $\Delta\psi$ is imposed experimentally (Lee *et al.*, 1982; Bennett & Spanswick, 1983; Blumwald & Poole, 1985b). It is difficult to use this calibration to determine absolute values of $\triangle\psi$ and $\triangle$pH created by a proton pump, mainly because it is unlikely that all the vesicles in a given preparation are behaving in an identical manner. The vesicles vary in size, and may also by chance have differing numbers of pump molecules or ion channels. Nevertheless, quantitative measurements of $\triangle$pH, $\triangle\psi$, and proton flux can be made in membrane vesicles, the values being placed on an arbitrary rather than an absolute scale.

Studies on membrane vesicles from plant cells have been reviewed by Sze (1985). In practice, tonoplast membranes prepared as described above usually do consist of sealed vesicles, which respond to added ATP + Mg^{2+} by acidification of their contents. These vesicles evidently contain proton pumps with their ATP-hydrolysing sites on the external surface, i.e. with the same orientation as in the tonoplast of the intact cell. It is not yet known whether inside-out vesicles are also present, since these would not be detected by the methods used.

An experiment on proton transport in tonoplast vesicles is shown in Fig. 3.1. On addition of Mg^{2+} in presence of ATP, rapid quenching of fluorescence from the weak base acridine orange indicates internal acidification of the vesicles. The initial rate of quenching should be proportional to the net flux of protons into the vesicles (Bennett & Spanswick, 1983). This is increased by Cl^- due to a direct action of Cl^- on the H^+-translocating ATPase (Bennett & Spanswick, 1983). When the negatively charged dye oxonol V is used to indicate $\Delta\psi$, rapid fluorescence quenching on addition of Mg^{2+} indicates the formation of an inside-positive potential by the H^+-ATPase. In this case, the rate of quenching is decreased by Cl^-, indicating that Cl^- is entering the vesicles in response to the potential and

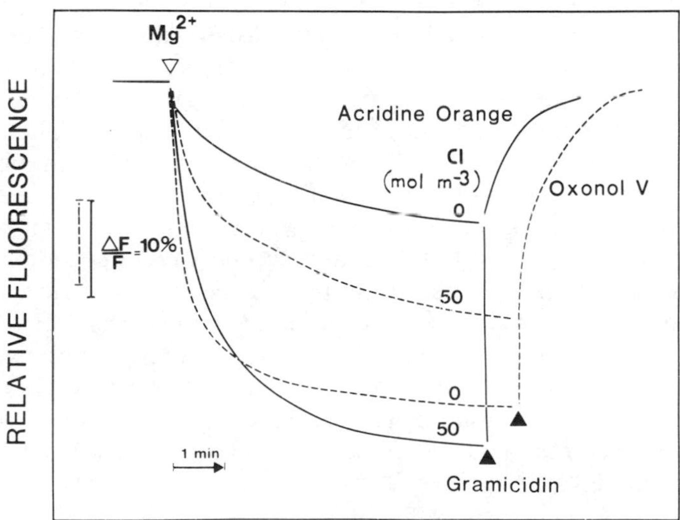

Fig. 3.1 Use of fluorescent dyes to detect changes in ΔpH and Δψ in tonoplast vesicles. ATP was present at the start, and proton pumping was initiated by addition of Mg^{2+}. Fluorescence quenching of acridine orange (solid lines) indicates acidification of vesicle contents, and that of oxonol V (broken lines) indicates formation of an inside-positive potential; 50 mol m^{-3} Cl^- was present from start where indicated. In absence of Cl^-, 25 mol m^{-3} SO_4^{2-} was included. Fluorescence quenching was reversed by gramicidin, which forms ion channels through the membranes. F, fluorescence. (From Blumwald & Poole, 1985b.)

thereby reducing its value. This can be confirmed by measuring $^{36}Cl^-$, uptake into the vesicles (Mettler *et al.*, 1982). Such experiments thus give information both on the properties of the proton pump and on the nature of ion channels across the tonoplast.

Plasma membrane preparations also appear vesicular in electron micrographs, but often do not seal well enough to permit detection of transport. The reasons for the poorer sealing of plasma membrane vesicles are not known. Nevertheless, some preparations do show internal acidification of vesicles on addition of ATP and Mg^{2+}. Plasma membrane vesicles responding in this way must be inside-out compared with the intact cell. Right-side-out vesicles are also formed from the plasma membrane, and these can be selectively purified by the phase partition method mentioned above (Sect. 3.3.4).

3.5 Proton pumps

Electrophysiological studies have shown that plant cells typically pump protons outward across the plasma membrane into the external solution or intercellular space, and also across the tonoplast into the vacuole. The proton pumps at the plasma membrane and tonoplast create gradients of pH and membrane potential which serve as driving forces for various secondary transport processes across these membranes. Vesicles derived from the plasma membrane or tonoplast transport protons and create a membrane potential when supplied with ATP + Mg^{2+}. The isolated membranes also hydrolyse ATP by enzymic reactions with properties corresponding to those of the proton pumps. Thus each membrane contains an H^+-transporting ATPase. However, the enzyme that performs this function at the tonoplast differs in molecular structure, properties and mechanism of action from its counterpart at the plasma membrane, and each of these enzymes differs from the 'energy-transducing' H^+-ATPases with 'F_1' and 'F_0' segments found in mitochondria, chloroplasts, and bacteria (Sze, 1984). Table 3.1 compares some of the structural properties of the three types of proton-pumping ATPase. These and other aspects of the plasma membrane and tonoplast H^+-ATPases are discussed below in Sects. 3.5.1 and 3.5.2.

3.5.1 Plasma membrane H^+-translocating ATPase

The plasma membrane ATPase was first identified as a K^+-stimulated enzyme with kinetics similar to the kinetics of K^+ uptake into plant roots. It is now known that the primary function of this enzyme is H^+ transport, as shown both *in vivo* (Spanswick, 1981; Marre & Ballarin-Denti, 1985)

Table 3.1 Subunit structure of three types of H^+-translocating ATPase found in plant cells

	Estimated sizes† of some subunits with defined functions, and of the entire enzymes (kDa)		
	Plasma membrane	**Tonoplast**	**F_1–F_0**
ATP-binding subunits	100	66–74 (89) 54–64 (64)	55–62 50–6
Phosphorylated subunits	100	None	None
DCCD-binding subunits	100	14–18 (19.5)	8
Functional size of enzyme	228	400–520	460

† Size ranges are the range of estimates reported by different investigators with different organisms. Values in parentheses are those reported for yeast. Estimates of functional size were made by radiation inactivation (target size) analysis. References in text, and Sussman & Slayman (1983); Vignais & Satre (1984); Briskin *et al.* (1985), Mandala & Taiz (1985); Uchida *et al.* (1985).

and in liposomes containing the purified enzyme (Serrano, 1984). Uptake of K^+, as well as the transport of various other ions and organic molecules, is directly or indirectly dependent on this enzyme, but it is still not known whether K^+ is carried by the ATPase itself. Although K^+ stimulates both H^+ transport and ATPase activity, it is not obligatory for either of these processes.

The structure and mechanism of the plant plasma membrane H^+-ATPase (Serrano, 1984) is similar to that of other 'plasma membrane type' ATPases, including the Na^+, K^+-ATPase of animal cells, the gastric H^+, K^+-ATPase, the sarcoplasmic reticulum Ca^{2+}-ATPase, and the K^+-transporting Kdp ATPase of *Escherichia coli* (Epstein, 1985). These enzymes have one major polypeptide 100 kDa or more in size, which spans the membrane and carries out both the hydrolytic and transport functions. Some of the pumps also have one or two other associated polypeptides. The gene for the major polypeptide has been cloned and sequenced in several cases, including yeast plasma membrane H^+-ATPase, which resembles that of higher plants, and strong homologies have been found (see Ch. 4; and Serrano *et al*, 1986). During the reaction cycle of these enzymes, the terminal phosphate of ATP is transferred to an aspartyl residue on the enzyme to form a transient phosphorylated intermediate, which is then hydrolysed to release inorganic phosphate. In the Na^+, K^+-ATPase, Na^+ stimulates phosphorylation and is transported during this step, while K^+ stimulates dephosphorylation as it in turn is transported by the enzyme. The action of K^+ on the plant enzyme is similar, in that K^+ stimulates the hydrolysis of the phosphorylated intermediate (Briskin & Poole, 1983). Transport ATPases which form a phosphorylated intermediate are typically inhibited by vanadate, which mimics the conformation of phosphate in its transitional state. In plants, vanadate inhibits the proton pump at the plasma membrane, but not those at the tonoplast, mitochondria or chloroplasts (Sze, 1984).

A number of other reagents inhibit the plasma membrane H^+-ATPase: these include the synthetic oestrogen diethylstilbestrol (DES), the lipophilic carboxyl reagent *N*, *N′*-dicyclohexylcarbodiimide (DCCD) and various sulphydryl reagents. These reagents also inhibit the other proton pumps to a greater or lesser extent. The fungal toxin FC (Marrè, 1979) is a specific activator of the plasma membrane proton pump, resulting in a wide range of secondary effects in plant cells, including hyperpolarization of the membrane potential, increased proton efflux, a transient increase in growth rate, increased transport of various other substances and opening of stomata. Although stimulation of proton pumping after incubation with FC has been demonstrated in isolated membrane preparations, there is no evidence that FC binds directly to the ATPase itself.

As might be expected for an enzyme playing a central role in plasma membrane energization, the plasma membrane H^+-ATPase is subject to regulation by a variety of factors. The most direct form of regulation is that exerted by the components of the proton gradient itself, namely the cytoplasmic and extracellular pH values and the membrane potential. Any change in these factors will tend to be compensated by a change in the rate of proton pumping. In this way, the rate of conversion of ATP energy into proton motive force by the H^+-ATPase will be adjusted to the demands of the various transport systems which utilize the energy of the proton gradient. In electrophysiological studies, the efflux of protons through the pump can be measured as an electric current. The relationship of the pump current to the membrane potential (I/V curve) can be plotted for various internal and external pH values to define exactly how the pump responds to these factors. Detailed studies of this kind have so far only been possible in certain types of cells, such as the giant cells of the algae *Nitella* and *Chara*, the hyphae of the fungus *Neurospora* and rhizoids of the liverwort *Riccia* (Spanswick, 1981; and Chs. 4 and 5).

The plasma membrane ATPase is also subject to regulation by a number of other factors. Thus it appears to be activated after a lag period by auxin (Bates & Goldsmith, 1983), it responds rapidly to various forms of environmental stress (Chastain & Hanson, 1982), and in the case of stomata (see Ch. 12) it is influenced by the various factors which regulate stomatal opening (light, ABA, etc.)

3.5.2 Tonoplast H^+-translocating ATPase

The H^+-translocating ATPase of the tonoplast (Sze, 1984, 1985) also serves to convert ATP energy into proton gradient energy, but as noted above, it differs markedly from the plasma membrane enzyme. The tonoplast ATPase is sensitive to anions rather than cations, being stimulated by Cl^-

and some other anions, but inhibited by NO_3^- It is not known to transport any ion except H^+. Partial purification of the tonoplast H^+-ATPase from various species and use of affinity labelling (Manolson *et al.*, 1985; Randall & Sze, 1986) has shown that it has a multimeric structure including two polypeptides of 70 kDa and 60 kDa which have nucleotide substrate binding sites, and a smaller polypeptide of about 16 kDa which binds DCCD. Other polypeptides are detectable in partially purified ATPase preparations (e.g. Marin *et al*, 1985) but in the absence of any identified function, it is not clear whether these could be additional components of the ATPase, or simply contaminants. These properties, and the lack of inhibition by vanadate, suggest a structure and mechanism somewhat resembling that of the mitochondrial type of ATPase. However, the tonoplast ATPase is insensitive to inhibitors of mitochondrial ATPase such as azide and oligomycin, and shows no antigenic cross-reactivity with mitochondrial ATPase (Bowman *et al.*, 1986).

The vacuole is one of the various organelles which are involved in membrane recycling and together form the endomembrane system of eukaryotic cells. These include the ER, Golgi apparatus, secretory vesicles, endocytotic and coated vesicles, endosomes and lysosomes. These endomembranes of plant and animal cells all appear to possess H^+-translocating ATPases which acidify their lumens. Present indications are that the H^+-ATPases of other endomembranes such as the Golgi apparatus of plants (Chanson & Taiz, 1985) and chromaffin granules of animals (Percy *et al.*, 1985) may resemble the tonoplast H^+-ATPase in structure and mechanism.

3.5.3 Tonoplast H^+-translocating pyrophosphatase

A pyrophosphatase found on the surface of isolated vacuoles and tonoplast vesicles has also been shown to function as a proton pump, acidifying the lumen and creating a membrane potential much like the tonoplast ATPase (Rea & Poole, 1985). The pyrophosphatase is cation sensitive, being stimulated by K^+ and inhibited by Na^+, and is insensitive to anions. It is highly specific for inorganic pyrophosphate (PP_i) as substrate. Depending on the tissue and on the conditions (e.g. Mg^{2+} concentration) it can be as active as the ATPase, and since its apparent K_m for PP_i can be as low as 15 mmol m^{-3} (Wang *et al.*, 1986) it should be amply supplied with substrate in the cell (cf. Smyth & Black, 1984). The importance of PP_i in plant energetics is also emphasized by the discovery of PP_i-dependent PFK as a control point for glycolysis, this enzyme being activated by the regulator compound fructose 2,6-bisphosphate (Smyth & Black, 1984).

The significance of two proton pumps operating in parallel at the tonoplast is not yet clear. However, this situation is not unique. An H^+-translocating ATPase and an H^+-translocating PPase occur in parallel in the thylakoids of *Rhodospirillum rubrum* (Baltscheffsky & Nyren, 1984) as well as in mitochondria and chloroplasts (Kulaev *et al.*, 1980), although in these cases the enzymes normally catalyse the reverse reactions, generating ATP and PP_i, respectively. Kulaev *et al.* (1980) have shown that the balance between ATP and PP_i production in mitochondria is correlated with the viscosity of the membrane lipids. In the case of the tonoplast, the H^+-ATPase and H^+-PPase may help to maintain the proton gradient under variable ionic conditions (NO_3^- or Na^+ inhibition) or variable supply of substrates (ATP or PP_i). As in the case of the plasma membrane ATPase, these tonoplast H^+ pumps are sensitive to the proton gradient and thus will tend to adjust their activity depending on the rate at which the proton gradient is dissipated by secondary transport mechanisms.

3.6 Other primary transport processes

Although proton pumps are usually the predominant energy transducers converting chemical energy into gradient energy at the plasma membrane and tonoplast, they are not the only enzymes which engage in such primary transport. Calcium-transporting ATPases appear to be associated with the plasma membrane (Dieter & Marmé, 1980) as well as with the endoplasmic reticulum (Buckhout, 1984; Bush & Sze, 1986) of plant cells. (The tonoplast also transports Ca^{2+}, but by means of a Ca^{2+}/H^+ antiport, see below.) These enzymes pump Ca^{2+} out of the cytosol and create a very large electrochemical gradient for Ca^{2+} across the plasma membrane. This makes the intracellular Ca^{2+} concentration very sensitive to perturbations of the membrane, and thus creates a system for regulating metabolism and development in response to extracellular signals (Hepler & Wayne, 1985). In marine algae such as *Acetabularia*, a Cl^- pump appears to replace the H^+ pump as the main energy transducer at the plasma membrane (see Ch. 5). Ion pumps driven by oxidation–reduction energy have also been proposed for the plant plasma membrane (see Ch. 1). Current evidence (Crane *et al.*, 1985) suggests that a redox system is present in the plasma membrane, and that it may play a role in transport. However, more evidence is needed to decide whether the redox system is regulatory in nature, or whether there is a direct utilization of redox energy for transport at the plasma membrane.

3.7 Secondary active transport at plasma membrane and tonoplast

Secondary active transport refers to solute transport which is energized by direct coupling to the flux of another solute (in plant cells usually H^+ moving down its electrochemical gradient (Ch. 1). *In vivo*, the 'downhill' direction for protons is inward across the plasma membrane and outward from the vacuole. Thus at either membrane, solutes may be moved into the cytoplasm by proton symport or out of the cytoplasm by proton antiport.

3.7.1 Characteristics of proton-coupled secondary transport

Proton-coupled transport shows a number of distinctive features (see e.g. Reinhold & Kaplan, 1984). In addition to the usual properties of carrier-mediated transport (saturation with increasing substrate concentration, substrate specificity, etc.) there should be a stoichiometric relationship between the substrate flux and its associated proton flux. This implies that a component of proton flux should show identical kinetics to the substrate flux with respect to substrate concentration, specificity, inhibition, etc. Also, the transport of substrate should be dependent on the proton chemical or electrochemical gradient (depending on whether the transporter carries a net charge), substrate transport should be reversed if the proton gradient is reversed, and it should be independent of other energy sources such as ATP. If the net charge of substrate plus protons transported is not zero, the transport kinetics should also be reflected by a component of electric current across the membrane, and by corresponding changes in membrane potential.

3.7.2 Proton-coupled transport at the plasma membrane

In practice, a detailed characterization of proton-coupled transport *in vivo* is difficult, not only because it is difficult to manipulate factors such as cytoplasmic pH, membrane potential and ATP levels independently, but also because of the tendency of the proton pumps to change their rates to compensate for any changes in ΔpH or $\Delta\psi$. Because of this, changes in proton flux and membrane potential following addition of a transport substrate may be only transitory. Changes in external pH due to proton fluxes may also be buffered by the cell wall, or confounded with pH changes caused by respiratory CO_2.

If the transport system carries a net charge, some of these difficulties can be avoided by studies of membrane current on addition of substrate under voltage clamp conditions (see Chs. 4 and 5 and Felle, 1983, for examples). This approach is limited to suitable types of cell and has not generally been possible in higher plants. Nevertheless, a considerable amount of information has been accumulated on proton-coupled influx of sugars and amino acids at the plasma membrane in many types of plant cells (Reinhold & Kaplan, 1984). For anion influx, anion–proton symports have been invoked, but apart from electrophysiological studies on *Chara* indicating a ($1Cl^-$: $2H^+$) symport (Sanders, 1980; Beilby & Walker, 1981), information on anion influx remains meagre. For Na^+ efflux, Ratner & Jacoby (1976) showed indirect evidence for Na^+/H^+ antiport in barley root tips. This antiport is probably electroneutral and therefore cannot be studied by electrophysiology. Activation of a plasma membrane K^+/H^+ antiport leading to cell death has been suggested as a basis for the hypersensitive reaction to pathogens, which limits the spread of disease within the plant (Atkinson *et al.*, 1985). If methods for obtaining sealed plasma membrane vesicles can be improved, this experimental system offers an important alternative approach for the study of these various transport mechanisms, as described below for tonoplast vesicles.

3.7.3 Proton-coupled transport at the tonoplast

As described in Sect. 3.4, sealed membrane vesicles are spontaneously formed from the tonoplast during membrane isolation, and a variety of methods are available for studying solute fluxes in such membrane vesicles under defined conditions. For example, the fluorescent weak base acridine orange accumulates in vesicles as the internal pH is decreased. This accumulation results in a quenching of fluorescence, which can be used to monitor ΔpH (Lee *et al.*, 1982). This method was applied by Blumwald and Poole (1985a) to characterize an Na^+/H^+ antiport at the tonoplast of red beet, a halophytic species in which the vacuolar accumulation of Na^+ is particularly important. The internal pH of tonoplast vesicles was adjusted to pH 6 by incubation of the membranes in a buffer of this pH. The membranes were then diluted into a buffer of pH 8, creating a ΔpH of 2 units across the membrane. In the presence of acridine orange, the fluorescence was immediately quenched and then recovered only very slowly, showing that the proton gradient remained essentially constant for a few minutes. On addition of Na^+, there was a rapid increase in fluorescence, indicating an Na^+-dependent efflux of H^+ from the vesicles (Fig. 3.2). Further experiments showed that reversal of the Na^+ gradient reversed the direction of the proton flux. The Na^+-dependent H^+ flux was not abolished by maintaining $\Delta\psi$ at zero by the presence of K^+ and valinomycin, and was therefore attributed to an electroneutral Na^+/H^+ antiport. The Na^+ dependent H^+ flux showed saturation with increasing Na^+ concentration and was inhibited by amiloride, a diuretic which inhibits Na^+/H^+ antiport in kidney epithelial cells (Kinsella & Aronson, 1981).

A Ca^{2+}/H^+ antiport has also been described in tonoplast vesicles (Schumaker & Sze, 1985; Bush & Sze, 1986). This antiport can be extremely fast (Blumwald & Poole, 1986) suggesting that the tonoplast plays an important role in controlling cytoplasmic Ca^{2+} levels.

The positive membrane potential maintained in plant vacuoles and tonoplast vesicles is sufficient to allow at least a two or three fold accumu-

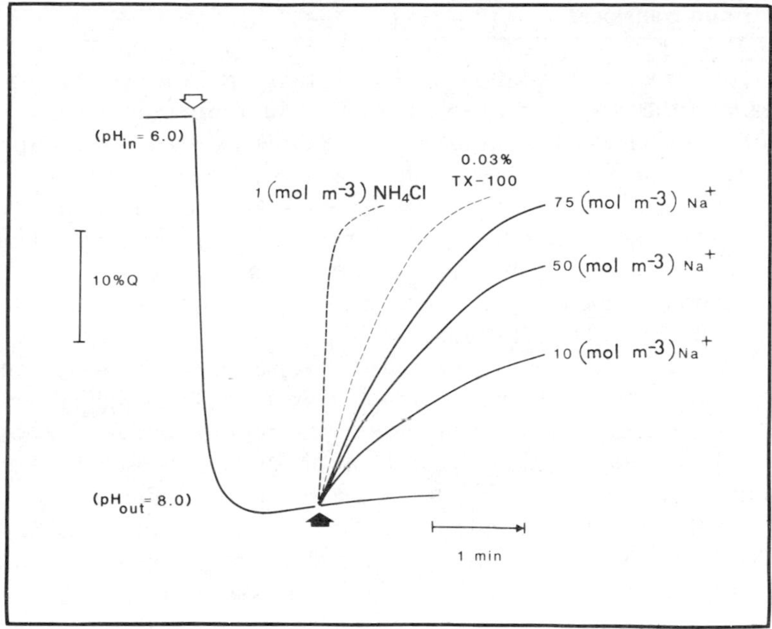

Fig. 3.2 Na^+-dependent H^+ efflux from tonoplast vesicles. Q, quenching of fluorescence from acridine orange. At the first arrow, tonoplast vesicles at pH 6 are added to a solution at pH 8, creating a ΔpH of 2 units. At the second arrow, addition of NH_4Cl or Triton X-100 to collapse ΔpH allows recovery of fluorescence to a level corresponding to zero ΔpH. Addition of Na^+ also allows recovery of fluorescence due to Na^+/H^+ antiport. (From Blumwald & Poole, 1985a.)

lation of anions from the cytoplasm (more for divalent ions), provided that a permeability mechanism (channel or uniport) is present in the membrane. Such passive permeability for Cl^- at the tonoplast is indicated by the observations that: (a)$^{36}Cl^-$ is accumulated in tonoplast vesicles in presence of ATP + Mg^{2+}, but this accumulation is prevented by protonophores (Mettler *et al.*, 1982); and (b) accumulation of Cl^- is accompanied by partial dissipation of $\Delta\psi$ but not ΔpH (Bennett & Spanswick, 1983; Blumwald & Poole, 1985b). In contrast to Cl^-. low concentrations of NO_3^- (insufficient to inhibit the tonoplast ATPase) dissipate both $\Delta\psi$ and ΔpH in tonoplast vesicles. This observation led to the suggestion (Blumwald & Poole, 1985b) that there is an NO_3^- : H^+ symport at the tonoplast which could enable the cell to retrieve this metabolically important ion from storage in the vacuole. Similar mechanisms may be involved in the transport of organic anions such as malate, which is subject to massive reversals of tonoplast net flux in Crassulacean acid metabolism (CAM) (Ch. 11) and in stomatal function (Ch. 12).

3.8 Group translocation

The term 'group translocation' refers to transport in which the substrate appears on the other side of the membrane in a different chemical form. Thus it is a chemical group, not an entire molecule, which is transported. Although there is evidence (Thom & Komor, 1984) for proton-coupled transport of glucose at the tonoplast, Thom & Maretzki (1985) have found evidence for a group translocation mechanism for sucrose accumulation in isolated vacuoles of sugar-cane. Addition of labelled UDPG resulted in accumulation of labelled sucrose and sucrose phosphate inside the vacuoles but not in the external medium. It is proposed that several associated enzymes at the tonoplast metabolize cytoplasmic UDPG and release sucrose and sucrose phosphate into the vacuole. Although a series of reactions is presumably involved, neither the addition of various supposed intermediates nor addition of sucrose or sucrose phosphate, with or without ATP + Mg^{2+}, promoted sucrose accumulation. The glucan synthase of the plasma membrane also involves group translocation, since UDPG must be supplied to the cytoplasmic face of the membrane in order to form β-glucans in the cell wall (Mueller & Maclachlan, 1983).

3.9 Ion channels and other uniports

Since lipid bilayers are essentially impermeable to most biologically important solutes, the passive permeability properties of membranes, just as much as their active transport properties, depend on the insertion of specific proteins, which act as selective channels through the membrane. Ion channels, which have been studied by electrophysiological methods in giant algal cells, have only recently become accessible to study in higher plant cells, through the use of the patch–clamp technique. These developments have been reviewed briefly by MacRobbie (1985) and at greater length by Takeda *et al.* (1985). In the patch–clamp technique, a patch of membrane is induced to seal tightly across the tip of a micropipette electrode. With appropriate circuitry, this enables ion currents to be detected even in a single channel formed by a protein molecule embedded in the membrane, and permits each conformational change resulting in channel opening or closing to be recorded as a separate event in time.

A distinction may be made between channels, in which the ion flux is so rapid that the opening and closing of single channels can be detected electrically, and carrier-mediated uniports which permit the passage of ions or molecules one at a time. An amine uniport has been identified at the plasma membrane, and although the transport of amines (especially NH_4^+) is rapid it appears to be carrier-mediated and competitive for NH_4^+ versus methylamine, with an apparent K_m of 2 mmol m^{-3} for NH_4^+ (Felle, 1980).

The first well defined example of an ion channel in higher plants is the demonstration of K^+-selective channels in the plasma membrane of stomatal guard cells (Schroeder *et al.*, 1984; and Ch. 12). It is likely that K^+ channels as well as anion channels will soon be detected in the tonoplast by these methods. The idea of cytoplasmic Ca^{2+} as a regulator of metabolism and development (Sect. 3.6) implies the existence of Ca^{2+} channels which can be opened transiently by external signals. (In practice, most types of channel have been found to operate intermittently and to be subject to various controlling factors.) In addition, there is evidence in various systems that developmental polarity is determined by the asymmetric distribution of ion channels around the cell (Hepler & Wayne, 1985). These new developments in membrane research are therefore likely to have far-reaching implications in plant biology.

In conclusion, Fig. 3.3 indicates some of the variety of mechanisms for solute transport currently under investigation at the plasma membrane and tonoplast.

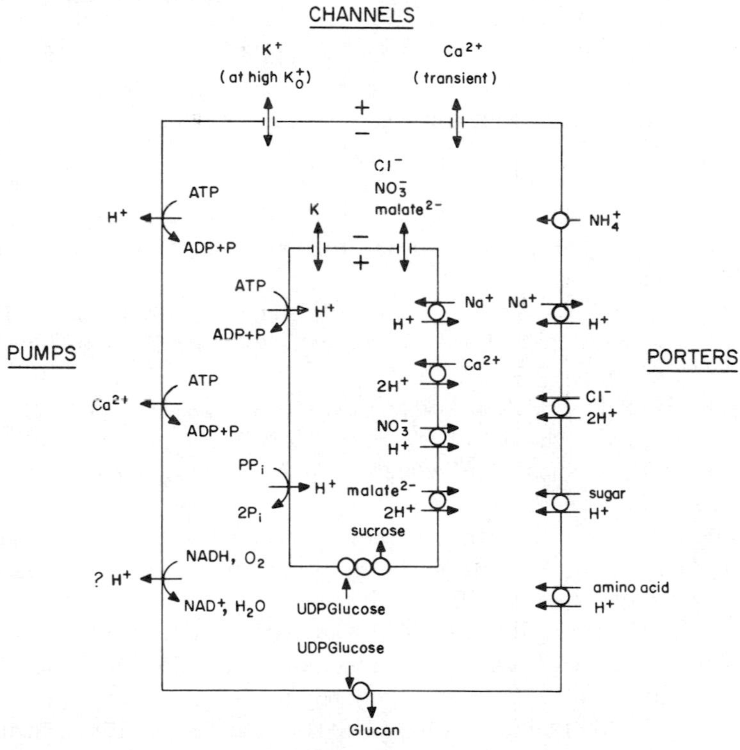

Fig. 3.3 Some transport systems identified or postulated at the plasma membrane and tonoplast of plant cells.

References

Aslam, M., Oaks, A. & Huffaker, R. C. (1976). Effect of light and glucose on the induction of nitrate reductase and on the distribution of nitrate in etiolated barley leaves. *Plant Physiology* **58**, 588–91.

Atkinson, M. M., Huang, J.-S. & Knopp, J. A. (1985). The hypersensitive reaction of tobacco to *Pseudomonas syringae* pv. *pisi*. Activation of a plasmalemma K^+/H^+ exchange mechanism. *Plant Physiology* **79**, 843–7.

Baltscheffsky, M. & Nyren, P. (1984). Membrane-bound inorganic pyrophosphatase. In Bolis, C. L., Helmreich, E. J. M. & Passow, H. (eds.). *Information and Energy Transduction in Biological Membranes*, pp. 199–207. New York, Alan R. Liss.

Bates, G. W. and Goldsmith, M .H .M. (1983). Rapid response of the plasma- membrane potential in oat coleoptiles to auxin and other weak acids. *Planta* **159**, 231–7.

Beilby, M. J. & Walker, N. A. (1981). Chloride transport in *Chara*. I. Kinetics and current–voltage curves for a probable proton symport. *Journal of Experimental Botany* **32**, 43–54.

Bennett, A. B., O'Neill, S. D. & Spanswick, R. M. (1984). H^+-ATPase activity from storage tissue of *Beta vulgaris*. I. Identification and characterization of an anion-sensitive H^+-ATPase. *Plant Physiology* **74**, 538–44.

Bennett, A. B. & Spanswick, R. M. (1983). Optical measurements of ΔpH and $\Delta\psi$ in corn root membrane vesicles: Kinetic analysis of Cl^- effects on a proton translocating ATPase. *Journal of Membrane Biology* **71**, 95–107.

Blumwald, E. & Poole, R. J. (1985a). Na^+/H^+ antiport in isolated tonoplast vesicles from storage tissue of *Beta vulgaris*. *Plant Physiology* **78**, 163–7.

Blumwald, E. & Poole, R. J. (1985b). Nitrate storage and retrieval in *Beta vulgaris*: Effects of nitrate and chloride on proton gradients in tonoplast vesicles. *Proceedings of the National Academy of Sciences*, USA **82**, 3683–7.

Blumwald, E. & Poole, R. J. (1986). Kinetics of Ca^{2+}/H^+ antiport in isolated tonoplast vesicles from storage tissue of *Beta vulgaris* L. *Plant Physiology* **80**, 727–31.

Booz, M. L. & Travis, R. L. (1980). Electrophoretic comparison of polypeptides from enriched plasma membrane fractions from developing soybean roots. *Plant Physiology* **66**, 1037–43.

Bowman, E. J., Mandala, S., Taiz, L. & Bowman, B. J. (1986). Structural studies of the vacuolar membrane ATPase from *Neurospora crassa* and comparison with the tonoplast membrane ATPase from *Zea mays*. *Proceedings of the National Academy of Sciences, USA* **83**, 48–52

Briskin, D. P. & Poole, R. J. (1983). Plasma membrane ATPase of red beet forms a phosphorylated intermediate. *Plant Physiology* **71**, 507–12.

Briskin, D. P., Thornley, W. R. & Roti-Roti, J. L. (1985). Target molecular size of the red beet plasma membrane ATPase. *Plant Physiology* **78**, 642–4.

Buckhout, T. J. (1984). Characterization of Ca^{2+} transport in purified endoplasmic reticulum membrane vesicles from *Lepidium sativum* L. roots. *Plant Physiology* **76**, 962–67.

Bush, D. R. & Sze, H. (1986). Calcium transport in tonoplast and endoplasmic reticulum vesicles from cultured carrot cells. *Plant Physiology* **80**, 549–555.

Campbell, N. A. & Garber, R. C. (1980). Vacuolar reorganization in the motor cells of *Albizzia* during leaf movement. *Planta* **148**, 251–5.

Chanson, A. & Taiz, L. (1985). Evidence for an ATP-dependent proton pump on the Golgi of corn coleoptiles. *Plant Physiology* **78**, 232–40.

Chastain, C. J. & Hanson, J. B. (1982). Control of proton efflux from corn root tissue by an injury-sensing mechanism. *Plant Science Letters* **24**, 97–104.

Crane, F. L., Sun, I. L., Clark, M. G., Grebing, C. & Low, H. (1985). Transplasma-membrane redox systems in growth and development. *Biochimica et Biophysica Acta* **811**, 233–64.

Dieter, P. & Marmé, D. (1980). Calmodulin activation of plant microsomal Ca^{2+} uptake. *Proceedings of the National Academy of Sciences, USA* **77**, 7311–314.

Dohrmann, U., Hertel, R., Pesci, P., Cocucci, S. M., Marrè, E., Randazzo, G. & Ballio, A. (1977). Localization of 'in vitro' binding of the fungal toxin fusicoccin to plasma membrane-rich fractions from corn coleoptiles. *Plant Science Letters* **9**, 291–9.

Epstein, W. (1985). The Kdp system: a bacterial K^+ transport ATPase. *Current topics in Membranes and Transport* **23**, 153–75.

Felle, H. (1980). Amine transport at the plasmalemma of *Riccia fluitans*. *Biochimica et Biophysica Acta* **602**, 181–95.

Felle, H. (1983). Driving forces and current–voltage characteristics of amino acid transport in *Riccia fluitans*. *Biochimica et Biophysica Acta* **730**, 342–50.

Flowers, T. J., Troke, P. F. & Yeo, A. R. (1977). The mechanism of salt tolerance in halophytes. *Annual Review of Plant Physiology* **28**, 89–121.

Gallagher, S. R. & Leonard, R. T. (1982). Effect of vanadate, molybdate and azide on membrane-associated ATPase and soluble phosphatase activities of corn roots. *Plant Physiology* **70**, 1335–40.

Goldsmith, M. H. M. (1982). A saturable site responsible for polar transport of indole-3-acetic acid in sections of maize coleoptiles. *Planta* **155**, 68–75.

Hall, J. L. & Moore, A. L, (eds.) (1983). *Isolation of Membranes and Organelles from Plant Cells*. London, Academic Press.

Hepler, P. K. & Wayne, R. O. (1985). Calcium and plant development. *Annual Review of Plant Physiology* **36**, 397–439.

Hodges, T. K., Leonard, R. T., Bracker, C. E. & Keenan, T. W. (1972). Purification of an ion-stimulated adenosine triphosphatase from plant roots: association with plasma membranes. *Proceedings of the National Academy of Sciences, USA* **69**, 3307–11.

Kinsella, J. L. & Aronson, P. S. (1981). Amiloride inhibition of the Na^+/H^+ exchanger in renal microvillus membrane vesicles. *American Journal of Physiology* **241**, F374–9.

Knoth, A. & Wiencke, C. (1984). Dynamic changes of protoplasmic volume and of fine structure during osmotic adaptation in the intertidal red alga *Porphyra umbilicalis*. *Plant, Cell and Environment* **7**, 133–19.

Kulaev, I. S., Mansurova, S. E., Burlakova, E. B. & Dukhovich, V. F., (1980). Why ATP instead of pyrophosphate? Interrelation between ATP and pyrophosphate production during evolution and in contemporary organisms. *BioSystems* **12**, 177–80.

Larsson, C. (1985). Plasma membranes. In Linskens, H. F. & Jackson, J. F. (eds.). *Modern Methods of Plant Analysis*, New Series, vol. 1, *Cell components*, pp. 85–104. Berlin, Springer-Verlag.

Lee, H. C., Forte, J. G. & Epel, D. (1982). The use of fluorescent amines for the measurement of pH_i: Applications in liposomes, gastric microsomes, and sea urchin gametes. In Nuccitelli, R. & Deamer, D. W. (eds.). *Intracellular pH: its Measurement, Regulation and Utilization in Cellular Functions*, pp. 135–60. New York, Alan R. Liss.

Leigh, R. A. & Branton, D. (1976). Isolation of vacuoles from root storage tissue of *Beta vulgaris* L. *Plant Physiology* **58**, 656–62.

Loomis, W. D. (1974). Overcoming problems of phenolics and quinones in the isolation of plant enzymes and organelles. In Fleischer, S. & Packer, L. (eds.). *Methods in Enzymology* **31A**, 528–44.

MacRobbie, E. A. C. (1985). Ion channels in plant cells. *Nature* **313**, 529.

Mandala, S. & Taiz, L. (1985). Partial purification of tonoplast ATPase from corn coleoptiles. *Plant Physiology* **78**, 327–33.

Manolson, M. F., Rea, P. A. & Poole, R. J. (1985). Identification of 3-*O*-(4-Benzoyl) benzoyl adenosine 5′-triphosphate- and *N,N′*-dicyclohexylcarbodiimide-binding subunits of a higher plant H^+-translocating tonoplast ATP-ase. *Journal of Biological Chemistry* **260**, 12273–9.

Marin, B., Preisser, J. & Komor, E. (1985). Solubilization and purification of the ATPase from the tonoplast of *Hevea*. *European Journal of Biochemistry* **151**, 131–40.

Marrè, E. (1979). Fusicoccin: a tool in plant physiology. *Annual Review of Plant Physiology* **30**, 273–88.

Marrè, E. & Ballarin-Denti, A. (1985). The proton pumps of the plasmalemma and the tonoplast of higher plants. *Journal of Bioenergetics and Biomembranes* **17**, 1–21.

Matile, P. (1978). Biochemistry and function of vacuoles. *Annual Review of Plant Physiology* **29**, 193–213.

Mettler, I. J., Mandala, S. & Taiz, L. (1982). Characterization of *in vitro* proton pumping by microsomal vesicles isolated from corn coleoptiles. *Plant Physiology* **70**, 1738–42.

Mueller, S. C. & Maclachlan, G. A. (1983). Radioautographic visualization of β-glucans formed by pea membranes from UDP-glucose. *Canadian Journal of Botany* **61**, 1266–75.

Percy, J. M., Pryde, J. G. & Apps, D. K. (1985). Isolation of ATPase I, the proton pump of chromaffin granule membranes. *Biochemical Journal* **231**, 557–64.

Pitman, M. G., Courtice, A. C. & Lee, B. (1968). Comparison of potassium and sodium uptake by barley roots at high and low salt status. *Australian Journal of Biological Science* **21**, 871–81.

Polonenko, D. R., & Maclachlan, G. A. (1984). Plasma membrane sheets from pea protoplasts. *Journal of Experimental Botany* **35**, 1342–9

Poole, R. J., Briskin, D. P., Kratky, Z. & Johnstone, R. M. (1984). Density gradient localization of plasma membrane and tonoplast from storage tissue of growing and dormant red beet. Characterization of proton transport and ATPase in tonoplast vesicles. *Plant Physiology* **74**, 549–56.

Quail, P. H. (1979). Plant cell fractionation. *Annual Review of Plant Physiology* **30**, 425–84.

Randall, S. K. & Sze, H. (1986). Properties of the partially purified tonoplast H^+-pumping ATPase from oat roots. *Journal of Biological Chemistry* **261**, 1364–71.

Ratner, A. & Jacoby, B. (1976). Effect of K^+, its counter anion, and pH on sodium efflux from barley root tips. *Journal of Experimental Botany* **27**, 843–52.

Rea, P. A. & Poole, R. J. (1985). Proton-translocating inorganic pyrophosphatase in red beet (*Beta vulgaris* L.) tonoplast vesicles. *Plant Physiology* **77**, 46–52.

Reinhold, L. & Kaplan, A. (1984). Membrane transport of sugars and amino acids. *Annual Review of Plant Physiology* **35**, 45–83.

Robinson, D. G. (1985). *Plant Membranes: Endo- and Plasma Membranes of Plant Cells*. New York, John Wiley.

Sanders, D. (1980). The mechanism of Cl^- transport at the plasma membrane of *Chara corallina*. I. Cotransport with H^+. *Journal of Membrane Biology* **53**, 129–41.

Scherer, G. F. E. & Morré, D. J. (1978). Action and inhibition of endogenous phospholipases during isolation of plant membranes. *Plant Physiology* **62**, 933–7.

Schroeder, J. I., Hedrick, R. & Fernandez, J. M. (1984). Potassium-selective single channels in guard cell protoplasts of *Vicia faba*. *Nature* **312**, 361–2.

Schumaker, K. S. & Sze, H. (1985). A Ca^{2+}/H^+ antiport system driven by the proton electrochemical gradient of a tonoplast H^+-ATPase from oat roots. *Plant Physiology* **79**, 1111–17.

Serrano, R. (1984). Plasma membrane ATPase of fungi and plants as a novel type of proton pump. *Current Topics in Cellular Regulation* **23**, 87–126.

Serrano, R., Kielland-Brandt, M. C. & Fink, G. R. (1986). Yeast plasma membrane ATPase is essential for growth and has homology with (Na^+ + K^+), K^+- and Ca^{2+}-ATPases. *Nature* **319**, 689–93.

Smyth, D. A. & Black, C. C., Jr (1984). Measurement of the pyrophosphate content of plant tissues. *Plant Physiology* **75**, 862–4.

Spanswick, R. M. (1981). Electrogenic ion pumps. *Annual Review of Plant Physiology* **32**, 267–89.

Stinissen, H. M., Peumans, W. J. & Chrispeels, M. J. (1985). Posttranslational processing of proteins in vacuoles and protein bodies is inhibited by monensin. *Plant Physiology* **77**, 495–8.

Sussman, M. R. & Slayman, C. W. (1983). Modification of the *Neurospora crassa* plasma membrane [H^+]-ATPase with *N,N′*-dicyclohexylcarbodiimide. *Journal of Biological Chemistry* **258**, 1839–43.

Sze, H. (1984). H^+-translocating ATPases of the plasma membrane and tonoplast of plant cells. *Physiologia Plantarum* **61**, 683–91.

Sze, H. (1985). H^+-translocating ATPases: advances using membrane vesicles. *Annual Review of Plant Physiology* **36**, 175–208.

Takeda, K., Kurkdjian, A. C. & Kado, R. T. (1985). Ionic channels, ion transport and plant cell membranes: potential applications of the patch–clamp technique. *Protoplasma* **127**, 147–62.

Thom, M. & Komor, E. (1984). H^+-sugar antiport as a mechanism of sugar uptake by sugarcane vacuoles. *FEBS Letters* **173**, 1–4.

Thom, M. & Maretzki, A. (1985). Group translocation as a mechanism for sucrose transfer into vacuoles from sugarcane cells. *Proceedings of the National Academy of Sciences, USA* **82**, 4697–701.

Uchida, E., Ohsumi, Y. & Anraku, Y. (1985). Purification and properties of H^+-translocating, Mg^{2+}-adenosine triphosphatase from vacuolar membranes of *Saccharomyces cerevisiae*. *Journal of Biological Chemistry* **260**, 1090–5.

Vignais, P. V. & Satre, M. (1984). Recent developments on structural and functional aspects of the F_1 sector of H^+-linked ATPases. *Molecular and Cellular Biochemistry* **60**, 33–70.

Wagner, G. J. & Seigelman, H. W. (1975). Large-scale isolation of intact vacuoles and isolation of chloroplasts from mature plant tissues. *Science* **190**, 1298–9.

Walker, R. R. & Leigh, R. A. (1981a). Characterisation of a salt-stimulated ATPase activity associated with vacuoles isolated from storage roots of red beet (*Beta vulgaris* L.) *Planta* **153**, 140–9.

Walker, R. R. & Leigh, R. A. (1981b). Mg^{2+}-dependent, cation-stimulated inorganic pyrophosphatase associated with vacuoles isolated from storage roots of red beet (*Beta vulgaris* L.) *Planta* **153**, 150–5.

Wang, Y., Leigh, R. A., Kaestner, K. H. & Sze, H. (1986). Electrogenic H^+-pumping pyrophosphatase in tonoplast vesicles of oat roots. *Plant Physiology* **81**, 497–502.

Yoshida, S. & Uemura, M. (1984). Protein and lipid compositions of isolated plasma membranes from orchard grass (*Dactylis glomerata* L.) and changes during cold acclimation. *Plant Physiology* **75**, 31–7.

Yoshida, S., Uemura, M., Niki, T., Sakai, A. & Gusta, L. V. (1983). Partition of membrane particles in aqueous two-polymer phase system and its practical use for purification of plasma membranes from plants. *Plant Physiology* **72**, 105–14.

4 Fungi

D. Sanders

4.1 Introduction

4.1.1 Organization of transport systems in fungi and plants

With taxonomic evidence now favouring the separation of fungi into an independent kingdom (Ainsworth, 1973), the inclusion of a chapter on fungal transport in a book devoted to plants might seem out of place. However, it is rapidly becoming clear that the strategies for organization and control of transport in the two groups bear marked similarities. These common features are most pronounced at the level of energy coupling. At the plasma and vacuolar membranes, a primary ATP-coupled pump translocates H^+ out of the cytoplasm to set up an electrochemical potential difference for H^+ ($\Delta\bar{\mu}_{H^+}$). Passive return flow of H^+ into the cytoplasm is then mediated by specific solute transport systems which have the effect of transducing the energy stored in $\Delta\bar{\mu}_{H^+}$ into an electrochemical potential gradient for the solute ($\Delta\bar{\mu}_S$) (cf. Ch.3).

Thus, the membranes of both fungi and plants might be said to operate a 'proton economy'. The similarities extend even to the level of protein chemistry, since although the pump which catalyses primary H^+ translocation at the plasma membrane is clearly very different in structure and reaction mechanism from its counterpart at the vacuolar membrane, the basic characteristics of each enzyme can be observed in fungi or in plants.

The analogous modes of organization of transport are especially impressive in the context of the vacuolar membrane because the morphological features and physiological function of the vacuoles of fungi and plants are markedly divergent. The higher plant cell normally possesses only a single vacuole which can account for as much as 95 % of the intracellular volume. The principal function of the plant vacuole appears to be to allow the cell to attain a large size without the associated problems: (a) of high energy consumption in synthesizing cytoplasmic consituents; and (b) of solute and gas diffusion through a considerable volume of cytoplasm. By contrast, most fungal cells contain several vacuoles, typically occupying in total less than half the intracellular volume. Their major function is normally the storage of amino acids and phosphate; growth in nitrogen-deficient conditions often results in the reduction of vacuolar volume.

4.1.2 Why study transport in fungi?

The fungi constitute a diverse kingdom. The amino acid sequence of cytochrome *c*, which is often used as a measure of evolutionary divergence, shows almost as much variation between the two Ascomycetes *Neurospora crassa* and the yeast *Saccharomyces cerevisiae* as between *N. crassa* and human cytochrome *c* (Heller & Smith, 1966). At the organismal level, the growth habit of the thallus is variable, taking the form of single plasmodial cells (e.g. the Myxomycota or slime moulds), or as single walled cells (e.g. yeasts), or as a filamentous mycelium (e.g. the Pyrenomycetes and Plectomycetes). Not surprisingly, this morphological heterogeneity is accompanied by considerable biochemical diversity: many yeasts, for example, are facultative anaerobes, while more advanced genera (e.g. *Neurospora*) tend to be obligately aerobic.

Only a small proportion of the large number of fungal species have been used for transport studies. Several criteria, most of which are not readily applicable to higher plants, can be listed to illustrate the utility of some fungi for fundamental research on the mechanism of membrane transport:

1. Cells can be grown axenically in liquid cultures of defined composition. Thus the nutritional status of experimental material can be carefully controlled.
2. The single cells harvested from liquid cultures are uniform. The absence of differentiation which accompanies tissue formation simplifies the interpretation of experimental results, since transport events monitored in the population can be assumed to represent activity of all the cells within it. Interpretational problems related to unstirred layers are also minimized through the use of single cells. Furthermore, the simple geometry of single cells and even hyphae allows electrophysiological data (e.g. membrane currents) to be quantified on a surface area basis, which is not possible in multicellular tissue.
3. The mean generation time of many species is less than 3 h. Thus, a large quantity of experimental material can be grown rapidly, with minimal resources. Besides conventional transport studies on intact cells, it is therefore also possible to produce within 1 d sufficient material for isolation of transport-competent membrane vesicles and purification of some transport proteins.
4. Some Ascomycetes are obtainable as highly inbred strains, which enables studies to be performed in the context of a uniform genetic

background. The elimination of genetic variability contributes substantially to consistent experimental results.

5. Transport systems can be distinguished on a genetic basis with mutants. General methods for the selection, screening and recovery of transport mutants are discussed by C. W. Slayman (1973). Many transport systems also exhibit derepressible or inducible activity, thereby enabling cells devoid of activity to be used as controls.
6. Haploidy dominates the life cycle and is a great aid to genetic manipulation, which is nowhere easier among eukaryotes than in *Saccharomyces cerevisiae*. Eventually, it will be possible to elucidate the structure/function relationships of transport proteins using such techniques as site-specific mutagenesis.
7. Cytoplasmic factors implicated in the control of transport can be investigated with high-time resolution by nuclear magnetic resonance because the cells are cytoplasm-rich and can be densely packed. Likewise, cytoplasmic impalement with ion-selective microelectrodes is relatively easy in some species.

The considerations listed above have resulted in a large body of work being performed on yeasts (particularly *Saccharomyces*, though the genera *Schizosaccharomyces, Candida and Rhodotorula* are also extensively used) and on *Neurospora*. Most work has focused on the primary H^+ pumps, and on the transport systems for organic nutrients, especially amino acids and carbohydrates. Among inorganic solutes, K^+ has received considerable attention. Cl^- which plays a major role in osmotic pressure generation in the vacuole of many higher plants, is physiologically much less important in the fungi, and research on its transport has therefore been rare (but cf. Miller & Budd, 1975, 1976).

4.1.3 Scope of this chapter

The literature on transport in fungi is vast. Rather than attempting to catalogue the many transport systems which have been worked on, this chapter will focus on recent work which has enhanced our fundamental understanding of the mechanism of transport, especially where such insights can potentially be extrapolated to plant cells. Useful reviews covering specific areas of fungal membrane transport in greater depth than is possible here are by Serrano (1984) on the plasma membrane proton pump; Goffeau & Slayman (1981) and Bowman & Bowman (1987) on fungal membrane ATPases; Eddy (1978) on H^+-coupled solute transport; Borst-Pauwels (1981) on ion transport in yeast; Eddy (1982) principally on amino acid and carbohydrate transport in yeast; Wolfinbarger (1980) on transport of amino acids. A general review on transport in fungi by Jennings (1976) gives much useful background of earlier studies.

4.2 Primary H^+ pumps

4.2.1 Plasma membrane H^+ pump: electrophysiological studies

4.2.1.1 In vivo evidence for H^+ pumping

Fungi were observed to acidify the external medium by Lavoisier almost 200 years ago (Amoury *et al.*, 1984), although it has not been until much more recently that mechanistic explanations for H^+ transport have been sought. Conway (1951) attempted to explain the H^+ efflux which occurs in yeast in response to K^+ addition in terms of a mobile redox carrier in the membrane. Metabolic reactions were proposed to reduce the carrier, with 4 H being translocated to the external surface of the membrane. Deposition of 4 H^+ would then leave the carrier negatively charged and able to transport 4 K^+ to the inside. Oxidized carrier could be regenerated by molecular oxygen. Some supportive evidence was forthcoming from the experiments of Conway & Kernan (1955) in which H^+ efflux was demonstrated to be a function of the redox potential of the medium. More recently, Crane *et al.* (1982) have demonstrated that *Saccharomyces cerevisiae* reduces external ferricyanide with a stoichiometric release of H^+. The failure of the cells to reduce exogenous cytochrome *c* led to the proposal that a transmembrane NADH dehydrogenase is responsible for the reduction. Although it is now clear that at least the major physiological component of H^+ efflux is mediated by an ATPase (Sect. 4.2.2), the possible existence of plasma-membrane-bound redox systems should not be overlooked.

Our current concept of the major pathway for H^+ transport owes much to the seminal work of C.L. Slayman on *Neurospora crassa*. It was found (Slayman, 1965a) that the membrane electrical potential difference ($\Delta\psi$) in hyphal cultures was more negative than −200 mV, considerably hyperpolarized in comparison with the equilibrium potentials of all ions. Furthermore, in the presence of Ca^{2+}, $\Delta\psi$ is relatively insensitive to variation of external [K^+] and [Na^+] but is, by contrast, very sensitive to external pH, especially in the range 3–6. These results, together with the observation that metabolic inhibitors abolish a large component of $\Delta\psi$ within seconds without appreciable resistance changes (Slayman, 1965b), pointed to the existence of an electrogenic ion pump, moving positive charge from the inside to the outside of the cell. Further work (reviewed by Slayman, 1970) revealed that activation of Na^+ efflux from Na^+-loaded cells by K^+ has little effect on $\Delta\psi$; that the anion influxes are insufficient to account for the current passed by the pump; and that net H^+ efflux parallels that change in $\Delta\psi$ effected by metabolic inhibitors. Taken together, these observations imply that the pumped ion is the proton.

Clearly some form of energy input to the pump is required in order that it might generate a membrane potential. Good circumstantial evidence that

ATP hydrolysis powers electrogenic H^+ efflux was forthcoming from experiments in which the time constant for cyanide-induced membrane depolarization was shown to be identical to that for depletion of ATP in similar conditions (Slayman *et al.*, 1970, 1973). Both events occur considerably more slowly than the cyanide-induced rate of pyridine nucleotide reduction, which militates against direct involvement of redox compounds in energization. Interestingly, there is a slight mismatch between membrane depolarization and ATP depletion during the first 5 s after cyanide application, with the larger changes occurring in the ATP level. By merging the two functions $\Delta\psi = f(t)$ and $[ATP] = f(t)$, it is possible to plot $\Delta\psi = f([ATP])$: the result is a good approximation to a rectangular hyperbola, with a small additional constant term representing the absence of electrogenic pumping. The resultant Michaelis constant for the electrogenic pump is 2.0 mol m^{-3}, in substantial agreement with *in vitro* estimates for the plasma membrane ATPase (Sect. 4.2.2.2).

It is important at this juncture to stress the importance of the electrophysiological approach in probing the *in vivo* properties of the plasma membrane H^+ pump. Simple measurements of H^+ efflux, conducted in dilute solutions with a pH stat, cannot be relied upon to give even a qualitative guide to proton pump activity because of the existence of alternative pathways for H^+ flow through the membrane. These pathways may comprise secondary H^+-solute symport systems (Sect. 4.3), as well as passive H^+ influx (uniport) under the influence of the large $\Delta\bar{\mu}_{H^+}$. In either case, some degree of $\Delta\psi$-sensitive H^+ influx is possible. Thus experiments in which the pump activity is assessed as the difference between H^+ efflux in the presence and absence of pump inhibitors are intrinsically unreliable because of the membrane depolarization which accompanies pump inhibition. One rather graphic instance of the disparity between net H^+ efflux and the pump current measured electrophysiologically occurs in *N. crassa* as a function of external pH: at pH 5.8, the measured H^+ efflux is only 20 % of that at pH 9 (Slayman, 1977), even though the pump current changes barely perceptibly over this pH range (Slayman & Sanders, 1984b). The apparent discrepancy might be accounted for by increased passive H^+ influx at the lower pH, or by effective buffering by excreted HCO_3.

4.2.1.2 Electrical kinetics of the plasma membrane H^+ pump

(a) Methods for deriving the pump current–voltage curve The arguments outlined above indicate the importance of direct measurements of pump current (which can be converted to the net proton flux through the pump by dividing by the Faraday constant, *F*). The high chitin content of the yeast cell wall confers considerable structural rigidity, making impalement with conventional glass micropipette electrodes difficult. Thus, although there are a few isolated reports of $\Delta\psi$ measurements in yeast with microelectrodes (Borst-Pauwels, 1981; Vacata *et al.*, 1981) the results are variable and no attempts have been made to assess membrane current. (The application of the patch–clamp technique to yeast suggests that some progress may be made quite soon in this regard (see Sect. 4.4.4).) To date, therefore, most electrophysiological measurements of *in vivo* H^+ pump activity have been made in *N. crassa*. However, an electrophysiological approach has also indicated the existence of an electrogenic pump in the slime mould *Physarum polycephalum* (Kuroda & Kuroda, 1981; Fingerle & Gradmann, 1982), in the marine fungus *Dendryphiella salina* (Brownlee, 1984) and in the water mould *Achlya bisexualis* (Kropf, 1986), the evidence principally resting on the sensitivity of $\Delta\psi$ to respiratory inhibitors.

In order to monitor the pump current, it is necessary to measure a membrane current–voltage (*I–V*) relationship. The plasma membrane is voltage-clamped with a feedback amplifier at a series of discrete membrane potentials, usually between −300 and 0 mV, and the current required to perform this clamp is recorded. Plotting these currents (normalized to membrane surface area) as a function of voltage yields the membrane *I–V* curve. The *I–V* curve represents the summed voltage response of all electrically active pathways in the membrane, and therefore the next step is to extract from the membrane *I–V* curve the specific contribution of the proton pump.

The dissection of the *I–V* curve can be performed in one of two ways. First, the membrane *I–V* relation can be measured initially in a control state and then again after rapid inhibition of the pump, for example with cyanide (Gradmann *et al.*, 1978; see Fig. 4.1,A). The difference between the two curves should then represent the *I–V* curve for the pump (Fig. 4.1,B). There are several disadvantages to this approach (Gradmann *et al.*, 1982): it yields only an empirical description of the pump *I–V* curve, without giving quantitative insight into the pump kinetics; the necessity to pulse with inhibitor limits the time resolution of any other experimentally induced change in pump activity to around 1 min, at best; and, most serious of all, the application of a reversible inhibitor can result not only in a scaling down of pump activity, but also in a shape change in the *I–V* curve. This last problem can potentially result in an *I–V* difference curve which does not accurately reflect the pump *I–V* curve if appreciable residual activity remains in the presence of the inhibitor (Chapman *et al.*, 1983), as appears to be the case for cyanide (Gradmann *et al.*, 1978).

Alternatively, the analysis can be initiated from the simple assumption that the pump behaves as a classical recycling carrier (Gradmann *et al.*, 1982). The rate constants for the carrier reaction cycle can then be extracted by fitting the available membrane *I–V* curve data to a reaction kinetic model for the pump, together with an additional function which describes the electrical activity of the residual membrane pathways. The carrier would be expected to exhibit discrete binding reactions for the pump substrates (ATP, H^+) as well as separate transmembrane reactions

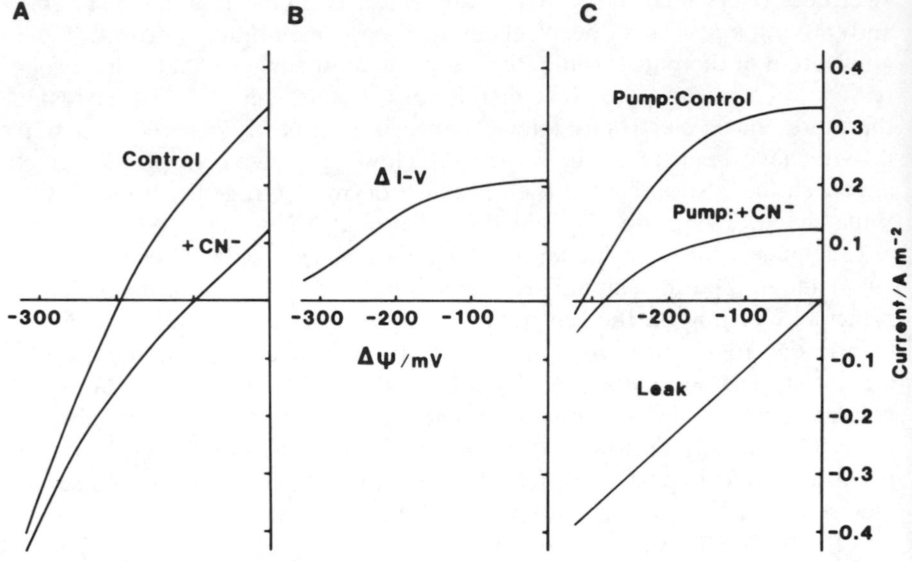

Fig. 4.1 Assessment of *in vivo* electrogenic pump activity in *N. crassa* from measurement of membrane current–voltage (*I–V*) relationships. A. Typical membrane *I–V* relationships measured in control conditions (external pH 5.8) and during metabolic restriction with CN^-. B. Estimation of pump *I–V* relationship as the difference between the two membrane *I–V* curves in A. C. Alternative method for estimation of pump *I–V* relationship from membrane *I–V* curves using the model in Fig. 4.2. The following parameter values (taken from Gradmann *et al.*, 1982: see Fig. 4.2, legend, and text for definitions) were used and substituted into Eq. (4.1) for calculation of all *I–V* curves: $n = 1$; $N = 10^{-8}$ mol m^{-2}; $K_{oi} = 350$ s^{-1} (control) and 130 s^{-1} (+ CN^-); $K_{io} = 2.4 \cdot 10^4$ s^{-1}; $k^{\circ}_{io} = 2.8 \cdot 10^6$ s^{-1}; $k^{\circ}_{oi} = 0.18$ s^{-1}; $G_L = 1.2$ S m^{-2}; $E_L = 0$ mV.

which serve to export H^+ and to return the unloaded H^+ binding site(s) to the inside. In fact, a two-state reaction kinetic model, in which all those reactions not translocating charge across the membrane are lumped together, can describe any given *I–V* curve (Hansen *et al.*, 1981; Fig. 4.2); the corollary of this statement is that a single *I–V* curve can be used only to specify the four reaction constants which link the two carrier states of the reduced model. Gradmann *et al.* (1982) assumed that the non-pump pathways, or ionic leaks, behave ohmically, in which case the overall membrane current can be described as a function of $\Delta\psi$ as

$$i = nFN \frac{K_{oi}k^{\circ}_{io}\exp(-F\Delta\psi/2RT) - K_{io}k^{\circ}_{oi}\exp(F\Delta\psi/2RT)}{k^{\circ}_{io}\exp(-F\Delta\psi/2RT) + k^{\circ}_{oi}\exp(F\Delta\psi/2RT) + K_{io} + K_{oi}} + G_L(\Delta\psi - E_L) \quad (4.1)$$

in which the reaction constants K_{oi}, K_{io}, k°_{io} and k°_{oi} are defined as in Fig. 4.2, n is the H^+/ATP stoichiometry, N is the total carrier density in

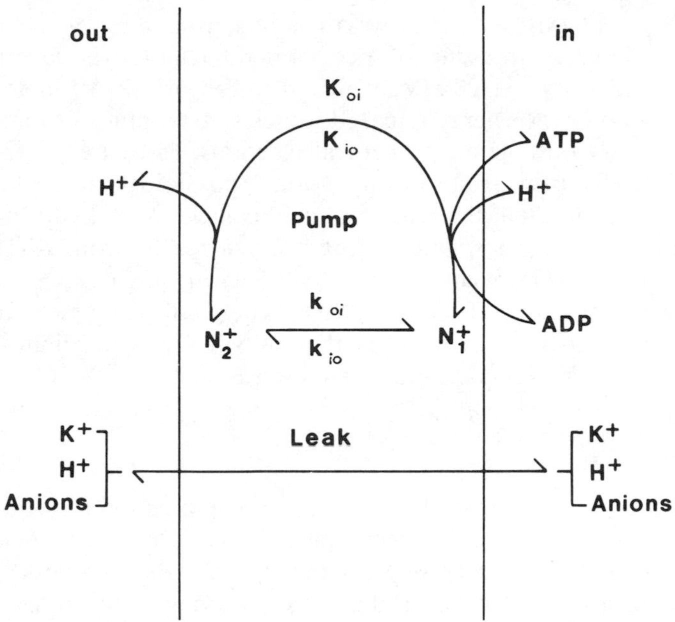

Fig. 4.2 Two-state reaction kinetic model of the electrogenic pump for dissection of membrane *I–V* relationships into pump and leak components. Voltage-insensitive reactions of the carrier (e.g. ligand binding) are subsumed into the two reaction constants K_{oi} and K_{io}. Transmembrane charge translocation reactions, designated k_{oi} and k_{io}, are explicit, with voltage sensitivity incorporated as $k_{oi} = \mathring{k}_{oi}$ $\exp(F\Delta\psi/2RT)$ and $k_{io} = \mathring{k}_{io} \exp(-F\Delta\psi/2RT)$. N_1^+ and N_2^+ are transition states of the carrier. Return current is carried by an ohmic 'leak' of unidentified ionic specificity.

the membrane (units, mol m^{-2}), G_L and E_L are the electrical conductance and equilibrium potential, respectively, of the leak, and R and T have their usual meanings.

(b)I–V relationship in standard conditions At the normal external pH of 5.8, the response of the pump current to $\Delta\psi$ shows several interesting features. As $\Delta\psi$ is lowered from −300 to around −200 mV, the current increases quasilinearly, but at voltages more positive than this, the pump becomes progressively less sensitive to $\Delta\psi$, until, at 0 mV, sensitivity is lost altogether (Fig. 4.1,B and C). This conclusion is independent of which analytical method is used (cf. Gradmann *et al.*, 1978, 1982). The response can be visualized as analogous to the Michaelis–Menten kinetics of an enzyme with respect to substrate concentration, with 'substrate concentration' in the present instance being $\Delta\psi$. The $\Delta\psi$-insensitive current is known as the 'saturation current'. Values of the reaction constants, derived from the modelling approach (assuming $N = 10^{-8}$ mol m^{-2} or $6 \cdot 10^3$ pump molecules μm^{-2} are as follows: $\mathring{k}_{io} = 2.8 \cdot 10^6\ s^{-1}$; $\mathring{k}_{oi} = 0.18\ s^{-1}$; K_{oi} =

350 s^{-1}; K_{io} = 2.4·10^4 s^{-1}. It is worth noting from these values that the major site of energy transduction occurs at the charge translocation step. At the normal $\Delta\psi$ of −200 mV, k_{io}/k_{oi} = 6·10^3, or −21 kJ mol^{-1} for the forward reaction. The energy released in this step is sufficient to overcome the slightly unfavourable electroneutral reactions, as well as to render the pump kinetically irreversible during normal physiological functioning. It has been speculated that the reaction might coincide with hydrolysis of the high-energy enzyme–phosphate complex which results from ATP binding and ADP release (Gradmann *et al.*, 1982; Sect. 4.2.2.2).

The reversal potential of the pump (E_P) is defined as that potential which must be applied to stall the pump at thermodynamic equilibrium. In terms of the reaction kinetic model, the definition is

$$E_P = \frac{RT}{nF} \ln \frac{\mathring{k}_{oi}\, K_{io}}{\mathring{k}_{io}\, K_{oi}} \tag{4.2}$$

Operationally, E_P can be read from the *I–V* curve as the intercept on the abscissa. Since it is a thermodynamic parameter, E_P is strictly independent of the particular kinetic response of the pump, though again both analytical methods are in agreement in predicting E_P as lying in the range −300 to −400 mV, though the precise value is rather uncertain because it lies outside the range of the experimentally attainable membrane *I–V* curve. The reversal potential can be used to determine the H^+/ATP stoichiometry because, with the pump at equilibrium, the change in free energy of the entire pump reaction (ΔG_P) is given by

$$\Delta G_P = (\Delta G_{ATP}/F) - n\Delta\bar{\mu}_{H^+} = 0 \tag{4.3}$$

where ΔG_{ATP} is the free energy of ATP hydrolysis. Expanding and rearranging,

$$n = \frac{\Delta G_{ATP}}{F\,[E_P + 59(pH_o - pH_c)]} \tag{4.4}$$

with pH_o and pH_c the external and cytoplasmic pH's, respectively. For $\Delta G_{ATP}/F$ = −520 mV (Slayman *et al.*, 1973), pH_c = 7.2 (Sanders & Slayman, 1982) and the minimum value of E_P = −300 mV, the maximum estimate of n is 1.36. This non-integral stoichiometry might be taken to indicate a mixed population of pumps–some functioning with n = 1, some with n = 2–or even a single population in which a pumped H^+ fails to dissociate from the carrier and returns via an internal leak. However, the uncertainty of the estimate for E_P means that a real value of −440 mV (required for n = 1.0) cannot be discounted, and it seems safest at present to conclude the n = 1. This interpretation is supported by the observation that Eq. (4.1) cannot generate sensible fits to the membrane *I–V* curve if $n > 1$ (D. Sanders, unpublished), and by the finding that the *in vitro* H^+/ATP stoichiometric ratio is very close to 1.0 (Sect.4.2.2.3). One firm conclusion, though, is that $n < 2$, since the value of E_P given by Eq. (4.4) for $n = 2$ (-177 mV) is more positive than the resting membrane potential.

The physiological pump current at pH 5.8 and $\Delta\psi = -200$ mV can be read straight from the *I–V* curve. Estimates lie in the range 0.12–0.25 A m^{-2} (Gradmann *et al.*, 1978, 1982; Sanders *et al.*, 1981; Slayman & Sanders, 1984a), equivalent to a net H^+ flux of 1.2–2.6 μmol m^{-2} s^{-1}. Since the H^+/ATP stoichiometry is estimated as 1, the proportion of cellular ATP production consumed by the pump can also be estimated (Gradmann *et al.*, 1978). For ATP synthesis at a rate of 3.7 μmol m^{-2} s^{-1} (C.L. Slayman, 1973) and a pump flux of 0.19 μmol m^{-2} s^{-1}, the H^+ pump would account for about half of total ATP consumption.

(c) Effect of inhibitors on the H^+ pump The effect of cyanide on the pump current-voltage curve, as estimated by reaction kinetic analysis, is rather straightforward (Gradmann *et al.*, 1982). The pump *I–V* curve scales down by a factor of almost 3 (Fig. 4.1,C) and there is only a small positive shift in reversal potential (as expected, since the term ΔG_{ATP} is dominated primarily by the standard free energy of ATP hydrolysis, with the mass action ratio having relatively little bearing). The effect is apparent kinetically simply as a reduction in the reaction constant K_{oi} which subsumes ATP binding. (ADP level is maintained almost constant in the presence of cyanide by adenylate kinase.)

The effects of the pump-specific inhibitor orthovanadate can be investigated in phosphate-starved cells (Kuroda *et al.*, 1980; Slayman & Sanders, 1984a), and turn out to be qualitatively similar to those of cyanide: vanadate inhibits the reloading reaction constant K_{oi} by a factor of 3, which is readily understood if the inhibitor competes for a low affinity ATP-binding site (Bowman & Slayman, 1979).

(d) Effects of pH on the H^+ pump The influence of $[H^+]$ on pump current is of particular interest for two reasons. First, it might be asked whether the kinetic effects of $[H^+]$ are compatible with the role of H^+ as substrate for the pump,i.e. does the pump current respond to variable $[H^+]$ with first-order kinetics, and if so, what are the respective pK_a's for the external and internal binding sites? Second, how do the effects of pH relate to the projected role of the proton pump as a primary regulator of cytoplasmic pH (Raven & Smith, 1976)?

The response of the pump *I–V* relation to changed external pH (pH_o) is markedly dependent on whether the shift is to the acid or the alkaline side of the control pH (5.8) (Slayman & Sanders, 1984a, b; Fig. 4.3). In the acid direction to $pH_o = 3.7$, the reversal potential moves 123 mV more positive (coinciding with the prediction of Eq. (4.4)) and the whole current–voltage curve shifts downward. By contrast, the effect of alkaline

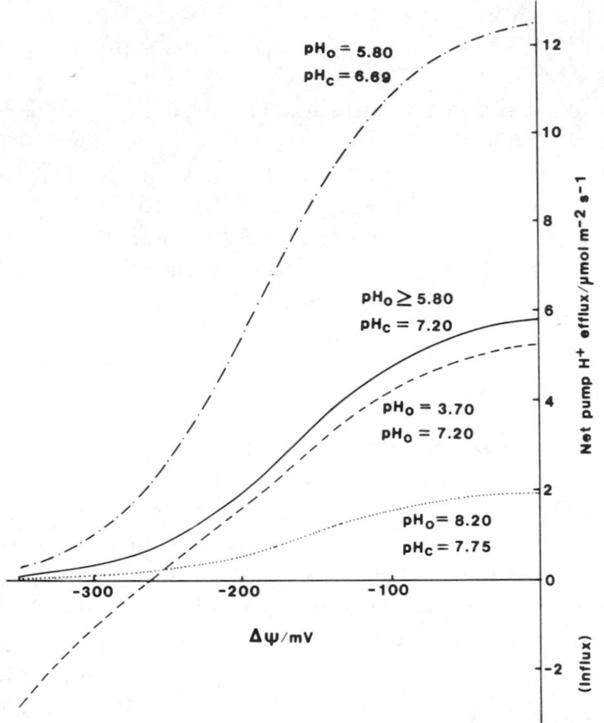

Fig. 4.3 pH-dependence of net H^+ transport by the electrogenic H^+ pump of *N. crassa*. Curves are derived from the model in Fig. 4.4, with the effects of pH incorporated simply via mass action of $[H^+]$ on the appropriate binding reactions, and the effects of $\Delta\psi$ according to the legend of Fig. 4. 2. The curves give a semi-quantitative description of the pH-dependence of the derived *I–V* curve for the electrogenic pump (see Slayman & Sanders, 1985).

pH_o is barely discernible. (The failure to detect the predicted 140 mV negative shift in E_P in the latter case might be rationalized on the basis that the kinetic characteristics of the pump result in only small currents at potentials more negative than $\simeq$ −350 mV–an interpretation which is supported by the sigmoid appearance of the curve at pH_o = 5.8.)

The cytoplasmic pH (pH_c) can be clamped to predetermined values, at constant pH_o with weak acids and bases. For example, addition of 5 mol m^{-3} butyric acid at pH_o = 5.8 results in a fall in pH_c to about 6.7 as the free acid permeates the plasma membrane and dissociates in the relatively alkaline cytoplasm (Sanders & Slayman, 1982). By contrast, a weak base with a pK_a above that of the cytoplasmic pH (e.g. procaine, pK_a = 9.1) will diffuse into the cell and associate with cytoplasmic H^+. For both acid and alkaline shifts of pH_c, significant changes in the pump *I–V* curve result (Sanders *et al.*, 1981; Slayman & Sanders, 1984b; Fig. 4.3). The primary changes are in the saturation currents (increasing for acid shifts,

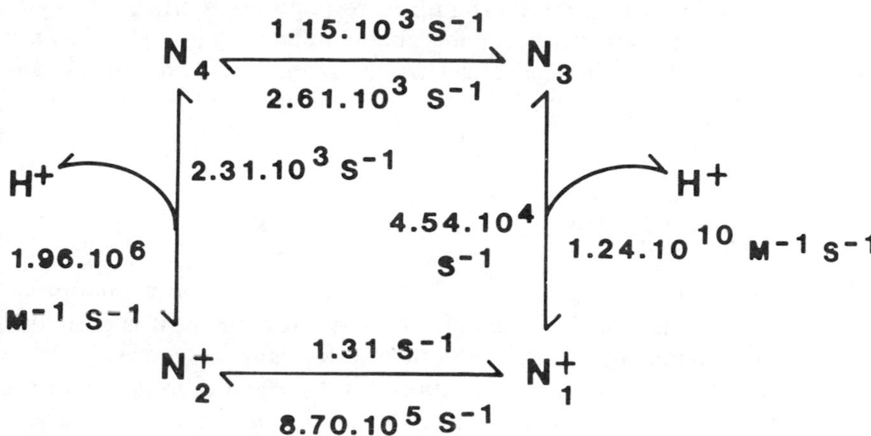

Fig. 4.4 Four-state reaction kinetic model for the *N. crassa* H^+ pump derived from pH-dependent behavior of the pump *I–V* relationship. Selective manipulation of pH_o and pH_c enables specific identification of the H^+ binding reactions: other voltage-insensitive pump reactions, which cannot be so identified, are lumped in the transition between carrier states 3 and 4. Values for the reaction constants (from Slayman & Sanders, 1985) are shown. The reaction constant from state i to state j is written as k_{ij} (see text).

decreasing for alkaline shifts), with comparatively minor effects on the reversal potential–as is to be expected for the relatively small pH_c changes.

Despite the seemingly disparate nature of the effects of pH_o and pH_c on the pump *I–V* curve, the results can be reconciled in a single kinetic scheme in which the two-state pump model is now expanded to four states, with the external and internal H^+ binding reactions explicit (Slayman & Sanders, 1984b, 1985; Fig. 4.4). Reasonable fits of the data for acid shifts in both pH_c and pH_o are generated by the model in which the effects of pH_o on the *I–V* curve are localized only to the external H^+ binding reaction (k_{42}) and the effects of pH_c are likewise restricted to k_{31}. The major conclusions from the fitting are: first, the relevant reaction constants k_{42} and k_{31}, which subsume $[H^+]$, vary in direct proportion to $[H^+]$, in apparent confirmation of the conclusion that $z = 1$, and suggesting that the effects of pH can be accounted for simply in terms of H^+ binding to and dissociating from the transport sites of the carrier; second, derivation of tentative values for the carrier reaction constants from the fits yield a pK_a for the internal H^+ binding site of 5.4, and for the external site, $pK_a = 2.9$. These estimates are significant from both mechanistic and physiological standpoints. They suggest that carrier recycling in the direction of

H^+ efflux is rate limited by binding of H^+ on the cytoplasmic side (recall $pH_c = 7.2$) but that external pH must become very acid before significant effects of H^+ pumping result. Moreover, the arrangement is physiologically meaningful because it allows the pump to respond appropriately to cytoplasmic pH changes without becoming unduly influenced by pH_o. This last finding explains the very small effect on the pump *I–V* curve of alkaline shifts in pH_o.

4.2.2 Plasma membrane H^+ pump: biochemical studies

4.2.2.1 Purification

Although electrophysiological studies on whole cells are most appropriate for investigating the electrical kinetics of the proton pump, it is clear that further characterization at the molecular level must rely on the more controlled conditions which can be attained by biochemical isolation of the pump. The major potential difficulty in the purification of plasma membranes is contamination by mitochondrial membranes which also possess a H^+ – ATPase (see Ch 2). Several strategies have now been successfully adopted to overcome this problem, and plasma membrane ATPase activity has been characterized from a wide variety of yeasts, including *Saccharomyces cerevisiae* (Willsky, 1979), *Schizosaccharomyces pombe* (Delhez *et al.*, 1977), *Candida tropicalis* (Blasco *et al.*, 1981), *C. albicans* (Hubbard *et al.*, 1986) *Metschnikowia reukaufii* (Aldermann & Höfer, 1984), *Debaryomyces hansenii* (Comerford *et al.*, 1985), as well as from *N. crassa* (E.J. Bowman *et al.*, 1981) and the slime mould *Dictyostelium discoideum* (Pogge-von Strandmann *et al.*, 1984).

One common method has been to stabilize the plasma membrane of spheroplasts with concanavalin A (Stroobant & Scarborough, 1979a; Bussey *et al.*, 1979; Hubbard *et al.*, 1986). Plasma membrane ghosts, rather than small vesicles, result from osmotic lysis, and these can be separated by low-speed or density-gradient centrifugation. Removal of extrinsic proteins and concanavalin A is then achieved with an NaCl plus α-methylmannoside wash. Alternatively, by subjecting spheroplasts to gentle lysis, the mitochondria remain intact and can be removed by low-speed centrifugation: the plasma membrane is then purified by density gradient centrifugation. Finally, if whole cells are used as starting material, the cell wall must be disrupted mechanically; intact cells are removed, and a total membrane fraction then isolated by differential centrifugation, with isopycnic sucrose gradient centrifugation used as a final purification step. Plasma membrane equilibrates at densities between 1.19 and 1.26 (Serrano, 1984).

The plasma membrane ATPase can be distinguished from mitochondrial ATPase by its lower pH optimum, its insensitivity to azide, and its sensitivity to orthovanadate. Typically, the specific activity of ATPase in plasma membrane preparations is in excess of 1 μmol mg protein^{-1} min^{-1}, though values as high as 25 μmol mg^{-1} min^{-1} have been reported for *Sch. pombe* (Dufour & Goffeau, 1980). Further purification of the ATPase from plasma membrane is achieved by solubilizing the enzyme with detergent (lysophosphatidylcholine, cholate + 45 % glycerol and Zwittergent-14 have all been used, but are species-specific in their utility: Serrano, 1984), followed by gradient density centrifugation in sucrose or glycerol. Aggregation of the ATPase causes it to enter the gradient ahead of other proteins.

The properties of the purified enzyme are similar to those of the ATPase in plasma membrane. M_r is estimated as 100 000 by sodium dodecylsulphate polyacrylamide gel electrophoresis (SDS–PAGE) (Dufour & Goffeau, 1978; Addison & Scarborough, 1981; B.J. Bowman *et al.*, 1981), and there is no apparent functional requirement for other polypeptides (Scarborough & Addison, 1984). The turnover number of the enzyme can be calculated (Bowman & Bowman, 1987) from the maximal specific activity of the purified enzyme (which approaches 100 μmol mg^{-1} min^{-1}: B.J. Bowman *et al.*, 1981), and is about 150 s^{-1} assuming the polypeptide functions as a monomer (see Sect. 4.2.2.4).

4.2.2.2 Mechanism of ATP hydrolysis

The enzyme exhibits a high degree of specificity for MgATP, which distinguishes it from the mitochondrial enzyme (Goffeau & Slayman, 1981). The K_m, in agreement with predictions from electrophysiological experiments, has generally fallen in the range 0.8–4.8 mol m^{-3} (Bowman & Bowman, 1987).

The finding that the ATPase is sensitive to orthovanadate with a K_i of 10^{-2} mol m^{-3} or less (Bowman & Slayman, 1979; Willsky, 1979; Dufour *et al.*, 1980; Borst-Pauwels & Peters, 1981) has suggested, by analogy with the ($Na^+ + K^+$)-ATPase of animal cell plasma membranes, that the reaction cycle involves formation of a phosphorylated transition state of the enzyme (E~P). The existence of E~P can be demonstrated directly: the enzyme is incubated with [γ-^{32}P]-ATP and the reaction then quenched rapidly with acid (Amoury *et al.*, 1980; Dame & Scarborough, 1980; Malpartida & Serrano, 1981a). Labelling of the 100 kDa polypeptide is shown by SDS–PAGE in acid conditions. The phosphorylation is complete within 100 ms (Amoury & Goffeau, 1982) and exhibits the characteristics of an acyl–phosphate bond (being hydroxylamine and alkali labile). After total protein hydrolysis in reducing conditions, the phosphate can be shown to be bound to an aspartyl residue (Dame & Scarborough, 1981; Amoury & Goffeau, 1982).

Specific details of that part of the reaction cycle which involves the formation of E ~ P have been investigated for the *Sch. pombe* enzyme with ^{18}O-labelled orthophosphate ([^{18}O]P_i) and [γ-^{18}O]ATP and have

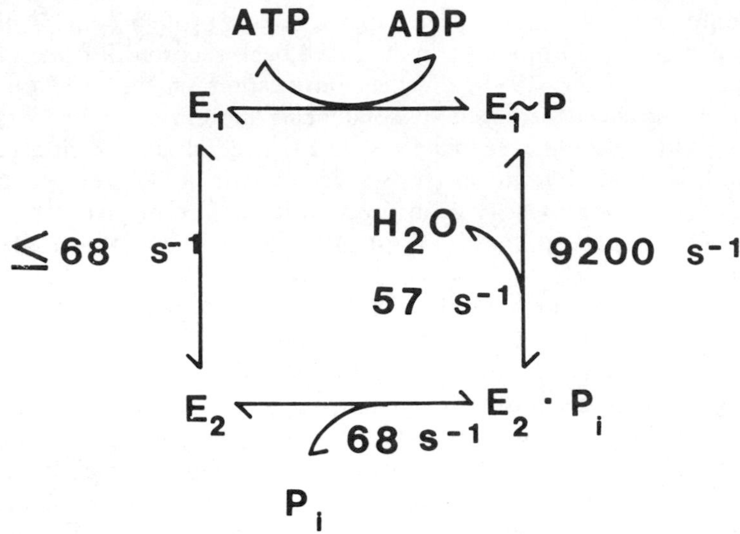

Fig. 4.5 Four-state reaction kinetic model for the *Sch. pombe* H^+-ATPase derived from ^{18}O-labelling (Amoury *et al.*, 1984). E_1 and E_2 represent two distinct conformations of the enzyme, and $E_1{\sim}P$ is the covalently phosphorylated state. Where values for the rate constants have been assigned, they are shown.

resulted in the conclusion that the hydrolytic reactions take the general form shown in Fig. 4.5 (Amoury *et al.*, 1982, 1984). The dilution of ^{18}O appearing in P_i after incubation with [γ-^{18}O]ATP implies that the hydrolytic reaction ($E{\sim}P \rightarrow E{\cdot}P_i$) is significantly reversible. The existence of two isomeric forms of the enzyme (E_1 and E_2) was deduced from the finding that the presence of ATP enhances exchange of ^{18}O between P_i and H_2O. Indeed, this latter observation implies rate limitation of the reaction cycle at the reaction $E_2 \rightarrow E_1$. Unidirectional rate constants for the hydrolytic reaction and for P_i release were determined by measuring the rate of ^{18}O exchange between [^{18}O]P_i and H_2O as a function of P_i concentration in the absence of nucleotides, and are shown in Fig. 4.5. The hydrolysis reaction is therefore exergonic, with a free energy change of -12.8 kJ mol^{-1}.

Brooker & Slayman (1982; 1983a, b, c) have characterized three discrete binding sites for nucleotides and Mg^{2+} on the *N. crassa* enzyme from the capacity of these ligands to protect the enzyme from inhibition by the sulphydryl reagent *N*-ethylmaleimide (NEM). The nucleotide binding site binds to the physiological substrate MgATP with an equilibrium constant (K_D) = 0.72 mol m^{-3}, and also binds to free ATP (K_D = 11.4 mol m^{-3}), MgADP (K_D = 0.18 mol m^{-3}) and free ADP (K_D = 1.5 mol m^{-3}), each of which acts as an inhibitor of ATP hydrolysis. The two additional binding sites are for free Mg^{2+}: binding to the high affinity site (K_D = 15 mmol m^{-3}) activates the enzyme and protects against NEM inhibition, while Mg^{2+}

bound to the low-affinity site (K_D = 3.5 mol m^{-3}) is inhibitory and enhances sensitivity to NEM. A model in which ligand binding to the three sites occurs randomly and reversibly is quantitatively capable of describing steady-state velocity-concentration plots for MgATP, including sigmoidicity which is observed at low MgATP concentration (Brooker & Slayman, 1983c). It might not, therefore, be necessary to propose that the enzyme cleaves ATP via an alternating site mechanism (cf. Bowman, 1983).

Binding of MgATP, P_i and vanadate to the enzyme are all pH-sensitive (Ahlers *et al.*, 1978; Borst-Pauwels & Peters, 1981; Amoury *et al.*, 1982) but it is not immediately obvious how to interpret these effects in terms of binding of the transported proton. However, it is possible, using the four-state model derived from electrophysiological analysis of the H^+ pump (Fig. 4.4), to predict the pH optimum of the ATPase *in vitro* (Slayman & Sanders, 1985). The predicted optimum is between pH 4 and pH 5, which contrasts markedly with the published optimum of pH 6.5 for the *N. crassa* ATPase (B.J. Bowman *et al.*, 1981). However, the measured pH optimum might well include an element of acid inactivation, and it is therefore noteworthy that the observed pH profile for *Sacch. cerevisiae* ATPase after correction for acid inactivation (Peters & Borst-Pauwels, 1979) is in much closer agreement with the calculated curve (Slayman & Sanders, 1985). The simple interpretation, then, is that the pH profile for ATP hydrolysis may, in part, reflect binding of H^+ to the transport site.

4.2.2.3 H^+ pumping

The vectorial, as well as the scalar, reactions of the ATPase can also be demonstrated *in vitro*, and the finding that the enzyme pumps protons constitutes definitive evidence that the ATPase is, in fact, the plasma membrane H^+ pump. Purified plasma membrane spontaneously forms vesicles, which, by freeze–fracture analysis of particle distribution on the P and E faces, are predominantly (87 %) everted in comparison with the orientation in the intact cell (Perlin *et al.*, 1984). Addition of MgATP to these preparations results in the formation of an inside acid pH gradient (ΔpH), which can be monitored as accumulation of weak bases in the vesicles (either by [^{14}C] labelling (Scarborough, 1980) or by concentration-dependent fluorescence quenching (Blasco & Gidrol, 1982; Perlin *et al.*, 1984), or as fluorescence quenching of intravesicular fluorescein-labelled dextran (Scarborough, 1980). Similar methods, as well as direct detection of extravesicular alkalinization on addition of ATP (Villalobo *et al.*, 1981), have been used to show that the purified ATPase also pumps H^+ to the vesicle interior if reconstituted into liposomes (Malpartida & Serrano, 1981b; Dufour *et al.*, 1982; Perlin *et al.*, 1984; Scarborough & Addison, 1984).

Charge translocation by the pump can be detected as the formation of an interior positive $\Delta\psi$ either in purified plasma membrane or proteoliposomes. Again either [^{14}C] or fluorescence techniques can be used. In the former case, $\Delta\psi$ is monitored as the accumulation of a permeant anion applied at low concentration (SCN^-: Scarborough, 1976), while in the latter, fluorescence of the $\Delta\psi$-sensitive dyes 1-anilinonaphthalene-8-sulphonic acid (Scarborough, 1976) or oxonol V (Perlin *et al.*, 1984) can be followed.

Protonophoric uncouplers are used to ensure that the putative changes in ΔpH and $\Delta\psi$ represent bona fide transport events, since these ionophores should completely short-circuit ATP-dependent H^+ translocation. Other ionophores can be employed to dissipate selectively either the ΔpH (e.g. nigericin) or $\Delta\psi$ (e.g. valinomycin) components of $\Delta\bar{\mu}_{H^+}$, and result in the conclusion that the two components are essentially interchangeable. For example, valinomycin stimulates the formation of ΔpH in reconstituted plasma membrane vesicles from *N. crassa* (Perlin *et al.*, 1984). Similar effects of monovalent anions (SCN^-, NO_3^-, Cl^-) have led to the general conclusion that these ions, by virtue of their membrane permeability, are also able to clamp $\Delta\psi$ close to zero.

The magnitude of the pump-generated $\Delta\bar{\mu}_{H^+}$ membrane vesicles is commonly found to be in the region of 120 mV–equivalent to a ΔpH of 2 units (Serrano, 1984). Since the measured plasma membrane $\Delta\bar{\mu}_{H^+}$ in intact cells is at least 2.5 times greater than this (Sanders & Slayman, 1982), the existence of a significant leak pathway for H^+ in membrane vesicles is clearly implied and its presence is manifested by the rapid relaxation of $\Delta\bar{\mu}_{H^+}$ after inhibition of the proton pumping (Perlin *et al.*, 1984). Thus the steady-state $\Delta\bar{\mu}_{H^+}$ is achieved when the proton fluxes through the pump and leak are equal and opposite, with the pump still poised far from equilibrium. Reference to Fig. 4.3 shows that only modest (< two fold) changes in pump activity would be predicted after dissipation of $\Delta\psi$ = 120 mV, and this, indeed is found to be the case in membrane vesicles after addition of monovalent anions (Perlin *et al.*, 1984). The interchangeability of ΔpH and $\Delta\psi$ in membrane vesicles can therefore be viewed as much a reflection of the ability of the H^+ leak to short-circuit pumping as of the factors which contribute to kinetic limitation of the pump itself.

It is possible to measure H^+/ATP stoichiometry for the proton pump despite the presence of the H^+ leak (Perlin *et al.*, 1986). After formation of a steady-state ΔpH in the presence of anions, the pump is rapidly turned off by complexing Mg^{2+} with EDTA. The rate of relaxation of acridine orange fluorescence is used to monitor relaxation of ΔpH, and the fluorescence signal is quantified by subjecting the vesicles to alkaline pH jumps of defined magnitude. The relaxation rate is 2.60 unit mg^{-1} min^{-1}. The intravesicular buffer capacity is estimated in separate experiments from the difference between the initial and final deflections in extravesicular pH when a vesicle suspension is pulsed with acid: this value was 0.26 μmol pH $unit^{-1}$ mg^{-1}. Finally, parallel determinations of ATP hydrolytic rate (0.55 $\mu mol\ mg^{-1}\ min^{-1}$) yield a H^+/ATP stoichiometry of 1.23. This value has to be corrected for the small proportion (13 %) of right-side-out vesicles, which are neither hydrolysing ATP nor pumping H^+, yet which contribute to the measured buffer capacity: after correction, values of H^+/ATP close to 1.0 are obtained, in agreement with the predictions from electrophysiological experiments (Sect. 4.2.1.2(b)).

4.2.2.4 Protein chemistry

Ultimate understanding of the mechanism by which the ATPase pumps H^+ will rely on a detailed molecular picture of the enzyme as it undergoes its reaction cycle. While this goal is still some years away, recent work has begun to address the issue of the chemistry of the protein.

(a) Functional molecular size A crucial first question concerns the functional molecular size of the pump: how many 100 kDa peptides constitute an operational pump unit? Reconstitution studies have revealed that maximal reactivation of solubilized *Sch. pombe* enzyme occurs at a ratio of 1 × 100 kDa polypeptide/vesicle, which appears to indicate that the functional pump unit is a monomer (Dufour & Tsong, 1981). However, arguments have been advanced on the basis of radiation inactivation studies (Bowman *et al.*, 1985; Bowman & Bowman, 1987) to suggest that the *N.crassa* enzyme functions as a dimer, and the matter is at present unresolved.

(b) Conformational changes during catalysis The kinetic evidence which supports the idea of a conformational change in the enzyme (Sect. 4.2.2.2) has been supported by structural studies with trypsin. The *N. crassa* ATPase is protected from tryptic cleavage into small fragments by concentrations of Mg^{2+} which would ensure occupation of the low-affinity (inhibitory) Mg^{2+} binding site, but not by concentrations in which only the Mg^{2+} activator site is occupied (Addison & Scarborough, 1982; Brooker & Slayman, 1983b). The inclusion of orthovanadate during incubation with trypsin markedly enhances the protective effect of Mg^{2+}, and this effect is apparent both in the ATPase activity and in the cleavage patterns identified by SDS–PAGE: the intact protein, with an apparent M_r = 105 000 degrades rapidly to an M_r = 95 000 form, then slowly to an M_r = 88 000 form (Addison & Scarborough, 1982). By contrast, orthovanadate alone has little effect on the control cleavage pattern. It is not yet possible to relate these results specifically to an E_1E_2-type model of the enzyme reaction cycle (Fig. 4.5) since the step at which low affinity Mg^{2+} binding takes place has not yet been identified, and, in particular, because similar cleavage patterns are identified in the presence of nucleotides.

(c) Chemical probes Some preliminary information on residues of the 100 kDa polypeptide which are essential for catalytic activity is available from studies with covalently binding probes. Di Pietro & Goffeau (1985) and Kasher *et al.* (1986) have demonstrated that the arginine reagents 2,3-butanedione and phenylglyoxal inhibit the ATPase in a nucleotide-protectable manner. [^{14}C]phenylglyoxal labelling reveals just one protectable labelling site per mol of ATPase. The results can be interpreted either as indicative of an essential arginine residue at the nucleotide-binding site, or as representing differential exposure of a residue elsewhere on the protein as a consequence of a nucleotide-binding-induced conformational change. Similar work with the sulphydryl reagent NEM and fluorescein isothiocyanate (FITC) has also demonstrated the existence of single, nucleotide-protectable cysteine (Brooker & Slayman, 1982, 1983a) and lysine (Hager & Slayman, 1986) residues, respectively.

In common with all classes of H^+-ATPase so far discovered, activity of the fungal plasma membrane ATPase is sensitive to the lipophilic carboxyl reagent *N,N′*-dicyclohexylcarbodiimide (DCCD) (Goffeau & Slayman, 1981). Hydrophilic carbodiimides are ineffective, which suggests the essential residue is embedded in the membrane bilayer (Sussman & Slayman, 1983). Labelling is, as might be expected, not affected by the presence of nucleotides.

(d) The primary amino acid sequence The structural genes for both the *Sacch. cerevisiae* and the *N. crassa* ATPases have been cloned by screening cDNA libraries in the expression vector λgt11 with affinity-purified antibody against the ATPase (Hager *et al.*, 1986; Serrano *et al.*, 1986). Subclones were then used to probe a genomic library for large overlapping restriction fragments. Further sub-cloning yields a 5 kilobase fragment which codes for the ATPase and contains some flanking sequences.

The deduced amino acid sequences of the yeast and *N. crassa* pumps exhibit a high degree of homology (74 %) and the proteins are of approximately equal size (yeast: 918 residues, M_r = 99 532; *N. crassa*: 920 residues, M_r = 99 886). Analysis of the primary sequences has resulted in several important conclusions concerning the structure and function of the protein.

First, the hydrophobicity profiles indicate that the polypeptide crosses the membrane ten (yeast) or eight (*N. crassa*; Fig. 4.6) times, but with only 4 % of the protein exposed to the external surface.

Second, both polypeptides show significant sequence homology with several other ion pumps: the (Na^+ + K^+)-ATPase of animal cell plasma membranes, the sarcoplasmic reticulum Ca^{2+}-ATPase, and (less so) the K^+-ATPase of *E. coli*. This result implies a common evolutionary origin, and confirms the conclusion from kinetic studies that all can be viewed as belonging to the same category of ATPase (the E_1E_2 category: Sect. 4.2.2.2). By contrast, no sequence homology has been detected with the H^+-ATPases of energy-coupling membranes (the F_0F_1 category, see Ch. 2).

Third, some of the active sites on the protein have been identified, partly by consideration of highly conserved regions within the E_1E_2 category and partly with the use of chemical probes. One region of striking homology is around the phosphorylation site (Asp ^{378}) in which all four eukaryotic ATPases share a common sequence of nine amino acids. This site occurs

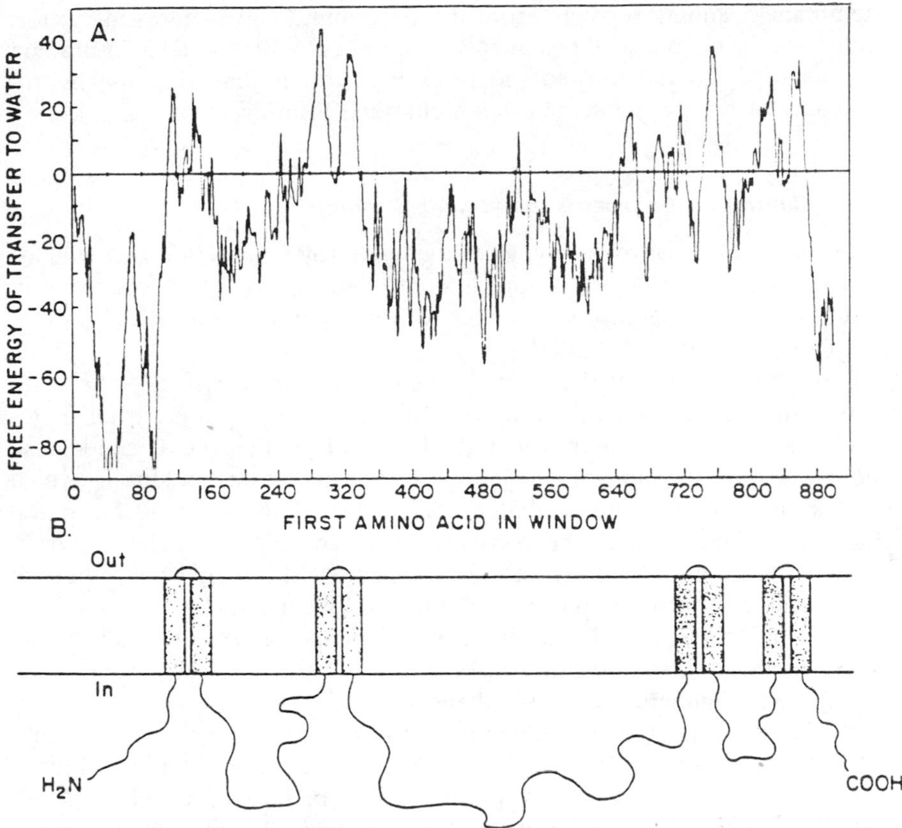

Fig. 4.6 A. Hydrophobicity profile of the *N. crassa* H^+-ATPase. The free energy (units: kcal mol^{-1}) for transfer of segments of twenty amino acids from a lipid phase to water is given as a function of the first amino acid in each window. B. Model, derived from A, of the probable transmembrane orientation of the enzyme, taking +20 kcal mol^{-1} as the threshold value for membrane partitioning of the hydrophobic segments. The N-terminus is placed on the cytoplasmic side of the membrane in accordance with the pattern for other membrane-spanning proteins. (From Hager *et al.*, 1986, with permission.)

on the large central hydrophilic domain (Fig. 4.6). Two other noteworthy homologous regions (Hager *et al.*, 1986) are those that might form part of the nucleotide binding site, as assessed by nucleotide-protectable labelling of animal E_1E_2 enzymes with FITC and 5′-(*p*- fluorosulphonyl) benzoyladenosine, respectively: residues 473–503 and 604–663, both also located on the central hydrophilic segment. The identity and location of the DCCD-binding carboxyl have been investigated by partial amino acid sequencing of the *N. crassa* enzyme labelled with [^{14}C]DCCD (M.R. Sussman *et al.*, 1987). The carbodiimide binds to a glutamyl carboxyl (Glu^{129}) which appears to lie slightly displaced from the centre of the first membrane-spanning segment from the *N*-terminal end. Interestingly, this region bears no particular sequence homology with the DCCD-binding domain of the 8 kDa proteolipid from F_0F_1, other than that the hydrophobicity of both regions indicates membrane location.

4.2.3 Control of the plasma membrane H^+ pump

There is ample electrophysiological evidence (Slayman, 1980) to suggest that the activity of the H^+ pump is controlled *in vivo* by mechanisms in addition to the straightforward effects of the pump reactants (ATP, $[H^+]_i$, membrane potential: Sect. 4.2.1).

Cyanide treatment of the *poky f* strain of *N. crassa*, which possesses a cyanide-insensitive electron transport chain, results in only partial metabolic restriction, with a transient depletion of ATP level. Cyanide also induces a train of damped oscillations in the membrane potential, which appear to emanate from oscillatory behaviour of the electrogenic pump and which cannot readily be correlated with changing ATP level (Gradmann & Slayman, 1975). Direct involvement of the membrane potential in the implied control loop can be eliminated by current clamp (which fails to elicit any change in frequency or phase of the oscillations: Gradmann & Slayman, 1975), and no concurrent oscillations in pH_c have been detected (D. Sanders, unpublished results).

Further investigations of pump current in *poky f* (Warncke & Slayman, 1980) have resulted in the conclusion that partial metabolic restriction by cyanide induces a change in pump stoichiometry from $1H^+/ATP$ to $2H^+/ATP$, with a consequent increase in efficiency: the difference between the *I–V* curve measured in the presence of cyanide alone and that in the presence of cyanide and salicylhydroxamic acid, which strongly depletes ATP level, indicates a reversal potential close to −200 mV, or about half the free energy of ATP hydrolysis.

Another clear example of control of the proton pump occurs after initiation of H^+-solute symport in *N. crassa*. The depolarization resulting from the onset of current through the H^+-glucose transport system is only transient, despite constancy of the sugar flux measured isotopically (Sanders & Slayman, 1984). This recovery in membrane potential results from activation of the proton pump, though there is no consistent drop in pH_i. Indeed, the physiological function of pump activation might be seen in terms of acute regulation of pH_i.

The biochemical basis of all these effects is unknown. Covalent phosphorylation of the H^+ ATPase by a membrane-bound protein kinase has been implied by the work of Sussman (1985) and McDonough & Mahler (1982) who found that [γ-^{32}P]-ATP labels serine residues on the protein in plasma membrane preparations. Nevertheless, if indeed the pump is kinase-regulated, other regulators must be invoked in some circumstances.

Serrano (1983) has demonstrated a reversible ten-fold activation of extractable plasma membrane ATPase activity from *Sacch. cerevisiae* incubated for 5 min with glucose. (Non-metabolizable, transportable glucose analogues are ineffective.) Although covalent modification can be invoked to explain these results, no change in membrane protein phosphorylation was detected.

4.2.4 Function of the plasma membrane H^+ pump

An operational plasma membrane proton pump is essential for growth (Serrano *et al.*, 1986), and several functions have been proposed for the role of the pump in cellular physiology. Most obviously, the pump generates a $\Delta\bar{\mu}_{H^+}$ for energization of secondary, H^+-coupled transport (Sect. 4.3). Another role is in regulation of cytoplasmic pH: to extrude from the cell those protons entering the cytoplasmic H^+ pool as a result of catabolic events or H^+-coupled transport. However, although the pump probably plays a crucial role in sustaining a neutral cytoplasmic pH during the life of the cell, and clearly responds to acute acidosis by increasing its activity, it is equally apparent that cytoplasmic pH can be controlled by other processes (perhaps metabolism itself) for periods of up to 1 h (Sanders & Slayman, 1982). Finally, a distinct role in raising cytoplasmic pH and consequent breaking of dormancy of yeast spores and activation of cell division in mature yeast is also possible (Barton *et al.*, 1980; Gillies *et al.*, 1981).

4.2.5 Vacuolar membrane H^+ pump

Like the plasma membrane H^+ pump, the primary electrogenic pump in the vacuolar membrane is an H^+-ATPase which pumps from the cytoplasm where its hydrolytic site is located. Structurally and mechanistically, however, it is clearly distinguishable from the plasma membrane H^+ pump, as well as from the F_0F_1 H^+-ATPase of energy-coupling membranes. The vacuolar ATPase of fungi appears, instead, to be member of a third category of H^+-ATPase, present in a wide variety of organelles including plant vacuoles (see Ch. 3), lysosomes, chromaffin granules, synaptosomes, clathrin-coated vesicles, platelet-dense granules and endosomes, Golgi and ER (Rea & Sanders, 1987). The function of the vacuolar H^+ pump is two-fold: to generate a $\Delta\bar{\mu}_{H^+}$ which enables H^+-gradient accumulation of vacuolar solutes, and to create an acidic environment in what can be regarded as a lysosomal compartment (Matile, 1978).

4.2.5.1 Purification

All experimental studies of the vacuolar H^+ pump have been performed on isolated vacuoles or vacuolar membranes because the organelles are essentially inaccessible *in vivo*. Intact vacuoles can be isolated either from spheroplasts by differential centrifugation (Kakinuma *et al.*, 1981; Bowman & Bowman, 1982; Okorokov & Lichko, 1983), or by disrupting intact cells in a glass-bead blender, again followed by differential centrifugation (Cramer *et al.*, 1983; E.J. Bowman, 1983). Osmotic lysis of the vacuoles, EDTA/EGTA washing and further centrifugation yields membranes with improved specific activity (0.7–3.3 μmol mg^{-1} min^{-1}) as intravacuolar proteins are removed (E.J. Bowman, 1983; Uchida *et al.*, 1985). So far, preparations of vacuolar membranes have been reported only from *N. crassa*, and from the yeasts *Sacch. cerevisiae* and *Sacch. carlsbergensis*.

Further purification of the ATPase is achieved by solubilizing with zwitterionic detergents and glycerol density gradient centrifugation (Lichko & Okorokov, 1984; Uchida *et al.*, 1985), resulting in specific activity of up to 18 μmol mg^{-1} min^{-1}. The purified enzyme has been reconstituted into liposomes by freeze–thaw (Lichko & Okorokov, 1984).

4.2.5.2 Hydrolytic properties

In comparison with the plasma membrane ATPase, the vacuolar ATPase has several distiguishing characteristics. The vacuolar enzyme appears to have a low substrate specificity, hydrolysing GTP at 70–80 % the rate of ATP (Kakinuma *et al.*, 1981; Bowman & Bowman, 1982; Uchida *et al.*, 1985). (However, there is some evidence, from inhibitor studies on the partially purified enzyme, that the GTPase activity might co-purify with the H^+-ATPase: (Lichko & Okorokov, 1985)). The pH optimum for the vacuolar ATPase is also rather higher than for the plasma membrane enzyme (6.9–7.0), and the K_m lower (0.2 mol m^{-3}). However, both enzymes show an absolute requirement for Mg^{2+}, and both are inhibited competitively by ADP (Uchida *et al.*, 1985).

The two enzymes can also be distinguished on the basis of their inhibitor sensitivities: the vacuolar ATPase is insensitive to orthovanadate (Kakinuma *et al.*, 1981; Bowman & Bowman, 1982), suggesting that no covalent phosphorylation occurs during the reaction cycle, and indeed, this has been confirmed directly (Lichko & Okorokov, 1985). By contrast, the vacuolar ATPase is inhibited by NO_3^- (albeit at considerably higher concentrations than its counterpart in plants), and is more sensitive than the plasma membrane enzyme to 7-chloro-4-nitrobenzo-2-oxa-1,3-diazole (NBD-Cl) and to 2′,3′-*O*-(2,4,6-trinitrophenyl)-adenosine 5′-triphosphate (TNP-ATP) (E.J. Bowman, 1983; Uchida *et al.*, 1985). Both enzymes are clearly distinguished from the mitochondrial F_0F_1 H^+-ATPase by their insensitivity to azide and oligomycin, and their relative sensitivity to quercetin. All three H^+-ATPases are inhibited by DCCD (E.J. Bowman, 1983; Uchida *et al.*, 1985).

4.2.5.3 H^+ pumping

Similar techniques to those used for the plasma membrane ATPase have enabled the demonstration that the vacuolar ATPase functions as an electrogenic H^+ pump (Kakinuma *et al.*, 1981; Okorokov & Lichko, 1983; Lichko & Okorokov, 1984). In vacuolar membrane vesicles, the overall $\Delta\bar{\mu}_{H^+}$ generated by the pump is higher than has been reported for plasma membrane vesicles (180 mV, with $\Delta\psi = +75$ mV and $\Delta pH = 1.7$ unit: Kakinuma *et al.*, 1981), indicating that the vacuolar vesicles might be sealed more tightly. There are no estimates yet available of H^+/ATP stoichiometry for the fungal vacuolar ATPase.

4.2.5.4 Protein chemistry

The inhibitor evidence suggests that the vacuolar ATPase might be distinct from both the plasma membrane and F_0F_1 enzymes, and studies on the purified protein confirm that this, indeed, is the case.

All studies on the purified enzyme suggest that the proton pump consists of at least three different polypeptides of M_r = 70 000, 60 000 and 15 000 (*N. crassa*: Bowman & Bowman, 1985); 89 000, 64 000 and 19 500 (*Sacch. cerevisiae*: Uchida *et al.*, 1985); 75 000, 62 000 and 16 000 (*Sacch. carlsbergensis*: Lichko & Okorokov, 1985). [A number of other polypeptides copurify with activity in *N. crassa* (52 kDa, 29 kDa and 27 kDa: Bowman *et al.*, 1986) and *Sacch. carlsbergensis* (14 kDa, 12 kDa and 9 kDa: Lichko & Okorokov, 1985), but two minor polypeptides of 55 kDa and 25 kDa in the *Sacch. cerevisiae* preparation can be eliminated by column chromatography without loss of ATPase activity, suggesting that these components are contaminants (Uchida *et al.*, 1985).]

The 70 kDa polypeptide of *N. crassa* exhibits nucleotide-protectable labelling with [^{14}C]NEM or [^{14}C]NBD-Cl, suggesting that this subunit possesses the nucleotide-binding site (Bowman & Bowman, 1985). The low M_r component can be labelled with [^{14}C]DCCD (E.J. Bowman, 1983; Uchida *et al.*, 1985; Lichko & Okorokov, 1985) and stains poorly with Coomassie blue. Both these characteristics are indicative of hydrophobicity, which might therefore point to a proton channel function for the small polypeptide analogous to the function of the 8 kDa DCCD-binding proteolipid of the F_0F_1 H^+-ATPase.

A clear distinction between the vacuolar ATPase and F_0F_1 has been obtained immunologically. There is no cross-reactivity between F_0F_1 antiserum and the components of the vacuolar ATPase, although cross-reactivity does exist between the 70 kDa polypeptide of *N. crassa* and antibody to its counterpart in the tonoplast ATPase of *Zea mays* (Uchida *et al.*, 1985; Bowman *et al.*, 1986).

The functional M_r of the *N. crassa* enzyme has been estimated by radiation inactivation as 530 000 (Bowman *et al.*, 1986). If it is accepted that the protein is comprised of just three discrete polypeptides, this result implies the presence of multiple copies of the subunits, and perhaps an analogous 'ball and stick' structure to F_0F_1 (Rea & Sanders, 1987). The determination of the primary structure of the vacuolar ATPase subunits is eagerly awaited, since the evolutionary relationship of the vacuolar and F_0F_1 H^+-ATPases is at present unclear.

4.3 Gradient-coupled transport

4.3.1 Accumulation of solutes by fungi: an overview

The ability of fungi to concentrate a wide variety of solutes from the external medium has been known for many years. The list includes monosaccharides (e.g. D-xylose and D-glucose: Kotyk & Höfer, 1965), disaccharides (e.g. α-thioethyl-D-glucoside: Okada & Halvorson, 1964), acidic and amino sugars (e.g. glucuronate: Niemietz & Höfer, 1984; glucosamine: Niemietz *et al.*, 1981), sugar alcohols (Klöppel & Höfer, 1976), a wide range of amino acids (Pall, 1971), purines (Reichert & Foret, 1977) and pyrimidines (Grenson, 1969), as well as the inorganic ions phosphate (Button *et al.*, 1973), sulphate (Roberts & Marzluf, 1971) and potassium (Boxman *et al.*, 1984). Solute accumulation of cations can, of course, be aided by the negative electrical potential across the plasma membrane – a circumstance which is considered explicitly in Sect. 4.4. Nevertheless, it is clear that even K^+ can be accumulated to a level beyond that possible were its uptake driven solely by the electrochemical potential difference (Boxman *et al.*, 1984).

Most of the transport systems operating in the range of low solute concentration, where accumulative transport can most clearly be demonstrated, exist in parallel to at least one other transport system with lower affinity for the same solute. The low-affinity systems are often constitutive, unable to concentrate solute and correspondingly less sensitive to metabolic inhibitors than are the high-affinity systems. The kinetic identification of multiple transport pathways can often be supplemented by manipulating growth conditions so that the presence of one of the systems is controlled. Typically (though not invariably) high-affinity uptake is manifested after a period of induction or derepression (incubation of cells in the presence or absence, respectively, of the solute). For example, in glucose-grown *N. crassa*, glucose is transported, but not accumulated, by a low-affinity system (K_m = 8–25 mol m^{-3}; Scarborough, 1970; Schneider & Wiley, 1971a, b). However, after 1–2 h in the absence of glucose, a second transport system for glucose (Glu II, K_m = 10–70 mmol m^{-3}) is derepressed. Glu II is able to concentrate non-metabolized analogues of glucose by a factor of up to 6000 (Slayman & Slayman, 1975).

Many transport systems exhibit specificities for more than one solute. This is exemplified not only with respect to non-metabolized analogues which have been used to probe the overall uptake capacity, but also for physiologically relevant solutes. Nowhere is the point clearer than in the case of amino acid transport. In yeast, a general amino acid permease, which is subject to catabolite repression during growth on NH_4^+, transports all of the common amino acids, and even the imino acid proline, albeit with low affinity (Grenson *et al.*, 1970; Darte & Grenson, 1975; Lasko & Brandriss, 1981; Grenson & Dubois, 1982). A similar transport system occurs on *N. crassa* (Pall, 1969; Rao *et al.*, 1975). However, other, more specific transport systems, most of which are not subject to repression by NH_4^+, are also present. In *Saccharomyces*, for example, additional independent systems with varying degrees of specificity exist for transport of lysine (Grenson, 1966), arginine (Grenson *et al.*, 1966), methionine, leucine, threonine (Gits & Grenson, 1967), histidine (Crabeel & Grenson, 1970), dicarboxylic amino acids (Darte & Grenson, 1975), glutamine (Grenson & Dubois, 1982) and proline (Magana-Schwencke & Schwenke, 1969). Most of these systems have been characterized genetically as well as kinetically.

Why more than one transport system for the same solute? For the glucose transport systems of *N. crassa*, the answer can probably be expressed in energetic terms: in cases where reduced carbon is plentiful (as it is in the normal growth medium) expenditure of energy in the accumulation of glucose is unnecessary, energy-dependent uptake is inhibited and carbohydrate enters the cell via a non-energy coupled transport system. Where reduced carbon in a form readily available for catabolism is scarce, glucose can nevertheless be accumulated by derepression of Glu II. Amino acids, however, constitute a contrasting case. The diversity of solute specificity in the case of the general amino acid permease endows it with the property of a scavenger for organic nitrogen, while the more specific systems have the ability to supply amino acids for more defined metabolic purposes. Accordingly, both the general and the specific systems have similarly low K_m's for some substrates (e.g. lysine: general permease. $K_m = 3$ mmol m^{-3}; lysine permease, $K_m = 25$ mmol m^{-3}) and correspondingly energy-dependent uptake of the single amino acid occurs on both transport systems (Eddy & Nowacki, 1971; Seaston *et al.*, 1973; Cockburn *et al.*, 1975).

4.3.2 Evidence for gradient-coupling at the plasma membrane

The findings that many solutes are accumulated against their electrochemical gradients led naturally to investigations into the mode of energy coupling. Over the last 15 years, it has become clear that accumulative uptake of the vast majority of solutes is coupled to the simultaneous uptake of protons. The energy for accumulation is derived ultimately from the primary electrogenic H^+ pump, with the resulting $\Delta\bar{\mu}_{H^+}$ powering the passive uptake of H^+ in association with the solute via a single carrier (Eddy, 1978).

4.3.2.1 Alkalinization of the external medium

Evidence for H^+-coupling of solute transport has been forthcoming from several approaches. Most commonly, the disappearance of protons from the external medium is followed with a pH electrode after introduction of the solute. Alkalinization concomitant with solute uptake has been detected for substrates of the glucose transport system of *Rhodotorula gracilis* (including xylose, galactose, glucosamine and glucuronate; Höfer & Misra, 1978; Niemietz *et al.*, 1981; Niemietz & Höfer, 1984), for α-glucosides in *Saccharomyces* (including sucrose, maltose and trehalose: Seaston *et al.*, 1973), for non-metabolized analogues of glucose in *N. crassa* (Slayman & Slayman, 1974) and for substrates of the general amino acid permease, as well as for some of the more specific amino acid transport systems, in *Saccharomyces* and *Candida* (Seaston *et al.*, 1973; Cockburn *et al.*, 1975; Eddy *et al.*, 1977). Accumulation of purines by *Sacch. cerevisiae* and phenol by the soil yeast *Trichospora cutaneum* are also coupled to H^+ uptake by the same criteria (Reichert & Foret, 1977; Mörtberg & Neujahr, 1985). Among inorganic anions, H^+-coupling has been proposed on the basis of transport-induced extracellular alkalinization for SO_4^{2-} uptake in *Saccharomyces* (Roomans *et al.*, 1979), for phosphate in *Saccharomyces* and in *Candida* (Cockburn *et al.*, 1975; Roomans & Borst-Pauwels, 1979; Eddy *et al.*, 1980), and for NO_3^- in *Candida* (Eddy & Hopkins, 1985). The co-transport of protons with solutes (symport) can often be shown to proceed in de-energized cells in which the ATP level is reduced to very low levels with metabolic inhibitors, thus ruling out the requirement for metabolism during transport. The absence of an additional energy supply for solute accumulation is also suggested by thermodynamic considerations (Sect. 4.3.3.2).

Although it is now possible to obtain accurate measurements of cytoplasmic pH with high time resolution, the high buffering capacity of the cytoplasm tends to obscure the acidification expected during symport (Ballarin-Denti *et al.*, 1984).

4.3.2.2 Changes in membrane electrical characteristics

A second method of detecting H^+-coupled transport is to monitor the accompanying electrical events across the plasma membrane. Thus, for neutral solutes, an inward (depolarizing) current must cross the membrane on addition of the solute if the solute and H^+ fluxes are coupled. For anionic solutes, an inward current will also flow if more than one H^+ accompanies each anion. Of course, the electrical measurement alone does not identify H^+ as the co-transported ion as such, although involvement of the proton can be surmised from ion-substitution experiments. Direct demonstrations of membrane depolarization resulting from onset of sugar or neutral amino acid transport have been reported for *N. crassa* and *Physarum polycephalum* (Slayman & Slayman, 1974; Fingerle & Gradmann, 1982; Sanders *et al.*, 1983).

For yeast, where classical electrophysiological methods cannot be used, membrane potential changes have to be monitored indirectly. Radiolabelled lipophilic cations and carbocyanine dyes have therefore been used. Solute uptake via putative symporters has yielded results consistent with membrane depolarization in *R. gracilis* (monosaccharides: Hauer & Höfer, 1978), in *Kluyveromyces marxianus* (= *Sacch. fragilis*) (fucose: Van den Broek *et al.*, 1982), *Sacch. carlsbergensis* (maltose; Eddy *et al.*, 1977) and *C. utilis* (amino acids and NO_3^-: Eddy & Hopkins, 1985). However, both classes of indirect probe can be used only qualitatively and with extreme caution, since partitioning into mitochondria (Korac & Varecka, 1981; Van den Broek *et al.*, 1982; Pena *et al.*, 1984) and, in the case of lipophilic cations, the possibility of uptake through specific carrier systems (Barts *et al.*, 1980; Eraso *et al.*, 1984) and non-specific binding (Vacata *et al.*, 1981; Boxman *et al.*, 1984) all result in considerable uncertainty surrounding a quantitative interpretation.

No mention has so far been made of cations, and in particular K^+. It has been known for many years that K^+-starved fungi exhibit high rates of K^+ influx in exchange for H^+ (Rothstein & Enns, 1946; Ryan & Ryan, 1972), or, in Na^+-loaded cells, in exchange for Na^+ (Kotyk & Kleinzeller, 1959; Slayman & Slayman 1968, 1970). Indeed, much of the early work on transport in fungi was devoted to cation exchange, and the stoichiometric nature of this process led to the notion that exchange occurs via a discrete transport system (Jennings, 1976). If the fluxes were tightly coupled and the exchange mediated by a single system, it is clear that transport would have to be energized, since, in the case of K^+/H^+ exchange, both ions move against their electrochemical potentials. But there exists the possibility that, rather than a chemical coupling of exchange, the fluxes are linked electrically with H^+ or Na^+ efflux acting to balance a depolarizing current carried by K^+. If coupling is electrically mediated, then simple measurement of membrane potential cannot distinguish between mechanisms of energization in which K^+ is co-transported with H^+ and those involving input of scalar free energy to non-coupled K^+ transport. (By contrast, membrane depolarization induced by transport of neutral and anionic solutes must represent obligatory coupling with another charged solute.)

Boxman *et al.* (1984) mooted the idea that K^+ uptake into K^+-starved yeast might be coupled to H^+ influx, and demonstrated that $\Delta\bar{\mu}_{H^+}$ would be energetically sufficient to drive stoichiometric H^+–K^+ symport. A direct demonstration of H^+–K^+ symport in *N. crassa* was forthcoming from a comparison of the K^+-induced membrane current with the net K^+ flux

(Rodriguez-Navarro *et al.*, 1986): the net inflow of positive charge was close to twice the net K^+ uptake. Observations of K^+/H^+ exchange can be accommodated satisfactorily by positing that the H^+ pump maintains electroneutrality during K^+ uptake by efflux of one equivalent of H^+ per equivalent of charge flow mediated by symport.

4.3.2.3. $\Delta\bar{\mu}_{H^+}$-driven solute accumulation in vesicles

All these kinetic methods for demonstration of H^+-coupled solute transport necessitate that the features particularly associated with coupling can be detected against either the background H^+ flux or the membrane electrical conductance. While techniques are available for minimizing such interference (inhibition of the H^+ pump in de-energized cells (Seaston *et al.*, 1973); lowering the membrane conductance by imposition of specific nutritional status (Slayman, 1980)), it is clear that H^+-coupled transport systems with low capacity might remain undetected. The problem can potentially be overcome with tightly sealed membrane vesicles (Franzuhoff & Cirillo, 1983a), since there the thermodynamic competence of $\Delta\bar{\mu}_{H^+}$ to drive accumulation of solute within the vesicle can be tested in controlled conditions. So far, plasma membrane vesicles have remained an under-exploited experimental system with respect to secondary transport. Nevertheless, vesicles (probably of plasma membrane origin) from *Phanerochaete chrysosporium* (white rot fungus) have been shown to accumulate glucose in response to either a pH or electrical potential gradient (Greene & Gould, 1984).

Membrane vesicles can, furthermore, be used to explore the mode of H^+-coupling of solutes normally exported from the cell, since, given a reasonable proportion of everted vesicles, the cytoplasmic side of the membrane is accessible. Ca^{2+} accumulation in plasma membrane vesicles from *N. crassa* is driven by an outwardly directed $\Delta\bar{\mu}_{H^+}$ (Stroobant & Scarborough, 1979b), which implies H^+/Ca^{2+} exchange. However, the measured cytoplasmic free calcium concentration (500 μmol m^{-3}: Miller *et al.*, 1987) is considerably lower than the K_m for the Ca^{2+} transport in vesicles ($\simeq$ 0.1 mol m^{-3}), and the physiological significance of H^+/Ca^{2+} antiport is not yet clear.

4.3.2.4 Involvement of other ions

While no evidence has been forthcoming for coupling of organic solute transport to ions other than H^+, the uptake of inorganic anions may exhibit some deviation from the general pattern. In yeast, three phosphate transport systems have been characterized kinetically. Low-affinity constitutive transport does not appear to be ion-coupled; a higher affinity system (K_m = 10 mmol m^{-3}) is probably H^+-coupled; and very high-affinity transport (K_m = 1 mmol m^{-3}) probably proceeds in association with Na^+ influx (Roomans & Borst-Pauwels, 1977, 1979). SO_4^{2-} accumulation by *Penicil-*

lium notatum is strongly promoted not only by external H^+, but also by Ca^{2+} (Cuppoletti & Segel, 1975). However, although SO_4^{2-} stimulates Ca^{2+} uptake, the apparent stoichiometric ratio $Ca^{2+} : SO_4^{2-}$ is only 0.23, and the possibility remains that Ca^{2+} acts rather non-specifically on the membrane surface potential to accelerate SO_4^{2-} influx (Roomans & Borst-Pauwels, 1979; Eddy, 1982).

4.3.3 Thermodynamics and the H^+: solute stoichiometric ratio (*n*)

4.3.3.1 Theory

The general reaction for symport at the plasma membrane can be written as

$$nH_o^+ + S_o \rightleftharpoons nH_c^+ + S_c \tag{4.5}$$

in which the subscripts o and c refer to the external and cytoplasmic compartment, respectively, n is the stoichiometric ratio of H^+ transported per S. The change in Gibbs' free energy for this reaction (left to right) is given as

$$\Delta G = n\Delta\bar{\mu}_{H^+} + \Delta\bar{\mu}_S \tag{4.6}$$

Equation (4.6) can be expanded and the result expressed in terms of the components of the respective electrochemical gradients as

$$\Delta G = (n + z)F\Delta\psi + RT\ln\frac{[H^+]_c^n[S]_c}{[H^+]_o^n[S]_o} \tag{4.7}$$

where z is the valence of S.

Thus, the value of n has a profound influence on the thermodynamic poise of the reaction, and the corresponding effect on the maximum attainable solute accumulation ratio can be found by setting Eq. (4.7) to 0 and rearranging, which yields

$$\frac{[S]_c}{[S]_o} = \frac{[H^+]_o^n}{[H^+]_c^n}\exp\left[\frac{-(n + z)\,F\Delta\psi}{RT}\right] \tag{4.8}$$

For a typical value of $\Delta\psi = -150$ mV and for $[H^+]_o/[H^+]_c = 100$, a unit increase in n generates an increase in the equilibrium solute accumulation ratio of 36 000-fold (Sanders & Rea, 1987). Evaluation of n is therefore essential if the thermodynamics of transport are to be understood.

4.3.3.2. Experimental determination of *n*

Either of the kinetic methods used to demonstrate the existence of H^+-coupled transport (Sect. 4.3.2.1 and 4.3.2.2) can legitimately be applied for measurement of n. The kinetic method involving measurement of solute-dependent H^+ uptake is most easily performed, and simply relies on measurement of an external pH change in conjunction with the buffer capacity of the external medium. Eddy and coworkers (Seaston *et al.*, 1973; Cockburn *et al.*, 1975; Brocklehurst *et al.*, 1977; Eddy *et al.*, 1977, 1980: Eddy & Hopkins, 1985) have measured n for a variety of symport systems in several yeasts. For the general amino acid permease of *Saccharomyces*, $n = 2$ for all substrates tested, whereas for the equivalent transport system in *Candida*, $n = 1$. The more specific amino acid permeases generally function with $n = 1$ in *Saccharomyces*. The L-glutamate permease is an exception. There, a higher H^+ : solute stoichiometry (= 2 or 3) ensures that transport is electrophoretic, and therefore that the membrane potential component in Eq. (4.8) contributes to the overall solute accumulation ratio. This pattern is reflected in the transport of other anions by *Saccharomyces*: for phosphate, $n = 2$ (Roomans & Borst-Pauwels, 1979) or 3 (Cockburn *et al.*, 1975); for NO_3^-, $n = 2$; for sulphate $n = 3$ (Roomans *et al.*, 1979). Transport of all these solutes is therefore electrophoretic. All carbohydrate transport systems so far investigated display a transport stoichiometry = 1 (Eddy, 1982), even though, in some cases, anionic substrates can be accepted as well as neutral ones (Niemietz & Höfer, 1984).

In the vast majority of cases, therefore, symport systems carry inward current, and therein lies a potential problem for determination of n by measurement of net H^+ uptake: the current through the symporter must be charge compensated by ion fluxes through other transport systems (Eddy, 1977), and if one of the ions involved is the proton, n will be underestimated. However, involvement of the electrogenic proton pump can be minimized in de-energized cells (e.g. Seaston *et al.*, 1973). Even the complete abolition of the background proton influx in such cells would not result in a significant underestimation of the H^+ flux through the symporter. In addition, the ΔpH, inside alkaline, of about 2 units will place the equilibrium potential for H^+ range too positive for net passive H^+ efflux. Therefore, in de-energized cells at least, the method can be regarded as reliable, although measurements of solute-dependent H^+ influx on energized cells may well result in underestimation of n. In most cases, a stoichiometric efflux of K^+ can be observed during symport (e.g. Eddy & Nowacki, 1971; Seaston *et al.*, 1973; Cockburn *et al.*, 1975; Reichert & Foret, 1977; Hauer & Höfer, 1982; Eddy & Hopkins, 1985), although this observation in itself does not eliminate the possibility of some charge compensation occurring via voltage-dependent H^+ pathways.

The second method for determination of n, and one which circumvents the problems associated with H^+ recirculation, is that in which the current through the symporter is measured directly in voltage–clamp experiments. A membrane I–V relationship is first obtained in the absence of solute, and a second I–V curve is then obtained as soon as possible after introduction of the solute. The difference in the membrane current at the resting potential in the presence of solute defines the current through the symporter (i_{sym}) and this can then be compared with the net solute flux (J_S), measured in parallel experiments, to yield n according to the relationship

$$n = -\left(\frac{i_{sym}}{FJ_S} + z\right) \tag{4.9}$$

Three estimates of n for symporters in *N. crassa* are available. For H^+–K^+ symport, $n = 1$ (Sect. 4.3.2.2), and for Glu II, a value of $n = 1$ can be derived from the I–V curves of Hansen & Slayman (1978) and the fluxes reported by Slayman & Slayman (1975), taking $1\ \mu A\ cm^{-2} = 2.39$ meq kg cell water^{-1} min^{-1}. For the general amino acid permease, $n = 2$ (Sanders *et al.*, 1983), as in *Saccharomyces*. The estimate for the general amino acid permease was independently confirmed by measuring the membrane depolarization in response to basic and neutral substrates: the ratio of the respective solute-induced shifts in $\Delta\psi$ was close to 1.5, and since the membrane can be taken to behave ohmically over the relevant voltage range, the value is consistent with $n = 2$.

In all cases for which reliable estimates of stoichiometry are available, the observed solute accumulation ratios do not exceed the limits dictated by Eq. (4.8). Thus, no input of free energy is required for solute accumulation apart from that donated by $\Delta\bar{\mu}_{H^+}$. Indeed, the observed steady-state accumulation ratios are frequently observed to fall well within the thermodynamic limits, indicating that the symport systems fail to attain equilibrium. For example, the general amino acid permease of *N. crassa*, operating in normal conditions (ΔpH = 1.4 unit (inside alkaline), $\Delta\psi = -150$ mV and $n = 2$), is capable of attaining an equilibrium solute accumulation ratio of $7.7 \cdot 10^7$ for neutral amino acids and $2.7 \cdot 10^{10}$ for basic amino acids (Sanders *et al.*, 1983). Thus, even at the low external solute concentration of 1 mmol m^{-3}, the cytoplasmic amino acid level would exceed 1 kmol m^{-3} if the system attained equilibrium. To appreciate why such unphysiological events do not occur, the kinetic properties of symport have to be explored.

4.3.4 Kinetics of H^+-coupled transport

4.3.4.1 Kinetics with respect to external solute concentration

Over fixed, low, concentration ranges, initial rates of solute and H^+ uptake, as well as i_{sym}, are commonly observed to exhibit saturable, Michaelis–Menten kinetics as a function of the external solute concentration (e.g. Slayman *et al.*, 1977). These kinetics are readily interpretable with conventional models of a recycling carrier, in which binding sites for H^+ and S are alternately exposed on each side of the membrane (Sanders *et al.*, 1984b; Fig. 4.7, A).

On the basis of the observation that the K_m for sorbose transport in *Kluyveromyces marxianus* is pH-insensitive, Van den Broek & Van Stev-

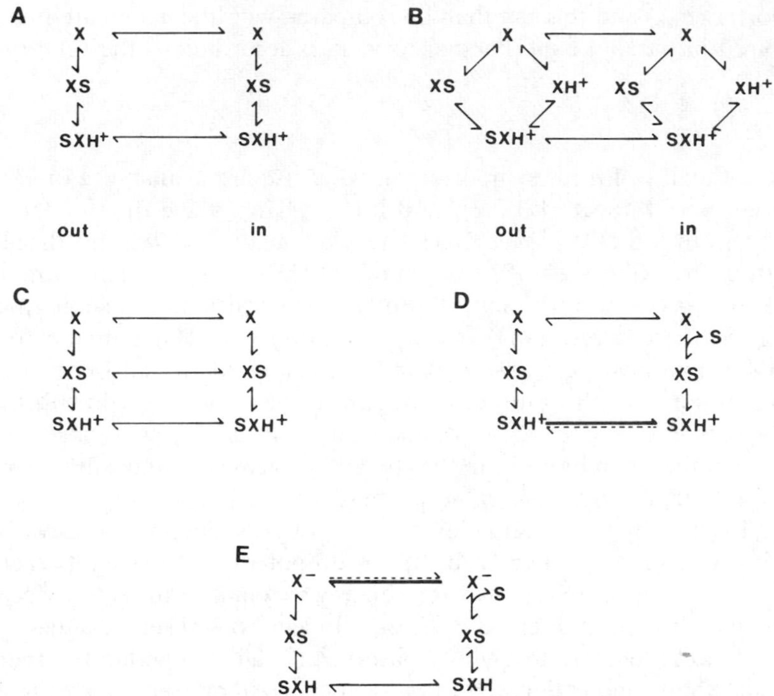

Fig. 4.7 Carrier models for symport of solute (S) and H^+ on a recycling carrier (X). A. Simple ordered binding model (first on-last off, with respect to S). Seven topological variants exist: three for other binding orders with charge translocation on the loaded form of the carrier, and another set of four with charge translocation on the unloaded form. B. Generalized model, with random ligand binding. C. Ordered binding model with 'slip': as S accumulates inside, efflux via the XS pathway prevents attainment of equilibrium with $\Delta\bar{\mu}_{H^+}$. D. Ordered binding model exhibiting transinhibition in physiological conditions. The negative membrane potential generates asymmetry in the charge translocation reaction. As internal solute concentration increases, carrier is progressively inactivated because internal binding sites become occupied by S. Note that the asymmetry of the charge translocation reaction prevents significant efflux. E. Transinhibition with a negatively charged carrier. The principles are as in D, though here, with $\Delta\psi$ generating asymmetry on the transmembrane reaction of the unloaded carrier, the steady state of near-zero net flux is attained primarily through an increase in unidirectional efflux. Kinetics of this type are common for transport of carbohydrates.

eninck (1980) have concluded that the carrier must exhibit random (rather than ordered) binding of the two ligands (Fig. 4.7, B). In *Penicillium notatum*, by contrast, the K_m but not the V_{max} for SO_4^{2-} transport is sensitive to external pH (Cuppoletti & Segel, 1975). This finding was taken to imply that H^+ binds to the carrier before SO_4^{2-}. However, both these conclusions are based on models which incorporate the (at present) unjustified assumption that translocation reactions of the carrier are rate limiting. Abandoning this assumption, it can easily be shown that both classes of kinetic response to external pH can be replicated by any of the four models for ordered ligand binding (Sanders *et al.*, 1984b). Thus the conclusions pertaining to binding order of ligands should be regarded as tentative.

As solute concentration is raised beyond levels which generate V_{max}, a further saturable phase often appears. The lower affinity phase might indicate one of several phenomena: (a) operation of a second discrete transport system (possibly not H^+-coupled); or (b) the ability of the symport system to catalyse the transport of solute uncoupled to that of protons (known as 'slip': Eddy, 1980; and Fig. 4.7, C); or (c) the random binding of the ligands to the carrier (Sanders, 1986). Possibility (a) can be distinguished from the other two by isolation of transport mutants for either of the two systems (e.g. Wiame *et al.*, 1985), by adjusting the growth conditions such that one of the phases fails to appear (e.g. Schnieder & Wiley, 1971a), and, more tentatively, by the failure of one of the kinetic phases to be inhibited by alternate substrates of the other (e.g. Pall, 1969) or by the widely differing sensitivity of the two phases to variation in external pH (e.g. Van den Broek & Van Steveninck, 1980).

Random ligand binding can be identified and distinguished from slip and from parallel functioning of a symporter and a uniporter if transport-associated currents are associated with both kinetic phases. The relevant experiments have not been performed. Hauer & Höfer (1982) have shown for *Rhodotorula gracilis* that H^+ : xylose stoichiometry falls and that the xylose-induced membrane depolarization is identified with high affinity transport. The authors argue for two separate routes for xylose entry–one H^+-coupled, the other not. However, the argument is based on the assumptions that the H^+ : sugar stoichiometry can be accurately assessed in energized cells (cf. Sect. 4.3.3.2) and that the membrane I–V curve for *R. gracilis* is linear: neither of these assumptions has been independently verified. Nevertheless, the observation that D-galactose interacts only weakly with the low-affinity kinetic phase, but is a competitive inhibitor of the high-affinity phase, tends to indicate that the kinetics arise from parallel operation of two discrete carriers (Alcorn & Griffin, 1978).

4.3.4.2 Electrical kinetics

The response of symport to the other component of the driving force, $\Delta\psi$, contrasts markedly with the response to variable ligand concentration. For both Glu II and the general amino acid permease of *N. crassa*, the activity of the transport system (i_{sym}) exhibits little sensitivity to $\Delta\psi$ over a large range from −300 to 0 mV (Hansen & Slayman, 1978; Sanders *et al.*, 1983). In other words, the symport behaves as a constant-current source over this range. Nevertheless, the result can readily be accommodated by conventional carrier models of the type shown in Fig. 4.7 (Hansen & Slayman, 1978): the equilibrium potential for symport is pushed far into the positive range by coupling to $\Delta\bar{\mu}_{H^+}$.

In the range of measureable potentials, therefore, the transport is effectively 'saturated' by voltage (just as it tends to saturate as a function of solute concentration). The H^+–K^+ symport system in *N. crassa* is, by contrast, extremely sensitive to $\Delta\psi$, increasing ten-fold for membrane hyperpolarization from -100 to -300 mV (Blatt *et al.*, 1987; Blatt & Slayman, 1987). Again, this result can be anticipated from a consideration of the reversal potential of the system, which is considerably more negative ($\simeq -50$ mV) than that for the amino acid and glucose transport systems because of high internal K^+ activity (Rodriguez-Navarro *et al.*, 1986).

4.3.4.3. The failure to attain equilibrium: transinhibition and slip

Why are the solute accumulation ratios predicted by Eq. (4.8) rarely attained? In a physiological sense, it is clearly essential to the survival of the cell that cytoplasmic solute concentrations are maintained below the levels predicted by Eq. (4.8), and a priori some kinetic control of transport might be expected. In this section, those features related directly to the carrier reaction scheme (i.e. the topology of the scheme itself and the cytoplasmic concentrations of the ligands) are considered. Other regulatory factors are considered in Sect. 4.3.5.

Studies on amino acid transport provided an early indication that control of solute transport is exercised directly and specifically by the intracellular concentration of solute (Crabeel & Grenson, 1970; Pall & Kelly, 1971; Kotyk & Rihova, 1972; Morrison & Lichstein, 1976). Transinhibition has been shown to be unrelated to *de novo* carrier synthesis (Crabeel & Grenson, 1970; Pall, 1971). Since efflux of amino acids is negligible (Seaston *et al.*, 1976), a simple scheme which has been proposed to account for these transinhibition effects (Pall, 1971) envisages a simple mass action effect of internal amino acid at the transport site, with the back reaction of the loaded carrier across the membrane occurring at a negligible rate. In terms of H^+-coupled transport, the effect of a negative membrane potential on distribution of the charged ternary complex would be expected to have exactly this effect (Fig. 4.7.D). Thus, as internal solute is accumulated, progressively more carrier is effectively removed from participation in the reaction cycle because its binding sites are occupied by cytoplasmic amino acid. Although transinhibition alone cannot reduce solute influx completely to zero until the transport system is at equilibrium, influx can be reduced to such an extent at driving forces well displaced from equilibrium that the resulting quasi-steady state of solute accumulation is experimentally indistinguishable from a genuine steady state (Sanders *et al.*, 1984b). Similar transinhibitory control by cytoplasmic $[H^+]$ is also implicated by the steep dependence (Hill coefficient = 3) of glycine influx in *Sacch. cerevisiae* on cytoplasmic pH (Ballarin-Denti *et al.*, 1984).

Carbohydrates, in complete contrast with amino acids, generally exhibit an approach to the steady state characterized principally by an increase in efflux. This general phenomenon is observed in plant cells as well as fungi (Sauer *et al.*, 1984). Nevertheless, if efflux and influx in the steady sate occur primarily by exchange (as is the case for the monosaccharide carrier of *R. gracilis*: Höfer, 1970; 1971a, b), the approach to a steady state can still be described in the same general terms as for amino acids, with the exception that the translocation reaction of the unloaded carrier is functionally irreversible (Sanders & Rea, 1987; Fig. 4.7,E). An exergonic translocation reaction involving reorientation of the free binding sites might be expected if this is the point in the carrier reaction cycle at which charge is carried across the membrane. Both the amino acid and carbohydrate classes of the reaction kinetic scheme have been analysed numerically, and can be shown, in simple conditions, to exhibit a net flux equal to 1 % of the control value (internal solute concentration = 0) when the internal solute concentration has risen to only 6 % or less of the equilibrium level (Sanders & Rea, 1987).

There are two caveats to the application of these simple transinhibition models to kinetic data. First, basic amino acids are sequestered in vacuoles (Sect. 4.3.6) which complicates the estimation of the concentration at the inner surface of the plasma membrane. Second, catabolite inactivation or repression is also likely to occur (Sect. 4.3.5). For example, Indge *et al.* (1977) have reported that glycine influx in nitrogen-starved *Sacch. uvarum* fails to autoregulate even when the intracellular concentration of glycine attains a value which would normally, in non-starved cells, be associated with transinhibition. The implication, therefore, is that metabolites of glycine, rather than glycine itself, are responsible for regulation in the case of non-starved cells. Similarly, Woodward & Kornberg (1981) have shown that recovery from transinhibition is prevented even after disappearance of the intracellular amino acid pool and requires protein synthesis. These observations imply the existence of additional controls but do not disprove kinetic models of transinhibition. Given that non-metabolized transport analogues are capable of transinhibition (Kotyk & Rihova, 1972; Cuppoletti & Segel, 1974), that the efficacy of a variety of amino acids to transinhibit the methionine transport system of *N. crassa* corresponds with their apparent affinity for transport (Pall, 1971), and that transinhibition has been directly demonstrated in at least one H^+-coupled transport system (Sanders, 1980), it seems simplest for the present to retain the kinetic approach whilst considering other controls which are also likely to be involved.

Slip reactions can also account for the maintenance of a steady state removed from equilibrium (Eddy, 1980, 1982). The steady state is characterized by influx via the symport mechanism balanced by efflux of solute down its electrochemical gradient on the non-protonated form of the carrier. (Alternatively, a tightly coupled symporter could operate in parallel with solute uniport to achieve the same effect.) Since the existence of slip has yet to be demonstrated directly (Sect. 4.3.4.1), the mechanism remains hypothetical, and its application to the control of amino acid transport is problematic given the virtual absence of efflux.

It might be asked why, in the case of the general amino acid permeases of *Neurospora* and *Saccharomyces*, has symport evolved with a stoichiometry of $2H^+$/amino acid when a unity stoichiometry is still capable of generating solute accumulation ratios in excess of 1000 with only half the energy dissipation. One speculation (Sanders *et al.*, 1983) is that because $n = 2$ poises the system far from equilibrium, transport will be rendered relatively insensitive to variation of driving force *per se*, thereby enabling an important nutrient scavenging system to operate in varied conditions of external pH and in conditions when the membrane is electrically depolarized. By contrast, the values of the rate constants describing carrier state transitions might well have evolved to yield sensitivity of transport to cytoplasmic ligand concentrations, thereby autoregulating the homeostatically relevant output parameters.

4.3.5 Extrinsic regulation

Extrinsic regulators are those which influence transport rate without themselves participating on the transport reaction (Sanders, 1984). Among the amino acid symporters, the general amino acid permease of *Sacch. cerevisiae* has been the subject of the most intensive study, and the mechanism of its regulation might serve as a model for the regulation of several other organic nitrogen uptake systems (Wiame *et al.*, 1985).

Ammonium-induced inhibition of transport by the general amino acid permease results from the action of two quite separate control systems coded for by distinct genes (Grenson, 1983a, b; Wiame *et al.*, 1985). The first, which results in reversible inactivation of the permease is achieved by a shift in the equilibrium between the activity of the *MUT2* and *MUT4* gene products, which inactivate the permease, with that of the *NPR1* gene product, which activates transport. Thus, NH_4^+ fails to inhibit pre-existing permease activity in *mut2* and *mut4* mutants, whereas in *npr1* mutants the permease is inactive, independently of the nutritional conditions. It is likely that NH_4^+ or one of its metabolic derivatives, represses synthesis of the *NPR1* product, which allows progressively more of the carrier to be locked in the inactive form. It is not yet known whether inactivation is achieved by covalent modification of the carrier protein, or by binding of a regulatory protein, or even by selective removal of permease from the plasma membrane (Wiame *et al.*, 1985). The second effect of NH_4^+ on the general amino acid permease is one of genuine repression of permease synthesis. The intracellular effector of this inhibitory mechanism is likely to be glutamine, a primary product of NH_4^+ assimilation.

Symport of α-glucosides with H^+ in *Sacch. cerevisiae* is likewise subject to two modes of inhibition by glucose (Peinado & Loureiro-Dias, 1986). Incubation of growing cells with glucose results in a reversible threefold rise of K_m for H^+-maltose symport. Reversal in the absence of glucose does not depend on protein synthesis, and is consistent with the notion that the permease is covalently modified. In parallel, however, glucose represses the synthesis of new permease (Holzer, 1976).

4.3.6 H^+-coupled transport at the vacuole

It has been clear for many years (Wiemken & Nurse, 1973; Weiss, 1973; Karlin *et al.*, 1976; Urech *et al.*, 1978) that one of the principal functions of fungal vacuoles is the storage of organic nitrogen (as basic amino acids) and of phosphate (as polyphosphate). One proposed mechanism for the accumulation of basic amino acids envisaged their retention by the large net negative charge on polyphosphates (Urech *et al.*, 1978): uptake would therefore occur essentially passively, driven by the Donnan potential set up by the polyphosphates. However, this hypothesis does not seem tenable, at least as the major factor in accumulation of basic amino acids, since the vacuolar pools of polyphosphate and basic amino acids are subject to independent regulatory control (Cramer *et al.*, 1980). And although studies with isolated vacuoles had demonstrated solute-specific pathways for uptake of amino acids (Boller *et al.*, 1975), it has not been possible to observe energy-dependent accumulation of amino acids in such preparations (Nakamura & Schlenk, 1974). However, with the development of assays for measurement of transport in vacuolar membrane vesicles (Ohsumi & Anraku, 1981), it is now apparent that a major route for uptake of amino acids is via H^+-coupled transport.

In a preliminary study, Ohsumi & Anraku (1981) demonstrated that right-side-out vacuolar membrane vesicles from *Sacch. cerevisiae* will accumulate arginine in response to activation of the vacuolar H^+ pump. Thus, provision of ATP leads to protonophore-sensitive arginine uptake. Amino acid uptake is also sensitive to inhibitors of the ATPase. The observations are consistent with uptake energized by $\Delta\bar{\mu}_{H^+}$, and therefore with H^+/amino acid antiport. This mechanism of energization could also function *in vivo*, since, in energized conditions, the intravacuolar pH is maintained 0.8–1.5 unit more acid than that of the cytoplasm (Nicolay *et al.*, 1982; Legerton *et al.*, 1983). [A ΔpH of 1.3 unit would be required to maintain the observed *in vivo* vacuolar accumulation ratio of 20 : 1 (Huber-Wälchi & Wiemken, 1979), assuming a H^+ : amino acid stoichiometry = 1 (Eq. (4.8).]

Subsequent kinetic studies have revealed the existence of as many as seven separate antiport systems for amino acids (Sato *et al.*, 1984a). Of the naturally occurring amino acids, ten are accumulated against a concentration gradient in vesicles. Arginine and lysine are transported with highest apparent affinities (K_m = 0.6 mol m^{-3}) as well as (in order of increasing K_m) phenylalanine, histidine, tryptophan, isoleucine, tyrosine, glutamine and leucine. The K_m's for transport of the neutral and aromatic amino acids are considerably higher than their presumed cytoplasmic concentrations for cells grown in standard medium, and this probably accounts for the relatively low vacuolar concentrations of these compounds in normal conditions. Asparagine is also accumulated in vesicles, but to a lesser extent than the other amino acids. Seven permeases were identified on the basis of mutual abilities to inhibit transport competitively: arginine, arginine/lysine/(ornithine/citrulline), histidine, phenylalanine/tryptophan, tyrosine, glutamine/asparagine and isoleucine/leucine. Curiously, histidine was observed to stimulate influx of arginine, and more detailed analysis revealed the existence of a discrete, high affinity (K_m (arg) = 0.1 mol m^{-3}) arginine/histidine exchange system, operating in the direction of arginine entry (Sato *et al.*, 1984b). The stimulatory effect of histidine on arginine uptake can therefore be ascribed to the prior entry of histidine.

Besides amino acids and phosphate, Ca^{2+} is also accumulated in yeast vacuoles (Eilam *et al.*, 1985). In vacuolar membrane vesicles, Ca^{2+} can, like amino acids, be accumulated significantly in response to H^+ pump activation, and again, H^+-coupled transport is implied (Ohsumi & Anraku, 1983). Antiport can be directly demonstrated as Ca^{2+}-dependent fluorescence quenching recovery of the ΔpH-indicating dye, quinacrine, in the presence of ATP. However, as with the plasma membrane H^+/Ca^{2+} antiporter (Sect. 4.3.2.3), the K_m for Ca^{2+} uptake (0.1 mol m^{-3}) is some 200 times the measured cytoplasmic concentration in fungi (Miller *et al.*, 1987), and the physiological relevance of the system has yet to be demonstrated.

4.4 Non-coupled solute transport

4.4.1 Glucose transport in Saccharomyces

Despite the undoubted importance of glucose in the metabolic reactions of yeast and many years of study of D-glucose transport, the mechanism by which monosaccharides enter *Saccharomyces* is still poorly understood and a matter of controversy. It seems clear that glucose uptake is not energized by coupling of transport to $\Delta\bar{\mu}_{H^+}$ (Brocklehurst *et al.*, 1977). Indeed, it is generally agreed that glucose (and its non-metabolized analogues) are not accumulated by *Saccharomyces* (Cirillo, 1968a; Heredia *et al.*, 1968; Eddy, 1982: Bisson & Fraenkel, 1983b).

Nevertheless, there is some evidence for the involvement of kinases during uptake, perhaps by a mechanism involving phosphorylation as an integral part of transport itself. Some evidence for a phosphorylation mechanism is the observation that labelled sugars appear to enter the intracellular sugar phosphate pool before their appearance in the intracellular pool of free sugar (Jaspers & Van Steveninck, 1975; Meredith & Romano, 1977; Franzuhoff & Cirillo, 1982). (Similar arguments have been proposed to account for galactose entry on the inducible galactose carrier: Van Steveninck & Dawson, 1968; Van Steveninck, 1972; but cf. Cirillo, 1968b; Kuo & Cirillo, 1970; Kotyk & Michaljanicova, 1974). However, the finding that the non-phosphorylated glucose analogue 6-deoxy-D-glucose, which competes for the same carrier, is nevertheless transported with high affinity argues against obligatory phosphorylation (Romano, 1982) and in favour of simple facilitated diffusion. Support for the existence of such a transport system has come from studies on plasma membrane vesicles in which the facilitated diffusion of glucose has been shown to exhibit similar kinetic properties to that of glucose transport in intact cells (Franzuhoff & Cirillo, 1983a). The transport system has been solubilized and reconstituted in liposomes (Franzuhoff & Cirillo, 1983b).

Firmer evidence for the involvement of kinases has been forthcoming from work on kinase mutants of *Sacch. cerevisiae* (Bisson & Fraenkel. 1983a). Constitutive transport of glucose exhibits biphasic kinetics, with K_m's of about 1 mol m^{-3} and 20 mol m^{-3}. In a mutant lacking hexokinases A and B, as well as glucokinase, the low K_m system is effectively absent. Introduction to the triple mutant of any one of the three kinases on a cloned wild type gene is sufficient to restore high affinity uptake. However, since the kinetics of 6-deoxy-glucose transport are also subject to modification by the presence of the kinases, the role of these enzymes must be unrelated to the metabolism of the sugar (Bisson & Fraenkel, 1983b). Perhaps the simplest interpretation at present is that the kinases modify transport kinetics by binding to the transport system. It seems possible that further studies on the reconstituted system (Franzuhoff & Cirillo, 1983b) might give further clues to this kinase-induced regulation of glucose transport, as well as uncovering the mechanistic basis of the Pasteur effect, which involves inhibition of glucose transport in aerobic conditions (Serrano & DelaFuente, 1974).

4.4.2 NH_4^+ uptake

Two (or possibly three) separate uptake systems exist for NH_4^+ in *Sacch. cerevisiae* (Dubois & Grenson, 1979). Methylamine is also transported by these permeases. The high-affinity–low-capacity system is repressed in the presence of glutamine, while the low-affinity–high-capacity system is repressed by NH_4^+. A high-affinity, derepressible NH_4^+ transport system has also been described in *Penicillium chrysogenum* (Hackette *et al.*, 1970).

The electrical characteristics of NH_4^+ transport have been investigated in *N. crassa* (Slayman, 1977). NH_4^+-induced membrane depolarization is elicited by nitrogen starvation, or alternatively, by NH_4^+ starvation in the presence of NO_3^-. The simplest interpretation of these results is that NH_4^+ uniport is derepressed in the absence of NH_4^+, although a symport mechanism with protons is not excluded. The time constant for the appearance of NH_4^+ transport activity is 15 min. As in yeast, two kinetic components can be characterized (K_m: 7 mmol m^{-3} and 8.5 mol m^{-3}), although, unlike the yeast permeases, there is no evidence that methylamine is a cosubstrate.

All the above evidence is compatible with carrier-mediated transport of NH_4^+, rather then diffusion of the free base through the bilayer: NH_4^+ transport is derepressible and may be lost through specific mutations (Dubois & Grenson, 1979), shows saturation kinetics and carries electrical current.

4.4.3 Efflux of organic anions

Most fungi excrete large amounts of Krebs' cycle acids, particularly when grown in carbon-rich conditions (Cochrane, 1958). Indeed, the phenomenon is exploited in the industrial production of citric acid by *Aspergillus* grown at low pH. However, the mechanistic aspects of the process are largely unstudied. The highly negative $\Delta\psi$ favours the passive efflux of anions, and it is tempting to speculate that this efflux might account for a large part of the background (non-pump) membrane electrical conductance in *N. crassa* (Sanders *et al.*, 1984a). However, coupling of organic anion efflux to that of protons is still possible. This mode of transport is exploited in *Escherichia coli*, where lactate efflux is coupled to an efflux of H^+ (Konings, 1985): the transport system therefore aids the generation of $\Delta\bar{\mu}_{H^+}$ and conserves energy which would be dissipated as heat were lactate simply allowed to diffuse from the cell.

4.4.4 K^+ channels

In addition to the H^+–K^+ symporter, which is derepressed only in special conditions of K^+ deficiency (Sect. 4.3.2.2), it has generally been considered likely that K^+ is also able to permeate the fungal plasma membrane via channels (uniport). A recent study by Gustin *et al.* (1986) has revealed the existence of K^+ channels in spheroplasts of *Sacch. cerevisiae* with the patch – clamp technique. In the whole cell mode, a depolarization-activated current was observed which could be abolished by tetraethylammonium (TEA) and Ba^{2+} – both identified as K^+ channel blockers in plant and animal systems. Isolated membrane patches exhibited conductance bursts characteristic of single-channel activity, with a slope conductance (5 °C) of 13 pS in the presence of 122 mol m^{-3} K^+ on both sides of the patch. In accord with the notion that these are K^+ channels, the unitary conductance is also blocked by TEA, and its I–V curve extrapolates to a reversal potential of 0 mV. Comparison of the single-channel and whole-cell currents yields an estimate of only from seven to ten channels per cell. However, since the channel does not appear to be active at hyperpolarizing (negative) potentials characteristic of intact cells, its physiological function is obscure.

4.5 Spatial heterogeneity of transport systems

Although membrane currents, taken over the plasma membrane as a whole in the steady state, must sum to zero, it is now apparent that transcellular flow of current can be detected in many eukaryotes (Harold *et al.*, 1985). Typically, current passes into the cell at the actual or presumptive growing tip, and this observation has led to some controversy concerning the role of the current in development: is the current *per se* involved in determining developmental events, or is the specific ion carrying the current more influential? A further possibility is that the current is nothing more than an interesting para-phenomenon which arises as a consequence of development and growth. Experiments with fungi have addressed some of these issues.

Inward currents at the hyphal tips of *Achlya* have been most thoroughly studied (Harold *et al.*, 1985), although transcellular currents have been detected macroscopically in a number of other genera (Stump *et al.*, 1980; Gow, 1984). In *Achlya bisexualis*, the tip current is abolished when amino acids are withdrawn from the medium, but not by any other ion substitutions (Kropf *et al.*, 1984), and it is therefore suggested that the current is carried by H^+-amino acid symport. The electrical circuit could be completed by an excess of proton pump activity further back along the hypha. Significantly, although inward current can predict the emergence of branches from the hypha, the polarity of the current at the tip of the hypha can reverse during branching, with no effect on tip growth (Kropf *et al.*, 1983). This observation suggests that the current *per se* plays no determinant role in growth. If indeed the current flows have any role in establishing polarity for growth, then this must arise from regional differences in ionic composition of the cytoplasm which result from differential activity of transport systems along the hypha.

Acknowledgements

I am grateful to M. R. Blatt, B. J. Bowman, K. M. Hager, D. S. Perlin and M. R. Sussman for communicating results prior to publication, and to P. A. Rea for critical comments on the manuscript. I am also indebted to A. A. Eddy, C. L. Slayman and C. W. Slayman for many enlightening conversations on transport in fungi.

References

Addison, R. & Scarborough, G. A. (1981). Solubilization and purification of the *Neurospora* plasma membrane H^+ ATPase. *Journal of Biological Chemistry* **256**, 13165–71.

Addison, R. & Scarborough, G. A. (1982). Conformational changes of the *Neurospora* plasma membrane H^+ ATPase during its catalytic cycle. *Journal of Biological Chemistry* **257**, 10421–6.

Ahlers, J., Ahr, E. & Seyfarth, A. (1978). Kinetic characterization of plasma membrane ATPase from *Saccharomyces cerevisiae*. *Molecular and Cellular Biochemistry* **22**, 39–50.

Ainsworth, G. C. (1973). Introduction and keys to higher taxa. In Ainsworth, G. C., Sparrow, F. K. & Sussman, A. S. (eds.). *The Fungi: An Advanced Treatise*, vol. 4, pp. 1–7. New York, Academic Press.

Alcorn, M. E. & Griffin, C. C. (1978). A kinetic analysis of D-xylose transport in *Rhodotorula glutinis*. *Biochimica et Biophysica Acta* **510**, 361–71.

Aldermann, B. & Höfer, M. (1984). Fractionation of membranes from *Metschnikowia reukaufii* protoplasts. Evidence for a plasma-membrane-bound ATPase. *Journal of General Microbiology* **130**, 711–723.

Amoury, A., Foury, F. & Goffeau, A. (1980). The purified plasma membrane ATPase of the yeast *Schizosaccharomyces pombe* forms a phosphorylated intermediate. *Journal of Biological Chemistry* **255**, 9353–57.

Amoury, A. & Goffeau, A. (1982). Characterization of the β-aspartyl phosphate intermediate formed by the H^+-translocating ATPase from the yeast *Schizosaccharomyces pombe*. *Journal of Biological Chemistry* **257**, 4723–30.

Amoury, A., Goffeau, A., McIntosh, D. B. & Boyer, P. D. (1982). Exchange of oxygen between phosphate and water catalysed by the plasma membrane ATPase from the yeast *Schizosaccharomyces pombe*. *Journal of Biological Chemistry* **257**, 12509–16.

Amoury, A., Goffeau, A., McIntosh, D. B. & Boyer, P. D. (1984). Contribution of ^{18}O technology to the mechanism of the H^+-ATPase from yeast plasma membrane. *Current Topics in Cellular Regulation* **24**, 471–83.

Ballarin-Denti, A., Den Hollander, J. A., Sanders, D., Slayman, C. W. & Slayman, C. L. (1984). Kinetics and pH- dependence of glycine–proton symport in *Saccharomyces cerevisiae*. *Biochimica et Biophysica Acta* **778**, 1–16.

Barton, J. K., Den Hollander, J. A., Lee, T. M., MacLaughlin, A. & Shulman, R. G. (1980). Measurement of the internal pH of yeast spores by ^{31}P nuclear magnetic resonance. *Proceedings of the National Academy of Sciences, USA* **77**, 2470–73.

Barts, P. W. J. A., Hoeberichts, J. A., Kaassen, A. & Borst-Pauwels, G. W. F. H. (1980). Uptake of the lipophilic cation dibenzyldimethylammonium into *Saccharomyces cerevisiae*. Interaction with the thiamine transport system. *Biochimica et Biophysica Acta* **597**, 125–36.

Bisson, L. F. & Fraenkel, D. G. (1983a). Involvement of kinases in glucose and fructose uptake by *Saccharomyces cerevisiae*. *Proceedings of the National Academy of Sciences, USA* **80**, 1730–34.

Bisson, L. F. & Fraenkel, D. G. (1983b). Transport of 6-deoxyglucose in *Saccharomyces cerevisiae*. *Journal of Bacteriology* **155**, 995–1000.

Blasco, F., Chapius, J. -P., & Giordani, R. (1981). Characterization of the plasma membrane ATPase (EC 3.6.1.3) of *Candida tropicalis*. *Biochemie* **63**, 507–14.

Blasco, F. & Gidrol, X. (1982). The proton-translocating ATPase of *Candida tropicalis* plasma membrane. *Biochemie* **64**, 531–6.

Blatt, M. R. Rodriguez-Navarro, A. & Slayman, C. L. (1987). Potassium–proton cotransport in *Neurospora*. II. Current–voltage analysis and evidence for coupling to $\Delta\bar{\mu}_{H^+}$, not ATP hydrolysis. *Journal of Membrane Biology*. In press.

Blatt, M. R. & Slayman, C. L. (1987). Role of 'active' potassium transport in the regulation of cytoplasmic pH by non-animal cells. *Proceedings of the National Academy of Sciences, USA* **84**, 2737–41.

Boller, T. P., Dürr, M. & Wiemken, A. (1975). Characterization of a specific transport system for arginine in isolated yeast vacuoles. *European Journal of Biochemistry* **54**, 81–91.

Borst-Pauwels, G. W. F. H. (1981). Ion transport in yeast. *Biochimica et Biophysica Acta* **650**, 88–127.

Borst-Pauwels, G. W. F. H. & Peters, P. H. J. (1981). Factors affecting the inhibition of yeast plasma membrane ATPase by vanadate. *Biochimica et Biophysica Acta* **642**, 173–81.

Bowman, B. J. (1983). Kinetic evidence for interacting active sites in the *Neurospora crassa* plasma membrane ATPase. *Journal of Biological Chemistry* **258**, 13002–7.

Bowman, B. J., Berenski, C. J. & Jung, C. Y. (1985). Size of plasma membrane H^+-ATPase from *Neurospora crassa* determined by radiation inactivation and comparison with the sarcoplasmic reticulum Ca^{2+}-ATPase from skeletal muscle. *Journal of Biological Chemistry* **260**, 8726–30.

Bowman, B. J., Blasco, F. & Slayman, C. W. (1981). Purification and characterization of the plasma membrane ATPase of *Neurospora crassa*. *Journal of Biological Chemistry* **256**, 12343–9.

Bowman, B. J. & Bowman, E. J. (1987). H^+-ATPases from mitochondria, plasma membrane and vacuoles of fungal cells. *Journal of Membrane Biology* **94**, 83–97.

Bowman, B. J. & Slayman, C. W. (1979). The effects of vanadate on the plasma membrane ATPase of *Neurospora crassa*. *Journal of Biological Chemistry* **254**, 2928–34.

Bowman, E. J. (1983). Comparison of the vacuolar membrane ATPase of *Neurospora crassa* with the mitochondrial and plasma membrane ATPases. *Journal of Biological Chemistry* **258**, 15238–44.

Bowman, E. J. & Bowman, B. J. (1982). Identification and properties of an ATPase in vacuolar membranes of *Neurospora crassa*. *Journal of Bacteriology* **151**, 1326–37.

Bowman, E. J. & Bowman, B. J. (1985). The H^+ translocating ATPase in vacuolar membranes of *Neurospora crassa*. In Marin, B. P. (ed.). *Biochemistry and Function of Vacuolar Adenosine-triphosphatase in Fungi and Plants*, pp. 132–41. Berlin, Springer-Verlag.

Bowman, E. J., Bowman, B. J. & Slayman, C. W. (1981). Isolation and characterization of plasma membranes from wild type *Neurospora crassa*. *Journal of Biological Chemistry* **256**, 12336–42.

Bowman, E. J., Mandala, S., Taiz, L. & Bowman B. J. (1986). Structural studies of the vacuolar membrane ATPase from *Neurospora crassa* and comparison with the tonoplast membrane ATPase from *Zea mays*. *Proceedings of the National Academy of Sciences, USA* **83**, 48–52.

Boxman, A. W., Dobbelmann, J. & Borst-Pauwels, G. W. F. H. (1984). Possible energization of K^+ accumulation into metabolizing yeast by the proton-motive force. Binding correction to be applied in the calculation of the yeast membrane potential from tetramethylphosphonium distribution. *Biochimica et Biophysica Acta* **772**, 51–7.

Brocklehurst, R., Gardner, D. & Eddy, A. A. (1977). The absorption of protons with α-methyl glucoside and α-thioethyl glucoside by the yeast N. C. Y. C. 240. *Biochemical Journal* **162**, 591–9.

Brooker, R. J. & Slayman, C. W. (1982). Inhibition of the plasma membrane [H^+]-ATPase of *Neurospora crassa* by *N*-ethylmaleimide. Protection by nucleotides. *Journal of Biological Chemistry* **257**, 12051–55.

Brooker, R. J. & Slayman, C. W. (1983a). [^{14}C]*N*-ethylmaleimide labeling of the plasma membrane [H^+]-ATPase of *Neurospora crassa*. *Journal of Biological Chemistry* **258**, 222–6.

Brooker, R. J. & Slayman, C. W. (1983b). Effects of Mg^{2+} ions on the plasma membrane [H^+]-ATPase of *Neurospora crassa*. I. Inhibition by *N*-ethylmaleimide and trypsin. *Journal of Biological Chemistry* **258**, 8827–32.

Brooker, R. J. & Slayman, C. W. (1983c). Effects of Mg^{2+} ions on the plasma membrane [H^+]-ATPase of *Neurospora crassa*. II. Kinetic studies. *Journal of Biological Chemistry* **258**, 8833–8.

Brownlee, C. (1984). Membrane potential components of the marine fungus *Dendryphiella salina* (Suth.) Pugh et Nicot. Possible involvement of calmodulin in electrophysiology and growth. *New Phytologist* **97**, 15–23.

Bussey, H., Saville, D., Chevallier, H. R. & Rank, G. H. (1979). Yeast plasma membrane ghosts. An analysis of proteins by two-dimensional gel electrophoresis. *Biochimica et Biophysica Acta* **553**, 185–96.

Button, D. K., Dunker, S. S & Morse, M. L. (1973). Continuous culture of *Rhodotorula rubra*: kinetics of phosphate–arsenate uptake, inhibition, and phosphate-limited growth. *Journal of Bacteriology* **113**, 599–611.

Chapman, J. B., Johnson, E. A. & Kootsey, J. M. (1983). Electrical and biochemical properties of an enzyme model of the sodium pump. *Journal of Membrane Biology* **74**, 139–53.

Cirillo, V. P. (1968a). Relationship between sugar structure and competition for the sugar transport system in bakers' yeast. *Journal of Bacteriology* **95**, 603–11.

Cirillo, V. P. (1968b). Galactose transport in *Saccharomyces cerevisiae* I. Non-metabolized sugars as substrates and inducers of the galactose transport system. *Journal of Bacteriology* **95**, 1727–1731.

Cochrane, V. W. (1958). *Physiology of Fungi*. New York, John Wiley.

Cockburn M., Earnshaw, P. & Eddy, A. A. (1975). The stoicheiometry of the absorption of protons with phosphate and L-glutamate by yeasts of the genus *Saccharomyces*. *Biochemical Journal* **146**, 705–12.

Comerford, J. G., Spencer-Phillips, P. T. H. & Jennings, D. H. (1985). Membrane-bound ATPase activity, properties of which are altered by growth in saline conditions, isolated from the marine yeast *Debaryomyces hansenii*. *Transactions of the British Mycological Society* **85**, 431–8.

Conway, E. L. (1951). The biological performance of osmotic work. A redox pump. *Science* **113**, 270–3.

Conway, E. L. & Kernan, R. P. (1955). The effect of redox dyes on the active transport of hydrogen, potassium and sodium ions across the yeast cell membrane. *Biochemical Journal* **61**, 32–6.

Crabeel, M. & Grenson, M. (1970). Regulation of histidine uptake by specific feedback inhibition of two histidine permeases in *Saccharomyces cerevisiae*. *European Journal of Biochemistry* **14**, 197–204.

Cramer, C. L., Ristow, J. L., Paulus, T. J. & Davis, R. H. (1983). Methods for mycelial breakage and isolation of mitochondria and vacuoles of *Neurospora*. *Analytical Biochemistry* **128**, 384–92.

Cramer, C. L., Vaughn, L. E. & Davis, R. H. (1980). Basic amino acids and inorganic polyphosphates in *Neurospora crassa*: independent regulation of vacuolar pools. *Journal of Bacteriology* **142**, 945–52.

Crane, F. L., Roberts, H., Linnane, A. W. & Löw, H. (1982). Transmembrane ferricyanide reduction by cells of the yeast *Saccharomyces cerevisiae*. *Journal of Bioenergetics and Biomembranes* **14**, 191–205.

Cuppoletti, J. & Segel, I. H. (1974). Transinhibition kinetics of the sulfate transport system of *Penicillium notatum*: analysis based on an Iso Uni Uni velocity equation. *Journal of Membrane Biology* **17**, 239–52.

Cuppoletti, J. & Segel, I. H. (1975). Kinetics of sulfate transport by *Penicillium notatum*. Interactions of sulfate, protons, and calcium. *Biochemistry* **14**, 4712–18.

Dame, J. B. & Scarborough, G. A. (1980). Identification of the hydrolytic moiety of the *Neurospora* plasma membrane H^+-ATPase and demonstration of a phosphoryl-enzyme intermediate in its catalytic metabolism. *Biochemistry* **19**, 2931–7.

Dame, J. B. & Scarborough, G. A. (1981). Identification of the phosphorylated intermediate of the *Neurospora* plasma membrane H^+-ATPase as β-aspartyl phosphate. *Journal of Biological Chemistry* **256**, 10724–30.

Darte, C. & Grenson, M. (1975). Evidence for three glutamic acid transporting systems with specialized physiological functions in *Saccharomyces cerevisiae*. *Biochemical and Biophysical Research Communications* **67**, 1028–33.

Delhez, J., Dufour, J. -P., Thines, D. & Goffeau, A. (1977). Comparison of the properties of plasma membrane-bound and mitochondria-bound ATPases in the yeast *Schizosaccharomyces pombe*. *European Journal of Biochemistry* **79**, 319–28.

Di Pietro, A. & Goffeau, A. (1985). Essential arginyl residues in the H^+-translocating ATPase of plasma membrane from the yeast *Schizosaccharomyces pombe*. *European Journal of Biochemistry* **148**, 35–9.

Dubois, E. & Grenson, M. (1979). Methylamine/ammonia uptake systems in *Saccharomyces cerevisiae*: multiplicity and regulation. *Molecular and General Genetics* **175**, 67–76.

Dufour, J. -P., Boutry, M. & Goffeau, A. (1980). Plasma membrane ATPase of yeast. Competitive inhibition studies of the purified and membrane-bound enzymes. *Journal of Biological Chemistry* **255**, 5735–41.

Dufour, J. -P. & Goffeau, A. (1978). Solubilization by lysolecithin and purification of the plasma membrane ATPase of the yeast *Schizosaccharomyces pombe*. *Journal of Biological Chemistry* **253**, 7026–32.

Dufour, J. -P. & Goffeau, A. (1980). Molecular and kinetic properties of the purified plasma membrane ATPase of the yeast *Schizosaccharomyces pombe. European Journal of Biochemistry* **105**, 145–54.

Dufour, J. -P., Goffeau, A. & Tsong, T. Y. (1982). Active proton uptake in lipid vesicles reconstituted with the purified yeast plasma membrane ATPase. *Journal of Biological Chemistry* **257**, 9365–71.

Dufour, J. -P. & Tsong, T. Y. (1981). Plasma membrane ATPase of yeast. Activation and interaction with dimyristoylphosphatidylcholine vesicles. *Journal of Biological Chemistry* **256**, 1801–8.

Eddy, A. A. (1977). Current status of the ion gradient hypothesis in some selected systems. In Kramer, M. & Lauterbach, F. (eds.). *Intestinal Permeation*, pp. 332–49. Amsterdam, Excerpta Medica.

Eddy, A. A. (1978). Proton-dependent solute transport in microorganisms. *Current Topics in Membrane Transport* **10**, 279–360.

Eddy, A. A. (1980). Slip and leak models of gradient-coupled solute transport. *Biochemical Society Transactions* **8**, 271–3.

Eddy, A. A. (1982). Mechanisms of solute transport in selected microorganisms. *Advances in Microbial Physiology* **23**, 1–78.

Eddy, A. A. & Hopkins, P. G. (1985). The putative electrogenic nitrate-proton symport of the yeast *Candida utilis*. Comparison with the systems absorbing glucose or lactate. *Biochemical Journal* **231**. 291–7.

Eddy, A. A. & Nowacki, J. A. (1971). Stoicheiometrical proton and potassium ion movements accompanying the absorption of amino acids by the yeast *Saccharomyces carlsbergensis*. *Biochemical Journal* **122**, 701–11.

Eddy, A. A., Philo, R., Earnshaw, P. & Brocklehurst, R. (1977). Some common aspects of active solute transport in yeast and mouse ascites tumour cells. *Federation of European Biochemical Societies Symposia* **42**, 250–60.

Eddy, A. A., Seaston, A., Gardner, D. & Hacking, C. (1980). The thermodynamic efficiency of cotransport mechanisms with special reference to proton and anion transport in yeast. *Annals of the New York Academy of Sciences, USA* **341**, 494–508.

Eilam, Y., Lavi, H. & Grossowicz, N. (1985). Cytoplasmic Ca^{2+} homeostasis maintained by a vacuolar Ca^{2+} transport system. *Journal of General Microbiology* **131**, 623–9.

Eraso, P., Mazon, M. J. & Gancedo, J. M. (1984). Pitfalls in the measurement of membrane potential in yeast cells using tetraphenylphosphonium. *Biochimica et Biophysica Acta* **778**, 516–20.

Fingerle, J. & Gradmann, D. (1982). Electrical properties of the plasma membrane of microplasmodia of *Physarum polycephalum. Journal of Membrane Biology* **68**, 67–77.

Franzuhoff, A. & Cirillo, V. P. (1982). Uptake and phosphorylation of 2-deoxy-D-glucose by wild-type and single-kinase strains of *Saccharo-*

myces cerevisiae. *Biochimica et Biophysica Acta* **688**, 295–304.

Franzuhoff, A. J. & Cirillo, V. P. (1983a). Glucose transport activity in isolated plasma membrane vesicles from *Saccharomyces cerevisiae*. *Journal of Biological Chemistry* **258**, 3608–14.

Franzuhoff, A. J. & Cirillo, V. P. (1983b). Solubilization and reconstitution of the glucose transport system from *Saccharomyces cerevisiae*. *Biochimica et Biophysica Acta* **734**, 153–9.

Gillies, R. J., Ugurbil, K., Den Hollander, J. A. & Shulman, R. G. (1981). ^{31}P NMR studies of intracellular pH and phosphate metabolism during cell division cycle of *Saccharomyces cerevisiae*. *Proceedings of the National Academy of Sciences, USA* **78**, 2125–9.

Gits, J. J. & Grenson, M. (1967). Multiplicity of the amino acid permeases in *Saccharomyces cerevisiae*. III. Evidence for a specific methionine-transporting system. *Biochimica et Biophysica Acta* **135**, 507–16.

Goffeau, A. & Slayman, C. W. (1981). The proton-translocating ATPase of the fungal plasma membrane. *Biochimica et Biophysica Acta* **639**, 197–23.

Gow, N. A. R. (1984). Transhyphal electrical currents in fungi. *Journal of General Microbiology* **130**, 3313–18.

Gradmann, D., Hansen, U. -P., Long, W. S., Slayman, C. L. & Warncke, J. (1978). Current–voltage relationships for the plasma membrane and its principal electrogenic pump in *Neurospora crassa*. I. Steady-state conditions. *Journal of Membrane Biology* **39**, 333–67.

Gradmann, D., Hansen, U. -P. & Slayman, C. L. (1982). Reaction-kinetic analysis of current–voltage relationships for electrogenic pumps in *Neurospora* and *Acetabularia*. *Current Topics in Membranes and Transport* **16**, 257–76.

Gradmann, D. & Slayman, C. L. (1975). Oscillations of an electrogenic pump in the plasma membrane of *Neurospora*. *Journal of Membrane Biology* **23**, 181–212.

Greene, R. V. & Gould, J. M. (1984). Electrogenic symport of glucose and protons in membrane vesicles of *Phanerochaete chrysosporium*. *Archives of Biochemistry and Biophysics* **228**, 97–104.

Grenson, M. (1966). Multiplicity of the amino acid permeases in *Saccharomyces cerevisiae*. II. Evidence for a specific lysine-transporting system. *Biochimica et Biophysica Acta* **127**, 339–49.

Grenson, M. (1969). The utilization of exogenous pyrimidines and the recycling of uridine-5′-phosphate derivatives in *Saccharomyces cerevisiae*, as studied by means of mutants affected in pyrimidine uptake and metabolism. *European Journal of Biochemistry* **11**, 249–60.

Grenson, M. (1983a). Inactivation–reactivation process and repression of permease formation regulate several ammonia-sensitive permeases in the yeast *Saccharomyces cerevisiae*. *European Journal of Biochemistry* **133**, 135–9.

Grenson, M. (1983b). Study of the positive control of the general amino-acid permease and other ammonia-sensitive uptake systems by the product of the *NPR1* gene in the yeast *Saccharomyces cerevisiae*. *European Journal of Biochemistry* **133**, 144–4.

Grenson, M. & Dubois E. (1982). Pleiotropic deficiency in nitrogen-uptake systems and derepression of nitrogen-catabolic enzymes in *npr-1* mutants of *Saccharomyces cerevisiae*. *European Journal of Biochemistry* **121**, 643–7.

Grenson, M., Hou, C. & Crabeel, M. (1970). Multiplicity of the amino acid permeases in *Saccharomyces cerevisiae*. IV. Evidence for a general amino acid permease. *Journal of Bacteriology* **103**, 770–7.

Grenson, M., Mousset, M., Wiame, J. -M. & Bechet, J. (1966). Multiplicity of the amino acid permeases in *Saccharomyces cerevisiae*. II. Evidence for a specific arginine-transporting system. *Biochimica et Biophysica Acta* **127**, 325–38.

Gustin, M. C., Martinac, B., Saimi, Y., Culbertson, M. R. & Kung, C. (1986). Ion channels in yeast. *Science* **233**, 1195–7.

Hackette, S. L., Skye, G. E., Burton, C. & Segel I. H. (1970). Characterization of an ammonium transport system in filamentous fungi with methyl ammonium-^{14}C as the substrate. *Journal of Biological Chemistry* **245**, 4241–50.

Hager, K. M., Mandala, S. M., Davenport, J. W., Speicher, D. W., Benz, E. J. & Slayman, C. W. (1986). Amino acid sequence of the plasma membrane H^+-ATPase of *Neurospora crassa*: deduction from genomic and cDNA sequences. *Proceedings of the National Academy of Sciences, USA*. **83**, 7693–7.

Hager, K. M. & Slayman, C. W. (1986). Molecular characterization of the plasma-membrane H^+ ATPase of *Neurospora crassa*. In Youvan, D. C. & Fevzi, D. (eds.). *Microbial Energy Transduction*, pp. 137–40. Cold Spring Harbor, Cold Spring Harbor Laboratory.

Hansen, U. -P., Gradmann, D., Sanders, D. & Slayman, C. L. (1981). Interpretation of current–voltage relationships for 'active' ion-transport systems. I. Steady-state reaction-kinetic analysis of Class-I mechanisms. *Journal of Membrane Biology* **63**, 165–90.

Hansen, U. -P. & Slayman, C. L. (1978). Current–voltage relationships for a clearly electrogenic cotransport system. In Hoffman, J.F. (ed.) *Membrane Transport Processes*, pp. 141–54. New York, Raven Press.

Harold, F. M., Kropf, D. L. & Caldwell, J. H. (1985). Why do fungi drive currents through themselves? *Experimental Mycology* **9**, 183–6.

Hauer, R. & Hofer, M. (1978). Evidence for interactions between energy-dependent transport of sugars and the membrane potential in the yeast *Rhodotorula gracilis (Rhodosporidium toruloides)*. *Journal of Membrane Biology* **43**, 335–49.

Hauer, R. & Höfer, M. (1982). Variable H^+/substrate stoichiometries in *Rhodotorula gracilis* are caused by a pH-dependent protonation of the carrier(s). *Biochemical Journal* **208**, 459–64.

Heller, J. & Smith, E. L. (1966). *Neurospora crassa* cytochrome *c*. II. Chymotryptic peptides, tryptic peptides, cyanogen bromide peptides and the complete amino acid sequence. *Journal of Biological Chemistry* **241**, 3165–80.

Heredia, C. F., Sols, A. & DelaFuente, G. (1968). Specificity of the constitutive hexose transport in yeast. *European Journal of Biochemistry* **5**, 324–29.

Höfer, M. (1970). Mobile membrane carrier for monosaccharide transport in *Rhodotorula gracilis*. *Journal of Membrane Biology* **3**, 73–82.

Höfer, M. (1971a). Transport of monosaccharides in *Rhodotorula gracilis* in the absence of metabolic energy. *Archives of Microbiology* **80**, 50–61.

Höfer, M. (1971b). A model of the monosaccharide uphill transporting cell membrane system in yeast. *Journal of Theoretical Biology* **33**, 599–603.

Höfer, M. & Misra, P. C. (1978). Evidence for a H^+-sugar symport in the yeast *Rhodotorula gracilis (glutinis)*. *Biochemical Journal* **172**, 15–22.

Holzer, H. (1976). Catabolite inactivation in yeast. *Trends in Biochemical Sciences* **1**, 178–81.

Hubbard, M. J., Surarit, R. & Sullivan, P. A. (1986). The isolation of plasma membrane and characterization of the plasma membrane ATPase from the yeast *Candida albicans*. *European Journal of Biochemistry* **154**, 375–81.

Huber-Wälchi, V. & Wiemken, A. (1979). Differential extraction of soluble pools from the cytosol and the vacuoles of yeast (*Candida utilis*) using DEAE-dextran. *Archives of Microbiology* **120**, 141–9.

Indge, K., Seaston, A. & Eddy, A. A. (1977). The concentration of glycine by *Saccharomyces uvarum*: role of the main vacuole and conditions leading to the explosive absorption of the amino acid. *Journal of General Microbiology* **99**, 243–55.

Jaspers, H. T. A. & Van Steveninck, J. (1975). Transport-associated phosphorylation of 2-deoxy-D-glucose in *Saccharomyces fragilis*. *Biochimica et Biophysica Acta* **406**, 370–85.

Jennings, D. H. (1976). Transport of fungal cells. In Lüttge, U. & Pitman, M. G. (eds.). *Encyclopedia of Plant Physiology*, New Series, vol. 2A, pp. 189–228. Berlin, Springer-Verlag.

Kakinuma, Y., Ohsumi, Y. & Anraku, Y. (1981). Properties of H^+-translocating adenosine triphosphatase in vacuolar membranes of *Saccharomyces cerevisiae*. *Journal of Biological Chemistry* **256**, 10859–63.

Karlin, J. N., Bowman, B. J. & Davis, R. H. (1976). Compartmental behavior of ornithine in *Neurospora crassa*. *Journal of Biological Chemistry* **251**, 3948–55.

Kasher, J. S., Allen, K. E., Kasamo, K. & Slayman, C. W. (1986). Characterization of an essential arginine residue in the plasma membrane H^+-ATPase of *Neurospora crassa*. *Journal of Biological Chemistry* **261**, 10808–13.

Klöppel, R. & Höfer, M. (1976). Transport and utilization of alditols in the yeast *Rhodotorula gracilis (glutinis)*. I. Constitutive transport of alditols. *Archives of Microbiology* **107**, 3229–34.

Konings, W. N. (1985). Generation of metabolic energy by end-product efflux. *Trends in Biochemical Sciences* **10**, 317–19.

Korac, L. & Varecka, L. (1981). Membrane potentials in respiring and respiration-deficient yeasts monitored by a fluorescent dye. *Biochimica et Biophysica Acta* **637**, 209–16.

Kotyk, A. & Höfer, M. (1965). Uphill transport of sugars in the yeast *Rhodotorula gracilis*. *Biochimica et Biophysica Acta* **102**, 410–22.

Kotyk, A. & Kleinzeller, A. (1959). Movement of sodium and cell volume changes in a sodium-rich yeast. *Journal of General Microbiology* **20**, 197–212.

Kotyk, A. & Michaljanicova, D. (1974). Nature of the uptake of D-galactose, D-glucose and α-methyl-D-glucoside by *Saccharomyces cerevisiae*. *Biochimica et Biophysica Acta* **332**, 104–13.

Kotyk, A. & Rihova, L. (1972). Transport of α-aminoisobutyric acid in *Saccharomyces cerevisiae*: feedback control. *Biochimica et Biophysica Acta* **288**, 380–9.

Kropf, D. L. (1986). Electrophysiological properties of *Achlya* hyphae: ionic currents studies by intracellular potential recording. *Journal of Cell Biology* **102**, 1209–16.

Kropf, D. L., Caldwell, J. H., Gow, N. A. R. & Harold, F. M. (1984). Transcellular ion currents in the water mold *Achlya*: amino acid proton symport as a mechanism of current entry. *Journal of Cell Biology* **99**, 486–96.

Kropf, D. L., Lupa, M. D. A., Caldwell, J. H. & Harold, F. M. (1983). Cell polarity: endogenous ion currents precede and predict branching in the water mold *Achlya*. *Science* **220**, 1385–7.

Kuo, S. -C. & Cirillo, V. P. (1970). Galactose transport in *Saccharomyces cerevisiae*. III. Characteristics of galactose uptake and exchange in galactokinaseless cells. *Journal of Bacteriology* **103**, 671–8.

Kuroda, H & Kuroda, R. (1981). Origin of the membrane potential in plasmodial droplets of *Physarum polycephalum*. Evidence for the existence of an electrogenic pump. *Journal of General Physiology* **78**, 637–55.

Kuroda, H., Warncke, J., Sanders, D., Hansen, U. -P., Allen, K. E. & Bowman, B. J. (1980). Effects of vanadate on the electrogenic proton pump in *Neurospora*. In Spanswick, R. M., Lucas, W. J.

& Dainty, J. (eds.). *Plant Membrane Transport: Current Conceptual Issues*, pp. 507–8. Amsterdam, Elsevier.

Lasko, P. F. & Brandriss, M. C. (1981). Proline transport in *Saccharomyces cerevisiae*. *Journal of Bacteriology* **148**, 241–7.

Legerton, T. L., Kanamori, K., Weiss, R. L. & Roberts, J. D. (1983). Measurements of cytoplasmic and vacuolar pH in *Neurospora* using nitrogen-15 nuclear magnetic resonance spectroscopy. *Biochemistry* **22**, 899–903.

Lichko, L. P. & Okorokov, L. A. (1984). Some properties of membrane-bound, solubilized and reconstituted into liposomes H^+-ATPase of vacuoles of *Saccharomyces carlsbergensis*. *FEBS Letters* **174**, 233–7.

Lichko, L. P. & Okorokov, L. A. (1985). What family of ATPases does the vacuolar H^+-ATPase belong to? *FEBS Letters* **187**, 349–53.

McDonough, J. P. & Mahler, H. P. (1982). Covalent phosphorylation of the Mg^{2+}-dependent ATPase of yeast plasma membranes. *Journal of Biological Chemistry* **257**, 14579–81.

Magana-Schwencke, N. & Schwencke J. (1969). A proline transport system in *Saccharomyces chevalieri*. *Biochimica et Biophysica Acta* **173**, 313–23.

Malpartida, F. & Serrano, R. (1981a). Phosphorylated intermediate of the ATPase from the plasma membrane of yeast. *European Journal of Biochemistry* **116**, 413–17.

Malpartida, F. & Serrano, R. (1981b). Purification of the yeast plasma membrane ATPase solubilized with a novel zwitterionic detergent. *FEBS Letters* **131**, 351–54.

Matile, P. (1978). Biochemistry and function of vacuoles. *Annual Review of Plant Physiology* **29**, 193–213.

Meredith, S. A. & Romano, A. H. (1977). Uptake and phosphorylation of 2-deoxy-D-glucose by wild type and respiration deficient baker's yeast. *Biochimica et Biophysica Acta* **497**, 745–59.

Miller, A. G. & Budd, K. (1975). Halide uptake by the filamentous ascomycete *Neocosmospora vasinfecta*. *Journal of Bacteriology* **121**, 91–8.

Miller, A. G. & Budd, K. (1976). Evidence for a negative membrane potential and for movement of Cl^- against its electrochemical gradient in the ascomycete *Neocosmospora vasinfecta*. *Journal of Bacteriology* **128**, 741–8.

Miller, A. J., Parsons, A., Jennings, I. R. & Sanders, D. (1987). Cytoplasmic free calcium in *Neurospora* and *Nitellopsis*: steady-state values and metabolically-induced transients. In Beilby, M. J., Smith, J. R. & Walker, N.A. (eds.). *Membrane Transport in Plants*. Canberra, Australian Academy. In press.

Morrison, C. E. & Lichstein, H. C. (1976). Regulation of lysine transport by feedback inhibition in *Saccharomyces cerevisiae*. *Journal of Bacteriology* **125**, 864–71.

Mörtberg, M. & Neujahr, H. Y. (1985). Uptake of phenol by *Trichospora cutaneum*. *Journal of Bacteriology* **161**, 615–19.

Nakamura, K. D. & Schlenk, F. (1974). Examination of isolated yeast cell vacuoles for active transport. *Journal of Bacteriology* **118**, 314–16.

Nicolay, K., Scheffers, W. A., Bruinenberg, P. M. & Kaptein, R. (1982). Phosphorus-31 nuclear magnetic resonance studies of intracellular pH, phosphate compartmentation and phosphate transport in yeasts. *Archives of Microbiology* **133**, 83–9.

Niemietz, C., Hauer, R. & Höfer, M. (1981). Active transport of charged substrates by a proton/sugar co-transport system. Amino-sugar uptake in the yeast *Rhodotorula gracilis*. *Biochemical Journal* **194**, 433–41

Niemietz, C. & Höfer, M. (1984). Transport of an anionic substrate by the H^+/monosaccharide symport in *Rhodotorula gracilis*: only the protonated form of the carrier is catalytically active. *Journal of Membrane Biology* **80**, 235–42

Ohsumi, Y. & Anraku, Y. (1981). Active transport of basic amino acids driven by a proton motive force in vacuolar membrane vesicles of *Saccharomyces cerevisiae*. *Journal of Biological Chemistry* **256**, 2079–82

Ohsumi, Y. & Anraku, Y. (1983). Calcium transport driven by a proton motive force in vacuolar membrane vesicles of *Saccharomyces cerevisiae*. *Journal of Biological Chemistry* **258**, 5614–17

Okada, H. & Halvorson, H. D. (1964). Uptake of α-thioethyl-D-glucopyranoside by *Saccharomyces cerevisiae*. II. General characteristics of an active transport system. *Biochimica et Biophysica Acta* **82**, 547–55

Okorokov, L. A. & Lichko, L. P. (1983). The identification of a proton pump on vacuoles of the yeast *Saccharomyces carlsbergensis*: ATPase is electrogenic H^+ translocase. *FEBS Letters*, **155**, 102–6.

Pall, M. L. (1969). Amino acid transport in *Neurospora crassa*. I. Properties of two amino acid transport systems. *Biochimica et Biophysica Acta* **173**, 113–27.

Pall, M. L. (1971). Amino acid transport in *Neurospora crassa* IV. Properties and regulation of a methionine transport system. *Biochimica et Biophysica Acta* **233**, 201–14.

Pall, M. L. & Kelly, K. A. (1971). Specificity of transinhibition of amino acid transport in *Neurospora*. *Biochemical and Biophysical Research Communications* **42**, 940–7.

Peinado, J. M. & Loureiro-Dias, M. C. (1986). Reversible loss of affinity induced by glucose in the maltose-H^+ symport of *Saccharomyces cerevisiae*. *Biochimica et Biophysics Acta* **856**, 189–92.

Pena, A., Uribe, S., Pardo, J. P. & Borbolla, M. (1984). The use of a cyanine dye in measuring membrane potential in yeast. *Archives of Biochemistry and Biophysics* **231**, 217–25.

Perlin, D. S., Kasamo, K., Brooker, R. J. & Slayman, C. W. (1984). Electrogenic H^+translocation by the plasma membrane ATPase of *Neurospora*. Studies on plasma membrane vesicles and reconstituted enzyme. *Journal of Biological Chemistry* **259**, 7884–92.

Perlin, D. S., San Francisco, M. J. D., Slayman, C. W. & Rosen, B. P. (1986). H^+/ATP stoichiometry of proton pumps from *Neurospora crassa* and *Escherichia coli*. *Archives of Biochemistry and Biophysics* **248**, 53–61.

Peters, P. H. J. & Borst-Pauwels, G. W. F. H. (1979). Properties of plasma membrane ATPase and mitochondrial ATPase of *Saccharomyces cerevisiae*. *Physiologia Plantarum* **46**, 330–7.

Pogge-von Strandmann, R., Kay, R. R. & Dufour, J. -P. (1984). An electrogenic proton pump in plasma membranes from the cellular slime mold *Dictyostelium discoideum*. *FEBS Letters* **175**, 422–8.

Rao E. Y. T., Rao, T. K. & De Busk, A. G. (1975). Isolation and characterization of a mutant of *Neurospora crassa* deficient in general amino acid permease activity. *Biochimica et Biophysica Acta* **413**, 45–51.

Raven, J. A. & Smith, F. A. (1976). Cytoplasmic pH regulation and electrogenic H^+ extrusion. *Current Advances in Plant Science* **8**, 649–60.

Rea, P. A. & Sanders, D. (1987). Tonoplast energization: two H^+ pumps, one membrane. *Physiologia Plantarum*. In press.

Reichert, U. & Foret, M. (1977). Energy coupling in hypoxanthine transport in yeast: potentiometric evidence for proton symport and potassium antiport. *FEBS Letters* **83**, 325–8.

Roberts, K. R. & Marzluf, G. A. (1971). The specific interaction of chromate with the dual sulfate permease systems of *Neurospora crassa*. *Archives of Biochemistry and Biophysics* **142**, 651–9.

Rodriguez-Navarro, A., Blatt, M. R. & Slayman, C. L. (1986). A potassium-proton symport in *Neurospora crassa*. *Journal of General Physiology* **87**, 649–74.

Romano, A. H. (1982). Facilitated diffusion of 6-deoxy-D-glucose in bakers' yeast: evidence against phosphorylation-associated transport of glucose. *Journal of Bacteriology* **152**, 1295–7.

Roomans, G. M. & Borst-Pauwels, G. W. F. H. (1977). Interaction of phosphate with monovalent cation uptake in yeast. *Biochimica et Biophysica Acta* **470**, 84–91.

Roomans, G. M. & Borst-Pauwels, G. W. F. H. (1979). Interaction of cations with phosphate uptake by *Saccharomyces cerevisiae*. Effects of surface potential. *Biochemical Journal* **178**, 521–7.

Roomans, G. M., Kuypers, G. A. J., Theuvenet, A. P. R. & Borst-Pauwels, G. W. F. H. (1979). Kinetics of sulfate uptake by yeast. *Biochimica et Biophysica Acta* **551**, 197–206.

Rothstein, A. & Enns, L. H. (1946). The relationship of potassium to carbohydrate metabolism in baker's yeast. *Journal of Comparative and Cellular Physiology* **28**, 231–52.

Ryan, J. P. & Ryan, H. (1972). The role of intracellular pH in the regulation of cation exchanges in yeast. *Biochemical Journal* **128**, 139–46.

Sanders, D. (1980). Control of Cl^- influx in *Chara* by cytoplasmic Cl^- concentration. *Journal of Membrane Biology* **52**, 51–60.

Sanders, D. (1984). Gradient-coupled chloride transport in plant cells. In Gerencser, G. A. (ed.). *Chloride Transport Coupling in Biological Membranes and Epithelia*, pp. 63–120. Amsterdam, Elsevier.

Sanders, D. (1986). Generalized kinetic analysis of ion-driven cotransport systems. II. Random ligand binding as a simple explanation for non-Michaclian kinctics. *Journal of Membrane Biology* **90**, 67–87.

Sanders, D., Ballarin-Denti, A. & Slayman, C. L. (1984a). The role of transport in regulation of cytoplasmic pH. In Cram, W. J., Janacek, K., Rybova, R. & Sigler, K. (eds.). *Membrane Transport in Plants*, pp. 303–8. Prague, Academia.

Sanders, D., Hansen, U. -P., Gradmann, D. & Slayman, C. L. (1984b). Generalized kinetic analysis of ion-driven cotransport systems: a unified interpretation of selective ionic effects on Michaelis parameters. *Journal of Membrane Biology* **77**, 123–52.

Sanders, D., Hansen, U. -P. & Slayman, C. L. (1981). Role of the plasma membrane proton pump in pH regulation in non-animal cells. *Proceedings of the National Academy of Sciences, USA* **78**, 5903–7.

Sanders, D. & Rea, P. A. (1987). H^+-coupled solute transport at the plasma membrane and tonoplast. In Beilby, M.J., Smith, J. R. & Walker, N. A. (eds.). *Membrane Transport in Plants*. Canberra, Australian Academy. In press.

Sanders, D. & Slayman, C. L. (1982). Control of intracellular pH. Predominant role of oxidative metabolism, not proton transport, in the eukaryotic microorganism *Neurospora*. *Journal of General Physiology* **80**, 377–402.

Sanders, D. & Slayman, C. L. (1984). Simultaneous measurement of cytoplasmic pH and membrane potential during H^+/sugar cotransport in *Neurospora*. In Cram, W. J., Janacek, K., Rybova, R. & Sigler, K. (eds.). *Membrane Transport in Plants*. pp. 341–2. Prague, Academia.

Sanders, D., Slayman, C. L. & Pall, M. L. (1983). Stoichiometry of H^+/amino acid cotransport in *Neurospora crassa* revealed by current–voltage analysis. *Biochimica et Biophysica Acta* **735**, 67–76.

Sato, T., Ohsumi, Y. & Anraku, Y. (1984a). Substrate specificities of active transport systems for amino acids in vacuolar–membrane vesicles of *Saccharomyces cerevisiae*. *Journal of Biological Chemistry* **259**, 11505–8.

Sato, T., Ohsumi, Y. & Anraku, Y. (1984b). An arginine/histidine exchange transport system in vacuolar-membrane vesicles of *Saccharomyces cerevisiae*. *Journal of Biological Chemistry* **259**, 11509–11.

Sauer, N., Komor, E. & Tanner, W. (1984). Regulation and characterization of two inducible amino acid systems in *Chlorella vulgaris*. *Planta* **159**, 404–10.

Scarborough, G. A. (1970). Sugar transport in *Neurospora crassa*. II. A second glucose transport system. *Journal of Biological Chemistry* **245**, 3985–7.

Scarborough, G. A. (1976). The *Neurospora* plasma membrane ATPase is an electrogenic pump. *Proceedings of the National Academy of Sciences, USA* **73**, 1485–8.

Scarborough, G. A. (1980). Proton translocation catalysed by the electrogenic ATPase in the plasma membrane of *Neurospora*. *Biochemistry* **19**, 2925–31.

Scarborough, G. A. & Addison, R. (1984). On the subunit composition of the *Neurospora* plasma membrane H^+-ATPase. *Journal of Biological Chemistry* **259**, 9109–14.

Schneider, R. P. & Wiley, W. R. (1971a). Kinetic characteristics of the two glucose transport systems in *Neurospora crassa*. *Journal of Bacteriology* **106**, 479–86.

Schneider, R. P. & Wiley, W. R. (1971b). Regulation of sugar transport in *Neurospora crassa*. *Journal of Bacteriology* **106**, 487–92.

Seaston, A., Inkson, C. & Eddy, A. A. (1973). The absorption of protons with specific amino acids and carbohydrates by yeast. *Biochemical Journal* **134**, 1031–43.

Seaston, A., Carr, G. & Eddy, A. A. (1976). The concentration of glycine by preparations of the yeast *Saccharomyces carlsbergensis* depleted of adenosine triphosphate: Effects of proton gradients and uncoupling agents. *Biochemical Journal* **154**, 669–76.

Serrano, R. (1983). In vivo glucose activation of the yeast plasma membrane ATPase. *FEBS Letters* **156**, 11–14.

Serrano, R. (1984). Plasma membrane ATPase of fungi and plants as a novel type of proton pump. *Current Topics in Cellular Regulation* **23**, 87–126.

Serrano, R. & DelaFuente, G. (1974). Regulatory properties of the constitutive hexose transport in *Saccharomyces cerevisiae*. *Molecular and Cellular Biochemistry* **5**, 161–71.

Serrano, R., Kielland-Brandt, M. C. & Fink, G. R. (1986). Yeast plasma membrane ATPase is essential for growth and has homology with ($Na^+ + K^+$), K^+- and Ca^{2+}-ATPases. *Nature* **319**, 689–93.

Slayman, C. L. (1965a). Electrical properties of *Neurospora crassa*. Effects of external cations on the intracellular potential. *Journal of General Physiology* **49**, 69–92.

Slayman, C. L. (1965b). Electrical properties of *Neurospora crassa*. Respiration and the intracellular potential. *Journal of General Physiology* **49**, 93–116.

Slayman, C. L. (1970). Movement of ions and electrogenesis in microorganisms. *American Zoologist* **10**, 377–92.

Slayman, C. L. (1973). Adenine nucleotide levels in *Neurospora*, as influenced by conditions of growth and metabolic inhibitors. *Journal of Bacteriology* **114**, 752–66.

Slayman, C. L. (1977). Energetics and control of transport in *Neurospora*. In Jungreis, A. M., Hodges, T. K., Kleinzeller, A. & Schultz, S. G. (eds.). *Water Relations in Membrane Transport in Plants and Animals*, pp. 69–86. New York, Academic Press.

Slayman, C. L. (1980). Transport control phenomena in *Neurospora*. In Spanswick, R. M., Lucas, W. J. & Dainty, J. (eds.). *Plant Membrane Transport*: Current Conceptual Issues, pp. 179–90. Amsterdam, Elsevier/North- Holland.

Slayman, C. L., Long, W. S. & Lu, C. Y.-H. (1973). The relationship between ATP and an electrogenic pump in the plasma membrane of *Neurospora crassa*. *Journal of Membrane Biology* **14**, 305–38.

Slayman, C. L., Lu, C. Y.-H. & Shane, L. (1970). Correlated changes in membrane potential and ATP concentrations in *Neurospora*. *Nature* **226**, 274–6.

Slayman, C. L. & Sanders, D. (1984a). Electrical kinetics of proton pumping in *Neurospora*. In Blaustein, M. P. & Lieberman, M. (eds.). *Electrogenic Transport: Fundamental Principles and Physiological Implications*, pp. 307–22. New York, Raven Press.

Slayman, C. L. & Sanders, D. (1984b). pH-dependence of proton pumping in *Neurospora*. In Forte, J. G., Warnock, D. G. & Rector, F. C. (eds.). *Hydrogen Ion Transport In Epithelia*, 47–56. New York, Wiley.

Slayman, C. L. & Sanders, D. (1985). Steady-state kinetic analysis of an electroenzyme. In Quinn, P. J. & Pasternak, C. A. (eds.). *The Molecular Basis of Movement Through Membranes*, pp. 11–29. London, Biochemical Society.

Slayman, C. L. & Slayman, C. W. (1968). Net uptake of potassium in *Neurospora*. Exchange for sodium and hydrogen ions. *Journal of General Physiology* **52**, 424–43.

Slayman, C. L. & Slayman, C. W. (1974). Depolarization of the plasma membrane of *Neurospora* during active transport of glucose: evidence for a proton-dependent cotransport system. *Proceedings of the National Academy of Sciences, USA* **71**, 1935–9.

Slayman, C. L., Slayman, C. W. & Hansen, U. -P. (1977). Current–voltage relationships for the glucose/H^+ cotransport system in *Neurospora*. In Thellier, M., Monnier, A., Demarty, M & Dainty, J. (eds.). *Transmembrane Ionic Exchanges in Plants*, pp. 115–22. Paris, CNRS.

Slayman, C. W. (1973). The genetic control of membrane transport. *Current Topics in Membrane and Transport* **4**, 1–174.

Slayman, C. W. & Slayman, C. L. (1970). Potassium transport in *Neurospora*. Evidence for a multisite carrier at high pH. *Journal of General Physiology* **55**, 758–86.

Slayman, C. W. & Slayman, C. L. (1975). Energy coupling in the plasma membrane of *Neurospora*: ATP-dependent proton transport and proton-dependent sugar transport. In Kaback, H. R., Neurath, H., Radda, G. K., Schwyzer, R. & Wiley, W. R. (eds.). *Molecular Aspects of Membrane Phenomena*, pp. 233–48. Berlin, Springer-Verlag.

Slayman, M. R., Strickler, J. E., Hager, K. M. & Slayman, C. W. (1987). Location of a dicyclohexylcarboiimide-reactive glutamate residue in the *Neurospora crassa* plasma membrane H^+-ATPase. *Journal of Biological Chemistry* **262**, 4569–73.

Stroobant, P. & Scarborough, G. A. (1979a). Large-scale isolation and storage of *Neurospora* plasma membranes. *Analytical Biochemistry* **95**, 554–8.

Stroobant, P. & Scarborough, G. A. (1979b). Active transport of calcium in *Neurospora crassa* plasma membrane vesicles. *Proceedings of the National Academy of Sciences*, USA **76**, 3102–6.

Stump, R. F., Robinson, K. R., Harold, R. L. & Harold, F. M. (1980). Endogenous electrical currents in the water mold *Blastocladiella emersonii* during growth and sporulation. *Proceedings of the National Academy of Sciences, USA* **77**, 6673–7.

Sussman, M. R. (1985). Protein kinase-mediated phosphorylation of the M_r = 100 000 plasma membrane H^+-ATPase of plants and fungi. *Plant Physiology* **77**, Suppl. 87.

Sussman, M. R. & Slayman, C. W. (1983). Modification of the *Neurospora crassa* plasma membrane [H^+]-ATPase with *N,N′*-dicyclohexylcarbodiimide. *Journal of Biological Chemistry* **258**, 1839–43.

Uchida, E., Ohsumi, Y. & Anraku, Y. (1985). Purification and properties of H^+-translocating, Mg^{2+}-adenosine triphosphatase from vacuolar membranes of *Saccharomyces cerevisiae*. *Journal of Biological Chemistry* **260**, 1090–5.

Urech, K., Dürr, M., Boller, T., Wiemken, A. & Schwenke, J. (1978). Localization of polyphosphate in vacuoles of *Saccharomyces cerevisiae*. *Archives of Microbiology* **116**, 275–8.

Vacata, V., Kotyk, A. & Sigler, K. (1981). Membrane potential in yeast cells measured by direct and indirect methods. *Biochimica et Biophysica Acta* **643**, 265–8.

Van den Broek, P. J. A., Christianse, K. & Van Steveninck, J. (1982). The energetics of D-fucose transport in *Saccharomyces fragilis*. The influence of the protonmotive force on sugar accumulation. *Biochimica et Biophysica Acta* **692**, 231–7.

Van den Broek, P. J. A. & Van Steveninck, J. (1980). Kinetic analysis of simultaneously occurring proton-sorbose symport and passive sorbose transport in *Saccharomyces fragilis*. *Biochimica et Biophysica Acta* **602**, 419–32.

Van Steveninck, J. (1972). Transport and transport-associated phosphorylation of galactose in *Saccharomyces cerevisiae*. *Biochimica et Biophysica Acta* **274**, 575–83.

Van Steveninck, J. & Dawson, E. C. (1968). Active and passive galactose transport in yeast. *Biochimica et Biophysica Acta* **150**, 47–55.

Villalobo, A., Boutry, M. & Goffeau, A. (1981). Electrogenic proton translocation coupled to ATP hydrolysis by the plasma membrane Mg^{2+}-dependent ATPase of yeast in reconstituted liposomes. *Journal of Biological Chemistry* **256**, 12081–7.

Warnke, J. & Slayman, C. L. (1980). Metabolic modulation of stoichiometry in a proton pump. *Biochimica et Biophysica Acta* **591**, 224–33.

Weiss, R. L. (1973). Intracellular localization of ornithine and arginine pools in *Neurospora*. *Journal of Biological Chemistry* **248**, 5409–13.

Wiame, J. -M., Grenson, M & Arst, H. N. (1985) Nitrogen catabolite repression in yeasts and filamentous fungi. *Advances in Microbial Physiology* **26**, 1–88.

Wiemken, A. & Nurse, P. (1973). Isolation and characterization of the amino acid pools located within the cytoplasm and vacuoles of *Candida utilis*. *Planta* **109**, 293–306.

Willsky, G. R. (1979). Characterization of the plasma membrane Mg^{2+}-ATPase from the yeast *Saccharomyces cerevisiae*. *Journal of Biological Chemistry* **254**, 3326–32.

Wolfinbarger, L. (1980). Transport and utilization of amino acids by fungi. In Payne, J. W. (ed.). *Microorganisms and Nitrogen Sources*, pp. 63–87. London, John Wiley.

Woodward, J. R. & Kornberg, H. L. (1981). Changes in membrane proteins associated with inhibition of the general amino acid permease of yeast (*Saccharomyces cerevisiae*). *Biochemical Journal* **196**, 531–6.

5 Algae

J. A. Raven

5.1 Introduction

The algae are a very diverse group of (predominantly) photolithotrophic Eukaryotes which have not attained to the reproductive complexity of the bryophytes (Bryophyta) or tracheophytes (Tracheophyta). This definition means that the prokaryotic Cyanobacteria and Prochloron (Chloroxybacteria) are excluded from the algae, but the organisms included have an enormous range of structural and chemical diversity, and can be either free-living or symbiotic with chemo-organotrophs (Fungi, Protozoa and Metazoa) (Bold & Wynne, 1985). Furthermore, they are found in many subaerial and soil habitats as well as in their 'usual' aquatic habitats.

The purpose of this chapter is as much to illuminate the potential of the algae for research into solute transport as to demonstrate the progress which has already been made. In particular, it seeks to balance the contribution to the solute transport properties of a given genotype which can be attributed to 'general' constraints (physical chemistry, and phylogeny, of membrane lipids and proteins) and 'specific' constraints (genotypic and phenotypic responses to the 'normal' habitat of the plant, with its complex of factors which determine the solute availability to each cell, e.g. bulk phase concentration of solute, unstirred layer effects and the extent to which solute acquisition and retention is a major contribution to inclusive fitness). Recent reviews of the topics mentioned in the present contribution may be found in Raven (1975, 1976, 1980, 1984a, 1986a,b, 1987) where more complete and equitable assessments of contributions (e.g. by Blinks, Dainty and MacRobbie) to the field will, it is hoped, be found.

5.2 Algal diversity

The higher taxa of the algae (divisions with the termination -phyta, classes with the termination -phyceae) are defined on the basis of their ultrastructure and biochemistry. It is important to note that the division- or class-specific combinations of characters are peripheral to the core of structure and function of photolithotrophic eukaryotes. For example, all algae have mitochondria, nuclei and (at least in photosynthetically competent cells) plastids, and their photolithotrophy is based on homologous structures and catalysts: examples are thylakoids; stroma; reaction centres and directly associated light-harvesting pigments for photoreactions one and two; the water dehydrogenase, cytochrome b_6–f, and ATP synthetase complexes; and RUBISCO (ribulose bisphosphate/carboxylase-oxygenase) and other PCRC (photosynthetic carbon reduction cycle) enzymes. The structures and catalysts whose combinations define the higher taxa of algae are, as far as the photolithotrophic apparatus is concerned, analagous rather than homologous; examples are the chromophores and proteins of the light-harvesting complexes; the catalysts of the cytochrome b_6–f complex – photoreaction one redox coupling (soluble cytochrome *c* and/or plastocyanin); and the number and properties of membranes around the plastid. The divisions and classes of the algae may be polyphyletic, at least as eukaryotic phototrophs, if the prokaryotic endosymbionts which gave rise to the plastids were already distinct in their light-harvesting pigments before becoming the plastids of the major 'pigment groups' of algae. At all events, very substantial parallel evolution has probably occurred in the algae to produce analagous structures and catalysts related to transport processes; examples are *inter*cellular symplastic transport conduits in multicellular organisms; rhizoids in macrophytes; and catalysts of primary active uniport at the plasmalemma.

In this context it is of interest to note that the 'higher' plants (Bryophyta and Tracheophyta) are all descendants of one of the (five?) classes, the Charophyceae, of one of the (twelve?) divisions, the Chlorophyta, of the algae. Accordingly, the great diversity of transport-related structures and catalysts in the algae is not surprising (see Raven, 1984a, 1987)..

Some kind of order in terms of nutrient availability to algae in their natural habitats can be seen if we consider the division by Luther (1949), Den Hartog & Segal (1964) and Raven (1981, 1984a) of phototrophs into planophytes (free in a water body), haptophytes (attached *to* substrate particles which are large relative to the organism) and rhizophytes (with a rhizoidal/rhizomatous portion embedded *in* a substrate whose particles are small relative to the organism).

Dealing first with the planophytes, numerically the most important subdivision are the planktophytes (= phytoplankters): microscopic, usually unicellular phototrophs which grow in the illuminated upper portion (euphotic zone) of most water bodies on earth. The smallest of the eukaryotic planktophytes (Micromonodophyceans such as *Micromonas*; Chlorophyceans such as the smaller strains of *Chlorella*) with volumes of less than 10^{-18} m^3 and effective spherical diameters of $\sim$1 μm are little bigger than the smallest prokaryotic (cyanobacterial) planktophytes (see Raven, 1986a). The largest unicellular planktophytes (phycoma stage of *Halosphaera* spp.; *Ethmodiscus rex*) have volumes of up to $\sim 10^{-9}$ m^{-3} and are relatively rare. The velocity of movement of these organisms relative to their immediate environment is of the order of <1 nm s^{-1} (the non-self-motile picoplankters of radius ~0.5 μm) to 0.01 m s^{-1} (the non-self-motile very largest cells radius ~500 μm, which are relatively uncommon) as a result of the sinking of dense cells, and by swimming of motile cells, at up to ~500 μm s^{-1}. Most planktophytes sink or swim at <100 μm s^{-1}. These velocities must be viewed in the context of the movement of bulk water, the spatial (and temporal) gradients of potentially growth-limiting nutrients, and the effect on the unstirred layers which can limit access of these nutrients to the cell surface. In 'mixed' waters the relative movements of portions of the water body can be of the order of m s^{-1} so that the efficacy of movement of planktophyte cells in moving the cell from 'low' to a 'high' nutrient environment are minimal. In 'stratified' waters, by contrast, with much smaller vertical movements and substantial nutrient solute gradients between, the achievable velocities of movement relative to the surrounding water due to swimming or, in the larger cells, by sinking, *are* significant in terms of moving cells into environments with high nutrient concentrations. Passive sinking achieves this in stratified waters to the extent that the ratio of photolithotrophic (mineral-nutrient using) to chemoorganotrophic (mineral-nutrient regenerating) activity decreases with depth, but at the expense of diminishing the photon availability for growth. Self-motile photolithotrophic cells can (granted the appropriate mixture of behavioural responses to photon flux density and nutrient concentration) maximize their access to photons and nutrients over a 24 h light–dark cycle in a stratified environment with largest photon availability high, and the greatest nutrient availability lower, in the water column (Raven & Richardson, 1984). Motility may also be important in moving to 'patches' of high nutrient availability (e.g. sites of intense chemo-organotrophic activity, such as at faecal pellets or 'marine snow').

It is essential to note that the movements which cells of the size (1–100 μm effective Stokes radius) considered here can make (i.e. at up to a few hundred μm s^{-1} have little impact on the effective thickness of external boundary layers which are, at least for the smaller cells, as wide as the cell radius. By a similar token, the velocity of cytoplasmic streaming (and other intracellular movements) i.e. $\leqslant$100 μm s^{-1}, are insufficient materially to aid the distribution of low M_r solutes within the cells (Nobel, 1983, p. 18; see also Gershon *et al.*, 1985). In view of the isotropy of the environment at the spatial scale of the organism there is no immediate reason why major fluxes other than *radial* fluxes of medium-derived solutes should be needed for those solutes although division of labour within the cell requires such movements.

Turning to benthic algae, an even wider size range is found here, with some unicells having a volume of only a few μm^3 while the largest benthic macroalgae can have volumes of up to 10^{-2} m^3 (e.g. the 10 m long acellular *Codium magnum*) or $10^{-1} m^3$ (e.g. the 50 m long multicellular *Macrocystis*: Levring *et al.*, 1969). Our scheme of life-forms categorizes benthic plants as haptophytes (plants growing on substrates of grain size *large* relative to the plant) or rhizophytes (plants growing with a rhizoid or 'rhizome' portions of the plant growing *in* a fine-grained (relative to the plant size) sediment); both *Codium* and *Macrocystis* are haptophytes.

The habitats of large haptophytes generally involve hydrodynamic regimes which prevent an accumulation of dead organic material around the photolithotrophs (Raven, 1981) so that, in the absence of a large standing crop of benthic, filter-feeding animals, such habitats are net sinks for dissolved plant nutrients. In stagnant water (not, as wc just saw, a likely long-term feature of most haptophyte habitats) we would envisage a continuous decline in nutrient concentration from the bulk phase value, down through the macrophyte bed, to the substratum. With unidirectional lateral watcr movement relative to the plant we would expect a downstream diminution in nutrient concentration which might be more significant than the vertical gradient which would, however, be expected to be more pronounced in the region of the substrate boundary layer. Jackson (1977) documents the vertical zonation of a nutrient in a Californian *Macrocystis* bed; the huge *Macrocystis* plants can span tens of metres of water depth with large external concentration differences. The inverse nutrient and light gradients in the *Macrocystis* bed appears to require substantial downward transport of photosynthate and upward transport of N within the plant; this can only be achieved at rates commensurate with the potential growth rate of the plant if *mass flow* is the means of intraplant transport (Nobel, 1983; Raven, 1984a, 1987). In addition to these requirements for intraplant transport that are related to spatial variation in resource availability over the thallus surface are those directly related to the separation of growth, storage and resource acquisition regions of the plant even with high and constant resource availability over the surface of the thallus (Raven, 1981, 1984a). For smaller haptophytes the distances involved, and required fluxes per unit TS area of thallus, are smaller than for the larger haptophytes, and over distances of up to a few tens of millimetres it is possible that diffusive symplastic transport is adequate (Nobel, 1983; Raven, 1984a, 1987).

For rhizophytes the hydrodynamic forces are frequently less vigorous relative to those exerted on haptophytes, and organic, mineralizeable material may accumulate in the sediment. This could make the sediment, with its gravitational input of organic matter, a net producer of mineral nutrients, i.e. organic input plus chemo-organotrophic activity might exceed the rate at which rhizoids and 'rhizomes' of rhizophytes can remove it, and CO_2, NH_4^+ and $H_2PO_4^-$ might be lost to the bulk water phase. At all events, we note that the nutrient gradients in the bulk phase for haptophyte communities may be reversed near the sediment in rhizophyte communities; that the rhizoids may be major sites of N and P acquisition for the whole plant; and that requirements for intraplant transport are, for a given plant size and degree of differentiation, at least as great as in haptophytes.

A point of difference between haptophytes and rhizophytes seems (Raven, 1981, 1984a) to relate to responses to low bulk aqueous phase potential flux of nutrients to the plant surface (a complex of low bulk phase concentration and and/or low water flow velocity relative to the plant surface), The haptophyte has no alternative (spatial) source of nutrients to the bulk phase, while the rhizophyte has the sediment, and, when the potential flux to the shoot is low, it frequently responds by increasing the number of (often colourless) hairs on the plant surface. These increase the surface area available for uptake and, especially, project this area through the 'bulk thallus' unstirred layer and may thus increase the achieved uptake per plant at low potential nutrient flux values.

Another aspect of the diversity of algae with respect to solute transport relates to variation in the *requirement* for net influx of nutrient solutes, and for net efflux of 'waste' solutes, during growth. Working initially in terms of plant *volume*, we note that the requirement for a nutrient during growth (mol nutrient (m^3 volume)$^{-1}$ s^{-1}) is the product of the specific growth rate of the organism (m^3 volume increase (m^3 volume)$^{-1}$ s^{-1}) and the volume-based nutrient content of the organism (mol nutrient (m^3 volume)$^{-1}$).

Specific growth rates at optimal resource availability (μ_{max}) vary substantially with algae taxon and with the size of the organism. Banse (1982) compares values for μ_{max} (normalized to 20 °C) for a number of taxa of the Bacillariophyta and Dinophyta, and finds that larger cells have lower μ_{max} values than do smaller cells, while, for a given size of cell, diatoms grow faster than dinophytes. Thus, if we relate μ_{max} to cell volume by an equation of the form $\mu_{max} = a$ (volume)^b, the data reviewed by Banse (1982) show that a is larger for diatoms than for dinoflagellates, and that the exponent b is negative, and is numerically smaller for these photolithotrophs than for chemo-organotrophs (where it is commonly $\leqslant -0.25$), being ~ -0.11 for diatoms and ~ -0.13 for dinoflagellates. The highest μ_{max} values (20 °C) for small algae are $\sim 3.10^{-5}$ s^{-1}, corresponding to a generation time of 6.4 h, for cells with a volume of $<30 \cdot 10^{-18}$ m^3. For larger algae the μ_{max} values are, understandably, less readily measured; data in Clendenning (1971a,b,c) for *Macrocystis* suggest a μ_{max} in nature (normalized to 20 °C assuming $Q_{10} = 2$) of 10^{-7} s^{-1}, i.e. a generation time of 80 d, for plants with a volume of $3 \cdot 10^{-3}$ m^3.

Nutrient content on a volume-of-organism basis varies with genotype and with the availability of the various resources required for growth. For growth at μ_{max} the second source of variation will be ignored since, by defition, the nutrient content must exceed the critical content (i.e. the minimum content at which the maximum growth rate is achieved), although there could still be 'luxury accumulation' of some nutrient(s). Variation with genotype relates to a number of factors; vacuolation tends to reduce the N and P content relative to that of C (in walls) and K^+, Na^+ and Cl^- (vacuolar osmotica): (see Raven, 1984a, Ch. 8).

Table 5.1 shows some elemental contents on a volume basis for contrasting algae, i.e. a very small Chlorococcalean (Chlorophyta; Chlorophyceae) unicells, such as *Nannochloris* or small-celled strains of *Chlorella*, and the largest Laminarian (Phaeophyta, Phaeophyceae), *Macrocystis* (Levring *et al.*, 1969). We note that, in addition to inorganic sources of C, N, S, P, K, Ca, Mg, Cl (and Na) given in Table 5.1, and of various trace elements (e.g. Fe, Mn, Cu, Zn), growth of these photolithotrophs needs an influx of O_2 (during the dark phase of the light-dark cycle) and of (effectively) H^+ or OH^- (during assimilation of NO_3^- and NH_4^+ respectively, as part of the pH-regulating processes of the organism (see Raven, 1984a, Ch. 1).

It will be seen from columns (1)–(3) of Table 5.1 that differences between nutrient element concentration between the small and the large algae are relatively small, with a *Chlorella* : *Macrocystis* ratio of 0.22 (K)–10.7 (N), with the exception of the lower values for Na (0.041) and Cl (0.039) for marine *Chlorella*. *Macrocystis* (and other marine macrophytes) has a much greater fraction of its total volume taken up by cell walls and other intercellular matrix material, and by vacuoles, than does *Chlorella* (see Ch. 6 and 8 of Raven, 1984a; Raven, 1986a, 1987). This explains the higher C : N and C : P ratios in *Chlorella* than in *Macrocystis*, since the cell wall has a very high C : N and C : P ratio relative to the cytoplasm, while the greater degree of vacuolation in *Macrocystis* than in *Chlorella* does not seem to permit greater luxury accumulation (as a fraction of the critical content) of C, N and P (Raven, 1984a, Ch. 8; see Redfield, 1958; Atkinson & Smith, 1983).

The differences in Na and Cl contents between *Chlorella* and *Macrocystis* may be attributable to three, partly interrelated causes; these are (1) the lower degree of vacuolation in *Chlorella* cells; (2) the higher fraction of cell walls in *Macrocystis* biomass: and [strongly interacting with (2)], (3) the exclusively marine habitat of *Macrocystis*.

The significance of the relatively small degree of vacuolation in *Chlorella* spp. ($\leq$0.1 of the intraplasmalemma volume) for the content of Na^+, Cl^- (and K^+) in the biomass relates to the concentration of free ions which the cytosol, and other 'N' phases (*sensu* Mitchell, 1979; see Schnepf, 1964, 1984; Raven & Smith, 1977) can tolerate. Raven (1984a; Table 7.1) tabulates 'typical' concentrations of solutes in 'N' phases (largely cytosol and stroma) of eukaryotes; for K^+, Na^+ and Cl^-, the '*whole organism*' values for *Chlorella* (row (1) of Table 5.1) are not high relative to the *range* quoted for the 'N ' phases of eukaryotes. However, the possibility that some solutes have been lost in the washing procedures (to which the

Table 5.1 Comparison of elemental content and net influx during growth in very small and in very large algae

Element	**(1) mol m^{-3}; 1–3 in** ***Chlorella*** **sp.†**	**(2) mol m^{-3} in** ***Macrocystis*** **sp.‡**	**(3) mol m^{-3}** ***Chlorella*****/ mol m^{-3}** ***Macrocystis*****§**	**(4) Influx, mol m^{-3} s^{-1} in** ***Chlorella*** **at μ_{max} ¶**	**(5) Influx mol m^{-3} s^{-1} in** ***Macrocystis*** **at μ_{max} ‖**
C	10 400	3900	2.67	$208 \cdot 10^{-3}$	$390 \cdot 10^{-6}$
N	1 198	112	10.7	$24 \cdot 10^{-3}$	$11.2 \cdot 10^{-6}$
S	47	44	1.1	$0.94 \cdot 10^{-3}$	$4.4 \cdot 10^{-6}$
P	114	16	7.1	$23 \cdot 10^{-3}$	$1.6 \cdot 10^{-6}$
K	91	413	0.22	$1.8 \cdot 10^{-3}$	$4.1 \cdot 10^{-6}$
Na	27	659	0.041	$0.54 \cdot 10^{-3}$	$66 \cdot 10^{-6}$
Ca	21	49	0.43	$0.42 \cdot 10^{-3}$	$49 \cdot 10^{-6}$
Mg	81	20	4.1	$1.6 \cdot 10^{-3}$	$2.2 \cdot 10^{-6}$
Cl	2 (13)	336	0.006 (0.039)	$0.04 \cdot 10^{-3}$ ($0.26 \cdot 10^{-3}$)	$34 \cdot 10^{-6}$

Table 5.1 (cont'd)

(6) mol m^{-3} s^{-1} *Chlorella*/mol m^{-3} s^{-1} *Macrocystis* ††	(7) Influx, mol m^{-2} s^{-1} in *Chlorella* at μ_{max} ‡‡	(8) Influx, mol m^{-2} s^{-1} in *Macrocystis* at μ_{max} §§	(9) mol m^{-2} s^{-1}/mol m^{-2} s^{-1} *Chlorella*/*Macrocystis* ¶¶
533	$35 \cdot 10^{-9}$	$106 \cdot 10^{-9}$	0.33
2143	$4 \cdot 10^{-9}$	$3 \cdot 1 \cdot 10^{-9}$	1.29
213	$0.16 \cdot 10^{-9}$	$1.2 \cdot 10^{-9}$	0.13
1438	$0.38 \cdot 10^{-9}$	$0.44 \cdot 10^{-9}$	0.86
43.9	$0.30 \cdot 10^{-9}$	$11.2 \cdot 10^{-9}$	0.027
8.2	$0.09 \cdot 10^{-9}$	$18.0 \cdot 10^{-9}$	0.005
85.7	$0.07 \cdot 10^{-9}$	$1.3 \cdot 10^{-9}$	0.054
727	$0.27 \cdot 10^{-9}$	$0.5 \cdot 10^{-9}$	0.54
1.18 (7.6)	$0.007 \cdot 10^{-9}$ ($0.043 \cdot 10^{-9}$)	$9.3 \cdot 10^{-9}$	0.00075 (0.0046)

† Computed from mean values for each element in Table 1.2 of Raven (1984a), assuming C = 125 kg (m^3 cell volume)$^{-1}$. Values are for *freshwater* strains grown in freshwater media. Marine *Chlorella salina* (Kirst, 1977) shows no greater cation or P concentrations, although Cl^- (brackets) is rather higher. Greenway and Šetter (1979), point out that the lengthy washing employed by Kirst (1977) may have underestimated intracellular ion concentrations, although they find that *C. emersonii* does not increase its intracellular K^+, Na^+ and Cl^- concentrations when grown on 355 mol m^{-3} NaCl relative to 1 mol m^{-3} NaCl medium (cf. Ahmad & Hellebust, 1984, 1985; Katz & Avron, 1985).

‡ From mean values for each element in Table 1.4 of Raven (1984a).

§ Column (1) divided by (2).

¶ From column (1), assuming a μ_{max} of $2 \cdot 10^{-6}$ s^{-1} (Table 4 of Raven, 1986a).

|| From column (2), assuming a μ_{max} of 10^{-7} s^{-1} (see text, and Clendenning, 1971a,b,c)

†† Column (4) divided by column (5).

‡‡ From column (5), assuming that the cells are at the lowest reported volume for *Chlorella* and *Nannochloris* cells; radius $0.5 \cdot 10^{-6}$ m, volume $0.524 \cdot 10^{-18}$ m^3, area $3.14 \cdot 10^{-12}$ m^2; $6 \cdot 10^6$ m^2 surface area (m^3 cell volume)$^{-1}$: see Table 4 of Raven (1986a).

§§ From column (6), assuming $3.66 \cdot 10^3$ m^2 thallus surface area (m^3 thallus volume)$^{-1}$; see Clendenning (1971a,b,c) and text of this paper.

¶¶ Column (7) divided by column (8).

microalgae, but not *Macrocystis* cells were subjected) cannot be ignored (see footnote[†] to Table 5.1). We also note (Table 7.1 of Raven, 1984a) that there is evidence from marine giant-celled algae consistent with higher total ion concentrations in 'N' phases than appear to be compatible (even allowing for the presence of enzyme-protective 'compatible solutes') with full activity of many eukaryotic enzymes. For ions such as Ca^{2+} the free 'N' phase concentration (or, at least, cytosol concentration) in eukaryotes is very low, <1 mmol m^{-3} (Table 7.1 of Raven, 1984a); however, the modified cytosol of sieve tube sap in both 'higher plants' and Phaeophyta (Sect. 5.7.5) may have free Ca^{2+} concentrations in excess of 1 mmol m^{-3} (Raven, 1986b; cf. Table 9.8 of Raven, 1984a).

Points (2) and (3) relate to the large extracellular volume of Donnan free space and water free space in *Macrocystis* together with its invariable marine habitat. The extent to which the cations in the Donnan free space represent the cations associated with the fixed negative charges when the walls were synthesized by exocytosis is not clear; especially significant for computations of the extent to which the production of cell wall fixed negative charge involves the influx at the plasmalemma of the cations associated with them in the wall is the degree to which —COOH groups produced by oxidation of C_6 of the hexoses are secreted as such, with external exchange of H^+ for Na^+, Mg^{2+}, etc. (see Raven & De Michalis, 1979; Ch. 1 of Raven, 1984a). At all events, the 'whole thallus' estimates of ion content in *Macrocystis* overestimates the need for transplasmalemma influxes of ions such as Na^+ and Cl^- to a greater extent than do estimates for *Chlorella* cell with their much lower fraction of wall volume and of fixed negative charge per unit wall volume (see Davison & Reed, 1985).

In terms of the requirement for transplasmalemma net solute influxes during growth [columns (4)–(9) of Table 5.1], we must further bear in mind that simply multiplying the cell content of the element (mol element (m^3 cell volume)$^{-1}$ by the specific growth rate (m^3 cell volume increase (m^3 cell volume)$^{-1}$ s^{-1} to obtain a volume-specific influx (mol element increase (m^3 cell volume)$^{-1}$ s^{-1}) gives a time-averaged influx over the 24 h light–dark cycle. The extent to which this procedure can underestimate *instantaneous* net influxes across the plasmalemma is probably greatest for inorganic C, where a net influx in the light period is replaced by a net efflux in the dark period (dark respiration). Accordingly, for growth in a 12 L : 12 D light–dark cycle, the instantaneous net inorganic C influx in the light must, for the same specific growth rate, be at least twice the values quoted in columns (4), (5), (7) and (8) of Table 5.1; even for a low dark respiration rate which is only 0.05 of the rate of net photosynthesis, the instantaneous mean inorganic C influx in the light phase must be 2.1 times the quoted values. This ratio would be even higher if the net efflux of soluble organic C from algae were taken into account, but very slightly reduced if the 'CAM- (Crassulacean acid metabolism: see Ch. 11) like' behaviour of certain Phaeophyta (Johnston, 1984; Raven *et al.*, 1985;

Johnston & Raven, 1986) extends (as it probably does *not*, since it has not thus far been reported for the Laminariales) to *Macrocystis*.

The values of volume specific nutrient element influxes during growth at μ_{max} are generally much higher for the small-celled *Chlorella* strains than for *Macrocystis* [column (6) of Table 5.1]; this is a result of the much higher specific growth rate of *Chlorella* ($2 \cdot 10^{-5}$ s^{-1}) than for *Macrocystis* (10^{-7} s^{-1}) outweighing, even for Na^+ and Cl^-, the higher volume-specific content of certain elements in *Macrocystis* than in *Chlorella* [column (3) of Table 5.1].

The much higher surface area per unit volume in the small strains of *Chlorella* ($6 \cdot 10^6$ m^{-1}) than in *Macrocystis* thalli ($3 \cdot 66.10^3$ m^{-1}) means that the net influx rates required to support growth at $2 \cdot 10^{-5}$ s^{-1} for *Chlorella* are less [with the exception (just!) of N] than those needed for growth of *Macrocystis* at 10^{-7} s^{-1} [columns (7)–(9) of Table 5.1]. Even allowing for the above-mentioned correction factor needed to take into account light–dark cycles, respiratory and organic C losses, the highest required net influx (C in *Macrocystis* in the light at ~250 nmol m^{-2} s^{-1}) is still substantially less than the maximum measured net photosynthetic values in *Macrocystis* fronds (5000 nmol m^{-2} s^{-1}; Table 5.7 of Raven, 1984a) at light saturation. This difference presumably reflects, *inter alia*, self-shading in whole plants of *Macrocystis* as well as any underestimation of μ_{max} from *in situ* measurements of growth.

In later sections of this article we shall explore the extent to which the intrinsic capacity for mediated transport per unit area of plasmalemma might restrict μ_{max} for algae, as well as the interaction of size and shape of algae with their hydrodynamic regime in determining their capacity to acquire nutrients from bulk media which have low concentrations of available forms of nutrient elements such as N and P.

5.3 Solute fluxes and unstirred layers

The previous section mentioned 'potential fluxes'. This concept (see Raven, 1984a) presents the maximum possible supply rate of dissolved nutrients to the plant surface in terms of mol m^{-2} s^{-1} which are analagous to those for the energy source, i.e. incident photosynthetically active radiation (mol photon m^{-2} s^{-1}). In both cases we note that the conversion of the potential flux to a flux absorbed by the plant involves a correction for the extent to which the plant acts as a perfect sink for solutes or photons which reach it.

For the solute fluxes we note that the potential flux can be derived from application of Fick's law (see Ch. 1) for the appropriate shape of the organism and the thickness of the unstirred layer; thus, for a planar plant surface (see Raven, 1984a):

$$J = \frac{D}{l} (C_b - C_s) \tag{5.1}$$

where J = solute flux (mol m^{-2} s^{-1}),
D = diffusion coefficient (m^2 s^{-1}),
l = effective unstirred layer thickness (m),
C_b = bulk phase solute concentration (mol m^{-3}),
C_s = plant surface solute concentration (mol m^{-3}).

J becomes the maximum potential flux, J_p, when $C_s = 0$, granted the given values of D, l and C_b. We note that the J_p values are, as defined here, based on the total plant surface area while for the potential (i.e. incident) photon flux is based (for a vector light field) on a *projected* area basis; for a strictly scalar light field the relevant area is (as for the solute fluxes) the *total plant surface area*.

For a given plant size and shape, the value of l depends on the velocity of water movement relative to plant. We have already seen that, for planktophytes, l is essentially independent of cell movement velocity relative to the bulk medium for achievable velocities of movement. For large benthic plants, we have seen that minimizing the unstirred layer thickness by water flow over the plant saturates at a few tenths of a metre per second, and that these sorts of velocities are more likely to occur for haptophytes than rhizophytes. For rhizophytes there is the additional possibility of nutrient acquisition by the rhizoids; in the sediment, the high C_b value, and the (often) large rhizoid area per unit biomass, is at least partially offset as far as l value is concerned by the essentially unstirred nature of the sediment interstitial water (Raven, 1984a).

In terms of the potential flux per unit volume of plant [mol (m^3 volume)$^{-1}$ s^{-1}] at a given cell quota for the nutrient (mol m^{-3}) we find that the small organisms are advantaged relative to larger ones.

We deal first with a 0.5 μm radius picoplankton cell with a 0.5 μm effective unstirred layer thickness under 'natural' conditions. This value of 0.5 μm is essentially unaltered by such movement as the organism can make relative to the medium, or by 'natural' turbulence, although a reduction can be effected by high rates of shearing in cultures, or by 'natural' movements of larger (10s of μm radius) cells (Munk & Riley, 1952; Pasciak & Gavis, 1974, 1975; Gavis, 1976; Purcell, 1977; Raven, 1984a, 1986b). With the 'normal' unstirred layer, the flux to the surface of a spherical cells is given by Raven (1986b):

$$J = \frac{3\,(C_b - C_s)\,D}{r^2} \tag{5.2}$$

where J = solute flux to cell surface [mol (m^3 cell volume)$^{-1}$ s^{-1}],
C_b = bulk phase solute concentration (mol m^{-3}),
C_s = surface solute concentration (mol m^{-3}),
D = solute diffusion coefficient (m^2 s^{-1}),
r = cell radius (m).

For $C_s = 0$, $C_b = 0.1$ mol m^{-3}, $D = 10^{-9}$ m^2 s^{-1}, $r = 0.5$ m, then $J =$ 1.2 mol m^{-3} s^{-1}. For P, granted reasonable assumptions about the cell P content (C : P at the 'Redfield ratio') and C/m^3, such a potential flux could support a specific growth rate of 0.0122 s^{-1}, equivalent to a generation time of a minute (Raven, 1986a). We thus see that, even granted a low C_b value, the potential P flux to the surface of a picoplankton organism is adequate to support growth rates more than 10^3-fold those observed at 20 °C (Raven, 1986b).

At the other extreme, even in turbulent flow at 50 mm s^{-1} (not always encountered *in situ*) the boundary layer thickness for a *Macrocystis* blade is 35 μm (Wheeler, 1980) giving an area-based potential flux [Eq. (5.1)] from $C_b = 0.1$ mmol m^{-3} of 2.86 nmol m^{-2} s^{-1}. For both sides of the blades (area 11 m^2) of a plant with a fresh weight of 3.055 kg (Clendenning, 1971a), the volume based flux (assuming a fresh weight of 3.055 kg corresponds to $2.8 \cdot 10^{-3}$ m^3 volume) is 11.3 mol P (m^3 volume)$^{-1}$ s^{-1}; if P is 8.78 mol per m^3 plant volume (Clendenning, 1971b; North, 1980), this corresponds to a specific growth rate of $1.3 \cdot 10^{-6}$ s^{-1}, i.e. a generation time of 6 d; for comparison, Clendenning (1971c) suggests only 2.5 turnovers of standing crop per year off California, so that P_i potential flux would overprovide P for growth by 24-fold. However, in view of the caveats entered earlier as to the *minimum* value of the unstirred layer thickness (35 μm) in that the water velocity might be lower than assumed, and might also be laminar; it is very likely that the *potential* flux of phosphate from a C_b of 0.1 mmol m^{-3} to a *Macrocystis* could be lower than the achieved, *in situ* growth rate reported earlier; since the surface of *Macrocystis* is not a 'perfect sink' for P_i, making 0.1 mmol m^{-3} phosphate a growth-limiting resource for this plant and, indeed, most large haptophytes.

A generalization of these conclusions is that even the *minimum* likely unstirred layer thickness for a benthic macrophyte imposes a larger potential constraint on growth rate than do the unstirred layers of smaller cells. Thus the maximum potential resource fluxes [mol (m^3 volume)$^{-1}$ s^{-1}] from a given C_b for the 0.5 μm radius picoplankter and the 3 kg fresh weight *Macrocystis* are in the ratio 100 000 : 1, while their specific growth rates (normalized to 20 °C) are in the ratio $3 \cdot 10^{-5} : 1 \cdot 10^{-7}$ or 300 : 1, an apparent advantage for the picoplankter of more than 300-fold. It appears that this sort of analysis contradicts the conclusions of Colinvaux (1980, pp. 72–8) as to the absence of macrophytes from the euphotic zone of the open ocean; Colinvaux opines that a floating macrophyte with minimal movement relative to the surrounding water could, by deploying a large surface area of rhizoids, compete well for nutrients with planktonic algae. Further investigation is required.

The discussion thus far has assumed that the diffusion coefficient for solute transport through the boundary layer is the same as that for free solution. It is unlikely that any cell has solute diffusion coefficients which are as high as those in free solution due to the presence of extracellular polymers. Even the so-called 'naked' algal cells (vegetative or reproductive) have a surface coating of acidic polysaccharides, probably analagous to the 'glycocalyx' of animal cells (Oliveira *et al.*, 1980; Brawley & Wetherbee, 1981; Yakote *et al.*, 1985) which decreases the diffusion coefficient for low M_r solutes by perhaps twofold for at least a few nanometres out from the plasmalemma. Cell walls are commonly at least 10 nm thick in the smallest walled algae cells and may be more than 10 μm thick in giant algae cells and on the outer surface of small-celled macrophytes; the turgor-resisting wall components may be supplemented by an outer layer of mucilage and/or a more ordered sporopollenin (lipid) or 'cuticle' (protein) layer (Raven, 1981, 1984a, 1986a, 1987). The cell wall probably also decreases the diffusion coefficient for nutrient solutes by at least twofold (Nobel, 1983; Raven, 1984a). We note that the effective pore radius of the sporopollenin layer in the walls of some algae apparently excludes $NADH^-$ (M_r = 664) but permits entry of malate^{2-} (M_r = 132); Raven (1984a, p. 483). This small effective pore radius would presumably impose a substantial constraint on the growth of any Vitamin B_{12} auxotrophs which had a cell wall sporopollenin layer since all known B_{12}-requiring organisms can use exogenous B_{12}-active molecules of $M_r > 1300$ (Dawson *et al.*, 1969; Provasoli & Carlucci, 1974). However, we note that none of the B_{12} auxotrophs listed by Provasoli & Carlucci (1974) are known to have a sporopollenin layer, and gram-negative prokaryotes which are B_{12} auxotrophs have a special mechanism for movement across the outer membrane since the pores in this membrane have effective radii little greater than those in sporopollenin (Raven, 1984a).

The sediment in which the rhizoids of rhizophytic algae dwell are (like cell walls) essentially unstirred and have a substantial fraction of the volume taken by solid particles. Thus, despite the sedimentation and chemo-organotrophic activity which regenerates nutrients, diffusion through the bulk phase can be slow. This effect is not so marked for NH_4^+ and P_i in anoxic sediments ($D_{NH_4^+} = 9.8 \cdot 10^{-10}$ m^2 s^{-1}; $D_{P_i} = 3.6 \cdot 10^{-6}$ m^2 s^{-1}; Krom & Berner, 1980a,b) than might be the case for oxygenated sediments where, while $D_{NO_3^-}$ might be quite high ($\sim 10 \cdot 10^{-10}$ m^2 s^{-1}), D_{P_i} is likely to be quite low with a low [P_i] (free) due to Fe_2O_3 complexing ($D_{P_i} \sim 10^{-13}$ m^2 s^{-1}; Nye & Tinker, 1977). An interesting possibility here is that the extension growth of algal rhizoids (30–90 nm s^{-1}; Mishra & Kefford, 1969; Sievers & Schroter, 1971) helps to tap undepleted nutrients in sediment water which would be out of range of diffusion to the surface of non-growing rhizoids (cf. Nobel, 1983). Once inside the rhizoid cytoplasm, cytoplasmic streaming at up to 100 μm s^{-1} can deliver nutrients to other parts of the plant (Raven, 1981, 1984a). Growth of nutrient uptake by rhizoids in anoxic sediments might be limited by O_2 supply from bulk water via the algae thallus (see Raven, 1981, 1984a). It is of interest in this context (see also Sec. 5.5.8) that the capacity for nutrient influx from low $\sim$ mmol m^{-3}) bulk phase concentrations (i.e. the sorts of concentrations which might be found for phosphate in an aerobic sediment) in roots of terrestrial plants are such that unstirred layers must be less than 20 μm thick if anything like their full transport capacity is to be expressed (Wild & Breeze, 1981). It would be of interest to see if that held for algal rhizoids (see Sect. 5.5.8).

Having considered the effects of unstirred layers and other *impediments* to solute supply from the bulk phase to the cell surface it behoves us to mention the *reverse* effect, i.e. the restriction of the loss of solutes from the outer surface of the plasmalemma to the bulk medium. This can be advantageous whenever the loss of a solute from the cell can be considered as disadvantageous to cell growth and survival. Imposition of unstirred layers and cell wall barriers to solute loss (in series with, and outside, the plasmalemma) reduces the net loss, granted constant *overall* driving forces between the cytosol and the bulk medium and constant plasmalemma properties (Raven, 1986a).

The propensity for net loss is reduced even further when there is a mediated accumulation mechanism for the leaking solute involved, or for some species which can be rapidly produced from the leaking species. Examples are the leakage of NH_3 from the cytosol of cells assimilating exogenous NO_3^-, urea or NH_4^+ (from low concentrations of NH_4^+ at low pH values where exogenous $[NH_3]$ is very low): see Raven (1980, 1986a); Kleiner (1985a,b), and of CO_2 from the cytosol of cells which are pumping HCO_3^- inwards at the plasmalemma (Raven, 1985b, 1986a). In the case of NH_3 leakage the scavenging of NH_4^+ requires the presence of an NH_4^+ accumulation mechanism (which may not otherwise be needed in a plant living on NH_3 or urea) but *not* of a catalyst of the intrinsically very rapid protonation of NH_3 to form NH_4^+; for the scavenging of HCO_3^- after CO_2 leakage, extracellular carbonic anhydrase is required to speed up the conversion of CO_2 to HCO_3^- (Raven, 1984b). The restriction of leakage by extracellular diffusion barriers is an essential component of the proposed mechanism of photosynthesis in internodal cells of certain ecorticate Characean species, whereby, at high bulk phase pH, HCO_3^- to CO_2 conversion is catalysed in 'acid bands' by high $[H^+]$ and carbonic anhydrase (for references and recent work, see Price *et al.*, 1985; Price & Badger, 1985) and of mechanisms [found in 'higher' (vascular) plants, but not yet sought in algae] for Fe acquisition based on extracellular production of H^+ and reductant (see Ch. 7 of Raven, 1984b).

To conclude this section it is necessary to reiterate (see Sect. 5.2) that unstirred layer effects are important inside, as well as outside, the plasmalemma. While cyclosis (at up to 100 $\mu m\ s^{-1}$) or other mechanisms of intracellular mass flow (at up to several hundred micrometres per second) is essential for symplastic movement of solutes at physiologically significant fluxes if the distances involved is more than a millimetre or so, fluxes from sources to sinks over tens or hundreds of micrometres can usually be accommodated by diffusion through 'unstirred' cytosol (Nobel, 1983; Raven, 1984a, 1986a,b, 1987).

5.4 Transport at the Plasmalemma

5.4.1 *Preamble*

Our consideration of the transport processes operating at the plasmalemma of algae cells begins by distinguishing between transport involving catalysis by membrane integral proteins ('mediated transport') and which involves a flux through the lipid component of the membrane ('lipid solution transport'). We then discuss individually the major categories of mediated transport (primary active transport, secondary active transport and passive uniport) and lipid solution transport, with respect to the solutes involved, the range of algae in which the system(s) have been demonstrated, and their characteristics in terms of energetics, maximum transport capacity and response to changing substrate (and other effector) concentrations.

5.4.2 *The distinction between 'mediated' and 'lipid solution' transport*

Experiments on whole cells cannot readily distinguish between mediated and lipid solution influx of solutes on the basis of apparent activation energy (Raven & Smith 1978; Homble, 1985), or, for metabolized solutes, saturation of influx at high external concentrations. 'Lipid solution' influx can show saturation kinetics related to the kinetics of an intracellular solute consuming reaction, and can have a temperature coefficient similar to that of mediated transport processes. There is no obvious mechanism for active transport in the absence of mediated transport, so that mediated (protein-related) transport can be inferred if the net transport occurs in the direction contrary to that predicted from the prevailing electrochemical potential gradient. When net transport is in the direction predicted by the electrochemical potential gradient, mediated transport is likely if the measured permeability coefficient is substantially in excess of that predicted for lipid solution transport, based on measurements on protein-free lipid bilayers.

On these criteria, mediated *and* active influx at the plasmalemma of algal cells is almost universally present for anions such as $H_2PO_4^-$, SO_4^{2-}, NO_3^- and Cl^-, and is also found, in some cases, for HCO_3^-; active efflux is found for H^+, Na^+ and Ca^{2+}. K^+ is rather tricky; no unanimity emerges from the numerous measurements made on a variety of organisms. It appears that active K^+ influx occurs in some organisms, while passive, mediated K^+ influx is adequate to account for net K^+ influx in algae with large, inside-negative ψ_{co} values. NH_4^+, arginine$^+$ and lysine$^+$ influx, may involve passive uniport, while in other cases active influx may occur (Walker *et al.*, 1979a; Walker *et al.*, 1979b; Sauer *et al.*, 1983). Passive mediated uniporters are known for H^+, Na^+, Ca^{2+} and Cl^- in addition to the K^+ and NH_4^+ already mentioned. Among neutral solutes, there are indications that active CO_2 influx can occur in some algae; other solutes bearing no net charge which at least some algae actively transport include urea, various hexoses, and certain neutral amino acids (Raven, 1980, 1984a; Sauer *et al.*, 1983). Lipid solution fluxes at the plasmalemma predominate for O_2 (in photosynthesis and respiration), CO_2 (for respiratory efflux and, in some cases for influx in photosynthesis) and NH_3 (at high pH and/or NH_4^+ concentrations). In the two latter cases we note that high concentrations of the substrate (inorganic C; NH_4^+ + NH_3) can repress the mediated, accumulative influx of HCO_3^-/CO_2 or NH_4 in organisms which possess such systems, and that 'lipid solution' efflux of CO_2 or NH_3 can short-circuit the accumulative influx and assimilation of, respectively, inorganic C (HCO_3^- and/or CO_2) and N sources such as NH_4^+, NO_3^- and urea. These various points (and more!) are discussed by Raven (1984a, Chs. 3, 5, 6, 7; 1986a,b,c). The following subsections of this chapter (5.4.3–5.4.8) deal with selected data which illustrate various points, and present a different, and more modern, perspective than those mentioned above.

5.4.3 Primary active transport at the algal plasmalemma

Primary (i.e. biochemically coupled) active ion influx in algae was first recognized by the occurrence of hyperpolarization of the inside-negative plasmalemma electrical potential difference to a value outwith the range permitted by diffusion potentials (Findlay & Hope, 1976; Raven, 1976, 1984a, 1986a). In practice, the most negative transplasmalemma diffusion potential is that for K^+. Evidence for such a hyperpolarization has come from experiments on Chlorophyta. Following the taxonomy of this division used by Raven (1987), these organisms include *Eremosphaera viridis* and *Hydrodictyon africanum* (Chlorophyceae), *Chara, Nitella, Lamprothamnium* and *Spirogyra* species (Charophyceae) *and Acetabularia* (Ulvophyceae). For the smaller-celled members of the Chlorophyceae for which hyperpolarization of ψ_{co} has been based on estimates using lipid-soluble cations it appears (Gimmler & Greenway, 1983; Ritchie, 1984) that quantitative conclusions must be drawn with great caution.

The increased inside-negatively of ψ_{co} could be accounted for by *either* an anion influx *or* a cation efflux. Manipulation of the extracellular medium is easier than is that of the cytosol, and the first well established correlation of the hyperpolarization with the availability of a certain ion was external Cl^- in *Acetabularia*. For the Characeae and *Hydrodictyon* the removal of external anions had little effect on the hyperpolarization, and active electrogenic efflux of a cation, probably H^+, was suggested.

More recent work on the giant-celled algae (Characeae, *Hydrodictyon*, *Acetabularia*) has emphasized the measurement of membrane conductance and voltage clamping technique with, in the case of the Characeae, a substantial contribution from internal perfusion, the use of these techniques (and especially the perfusion procedure) being facilitated by the very large cell sizes (volumes up to 10^{-7} m^3) of the organisms under study. The work on *Acetabularia* (summarized by Gradmann *et al.*, 1982a; Gradmann *et al.*, 1982b; Tittor *et al.*, 1983; Goldfarb *et al.*, 1984a,b) shows that the Cl^- influx pump is vanadate-sensitive and ATP-powered, using 1 mol ATP to pump 2 mol Cl^-; manipulation of driving forces on Cl^- across the plasmalemma permits reversal of the pump, with net ATP synthesis accompanying downhill Cl^- entry through the pump. Both tracer ($^{36}Cl^-$) influx and voltage clamp measurements are consistent with a maximum Cl^- influx capacity in sea water medium of up to 10 μmol m^{-2} s^{-1} (see also vibrating probe estimates of current efflux: Bowles & Allen, 1984); with a Cl^- pump density of 50 nmol m^{-2}, a maximum specific reaction rate of 200 s^{-1} is required. We note that the quoted (Tittor *et al.*, 1983) pump site density of 50 nmol m^{-2} (cf. the 'small particle' density in freeze-etching of 100 nmol m^{-2}) requires, with an M_r $\approx 10^5$ for a vanadate-sensitive ion-transporting ATPase (Goffeau & Slayman, 1981), 5 mg of pump protein per square metre of plasmalemma. Even if the pump molecules are cylinders 15 nm long (twice the thickness of a lipid bilayer) oriented at right angles to the membrane surface and with a density of 1300 kg m^{-3}, they would occupy 0.257 of the membrane surface. We see that the measured active Cl^- fluxes in *Acetabularia* are very high, and demand a high porter density fraction of membrane devoted to porters, and porter-specific reaction rates, relative to many other membranes (see Ch. 3 of Raven, 1984a). A complication in interpreting quantitative relationships of active Cl^- influx (and for active H^+ efflux in Characeae) is the localization of pump sites (Bowles & Allen, 1984; Nawata, 1984).

The internodal cells of ecorticate Characeans have also been subjected to very substantial recent studies which combine perfusion, voltage clamp and net flux measurements in investigating the nature of the primary active transport mechanism. A very important recent finding is that of Takeshige *et al.* (1985) who found that, with a stirred medium at pH 8.2 (minimizing extracellular H^+ accumulation and passive H^+ re-entry), net ATP-dependent H^+ efflux and ATP-dependent positive charge efflux in a voltage clamp were very similar (19–20 mA m^{-2}, equivalent to 197–207 nmol H^+ m^{-2} s^{-1}) in perfused *Nitellopsis obtusa* internodal cells. This finding is very important evidence in favour of the electrogenic pump of the Characeae being an H^+ efflux pump, although Takeshige *et al.* (1985) point out that the chemical–electrical parallelism must be tested under a variety of conditions. Lühring & Tazawa (1985) found ATP-dependent H^+ effluxes in perfused *Chara australis* (= *C. corallina*) of 231–251 nmol m^{-2} s^{-1}, while Fujii *et al.* (1979) found an ATP-dependent outward current of some 25 mA m^{-2} (= 259 nmol m^{-2} s^{-1}) for *Chara australis* under similar conditions; again, the agreement is good (see also Smith & Walker, 1981). We note that the work of Fujii *et al.* (1979) was aimed at measuring the effects of changes in external and cytoplasmic pH on pump activity; with respect to *cytoplasmic* pH they found that *maximum* pump activity (25 mA m^{-2}) occurred at a cytoplasmic pH of 6.9, with a decline to 14 mA m^{-2} at pH 8.7, and to 6 mA m^{-2} at pH 5.1. Since the cytoplasmic pH of *Chara corallina* (= *C. australis*) is always above pH 6.9 in intact cells (Smith & Raven, 1979), and is regulated at pH 7.7–7.8 in perfused cells (Lucas & Shimmen, 1981) we note that the *increase* in H^+ efflux pump activity as cytoplasmic pH *decrease* is in at least semiquantitative agreement for this response playing a role in cytoplasmic pH regulation (see Raven, 1984b, 1985a). The increase in pumping rate as H^+ activity increases from 2 mmol m^{-3} to 12.6 mmol m^{-3}, i.e. from 14 mA m^{-2} to 25 mA m^{-2} (with a pumping rate of 20.5 mA m^{-2} at an H^+ activity of 16 mmol m^{-3}) (Fig. 9 of Fujii *et al.* 1979), does not fit a simple Michaelis–Menten relationship for H^+ efflux as a function of free H^+ concentration in the cytosol, and suggests a low $K_{\frac{1}{2}}$ value (<2 mmol H^+ m^{-3}) for the pump.

A point disputed for several years is the H^+ : ATP ratio for the *Chara corallina* plasmalemma H^+ pump (e.g. Walker & Smith, 1975; Smith & Walker, 1971; Lucas, 1985); it is likely that (as in the Ascomycete *Neurospora crassa*) the *Chara* plasmalemma H^+ pump can operate with either 1 or 2 H^+ expelled per ATP hydrolysed (Smith, 1984a, Takeuchi *et al.*, 1985).

A complicating feature of the ecorticate Characean internode is the 'banding' phenomenon. This is an elaboration of the current circulation in *Acetabularia* where (in uninjured cells) the current enters, at a high flux density, in the rhizoid and leaves, at a lower flux density, over the 'shoot' part of the organism; in the Characean internode of ~100 mm length there are perhaps from five to ten circumferential 'acid bands' which are sites of (positive) current exit and as many intercalated 'alkaline bands' through which current enters the cell. The active H^+ efflux alluded to above takes place in the 'acid zones' of 'banded' cells. Vibrating probe measurements on 'banded' cells show that the *peak* outward current in the 'acid bands' was 100–300 mA m^{-2} equivalent to an H^+ efflux of 1.04–3.11 μmol m^{-2} s^{-1} (Walker & Smith, 1977; Lucas & Nucitelli, 1980; cf. Walker *et al.*, 1980; Lucas *et al.*, 1983; Nawata, 1984). In the experiments of Lucas & Nucitelli (1980) the 'acid bands' occupied more than half of the cell surface; this may not always be the case (Smith, 1985). The nature of the inward current in the 'alkaline bands' will be discussed later. *Inter alia* the banding phenomenon is related to the 'use' of exogenous HCO_3^- in photosynthesis (see Lucas, 1975; Smith, 1985; Walker, 1985). In the freshwater giant-celled alga *Hydrodictyon africanum* (Chlorophyceae) with an electrogenic mechanism apparently similar to that of the Characeans, there does not seem to be an analogue of the 'banding' phenomenon on its spherical cells (Walker, 1985, citing unpublished data of G. P. Findlay & N. A. Walker). Circulating electric currents have also been reported for other freshwater algae (e.g. the Charophyceans *Closterium* and *Micrasterias*: Troxell *et al.*,

Table 5.2 Examples of primary active transport at the plasmalemma of algal cells

Organism[†]	Solute	Evidence	Capacity μ mol m^{-2} s^{-1}	Reference‡
Chara corallina internodal cells	Proton efflux	Peak current efflux in 'acid bands'	1–3 (based on 'acid bands' only)	1
Corallina internodal cells	Proton efflux	ATP-dependent conductance; ATP-dependent net H^+ efflux	~0.25 (based on whole cell surface area)	2
Nitellopsis obtusa internodal cells	Proton efflux	ATP-dependent conductance; ATP-dependent net H^+ efflux	~0.2	3
Hydrodictyon africanum cells	Proton efflux (?)	'Metabolic' conductance	0.008	4
Acetabularia mediterranea cells	Chloride influx	Metabolic conductance	⩽ 30	5
		Tracer Cl^- influx	⩽ 10	6
Halicystis parvula cells	Chloride influx	Short-circuit current = net short-circuit Cl^- influx	6–10	7
H. ovalis cells	Chloride influx	A portion of the short-circuit current is accounted for by net short-circuit Cl^- influx	~3	8
H. ovalis cells	Sodium efflux	A portion of the short-circuit current is accounted for by net short-circuit Na^+ efflux	~0.6	9

† Of the species used, only *Chara corallina* shows the 'banding' phenomenon. *Nitellopsis obtusa* does not have 'banded' internodal cells. Some degree of spatial inhomogeneity of primary electrogenic ion fluxes occurs in *Acetabularia* (and, probably, in the other rhizoid-bearing cells, i.e. *Halicystis*, but not in *Hydrodictyon* (see Ch. 7 of Raven, 1984a, Sect. 5.4.3).

1. Lucas & Nucitelli (1980); Lucas *et al.*, (1983); Walker *et al.* (1980); Nawata (1984)
2. Lühring & Tazawa (1985); Fujii *et al.*, (1979)
3. Takeshige, *et al.* (1985)
4. Findlay, (1982)
5. Gradmann, (1975)
6. Gradmann, *et al.*, (1982b) Saddler (1970a, b)
7. Graves & Gutknecht, (1977a, b)
8. Blount & Levedahl, (1960)
9. Blount & Levedahl (1960)

1985); their relationship to primary active H^+ efflux, and to HCO_3^- use, is not yet defined.

The occurrence of these primary active electrogenic ion transport mechanisms at the plasmalemma of *Acetabularia*, Characeans, *Hydrodictyon, Spirogyra, Eremosphaera* and (probably) a number of other freshwater Chlorophyte algae (see Ch. 7 of Raven, 1984a; Raven, 1986a; cf. Tromballa, 1980, 1981) is not always reflected in a ψ_{co} more negative than the ψ_{K^+}. (Saddler, 1970a,b; Walker, 1980; Ch. 7 of Raven, 1984a). Treatments such as increased external [K^+], decreased external [Ca^{2+}], darkness (temporarily), by passage of an action potential, and (for the tropical and subtropical *Acetabularia*) low temperatures depolarize ψ_{co} to a value equal to, or more positive, than ψ_{K^+}. The extent to which these depolarizations reflect changes in pump activity (mol ion transported m^{-2} s^{-1}) rather than increased short-circuiting of the pump by downhill ion fluxes varies with the treatments which lead to the depolarization and, in some cases, is not known (e.g. Bisson, 1984; Smith, 1984a,b; Walker, 1980; Beilby, 1985; Kishimoto *et al.*, 1985; Smith & Gibson, 1985). There are clearly energetic constraints on the extent to which the active ion flux can occur in cells in the dark when only oxidative (and glycolytic) phosphorylation, with a limited capacity relative to photosynthetic phosphorylation, is available to power ATP-requiring processes.

A substantial number of algae for which the ψ_{co} has been measured have *never* been shown to have ψ_{co} more negative than ψ_{K^+} (Ch. 7 of Raven, 1984a). In the case of *Halicystis* and *Bryopsis* (two marine giant-celled members of the Ulvophyceae which are in the order Codiales rather than the Dasyclydales to which *Acetabularia* belongs) the removal of external Cl^- leads, as in *Acetabularia*, to a substantial depolarization of ψ_{co}; however, the 'plus Cl^-' ψ_{co} is more positive than ψ_{K^+}, so that the resulting 'minus Cl^-' ψ_{co} is much more positive than ψ_{K^+}. Although much remains to be discovered about the factors determining ψ_{co} in *Halicystis* and *Bryopsis*, it would appear that there *is* a primary active, electrogenic Cl^- influx pump in these genera. For other algae we have no such electrical evidence for the occurrence of primary active ion fluxes at the plasmalemma; in these cases there clearly *must* be one or more primary active transport processes, in view of the substantial evidence for active transport of many solutes, and the clear requirement for a biochemical energy input to one or more of these solute transport reactions, rather than energization from some downhill solute flux which is not related to a biochemical energy input. Primary active transport of a neutral solute, or of ions via electroneutral symport or antiport, would have no direct electrical signature. Such primary active transporters can be sought *inter alia* by characterizing solute stimulated plasmalemma ATPases. Evidence for ATPases stimulated by $K^+ \pm Na^+$, Ca^{2+}, and NO_3^- have been reported for plasmalemma-enriched fractions of various algae (see Raven, 1976, 1980, 1984a; Gross & Marme, 1978). We note that for *Halicystis ovalis* (Blount & Levedahl, 1960), some

0.392 (~64 nmol m^{-2} s^{-1}) of the short-circuit current was accounted for by net Na^+ efflux with, in a different experiment, a further 0.576 (~3240 nmol m^{-2} s^{-1}) of the current accounted for by net Cl^- influx. These data are consistent with primary active, electrogenic Na^+ efflux at the plasmalemma of *Halicystis ovalis*. However, comparable experiments on *Halicystis parvula* (Graves & Gutknecht, 1977a,b) showed that all of the short-circuit current could be accounted for by net Cl^- influx.

5.4.4 Secondary active transport at the algal plasmalemma

It is currently fashionable (following Smith, 1970) to regard many of the active solute fluxes at the plasmalemma of algal cells as being catalysed by secondary active transport mechanisms (e.g. Raven, 1984a; Sanders, 1984; Lucas, 1985). An essential coupling element is the 'working ions' whose active efflux (cations) or influx (anions) generates an electrochemical potential gradient for those ions, thus permitting the exergonic 'downhill' backflux of the working ions to be coupled (via an appropriate co-transporter) to the 'uphill' (endergonic) flux of a co-transported solute (see also Chs. 1, 3). Identification of a given active solute flux as being the result of secondary active transport involves demonstrating that the active solute flux requires the simultaneous downhill flux of the 'working ion'; that the downhill flux of the 'working ion' is (reciprocally) stimulated by the presence, and further, the transmembrane endergonic flux, of the co-transported solute; and that the stoichiometry of the downhill flux of the 'working ion' and the uphill flux of the co-transported solute, together with the electrochemical potential differences for the 'working ion' and the 'co-transported solute' make the coupled fluxes of 'working ion' and 'co-transported solute' exergonic and thus not requiring any additional energy input. Of course, the ultimate test of the occurrence of such a secondary active transport system in the plasmalemma of an algal cell would be the isolation, purification and characterization of the porter concerned, with its reconstitution into otherwise protein-free lipid bilayers culminating in the demonstration of a downhill flux of the 'working ion' coupled, with no other energy supply available, to the uphill flux of the co-transported solute.

None of the algal systems so far examined (see Tables 5.3 and 5.4 and Ch. 7 of Raven, 1984a) has yet achieved the level of resolution indicated in the final sentence of the last paragraph, although there has been substantial progress on identification and purification of the *Chlorella* glucose porter (Sauer & Tanner, 1984). Only two, the *Chara* 2 H^+ : 1 Cl^- symporter (Sanders, 1980, 1984; Beilby & Walker, 1981) and the *Chlorella* 1 H^+ : 1 glucose symporter (Komor & Tanner, 1980) can be said to have satisfied the criteria outlined in the penultimate sentence of the last paragraph. Even so, the *Chlorella* example relies, for the testing of consistency

Table 5.3 Well authenticated examples of ATP-independent proton co-transport (symport or antiport) at the plasmalemma of freshwater algal cells†

Organism	Co-transport (symport or antiport)	Evidence	Reference‡
Chlorella vulgaris	Glucose symport	Initial 1 H^+ : 1 glucose stoichiometry upon addition of glucose; capacity of 200 nmol glucose m^{-2} s^{-1}	1
	Nitrate symport	Initial 1 H^+ : 1 NO_3^- stoichiometry upon addition of nitrate (not confused with subsequent pH-stat net H^+ influx!)	2
Chara corallina	Chloride symport	Voltage clamp fluxes suggest 2 H^+: 1 Cl^-	3

† Based on Table 7.5 of Raven (1984a)

1. Komor & Tanner (1980); Raven (1984a)
2. Schlee *et al.* (1985)
3. Smith (1970); Sanders (1980); Beilby & Walker (1981); Sanders *et al.* (1985)

of the magnitude (J mol^{-1}) of the 'driving' ($\Delta\mu_{H^+}$) and the 'driven' ($\Delta\mu_{glucose}$) processes with the observed stoichiometry of 1 H^+ : 1 glucose, on the use of a 'lipid-soluble cations' measure ψ_{co}; the quantitative use of this technique in eukaryotic cells is in some doubt (Gimmler & Greenway, 1983; Astle & Rubery, 1984; Ritchie, 1984).

For other cases in which H^+ co-transport has been invoked for the algae plasmalemma, it is not established that the active transport process could be entirely energized by the $\Delta\mu_{H^+}$. This applies to the 1 H^+ : 1 NO_3^- symport in *Chlorella vulgaris* (Schlee *et al.*, 1985); the (possible) H^+ : Cl^- symport in *Chlorella fusca* (Doblinger & Tromballa, 1982) if H^+ : Cl^- is <2.0 at external pH values above 6.5; the (possible) H^+ : Na^+ antiport seen in *Chlorella fusca* cells loaded with Na^+ at high pH and then returned to a lower external pH in the absence of external K^+ (Tromballa, 1974); the (?) 2 H^+ : 1 NO_3^- symport in *Hydrodictyon africanum* (Raven & De Michelis, 1979; Raven, unpublished); the H^+ : Na^+ antiport in *Dunaliella salina* (Kaplan & Schreiber, 1981); and the (?) 2 H^+ : 1 Cl^- symporter in *Scenedesmus obliquus* (Findenegg, 1974a,b; Sanders, 1984).

In the case of H^+-coupled co-transports in *Chara, Hydrodictyon* and *Chlorella*, the $\Delta\mu_{H^+}$ which (at least partially) energizes the processes is generated by primary active transport of H^+ (Sect. 5.4.3). For *Dunaliella* it is likely that the H^+ : Na^+ antiporter generates a $\Delta\bar{\mu}_{H^+}$ using a $\Delta\bar{\mu}_{Na^+}$ generated by primary active Na^+ efflux, or from some secondary active transport not involving H^+ as the 'working ion' (see discussion of *Acetabularia* below). A probable secondary active transport process at the

Table 5.4 Examples of effects of external Na^+ on the flux of various solutes at the plasmalemma of some marine algal cells: from Table 7.6 of Raven (1984a) except where otherwise indicated

Organism	'Co-transported' solute	Sym– or anti-port	Dependence on external Na^+
Cyclotella cryptica	Glucose	Symport	Total
	Glutamate	Symport	Total
	Alanine	Symport	Total
	Arginine	Symport	Total
	Methylammonium	Symport	Total
	KCl	Symport	Partial
Phaeodactylum tricornutum	Methylammonium	Symport	Total
	Nitrate	Symport	Total
	Urea	Symport	Total
	Guanine	Symport	Total
	Inorganic carbon†	Symport	Partial‖
	Arginine‡	Symport	Total
	Lysine‡	Symport	Total
Nitzschia alba	Silicic acid	Symport	Partial
Chlamydomonas pulsatilla§	Methylammonium	Symport	±None
	Phosphate	Symport	Partial
	Inorganic carbon	Symport	None ‖
Dunaliella tertiolecta§	Methylammonium	Symport	Total
	Phosphate	Symport	Total
	Inorganic carbon	Symport	None ‖
Dunaliella salina	Protons	Antiport	Total?
Lamprothamnium papulosum (euryhaline)	Sucrose¶	Symport	Partial
	glucose¶	Symport	Partial

† Rees (1984).
‡ Flynn & Syrett (1985).
§ Hellebust (1985
¶ Kirst & Bisson (1983).
‖ The Na^+ stimulation in *Phaeodactylum tricornutum* was found only in cells grown at air-equilibrium CO_2 levels (i.e. with an operational inorganic carbon concentrating mechanism) and was much more marked at rate-limiting than at saturating inorganic C concentrations. The *Chlamydomonas pulsatilla* and *Dunaliella tertiolecta* cells were grown at air levels of CO_2 which would (certainly in *Dunaliella*) de-repress an inorganic carbon concentrating mechanism; the effect of Na^+ on photosynthesis was tested at a saturating inorganic C concentration (see Aizawa *et al*, 1985).

Chara corallina plasmalemma for which the 'working ion' has not yet been identified, but could well be H^+, is the efflux of Ca^{2+}. Lühring & Tazawa (1985) found that the $^{45}Ca^{2+}$ efflux across the plasmalemma of perfused cells was ATP-independent, and was larger (~10 nmol m^{-2} s^{-1}) than the resting influx in non-perfused cells (~1 nmol m^{-2} s^{-1}). Since the $\Delta\mu_{Ca^{2+}}$

predicts net passive Ca^{2+} *influx*, the ATP-independent net Ca^{2+} efflux is clearly active, and presumably occurs via secondary active transport. The affinity for free Ca^{2+} seems to be high, with a $K_{\frac{1}{2}}$ of ~1 mmol Ca^{2+} m^{-3} (Lühring & Tazawa, 1985).

For Na^+ as putative 'working ion' the situation is even less satisfactory than is found for H^+. Table 7.6 of Raven (1984a) lists some possible Na^+ co-transport reactions; further examples may be found in Raven (1984) and in Table 5.4. However, the evidence for Na^+ co-transport at the algal plasmalemma rests solely on the stimulation of (usually tracer) influx of various solutes by the presence of exogenous Na^+, so that neither of the basic criteria listed in the penultimate paragraph (mutual stimulation of fluxes of the 'working ion' and the co-transported solute, and energetic explicability of the co-transported system without the need to invoke additional energy inputs) have been met for Na^+. Raven (1984a, 1987) points out that, for the rapidly growing 'ruderal' macrophytes *Porphyra* and *Enteromorpha*, where Na^+ fluxes and $\Delta\bar{\mu}_{Na^+}$ have been well studied, the measured (downhill) tracer Na^+ influx is not quantitatively adequate for co-transport of major nutrient solutes (e.g. NO_3^-, $H_2PO_4^-$) during rapid growth, while, for *Dunaliella*, the tracer Na^+ influx is so large relative to net, growth-related influxes of major nutrient ions that stimulation of tracer Na^+ influx would (granted Na^+ : co-transported solute ratios not more than five-fold the minimum ratio needed to avoid the necessity to invoke an extra energy input) be difficult to detect. We can, then, conclude that the potential advantage of being able to measure bidirectional (tracer) fluxes for Na^+, but only net flux for H^+, at the algae plasmalemma has not thus far been realized in the context of co-transport processes. On a more positive note, the data to hand on the concentration-dependence of the Na^+ stimulation of transport of putatively co-transported solutes in a number (mainly marine diatoms) of algae do form a useful base for further investigations.

For certain marine algae there is, accordingly, evidence *consistent* with an involvement of Na^+ as 'working ion' in co-transport reactions at the plasmalemma; one of the many 'unknowns' here is the energy source for the production of the $\Delta\bar{\mu}_{Na^+}$ which drives the putative co-transport system. We have seen (Sect. 5.4.3) that the evidence for primary active Na^+ efflux in algae is not overwhelming, so there is a possibility that any co-transport processes at the algal plasmalemma with Na^+ as working ion are themselves dependent on some other secondary active transport process which is responsible for active Na^+ efflux with some 'working ion' which is itself subject to *primary* active transport. H^+ does not seem to be a very likely 'working ion' which undergoes primary active transport in many marine algae (Raven & Smith, 1980a); Cl^- is a possibility in at least certain Dasyclydales and Codiales (Chlorophyta : Ulvophyceae), with downhill Cl^- efflux coupled, *ex hypothesi*, to uphill Na^+ efflux (see Sect. 5.4.3, and Raven, 1984a, 1987).

Our conclusion must be that critical data are not available fully to substantiate many of the claims that co-transport occurs at the algal plasmalemma. However, in view of the (probably) restricted number of solutes involved in primary active transport at any one membrane (Raven & Smith, 1980b, 1982), it is likely that many of the solutes which are actively transported at the algal plasmalemma are moved by secondary active transport reactions. Raven & Smith (1980b) and Raven (1980, 1984a) point out that, in the long term, the working ion flux downhill via co-transporters, uniporters and leaks must equal the uphill flux via (usually primary) active transport of that ion (if any corrections for net influxes or efflux of H^+ required for pH homoiostasis, and net influxes of H^+, Na^+ and Cl^- needed for growth, are allowed for). By 'long term' we mean only fractions of a second in terms of the depolarizing effect of downhill, electrogenic fluxes of the working ion, short-circuiting primary active, electrogenic fluxes of that ion. Since the total capacity of secondary active transport *plus* uniport systems substantially exceeds that of primary active transport in terms of backflux of the working ion and of capacity for charge transfer across the plasmalemma, as well as giving net nutrient influxes in excess of those needed for steady-state growth (Raven, 1984a), regulation of the expressed activity of these co-transporters and passive uniporters is clearly required; similarly, regulation of co-transport is needed under conditions of low energy supply (see Raven, 1980, 1984a).

The large combined capacity of the co-transporters does not necessarily mean that they pre-empt a larger fraction of the plasmalemma proteins than do primary active porters, since their specific reaction rate [mol solute moved (g protein)$^{-1}$ s^{-1}) may well be higher than that of the primary active porters (Raven & Smith, 1980b; Raven, 1984a).

5.4.5 Mediated, passive uniport at the algal plasmalemma

Chapter 7 of Raven (1984a) discusses *inter alia*, examples of passive uniporters in the algal cell plasmalemma; Table 7.7 of Raven (1984a) lists some examples of these uniporters, with some of their properties. Table 5.5 of the present discourse is (except where otherwise indicated) taken from Table 7.7 of Raven (1984a). We see that the combined transport capacity of the uniporters can, in *Chara corallina*, exceed the capacity of the active H^+ efflux pump, so that in the steady state the sum of the uniporters must be running well below their capacity. For Cl^- and Ca^{2+} the action potential values of the uniport charge entry occurs over for, at most, a few per cent of the organism's life. With H^+, the constraint is spatial, less than half of the plasmalemma area of ecorticate Characean cells is occupied by the 'alkaline bands' whose high extracellular pH values are caused by, and perpetuate, the 'high P_{H^+}' state of the plasmalemma. For K^+, the voltage dependence of the porters (reduced capacity at more negative values of ψ_{co}) reduced the extent of short-circuiting of the H^+ pump in the 'pumping state' of *Chara corallina*. In the case of NH_4^+ the fraction of the uniporter capacity which is expressed is regulated by the N status of the plant, with N-deprived plants exhibiting the full capacity and a decreasing fraction of the capacity being shown by progressively more N-satiated organisms. Perhaps the very high NH_4^+-saturated capacity for NH_4^+ transport is related to maximizing the transport ability of the plasmalemma at low ambient NH_4^+ concentrations, granted the measured half-saturation value (corrected for unstirred layer effects) for the NH_4^+ porter of 3 mmol m^{-3}.

Most of the data in Table 5.5 relate to giant-celled algae with cell volumes in excess of 1 mm^3 and, granted a 10 μm thick layer of cytoplasm, 10^5 m^2 per m^3 of cytoplasm. For a very small eukaryote (the 1 μm diameter sphere of Table 5.1) the corresponding surface area is $6 \cdot 10^6$ m^2 per m^3 of cytoplasm. This means that the small cell has only one-sixtieth as much cytoplasm, and hence one-sixtieth as much catalytic capacity for energy transduction (under high-light conditions) or pigment for photon absorption (under low-light conditions) per unit plasmalemma area. For constant plasmalemma permeabilities, the small cell is disadvantaged in terms of the cost of maintaining intracellular solute concentrations in the face of passive uniport at the plasmalemma. At high photon flux densities a (small) decrement in growth rate is involved in providing extra catalytic machinery to permit leaked solutes to be replaced (by synthesis, e.g. compatible solutes) or pumped in (e.g. Cl^-) or out (e.g. Na^+) (Raven, 1986a). At low photon flux densities, the decreased 'package effect' in the smaller cells does not completely make up for larger area of plasmalemma per unit of photosynthetic pigment, so that a greater fraction of photons incident on the smaller cell must be used in resynthesis, reaccumulation or expulsion of uniported solutes (Raven, 1986a). It might, then, be envisaged that mediated passive uniport would be more of a selective liability to small than to large algae cells; the data in Table 5.5 are in accord with a lower uniport capacity for K^+ and Na^+ in small than in large-celled algae.

5.4.6. 'Lipid solution' passive uniport at the algal plasmalemma

The energetic, and resource-retention, problems associated with *mediated* passive uniport (Sect. 5.5.5) could be eliminated by removing from the membrane, or deactivating, the uniporters, although either of these courses of action (especially the former) could have undesirable side-effects. Even with mediated passive uniport eliminated there would still be a potential for passive uniport via a 'lipid solution' mechanism. The permeability coefficients for important solutes in lipid bilayers with a composition similar to that of the plasmalemma lipids varies from some

Table 5.5 Some characteristics of mediated uniporters at the plasmalemma of algal cells: from Table 7.7 of Raven (1984a) except where otherwise indicated

Organism	Uniport solute	Maximum capacity for uniport of solute at algal plasmalemma (μmol m^{-2} s^{-1})	Maximum permeability coefficient for uniport of solute at algal plasmalemma (m s^{-1})	Permeability coefficient for uniport of solute in lipid bilayers (m s^{-1})‡
Chara corallina	NH_4^+	2.0	$7 \cdot 10^{-4}$	$(\sim 10^{-12}?)$‡
	K^+	0.57	$2 \cdot 10^{-8}$	$\sim 10^{-12}$‡
		0.03 (resting)	10^{-11}–10^{-10} (resting)	
	Cl^-			$\sim 10^{-12}$‡
		2.3 (action potential)	10^{-6}	
	Ca^{2+}	0.00 (resting)†	$3 \cdot 10^{-10}$ (resting)†	$(\sim 0.5 \cdot 10^{-12}?)$‡
		0.35 (action potential)†	10^{-7} (action potential)†	
	H^+	2.0 ($pH_0 \geqslant 10.0$)	2.0 ($pH_0 \geqslant 10.0$)	10^{-3} ($pH \sim 10$)§
		0.1 ($pH_0 \simeq 4.0$)	10^{-6} (pH_0 5.0–7.0)	$3 \cdot 10^{-7}$ ($pH \sim 7$)§
	Na^+	$\geqslant 0.02$	10^{-9}	$\sim 10^{-12}$‡
Acetabularia mediterrenea		0.14 (resting $\psi_{co} = -180$ mV)		
	K^+	2.0 (resting $\psi_{co} = -50$ mV)		$\sim 10^{-12}$
		10.0 (action potential)		
		2.0 (resting, $\psi_{co} = -180$ mV)		$\sim 10^{-12}$‡
	Cl^-	10.0 (action potential)		
Griffithsia pulvinata	K^+	3.8 (resting)	$3 \cdot 10^{-7}$	$\sim 10^{-12}$‡
	Na^+	0.15	$3 \cdot 10^{-9}$	$\sim 10^{-12}$‡
Chlorella pyrenoidosa	K^+	0.01	$2 \cdot 10^{-10}$	$\sim 10^{-12}$‡
	Na^+	0.002	$2 \cdot 10^{-11}$	$\sim 10^{-12}$‡

† Hayama *et al*. (1979); Lühring & Tazawa (1985).
‡ Raven (1984a); Raven (1986a).
§ Gutknecht (1984)

10^{-3} m s^{-1} for CO_2 and O_2 down to 10^{-8}m s^{-1} for urea and glycerol and 10^{-12} m $s^{-1}$for K^+, Na^+, Cl^-, hexoses and hexitols (see Table 3.5 of Raven, 1984a). For H^+ there has been much disputation (see Ch. 3 of Raven, 1984a); recently, clarification has been provided by Gutknecht (1984). It was found that while the H^+/OH^- conductance of the membrane was essentially constant in the pH range 1.6–10.5, the variations in $[H^+]$ and $[OH^-]$ with pH means that P_{H^+/OH^-} varies from 10^{-3} m s^{-1} at pH 2.0 and pH 10.0 to some 3.10^{-7} m s^{-1} at pH 7.0 (Fig. 6 of Gutknecht, 1984). We note that, while Gutknecht (1984) cautions that such agreements may be coincidental, the P_{H^+/OH^-} of artificial bilayers at near-neutral pH values *is* very similar to 'background' P_{H^+/OH^-} of many 'real' membranes at near-neutral pH values (p. 111 of Gutknecht, 1984).

Comparing the P_{solute} values for bilayers with those of mediated uniporters we find that, for ions (Table 5.5), the lowest reported values for mediated uniport at the algae plasmalemma suggest that the catalysed passive uniport only increases the lipid-solution permeability by about an order of magnitude for Na^+ and Cl^- and by two orders of magnitude for K^+, while P_{H^+} for the algal plasmalemma at near-neutral pH values is little larger than is P_{H^+} for lipid membranes. By contrast, the peak P_{solute} values in action potentials (for Cl^-, K^+ and Ca^{2+}) and during rapid NH_4^+ or K^+ entry are up to $7 \cdot 10^8$ times the lipid solution ionic permeability.

For non-electrolytes passive uniport permeability coefficients for the algal plasmalemma may be similar to those predicted from work on lipid bilayers, or slightly (up to ten times) higher, indicating some additional passive, mediated uniport; examples are glycerol and urea in members of the Characeae and in *Valonia* (see Gutknecht, 1968; Raven 1984a; Wilson, 1985). However, in a number of cases the permeability of the algal plasmalemma appears to be substantially *lower* than is consistent with the data from lipid bilayers. Perhaps the best documented case is that of glycerol in *Dunaliella* spp. (see discussion by Raven, 1984a, pp. 346–68). Glycerol is the major compatible solute in this halotolerant algal genus; leakage of glycerol from the very high cytosol concentrations (up to several kmol m^{-3}) to glycerol-free seawater is minimized by a $P_{glycerol}$ of some 10^{-12} m s^{-1}, i.e. some six orders of magnitude less than the 'normal' value, and similar to the permeability of lipid bilayers to glucose or mannitol. Less well quantified is the low P_{CO_2} (relative to that for lipid bilayers) found for the plasmalemma of algae (including *Dunaliella*) with an operative 'CO_2 concentrating mechanism' based on active inorganic carbon transport at that membrane; here the P_{CO_2} of as little as 10^{-6} m s^{-1} is four (three?) orders of magnitude less than that for lipid bilayers or, probably, than P_{CO_2} for the plasmalemma of algae with diffusive CO_2 entry for photosynthesis (Raven, 1984a, 1985b; Raven & Lucas, 1985). How these very low permeabilities are achieved, and, for CO_2, how the same genotype can so dramatically change its P_{CO_2} between the 'CO_2-accumulating' and the 'CO_2-diffusion' phenotypes, is not known (Raven, 1984a).

5.4.7 The extent and significance of passive uniport ('leakage' and 'mediated') at the algal plasmalemma

The extent of leakage has important repercussions for the efficiency of resource use by algae. This is certainly true for energy, carbon and nitrogen. In the case of energy it is clear that an additional energy (photon) input is required, relative to the situation for a hypothetical 'leak-proof' cell, to replace (by synthesis) organic solutes which are lost by leakage and to power active transport which permits homoiostasis in the face of leakage influxes (of H^+, Na^+ and Ca^{2+}) and effluxes (e.g. of CO_2 in organisms with a 'CO_2 concentrating mechanism'; NH_3 in organisms using NH_4^+, NO_3^- or urea as exogenous N source; HNO_2 in organisms using NO_3^- or NO_2^- exogenous N sources; and Cl^-): see Raven, 1980, 1984a, 1986a; Anderson & Roels, 1981; Kleiner, 1985a, b).

More obviously, perhaps, are the actual loss of C and N which the above-mentioned C and N fluxes imply, and which constrain the rate at which C and N net acquisition can proceed from low exogenous levels of inorganic C for inorganic N relative to that found in an otherwise comparable cell which is 'leak-proof'. These 'leakage' phenomena add to the incident photon flux density required to maintain the biomass of a non-growing organism, or to achieve a given growth rate, and increase both the external inorganic carbon concentration required to give a growth rate, and the inorganic carbon compensation concentration, relative to what would be the case for a comparable leak-proof cell. Furthermore, leakage would be expected to yield a finite 'compensation concentration' for exogenous N sources which would be absent for a leak-proof organism.

Three further points on leakage remain; these relate to cell size effects, to the distinction between leakage and passive uniport, and the role of these two phenomena in algal symbioses with chemo-organotrophs.

Leakage expressed on an element-specific basis is clearly cell-size-dependent. For a spherical cell, with given transplasmalemma solute concentration differences, and plasmalemma permeability, the leakage efflux of C-containing solutes (proportional to surface area) decreases on the basis of cell C content (proportional, for a non-vacuolate cell, to cell volume) in inverse proportion to cell radius (see Raven, 1987). However, this apparent advantage of larger cells is substantially offset by their vacuolation. If the thickest mean cytoplasmic layer thickness in a 5 mm radius giant cell is 10 μm, and the cell wall is also 10 μm thick (see Raven, 1984a, Ch. 9), and all of the cell C is present in these two fractions at a concentration of 125 kg m^{-3}, then C per m^2 of plasmalemma area is 208 mmol C m^{-2}. This contrasts with 1.74 mmol C m^{-2} for a 0.5 μm radius cell, also with 125 kg C m^{-3}. Thus the ratio (C-specific leakage in a 0.5 μm radius cell C-specific leakage in a 500 μm radius cell) is 120 rather than the ratio of 1000 predicted for cells with a 1000-fold ratio of radii and an identical volume-based C content (10^9 times longer in the longer cell)).

In view of the difficulties mentioned in Sect. 5.4.2, and the discrepancies mentioned above for glycerol and CO_2 between 'lipid solution' predictions of P_{solute} and observed values, distinguishing between leakage and mediated uniport is not always easy. This distinction is relevant to the ecologically important, and contentious, issue of the fraction of primary productivity of algae which is lost as soluble organic material (Mague *et al.*, 1980; Carlson & Carlson, 1984) since 'lipid-solution' leakage presumably represents the minimum level of loss upon which mediated uniport is superimposed. In view of the range of soluble organic compounds lost from algal cells (Hellebust, 1974) and the unknown concentration differences across the plasmalemma for individual solutes, it is impossible to see how the observed loss rates agree with predictions from a 'lipid solution' mechanism for the effluxes. However, it is possible to make a few generalizations. One is that the specific organic C efflux rate (mol organic C lost (mol cell C)$^{-1}$ s^{-1}) is positively correlated with the specific growth rate (mol C gain (mol cell C)$^{-1}$ s^{-1}) when specific growth rate is altered by either photon supply (Verity, 1981) or by nitrate supply (Laws & Wong, 1978). It is unlikely that a similar growth rate dependence occurs for the cytosol concentrations of the effluxing organic solute (Ch. 3 and Table 8.1 of Raven, 1984a; Dortch *et al.*, 1984); rather, a *relative* constancy of the concentrations would be expected. This means that *neither* a 'lipid solution' efflux (directly proportional to internal concentration with negligible external organic solute) *nor* a (saturable) mediated uniport efflux (directly proportional to internal concentration when these are relatively low; independent of internal concentration at higher internal concentrations) can explain the observed *direct* correlation between specific excretion rate and specific growth rate (cf. Huntsman, 1972). A second is that the observed increase in the ratio specific excretion rate/specific growth rate at lower growth temperatures (Verity, 1981) is in accord with the notion of 'lipid solution' efflux *if* it is accepted that this process is less temperature-dependent than is the generality of processes which limit metabolism and growth; however, this assumption needs careful scrutiny (see Sect. 5.4.2). Clearly compartmental analysis could be helpful in defining the mechanism of efflux.

Turning finally to the interaction between algae and chemo-organotrophs (and, more generally, to that between differentiated cells with different resource-acquisition capacities in multicellular algae), it is clear that *inter*-cellular transport is required to move oganic solutes from photosynthetic to chemo-organotrophic cells and, in some cases, inorganic solutes from chemo-organotrophic cells to photosynthetic cells. With the exception of certain Rhodophyta parasitizing other Rhodophyta, the symbiotic (*sensu lato*) interactions do not involve structures which could permit solute transport between cells without crossing a plasmalemma which is differentially permeable to low M_r ($< 10^3$) solutes (see Ch. 9 of Raven, 1984), and so the solute transfer must relate to flux out of the 'donor' cell across a differentially permeable plasmalemma. For the non-symbiotic differentiated plants there are almost invariably plasmodesmata (Chlorophyta,

Phaeophyta) or 'pit connections' (Rhodophyta) between cells which could bypass the need for transplasmalemma fluxes for intercellular fluxes of low M_r solutes (see Ch. 9 of Raven, 1984a).

5.4.8 Unstirred layers revisited: constraints on usable combinations of maximum transport capacity and substrate affinity for transport systems

We have already discussed some aspects of unstirred layer influxes on solute fluxes (Sect. 5.3). The point we wish to emphasize here is the extent to which high-capacity, high-affinity mediated transport processes can lead to diffusive limitation under conditions of low bulk phase solute concentration. Pasciak & Gavis (1974; cf. p. 167 of Gavis, 1976) in their Tables 1 and 2 indicate that transport limitation by unstirred layers could limit uptake from low concentrations of NO_3^- in the non-motile cells of *Coscinodisus lineatus* (a diatom with an effective spherical radius of 25 μm, an apparent $K_{\frac{1}{2}}$ of 2.8 mmol NO_3^- m^{-3} and a maximum uptake capacity of 2.5 fmol NO_3 $cell^{-1}$ s^{-1} or 38 mmol (m cell volume)$^{-1}$ s^{-1}) and *Rhizosolenia robusta* (a diatom with an effective spherical radius of 42 μm, an apparent $K_{\frac{1}{2}}$ of 9.3 mmol NO_3^- m^{-3} and a maximum uptake capacity of 6.3 fmol $cell^{-1}$ s^{-1} or 20 mmol (m^3 cell volume)$^{-1}$ s^{-1}). The dinoflagellate *Gymnodinium splendens* swims at 200 μm s^{-1}; despite this motility there is still transport limitation by unstirred layers for these cells with an effective spherical radius of 20 μm, an apparent $K_{\frac{1}{2}}$ of 15 mmol m^{-3}, and a maximum uptake capacity of 11 fmol $cell^{-1}$ s^{-1} or 328 mmol (m^3 cell volume)$^{-1}$ when NO_3^- is being taken up from low external concentrations. We note that these examples of transport limitation by unstirred layers refer to cells which, by planktophyte standards, are quite large (Sournia, 1982); Tables 1 and 2 of Pasciak & Gavis (1974) cite thirteen more examples for which limitation by unstirred layers is negligible or, at least, much less significant than for the three cases cited above; effective spherical radii for these cases range from 2 μm to 75 μm with $K_{\frac{1}{2}}$ for NO_3^- uptake varying from 0.19 to 7.5 mmol m^{-3}, and maximum uptake capacity of 0.0014–0.35 fmol $cell^{-1}$ s^{-1} or 0.2–40 mmol (m^3 cell volume)$^{-1}$ s^{-1}, with the *highest* values for the *smaller* cells. Studies on phosphate influx *in situ* in estuarine phytoplankton are also consistent with unstirred layer limitation (Friebele *et al.*, 1978).

Malone (1980) provides a plot of the apparent $K_{\frac{1}{2}}$ of NO_3^- influx for a variety of diatoms with mean spherical diameters of 3–200 μm^3 (values of $14 \cdot 10^{-18}$ m^3–$4.2 \cdot 10^{-12}$ m^{-3}). While the variation in $K_{\frac{1}{2}}$ at a given cell size is quite large, there is a pronounced tendency for the larger $K_{\frac{1}{2}}$ values to be associated with the larger cell sizes. This could be due to an increased 'real' $K_{\frac{1}{2}}$ value for the NO_3^- porter *or* a greater impact of unstirred layers for the larger cells. It is significant in this respect that data analysed by Banse (1982) show μ_{max} (at 20 °C) for the smallest cells considered by Malone (1980) is $3 \cdot 10^{-5}$ s^{-1} and, for the largest cells, $1 \cdot 10^{-5}$ s^{-1}. Steady-state NO_3^- influxes needed to provide N for growth at this rate [assuming

1 mol N (m^3 cell volume)$^{-1}$ for small cells and 0.1 mol N (m^3 cell volume)$^{-1}$ for large cells with a large fraction of vacuole in the cell volume] at an external concentration equal to *twice* the $K_{\frac{1}{2}}$ are [Eq. (5.2)], for the smallest cells, just over one-tenth of the potential flux and, for the largest cells, almost half of that potential flux. This somewhat different approach supports the conclusions of Pasciak & Gavis (1974, 1975) and of Gavis (1976); however, the work of Smith & Kalff (1982) on phosphate influx characteristics of planktophytes of various sizes did not indicate large unstirred layer effects.

The ability of planktophyte cells to benefit from pulses of high bulk-phase nutrient concentrations, superimposed on a low, growth-limiting 'background' level, by 'luxury accumulation' (see Raven, 1976) might also be limited by unstirred layer effects, at least in the context of the kinetics of the plasmalemma transport systems (see Parslow *et al.*, 1985).

Similar constraints apply to the kinetics of an enzyme such as RUBISCO when it is functioning in a cell which lacks a 'CO_2 concentrating mechanism' the extent to which a high activity (per unit cell volume) of a RUBISCO with a low $K_{\frac{1}{2}(CO_2)}$ could function at low bulk phase CO_2 concentrations can be severely constrained by diffusion through unstirred layers outside cells, not to mention cell membrane and intervening cytosol and stroma (see MacFarlane & Raven, 1985). The supply to RUBISCO of CO_2 entering the cell by diffusion can be increased by the presence of carbonic anhydrase which overcomes the low rate constant for uncatalysed CO_2/HCO_3^- interconversion and permits HCO_3^- to act as a 'carrier' for inorganic C in the phases in which the enzyme occurs (see Raven & Glidewell, 1981). Raven (1977, 1987) has also emphasized the significance of the use of 'buffers' [H^+] and 'chelaters' [Ca^{2+}] in fluxes through unstirred cytoplasm of these solutes whose free concentration is only 0.1 mmol m^{-3}. In these cases the intrinsic rate of H^+ and Ca^{2+} association and dissociation are high, the increased flux capacity through the cytosol which this buffering/chelating permits is important in permitting large mediated effluxes of these cations (especially of H^+) from the cytosol. The $K_{\frac{1}{2}}$ values for mediated, active efflux of these cations appears to be low (see Sect. 5.4.3 and 5.4.4), in contrast to that for Na^+ efflux which is in excess of 10 mol m^{-3} in *Chlorella* (see Shieh & Barber, 1971; Tromballa, 1974). It may be significant that Na^+ binding to soluble cell components is much less than is that for H^+ or Ca^{2+}. Chelated forms of essential trace metals, such as Mn and Fe, may be important in moving these metals from the bulk phase to the outer surface of the plasmalemma where the porters have very high affinities for the substrate forms of the metals (see Sunda & Huntsman, 1985), however, the potential flux of free Mn [Eq. (2)] from a bulk phase concentration of free Mn equal to the $K_{\frac{1}{2}}$ value of 10^{-4} mol Mn m^{-3} is much higher than the observed flux, so unstirred layers are not limiting (computations based on data of Sunda & Huntsman, 1985). For Mn and Fe in the cytosol, the concentration of free ions is probably very low (Williams, 1982).

For benthic macrophytes the role of unstirred layers in restricting solute supply to the plant surface, and its significance for the measurement of the kinetics of membrane porters (and, for 0_2 and CO_2 with diffusive entry, the kinetics of cytochrome oxidase and the carboxylase function of RUBISCO) has been analysed by (*inter alia*) Walker *et al.* (1979a), Smith & Walker (1980), Wheeler (1980), Raven (1981, 1984a) and MacFarlane (1985). Wallentinus (1984) and Raven (1981, 1984a) have attempted to relate uptake kinetics (Wallentinus, 1984) and unstirred layers (Raven, 1981, 1984a) to life- and growth-forms of macroalgae; much further work in this area is needed. A start has been made by Williams (1984) and Williams & Fisher (1985) who have demonstrated the importance of rhizoids in the N nutrition of the rhizophyte *Caulerpa cupressoides*.

5.5 Transport at the Tonoplast

5.5.1 Preamble

Clarkson & Lüttge (1984) have recently evaluated the techniques available, and the data which they have yielded, as to the functioning of the plant cell tonoplast (see Ch. 3). The only area in which algae are favoured experimental material for those interested in tonoplasts *per se* is in the direct estimation of vacuolar contents (by microsurgery), transtonoplast fluxes (using microsurgery, and perfusion) and transtonoplast potential differences (micropipettes) and tonoplast conductances (micropipettes) in giant cells. Algae are conspicuously missing from the recent book edited by Marin (1985) on vacuolar ATPases of fungi and plants (but see Moriyasu *et al.*, 1984a, b). In our brief consideration of transport at the tonoplast of algal cells (for a more detailed analysis, see Raven, 1984a) we shall consider the systems in the same order as for the plasmalemma in Sect. 5.4.

5.5.2 The distinction between different modes of transport

Using criteria outlined for the plasmalemma (Sect. 5.4.2), we can identify H^+, Ca^{2+} and Na^+ as the ions which are, apparently, universally subject to active transport from cytoplasm to vacuole in algae. In most cases it appears that Cl^-, NO_3^-, $H_2PO_4^-$ and SO_4^{2-} are also actively transported into the vacuole, although in those Chlorophyta/Ulvophyceae which have a large positive (~80 mV) of ψ_{vc} this is not well established. In members of this 'high positive ψ_{vc}' group with high vacuolar K^+ concentrations, K^+ is also actively transported from cytoplasm to vacuole; this may be the case in some other algae with less positive ψ_{vc}, such as *Griffithsia* (Rhodophyta). An active K^+ flux from vacuole to cytoplasm may be needed in algae with high vacuolar Na^+/K^+ ratios and small values of ψ_{vc}. How organic solutes are distributed between cytoplasm and vacuole, and which *must* be actively transported to explain this distribution, are both largely unknown. Storage of NH_4^+ in algal vacuoles *might* occur via an 'acid trap' mechanism, with lipid-solution influx of NH_3 from the alkaline cytoplasm to the acid vacuole (see Walker *et al.*, 1979b; Raven, 1980, 1984a; Sakano & Tazawa, 1984).

5.5.3 Primary active transport at the algal tonoplast

On general comparative grounds we might expect a major primary active transport system at the algal tonoplast to involve an electrogenic H^+-pumping ATPase such as occurs in higher plants (Marin, 1985); such a phylogenetic argument is strongest for the Charophyceae which are the closest extant algal relatives of the higher plants (Raven, 1987). It is for the Characeae (member of the Charophyceae) for which most evidence is available on tonoplast H^+ transport in algae. Moriyasu *et al.* (1984a, b) have shown, from measurements of the rate of recovery (by downhill H^+ flux from vacuole to cytoplasm) of the vacuolar pH after perfusion with a more acid solution than the normal vacuolar solution, that the tonoplast P_{H^+} of *Chara corallina* is $3 \cdot 10^{-5}$ m s^{-1}. This implies a normal active H^+ flux from cytoplasm to vacuole of 300 nmol H^+ m^{-2} s^{-1} to balance the downhill flux from vacole to cytoplasm. This active flux appears to be electrogenic (Moriyasu *et al.* 1984a, b). Moriyasu *et al.* (1984b) showed that acidification of the interior of *Chara corallina* tonoplast vesicles required external ATP, but provided no data on the electrical PD response to ATP such as is needed to verify the occurrence of an electrogenic ATP-driven pump.

Of the other actively transported solutes mentioned in Sect. 5.5.2, electrical evidence is available which bears on the primary nature of active transport of Cl^- and of K^+ from cytoplasm to vacuole in giant-celled algae. For Cl^-, Blount & Levedahl (1960) suggested that the fraction of the short-circuit current (positive charge flow from vacuole to bathing solution) which could be attributed to net Cl^- influx was probably due to active electrogenic Cl^- influx at the tonoplast of *Halicystis ovalis*; however, later work on *H. parvula* and on *Acetabularia* has been interpreted in terms of such an influx at the plasmalemma (see Sect. 5.4.3). For K^+, the short-circuit current (positive charge flow from bathing medium to vacuole) in *Valonia* could be related to a primary, electrogenic active K^+ flux from cytoplasm to vacuole (Hastings & Gutknecht, 1976).

5.5.4 Secondary active transport at the algae tonoplast

Aside from the bland generalization that, on comparative grounds, the majority of active transport processes at the algal tonoplast are likely to be secondary (Raven & Smith, 1980b), little can be said.

5.5.5 Mediated passive uniport at the algae tonoplast

The scant evidence here comes from studies of action potentials at the tonoplast of intact Characean internode cells (which show that variations in the specific reaction rates of ion uniporters for one or more of the ions which are far from thermodynamic equilibrium across the tonoplast: see Findlay & Hope, 1976). While the measurement of resting electrical properties of the Characean tonoplast as a function of changes (by perfusion) in vacuolar composition clearly shows that ion fluxes (excluding primary active transport) occur at faster rates than 'lipid solution' can accommodate (Moriyasu *et al.*, 1984a, b), it does not clearly distinguish mediated uniport from electrogenic secondary active transport (Sect. 5.5.4).

5.5.6 'Lipid solution' passive uniport at the algae tonoplast

Evidence cited in Sect. 5.5.3. suggests that P_{H^+} at the tonoplast of *Chara corallina* is $3 \cdot 10^{-5}$ m s^{-1}; this is compatible with the notion that the H^+ flux could be a 'lipid solution' flux, since at the low pH_v values used in the experiment of Moriyasu *et al.* (1984a) the 'lipid solution' P_{H^+} value could be of about this magnitude (Gutknecht, 1984). However, the data do not rule out a contribution from mediated uniport, or co-transport, H^+ fluxes.

5.5.7 The extent and significance of passive uniport ('leakage' and 'mediated') at the algal tonoplast

The plasmalemma and the tonoplast are both membranes across which passive uniport losses of solute from the cytoplasm, and gains of solute by the cytoplasm, can occur, although clearly the topological disposition of the vacuole(s) and its restricted volume mean that its capacity as a sink for cytoplasmic solutes is restricted relative to that of the medium, while its capacity to act as a source of solutes for the cytoplasm is predicated on prior solute entry via the cytoplasm. Granted the presence of a higher concentration of (absolutely impermeant) macromolecules in the cytoplasm than the vacuole, the long-term effect of inhibition of the transport processes which normally reverse the uniport fluxes would be to increase the cytoplasmic volume at the expense of that of the vacuole (see Ch. 8 of Raven, 1984a). It is likely that uniport of solutes at the tonoplast has much less effect on nutrient use efficiency at low nutrient concentrations in the bulk medium than does uniport at the plasmalemma (nutrients are not lost from the cell if they move into the vacuole), while the energetic impact of tonoplast uniport might approach that of plasmalemma uniport (on a membrane area basis) for algal performance at low photon flux densities (see Sect. 5.4.7). In relation to the effect of cell size on the energy cost (expressed as a specific maintenance rate, mol C lost (mol cell C)$^{-1}$ s^{-1}) of maintaining the *status quo* in the face of membrane uniport, we note that while the tonoplast area of our small (0.5 μm radius) alga may be 0.35 or less of that of the plasmalemma, the ratio of areas for a giant algal cell may be 1.0 (see Raven, 1980 and Ch. 8 of Raven, 1984a). This means that the area of 'uniporting' plasmalemma *plus* tonoplast would be $\leqslant$776 m^2 (mol cell C)$^{-1}$ for the 0.5 μm, radius cell and 9.6 m^2 (mol cell C)$^{-1}$ for the 500 μm radius cell, as compared to 575 m^2 (mol cell C)$^{-1}$ and 4.8 m^2 (mol cell C)$^{-1}$, respectively, if only plasmalemma area is considered (see Sect. 5.4.7).

5.5.8 Unstirred layers internalized: effects on transtonoplast fluxes

Cytoplasmic streaming, together with frictional interactions between the tonoplast and the vacuolar solution, would help to mix the vacuolar contents, but would still leave a very substantial boundary layer at the observed velocities of cytoplasmic streaming ($\leqslant$100 μm s^{-1}); in the absence of cytoplasmic streaming, no such mixing would occur. The effects on transtonoplast fluxes would not, however, be very great, even in giant cells; with mediated net transtonoplast fluxes of <1 μmol m^{-2} s^{-1} and vacuolar concentrations of >10 mol m^{-3} for most major solutes (Ch. 8 of Raven, 1984a) decrements, or increments of <10 % of the vacuolar concentration would be expected even if there were an unstirred layer of 1 mm. Raven (1977, 1986c) discusses the role of buffers and chelates in cytosol fluxes of H^+ and Ca^{2+} whose free ion concentrations are only 0.1 mol m^{-3}. However, for the estimation of $P_{\text{tonoplast}}$ for rapidly permeating solutes such effects would be very important (cf. Walker *et al.*, 1979b).

5.6 Transport by Membranes of other Algal Organelles

5.6.1 Preamble

The book edited by Wiessner *et al.* (1984) surveys the range of compartments found in algal cells and their interactions. Very little is known about the solute relations of these compartments for most algae; the main purpose of this section is to direct the reader to source materials of what is known, or of what should, and can, be investigated. General concepts of algal cell compartments are discussed by Schnepf (1964, 1984), Mitchell (1979) and Raven (1984a).

5.6.2 Contractile vacuoles

These, or their functional equivalents, are important in cell-volume regulation in wall-less cells (including apparently walled, but flagellate, cells): Raven (1982, 1984a); Hausmann & Patterson (1984).

5.6.3 Microbodies

Stabenau (1984) discusses the role of these organelles, in conjunction with other organelles (mitochondria, plastids) in prosecuting the photorespiratory carbon reduction cycle (PCOC) in photolithotrophic algae. The details of the pathway, the extent of microbody involvement, and the quantitative significance of the pathway (C flux as a fraction of net C fixation) vary genotypically and phenotypically in the algae (Collins & Merrett, 1975; Raven, 1984a; Stabenau, 1984; see Sect. 5.6.4). It is likely that the microbody membrane is permeable (probably via protein pores such as are found in the outer mitochondrial membrane, and in the outer membrane of the 'normal' two-membraned plastid envelope) to solutes of M_r 500–1000, i.e. the substrates and products of the PCOC reactions occurring in the microbodies appear to cross the microbody membrane by a non-specific uniport (Raven, 1984a).

5.6.4 Mitochondria

Algal mitochondria (Lloyd, 1974a, b; Lloyd & Turner, 1980; Raven, 1984a) *generally* have properties more similar to those of higher plant and fungal mitochondria rather than those of metazoans (see Ch. 2). However, relatively few data are available on the H^+ fluxes associated with redox H^+ pumps, F_0F_1 ATP synthetase, and phosphate and organic anion co-transport, and more generally on the influx of substrates and efflux of products, in algal mitochondria. The significance for transport processes of the differences in crista structure between different major taxa of algae (see Cavalier-Smith, 1982; Stewart & Mattox, 1984) is unclear. An apparently unrelated phyletic difference relates to the occurrence of PCOC enzymes in mitochondria; the Charophyceae (like 'higher' plants) seem to have only the glycine–serine step in their mitochondria, while *Euglena* and the Chlorophyceae have the complete glycollate to glycerate sequence in *their* mitochondria, although *Euglena* also seems to have glycollate dehydrogenase in its microbodies (Collins & Merrett, 1975; Stabenau, 1984). The implications of this for transport at the inner mitochondrial membrane have not been explored.

5.6.5 Plastids

Isolation of plastids from algae in a state which permit them to fix CO_2 at light and inorganic C saturation at the rate (per unit pigment) which is found in the intact organism is only *routinely* possible for various coenocytic members of the Chlorophyta and (*Griffithsia*) Rhodophyta which have discoid, 'higher-plant-like' plastids (Raven, 1984a). For other organisms (*Euglena*; *Chlamydomonas*) the isolation of plastids with even *half* of the *in vivo* CO_2 fixation capacity has only just become possible (Raven, 1984a).

The transport properties of the outer membranes of the most intact of the algal plastids thus far isolated have not been examined in as much detail as have those of several 'higher' terrestrial plants. Raven (1984a, 1987) reviews the available data, which show that, in the Chlorophyta, the most 'higher-plant-like' of the organisms looked at are in the Chlorophyceae, while plastids of at least some Ulvophyceae seem to have an additional hexose-P porter; alas, no data are available for the Charophyceae, the class of Chlorophyta closest to the 'higher' plants. The Chlorophyta (and Rhodophyta) have, like the 'higher' plants, two plastid envelope membranes (see Ch. 2). Other algae have an additional membrane around their plastids (Euglenophyta, most Dinophyta), or have two extra membranes ('chloroplast endoplasmic reticulum', or some related structure, in Chlorarachniophyta, and in 'Chromophyte' algae other than the Dinophyta): see Raven (1984a, 1987). The only algae with more than two plastid envelope membranes which has been investigated with respect to transport across the envelope is *Euglena* (see Raven, 1984a); here the departure from the higher plant paradigm is more profound than is the case for the Chlorophyta.

An important question, which is still unresolved, relates to the location of the 'pump' for the inorganic carbon accumulation mechanism, based on active transport of CO_2 and/or HCO_3^-, which occurs in many algae grown at low CO_2 concentrations (Raven, 1985b). The two likely locations are the plasmalemma or the innermost membrane of the plastid envelope (Beardall & Raven, 1981). Experiments with Chlorophycean *Chlamydomonas reinhardtii* treated with plasmalemma-permeabilizing agents suggest that the plasmalemma is the pump site in this organism (Marcus *et al.*, 1984) since interference with plasmalemma integrity abolishes the 'CO_2 concentrating mechanism' (cf. Moroney *et al.*, 1985). Furthermore, plastids isolated from the marine Ulvophycean *Udotea* sp. show substantial inhibition of CO_2 fixation from 'air equilibrium' CO_2 levels by 21 % O_2 relative to the rate in low O_2, whereas the intact thallus shows much less O_2 inhibition (Bowes, 1985). *If* the avoidance of O_2 inhibition in the intact thallus is the result of active inorganic C transport (as opposed to some C_4-like biochemistry), then the results of Bowes (1985) suggest a plasmalemma location for this pump. It is important, in further investigations of this problem which involve comparisons of responses of isolated plastids relative to whole cells or thalli to $[CO_2]$, $[HCO_3^-]$ and $[O_2]$, to bear in mind that unstirred layer effects on kinetics will be more important for whole cells or thalli than for isolated plastids since the latter are smaller and, per unit pigment, have more area exposed to the surface *and* a thinner unstirred layer (cf. Sect. 5.3). It is possible that parts of the difference in $K_{\frac{1}{2}(CO)_2}$ noted by Tsuzuki *et al.* (1985) between photosynthesis by protoplasts (*relatively* high $K_{\frac{1}{2}}$) and plastids (*relatively* low $K_{\frac{1}{2}}$) of C_3 terrestrial plants can be attributed to this cause.

Burrowing deeper into the plastids, little is known *directly* of H^+/e^- and H^+/ATP ratios, and the magnitude of $\Delta\ \bar{\mu}_{H^+}$, for thylakoid reactions in algae, although what data are available on isolated plastids and thylakoids show that the stoichiometries and magnitudes involved are likely to be similar to those of higher plants (see Ch. 3 and 4 of Raven, 1984a). Certainly the photon yields of photosynthesis and growth in intact algae (see Geider, *et al.*, 1985, 1986; Luning & Dring, 1985) demand that the H^+/e^- and ATP/H^+ ratios, and the $\Delta \bar{\mu}_{H^+}$ across the thylakoid membrane, are no lower than in higher plants.

5.7 Long-distance transport in algae

5.7.1 Preamble

Long-distance transport in algae has recently been dealt with at some length by the present author (Raven, 1981, 1984a, 1987); only a few points, mainly of particular relevance to membrane transport processes, will be mentioned here.

5.7.2 The requirement for long-distance transport

We have already seen (Sect. 5.2, 5.3) that spatial separation of sites of resource acquisition, resource storage, and resource use in growth and maintenance, by more than a few millimetres requires that the chemical substances bearing the resources shall move other than by diffusion (see Raven, 1981, 1984a, 1986b, 1987).

5.7.3 Mechanisms of long-distance transport: Cyclosis

The first mechanism to be considered is cyclosis, based on the interaction of actin attached to some skeletal element, with myosin ATPase; the myosin-containing portion of the cytoplasm moves relative to the anchored actin. This cyclosis is most effective in long-distance transport if there are no diffusive steps, e.g. through 'classical' plasmodesmata, along the path from source to sink. This is achieved in the more differentiated coenocytic members of the Ulvophyceae and the Tribophyta, by having *no* cross-walls. In the Characeae (Charophyceae) there are relatively few arrays of plasmodesmata in the plant axes, with most of the length (and volume) of the plant occupied by giant internodal cells; thus a metre-long plant of a Characean may have thirty plasmodesmatal arrays, while a metre length of 50 μm diameter cells has 20 000 arrays of plasmodesmata.

Cyclosis can occur at up to 100 $\mu m\ s^{-1}$; the streaming cytoplasm moving in the 'source' to 'sink' direction may occupy 0.02 of the transectional area of a Characean axis; and the concentration of mobile (through plasmodesmata) solutes can be 85 $mol\ N\ m^{-3}$ and 296 $mol\ C\ m^{-3}$ (higher values than those suggested by Raven, 1984a, p. 540; based on 10 $mol\ m^{-3}$

sucrose in addition to the 176 mol organic C in amino acids computed from data in Sakano & Tazawa, 1984). The C flux (source to sink) is, then $10^{-4} \times 296 \times 0.02$ or $5.92 \cdot 10^{-4}$ mol C (m^2 plant axis TS area)$^{-1}$ s^{-1}. Since a 1 mm diameter axis has a circumference of $4 \cdot 10^3$ m per m^2 transectional area, a mean net photosynthetic rate over 24 h of 10^{-6} mol C (m^2 plant area)$^{-1}$ s^{-1} (Raven, 1981) means that the photosynthesis of 592 m^2 plant area is needed to supply the mobile C to 1 m^2 of plant axis, contained in a length of 0.148 m or 14.8 % of the axis length of 1 m long plant. We can see that the plant surface area needed for loading the long-distance transport pathway with photosynthate is 592 times the axis area, or 29 600 times the area of streaming cytoplasm moving from source to sink. Recent work on long-distance transport in Characeans is cited by Raven (1984a, 1987), Andrews *et al.* (1984) and Box *et al.* (1984).

5.7.4 Mechanisms of long-distance transport: movement of organelles along microtubules

It has recently become evident that 'fast axonal transport' involves movement of organelles along microtubules with mechanochemical transduction via an ATPase ('kinesin') other than the 'normal' (e.g. in flagella) microtubule-associated dynein and the actomyosin system; the 'kinesin' system also seems to be at least part of the mechanism of plastid movement in some coenocytic members of the Ulvophyceae (see Menzel *et al.*, 1985; Vale *et al.*, 1985; Raven, 1987). It is possible that movement of plastids containing starch, protein and phosphorus could supplement or even replace, cytoplasmic streaming as a mechanism of long-distance movement of resources in these plants. Quantitation of this mechanism is not easy, although it is clear that the velocity of movement is an order of magnitude less than the 10 μm s^{-1} found for cytoplasmic streaming in many Ulvophyceae. We note that this mechanism for long-distance resource transport has implications for fluxes across the plastid envelope of these Ulvophycean algae.

5.7.5 Mechanisms of long-distance transport: phloem-type mass flow of solution

This type of transport seems to be limited to certain of the larger members of the Phaeophyta (many Laminariales and Fucales; probably also Durvilleales and some Desmarestiales: Ch. 9 of Raven, 1984a). Here mass flow of a concentrated solution, essentially modified cytosol, with up to 5 kmol organic C m^{-3} (Table 9.8 of Raven, 1984a) typically occurs through elongate cells with enlarged plasmodesmata (<6 μm diameter) in their end walls (Table 9.6 of Raven, 1984a) at velocities up to 200 μm s^{-1} (Table 9.7 of Raven, 1984a). This combination of concentration and velocity yields organic C fluxes[1] of 1 mol (m^2 transectional area of transport pathway)$^{-1}$ s^{-1} or, with the transport pathway occupying only 0.01 of the transectional area of the plant, an organic C flux of 10^{-2} mol (m^2 transectional area of the plant)$^{-1}$ s^{-1}. We note that this flux can be unidirectional (unlike cytoplasmic, streaming, which is recirculatory) with a net flux of solutes *and* solvent from 'source' to 'sink'.

How do these long-distance fluxes relate to transmembrane fluxes? The net photosynthesis over 24 h of ~ 1 mol C fixed (m^2 blade surface)$^{-1}$ s^{-1} for a large phaeophyte such as *Macrocystis* (p. 513 of Raven, 1984a) is some 10^{-4} of the axial organic C flux quoted above, i.e. 10^4 m^2 of photosynthetic area is needed to provide photosynthate to 1 m^2 transectional area of translocating axis *if* the translocatory flux is working at its full potential. For at least part of the path from photosynthetic cells to 'phloem-like' cells in large Phaeophyta there seems to be a quantitatively adequate provision of plasmodesmata to support the required C flux via a symplastic route (Schmitz & Kuhn, 1982; Raven, 1984a, p. 513), although a role for an apoplastic route over part of the loading pathway has not been excluded (cf Ch. 8). Even more dubiety surrounds the unloading mechanism in these plants. A final membrane-related point concerns the mechanism of energization of the mass flow: Raven (1984a, pp. 530–5) suggests that the evidence for pressure flow in such Phaeophytes as *Macrocystis* is at least as compelling as for many vascular plants (see Ch. 8, pp. 306). If this is the mechanism it gives transmembrane fluxes a key role in not only loading the solutes but also in bringing about their longitudinal movement.

5.7.6 Conclusions

While long-distance *apoplastic* transport by non-diffusive mechanisms is, apparently, not important in algae, they show a wider range of mechanisms for long-distance *symplastic* movement by non-diffusive mechanisms than do higher plants (see Ch. 8).

5.8 Future developments

In phylogenetic terms this question has been answered for the algal class Charophyceae; they are the closest algal relatives of the 'higher' plants (Bryophyta and Tracheophyta). The algae lack some membrane transport phenomena related to the homoiohydric state (e.g. the transcellular or 'epithelial' phenomena associated with xylem loading and salt glands and the turgor-control mechanisms associated with stomatal regulation and

[1] Note that the value of 7.4 mol organic C m^{-2} s^{-1} on line 27, p. 525 of Raven (1984a) should be 0.74 mol organic C m^{-2} s^{-1}.

pulvinar function) but have other transport phenomena not obviously shared by 'higher' plants (see Sect. 5.7.3 and 5.7.4). Aside from these algal 'specialities', we can point to further potential for a genetic approach to transport problems in, for example, *Chlamydomonas* and *Acetabularia*, the utility of microalgae as 'microbes' and of giant algal cells as 'giant axons' and 'giant muscle fibres' in the use of techniques from microbial and metazoan studies to further our understanding of transport in phototrophs, and the intrinsic interest of such a taxonomically, morphologically and ecologically varied group of organisms, as justification for further work on algal transport processes.

Acknowledgements

SERC and NERC have funded work in algal transport and energy-transduction processes in the author's laboratory; Dale Sanders and Mazashi Tazawa kindly furnished unsolicited, but very welcome, re- and pre-prints.

References

Ahmad, I. & Hellebust, J. A. (1984). Osmoregulation in the extremely euryhaline marine microalga *Chlorella autotrophica*. *Plant Physiology* **747**, 1010–15.

Ahmad, I. & Hellebust, J. A. (1985). Osmoregulation in the euryhaline flagellate *Brachiomonas submarina* (Chlorophyceae). *Marine Biology* **87**, 245–50.

Aizawa, K., Nakamura, Y. and Miyachi, S. (1985). Variation of PEPc activity in *Dunaliella* associated with changes in atmospheric CO_2 concentration. *Plant and Cell Physiology* **26**, 1199–203.

Anderson, S. M. & Roels, D. A. (1981). Effects on light intensity on nitrate and nitrite uptake and excretion by *Chaetoceros curvisetus*. *Marine Biology* **62**, 257–61.

Andrews, M., Box, R., Fyson, A. & Raven, J. A. (1984). Source–sink characteristics of carbon transport in *Chara hispida*. *Plant, Cell and Environment* **7**, 683–7.

Astle, M. C. & Rubery, P. H. (1984). Evaluation of $TPMP^+$ cation as a probe of membrane potential in suspension-cultured runner bean cells. *Plant Science Letters* **36**, 43–49.

Atkinson, M. J. & Smith, S. V. (1983). C : N : P ratios of benthic marine plants. *Limnology and Oceanography* **28**, 568–74.

Banse, K. (1982). Cell volumes, maximal growth rates of unicellular algae and ciliates, and the role of ciliates in the marine pelagial. *Limnology and Oceanography* **27**, 1059–71.

Beardall, J. & Raven, J. A. (1981). Transport of inorganic carbon and the

'CO_2 concentrating mechanism' in *Chlorella emersonii* (Chlorophyceae). *Journal of Phycology* **17**, 134–41.

Beilby, M. J. (1985). Potassium channels at *Chara* plasmalemma. *Journal of Experimental Botany* **36**, 228–39.

Beilby, M. J. & Walker, N. A. (1981). Chloride transport in *Chara*. I. Kinetics and current–voltage curves for a probable proton symport. *Journal of Experimental Botany* **32**, 43–54.

Bisson, M. A. (1984). Calcium effects on electrogenic pump and passive permeability of the plasma membrane of *Chara corallina*. *Journal of Membrane Biology* **81**, 59–67.

Blount, R. W. & Levedahl, B. H. (1960). Active Na^+ and Cl^- transport in the single celled marine alga *Halicystis ovalis*. *Acta Physiologia Scandinavica* **49**, 1–9.

Bold, H. C. & Wynne, M. J. (1985). *Introduction to the Algae. Structure and Reproduction*. Englewood Cliffs, NJ, Prentice-Hall.

Bowes, G. (1985). Pathways of CO_2 fixation by aquatic organisms. In Lucas, W. J. & Berry, J. A. (eds.). *Inorganic Carbon Uptake by Aquatic Photosynthetic Organisms*, pp. 187–211. Rockville, Md., American Society of Plant Physiologists.

Bowles, E. A. & Allen, N. S. (1984). Steady currents go through *Acetabularia*: a vibrating probe analysis. *Biological Bulletin* **167**, 501–4.

Box, R., Andrews, M. & Raven, J. A. (1984). Intracellular transport and cytoplasmic streaming in *Chara hispida*. *Journal of Experimental Botany* **35**, 1016–21.

Brawley, S. M. & Wetherbee, R. (1981). Cytology and ultrastructure. In Lobban, C. S. & Wynne, M. J. (eds.). *The Biology of Seaweeds*, pp. 248–99. Oxford, Blackwell.

Carlson, D. J. & Carlson, M. L. (1984). Re-assessment of exudation by fucoid macroalgae. *Limnology and Oceanography* **29**, 1077–87.

Cavalier-Smith, T. (1982). The origin of plastids. *Biological Journal of the Linnean Society* **17**, 289–306.

Clarkson, D. T. & Lüttge, U. (1984). Mineral nutrition: vacuoles and tonoplasts. *Progress in Botany* **46**, 56–67.

Clendenning, K. A. (1971a). Photosynthesis and general development in *Macrocystis*. *Beihefte zur Nova Hedwigia* **32**, 169–90.

Clendenning, K. A. (1971b). Gross composition of kelp. *Beihefte zur Nova Hedwigia* **32**, 197–209.

Clendenning, K. A. (1971c). Organic productivity in kelp areas. *Beihefte zur Nova Hedwigia* **32**, 259–63.

Collins, N. & Merrett, M. J. (1975). The localization of glycolate-pathway enzymes in *Euglena*. *Biochemical Journal* **148**, 321–8.

Colinvaux, P. (1980). *Why Big Fierce Animals are Rare*. Harmondsworth, Penguin.

Davison, I. R. & Reed, R. H. (1985). Osmotic adjustment in *Laminaria digitata* (Phaeophyta) with particular reference to seasonal changes in internal solute concentration. *Journal of Phycology* **21**, 41–50.

Dawson, R. M. C., Elliott, D. C., Elliott, W. H. & Jones, K. M. (1969). *Data for Biochemical Research*. Oxford, Clarendon.

Den Hartog, C. & Segal, S. (1964). A new classification of the water plant communities. *Acta Botanica Neerlandica* **13**, 367–93.

Doblinger, R. & Tromballa, H. W. (1982). The effect of glucose on chloride uptake by *Chlorella*. I. General characteristics of chloride transport. *Planta* **156**, 10–15.

Dortch, Q., Clayton, J. R., Jr., Thoreson, S. S. & Ahmed, S. I. (1984). Species differences in accumulation of nitrogen pools in phytoplankton. *Marine Biology* **81**, 237–50.

Findenegg, G. R. (1974a). Beziehungen zwischen carboanhydrase-aktivitat und aufnahme von HCO_3 und Cl^- bei der photosynthese von *Scenedesmus obliquus*. *Planta* **116**, 123–131.

Findenegg, G. R. (1974b). Carbonic anhydrase and the driving force for light-dependent uptake of Cl^- and HCO_3 by *Scenedesmus*. In Zimmermann, U. & Dainty, J. (eds.). *Membrane Transport in Plants*, pp. 192–196. Berlin, Springer-Verlag.

Findlay, G. P. (1982). Electrogenic and diffusive components of the membrane of *Hydrodictyon africanum*. *Journal of Membrane Biology* **68**, 179–89.

Findlay, G. P. & Hope, A. B. (1976). Electrical properties of cells: methods and findings. In: Lüttge, U. & Pitman, M. G. (eds.). *Encyclopedia of Plant Physiology*, New Series, vol. 2A, pp. 53–92. Berlin, Springer-Verlag.

Flynn, K. J. & Syrett, P. J. (1985). The uptake of nitrogenous compounds by *Phaeodactylum tricornutum* with particular reference to lysine uptake. *British Phycological Journal* **20**, 185.

Friebele, E. S., Correll, D. L. & Faust, M. A. (1978). Relationship between phytoplankton cell size and the rate of orthophosphate uptake: *in situ* observations on an estuarine population. *Marine Biology*. **45**, 39–52.

Fujii, S., Shimmen, T. & Tazawa, M. (1979). Effect of intracellular pH on the light-induced potential change and electrogenic activity in tonoplast – free cells of *Chara australis*. *Plant and Cell Physiology* **20**, 1315–28.

Gavis, J. (1976). Munk and Riley revisited: nutrient diffusion transport and rates of phytoplankton growth. *Journal of Marine Research* **34**, 161–79.

Geider, R. J., Osborne, B. A. & Raven, J. A. (1985). Light dependence of growth and photosynthesis in *Phaeodactylum tricornutum*. *Journal of Phycology* **21**, 609–19.

Geider, R.J., Osborne, B. A. & Raven, J. A. 1986). Growth, photosynthesis and maintenance metabolic costs in the diatom *Phaeodactylum tricornutum* grown at very low light levels. *Journal of Phycology* **22**. In press.

Gershon, N. D., Porter, K. R. & Trus, B. L. (1985). The cytoplasmic matrix: its volume and surface area and the diffusion of molecules through it. *Proceedings of the National Academy of Sciences, USA* **82**, 5030–4.

Gimmler, H. & Greenway, H. (1983). Tetraphenylphosphonium (TPP^+) is not suitable for the assessment of electrical potentials in *Chlorella emersonii. Plant, Cell and Environment* **6**, 739–44.

Goffeau, A. & Slayman, C. W. (1981). The proton-translocating ATPase of the fungal plasmalemma membrane. *Biochimica et Biophysica Acta* **639**, 197–224.

Goldfarb, V., Sanders, D. & Gradmann, D. (1984a). Phosphate relations in *Acetabularia*: phosphate pools, adenylate kinetics, and ^{32}P influx kinetics. *Journal of Experimental Botany* **35**, 626–44.

Goldfarb, V., Sanders, D. & Gradmann, D. (1984b). Reversal of electrogenic Cl^- pump in *Acetabularia* increases level and ^{32}P labelling of ATP. *Journal of Experimental Botany* **35**, 645–58.

Gradmann, D. (1975). Analog circuit of the *Acetabularia* membrane. *Journal of Membrane Biology* **25**, 183–208.

Gradmann, D., Hansen, V–P. & Slayman, C. L. (1982a). Reaction-kinetic analysis of current–voltage relationships for electrogenic pumps in *Neurospora* and *Acetabularia. Current Topics in Membranes and Transport* **16**, 257–76.

Gradmann, D., Tittor, J. & Goldfarb, V. (1982b). Electrogenic Cl^- pump in *Acetabularia. Philosophical Transactions of the Royal Society, London* B **299**, 447–57.

Graves, J. S. & Gutknecht, J. (1977a). Chloride transport and the membrane potential in the marine alga, *Halicystis parvula. Journal of Membrane Biology* **36**, 65–82.

Graves, J. S. & Gutknecht, J. (1977b). Current voltage relationships and voltage sensitivity of the chloride pump in *Halicystis parvula. Journal of Membrane Biology* **36**, 83–95.

Greenway, H. & Setter, T. L. (1979). Na^+, Cl^- and K^+ concentrations in *Chlorella emersonii* exposed to 100 and 335 mM NaCl. *Australian Journal of Plant Physiology* **6**, 61–7 (corrigendum on p. 571).

Gross, J. & Marme, D. (1978). ATP-dependent Ca^{2+} uptake into plant membrane vesicles. *Proceedings of the National Academy of Sciences, USA* **75**, 1232–6.

Gutknecht, J. (1968). Permeability of *Valonia* to water and solutes: apparent absence of aqueous membrane pores. *Biochimica et Biophysica Acta* **163**, 20–9.

Gutknecht, J. (1984). Proton/hydroxide conductance through lipid bilayer membranes. *Journal of Membrane Biology* **82**, 105–12.

Hastings, D. F. & Gutknecht, J. (1976). Ionic relations and the regulation of turgor pressure in the marine alga, *Valonia macrophysa. Journal of Membrane Biology* **28**, 263–76.

Hausmann, K. & Patterson, D. J. (1984). Contractile vacuole complexes in algae. In Wiessner, W., Robinson, D. G. & Starr, R. C. (eds.).

Compartments in Algal Cells and their Interaction, pp. 134–46. Berlin, Springer-Verlag.

Hayama, T., Shimmen, T. & Tazawa, M. (1979). Participation of Ca^{2+} in cessation of cytoplasmic streaming induced by membrane excitation in Characean cell internodes. *Protoplasma* **99**, 305–21.

Hellebust, J. A. (1974). Extracellular products. In Stewart, W. D. P. (ed.). *Algal Physiology and Biochemistry*, pp. 838–63. Oxford, Blackwell.

Hellebust, J. A. (1985). A comparative study of sodium and osmotic requirements for growth and nutrient uptake of two related green flagellates, *Dunaliella tertiolecta* and *Chlamydomonas pulsatilla*. *Archives of Microbiology*. **143**, 11–14.

Homble, F. (1985). Effect of sodium, potassium, calcium, magnesium and tetraethylammonium on the transient voltage response to a galvanostatic step and the temperature on the steady membrane conductance of *Chara corallina*: a further evidence for the involvement of potassium channels in the fast time variant conductance. *Journal of Experimental Botany* **36**, 1603–11.

Huntsman, S. A. (1972). Organic excretion by *Dunaliella tertiolecta*. *Journal of Phycology* **8**, 59–63.

Jackson, G. A. (1977). Nutrients and production of giant kelp, *Macrocystis pyrifera*, off Southern California, *Limnology and Oceanography* **22**, 979–95.

Johnston, A. M. (1984). The assimilation of inorganic carbon by *Ascophyllum nodosum* (L.) Le Jolis. Ph.D. Thesis, University of Dundee.

Johnston, A. M. & Raven, J. A. (1986). Dark fixation studies on the intertidal macroalga *Ascophyllum nodosum* (Phaeophyta). *Journal of Phycology* **22**, In press.

Kaplan, A. & Schreiber, V. (1981). Light-induced proton gradient formation in intact cells of *Dunaliella salina*. *Plant Physiology* **68**, 236–9.

Katz, A. & Avron, M. (1985). Determination of intracellular osmotic volume and sodium concentration in *Dunaliella*. *Plant Physiology* **78**, 817–20.

Kirst, G. O. (1977). Ion composition of unicellular marine and freshwater algae, with special reference to *Platymonas subcordiformis* cultivated in media with different osmotic strengths. *Oecologia* **28**, 177–90.

Kirst, G. O. & Bisson, M. A. (1983). Sugar uptake in a euryhaline charophyte, *Lamprothuminium papulosum. Zeitschrift fur Pflanzenphysiologie* **111**, 105–14.

Kishimoto, V., Takeuchi, Y., Ohkawa, T. & Kami-ike, N. (1985). A kinetic analysis of the electrogenic pump of *Chara corallina*. III. Pump activity during action potential. *Journal of Membrane Biology* **86**, 27–36.

Kleiner, D. (1985a). Energy expenditure for cyclic retention of NH_3/NH_4^+

during N_2 fixation by *Klebsiella pneumoniae*. *FEBS Letters* **187**, 237–9.

Kleiner, D. (1985b). Bacterial ammonium transport. *FEMS Microbiology Reviews* **32**, 87–100.

Komor, E. & Tanner, W. (1980). Proton cotransport of sugars in plants. In Spanswick, R. M., Lucas, W. J. & Dainty, J. (eds.). *Plant Membrane Transport: Current Conceptual Issues*, pp. 247–57. Amsterdam, Elsevier.

Krom, M. D. & Berner, R. A. (1980a). The diffusion coefficients of sulfate, ammonium and phosphate ions in anoxic marine sediments. *Limnology and Oceanography* **25**, 327–37.

Krom, M. D. & Berner, R. A. (1980b). Adsorption of phosphate in anoxic marine sediments. *Limnology and Oceanography* **25**, 797–806.

Laws, E. A. & Wong, D. C. L. (1978). Studies of carbon and nitrogen metabolism by three marine phytoplankton species in nitrate-limited continuous culture. *Journal of Phycology* **14**, 406–16.

Levring, T., Hoppe, H. A. & Schmid, O. J. (1969). *Marine Algae, a Survey of Research and Utilization*. Hamburg, Cram, De Gruyter & Co.

Lloyd, D. (1974a). *The Mitochondria of Microorganisms*. London, Academic Press.

Lloyd, D. (1974b). Dark respiration. In Stewart, W. D. P. (ed.). *Algal Physiology and Biochemistry*, pp. 505–29. Oxford, Blackwell.

Lloyd, D. & Turner, G. (1980). Structure, function, biogenesis and genetics of mitochondria. *Symposium of the Society for General Microbiology* **30**, 143–80.

Lucas, W. J. (1982). Mechanism of acquisition of exogenous bicarbonate by internodal cells of *Chara corallina*. *Planta* **156**, 181–92.

Lucas, W. J. (1985). Bicarbonate utilization by *Chara*: a re-analysis. In Lucas, W. J. & Berry, J. A. (eds.). *Inorganic Carbon Uptake by Aquatic Photosynthetic Organisms*, pp. 229–54. Rockville, Md., American Society of Plant Physiologists.

Lucas, W. J., Keifer, D. W. & Sanders, D. (1983). Bicarbonate transport in *Chara corallina*: evidence for cotransport of HCO_3^- with H^+. *Journal of Membrane Biology* **73**, 263–74.

Lucas, W. J. & Nucitelli, R. (1980). HCO_3^- and OH^- transport across the plasmalemma of *Chara*. Spatial resolution obtained using extracellular vibrating probe. *Planta* **150**, 120–31.

Lucas, W. J. & Shimmen, T. (1981). Intracellular perfusion and cell centrifugation studies on plasmalemma transport processes in *Chara corallina*. *Journal of Membrane Biology* **58**, 227–37.

Lühring, H. & Tazawa, M. (1985). Effect of cytoplasmic Ca^{2+} on the membrane potential and membrane resistance of *Chara* plasmalemma. *Plant and Cell Physiology* **26**, 635–46.

Luning, K. & Dring, M. J. (1985). Action spectra and spectral quantum yield of photosynthesis in marine macroalgae with thin and thick thalli. *Marine Biology* **87**, 119–29.

Luther, H. (1949). Vorschlag zu einer okologischen Grundeinteilung der Hydrophytem. *Acta Botanica Fennica* **44**, 1–15.

MacFarlane, J. J. (1985). Diffusion, boundary layers and the uptake of nutrients by aquatic macrophytes. Ph.D. Thesis, University of Adelaide.

MacFarlane, J. J. & Raven, J. A. (1985). External and internal CO_2 transport in *Lemanea*: interactions with the kinetics of ribulose bisphosphate carboxylase. *Journal of Experimental Botany* **36**, 610–22.

Mague, T. H., Friberg, E., Hughes, D. J. & Morris, I. (1980). Extracellular release of carbon by marine phytoplankton: a physiological approach. *Limnology and Oceanography* **25**, 262–79.

Malone, T. (1980). Algae size. In Morris, I. (ed.). *The Physiological Ecology of Phytoplankton*, pp. 433–463. Oxford, Blackwell.

Marcus, Y., Volokita, M. & Kaplan, A. (1984). The location of the transporting system for inorganic carbon and the nature of the form translocated in *Chlamydomonas reinhardtii*. *Journal of Experimental Botany* **35**, 1136–44.

Marin, B. P. (ed) (1985). *Biochemistry and Function of Vacuolar Adenosine-triphosphatase in Fungi and Plants*. Berlin, Springer-Verlag.

Menzel, D., Bankston, L. A. & Schliwa, M. (1985). Chloroplast movement in the siphonous green alga *Bryopsis* requires both microtubules and actin filaments. *Journal of Cell Biology* **101**, 399a.

Mishra, A. K. & Kefford, N. P. (1969). Developmental studies on the coenocytic alga *Caulerpa sertularioides*. *Journal of Phycology* **5**, 103–9.

Mitchell, P. (1979). Direct chemiosmotic ligand conduction mechanisms in proton-motive complexes. In Lee, C. P., Schotz, G. & Ernster, L. (eds.). *Membrane Bioenergetics* pp. 361–72. Reading, Mass., Addison-Wesley.

Moriyasu, Y., Shimmen, T. & Tazawa, M. (1984a). Vacuolar pH regulation in *Chara australis*. *Cell Structure and Function* **9**, 225–34.

Moriyasu, Y., Shimmen, T. & Tazawa, M. (1984b). Electric characteristics of the vacuolar membrane of *Chara* in relation to pH_v regulation. *Cell Structure and Function* **9**, 235–46.

Moroney, J. V., Husic, H. D. & Tolbert, N. E. (1985). Effects of carbonic anhydrase inhibitors on inorganic carbon accumulation by *Chlamydomonas reinhardtii*. *Plant Physiology* **79**, 177–83.

Munk, W-H. & Riley, G. A. (1952). Absorption of nutrients by aquatic plants. *Journal of Marine Research* **11**, 215–40.

Nawata, T. (1984). A simple method of making a vibrating probe system. *Plant and Cell Physiology* **25**, 1089–94.

Nobel, P. S. (1983). *Biophysical Plant Physiology and Ecology*. San Francisco, Calif., W. H. Freeman.

North, W. J. (1980). Trace metals in giant kelp, *Macrocystis*. American *Journal of Botany* **67**, 1097–101.

Nye, P. H. & Tinker, P. B. (1977). *Solute Movement in the Soil–Root System*. Oxford, Blackwell.

Oliveira, L., Bisalputra, T. & Antia, N. J. (1980). Ultrastructural observations on the surface coat of *Dunaliella tertiolecta* from staining with cationic dyes and surface treatments. *New Phytologist* **85**, 385–392.

Parslow, J. S., Harrison, P. J. & Thompson, P. A. (1985). Interpreting rapid changes in uptake kinetics in the marine diatom *Thalassiosira pseudonana* (Hustedt). *Journal of Experimental Marine Biology and Ecology* **91**, 53–64.

Pasciak, W. J. & Gavis, G. (1974). Transport limitation of nutrient uptake in phytoplankton. *Limnology and Oceanography* **19**, 881–8.

Pasciak, W. J. & Gavis, G. (1975). Transport limited nutrient uptake rates in *Ditylum brightwellii*. *Limnology and Oceanography* **20**, 604–17.

Price, G. D. & Badger, M. R. (1985). Inhibition by proton buffers of photosynthetic utilization of bicarbonate in *Chara corallina*. *Australian Journal of Plant Physiology* **12**, 257–67.

Price, G. D., Badger, M. R., Bassett, M. E. & Whitecross, M. I. (1985). Involvement of plasmalemmasomes and carbonic anhydrase in photosynthetic utilization of bicarbonate in *Chara corallina*. *Australian Journal of Plant Physiology* **12**, 241–56.

Provasoli, L. & Carlucci, A. F. (1974). Vitamins and growth regulators. In Stewart, W. D. P. (ed.). *Algae Physiology and Biochemistry*, pp. 741–87. Oxford, Blackwell.

Purcell, E. M. (1977). Life at low Reynolds number. *American Journal of Physics* **45**, 3–11.

Raven, J. A. (1975). Algal cells. In Baker, D. A. & Hall, J. L. (eds.). *Ion Transport in Plant Cells and Tissues*, pp. 125–60. Amsterdam, North-Holland.

Raven, J. A. (1976). Transport in algal cells. In Lüttge, U. & Pitman, M. G. (eds.). *Encyclopedia of Plant Physiology*, New Series, vol. 2A, pp. 129–8. Berlin, Springer-Verlag.

Raven, J. A. (1977). H^+ and Ca^{2+} in phloem and symplast: relation of relative immobility of the ions to the cytoplasmic nature of the transport paths. *New Phytologist* **79**, 465–80.

Raven, J. A. (1980). Nutrient transport in microalgae. *Advances in Microbial Physiology* **21**, 47–226.

Raven, J. A. (1981). Nutritional strategies of submerged benthic plants: the acquisition of C, N and P by rhizophytes and haptophytes. *New Phytologist* **88**, 1–30.

Raven, J. A. (1982). The energetics of freshwater algae: energy requirements for biosynthesis and volume regulation. *New Phytologist* **92**, 1–20.

Raven, J. A. (1984a). *Energetics and Transport in Aquatic Plants*. New York, Alàn R. Liss.

Raven, J. A. (1984b). The role of membranes in pH regulation: implications for water use efficiency of higher plant growth with nitrate as nitrogen source. In: Baudet, A. M., Alibert, G., Mariga, G. & Lea, P. J. (eds.). *Annual Proceedings of the Phytochemical Society of Europe*, vol. 24, pp. 89–98. Oxford University Press.

Raven, J. A. (1985a). pH regulation in plants. *Science Progress (Oxford)* **69**, 495–509.

Raven, J. A. (1985b). The CO_2 concentrating mechanism. In Lucas, W. J. & Berry, J. A. (eds.). *Inorganic Carbon Uptake by Aquatic Photosynthetic Organisms*, pp. 67–78. Rockville, Md., American Society of Plant Physiologists.

Raven, J. A. (1986a). Physiological consequences of extremely small size for autotrophic organisms in the sea. In Platt, T. R. and Li, W. K. W. (eds). *Photosynthetic Picoplankton. Canadian Bulletin of Fisheries and Aquatic Science* **214**, 1–70.

Raven, J. A. (1986b). Long distance transport of calcium. In Trewavas, A. (ed.). *Molecular and Cellular Aspects of Calcium in Plant Development*, pp. 241–50. New York, Plenum Press.

Raven, J. A. (1987). Biochemistry, biophysics and physiology of chlorophyll *b*-containing algae: implications for taxonomy and phylogeny. *Progress in Phycological Research* **5**. 1–121.

Raven, J. A. & De Michelis, M. I. (1979). Acid–base regulation during nitrate assimilation in *Hydrodictyon africanum*. *Plant, Cell and Environment* **2**, 245–57.

Raven, J. A. & Glidewell, S. M. (1981). Processes limiting photosynthetic conductance. In Johnson, C. B. (ed.). *Physiological Processes Limiting Plant Productivity*, pp. 109–36. London, Butterworths.

Raven, J. A. & Lucas, W. J. (1985). The energetics of carbon acquisition. In Lucas, W. J. & Berry, J. A. (eds). *Inorganic Carbon Uptake by Photosynthetic Organisms*, pp. 305–24. Rockville, Md., American Society of Plant Physiologists.

Raven, J. A., Osbońe, B. A. & Johnston, A. M. (1985). Uptake of CO_2 by acquatic vegetation. *Plant, Cell and Environment* **8**, 417–25.

Raven, J. A. & Richardson, K. (1984). Dinophyte flagella: a cost–benefit analysis. *New Phytologist* **98**, 259–76.

Raven, J. A. & Smith, F. A. (1977). Characteristics, functions and regulation of active proton extrusion. In Marrè, E. & Ciferri, O. (eds.). *Regulation of Cell Membrane Activity in Plants*, pp. 25–40. Amsterdam, Elsevier.

Raven, J. A. & Smith, F. A. (1978). Effect of temperature on ion content, ion fluxes and energy metabolism in *Chara corallina*. *Plant, Cell and Environment* **1**, 231–8.

Raven, J. A. & Smith, F. A. (1980a). Intracellular pH regulation in the giant-celled marine alga *Chaetomorpha darwinii*. *Journal of Experimental Botany* **31**, 1357–71.

Raven, J. A. & Smith, F. A. (1980b). The chemiosmotic approach. In Spanswick, R. M., Lucas, W. J. & Dainty, J. (eds.). *Plant Membrane Transport: Current Conceptual Issues*, 161–78. Amsterdam, Elsevier.

Raven, J. A. & Smith, F. A. (1982). Solute transport at the plasmalemma and the early evolution of cells. *Biosystems* **15**, 13–26.

Redfield, A. C. (1958). The biological control of chemical factors in the environment. *American Scientist* **46**, 204–21.

Rees, T. A. V. (1984). Sodium dependent photosynthetic oxygen evolution in a marine diatom. *Journal of Experimental Botany* **35**, 332–7.

Ritchie, R. J. (1984). A critical assessment of the use of lipophilic cations as membrane potential probes. *Progress in Biophysics and Molecular Biology* **43**, 1–32.

Saddler, H. D. W. (1970a). The membrane potential of *Acetabularia mediterranea*. *Journal of General Physiology* **55**, 802–21.

Saddler, H. D. W. (1970b). The ionic relations of *Acetabularia mediterranea*. *Journal of Experimental Botany* **21**, 345–59.

Sakano, K. & Tazawa, M. (1984). Intracellular distribution of free amino acids between the vacuolar and extravacuolar compartments in internodal cells of *Chara australis*. *Plant and Cell Physiology* **25**, 1477–86.

Sanders, D. (1980). The mechanism of Cl^- transport at the plasma membrane of *Chara corallina*. I. Cotransport with H^+. *Journal of Membrane Biology* **53**, 129–41.

Sanders, D. (1984). Gradient-coupled chloride transport in plant cells. In Gerencser, G. A. (ed.). *Chemical Transport Coupling in Biological Membranes and Epithelia*, pp. 63–110. Amsterdam, Elsevier.

Sanders, D., Smith, F. A. & Walker, N. A. (1985). Proton/chloride cotransport in *Chara*: mechanism of enhanced influx after rapid external acidification. *Planta* **163**, 411–18.

Sauer, N., Komor, E. & Tanner, W. (1983). Regulation and characterization of two inducible amino-acid transport systems in *Chlorella vulgaris*. *Planta* **159**, 404–10.

Sauer, N. & Tanner, W. (1984). Partial purification and characterisation of inducible transport proteins of *Chlorella*. *Zeitschrift fur Pflanzenphysiologie* **114**, 367–76.

Schlee, J., Cho, B-H. & Komor, E. (1985). Regulation of nitrate uptake by glucose in *Chlorella*. *Plant Science* **39**, 25–30.

Schmitz, K. & Kuhn, R. (1982). Fine structure, distribution and frequency of plasmodesmata and pits in the cortex of *Laminaria hyperborea* and *Laminaria saccharina*. *Planta* **154**, 385–92.

Schnepf, E. (1964). Zur Feinstruktur von *Geosiphon pyriforme*. *Archiv fur Mikrobiologie* **49**, 112–31.

Schnepf, E. (1984). The cytological viewpoint of functional compartmentation. In Wiessner, W., Robinson, D. G. & Starr, R. C. (eds.). *Compartments in Algal Cells*, pp. 1–10. Berlin, Springer-Verlag.

Shieh, Y. J. & Barber, J. (1971). Intracellular sodium and potassium concentrations and net cation movements in *Chlorella pyrenoidasa*. *Biochimica et Biophysica Acta* **233**, 594–603.

Sievers, A. & Schroter, K. (1971). Versuch einer Kausalanalyse der geotropischen Reaktionskette in *Chara*–rhizoide. *Planta* **96**, 339–53.

Smith, F. A. (1970). The mechanism of chloride transport in Characean cells. *New Phytologist* **69**, 903–17.

Smith, F. A. (1984a). Regulation of cytoplasmic pH of *Chara corallina* in the absence of external Ca^{2+}; its insignificance in relation to the activity and control of the H^+ pump. *Journal of Experimental Botany* **35**, 1525–36.

Smith, F. A. (1984b). Regulation of the cytoplasmic pH of *Chara corallina*: response to changes in external pH. *Journal of Experimental Botany* **35**, 43–56.

Smith, F. A. (1985). Biological occurrence and importance of HCO_3^--utilizing systems: macroalgae (Charophytes). In Lucas, W. J. & Berry, J. A. (eds.). *Inorganic Carbon Uptake by Aquatic Photosynthetic Organisms*, pp. 111–124. Rockville, Md., American Society of Plant Physiologists.

Smith, F. A. & Gibson, J. L. (1985). Effect of cations on the cytoplasmic pH of *Chara corallina*. *Journal of Experimental Botany* **36**, 1331--140.

Smith, F. A. & Raven, J. A. (1979). Intracellular pH and its regulation. *Annual Review of Plant Physiology* **30**, 289–311.

Smith, F. A. & Walker, N. A. (1980). Photosynthesis by aquatic plants: effects of unstirred layers in relation to assimilation of CO_2 and HCO_3^- and to carbon isotope discrimination. *New Phytologist* **86**, 245–59.

Smith, P. T. & Walker, N. A. (1981). Studies on the perfused plasmalemma of *Chara corallina*: Current–voltage curves: ATP and potassium dependence. *Journal of Membrane Biology* **60**, 223–6.

Smith, R. E. M. & Kalff, J. (1982). Size-dependent phosphorus uptake kinetics and cell quota in phytoplankton. *Journal of Phycology* **18**, 275–84.

Sournia, A. (1982). Form and function in marine phytoplankton. *Biological Reviews* **57**, 347–94.

Stabenau, H. (1984). Microbodies in different algae. In Wiessner, W., Robinson, D. G. & Starr, R. C. (eds.). *Compartments in Algal Cells and their Interactions*, pp. 183–90. Berlin, Springer-Verlag.

Stewart, K. D. & Mattox, K. R. (1984). The case for a polyphyletic origin of mitochondria: morphological and molecular consideration. *Journal of Molecular Evolution* **21**, 54–7.

Sunda, W. G. & Huntsman, S. A. (1985). Regulation of cellular manganese and manganese transport rates in the unicellular alga *Chlamydomonas. Limnology and Oceanography* **30**, 71–80.

Takeshige, K., Shimmen, T. & Tazawa, M. (1985). Electrogenic pump current and ATP-dependent H^+ efflux across the plasmamembrane of *Nitellopsis obtusa. Plant and Cell Physiology* **26**, 661–8.

Takeuchi, Y., Kishimoto, V., Ohkaea, T. & Kami-ike, N. (1985). A kinetic analysis of the electrogenic pump of *Chara corallina*. II. Dependence of the pump activity on external pH. *Journal of Membrane Biology* **86**, 17–26.

Tittor, J., Hansen, V-P. & Gradmann, D. (1983). Impedance of the electrogenic Cl pump in *Acetabularia*: electrical frequency entrainments, voltage-sensitivity, and reaction kinetic interpretation. *Journal of Molecular Biology* **75**, 129–40.

Tromballa, H. W. (1974). Der Einfluss de pH-Werts auf Aufnahme und Abgahbe von Natrium durch *Chlorella. Planta* **117**, 339–48.

Tromballa, H. W. (1980). Electrogenicity of potassium transport in *Chlorella. Zeitschrift für Pflanzenphysiologie* **96**, 12–133.

Tromballa, H. W. (1981). The effect of glucose on potassium transport by *Chlorella fusca. Zeitschrift für Pflanzenphysiologie* **105**, 1–10.

Troxell, C. L., Scheffey, G. E. & Pickett-Heaps, J. D. (1985). Ionic currents are associated with primary cell wall expansion in the green algae desmids *Microsterias* and *Closterium. Journal of Cell Biology* **101**, 362a.

Tsuzuki, M., Miyachi, S. & Edwards, G. E. (1985). Localization of carbonic anhydrase in mesophyll cells of terrestrial plants in relation to CO_2 assimilation. *Plant and Cell Physiology* **26**, 881–91.

Vale, R. D., Schnapp, B. J., Reese, T. S. & Sheetz, M. P. (1985). Movement of organelles along actin filaments dissociated from the cytoplasm of the squid giant axon. *Cell* **40**, 449–54.

Verity, P. G. (1981). Effects of temperature, irradiance and daylength on the marine diatom *Leptocylindrus danicus* Cleve. II. Excretion. *Journal of Experimental Marine Biology and Ecology* **55**, 159–69.

Walker, N. A. (1980). The transport systems of charophyte and chlorophyte gaint algae and their integration into modes of behaviour. In Spanswick, R. A., Lucas, W. J. & Dainty, J. (eds.). *Membrane Transport in Plants: Current Conceptual Issues*, pp. 287–304. Amsterdam, Elsevier.

Walker, N. A. (1985). The carbon species taken in by *Chara*: a question of unstirred layers. In Lucas, W. J. & Berry, J. A. (eds.). *Inorganic Carbon Uptake by Aquatic Photosynthetic Organisms*, pp. 31–7. Rockville Md, American Society of Plant Physiologists.

Walker, N. A., Beilby, M. J. & Smith, F. A. (1979a). Amine uniport at the plasmalemma of charophyte cells. I. Current–voltage curves, saturation kinetics, and effects of unstirred layers. *Journal of Membrane Biology* **49**, 21–56.

Walker, N. A. & Smith, F. A. (1975). Intracellular pH in *Chara corallina* measured by DMO distribution. *Plant Science Letters* **4**, 125–32.

Walker, N. A. & Smith, F. A. (1977). Circulating electric currents between acid and alkaline zones associated with HCO_3^- assimilation in *Chara*. *Journal of Experimental Botany* **28**, 1190–206.

Walker, N. A., Smith, F. A. & Beilby, M. J. (1979b). Amine uniport at the plasmalemma of charophyte cells. II. Ratio of matter to charge transported and permeability of free base. *Journal of Membrane Biology* **49**, 283–96.

Walker, N. A., Smith, F. A. & Cathers, I. R. (1980). Bicarbonate assimilation by fresh-water charophytes and higher plants. I. Membrane transport of bicarbonate ions is not proven. *Journal of Membrane Biology* **57**, 51–8.

Wallentinus, I. (1984). Comparison of nutrient uptake rates for Baltic macroalgae with different thallus morphologies. *Marine Biology* **80**, 215–25.

Wheeler, W. N. (1980). Effect of boundary layer transport on fixation of carbon by the giant kelp *Macrocystis pyrifera*. *Marine Biology* **56**, 103–110.

Wiessne, W., Robinson, D. G. & Starr, R. C. (1984) (eds.). *Compartments in Algal Cells and their Interaction*. Berlin, Springer-Verlag.

Wild, A. & Breeze, V. G. (1981). Nutrient uptake in relation to growth. In Johnson, C. B. (eds.). *Physiological Processes Limiting Plant Productivity*, pp. 331–44. London, Butterworths.

Williams, R. J. P. (1982). Free manganese (II) and iron (II) cations can act as intracellular cell controls. *FEBS Letters* **140**, 3–10.

Williams, S. L. (1984). Uptake of sediment ammonium and translocation in a marine macroalga *Caulerpa cupressoides*. *Limnology and Oceanography* **29**, 374–9.

Williams, S. L. & Fisher, T. R. (1985). Kinetics of nitrogen-15 labelled ammonium uptake by *Caulerpa cupressoides* (Chlorophyta). *Journal of Phycology* **21**, 287–96.

Wilson, M. R. (1985). A study of urea transport and metabolism in *Chara australis*. Ph.D. Thesis, University of Sydney.

Yakote, M., Honjo, T. & Asakawa, M. (1985). Histochemical demonstration of a glycocolyx on the cell surface of *Heterosigma askashiwo*. *Marine Biology* **88**, 295–9.

Note added in proof: Since this chapter was written, two papers relevant to the transport properties of (non-algal) microbodies have appeared. These are: La barca, P., Wolff, D., Soto, V., Necochea, C. & Leighton, F. (1986). *Journal of Membrane Biology* **94**, 285–91, and Dauma, A. C., Veenhuis, M., Sulter, G. J. & Harder, W. (1987). *Archives of Microbiology* **94**, 285–91. Furthermore, data suggesting that active inorganic carbon accumulation occurs at the plastid envelope of *Chlamydomonas reinhardtii* has appeared: Moroney, J. V., Kitayama, M., Togasaki, R. K. and Tolbert, N. E. (1987) *Plant Physiology* **83**, 460–3. These papers are important collateral reading for Sections 5.6.3 and 5.6.5.

6 Some transport properties of cells within tissues

A. D. Tomos and R. G. Wyn Jones

6.1 Compartments in plants

Between the fully reductionist approach of traditional biochemistry and whole plant analysis lies an area of study that is dominated by the concept of compartments and vectorial events. These extend in scale from the subcellular 'micro-domains' on the surface or interiors of macromolecular structures, of which the active sites of enzymes may serve as examples, to macro-divisions concerning the major organs of plants. In between, and the subject of this chapter, are the cells themselves, their subcellular organelles and endomembrane systems, which give rise to the three major compartments that are characteristic of plants, namely the cell wall, the cytoplasm and the vacuole. These in turn lead to the apoplast, being the continuum of the cell walls of individual cells; the symplast, combining the cytosolic compartments; and the vacuoles (Münch, 1930; and see Ch. 7). Despite their confinement to single cells the properties of the vacuoles as a population are central to the understanding of how multicellular plants work (Leigh & Wyn Jones, 1986). Similarly we need to consider not only the chemical and physical characteristics of cell walls and cytosols (initially neglecting cytosolic organelles) but also the vectorial properties of the transport of water, ions and organic solutes across three-dimensional matrices made up from these components. In photosynthetic tissue the chloroplasts often dominate the volume of the cytosol. In this chapter, cytosol–chloroplast interaction will not be considered as they are the subject of Ch. 2.

6.2 Of uniformity and gradients

The majority of plant physiological and biochemical measurements are performed on bulked tissue material ('grind and find'!). Successful interpretation of such data depends on uniformity or smooth gradients of cell parameters across the tissues concerned. In interpreting any physiological or chemical data derived from a tissue or organ, the resolution of those data must be at least of the order of the scale of variation within that tissue. Tissue-averaged data may well provide solutions to some questions, but equally they may be totally misleading. A recent example of the latter from this laboratory involved the determination of wheat root turgor pressures (Jones *et al.*, 1987). A 'conventional' method involving measurements of osmolality of extracted tissue sap (e.g. used in Tomos *et al.*, 1984) suggested that treatment of the tissue with abscisic acid (ABA) had no effect on turgor pressure, since little or no change in osmolality was observed. Detailed measurement with the pressure probe (Hüsken *et al.*, 1978), on the other hand, revealed that dramatic changes in turgor pressures were occurring but that different regions of the tissue behave in diametrically opposite ways. These involved changes in osmotic potential that approximately cancelled out when measured with insufficient resolution (Fig. 6.1).

Within a group of cells, uniformity is generally assumed and each cell is taken to be behaving like its neighbours in the tissue or organ, or at least within a narrow Gaussian distribution of properties. In some aspects of multicellular physiology such as stomatal and secretory gland function, however, such an assumption cannot be made and it is not surprising that exciting new techniques of studying plant cells have been developed in these fields. The techniques of 'single cell biochemistry' developed by Lowry, Outlaw and colleagues (e.g. Lowry & Passoneau, 1972) will contribute enormously to our understanding as more laboratories acquire the equipment and patient expertise to apply them to a greater range of questions. The work of Bowling and colleagues (e.g. Bowling, 1972, 1973a, b; Penny & Bowling, 1974; Kelday & Bowling, 1980) has indicated the possibilities innate in the use of microelectrode technology for studying local gradients of various types within the tissues of higher plants. Similarly X-ray microanalysis (see Flowers & Läuchli, 1983) is beginning to be a powerful quantitative technique. Micro-immunological techniques will also contribute considerably in this field. Finally we believe that the use of the pressure probe (Hüsken *et al.*, 1978), when used in conjunction with other techniques, will provide valuable information on solute transport as well as water relations parameters at this cellular level of resolution.

6.2.1 Water relations – a model for uniformity?

While this book is particularly concerned with solute transport there are important reasons for drawing water relations into the discussion. Primarily this is because a major proportion of plant solute transport is dictated by its implication for water relations. Additionally, much can be learned of relevance to solute transport from concepts developed from the successful

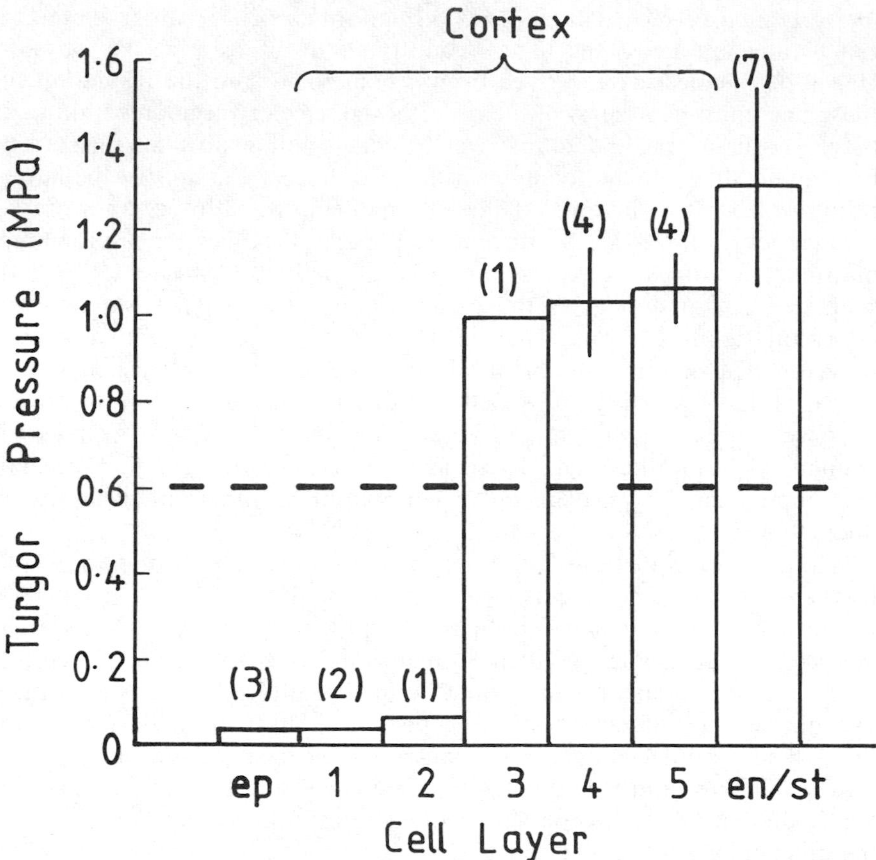

Fig. 6.1 Heterogeneity in cell properties is only evident when measured at sufficient resolution. Cell turgor pressures of wheat root cortical cells following treatment with 0.025 mol m^{-3} ABA show a step midway across the cortex. A method based on psychrometry of the expressed sap gave a tissue averaged value indistinguishable from the (uniform) pressure value of untreated roots (– – – – – –)(ep, epidermis; en/st, endodermis/stele). (Jones, H., 1986; PhD Thesis, University of Wales.)

application, since the early 1960s, of elementary thermodynamics to the water relations of plant tissues; much of this was in response to the work of Dainty (e.g. Dainty, 1963). This is not to imply that solute relations have not been dealt with as 'rigorously' as water relations but several features make the approach to water relations simpler than that for solutes. The most significant of these is that active transport of water across single membranes is unlikely. Where it may occur across composite membranes (Ginsburg, 1971; Ginsburg & Ginzburg, 1971), it can be readily described by relatively simple thermodynamic parameters. The second most important factor is that water permeates most (if not all) plant structures.

These two features together make the properties of water across a tissue uniform or close to it. This has allowed the successful application of non-equilibrium thermodynamics to plant water relations (unlike its classical cousin, such thermodynamics allow for small deviations from equilibrium: see Zimmermann & Steudle, 1978; Nobel, 1983).

The implications of data from solute relations experiments must therefore be consistent with realistic water relations parameters. This had not always been so. Indeed, in many cases, the water relations parameters (especially turgor pressure which can now be measured at single cell resolution in intact tissue with the pressure probe) may prove to be useful cross-checks for the validity of analytic data and even for theories concerning solute relations. Equally, all water relations models must also be consistent with demonstrable solute relations. The implications of these statements have not, as yet, been fully exploited.

6.2.2 Can solute relations of tissues be modelled?

By analogy with the water flow model of Molz & Ferrier (1982), Fig. 6.2 illustrates the interrelationship between the three compartments under consideration. Each of the pathways has its own specific physical prop-

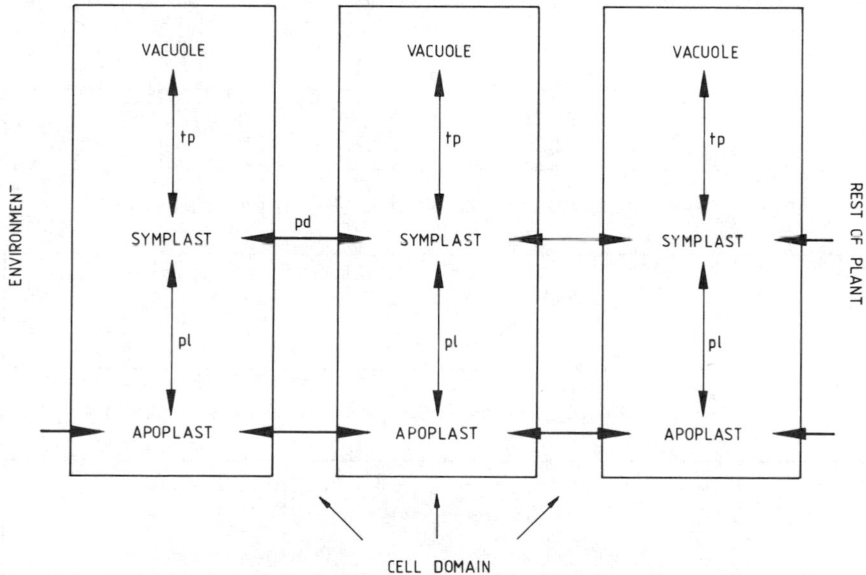

Fig. 6.2 Scheme of transport interrelations between tissue compartments. Thin arrows represent membrane transport, thick lines, that through apparent continua. Each solute, as well as water, will have a different driving force and resistance for each pathway. Heterogeneity of cell contents will be based on the individual cell domains within the continua (pl, plasmalemma; tp, tonoplast; pd, plasmodesmata).

erties. Fluxes of solutes and water along each will vary according to their individual driving forces and resistances and also may be influenced by each other. For example, bulk water flow through a pathway is highly likely to carry with it dissolved solutes. This mutual influence is a gross example of the Onsager cross-coefficients included in the non-equilibrium thermodynamic treatment of such systems (Zimmermann & Steudle, 1978; Nobel, 1983). Other more subtle non-zero values for these would include descriptions of coupled active transport across membranes (Wilson & Westerhoff, 1982; Westerhoff *et al.*, 1984).

6.2.3 Micropartition chromatography?

At what level of resolution do we require to measure the properties of a tissue behaving according to Fig. 6.2? If the pathways are uniform, flow across such a system would resemble the development of partition chromatography with the familiar partition between two phases, as in paper or ion exchange chromatography (one mobile and one stationary), replaced by one stationary phase (the vacuoles) and two mobile phases (the symplast and apoplast) each in turn partitioning solutes between them (Fig. 6.3). (Each cell domain of Fig. 6.2 would be expected to represent a single theoretical plate: see e.g. Marshall, 1978.) The apoplast, which would not be able to partition directly with the fixed vacuoles, has itself

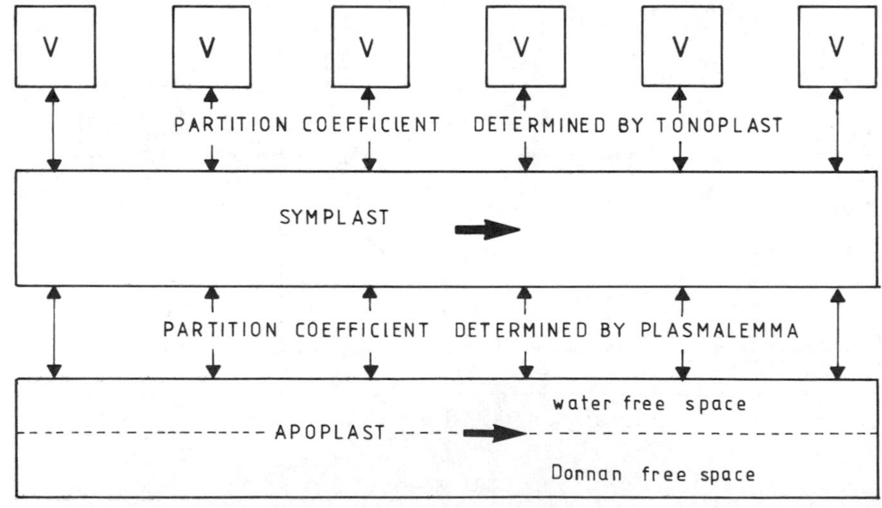

Fig. 6.3 Scheme of transport interrelations of uniform tissue based on a partition chromatography model. The vacuoles, V, represent a stationary phase, while apoplast and symplast are mobile phases. Partition between the phases will depend on membrane transport properties. The apoplast will also have many of the properties of an ion exchanger due to its Donnan free space.

the potential of forming two phases by virtue of its Donnan and water free spaces (Briggs *et al.*, 1961; Demarty *et al.*, 1980). The rate of passive diffusion and active pumping of solutes across membranes would correspond to the affinity partitioning of chromatography. However, here it must be remembered that the general membrane transport parameters are likely to be under dynamic biochemical and not purely passive physical control.

Movement of the two mobile phases (apoplast and symplast) ought not to be complex. In the apoplast movement would be dominated by bulk water flow driven by hydrostatic (transpiration) gradients. Osmotic flow will be minimal due to the assumed low reflection coefficient of the pathway (Steudle, 1985) and solute diffusion insignificant over intercellular distances (Nobel, 1983). In this context, the situation encountered by the widely studied root apoplast in aerated hydroponic culture, in which the dimensions of unstirred layers are greatly reduced, may under some circumstances be misleading. An example of the importance of unstirred layers is seen in considerations of iron uptake by maize roots (Römheld & Marschner, 1986). The short-term control of flow through this pathway will be dominated by the activity of the stomata. It will be modulated on an extended time basis by the resistances to flow as a result of cell wall modifications (e.g. Casparian strips, see Ch. 7) and pathway tortuosity. In non-transpiring tissue, such as storage roots, apoplasmic flow may prove to be very slow.

The flow rates of individual solutes in the apoplast will be influenced by the properties of the Donnan phase and may be retarded relative to the water flow. Due to charge repulsion by the fixed anionic charges of the pectin components in the wall, anions will tend to diffuse more slowly than cations through the Donnan space of the walls e.g. Cl^- has only 30 % of the mobility of Rb^+ through the sugar-beet apoplast (Aikman *et al.*, 1980). Conversely, due to chelating effects, Ca^{2+} will be bound strongly to (Demarty *et al.*, 1978) and move slowly through the same space. Since wall fixed charges are saturated by low Ca^{2+} concentrations, monovalent cations such as K^+ and Na^+ will normally be mobile in the apoplast, as will simple uncharged organic molecules. However, Richter & Ehwald (1983) have shown that the apoplast of sugar-beet tap-root has a rather high resistance to sucrose movement and question whether this is consistent with apoplasmic pathways of sucrose transport in this tissue.

Movement in the symplast is not so well characterized. Tyree (1970) proposed that diffusion alone would account for observed rates of solute flux through plasmodesmata and possibly even across the individual cytoplasms. It seems likely, however, that metabolically driven cyclosis might also maintain uniform concentrations in the individual cytoplasms, although Spanswick (1976) presents clear examples where cyclosis does not play a role. Discrimination by different flow rates between different solutes in the cytoplasms of individual cells is therefore unlikely.

For bulk flow through the plasmodesmata the reader is referred to Ch. 7. Here solute discrimination is theoretically possible. In mammalian cell gap-junctions (structures analogous to plasmodesmata), for example, Donnan charge repulsion appears to play a role in preventing intercellular passage of anions (Flagg-Newton *et al.*, 1979). In plants, however, charge appears to have no effect on qualitative transfer of solutes; at least through the plasmodesmata of nectary trichomes of *Abutilon thomsonii* (Terry & Robards, unpublished) whereas polarity may play a role. Some dynamic control may also occur at this point. Ca^{2+}, Mg^{2+} and Sr^{2+} have been shown to inhibit reversibly intraplasmodesmatal transport in *Elodea* (*Egeria*) *densa* (Erwee & Goodwin, 1983). Gunning & Overall (1983) point out that under conditions in which diffusive gradients and bulk flow are in opposite directions, bidirectional flow through individual pores would be expected to occur. Another factor about the passage of solutes through plasmodesmata is worth considering. The enormous surface to volume ratio at this point will greatly facilitate non-specific diffusive 'leakage' of solutes into the apoplast. It may be that this does not occur due to the plasmodesmatal membrane, or the cell wall immediately adjacent to it, being unusually impermeable. Linked with a solute-retrieval system, however, as proposed for sucrose and glucose by Maynard & Lucas (1982) in the preceding cell, this would act as a system to select solutes for symplasmic movement.

In our consideration of tissue uniformity, however, we require to know not only what controls the pathways described in Fig. 6.2, but whether the control mechanisms operate uniformly across the tissue. In some cases this can be shown clearly *not* to be the case. Where it is modified by the Casparian strip or other hydrophobic encrustrations (in roots, leaves, etc.), the apoplast has a very high resistance to hydrophilic substances. Similarly, plasmodesmatal connections are not universal. In *Commelina cyanea* leaves, the cells of the mesophyll (both spongy and palisade), epidermis (with the notable exception of stomatal guard cells) and even vascular tissue, all appear to be interconnected by open plasmodesmata (Erwee *et al.*, 1985). On the other hand, in *Elodea* (*Egeria*) *densa* distinct domains of continuous symplast are separated by visible, but apparently closed, plasmodesmata (Erwee & Goodwin, 1985). The configurations available due to tissue architecture are evidently numerous.

What, however, of the majority of uniform tissues (e.g. parenchyma) that appear at least to have no specific architectural features that create inhomogeneity in the pathways of Figs. 6.2 and 6.3? What data do we have to test the functional uniformity of such tissues, either in the form of solute concentration gradients consistent with the application of the 'chromatographic' scenario detailed above, or in the form of diffusion gradients in each of the 'open' continua?

6.2.3.1 Symplasmic uniformity

Elsewhere arguments have been presented for a large degree of cytoplasmic ionic homeostasis on a whole tissue basis (Wyn Jones *et al.*, 1979; Leigh & Wyn Jones, 1986). There is evidence that mechanisms exist for controlling the cytoplasmic concentrations of H^+, Na^+ and Cl^- (Wyn Jones *et al.*, 1979; Flowers & Läuchli, 1983; Wyn Jones & Pollard, 1983) and P_i (Lee & Ratcliffe, 1983) within relatively narrow ranges when studied at cell level. It has been argued that this is necessary for the biochemical processes of the compartment (Wyn Jones & Pollard, 1983). K^+ homeostasis has been singled out as being particularly significant for a number of processes including enzyme activation (RNA translation), stabilization of native enzyme (and possibly membrane) conformation, cytoplasmic volume regulation, transmembrane energy conservation and pH regulation. Mg^{2+} levels are probably also tightly controlled as the ion is an important cofactor, e.g. MgATP. The free cytoplasmic Ca^{2+} levels are low, a characteristic which may be related to the role of this ion as a second messenger as in animal cells (Hanson, 1984). The hypothesis of cytoplasmic ionic homeostasis would suggest that the symplast should be uniform across tissues, and indeed across whole plants. This appears not to be consistent with the implied diffusion gradients that drive solute flow across this compartment. However, in his rigorous theoretical treatment of the symplasmic gradients required to drive KCl flux across onion root cortices, Tyree (1970) proposed that gradients of the order of 1 mol m^{-3} would be sufficient. Clearly such a gradient against a background of from 100 to 150 mol m^{-3} K^+ (a reasonable estimate for cytosolic K^+ concentrations) is unlikely to be important.

Testing this, however, is technically very difficult. Owing to its generally small dimensions, the symplast is the most difficult compartment to analyse and reliable data are limited. The variation in X-ray microprobe analytical values for the major ions in the cytoplasms of *Atriplex spongiosa* (grown in 400 mol m^{-3} NaCl) expressed as a standard deviation is of the order of ± 54 % for K^+ in bundle sheath cells (Storey *et al.*, 1983a) and ± 49 % for K^+ in root tips (Storey *et al.*, 1983b). (These authors do not note the spatial distribution of their data.) Given the technical problems, these data are insufficient to determine uniformity which can be neither excluded nor concluded. On the other hand, Harvey and Thorpe (1986) report a range of from 32 to 3502 mol m^{-3} in the cytoplasms of wheat mesophyll cells and comment specifically on the absence of any spatial relationship for this range. They had previously shown similar data for maize roots (Harvey, 1984). These data strongly imply a total absence of uniformity (and hence homeostasis) at the cellular level. It is worth noting, however, that such enormous ranges would have significant implications with regards to the cellular water relations (especially turgor pressure), and some doubt must be cast on their validity. The authors do themselves point out that the higher concentrations reported are too high to be compatible with biochemical activity.

It must be said, therefore, that symplasmic profiles remain to be described either within an apparent continuum, or across symplasmic domain boundaries such as the *Elodea* epidermis/mesophyll boundary, where concentration steps might be expected to occur.

A feature of cytoplasmic uniformity that has not to our knowledge been considered, concerns the issue of cytoplasmic volume control. The differential gradient of indiffusible, predominantly anionic, protein across the tonoplast will result in the tendency of small molecular weight ions to approach an unequal distribution across that membrane, a Donnan equilibrium (Briggs *et al.*, 1961; and see Ch. 1). If this were allowed to occur, water would be drawn into the cytoplasm from the vacuole. That this does not occur is due to constant active pumping of the diffusible solutes back into the vacuole. Conversely, under condition of water stress, organic non-toxic 'compatible solutes' (e.g. glycinebetaine, proline, sorbitol, etc.) are accumulated in the cytoplasm to prevent its desiccation (see Wyn Jones *et al.*, 1977a, b; Wyn Jones, 1984). This would suggest a degree of cytoplasmic volume regulation and the integration of transport process of the tonoplast and plasmalemma. It may be argued (R. A. Leigh, personal communication) that cytoplasmic volume regulation is the underlying feature of a range of compartmentation phenomena. With regard to a tissue, therefore, it might be pertinent to ask whether this feature is expressed as regulation at the single cell level, or is the 'symplasmic volume' as a whole regulated in some way?

6.2.3.2 Vacuolar uniformity

If symplasmic conditions were to prove to be uniform (or follow shallow gradients), the 'chromatographic' approach would suggest similar behaviour across the 'stationary' vacuolar phase. After all, *all* vacuolar solutes presumably originate in the cytoplasm. Solutes with very high 'partition coefficients' with respect to the vacuolar lumen would accumulate in vacuoles at the entry points to the symplasm (e.g. endodermis).

Vacuolar contents are relatively easier to quantify than those of the cytoplasm. Yet here again, there is no consensus of opinion. Microsampling of vacuolar sap (Kelday & Bowling, 1980) and microelectrode measurements (Bowling, 1972; Dunlop, 1973) have demonstrated an absence of K^+ and Cl^- gradients across the roots of *Commelina communis*, maize, sunflower and rye grass. More recently, however, van Steveninck *et al.* (1982a, b, c) using X-ray microanalysis have shown a consistent decrease of Cl^- content and an increase in phosphorus and sulphur in a centripetal direction in roots of *Lupinus luteus* grown under saline conditions. In contrast, these workers also report *random* tangential cell to cell inhomogeneity as described briefly earlier (van Stevenink *et al.*, 1980). Due to the random microheterogeneity van Steveninck *et al.* (1982a) conclude that vacuolar content does not reflect symplasmic content. As for their data for the cytoplasm, the data presented by Harvey & Thorpe (1986) indicate large heterogeneity in wheat leaf mesophyll vacuoles. If we were to assume a uniform symplasmic composition these findings indicate that individual vacuoles must be behaving individually and differently against such a uniform background.

How, and indeed whether, such gradients are controlled at a cellular level is relevant to future developments in genetic engineering in plants. Optimal transport proteins, for example, may well be present in certain cases, but their distribution may be suboptimal to meet particular environmental requirements. An example would be the control of Na^+ or Cl^- transport in halophytes (see Ch. 10). Microheterogeneity or not, the scatter in the data of Storey *et al.* (1983a, b) for *Atriplex* roots and leaves is no greater than that for the tentatively 'uniform' cytoplasm.

Under these circumstances, how could the plant control a population of 'independently-minded' vacuoles? Several parameters could integrate the behaviour of compartments that are not physically contiguous such as the vacuoles of a tissue. One important mechanism is likely to involve plant growth substances. A classical example of this is the Cholodny–Went theory of light-induced phototropism (see critiques in Digby & Firn, 1976; Trewavas, 1981) in which it is proposed that the light stimulus sets up a gradient of growth substance concentration across the organ. This in turn generates a gradient of differential growth responses resulting in the bending of the tissue. Other workers (e.g. Trewavas, 1981) would suggest that gradients of *sensitivity* across tissue are responsible for such behaviour. The controversy is in several ways central to the discussion of the basis of uniformity within tissues, as it distinguishes between endogenous and exogenous influences on the functioning of individual cells. Microheterogeneity of vacuolar contents despite a uniform apoplast or symplast would, for example, favour the endogenous interpretation of the control of function. The reader is referred to Trewavas and Cleland (1983) for an outline of the argument.

A second, less extensively studied, factor that would orchestrate isolated compartments is turgor pressure. In general, turgor pressure appears to be maintained throughout changes in water potential of the tissue. (Although it is not as yet possible to distinguish between turgor or volume as the crucial parameter as they are dependent variables.) This is observed, for example, in the case of *Suaeda maritima* (Clipson *et al.*, 1985) where constancy of turgor is accompanied by changes in osmotic potential equivalent to the changes in external osmotic (water) potential. Such a process has been observed in many tissues during salt and water stress. Recently, a study with excised beet tap-root tissue (Perry *et al.*, 1987) has observed the phenomenon *in vitro*. In this study, excised tissue was bathed in a series of mannitol solutions of varying strengths. Bulk changes in osmotic

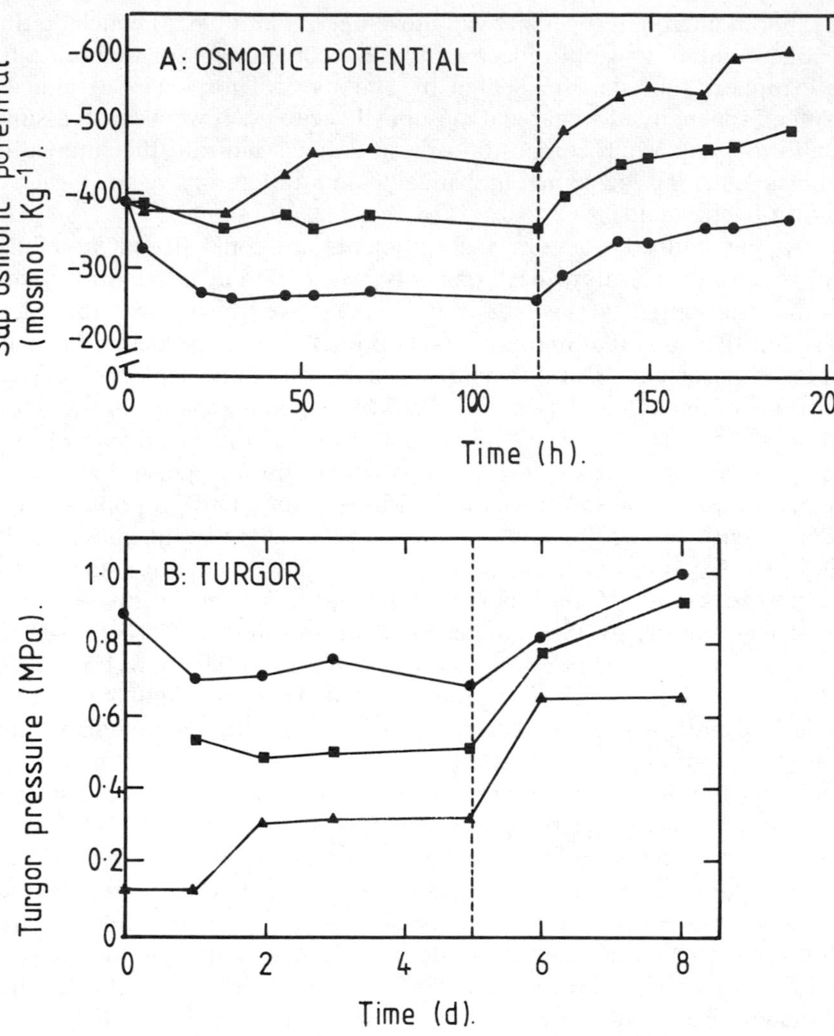

Fig. 6.4 Osmotic and turgor pressure adjustment of excised red beet tap-root tissue bathed in three concentrations of mannitol (0, 200 and 400 mol m^{-3}. ●, ■, and Δ, respectively) ('open system' cf. Fig. 6.5). At day 5 NaCl and KCl were added to the solutions. Osmotic potentials were measured psychrometrically, turgor pressures with the pressure probe. (Perry *et al.*, 1987.)

potential were observed that indicate a significant degree of turgor regulation (Fig. 6.4). Furthermore, the mid-point of the stable turgor pressures reached correspond to the turgor pressure observed in the intact plant (Leigh & Tomos, 1983).

If salt solutions are subsequently added to the bathing medium the maintenance of turgor is lost. The salts are taken up and turgor rises. The significance of this in terms of turgor pressure control is as yet not understood. Interestingly, this absorption of salt is accompanied by a loss of sugars that were maintained during the constant turgor pressure phase. It is as if the residual sugars, used as osmotica to maintain turgor, cannot be fully mobilized and exported from the cell until *alternative* solutes are available. Similar linking of osmotic solutes have been observed for salts and sugars in barley (Pitman *et al.*, 1971), for NO_3^- and sugars in ryegrass (Veen & Kleindorst, 1985), and for K^+ and sugars in carrot (Steingröver, 1983).

Under such circumstances it could be envisaged that the availability of alternative solute determines the ease of availability of stored nutrient that occur in osmotically significant quantities (R. A. Leigh, personal communication). Another example of this is the nature of the osmotica used by growing root cortical cells. In *excised* tissue salts are absorbed from the medium to maintain turgor-generating osmotic potentials. In intact plants, however, sugars from the shoot are used despite the ready availability of the energetically less demanding supply of salts from the medium (Pitman *et al.*, 1971; Jones *et al.*, 1987).

Such selection of vacuolar osmotica would be expected to operate largely at the level of the individual cell. Homogeneity or not of vacuolar content, therefore, would once again be a function of the response to turgor pressure changes of the transport systems involved.

6.2.3.3 Apoplast uniformity

Here we have almost no data available. Heterogeneity normal to and parallel with the plasmalemma are both possible. Walls are multilaminate structures that reflect cell differentiation (Northcote, 1985). This factor has been considered extensively at the chemical level extending to its association with gene expression (Jones & Northcote, 1981). At a more subtle, but equally important, physiological level few data have been presented, once again because of the formidable technical problems involved. Distribution of Ca^{2+} throughout the apoplast is especially important as it will influence the movement of both anions and cations (Demarty *et al.*, 1980). The behaviour of the internal cell wall surface as a 'partition interface' adjacent to the plasmalemma will certainly influence the ionic environment of the membrane as a result of apoplast diffusion potentials, local pH and concentrations of free and ionized fixed groups (Sentenac & Grignon, 1981).

The distribution of unmethylated pectin and other charged polymers across the apoplast might be an example of a molecule expected to generate heterogeneity in the physiological properties of the apoplast. Staining techniques for the investigation of this are available (e.g. ruthenium red). Preliminary data by Saftner and Wyse (personal communication) suggest that systematic gradients in charge density may indeed occur in sugar-beet tap-root tissue. Significantly in this case, areas rich in material staining with ruthenium red also appeared rich in K^+. This is consistent with charge balance, but is very unexpected in the context of Ca^{2+}/K^+ exchange and implies cation binding sites highly specific for K^+ or otherwise the exclusion of Ca^{2+} from these areas in the wall. The observations of Demarty *et al.* (1978) would suggest that the fixed wall anionic groups would be saturated with available Ca^{2+}. A study of the distribution of this ion in walls may well be overdue in this respect. The involvement of Ca^{2+} in wall loosening (Cleland & Rayle, 1977) may also arise from ion exchange requirements. It has been shown, however, that wall loosening by simple modulation of Ca^{2+} cross-links is not the role of Ca^{2+} in this respect (Cleland & Rayle, 1977).

Evidence that cell wall physical properties can play an active role in physiological processes is provided by the dramatic contractile properties of some fungal spore cell walls (see Ellar, 1978) where the cortex of the wall has been proposed as an osmoregulatory organ directly or as a structure that expands hydraulically into the protoplast to dehydrate it by reverse osmosis. Such a major role for the wall has yet to be shown for a higher plant.

The solute content of the apoplast is also likely to play a role in the behaviour of membrane solute transport involving the turgor regulation of cells in a tissue. Osmotic adjustment leading to turgor maintenance need not involve only the osmotic potential of the protoplast. Cram (1976) pointed out that the flux of solutes across the plasmalemma (e.g. from protoplast to cell wall) will result in a much larger relative change in the *wall* osmotic potential (hence water potential) than in that of the protoplast. This is because of the very much smaller relative volume of the wall. Since turgor pressure is governed by the *differential* osmotic potential across the plasmalemma, this process could operate as an amplifying system for modulating turgor pressure. Note that this process allows modulation of turgor pressure with no change in the mean osmotic potential.

Behaviour consistent with this hypothesis has been observed for excised red beet tap-root tissue (Tomos, unpublished). When immersed in silicone oil, excised beet discs initially have a low turgor pressure when measured with the pressure probe (Fig. 6.5). This is due to the cell sap released from the broken cells flooding the extracellular spaces. However, the turgor immediately begins to rise reaching a plateau after about 2 h. This preparation, with its low volume of 'apoplast' relative to protoplast resembles the *in vivo* situation far more than does one in which excised tissue is bathed in a large external aqueous volume (e.g. Perry *et al.*, 1987) which also undergoes turgor regulation, but over a much longer time scale (Fig. 6.4).

In this context, however, it must be mentioned that when turgor pressure is modified *in vivo* by promoting transpiration, stable turgor press-

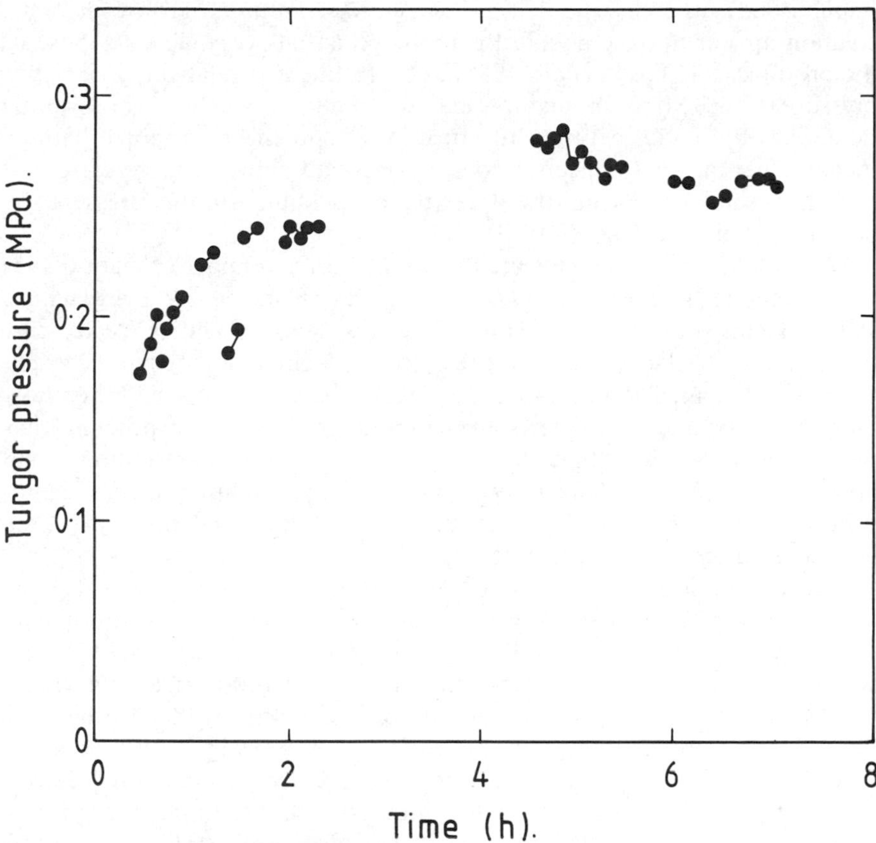

Fig. 6.5 Turgor pressure adjustment of excised red beet tap-root tissue bathed in silicone oil to reduce extracellular aqueous volume ('closed system' cf. Fig. 6.4). The turgor pressure increased much more rapidly than in the equivalent experiment illustrated in Fig. 6.4. This behaviour would be expected if a wall-protoplast shunt was operating (see text). Points linked represent measurments on a single cell.

ures established after 30–60 min are *not* adjusted back to any 'pre-set' level as the result of this latter process (Fig. 6.6; Palta, Wyn Jones & Tomos, 1987).

The operation of a wall–protoplast shunt depends on the presence of a significant osmotic potential in the cell wall. It appears that wall solute composition varies greatly between species. One rapid way of measuring this is to compare the turgor pressure of cells under non-transpiring conditions (where apoplast hydrostatic pressure should rise to atmospheric) with the osmotic potential of the cell protoplast (determined either by means of expressed cell sap or by measurement of the turgor pressure of cells in tissue infiltrated with solutions of known osmotic potential).

Using these criteria, the leaves of hydroponically grown wheat seedling have only a small osmotic component in the leaf cell apoplasts (Tomos, unpublished) whereas the leaves of *Suaeda maritima* grown in NaCl solution appear to have apoplast osmotic potentials very close to those of the protoplasts (Clipson *et al.*, 1985). This results in a relatively low turgor pressure (<0.1 MPa) in mature leaves despite a much larger osmotic potential differential between the protoplast and the hydroponic bathing media. Beet tap-root appears to be intermediate between these extremes, the wall solute reducing the 'maximum' possible turgor pressure by 30–50 % (Leigh & Tomos, 1983).

Wall solutes have been indicated in an increasing number of plant tissues using other techniques, e.g. sugar-beet leaves (Kursanov & Brovchenko, 1970), stems of *Pisum sativum, Cucumis sativus* and *Glycine max* (Cosgrove & Cleland, 1983) and developing legume seed coats (Patrick, 1984; Gifford & Thorne, 1985; Wolswinkel *et al.*, 1986 and references therein), a significant and in some cases large wall solute potential has indeed been detected. It is in such systems that the contribution of an apoplast not freely exchanging with an external medium becomes significant to turgor regulation. In other systems, e.g. wheat seedling leaves, the wall solute potential is negligible.

In plants such as *Suaeda* this large osmotic potential in the wall illustrates that some plants adjust osmotically to maintain a *lower* turgor than is maximally possible. This is the converse of the situation in water stress, where maintaining turgor of any kind could be a problem for the plant. In *Beta* tap-root a critical role for turgor reduction by this mechanisms might be that storage to high concentration of osmotically active molecules (sucrose) and lead to turgor pressures that are too high for the strength of the cell wall. This is put in context when it is realized that turgor pressures up to about 1 MPa are typical for plant cells. This pressure is sustained by a 1 μm thick wall. (The tyre of a motor car uses much more material to sustain pressures of only 0.15 MPa.) Cell walls are robust, therefore, but they have their limits. Salt-acclimatized *Suaeda maritima* cells burst when their pressures are artificially increased towards the value dictated by their internal osmotic potential (about −3 MPa); this is achieved by infiltrating leaves with distilled water (Tomos & Clipson, unpublished). By maintaining a significant wall solute potential this problem can be alleviated.

This role of wall solutes as a regulator in turgor adjustment is a feature only of cells within tissues. It provides the possibility of each cell *individually* monitoring constant turgor despite different osmotic potentials. This might explain the conflict between the observation described above, of adjacent cells with different solute contents, and a presumed continuity of symplast through the plasmodesmata, which presumably must result in uniform turgor pressure across the symplast and vacuoles. Figure 6.7 illustrates this uniformity of turgor pressure between adjacent epidermal cells around stomatal complexes of *Tradescantia virginiana* measured with the pressure probe (Tomos & Zimmermann, unpublished).

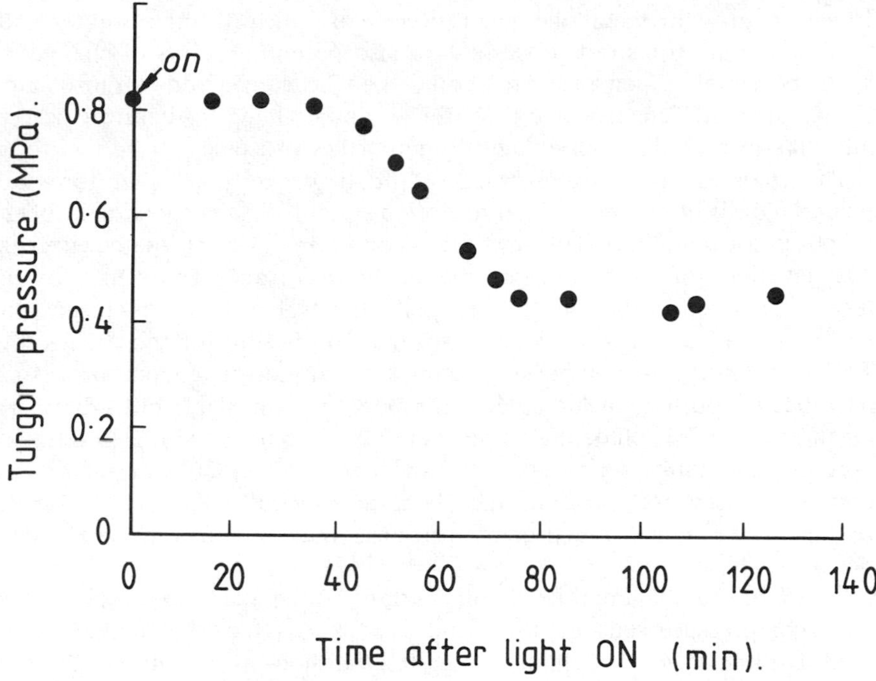

Fig. 6.6 Turgor pressure changes in intact sugar-beet tap-root cells induced by increasing transpiration. After about 80 min a steady state is reached. The absence of turgor adjustment is in contrast with what would be expected from Fig. 6.5. (Palta, Wyn Jones & Tomos, 1987.)

6.3 Turgor pressure and solute transport processes

Since water is present as an unbroken continuum through the entire plant body, changes in water potential will propagate through the plant, at speeds and dissipation rates dependent mainly on the pathway of propagation. For example, it is likely that movement of a hydrostatic gradient of changing water potential along the bulk, uninterrupted aqueous phase of xylem vessels will be at the speed of sound (i.e. 'instantaneous'), while passage through tissues will be slower. The *slowest* case is likely to be via the cell membrane in series as described by Philip (1958). This pathway has been investigated for leaves of *Tradescantia* (Tomos *et al.*, 1981), where it is of relatively low resistance. Changes in water potential would result in rapid turgor pressure changes in the cells they involve as the equilibration time of water potential across each membrane is measured in seconds (Hüsken *et al.*, 1978).

The initial change in water potential can be generated in several ways, the most dramatic being a change in the environmental water potential.

However, growth, metabolic generation of osmotically active solutes and transmembrane transport processes are also potential points of change in water potential. (Such processes could then be considered 'turgor-genic' by analogy with 'electrogenic'). Water potential changes, therefore, carry information regarding physiologically important processes.

The changes in turgor pressure of the tissue cells are therefore the potential basis of an information reception system [this has previously been proposed for phloem by Huisinga (1979) and Lang (1983)], the occurrence and sensitivity of which will depend on the occurrence and sensitivity of turgor-dependent processes in the cells themselves. In this respect it represents a situation analogous to that proposed for hormones by Trewavas (1981). An increasing number of turgor-dependent processes have been reported, including K^+ transport in giant algal cells (Zimmermann *et al.*, 1976), phloem loading (Smith & Milburn, 1980), solute transport in leguminous seed coats (Wolswinkel *et al.*, 1986 and references therein) and sucrose uptake in sugar-beet tap-root cells (Wyse *et al.*, 1986). In this last example the primary response may be on a plasmalemma proton ATPase.

At a more fundamental level, very little is known of the molecular basis of turgor pressure sensing. Benz and coworkers (Benz & Zimmermann, 1983; Büchner *et al.*, 1985) have suggested that non-protein mobile charges associated with proteins within *both* the plasmalemma and tonoplast may play a role. This work is currently restricted to giant algal cells, the link between such processes and solute transport in higher plants remains to be demonstrated.

Turgor regulation has been well studied at the gross physiological level in the field of water and solute stress (e.g. Clipson *et al.*, 1985; and see Ch. 10), and this work is being extended to excised tissues (Perry *et al.*, 1987). The initial signal in this process could either be a response to turgor pressure changes, cell volume changes or to osmotic potential changes. Experiments with penetrant solutes tend towards the conclusion that these processes respond to changes either in turgor or volume rather than changes in osmotic potential. Since it is technically extremely difficult to distinguish between turgor or volume control at present we shall probably have to wait for the identification of the receptors and then determine whether turgor or volume is the crucial parameter.

We have here, however, an apparent dilemma. How can turgor pressure changes convey to a cell information if the same cell possesses the machinery of turgor regulation? The answer to this question is beyond the scope of this current chapter, and much more information is required regarding the various feedback loops involved before realistic models can be proposed.

6.3.1 Self-consistency of data

The interdependence of solute and water relations gives rise to a number of useful relationships that provide valuable potential checks on the validity of chemical, physiological and biophysical data on both unicellular and multicellular plants. With the continuing diversification of approaches of studying plant transport and solute relations it is important to take advantage of these checks, and of others such as the necessity of charge balance in any cellular (or tissue) compartment. (The charge imbalance required to maintain physiologically relevant potential differences across membranes is negligible in this context.)

Two aspects of water–solute relations are applicable here. First, proposed solute concentrations for each compartment must be consistent with the measured osmotic potentials of the compartments. Secondly, the difference between the proposed protoplast osmotic potential and wall water potential must be consistent with measured turgor pressures.

In some experimental material, such as roots grown under hydroponic conditions or excised and washed storage tissue discs, the immediate environment of each protoplast is in sufficient close contact with the bathing medium to be at, or very close to, water potential equilibrium with it. As hydrostatic pressure in the bathing medium will always be at atmospheric pressure, the water potential of this medium will be equivalent to its osmotic potential and hence easily quantified. The influence of surface–water interactions (i.e. true matrix effects) and Donnan equilibria set up *within* the cell wall will be restricted to the interior of the wall, i.e. the protoplast and bathing medium will be at thermodynamic equilibrium regardless of the properties of the intervening wall.

In this context it can initially be assumed that the solute–membrane reflection coefficients that relate osmotic and hydrostatic components of molecules across membranes (see Zimmermann & Steudle, 1978) have a value approaching unity, an assumption that appears to be justified for most osmotically important solutes in tissues, e.g. organic solutes (Tyerman & Steudle, 1982) and K^+ (Hastings & Gutknecht, 1978). Methods of assessing values for these constants are available (Tyerman & Steudle, 1982).

Cell turgor pressure (P), water potential (ψ) and osmotic potential (Ψ_π) are related by the equation:

$$\Psi = P + \Psi_\pi^i \tag{6.1}$$

thus independent determination of P by the pressure probe and intra- (Ψ_π^i) and extra-cellular (Ψ_π^o) osmotic potentials by a freezing point depression or equilibrium vapour pressure osmometer *must* produce mutually consistent results.

Since the commonly used estimate of (Ψ_π^i) as being similar to the osmotic potential of disrupted tissue cannot avoid the dilution of the protoplasmic

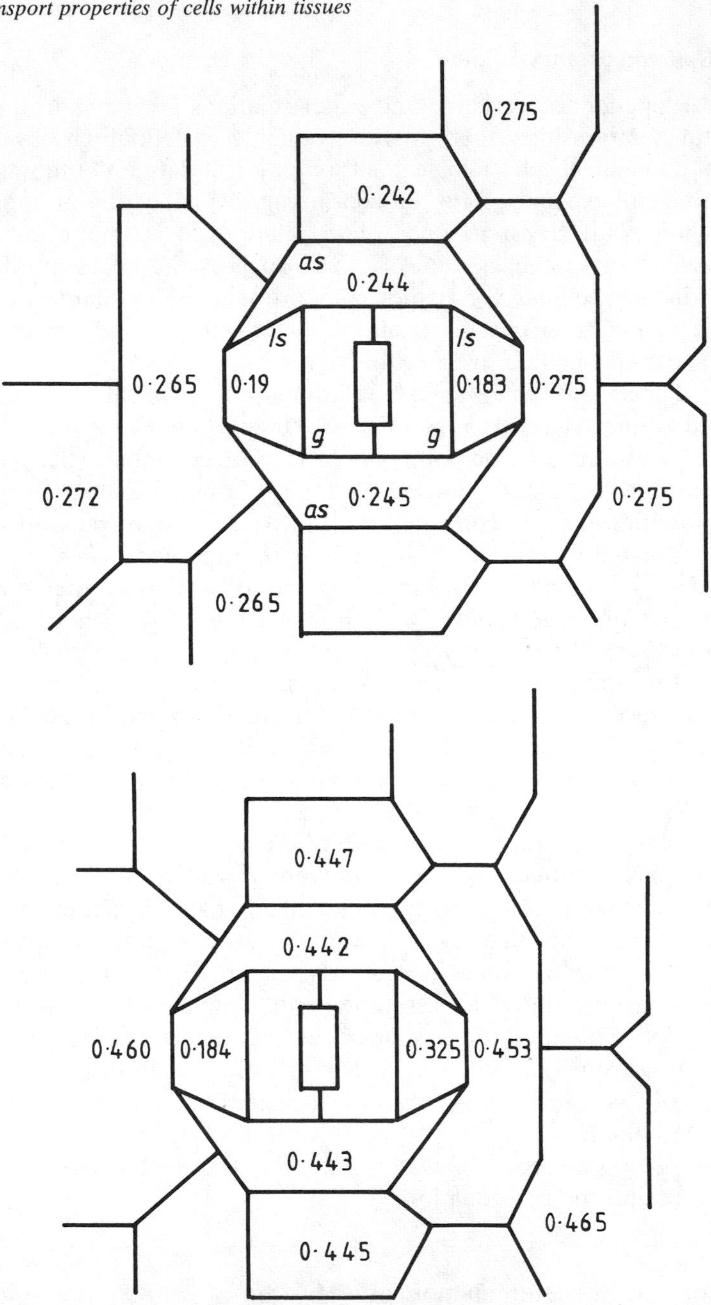

Fig. 6.7 Tissue turgor pressure uniformity is illustrated by pressure probe measurements of individual cells surrounding the stomatal complex of *Tradescantia virginiana*. Two different plants are illustrated (pressures in MPa.) Epidermal cell turgor pressures are indistinguishable within experimental error. Apical subsidiary cells (as) have slightly, and lateral subsidiary cells (ls) have considerably lower pressures than the epidermal cell (Tomos & Zimmermann, unpublished).

sap by the wall volume the accurate determination of the osmotic discrepancy allows an estimation of the wall volume (Tomos *et al.*, 1984) which in turn can be cross-related to an independent measurement of free space by non-permanent solutes. A major discrepancy between cell turgor (pressure probe) and the anticipated turgor based simply on (Ψ_{π}^{i}) alone in other tissues indicates a significant decrease in wall water potential either because of solute accumulation or transpirationally induced tension.

An indication of the relative contributions of solute and hydrostatic potentials can then be obtained either by eluting the solute from the walls (Cosgrove & Cleland, 1983; Leigh & Tomos, 1983) or by stopping transpiration in some way and observing the increase in turgor pressure on the hydrostatic tension in the wall due to transpiration decays.

The realization that wall solutes are crucial to interpretation of pressure bomb results was underlined by Scholander *et al.* (1964), but is sometimes ignored, and may be the basis of some errors in the reported descriptions of tissue water relations (Cosgrove & Cleland, 1983).

Cross-relating different analytical approaches becomes an even more powerful tool if chemical (solute) composition is related to osmotic potential as well as turgor. In any given tissue it should be possible to account for the measured osmotic potential from the sum of the osmotic concentrations of the individual organic and inorganic solutes. For the total osmolality to be computed, the osmotic coefficients and degree of dissociation of the individual solutes should be known. No information on the exact osmotic coefficients applicable within cells is available but since the water-filled vacuole accounts for some 90–95 % of the sap, standard chemical potential tables (e.g. Robinson & Stokes, 1959) can be used with some confidence for this compartment. It is interesting that some biologically important solutes (e.g. potassium malate and malic acid) have not been studied by the physical chemists and knowledge of their basic physical chemistry awaited the plant biologist (Smith & Lüttge, 1985; see Ch. 11).

Even where a complete balance is not available an approximate assessment of the contributions of individual solutes to the osmotic potential of tissues can guard against gross error (cf. Shephard & Bowling, 1979) and the practice of recording solute concentration only on a per unit dry weight basis should be discouraged (Leigh & Johnston, 1983). The same fundamental physical parameters should also be considered at the cellular and subcellular level, but unfortunately the lack of reliable analytical data on higher plants is pervasive as emphasized above. Cytoplasm and vacuole must have identical water potentials, as only a negligible pressure differential can be set up across the tonoplast. This necessitates equality of osmotic potentials.

Data derived from X-ray microprobe analysis are difficult to analyse in this context, however, as ions bound to these Donnan phases in which physical stresses are set up within a structure (e.g. mutual repulsion of linked fixed charges in a protein or polysaccharide chain) will not contribute to the bulk water potential as the stresses will balance the effect of some of the charged solute. The consignment of an osmotic role to elements seen by the X-ray microprobe analysis are far from straightforward, hence the importance of cross-correlating the various techniques. In animal systems these problems, however, seem to be small (Hoffmann, 1977).

Charge balance within the narrow limits set by the membrane potentials must also be achieved in both compartments. The extensive data available (e.g. of Kirkby & Knight, 1977) showing the role of organic acids and pectic acids in charge balance refer to the vacuolar and apoplast compartments, respectively. No data exist on charge balance in the cytosol (symplasm) of higher plants and the best estimates can only be derived from data obtained from animal cells. These indicate that the net anion charges on proteins and of phosphate residues (HPO_4^{2-} and $H_2PO_4^-$), (both inorganic and organic) are the main contributors to the anionic charge while K^+ is the dominant cation (Wyn Jones & Pollard, 1983).

6.3.1.1 Air spaces in tissues

Air spaces have not been considered in this discussion. This could possibly be a mistake as gradients of gaseous compounds, especially oxygen and CO_2, across tissues may play a role in their integration. In this context, Byott (1976) has suggested a significance in the difference in extent of air space in C_3 and C_4 plants, although Bowling (1973a) showed only relatively small gradients of oxygen tension across the cortices of intact maize root (see Ch. 7). Similar observations have been reported for potato tubers (Burton, 1950). The observation of Burström (1959), that root intercellular spaces appear to be full of pure carbon dioxide, also suggests little possibility of gradients.

6.4 Uptake kinetics and the compartmentation model

While information regarding solute–membrane transport is scarce at the level of the single cell, much information regarding the behaviour of entire tissues is available. However, the interpretation of such data has proved an obstacle.

In an equivalent chapter in a previous volume, Wyn Jones (1975) dealt with ion uptake by excised roots and dwelt at some length on the vexed question of the interpretation of the kinetic parameters (referred to as 'isotherms') derived from the concentration dependence of the ion and solute uptake into roots and other tissues. The original arguments, represented primarily by the views of Epstein and co-workers (Epstein *et al.* 1963; Epstein, 1976) and Laties (1969), revolved around the interpretation of two major phases of uptake, one in the low (10^{-3}–1 mol m^{-3}) and the other in the higher (1–100 mol m^{-3}) external concentration ranges. Borstlap (1983) has suggested that many of the published isotherms are best described by the sum of two Michaelis–Menten terms and a linear component (see Ch. 1). A more complex multiphasic interpretation has been strongly and consistently advocated by Nissen (1974) and Nissen & Nissen (1983).

It is not our intention in this chapter to reopen these discussions but to underline certain conceptual changes that have taken place in the last 10 years, particularly those arising from the solute compartmentation model (Leigh & Wyn Jones, 1984). In all the interpretations the isotherms in the low concentration range are ascribed to a carrier (porter) in the plasmalemma with a high affinity for the transported solute. Laties (1969) and Epstein and co-workers (Epstein *et al.*, 1963; Epstein 1976), however, differed in the cytological location of the hypothetical low-affinity carrier. The former favoured the tonoplast and the latter the plasmalemma. Laties (1969) specifically saw the cytoplasm being entirely filled with external K^+ via the high-affinity carrier before a low-affinity carrier on the tonoplast could be detected. Clearly this concept has been rendered obsolete by evidence for the preferential, homeostatic regulation of a high cytoplasmic K^+ level at all times (e.g. Pitman *et al.*, 1981; Leigh & Wyn Jones, 1984, 1986). Similarly, the models proposed in the late 1970s to account for the feedback regulation of K^+ influx by the cytoplasmic K^+ concentration (e.g. Glass, 1976; Jensen & Pettersson, 1978) have had to be altered to take into account cytoplasmic homeostatis (see Glass & Siddiqi, 1984). It remains, however, to be shown how the uptake of solutes into the vacuole is in fact controlled by external concentrations when the two compartments are separated by the 'insulating' layer of a homeostatic cytoplasm.

In Wyn Jones (1975) the evidence for a major diffusive or exchange component in K^+ uptake in the high-concentration range was pointed out. Cheeseman & Hanson (1979) expanded on the pioneering electrophysiological work of Mertz & Higinbotham (1974) and showed conclusively that, in the upper range, K^+ uptake was energetically downhill and thus not comparable to the high-affinity energy-dependent uptake in the micromolar range. Despite the reservations of Epstein (1976), diffusion, either facilitated or simple, is almost certainly a major factor producing the uptake kinetics observed in the high-concentration range (cf. Epstein, 1976). Thus, apart from any change in ligand binding due to positive or negative cooperativity, factors such as membrane potential and thermodynamic driving force are changing significantly over the four or five decades of concentrations considered in the various theories concerning concentrations isotherms. Given those and other limitations, it is not readily apparent that such uptake kinetics will in themselves advance our understanding of either uptake of ions or organic solutes into simple cells or complex tissues (see also Cram, 1974).

6.5 Conclusions

Until recently the technical difficulties of measuring the behaviour of single cells within tissue has necessitated the consideration of such tissue as uniform. Thus their component cells have of necessity been described in averaged terms. The presence of an apoplast that can not only be separated from the environment by resistance to diffusion and bulk flow, but can also be subdivided into domains by both active and passive processes, greatly extend the options available to the plant in organizing both its solute and water relations. It may well prove that tissues are very uniform, and that they may be considered equivalent to a collection of unicellular algae. But at present, what little data are available seem to indicate a certain heterogeneity of all cells within tissue. As we have discussed, this leads to the fundamental question. First, how does the plant orchestrate and control such a system and, secondly, is the heterogeneity of specific advantage to the plant, and hence indirectly to the physiologist looking for characters he can pass on to the genetic engineer? We feel that the detailed techniques required to answer these questions are becoming available, and that much progress in understanding the full implications for multicellular tissues will be made.

Acknowledgements

The authors thank Dr R. A. Leigh for helpful discussions and Mrs N. Parry for her patient typing during preparation of the manuscript.

References

Aikman, D. P., Harmer, R. & Rust, T. S. O. (1980). Ion movement through the apoplast of *Beta vulgaris* and *Zea mays*. In Spanswick, R. M., Lucas, W. J. and Dainty, J. (eds.). *Plant Membrane Transport: Current Conceptual Issues*, pp. 433–4. Amsterdam, Elsevier/North Holland.

Benz, R. & Zimmermann, U. (1983). Evidence for the presence of mobile charges in the cell membrane of *Valonia utricularis*. *Biophysical Journal* **43**, 13–26.

Borstlap, A. C. (1983). The use of model-fitting in the interpretation of 'dual' uptake isotherms. *Plant, Cell and Environment* **6**, 407–16.

Bowling, D. J. F. (1972). Measurements of profiles of potassium activity and electrical potential in the intact root. *Planta* **108**, 147–51.

Bowling, D. J. F. (1973a). Measurement of a gradient of oxygen partial pressure across the intact root. *Planta* **111**, 323–8.

Bowling, D. J. F. (1973b). A pH gradient across the root. *Journal of Experimental Botany* **24**, 1041–45.

Briggs, G. E., Hope, A. B. & Robertson, R. N. (1961). *Electrolytes and Plant Cells*. Philadelphia, Penns., Davis F. A. Company.

Büchner, K.-H., Rosenheck, K. & Zimmermann, U. (1985). Characterisation of the mobile charges in the membrane of *Valonia utricularis*. *Journal of Membrane Biology* **88**, 131–7.

Burström, H. (1959). Growth and formation of intercellularies in root meristems. *Physiologia Plantarum* **12**, 371–85.

Burton, W. G. (1950). Studies on the dormancy and sprouting of potatoes. I. The oxygen contents of the potato tuber. *New Phytologist* **49**, 121–34.

Byott, G. S. (1976). Leaf air space systems in C_3 and C_4 species. *New Phytologist* **76**, 295–9.

Cheeseman, J. M. & Hanson, J. B. (1979). Energy-linked potassium influx as related to cell potential in corn roots. *Plant Physiology* **64**, 842–5.

Cleland, R. E. & Rayle, D. L. (1977). Reevaluation of the effect of calcium ions on auxin-induced elongation. *Plant Physiology* **60**, 709–12.

Clipson, N., Tomos, A. D., Flowers, T. J. & Wyn Jones, R. G. (1985). Salt tolerance in the halophyte *Suaeda maritima* L. Dun. The maintenance of turgor pressure and water potential gradients in plants growing at different salinities. *Planta* **165**, 392–6.

Cosgrove, D. J. & Cleland, R. E. (1983). Osmotic properties of pea internodes in relation to growth and auxin action. *Plant Physiology* **72**, 332–8.

Cram, W. J. (1974). Influx isotherms – their interpretation and use. In Zimmermann, U. and Dainty, J. (eds.). *Membrane Transport in Plants*, pp. 334–7. Berlin, Heidelberg, New York, Springer-Verlag.

Cram, W. J. (1976). Negative feedback regulation of transport in cells. The maintenance of turgor, volume and nutrient supply. In Lüttge, U. and Pitman, M. G. (eds.). *Encyclopedia of Plant Physiology*, New Series, vol. 2B, pp. 284–316. Berlin, Heidelberg, New York, Springer-Verlag.

Dainty, J. (1963). Water relations of plant cells. *Advances in Botanical Research* **1**, 279–326.

Demarty, M., Morvan, C. & Thellier, M. (1978). Plant cell walls. Exchange properties of isolated cell walls of *Lemna minor*. *Plant Physiology* **62**, 477–81.

Demarty, M., Ripoll, C. & Thellier, M. (1980). Ion exchange in plant cell walls. Spanswick, R. M., Lucas, W. J. and Dainty, J. (eds.) *Plant Membrane Transport: Current Conceptual Issues*. pp. 33–47. Amsterdam, Elsevier/North Holland.

Digby, J. & Firn, R. D. (1976) A critical assessment of the

Cholodny–Went theory of shoot geotropism. *Current Advances in Plant Science* **8**, 953–60.

Dunlop, J. (1973). The transport of potassium to the xylem exudate of ryegrass. I. Membrane potentials and vacuolar K^+ activities in seminal roots. *Journal of Experimental Botany* **24**, 995–1002.

Ellar, D. J. (1978). Spore specific structures and their function. *Symposium of the Society of General Microbiology* **28**, 295–325.

Epstein, E. (1976). Kinetics of ion transport and the carrier concept. In Lüttge, U. and Pitman, M. G. (eds.). *Encyclopedia of Plant Physiology*, New Series, vol. 2B, pp. 70–94. Berlin, Heidelberg, New York, Springer-Verlag.

Epstein, E., Rains, D. W. & Elzam, O. E. (1963). Resolution of dual mechanisms of potassium absorption by barley roots. *Proceedings of the National Academy of Sciences*, USA **49**, 684–92.

Erwee, M. G. & Goodwin, P. B. (1983). Characterisation of the *Egeria densa* Planch. leaf symplast. Inhibition of the intercellular movement of fluorescent probes by group II ions. *Planta* **158**, 320–8.

Erwee, M. G. & Goodwin, P. B. (1985). Symplast domains in extrastelar tissues of *Egeria densa* Planch. *Planta* **163**, 9–19.

Erwee, M. G., Goodwin, P. B. & van Bel, A. J. E. (1985). Cell–cell communication in the leaves of *Commelina cyanea* and other plants. *Plant, Cell and Environment* **8**, 173–8.

Flagg-Newton, J., Simpson, I. & Loewenstein, W. R. (1979). Permeability of the cell-to-cell membrane channels in mammalian cell junctions. *Science* **205**, 404–7.

Flowers, T. J. & Läuchli, A. (1983). Sodium versus potassium: Substitution and compartmentation. In Läuchli, A. and Bieleski, R. L. (eds.). *Encyclopedia of Plant Physiology*, New Series, vol. 15B, pp. 651–81. Berlin, Heidelberg, New York, Springer-Verlag.

Gifford, R. M. & Thorne, J. H. (1985). Sucrose concentration of the apoplastic interface between seed coat and cotyledons of developing soybean seeds. *Plant Physiology* **77**, 863–8.

Ginsburg, H. (1971) Model for iso-osmotic water flow in plant roots. *Journal of Theoretical Biology* **32**, 147–58.

Ginsburg, H. & Ginzburgh, B. Z. (1971). Evidence for active water transport in a corn root preparation. *Journal of Membrane Biology* **4**, 29–41.

Glass, A. D. M. (1976). Regulation of potassium absorption in barley roots: An allosteric model. *Plant Physiology*. **58**, 33–7.

Glass, A. D. M. & Siddiqi, M. Y. (1984). The control of nutrient uptake rates in relation to the inorganic composition of plants. In Tinker, P. B. & Läuchli, A. (eds.). *Advances in Plant Nutrition*, vol. 1, pp. 103–147. New York, Praeger.

Gunning, B. E. S. & Overall, R. L. (1983). Plasmodesmata and cell-to-cell transport in plants. *BioScience* **93**, 260–5.

Hanson, J. B. (1984). The function of calcium in plant nutrition. In

Tinker, P. B. & Läuchli, A. (eds.). *Advances in Plant Nutrition*, vol. 1, pp. 149–208. New York, Praeger.

Harvey, D. M. R. (1984). The effect of salinity on ion distributions in wheat and maize cells. In Cram, W. J., Janaček, K., Rybova, R. & Sigler, K. (eds.). *Membrane Transport in Plants*, pp. 495–496. Prague, Academia.

Harvey, D. M. R. & Thorpe, J. R. (1986). Some observations on the effects of salinity on ion distributions and cell ultrastructure in wheat leaf mesophyll cells. *Journal of Experimental Botany* **37**, 1–7.

Hastings, D. F. & Gutknecht, J. (1978). Potassium and turgor pressure in plants. *Journal of Theoretical Biology* **73**, 363–6.

Hoffman, E. K. (1977). Control of Cell Volume. In Gupta, B. J., Moreton, R. B., Oschman, J. L. & Wall B. J. (eds.). *Transport of Ions and Water in Animals*, pp. 285–332, London, Academic Press.

Huisinga, B. (1979) Control of loading and unloading by turgor regulation in long distance transport. *Acta Botanica Neerlandica* **28**, 67–72.

Hüsken, D., Steudle, E. & Zimmermann, U. (1978). Pressure probe technique for measuring water relations of cells in higher plants. *Plant Physiology* **61**, 158–63.

Jensen, P. & Pettersson, S. (1978). Allosteric regulation of potassium uptake in plant roots. *Physiologia Plantarum* **30**, 24–9.

Jones, D. H. & Northcote, D. H. (1981). Induction by hormones of phenylalanine ammonium-lyase in bean cell suspension culture. Inhibition and superinduction by actinomycin D. *European Journal of Biochemistry* **116**, 117–25.

Jones, H., Leigh, R. A., Tomos, A. D. & Wyn Jones, R. G. (1987). The effect of abscisic acid on cell turgor pressures, solute content and growth of wheat roots. *Planta*. **170**, 257–62.

Kelday, L. S. & Bowling, D. J. F. (1980). Profiles of chloride concentration and potential difference in the root of *Commelina communis*. *Journal of Experimental Botany* **31**, 1347–55.

Kirkby, E. A. & Knight, A. H. (1977). Influence of the level of nitrate nutrition on ion uptake and assimilation, organic acid accumulation and cation-anion balance in whole tomato plants. *Plant Physiology* **60**, 349–53.

Kursanov, A. L. & Brovchenko, M. I. (1970). Sugars in the free space of leaf blades: their origin and possible involvement in transport. *Canadian Journal of Botany* **48**, 1243–50.

Lang, A. (1983) Turgor regulated translocation. *Plant, Cell and Environment* **6**, 683–9.

Laties, G. G. (1969). Dual mechanism of salt uptake in relation to compartmentation and long-distance transport. *Annual Review of Plant Physiology* **20**, 89–116.

Lee, R. B. & Ratcliffe, R. G. (1983). Phosphorus nutrition and the intracellular distribution of inorganic phosphate in pea root tips: A quantitative study using ^{31}P-NMR. *Journal of Experimental Botany* **34**, 1222–44.

Leigh, R. A. & Johnston, A. E. (1983). Concentration of potassium in the dry matter and tissue water of field-grown spring barley and their relationships to grain yield. *Journal of Agricultural Science* **101**, 675–85.

Leigh, R. A. & Tomos, A. D. (1983). An attempt to use isolated vacuoles to determine the distribution of sodium and potassium in cells of storage roots of red beet (*Beta vulgaris* L.) *Planta* **159**, 469–75.

Leigh, R. A. & Wyn Jones, R. G. (1984). A hypothesis relating critical potassium concentration for growth to the distribution and functions of this ion in the plant cell. *New Phytologist* **97**, 1–13.

Leigh, R. A. & Wyn Jones, R. G. (1986). Cellular compartmentation in plant nutrition: The selective cytoplasm and the promiscuous vacuole. In Tinker, B. & Läuchli, A. (eds.). *Advances in Plant Nutrition*, vol. 2, pp. 249–79. New York, Prager Press.

Lowry, O. H. & Passoneau, J. W. (1972). *A Flexible System of Enzymatic Analysis*. London, Academic Press.

Marshall, A. G. (1978). *Biophysical Chemistry. Principles, Techniques and Applications*. Chichester, John Wiley.

Maynard, J. W. & Lucas, W. J. (1982). Sucrose and glucose uptake into *Beta vulgaris* leaf tissues. A case for general (apoplastic) retrieval systems. *Plant Physiology* **70**, 1436–43.

Mertz, S. M., Jr. & Higinbotham, N. (1974). The cellular electro-potential isotherm as related to the kinetic K^+ absorption isotherm in low-salt barley roots. In Zimmermann, U. & Dainty, J. (eds.). *Membrane Transport in Plants*, pp. 343–6. Berlin, Heidelberg, New York, Springer-Verlag.

Molz, F. J. & Ferrier, J. M. (1982). Mathematical treatment of water movement in plant cells and tissue: a review. *Plant, Cell and Environment* **5**, 191–206.

Münch, E. (1930). *Die Stoffbewegung in der Pflanze*. Jena, Fischer.

Nissen, P. (1974). Uptake mechanisms: inorganic and organic. *Annual Review of Plant Physiology* **25**, 53–79.

Nissen, P. & Nissen, O. (1983). Validity of the multiphasic concept of ion absorption in plants. *Physiologia Plantarum* **57**, 46–56.

Nobel, P. S. (1983). *Biophysical Plant Physiology and Ecology*. San Francisco, Calif., W. H. Freeman and Co.

Northcote, D. H. (1985). Control of cell wall formation during growth. In Brett, C. T. & Hillman, J. R. (eds.). *Biochemistry of Plant Cell Walls*. Society for Experimental Biology Seminar Series, vol. 28, pp. 177–98. Cambridge University Press.

Palta, J. A., Wyn Jones, R. G., & Tomos, A. D. (1987) Leaf diffusive conductance and tap root cell turgor pressure in sugarbeet. *Plant, Cell and Environment*. In press.

Patrick, J. W. (1984). Photosynthate unloading from seed coats of *Phaseolus vulgaris* L. Control by tissue water relations. *Journal of Plant Physiology* **115**, 297–310.

Penny, M. G. & Bowling, D. J. F. (1974). A study of potassium gradients in the epidermis of intact leaves of *Commelina communis* L. in relation to stomatal opening. *Planta* **119**, 17–35.

Perry, C. A., Leigh, R. A., Tomos, A. D., Wyse, R. E. & Hall, J. L. (1987). The regulation of turgor pressure during sucrose mobilisation and salt accumulation by storage root tissue of red beet. *Planta* **170**, 353–61.

Philip, J. R. (1958). Osmosis and diffusion in tissues: Half tissues and internal gradients. *Plant Physiology* **33**, 275–8.

Pitman, M. G. & Cram, W. J. (1977). Regulation of ion content in whole plants. In *Integration of Activity in the Higher Plant*. Society for Experimental Biology Symposium, vol. 31, pp. 391–424. Cambridge University Press.

Pitman, M. G., Läuchli, A. & Stelzer, R. (1981). Ion distribution in roots of barley seedlings measured by electron probe X-ray microanalysis. *Plant Physiology* **68**, 673–9.

Pitman, M. G., Mowat, J. & Nair, H. (1971). Interactions of processes for accumulation of salt and sugar in barley plants. *Australian Journal of Biological Science* **24**, 618–31.

Richter, E. & Ehwald, R. (1983). Apoplastic mobility of sucrose in storage parenchyma of sugar beet. *Physiologia Plantarum* **58**, 263–8.

Robinson, R. A. & Stokes, R. H. (1959). *Electrolyte Solutions*, 2nd edn. London, Butterworths.

Römheld, V. & Marschner, H. (1986). Mobilisation of iron in the rhizosphere of different plant species. In Tinker, B. and Läuchli, A. (eds.) *Advances in Plant Nutrition*, vol. 2, pp. 155–204. New York, Praeger Press.

Scholander, P. F., Hammel, H. T., Hemmingsen, E. A. & Bradstreet, E. D. (1964). Hydrostatic pressure and osmotic potentials in leaves of mangroves and some other plants. *Proceedings of the National Academy of Sciences, USA* **52**, 119–125.

Sentenac, H. & Grignon, C. (1981). A model for predicting ionic equilibrium concentrations in cell walls. *Plant Physiology* **68**, 415–19.

Shephard, U. H. & Bowling, D. J. F. (1979). Sodium fluxes in roots of *Eleocharis uniglumis*, a brackish water species. *Plant, Cell and Environment* **2**, 123–30.

Smith, J. A. C. & Lüttge, U. (1985). Day–night changes in leaf water relations associated with the rhythm of crassulacean acid metabolism in *Kalanchoë daigremontiana*. *Planta* **163**, 272–82.

Smith, J. A. C. & Milburn, J. A. (1980). Phloem turgor and the regulation of sucrose loading in *Ricinus communis* L. *Planta* **148**, 42–8.

Spanswick, R. M. (1976). Symplasmic transport in tissues. In Lüttge, U.

& Pitman, M. G. (eds.). *Encyclopedia of Plant Physiology*, New Series, vol. 2B, pp. 35–53. Berlin, Heidelberg, New York, Springer-Verlag.

Steingröver, E. (1983). Storage of osmotically active compounds in the taproot of *Daucus carota* L. *Journal of Experimental Botany* **34**, 425–33.

Steudle, E. (1985). Water transport as a limiting factor in extension growth. In Baker, N. R., Davies, W. J. & Ong, C. K. (eds.). *Control of Leaf Growth*. Society of Experimental Biology Seminar Series, vol. 27, pp. 35–56. Cambridge University Press.

Storey, R., Pitman, M. G., Stelzer, R. & Carter, C. (1983a). X-Ray microanalysis of cells and cell compartments of *Atriplex spongiosa*. I. Leaves. *Journal of Experimental Botany* **34**, 778–94.

Storey, R., Pitman, M. G., Stelzer, R. & Carter, C. (1983b). X-Ray microanalysis of cell compartments of *Atriplex spongiosa*. II. Roots. *Journal of Experimental Botany* **34**, 1196–206.

Tomos, A. D., Leigh, R. A., Shaw, C. A. & Wyn Jones, R. G. (1984). A comparison of methods for measuring turgor pressures and osmotic pressures of red beet storage tissue. *Journal of Experimental Botany* **35**, 1675–83.

Tomos, A. D., Steudle, E., Zimmermann, U. & Schulze, E.-D. (1981). Water relations of leaf epidermal cells of *Tradescantia virginiana*. *Plant Physiology* **68**, 1135–43.

Trewavas, A. J. (1981). How do plant growth substance work? *Plant, Cell and Environment* **4**, 203–8.

Trewavas, A. J. & Cleland, R. E. (1983). Is plant development regulated by changes in the concentration of growth substance or by change in the sensitivity to growth substances? (Discussion forum.) *Trends in Biochemical Sciences* **8**, 354–7.

Tyerman, S. D. & Steudle, E. (1982). Comparison between osmotic and hydrostatic water flows in a higher plant cell. Determination of hydraulic conductivities and reflection coefficients in isolated epidermis of *Tradescantia virginiana*. *Australian Journal of Plant Physiology* **9**, 461–80.

Tyree, M. T. (1970). The symplast concept. *Journal of Theoretical Biology* **26**, 181–214.

van Steveninck, R. F. M., van Steveninck, M. E. & Läuchli, A. (1982a). Profiles of chloride content of vacuoles in lupin root cells as shown by electron probe X-ray microanalysis. *Zeitschrift für Pflanzenphysiologie* **108**, 215–22.

van Steveninck, R. F. M., van Steveninck, M. E. Stelzer, R. & Läuchli, A. (1980). Electron probe X-ray microanalysis of ion distribution in *Lupinus luteus* L. seedlings exposed to salinity stress. In Spanswick, R. M., Lucas, W. J. & Dainty, J. (eds.). *Plant Membrane*

Transport: Current Conceptual Issues, pp. 489–90. Amsterdam, Elsevier/North Holland.

van Steveninck, R. F. M., van Steveninck, M. E., Stelzer, R. & Läuchli, A. (1982b). Studies on the distribution of Na and Cl in two species of lupin (*Lupinus luteus* and *Lupinus angustifolius*) differing in salt tolerance. *Physiologia Plantarum* **56**, 465–73.

van Steveninck, R. F. M., van Steveninck, M. E., Stelzer, R. & Läuchli, R. (1982c). Variations in vacuolar solutes of *Lupinus luteus* L. leaf tissue shown by electron probe X-ray microanalysis. *Zeitschrift für Pflanzenphysiologie* **107**, 91–5.

Veen, B. W. & Kleindorst, A. (1985). Nitrogen accumulation and osmotic regulation in Italian ryegrass (*Lolium multiflorum* Lam). *Journal of Experimental Botany* **36**, 211–18.

Westerhoff, H. V., Melandri, B. A., Venturoli, G., Azzone, F. G. & Kell, D. B. (1984). A minimal hypothesis for membrane-linked energy transduction. *Biochimica et Biophysica Acta* **768**, 257–92.

Wilson, D. F. & Westerhoff, H. V. (1982). Should irreversible thermodynamics be applied to metabolic systems? (Discussion forum.) *Trends in Biochemical Science* **7**, 275–9.

Wolswinkel, P., Kraus, E. & Ammerlaan, A. (1986). Effect of the osmotic environment on the balance between uptake and release of sucrose and amino acids by the seed coat and cotyledons of developing seed of *Pisum sativum*. *Journal of Experimental Botany* **37**, 1462–71.

Wyn Jones, R. G. (1975). Excised roots. In Baker D. A. & Hall, J. L. (eds.). *Ion Transport in Plant Cells and Tissues*, pp. 193–230. Amsterdam, Elsevier/North Holland.

Wyn Jones, R. G. (1984). Phytochemical aspects of osmotic adaptation. In Timmermann, B. N., Steelink, C. & Loewus, F. A. (eds.). *Recent Advances in Phytochemistry*, vol. 18, *Phytochemical Adaptations to Stress*, pp. 55–78. New York, Plenum Press.

Wyn Jones, R. G., Brady, C. J. & Speirs, J. (1979). Ionic and osmotic regulation in plants. In Laidman, D. L. & Wyn Jones, R. G. (eds.). *Recent Advances in the Biochemistry of Cereals*, pp. 63–104. London, New York, Academic Press.

Wyn Jones, R. G. & Pollard, A. (1983). Proteins, enzymes and inorganic ions. In Läuchli, A. & Bieleski, R. L. (eds.). *Encyclopedia of Plant Physiology*, New Series, vol. 15B, pp. 651–81. Berlin, Heidelberg, New York, Springer-Verlag.

Wyn Jones, R. G., Storey, R., Leigh, R. A., Ahmad, N. & Pollard, A. (1977b). A hypothesis on cytoplasmic osmoregulation. In Marrè, E. & Ciferri, O. (eds.). *Regulation of Cell Membrane Activities in Plants*, pp. 121–36. Amsterdam, North Holland.

Wyn Jones, R. G., Storey, R. & Pollard, A. (1977a). Ionic and osmotic regulation in plants particularly halophytes. In Thellier, M.,

Monnier, A., Demarty, M. & Dainty, J, (eds.). *Transmembrane Ionic Exchange in Plants*, pp. 537–44. Paris, Colloques Internationaux CNRS.

Wyse, R. E., Zamski, E. & Tomos, A. D. (1986). Effect of turgor on the kinetics of sucrose uptake. *Plant Physiology* **81**, 478–81.

Zimmermann, U. & Steudle, E. (1978). Physical aspects of water relations of plant cells. *Advances in Botanical Research* **6**, 45–117.

Zimmermann, U., Steudle, E. & Lelkes, P. I. (1976). Turgor pressure regulation in *Valonia utricularis*. Effect of cell wall elasticity and auxin. *Plant Physiology* **58**, 608–13.

7 Movements of ions across roots

D. T. Clarkson

7.1 Introduction

The simplest way in which we can begin to consider ion transport across a root is to reduce its structural complexity to that of an epithelium separating two inorganic salt solutions (Fig. 7.1); the solution on the inside is invariably more concentrated than that on the outside in natural conditions. The basic requirements of this model are that ions should be absorbed at the outer surface, by the energy-dependent processes considered in earlier chapters, and discharged into the xylem at the inner surface. For most of its length the xylem comprises dead cells linked into conduits with few transverse walls, a design which minimizes axial resistance to flow. The model must contain two important structural features if fluxes across the layer are to be ion selective. The transport mechanisms of the plasmalemma on the outer and inner surfaces must be different since they facilitate net influx and net efflux of ions, respectively, and secondly, direct flows of ions from the outside to the inside via the cell wall must be minimized by some kind of apoplastic barrier.

We will begin by comparing this simple model with the actual structure (Fig. 7.2) of a young, recently matured part of a root of *Hordeum vulgare* (barley). The outer face of the model is equivalent to the epidermal layer of the root which may or may not have radial projections caused by root hairs; the inner face is equivalent to the boundary between stelar parenchyma cells and the xylem vessels. The flow of ions and water through the cell walls is restricted by the Casparian band in the radial and transverse walls of the endodermis and the attachment of the plasmalemma to that structure. Thus a model which contains three cells in series is more realistic than the one we started with (Fig. 7.3). In the more refined model, provision is made for cytoplasmic continuity between the three cells. This creates not only a pathway by which materials can move (the symplast) but also an opportunity for the cells to communicate and coordinate their

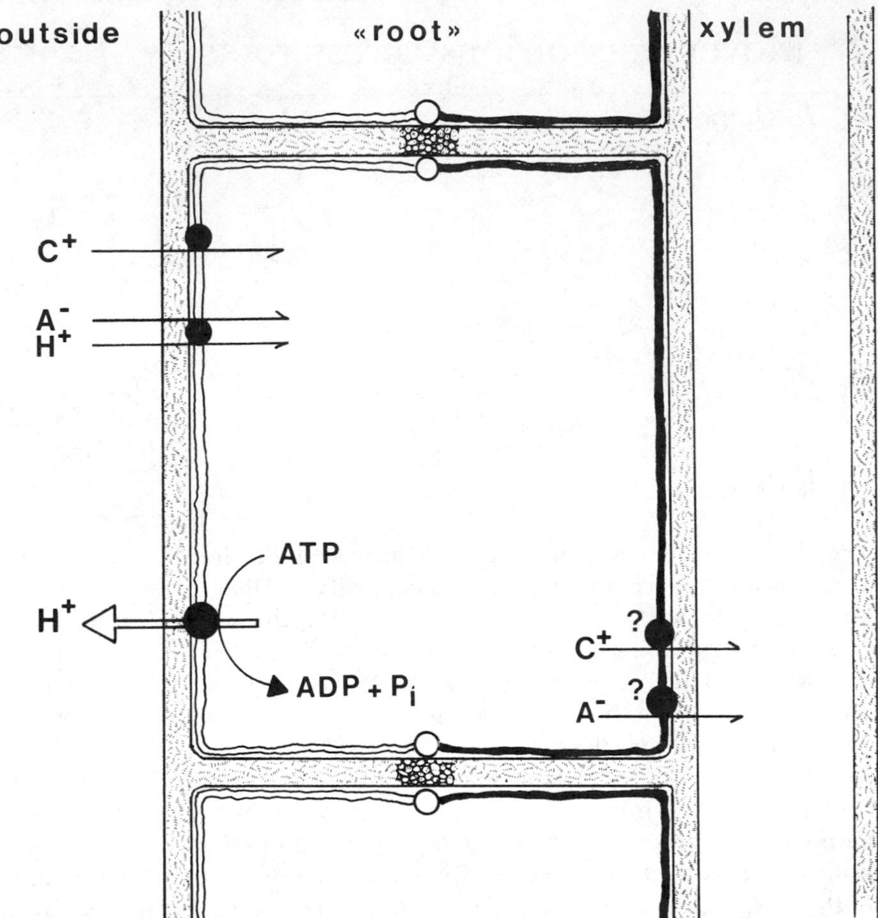

Fig. 7.1 The simplest view of a root as an epithelium. Primary active H^+ transport on the outside surface provides driving forces for C^+ and A^- uptake via specific carrier proteins. At the xylem surface there is a net export of C^+ and A^- by mechanisms which are largely unknown. The plasma membrane on the outside and xylem side must have different properties – this is emphasized here by the different weight given to the lines. Since cell walls are porous, some osmotic barrier needs to be in place to prevent back-flow of ions from the relatively concentrated xylem sap to the outside solution which is usually more dilute.

activities. It is only in highly specialized cell types, e.g. guard cells (see Ch. 12), that such cytoplasmic continuity is lacking. The structures involved are plasmodesmata; a résumé of their structure in higher plants will be given later. The three-cell model also makes it somewhat easier to envisage how the polarity, an essential feature of Fig. 7.1, can be brought about since the outer and inner cells exist in different environments; it is possible that they have different metabolic and structural characteristics

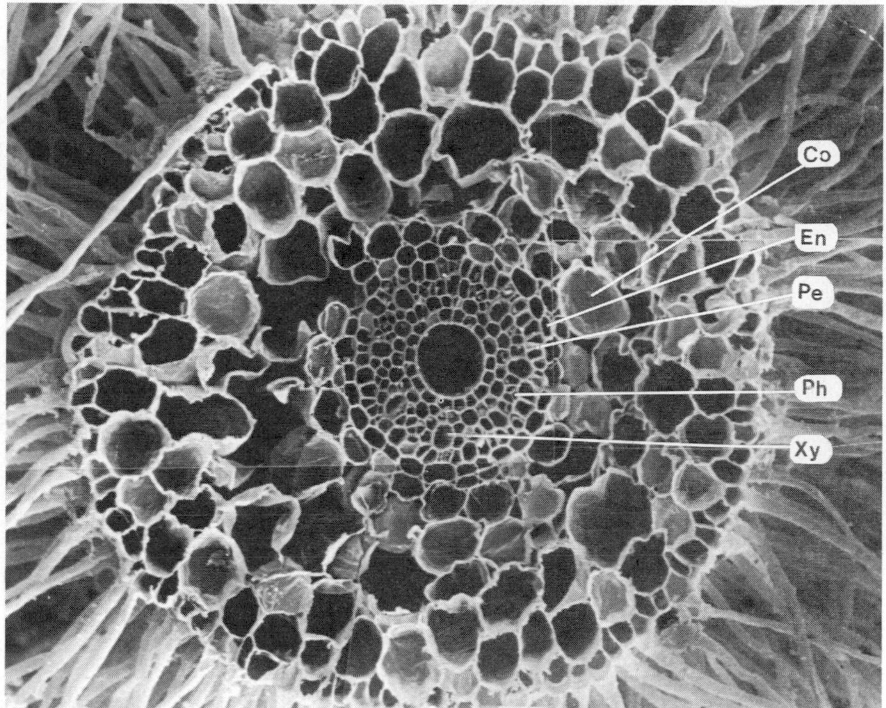

Fig. 7.2 Transverse fracture of a seminal axis of barley. The root is surrounded by many root hairs which are as long as the root diameter. The cross walls of some cortical cells, Co, are seen in this fracture – the fragmentation of some cells in the inner cortex is accidental. The endodermis, En, separates the cortex from the stelar tissue. Groups of phloem, Ph, alternate with metaxylem elements, Xy. Lateral roots are initiated from the pericycle, Pe. (Photographs courtesy of Dr A. W. Robards, University of York.

which might interact with transport processes across the plasmalemma. The model shows that there are two ways for a solute to reach the endodermal cell, via an extracellular route, which bypasses cell 1, or an intracellular route via the plasmodesmata; these two pathways will be referred to as the 'apoplast' and the 'symplast', respectively. Each cell contains a vacuole, a large compartment limited by the tonoplast. Vacuoles can act as storage pools in which osmotically active substances in excess of current metabolic requirements may be sequestered, leaving the cytoplasm with optimum or near-optimum levels of required nutrients. Thus, Na^+, Cl^- and most of the cellular free Ca^{2+} are sequestered in the vacuole to prevent the damage they can cause to metabolic processes; nutrients such as NO_3^-, $H_2PO_4^-$ or SO_4^{2-} surplus to requirements may be stored or withdrawn when necessary. The vacuoles of adjacent cells are not linked directly as is the cytoplasm and thus they are not part of the symplast. Ions flowing along the symplast

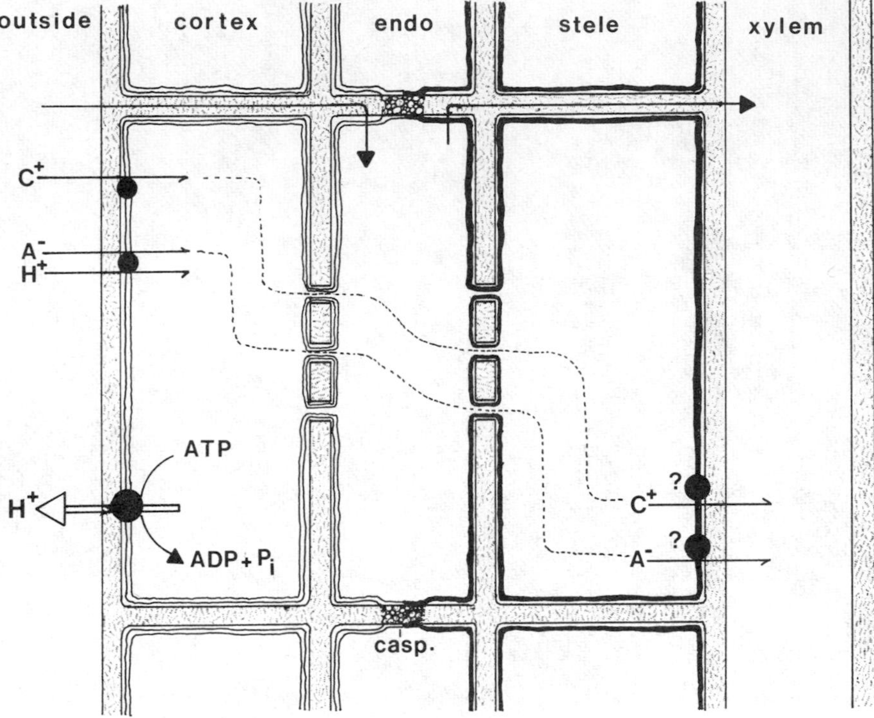

Fig. 7.3 The elaboration of the root model by the addition of further tissues adds no conceptual difficulties greater than those in Fig. 1. Cells are connected via plasmodesmata, the pores of which are lined by plasma membrane. The osmotic barrier is provided by the endodermis and its Casparian band (see p. 258). This tissue now takes on an epithelial character since the properties of its plasma membrane differ on the two sides (see p. 272). Solutes can reach the endodermis either by movement through the porous cell wall (the apoplast) or via the plasmodesmata in the cortex-endodermis junction (the symplast); similar alternative pathways exist on the stelar side of the endodermis.

pathway may be diverted into, or joined by fluxes from the vacuoles. Clearly, the magnitude of the flux of a given ion in the symplast relative to the fluxes across the tonoplast will influence the extent to which ions recently absorbed by cell 1 are diluted by those previously located in vacuoles. Whenever parallel pathways exist, the flow of a given material along them will be determined by the relative resistances they offer. As is frequent in analogies of this kind, the word 'resistance' is very broadly defined and may include such factors as the probability that an ion will bind to a vacant carrier site and be transported into a cell, relative to the probability that it will be carried by mass flow of water into the apoplast. I have chosen this particular example because of revived interest in the speculation about whether ion absorption by roots occurs exclusively at the periphery. I hope to show, in the course of this chapter, that this exclusive view, like so many cut-and-dried statements about complex systems, is too simplistic and applies only to certain ions.

A root system of a growing plant constantly produces new cells and, in the older zones of the root axis, the primary structure seen in Fig. 7.2 gives way to secondary developments which may influence the capability of the root tissues to absorb ions. One might envisage both declining metabolic activity and changes in structure as factors here. I hope to show that developmental changes have a greater influence on the apoplast pathway than on the symplast and that this can be used to distinguish those nutrients which move predominantly by one of the two pathways.

It will become clear to the reader that little is known about the events which occur on the inside of the endodermis particularly the factors regulating the unloading of ions from the symplast of cell 3 (Fig. 7.3) into the xylem. This ignorance leaves us in a poor position to explain a readily observable physiological phenomenon, viz. that the release of ions into the xylem and even the intensity of the initial absorption process in cell 1 can be strongly influenced by the 'demand' for nutrients created by shoot growth; some examples of this will be given later.

7.2 Resistance to ion movement in the root approaches and the apoplast

Root surfaces are not comprised of naked, porous cell walls which give relatively unhindered access to the epidermal protoplasts from the outside solution; several types of barrier intervene.

7.2.1 Mucigel

The epidermal surface is usually invested with polysaccharide gel material (known generically as 'mucigel') some of plant and some of microbial origin (Foster, 1981). External to this pellicle there may be an extracellular layer of cuticle-like material (Scott, 1963). It is difficult to be certain how completely these layers invest the root surface; they seem readily disrupted and cracked by the preparative techniques for electronmicroscopy. Flaws or cracks occurring in nature would be pathways of lower resistance for diffusion or mass flow of ions and would represent preferred entry sites. Root hairs project from the root surface, often for considerable distances into the surrounding medium (Itoh & Barber, 1983; Clarkson, 1985). In soil they are strategically placed to absorb ions whose movement to the root surface is limited by diffusion or ions which are very dilute. The surface of root hairs is also covered with mucigel and a cuticular layer of uncertain composition has been found to cover most of the root surface of the few species which have been studied (Scott, 1963; Cailloux, 1972).

7.2.2 Unstirred layers

All parts of the root surface will be surrounded by an unstirred layer of solution. Nye & Tinker (1977) estimated its thickness to lie in the range from 10 to 100 μm for roots in a flowing culture solution and it would probably be greater in the static solution in the soil. At most, therefore, the layer will be wider than the radial dimension of an epidermal cell. Diffusion of ions through this layer may limit the rate at which they can be absorbed from the external medium (Fig. 7.4). The effect will be greatest for ions such as nitrate which are absorbed rapidly in large quantities. Pitman (1982) estimated that the potential rate of NO_3^- uptake from a solution of 10 mol m^{-3} NO_3^- would be reduced by 14 % and 40 %, respectively by unstirred layers of 10 and 100 μm thickness respectively. Figure 7.4 shows that this is because the concentration at the root surface is lower than in the surroundings; it is evident that the lower the external solution concentration the greater the depletion will be. When plants are transpiring, mass flow of solution towards the root surface can counteract to some extent the depletion illustrated in Fig. 7.4, but the effect will only be significant where the external concentration is relatively high.

7.2.3 Cell wall pores

As indicated in the model (Fig. 7.3) there are two pathways, in theory at least, that can be followed once ions have negotiated the surface barriers. They may move directly through the wall, interact with a carrier site in the plasmalemma of cell 1 or they may move in the plane of the wall in pores which may also conduct a bulk flow of water in a transpiring plant. In either case the pores in the wall are provided by the interstices of bundles of cellulose microfibrils which are disposed at angles to one another to create a three-dimensional meshwork. The pores, especially those in the plane of the wall, are tortuous so that passage though them involves a longer journey than the linear distance between one cell and another. There have been relatively few attempts to determine the porosity of primary cell walls. The results in Table 7.1 are from some ingenious experiments by Carpita *et al.* (1979) who used osmotica of increasing molecular size to determine average pore diameters in the walls of *Raphanus sativus* (radish) root hairs and palisade cells from *Xanthium* (cocklebur). Small osmotica penetrate the wall pores and the protoplasts shrink as in the normal pattern of plasmolysis. When hypertonic solutions of large molecular weight osmotica were used, the cells lost water but the protoplast did not shrink, rather the cell buckled and deformed in a

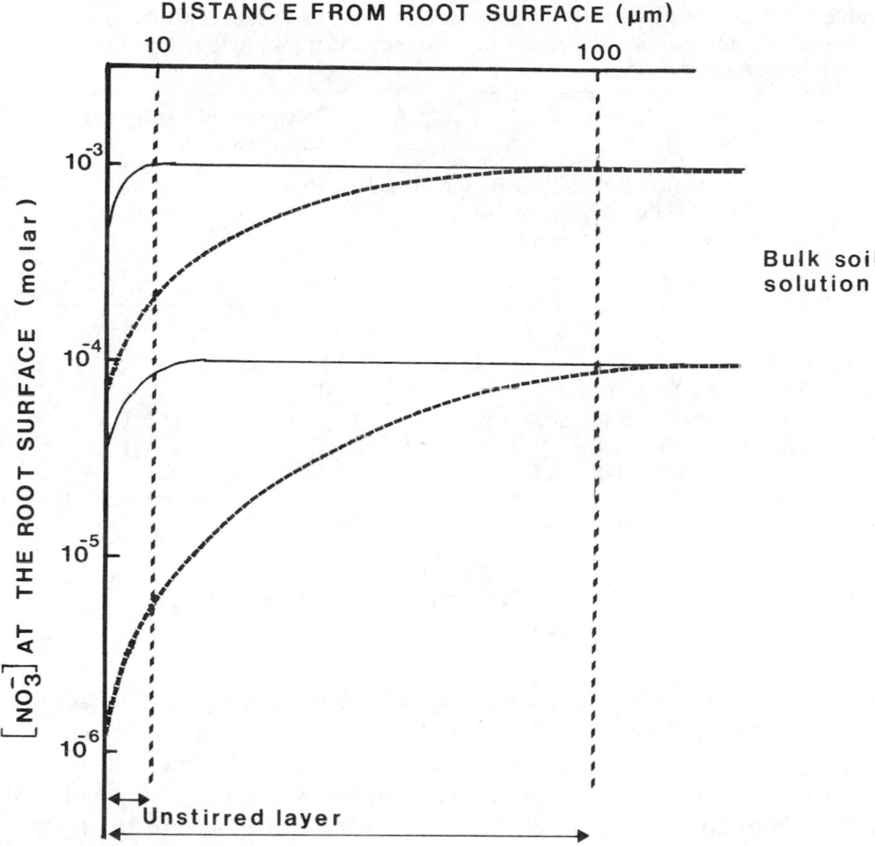

Fig. 7.4 Diagram to illustrate the influence of thickness of an unstirred layer on the concentration of NO_3^- at the surface of a root. A layer 10 μm in thickness results in a very substantial reduction in NO_3^- concentration if roots are absorbing at a normal rate (*c*.5 mol g^{-1} fresh wt h^{-1} and the NO_3^- in the soil has a diffusion coefficient of about 10^{-6} cm^2 s^{-1}. As the bulk solution concentration falls, the influence of the unstirred layer becomes more pronounced – an increase in absorption rate would have the same effect. (Based on calculations in Pitman, 1982.)

process known as 'cytorrhysis'. This happened because the molecules could not penetrate the wall. By employing a series of polyethylene glycols of increasing molecular size they found that, at around 1000–1600 *MW*, cells first showed cytorrhysis followed by plasmolysis. This result is compatible with there being a small number of large pores in the wall which slowly allowed the osmoticum to penetrate the space between the wall and the protoplasts. Above a molecular diameter of 3.8 nm only cytorrhysis was seen. This sets the exclusion limit somewhere between 3–3.5 nm. Thus, movement in the cell wall from the root surface to the next cortical cell might involve passage through a system of pores as much as 100 000 nm

Table 7.1 Effects of osmotica on increasing molecular weight and size on the response of root hairs of *Raphanus sativus* and leaf mesophyll cells of *Xanthium* to hypertonic solutions†

Osmoticum	Molecular Weight	Approx. molecular diameter (nm)	Response to hypertonic solutions	
			Root hair	Mesophyll
Mannitol	182	0.8	P‡	P
Sucrose	342	1.0	P	P
Polyethylene glycol (PEG)				
PEG 600	570–630	2.9	P	P
PEG 1000	950–1050	3.6	P + C§	C ¶ P
PEG 1540	1300–1600	3.8	C	C ¶ P
PEG 4000	3000–3700	4.5	C	C

† Data from Carpita *et al.* (1979).
‡ Plasmolysis.
§ Cytorrhysis.
¶ Transient cytorrhysis giving way to plasmolysis.

long and 3.5 nm wide. The diffusive resistance of such a pathway has been estimated to be 100–1000 times greater, for univalent cations, and ten times greater, for univalent anions than an equivalent thickness of solution (Walker & Pitman, 1976). The pores themselves are lined with material with a considerable cation exchange capacity which is due to the dissociation of fixed carboxyl groups of the galacturonic and glucuronic acid components of pectins and hemicelluloses. These negatively charged sites restrain the movement of cations and greatly complicate calculations of the rates of diffusion in the anoplast. In most circumstances the cation exchange capacity of roots is saturated, a majority of sites being occupied by calcium. The movement of a given ion through the apoplast is akin to that through a saturated chromatography column; its progress will be slowed down by repeated association and dissociation with exchange sites. If, however, the column is saturated, the entry of a new ion at one end must displace another ion at the other. This notion raises the question as to where the apoplast ends. The conventional answer to this question is a short one; the endodermis.

7.2.4 The endodermis

We need to appreciate two structural properties of this tissue to see how it works, and the timing of their development to appreciate their physiological significance. In the radial and cross-walls of fully expanded endo-

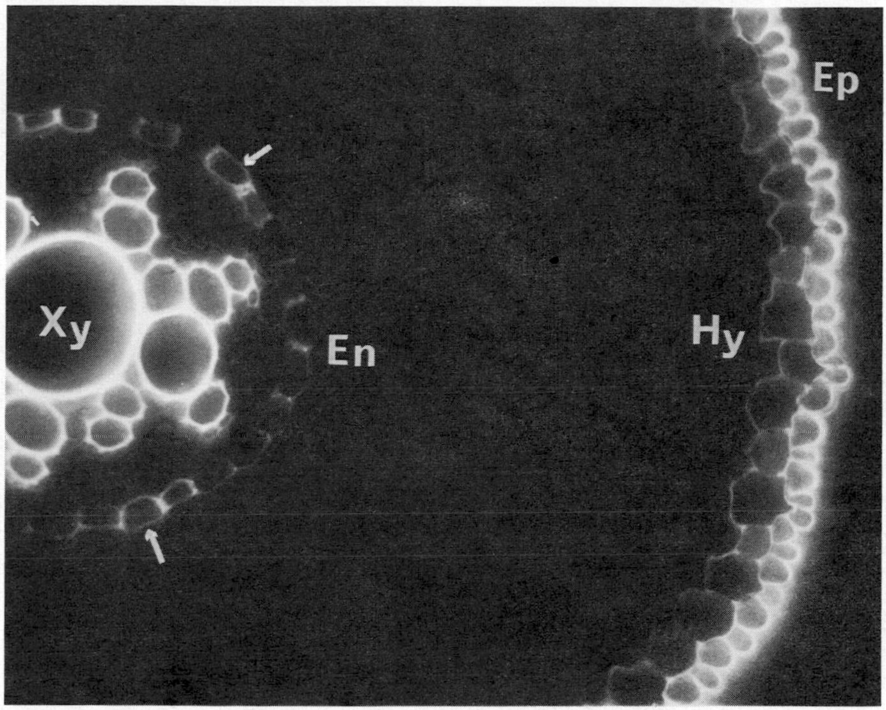

Fig. 7.5 Autofluorescent walls in a transverse section of onion root cut approximately 100 mm from from the root tip. The cell layer immediately below the brightly fluorescing epidermis is the hypodermis (see p. 278). The endodermis, En, is seen to have fluorescent material in all of its radial walls; this is where the Casparian band is formed. At this stage of development some of the endodermal cells have additional suberin in their walls (arrow) (see p. 276). The lignin of the xylem vessels, Xy, fluoresces very brightly. (Photograph kindly provided by Dr C. A. Peterson.)

dermal cells there is a region (see Fig. 7.5), described as the Casparian band, which stains positively for suberin and lignin, or lignin precursors (Van Fleet, 1961; Scott & Peterson, 1979b). These substances are interpolated in the meshwork of the fibrillar wall components, they are hydrophobic and probably create a high resistance to the movement of materials through the wall (Clarkson & Robards, 1975). The plasmalemma of the endodermal protoplast becomes firmly attached to the Casparian band (Bonnett, 1968; Robards *et al.*, 1973; Scott & Peterson, 1979a), so firmly in fact that, when endodermal cells are plasmolysed, the protoplast may shrink to form a septum stretched across the lumen of the cell (Bryant, 1934).

Consider now the situation confronting water, solutes and suspended material which has moved radially in the cortical apoplast (Fig. 7.6).

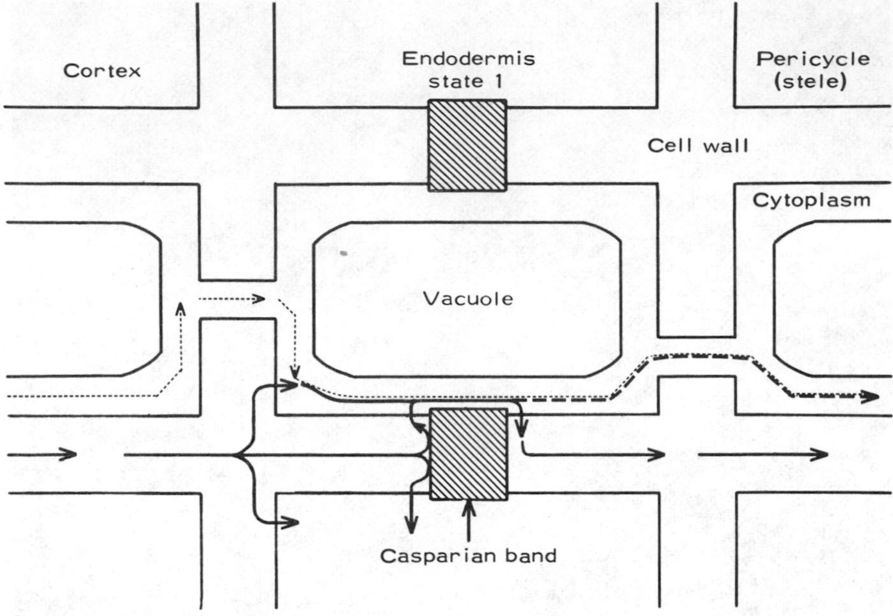

Fig. 7.6 The influence of the Casparian band in the state 1 endodermis on the flow of solutes and water via the symplast (dotted line) and the apoplast (solid line).

Further progress through the Casparian band is likely to be very slow (Nagahashi *et al.*, 1974) and movement over its surface, to circumvent the resistance, is prevented by the attachment of the plasmalemma. Clearly these features leave passage across the plasmalemma of the endodermal protoplast as the only plausible pathway. Suspended materials, such as colloidal lanthanum, can sometimes be seen accumulating on the cortical side of the Casparian band but not internal to it.

The developmental event which is closely correlated with the appearance of the Casparian band is the maturation of the protoxylem; this is reported in *Hordeum vulgare* (Robards *et al.*, 1973) and *Zea mays* (Haas & Carothers, 1975). Xylem maturation involves the death of the protoplast and evidently the loss of selectivity over what can enter the cell lumen. This development makes it imperative to introduce some barrier into the apoplast, otherwise, scarcely modified soil solution, with its incorrect ratios of nutrient ions, toxins and suspended material would be drawn by transpiration into the vessels. In essence, when the xylem cells die the selectivity function, formerly discharged by their plasmalemmas, is transferred to the endodermis.

Within the stele, the walls of the parenchyma and other cells may have pores so that materials may move through a stelar apoplast before entering its most prominent component, the xylem. The endodermis undergoes further changes which may influence the movement of water and some solutes into the stele; these will be considered later. In the recently matured zone of the root the endodermis seems likely to be the place where symplastic and apoplastic flows of materials combine, at least for a short distance. The plasmalemma of endodermal cells from *Z. mays* (Robards *et al.*, 1980) has a far higher frequency of intramembrane particles (IMPs), revealed by freeze–fracture electron microscopy, than membranes of the cortex of epidermal region (Table 7.2). In other membrane systems there seems to be a positive correlation between IMP frequency and the intensity of material fluxes across them (Branton & Deamer, 1972). Thus, the morphology of endodermal membranes is compatible with high fluxes of materials from the apoplast.

Table 7.2 Intercalated membrane particle (IMP) frequency on plasmalemma fracture faces from endodermal and cortical cells in seminal roots of *Zea mays* (maize)†

Fracture face	Cell type	IMP frequency (μm^{-2} ± SEM)
P (cytoplasmic side)	Endodermis	2500 ± 110
	Cortex	1140 ± 65
E (apoplast side)	Endodermis	1295 ± 25
	Cortex	350 ± 20

† Adapted from Robards *et al.* (1980).

7.3 The symplast

Viewed simply, the symplast is equivalent to the cytoplasm of all living cells linked by plasmodesmata. Some current ideas and controversies about plasmodesmata are sketched in Fig. 7.7. Compelling evidence is now available that solutes do actually move through plasmodesmata in higher plant cells. Molecules of up to 650 Da can move freely from an injected cell into neighbouring cells in the leaf epidermis of *Egeria densa*, but the exclusion limit is lower, *c.* 376 Da, in roots of this species (Goodwin, 1983; Erwee & Goodwin, 1985). In this work small peptides were labelled with fluorescein and injected into cells; the peptides were neither metabolized nor showed any leakage outward across the plasmalemma. The most plausible pathway for their rapid spread to cells neighbouring the injection site is via the plasmodesmata. Earlier it had been shown that electrical conductivity between cells was high, relative to the conductance of the plasmalemma (Spanswick, 1972). Electric current entering one cell caused changes in the membrane potential in neighbouring cells, from which the extent of electrical coupling could be calculated. Between 5 and 10 % of the current leaving the injected cell would have moved to the nearest

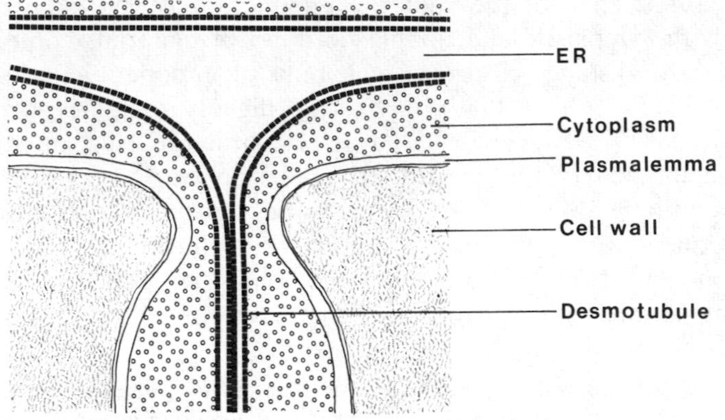

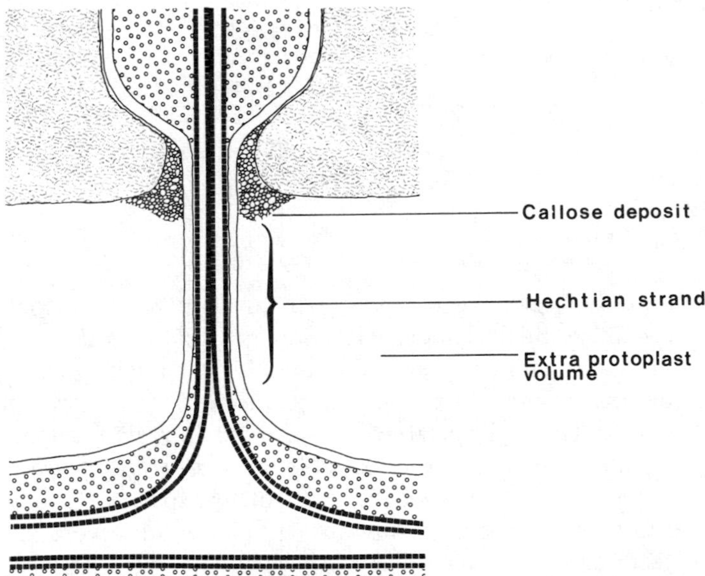

Fig. 7.7 A recent interpretation of the structure found in the pore of plasmodesmata. The central structure, the desmotubule, is formed from compressed and modified endoplasmic reticulum envelope membrane. The inner polar regions of the envelope membrane become compressed to form a cylindrical array of associated phospholipid head groups surrounded by the outer half of the bilayer. The molecular justification of this design is discussed in Overall *et al* (1982). The desmotubule is surrounded by an annulus of cytoplasm contained within a plasmalemma-lined cavity in the wall. In the lower half of the figure two aspects of plasmolysis are depicted. The neck region of the cavity becomes occluded by callose (Drake *et al.*, 1978) and the desmotubule extends to form the so-called Hechtian strand (Burgess, 1971). The plasmalemma remains closely appressed to the desmotubule and continuity between cells is lost.

neighbours. When it is recalled that the plasmodesmata occupy no more than 1–2 % of the plasmalemma area it can be appreciated that electrolyte flows more readily through the cell junctions than across an equivalent area of membrane to the outside. More recently it has been shown that electrical coupling between cell in roots of *Azolla* is directly dependent on the number of plasmodesmata linking them (Overall & Gunning, 1982).

The nature of the channel, or channels, provided by plasmodesmata is still far from clear. Recently there has been a new interpretation of the osmiophilic structures in the centre of the plasmodesmatal canals. Formerly these were thought to constitute a central channel, or desmotubule, which provided continuity between the ER in adjacent cells (Robards, 1975). In this interpretation the appearance of a central rod of dark-staining material had never been satisfactorily explained and remained the Achilles heel of Robards's hypothesis. The more recent view of Overall *et al.* (1982) is that the central structure is comprised of tightly appressed ER envelope membrane trapped in the wall during cell plate formation at the end of mitosis (Hepler, 1982). If this is correct the central stucture cannot be thought of as a channel connecting adjacent cell compartments. Work with fluorescent peptides tells us something about the nature of the channel which does exist. Since fluorescent peptides do not appear to cross membranes it seems most unlikely that their movement from cell to cell could have been via the desmotubule, even if this structure really exists. It is much more likely that they pass through the annulus of cytoplasm at the neck of the plasmodesmatal canal shown on the upper half of Fig. 7.7. Goodwin (1983) estimates the diameter of the 356 Da peptide which diffuses between root cells in *E. densa* as approximately 1.2 nm. This is appreciably larger than the diameters of hydrated monovalent cations and anions and many of the smaller metabolites, e.g. ATP, and thus plasmodesmata could provide a channel through which electrolytes might move.

The idea of a desmotubule coupling the ER between cells was attractive to physiologists because it offered a means by which flows in opposite directions might be accommodated within a single plasmodesma and because various pieces of circumstantial evidence suggest that the symplast might be a compartment within the cytoplasm, not the whole of the cytoplasm. The kinetics of K^+ uptake and movement into the xylem of low-salt barley plants were such that Hooymans (1974) concluded that there must be a separate compartment within the cytoplasm where a portion of the K^+ was sequestered. During the first few hours of K^+ uptake a relatively small compartment, represented by the root symplast was filled and movement into the xylem or into the vacuole was delayed until this phase neared completion. Hooymans (1976) also found that K^+ fluxes to the xylem, relative to those into the vacuole, could be altered without there being any measurable change in the cytoplasmic $[K^+]$. The results might be interpreted by an assumption that there is a compartment within the cytoplasm through which K^+ passes on its way to the xylem. There is other evidence which points to a similar conclusion. Rufty *et al.* (1986) found that when roots of *Z. mays* were treated with a low [NO_3^-], after preliminary growth in NO_3^- free medium, nitrate reductase (NR) was induced only in cells at the root periphery (i.e. where the NO_3^- was probably absorbed). No NR was induced in the cortex or the stele even though an appreciable fraction of the NO_3^- passed across the root to the xylem. This movement would have occurred in the symplast and yet it failed to induce NR in the cells through which it passed. Evidently either the NO_3^- or the NR must be sequestered in some way. If roots were treated with a high [NO_3^-], NR was induced in both peripheral, cortical and stelar cells. This suggests that NO_3^- must enter from the apoplast and mix with the whole cytoplasm if NR is to be induced and that the symplasmic component is something separate from the whole cytoplasm. The ER has been a frequent candidate in speculation about this separate compartment, especially since it can be shown to have ion-sequestering properties, e.g. for Ca^{2+} (Marmé, 1983).

The results of plasmolysis experiments might be taken to weaken the case for the ER as a component of the symplast. A number of authors have shown that even mild plasmolysis can severely disrupt transport via the symplast (e.g. Falk *et al.*, 1965; Van Iren & Boers-van der Sluijs, 1980) and electrical coupling (Drake *et al.*, 1978). It has been shown that such treatments do not necessarily break plasmodesmatal connections in the way that the authors assumed. Burgess (1971) showed that the desmotubule can extend considerably and that shrunken protoplasts remain connected via the so-called Hechtian strand. These strands appear to be surrounded by tightly appressed plasmalemma; I have tried to represent this state in the lower half of Fig. 7.7. It is also evident that plasmolysis promoted callose formation, which becomes very prominent around the necks of the plasmodesmatal canal, occluding structures which had been disrupted and constricting those which had stretched (Drake *et al.*, 1978). Thus, plasmolysis disrupts the cytoplasmic pathway by constricting the annulus of cytoplasm around the plasmodesmatal neck region; it does not seem to disrupt the connection between the ER of neighbouring cells, yet symplasmic transport was largely eliminated. It should be remembered, however, that plasmolysis has been found to have catastrophic effects on the cell membrane potential in *Avena* coleoptiles (Drake *et al.*, 1978) and cultured cells of *Acer psuedoplatanus* (Rona *et al.*, 1980) and also greatly decreased P_i uptake by roots of *Z. mays* (Attia & Jeanjean, 1983). It may be, therefore, that primary transport processes are badly disturbed in plasmolysed cells and that disturbances in symplasmic transport follow from this rather than because the pathway is disrupted.

If there is no desmotubule with a conductive function one can still envisage how the central rod of trapped membrane could be involved in regulating flow via the annulus of cytoplasm surrounding it. Erwee & Goodwin (1983) showed that when Ca^{2+} or Mg^{2+} was injected into cells of *E. densa* along with fluorescent peptides there was no diffusion of label from the injected cell for 30–40 min; the elevated cytosol concentration of group II ions evidently restricted some part of the plasmodesmatal pathway. Results of this kind encourage the speculation that there is some sphincter-like control over the annulus of cytoplasm at the necks of the plasmodesmata and that, when the plasmalemma is closely appressed to the desmotubule, symplasmic continuity is shut down (Olesen, 1980). There is certainly considerable variation in the exclusion limit for fluorescent peptides within different tissues in *E. densa* and especially between certain cell types (Erwee & Goodwin, 1985). In the roots of this species there was poor conductance of a 356 Da molecule between the epidermis and the cortical layer beneath it, and between the root cap and the apical meristem, even though there were abundant plasmodesmata which did not appear to differ in structure from those where the exclusion limit was much larger. These observations lead the authors to propose that there are domains within the symplast; this is perhaps rather premature because the exclusion limits against which these domains are categorized are appreciably larger than the ions and metabolites which are the major constituents of symplasmic flows. If there is sphincter-like control it might be exercised by some protein or by growth of callose around the neck of the canal. Callose deposition can be triggered by elevated cytosol free Ca^{2+} (Kauss *et al.*, 1983). It should be noted, however, that marked increases in plasmodesmatal resistance can be induced by treatment with NaN_3 which did not promote callose formation on oat coleoptiles (Drake, 1979).

7.3.1 Distribution of plasmodesmata

Comparative anatomy of tissues and species suggests that there is a correlation between the number of plasmodesmata which link cells and the fluxes of materials which occur between them. Thus, large numbers are seen in the stalk cells of salt glands (Thomson & Liu, 1967; see Ch. 13), extrafloral nectaries (Gunning & Hughes, 1976; see Ch. 14) and companion cell/sieve tube junctions (Warmbrodt, 1985; see Ch. 8). In the barley root (Table 7.3) it was found that there were twice as many plasmodesmata at the junction between the endodermis and the pericycle than on the junction between the endodermis and the cortex (Robards *et al.*, 1973). One might predict the main direction of symplast flows from maps of plasmodesmatal frequency (e.g. Warmbrodt, 1985). Because of the arduous nature of the work, few such maps have been produced. This approach has been used to see if root hairs are the major loading point of the symplast (Fig. 7.8). Were this to be so, root hair-bearing epidermal cells (= trichoblasts) would have greater frequencies of plasmodesmata in the junction with the cortex than epidermal cell lacking hairs (= atrichoblasts). This was found to be the case with *Trianea bogotensis* (a small

Table 7.3 Frequency of plasmodesmata in walls of cortical, endodermal and pericycle cells in the seminal root axis of *Hordeum vulgaris* (barley)†

Junction	Radial distribution	Orbital distribution
Cortex/cortex	0.28	0.24
Cortex/endodermis	0.37	–
Endodermis/endodermis	–	0.32
Endodermis/pericycle	0.75	–
Pericycle/pericycle	–.	0.54
Pericycle/stelar parenchyma	0.25	–

† Data compiled by Jackson & Robards (unpublished) and Robards *et al.* (1973).

floating pond weed) (Vakhmistrov & Kurkova, 1979) but not in *Raphanus sativus* (radish) (Vakhmistrov *et al.*, 1981). This result tends to confound one's intuition since one would have imagined that the latter species, growing in soil and being therefore dependent on diffusion of nutrients to the root surface, to be more likely to attach special significance to its root hairs.

Moving from the periphery towards the stele, the surface area of each rank of cells becomes smaller. Thus if the symplast flux is to be carried at a relatively constant rate per plasmodesma, the number of plasmodesmata per unit area must rise. It is to be expected, therefore, that plasmodesmatal frequencies will be greater on the inner faces of the cortex than near the root periphery. The marked increase in frequency between the outer and inner tangential walls of the endodermis of barley (Table 7.3) is much larger than one would expect from geometrical considerations and implies that the flux leaving the endodermal cell is greater than that which arrives at its outer surface, perhaps because of the combination with apoplastic flows described earlier.

7.3.2 Cytoplasmic streaming

The distribution of materials within the cytoplasm must be influenced by cytoplasmic streaming, a process dependent on the interaction of free Ca^{2+} and protein filaments resembling F-actin and myosin (Williamson & Ashley, 1982; Hepler & Wayne, 1985). The cytoplasm does not move *en masse*, there are mobile and immobile domains which enhance the mixing effect of the streaming. In *Lycopersicon esculentum* (tomato), rates of streaming of from 2–6 $\mu m\ s^{-1}$ were measured in foliar trichomes (Patterson & Graham, 1977). It is reasonable to ask what influence streaming has on rates of movement through the symplast. The drug cytochalasin B inhibits the aggregation of actin microfilaments into bundles, inhibiting

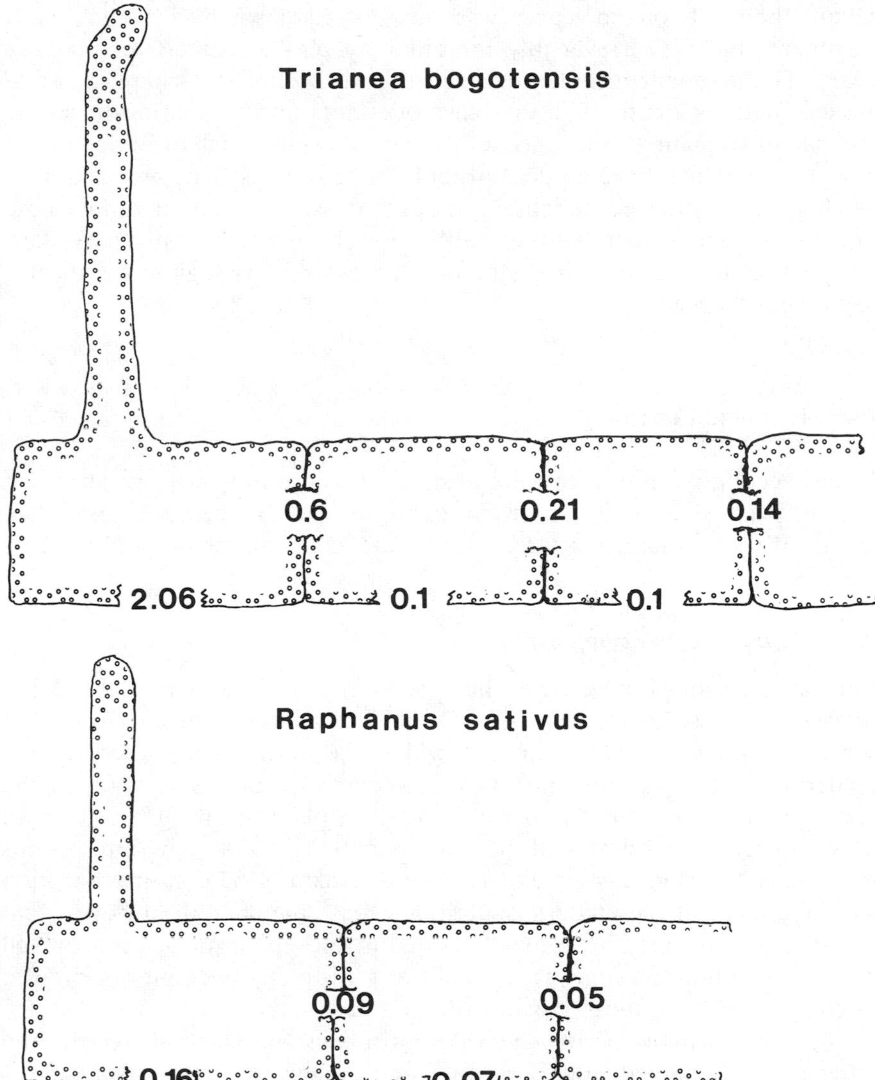

Fig. 7.8 Frequency of plasmodesmata in epidermal cells in relation to root hairs. In *Trianea* there are very large numbers of plasmodesmata linking the junction of a trichoblast with the underlying cortex, and also some suggestion of a tangential gradient of plasmodesmatal frequency in the radial walls of epidermal cells. The above arrangement is far less marked in *Raphanus*.

streaming completely, and yet this treatment had no influence at all on the movement of ^{86}Rb-labelled K^+ into the xylem sap of detached roots of tomato, maize and *Cucumis sativus* (cucumber) (Glass & Perley, 1979). Similarly, the movement of disodium fluorescein between cells in multicellular hairs of tomato leaves was not slowed down by cytochalasin B treatment, but was irreversibly inhibited by plasmolysis (Barclay *et al.*, 1982). The authors concluded that movement through the symplast is rate limited by transport through the plasmodesmata and that diffusion, in the absence of streaming, could sustain the flux. In cells which have open plasmodesmata, which have no desmotubule or analogous structure, and have much lower resistance, the inhibition of cytoplasmic streaming reduced the rate of Cl^- movement between cells, e.g. *Chara* (Bostrom & Walker, 1976). It should be noted that streaming rates are very high in these cells, being about 100 $\mu m\ s^{-1}$.

7.4 Sites of ion uptake

In this section we return to the question of where ions are first absorbed in their passage across the root and consider how uptake varies along the length of roots where structure and metabolic activity change with age.

7.4.1 Entry into the symplast

Cells at the root periphery are the first to encounter ions brought to the surface by mass flow or diffusion. If the rate at which ions arrive at the surface is slower than the rate at which they can be absorbed by the epidermal cells, it seems unlikely that they will progress very far in the apoplast. It can be argued reasonably that, for plants growing in soil where most nutrients are dilute and have their mobility reduced by interactions with the soil exchange complex (see Nye & Tinker, 1977), most absorption will take place at the epidermis. It should be remembered, however, that roots in the soil may have many lesions in the epidermis caused by soil bacteria and fungi; soil-grown roots have a distinctly decrepit appearance in comparison with those grown in laboratory media.

The long tortuous pores which comprise the apoplast will restrict the movement of charged solutes as discussed earlier, and it has been calculated that, even in young tissues close to the root tip, diffusion of dilute salt solutions into the free space is slower than uptake into the cells (Bange, 1973; Ehwald *et al.*, 1973). These authors attribute the high-affinity transport mechanism for ions (the system I of Epstein, see Ch. 6) to uptake by the epidermal cells and the lower affinity uptake observed at higher concentrations (system II) to diffusion into the apoplast to more deeply located absorption sites within the tissue. There are a number of reasons, not least the fact that system I and system II absorption can be observed in single cells or very small cell aggregates, for challenging this point of view. It is quite clear, however, that ions from more concentrated solutions, say greater than 1 mol m^{-3}, do readily penetrate the apoplast

(= free space) from which they can be rapidly exchanged with half-times in the order of from 1–2 min (Walker & Pitman, 1976). Examination of the kinetics of K^+ uptake by excised segments of maize roots showed that, over a wide range of concentrations there was a component which saturated in the 0.1–0.5 mol m^{-3} range, which is equivalent to the high-affinity system reported by other authors, and a first-order linear component most evident at higher concentrations (Kochian & Lucas, 1983). At lower concentrations of K^+ (0.2 mol m^{-3}) these authors propose that most, if not all, K^+ is absorbed by the epidermis plus hypodermis. The evidence for this view comes from short-term pretreatment of the roots with two sulphydryl group inhibitors, *n*-ethyl maleimide and *p*-chloromercuribenzene sulphonic acid (PCMBS). The latter compound was labelled with ^{203}Hg and its distribution in the tissue observed by autoradiography. There was a much higher intensity of silver grains over the epidermal cells than elsewhere in sections of freeze-substituted root treated for 1 min with ^{203}Hg-PCMBS. This peripheral distribution of inhibitor was associated with a 70–85 % reduction in $^{86}Rb(K^+)$ uptake by the tissue over a 10 min period. The authors point out that this period would have been ample for K^+ to have diffused into the apoplast and that the hypodermis (see Sect. 7.4.3.3) did not constitute any barrier to diffusion. They conclude that K^+ uptake capability in the cortical cells must have been weakly developed. There is, however, difficulty in deciding, on the basis of autoradiographs alone how much of a given intensity of labelling is enough to have an effect. Kochian and Lucas's autoradiographs show that there is some ^{203}Hg-PCMBS in the cortex; about 25 % of that in the epidermis. How can one be sure that this material is not sufficient to inhibit K^+ transport by cortical cells and the higher concentration at the root periphery is not merely 'over-kill'? It is significant that protoplasts isolated from the cortex did have K^+ uptake capability the kinetics of which showed the same characteristics seen in root segments, although the rates of transport were lower. The lower rates may have been caused by plasmolysis which occurs during protoplast isolation (Rona *et al.*, 1980; Lefebvre & Clarkson, 1984).

The roots of some plants can adapt their structure so as to provide large air channels in the cortex when the root environment becomes depleted of oxygen. The channels (aerenchyma) are formed after the death of a large proportion of the cells in the cortex (Fig. 7.9). This process occurs in maize and provided an opportunity to see if large-scale disruption of the cortex disturbed ion uptake and transport to the stele. Drew *et al.* (1980) found no significant difference in the radial movement of $^{86}Rb(K^+)$ into the xylem in aerenchymatous and non-aerenchymatous roots of similar age. The loss of cortical cells does more than reduce the numbers of cells potentially available for uptake, since it also reduces the size of the symplastic pathway from the peripheral cell layers to the inner cortex. Perhaps the most surprising feature of these results, and those of Drew and Saker (1986), was that the loss of most of the plasmodesmatal connec-

Fig. 7.9 Transverse fracture of a frozen adventitious root of *Zea mays* treated for several days in solution with only 1 % O_2. Note the extensive collapse of the cortex to form air spaces and the narrow bridges of remaining living cells. Despite the reduction in both symplast and apoplast, radial movement of ions in roots such as these is comparable with normal roots (Drew & Saker, 1986). (Photograph kindly provided by Dr M. C. Drew.)

tions between the epidermal layer and the endodermis had no influence on the flux of ions across the root. This result suggests either that the capacity of the symplast is much larger than the fluxes it carries in most circumstances, or that the capacity is not rate limiting in radial ion movement. The limitation might be, say, in release from the xylem parenchyma (see Sect. 7.5)

Perhaps the most radical proposal concerning the transport properties of the cortical cells has come from the work of Van Iren & Boers van der Sluijs (1980) who found that cortical cells in plasmolysed roots absorbed very little $^{86}Rb(K^+)$ in comparison with epidermal cells and that little, if any, ^{86}Rb was found within the stele. Again the evidence was autoradiographic and no quantitation of the grain distribution was presented. The argument turned on the fact that ^{86}Rb was taken up by plasmolysed cortical cell initials in the meristematic zone, and they concluded that cortical cells lost the capacity for ion transport as they matured. Leaving aside technical criticism, this seems a most curious proposal. All available evidence suggests that the membrane potential in cortical cells differs little from that in the epidermis (e.g. Dunlop & Bowling, 1970), and values for the resting potential are generally reckoned to depend partially on the electrogenic activity of a proton pump in the plasmalemma. This pump controls not only the membrane potential but regulates the pH of the cytoplasm (see Ch. 3). These considerations make it seem probable that plasmalemmas in cortical cells possess the primary active transport mechanism. Other considerations suggest that they are likely to control their internal calcium concentration by active pumping back into the apoplast. A further process which is likely to occur is the efflux of ions from the cortical cell protoplasts into the free space, especially true for the anions (e.g NO_3^-), which are at electrochemical potentials which arc vcry much higher in the cytoplasm than outside. It is difficult to imagine that the system could change so as to lose its influx mechanisms and thus be unable to retrieve the inevitable loss of ions by efflux, whilst retaining primary active transport.

In summary, it seems likely that most of the ions absorbed from dilute solutions in the system I range of concentrations will be transported across the plasmalemmas of epidermal cells. At higher concentrations ion movements into the apoplast present opportunities for cortical cells to act as entry points into the symplast flow.

7.4.2 Radial movement of calcium

Most evidence leads to the conclusion that the activity of free Ca^{2+} in the cytoplasm is kept between 10^{-4} and 10^{-3} mol m^{-3} by a combination of internal sequestration in the ER and other organelles and efflux pumping across the plasmalemma (Hepler & Wayne, 1985). Thus, the quantity of Ca^{2+} available for long-distance transport in the symplast is small and quite inadequate to supply the needs for Ca^{2+} in the shoot apoplast which will be in the mol m^{-3} range (Clarkson, 1984; Hanson, 1984). Nothing suggests that Ca^{2+} does not move via the symplast but it is certain that the amount moved is small. These considerations make the apoplast the preferred pathway for the movement of Ca^{2+} across the root. It will be shown later that the movement of Ca^{2+} into the xylem is strongly influenced by anatomical changes in the apoplast which have little or no effect on the radial movement of symplastically transported ions. The conceptual difficulty comes in trying to understand how Ca^{2+} fluxes across the endodermis are managed. As discussed earlier, apoplastic flow of Ca^{2+} would be diverted across the plasmalemma of the endodermis on its outer half; this entry is likely to be through a channel and would be passive in nature. So as to prevent an excessive accumulation of Ca^{2+} in the cytosol, it would be necessary to have Ca^{2+} efflux pumps operating on the inner half of the endodermis at a rate more or less equivalent to the passive entry

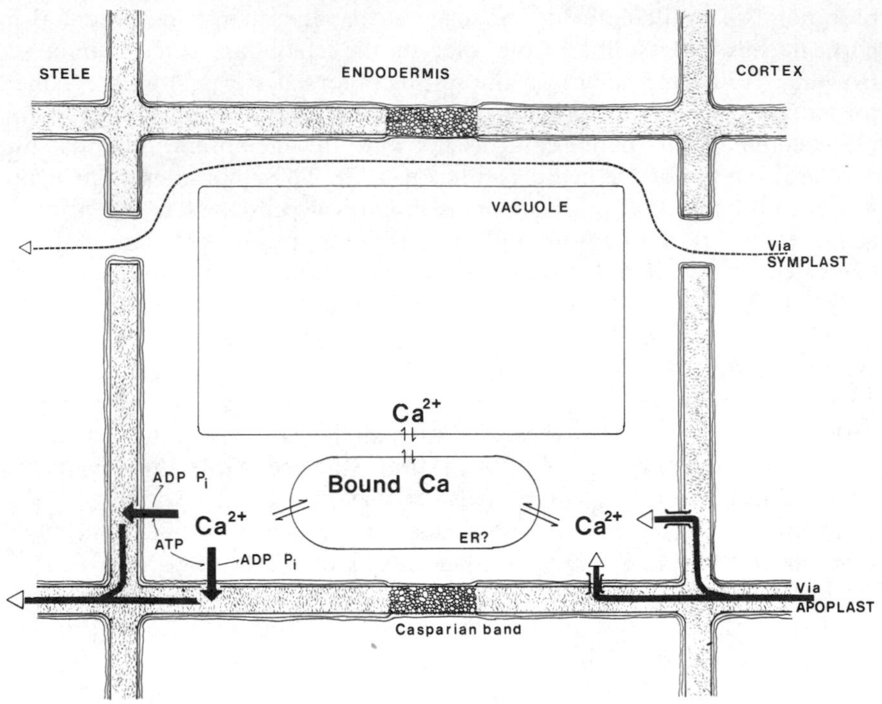

Fig. 7.10 Diagram which emphasizes the possible epithelial-like properties of the endodermis with respect to Ca^{2+} transport. It is envisaged that the plasma membrane on the cortical side of the Casparian band admits Ca^{2+} from the apoplast via channels. The rate at which Ca^{2+} is removed from the cytosol by active transport (Ca^{2+} transporting ATPase) and influx via the channels is closely coupled via some monitor of free Ca^{2+} in the cytosol (e.g. Ca^{2+} – calmodulin regulation of the putative Ca^{2+}-pump). There may be some symplastic transport of Ca^{2+} but this is small due to the low activity of free Ca^{2+} in the cytoplasm (*c.* 10^{-4} mol m^{-3}). (Adapted from Clarkson, 1984.)

(Fig. 7.10). Epithelial transport of Ca^{2+} which conforms precisely to this arrangement, has been reported in the small intestine (Nellans & Popovitch, 1981) and kidney cortex (Gmaj *et al.*, 1979) of the rat. The quantitative aspects of this model are discussed in Clarkson (1984). The movement from the inner surface of the endodermis to the xylem vessels would be through the stelar apoplast.

7.4.3 Ion uptake/release along roots

A number of approaches have been taken to study the uptake and release of ions along the length of roots. Some give more precise information than others but all of them show that the capability for ion transport is widely distributed over the root surface and by no means restricted to absorbing zones associated with root apices.

7.4.3.1 Whole root scanning

One of the early attempts to visualize the distribution of ion uptake capacity made use of short-term exposure of roots to radiolabelled ions followed by scanning using a modified chromatogram scanner (Bowen & Rovira, 1967). These showed widespread ion absorption of ^{32}P, ^{36}Cl and ^{86}Rb. Although adsorbed ions could be removed by exchange from the free space, no direct assessment could be made of the radial movement of ions from the uptake sites into the xylem. The techniques also tend to confound uptake by lateral roots and the axis from which they arise.

Scanning of a novel kind has been developed by Marschner *et al.*, (1982) to observe the distribution of net proton release by roots. Various techniques were used to apply liquid agar or sheets of solidified agar or agarose (Haussling *et al.*, 1985) containing 0.006 % bromocresol purple, to roots grown in solution culture or in soil (Römheld *et al.*, 1984). The release of protons is seen as a change in the colour of the dye from purple to yellow. Very marked changes in colour can be seen after quite short periods of time (from 10–20 min) in roots which are iron-deficient; somewhat longer periods of from 0.5–2 h may be required in other circumstances. There are numerous matters which can be explored using this interesting technique, e.g. variations between species, between different sources of nitrogen or between roots grown in soil or solution culture. The examples in Fig. 7.11 are from roots of *Helianthus annuus* (sunflower) and show that the pattern of net proton release varied enormously with the nutritional status/regime to which the plants had been subjected. Only a small net acidification was recorded in the extreme basal zone of main axes when the plants were supplied with a complete culture solution containing NO_3^- as the dominant anion. The substitution of K_2SO_4 for KNO_3 resulted in the more rapid absorption of K^+ than the anion and a balancing counterflow of protons (Hiatt, 1967). It is interesting that all parts of the root released protons indicating, perhaps, that excess K^+ absorption occurs over the whole root surface. By contrast, the induction of proton pumping by iron stress is confined to the apical parts of axes and lateral branches. This is part of the response of 'iron-efficient' plants to Fe-deficiency and is associated with the formation of transfer cells in the epidermis of many dicot species (Römheld & Kramer, 1983). The greatly increased surface area of plasmalemma, which is conformable with the complex labyrinth of wall ingrowths, is associated with an enormous H^+ efflux, especially if this is expressed on the basis of tissue weight; Römheld *et al.* (1984) estimated it to be between 80 and 150 μmol H^+ g^{-1} fresh wt h^{-1}. Such fluxes are comparable in size to those which can be measured between the stelar parenchyma and the xylem (see later).

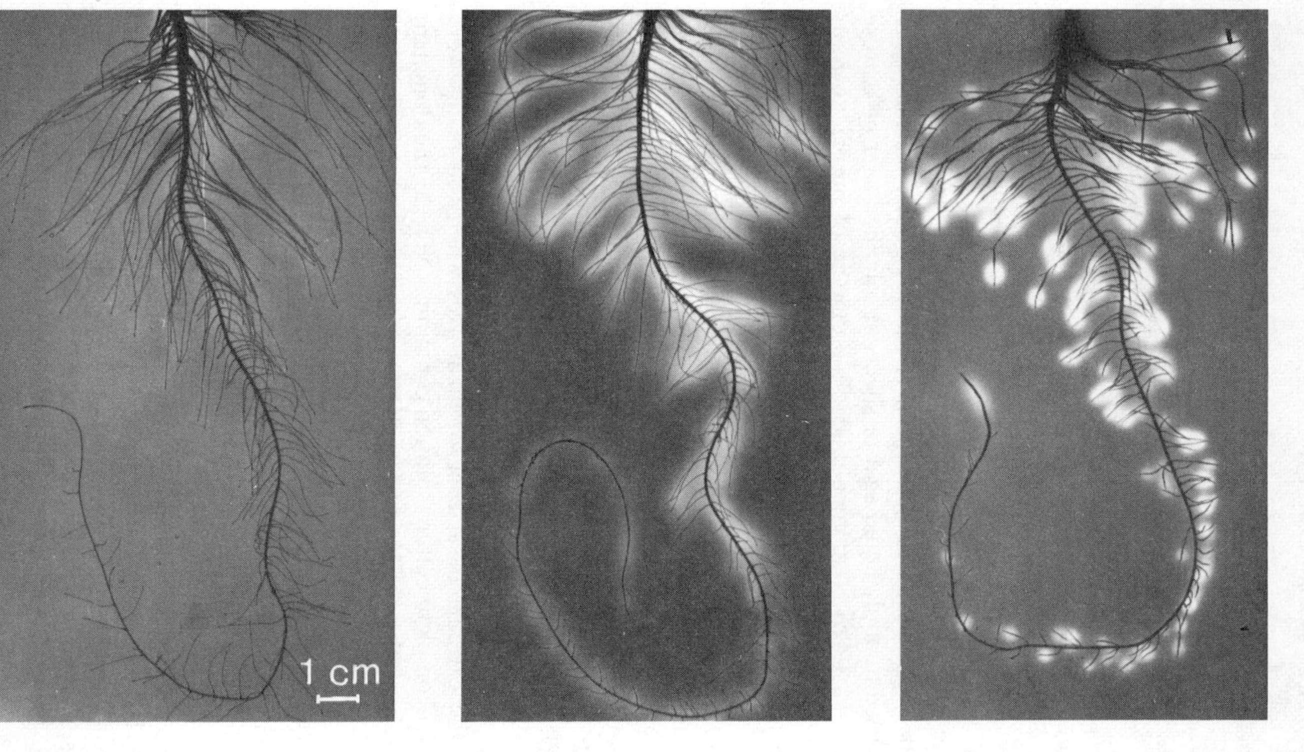

Fig. 7.11 Different patterns of proton release from sunflower roots dependent on the composition of the culture solution or the iron status of the plants. The roots are covered with a dilute agar gel containing 0.006 % bromo cresol purple. Acid production is associated with the white areas on these prints. The original colour change was from purple to yellow. (Photographs kindly provided by Dr V. Römheld.)

7.4.3.2 Measurements by electrodes close to the root surface

Small electric fields are generated by currents of ions passing through the root and the solution bathing it. Jaffe and Nuccitelli (1974) developed a vibrating probe electrode to detect such fields. The tip is vibrated through a defined amplitude and the tip measures the voltage at the extremes of the vibration amplitude (usually set at 20–25 μm and at 90° to the root surface). While there is some difficulty in relating quantitatively the observed voltage difference to an equivalent electric current (see Lucas & Kochian, 1985) there is no doubt that the technique can provide a new insight into the location and intensity of primary electrogenic transport mechanisms of the root. Since it is non-invasive, measurements can be made over long periods of time and the response of the root to changing conditions observed. Weisenseel *et al.* (1979) showed that a steady inward electrical current could be measured in the elongation zone and at the tips of root hairs of barley roots and appeared to re-enter the medium from the epidermal surface of expanded cells. Using a technique similar to that described above (pH indicator-containing agar) they deduced that these currents represented net proton movements. A slightly different pattern, of inflowing current at the tip and outflowing current associated with emerging root hairs, was seen in roots of *Trifolium repens* (clover) by Miller *et al.* (1986).

A more recent development makes use of ion-selective microelectrodes which can be placed at variable distances from the root surface by micromanipulators. If convective movements of ions to the surface are minimized by increasing the viscosity of the medium it is possible to measure localized depletion profiles at different points along the root. A study of K^+ uptake by maize roots shows a zone of high absorption associated with the root apex followed by a trough in the zone 2–4 cm from the tip and a distinct rise in the zone 5–6 cm from the tip. This pattern seemed to match that of net acidification of the rhizosphere solution (Kochian & Lucas, 1985). The second peak in K^+ uptake may be associated with the maturation of the xylem vessels and a greater radial flux of K^+ in the symplast.

7.4.3.3 Segment labelling techniques

If a short length of an intact root is isolated from the rest of an axis or branch, by enclosing it in some kind of chamber, and then treated by labelled ions one can measure both the uptake by the segment and radial movement to the stele. The latter can be deduced from the labelled ions translocated to other parts of the plant in the xylem or phloem. In this respect segment labelling techniques are more informative than simple measurements of initial absorption because it is possible to make correlations between developments in root anatomy and the radial transport of ions. When Ca^{2+} and $H_2PO_4^-$ were presented to the main axis of barley

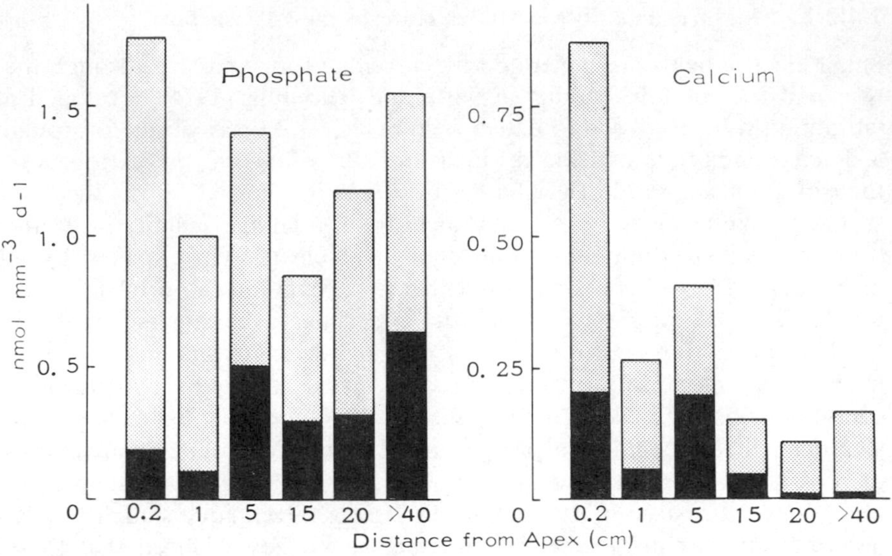

Fig. 7.12(a) Histograms illustrating absorption (entire column) and translocation (lower section) of phosphate and Ca^{2+} determined 3.5 mm regions of intact seminal axes of barley plants about 3 weeks old. Uptake measured over 24 h at 20 °C; external concentrations 3 mmol m^{-3} KH_2PO_4 (^{32}P-labelled) 1.25 mol m^{-3} $CaCl_2$ (^{85}Sr-labelled).

at different distances from the tip of a main axis (Fig. 7.12), it became evident that at around 12–18 cm from the tip the movement of Ca^{2+} into the xylem was curtailed while that of phosphate was unchanged (see Clarkson, 1974; Russell & Clarkson, 1975). Even at the base of roots more than 50 cm long, a high proportion of the phosphate absorbed was translocated to other parts of the plant, mainly to the shoot. Behaviour of a similar kind was seen when Ca^{2+} and K^+ uptake and translocation were compared in segments of intact roots of *Cucurbita pepo* (marrow) (Harrison-Murray & Clarkson, 1973). In both barley and marrow the decline in Ca^{2+} translocation was correlated with the appearance of suberin lamellae in the walls of endodermal cells and was matched by an equivalent decrease in water flow across the root (Fig. 7.13). During suberization the strong contact between the Casparian band and the plasmalemma of the endodermal cell is broken and suberin is deposited over the entire wall surface of the cells. The anatomical features of this development are discussed in Robards *et al.* (1973) and Clarkson & Robards (1975); in summary, it results in a layer, which most people have assumed to be hydrophobic, which reduces access to the endodermal plasmalemma from the apoplast. Such secondary developments place a further restriction on materials which move across the root largely via the apoplast; the parallel behaviour of calcium and water is significant. Suberin is laid around the

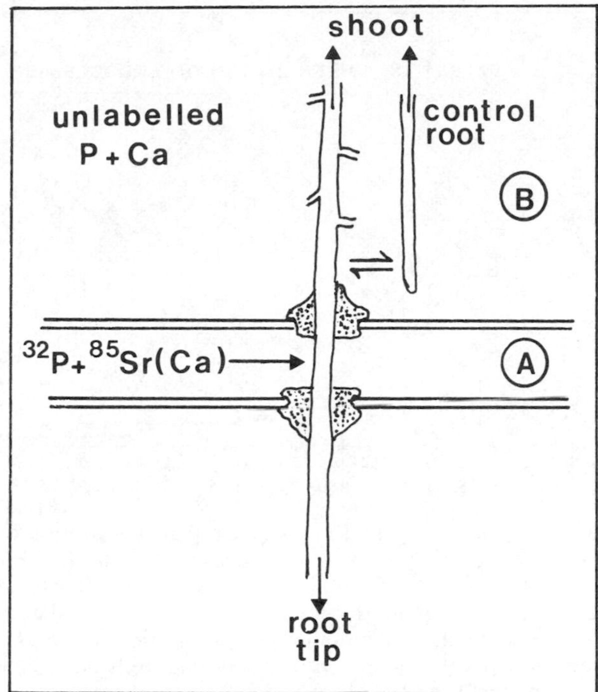

Fig. 7.12(b) Simple principle involved in feeding labelled nutrients to selected locations on intact roots. The root is sealed into a slit cut across the diameter of the tube A through which aerated, labelled culture solution is pumped. The remainder of the root system and the roots of control plants are treated with unlabelled solution in B. Any leakage of radioactivity from A or exchange from the treated root may be reabsorbed by the remaining roots. The control measures this process and its radioactivity can be deducted from the treated plant.

plasmodesmata in the endodermal walls and does not block the progress of symplastically transported ions as they move into and out from the endodermis. These observations suggest that ion supplies to the shoot can be derived from the extensive absorption over the root surface. Ca^{2+} (for the reasons outlined earlier), possibly Mg^{2+} (Ferguson & Clarkson, 1976b) and Fe (for largely biochemical reasons) seem to be the major nutrients whose supply to the shoot is restricted to part of the root system in barley and probably in many other species.

There are some species, however, where radial transport of all nutrients seems to be restricted to a short apical zone which ends shortly after cell elongation has been completed. In *Carex arenaria* (sand sedge) the cortical layers immediately below the epidermis become suberized and lignified to form a multilayered hypodermis (Robards *et al.*, 1979). This structure seems to block radial movement of K^+, $H_2PO_4^-$ and Ca^{2+} even though the epidermis has numerous root hairs. There would appear to be neither

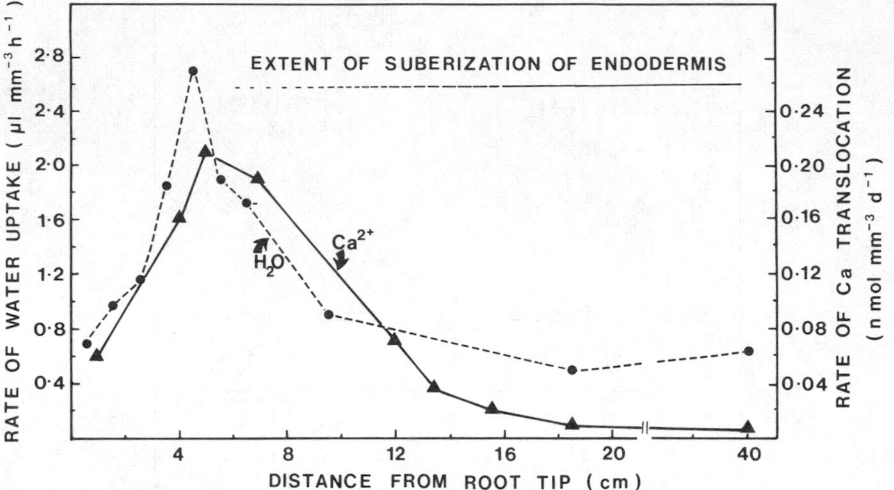

Fig. 7.13 Comparison of the influence of endodermal suberization on the uptake of water and the translocation of Ca^{2+} from short lengths (3.5 mm) of intact seminal root axes of barley. Ca^{2+} uptake measured in a system similar to that outlined in Fig. 7.12; water uptake by micropotometry as described in Sanderson (1983). The dotted line describing endodermal suberization indicates that between 6 and 20 cm from the tip only some of the endodermal cells have suberin lamellae in their walls. At greater distances all of the cells are suberized.

apoplast nor symplast continuity through this hypodermis; if it is removed by microdissection, ions move readily through the remaining unsuberized cortical cells and cross the endodermis. There has been considerable interest in the hypodermis lately, particularly as to whether it is endodermis-like in its properties. Some of the earlier discussion about whether or not cortical cells participate in ion absorption has clearly contributed to this revived interest. In a number of species, e.g. *Allium cepa* (onion) (Peterson *et al.*, 1978) and maize (Peterson & Perumalla, 1984; Clarkson *et al*, 1985), the radial walls of the hypodermis (see Fig. 7.5) have depositions of suberin and lignin precursors at short distances from the root tip (e.g. 50 mm in maize). Peterson *et al.* (1982) and Peterson & Perumalla (1984) described this development as a Casparian band and have briefly reported the attachment of the plasmalemma to it in onion roots (Peterson & Emanuel, 1983). In the roots of maize, segment-labelling experiments did not detect any change in the movement of Ca^{2+} across the cortex after this feature developed, although there was a decline in Ca^{2+} transport in more basal zones where the endodermis became suberized (Ferguson & Clarkson, 1976a). It would appear, therefore, that the suberized hypodermis in the intact maize root is not necessarily a barrier to apoplastic movement of either Ca^{2+} or water (Stephens, 1981) in the early stages of its development. Recently it has been shown that in the basal zone of the root the apoplast permeability of the hypodermis may decrease thirty-fold if it is exposed to damp air rather than being kept immersed in solution (Clarkson *et al.*, 1987). This change was not accompanied by any visible alteration of the walls or the suberin lamellae in them and may represent some change in the chemical state of the suberin polymers or the deposition of wax in response to increased oxygen supply. Thus the evidence suggests that the hypodermis is not always a barrier to apoplast movement of ions and water even when electron microscopy reveals the presence of suberin lamellae in its radial walls; the evidence of Peterson *et al.* (1982) is consistent with the exclusion of solutes of large molecular weight from entry into the apoplast by the thickenings in the hypodermal walls.

7.4.4 Lateral roots

Most of the total length of a root system is made up from its lateral branches. While their diameters may be much less than the axis from which they arise, they are usually similar to the main axis in their general anatomy. On the basis of weight (volume) ion uptake by lateral roots of barley was comparable to that by main axes (Russell & Sanderson, 1967). On the basis of their surface area, however, it seems probable that in soil most of the nutrient intake by the plant will occur through them. The same anatomical correlations between endodermal suberization and Ca^{2+} movement into the stele were seen in primary laterals as described above for main axes of barley (Robards *et al.*, 1973).

During their development, lateral roots cause a hiatus in the structure of the endodermis of the main axis. They arise from meristematic initials in the pericycle; as they divide, a dome of cells is produced and the endodermis divides to form the epidermal layer of the emerging lateral (McCully, 1975). At the junction of the lateral and the endodermis there is a ring of cells which lack Casparian bands (Dumbroff & Peirson, 1970) and which seem to permit apoplast continuity between the cortex and the stele. Apoplastic tracers were seen to enter the stele at this point (Peterson *et al.*, 1981) and water uptake in the zone of lateral emergence on barley roots was higher than elsewhere and very responsive to changes in transpiration rate (Sanderson, 1983). The quantitative significance of this apoplastic 'leak' in the overall uptake of water and nutrients has not been worked out.

7.4.5 Mycorrhizae

It is a fact, much ignored by laboratory investigators, that roots in soil are invariably parasitized by mycorrhizal fungi. There appear to be only two families of plants in which this is not so, the Cruciferae and the Chenopodiaceae. Fungal hyphae extend for great distances from the surface of parasitized roots and may have access to sources of nutrient which the root itself cannot reach. In the case of phosphate it is demonstrated that the fungi absorb the ions, translocate them within the hyphae and release them into the host tissue. These basic features are found in the several types of mycorrhizae but they have been most intensively studied in the ecto- or vesicular–arbuscular (V–A) mycorrhizae which are found in herbaceous plants and some shrubs and trees. There is a burgeoning literature on the mycorrhizal symbiosis, for symbiosis it is. In exchange for carbon substrates, the fungus effectively forages for nutrients and may be absolutely vital in obtaining supplies of slowly diffusing ions such as phosphate, zinc and copper. The price exacted by the fungus has been estimated as equivalent to an increase of root respiration of 74 % in *Vicia faba* (Pang & Paul, 1980) and 18 % in *Allium porrum* (Snellgrove *et al.*, 1982); these increases imply that an increased proportion of the carbon fixed by the plant is consumed in the root system.

The nature of the interchange of phosphate and sugar between host and fungus is relevant to this chapter because, in the real world, most of the phosphate entering the cortex of a root probably depends on this process, particularly in the more mature portions of axes and laterals. Figure 7.14 shows that the internal hyphae of the fungus grow between the host cells in the apoplast. At intervals, throughout the cortical volume, short branches penetrate the cell wall and ramify to form a highly lobed structure known as the 'arbuscule'. The host plasmalemma becomes conformable with these arbuscular branches and its surface area greatly increases (Cox & Tinker, 1976; Toth & Miller, 1984) much as it does in transfer cells. Between the plasmalemmas of host and arbuscule there is a thin layer of highly modified 'wall' materials and a space known as the 'matrix' (see Fig. 7.15). Phosphate absorbed by the hyphae is polymerized into polyphosphate granules which move by cytoplasmic streaming along the external hyphae and enter the arbuscules (Cox *et al.*, 1980). Polyphosphatases, localized in the arbuscules (Capaccio & Callow, 1982), break down the granules producing a high local P_i concentration. P_i moves into the matrix (mechanism unclear) and is absorbed across the host plasmalemma which has strongly developed H^+-ATPase activity (Marx *et al.*, 1982). Sugar released from the host (probably by simple diffusion) may be absorbed by the hyphae which could make use of the proton motive force created by the H^+-pumps of the host in the matrix. Thus, cortical cells with arbuscules are loaded directly with phosphate, and any other ions which may be exchanged, which then move symplastically to the xylem. There is evidence that hyphae can also transfer SO_4^{2-}, NO_3^+ and NH_4^+ to the root (see Clarkson, 1985 for references), but the quantitative significance of this transfer is less than it is for phosphate since these ions are much more mobile in the soil and diffuse readily towards the root surface.

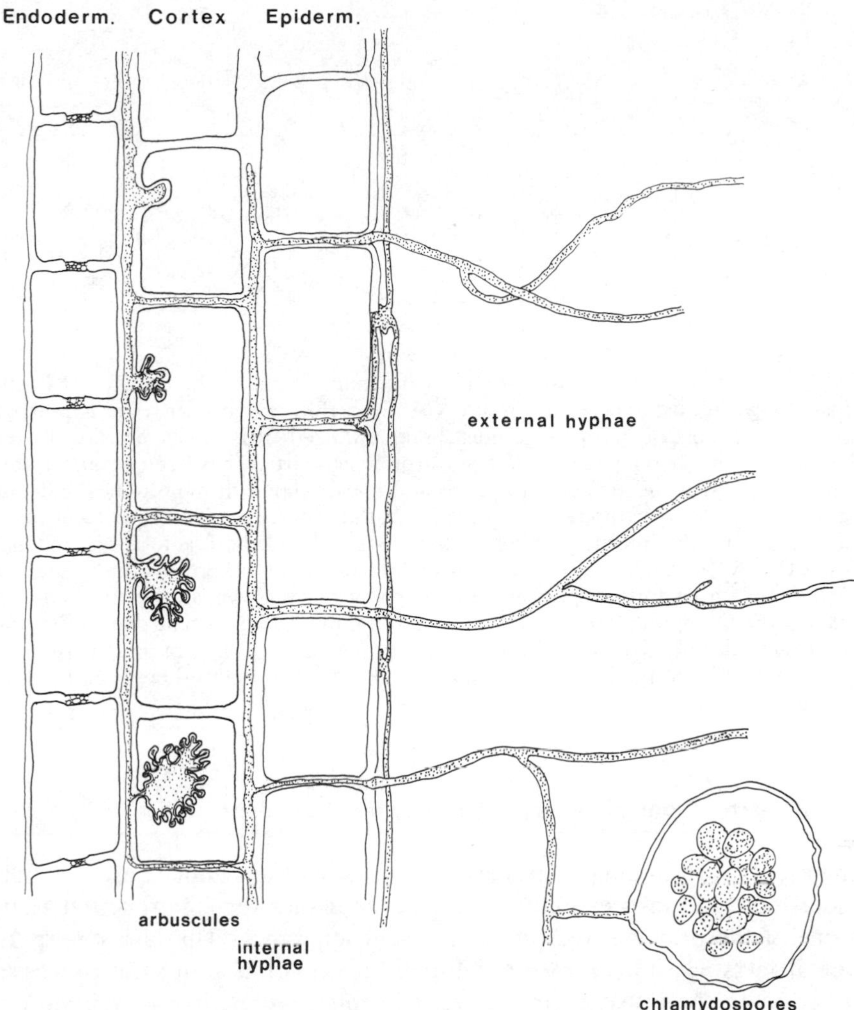

Fig. 7.14 Some general features of the association of vesicular–arbuscular mycorrhizal (VAM) fungus and a plant root. Hyphae running parallel to the root surface develop appressoria which penetrate *between* the epidermal cells. The internal hyphae are restricted to the apoplast–intrusions of hyphae into the cortical cells are always surrounded by the host plasmalemma – there is no direct contact between host and fungal cytoplasms. By repeated branching the surface area between arbuscule and host plasmalemma is greatly expanded. External hyphae are occasionally branched and extend for relatively great distances into the soil. There is strong evidence of adjacent plants being linked by a network of external hyphae. The internal hyphae do not penetrate the endodermis and are thus excluded from the stele. The fungus propagates via its chlamydospores.

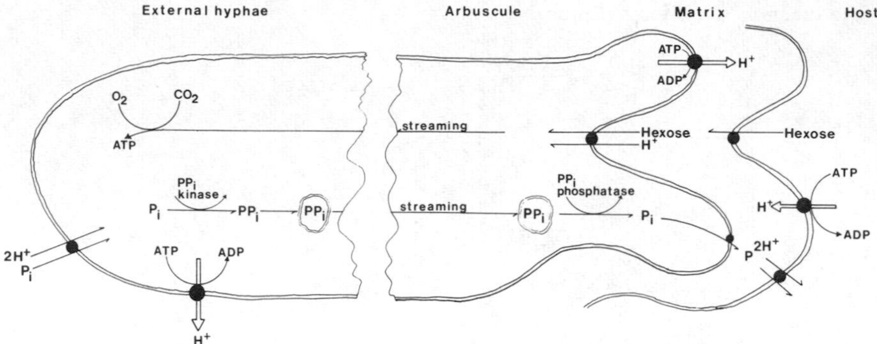

Fig. 7.15 A scheme to illustrate some of the main events in the transfer of P_i and sugars between host and fungus in a VAM mycorrhiza association. The hypothetical nature of most of this scheme must be emphasized. It is likely, but not proved that the external hyphae absorb P_i via a proton symport. It has been demonstrated that P_i is polymerized to polyphosphate and encapsulated in membranes and that these are carried by cytoplasmic streaming to the arbuscules. The activity of phosphatase (PP_i)in the internal hyphae is demonstrated and there is indirect evidence of the H+-ATPase in the arbuscule plasmalemma. It is not known whether hexose uptake is via a proton symport and the mechanism for P_i release by the fungus is unknown. Both plasmalemmata are invested in thin 'wall' material, that of the host being quite different in composition from normal walls. The protons, hexose and Pi are released into the matrix – a space of unknown properties – between the host and fungus.

7.5 Release of ions at the symplast – xylem boundary

The events surrounding the release of ions from the symplast are not fully understood; the inaccessibility of the stelar tissues for experimental study is one of the principal reasons for present ignorance. The lack of experimental facts has not, however, inhibited speculation about the processes involved. In the introductory remarks in this chapter, it was pointed out that the cells of the stele must preside over a net efflux of ions across the plasmalemma. Either the conditions affecting the plasmalemma or the transport mechanisms it contains must be quite different from those of the epidermal – cortical plasmalemmas. The hypotheses explaining release into the xylem cover a wide spectrum of ideas, from those which envisage the process as merely uncontrolled leakage to those which imply secretion mechanisms. As with most things the truth probably lies between the extremes.

7.5.1 Passive release

An early view of the process had, as its central idea, that cells within the stele existed within an oxygen-depleted environment since they were most distant from the external source of dissolved oxygen. In advancing this theory, Crafts and Broyer (1938) observed that, in anoxic conditions, roots leaked ions to the external medium. The combination of an inwardly directed flow of water and this leakage would carry ions into the xylem. Actual measurement of the partial pressure of oxygen in the core of roots is not an easy matter, especially since the insertion of any detector or incisions in the root may quickly alter the oxygen status. It was possible to insert oxygen-sensitive microelectrodes into primary lateral roots of sunflower (Bowling, 1973). These roots had a diameter of 0.5 mm and it was found that the oxygen diffusion into the stele was rapid so that a depletion of only 8 % was found between the xylem parenchyma and the surrounding solution. The technical *tour de force* involved in this work does not appear to have been repeated with other species or thicker roots. Recent technical advances in perfusing xylem vessels present an opportunity to re-examine this matter (see below).

During their maturation, xylem cells die and release their contents into the developing vessels behind them. Although this makes some contribution of ions it is most improbable that xylem maturation could be the major source of the contents of the xylem sap as was suggested by Hylmö (1953) and Higinbotham *et al.* (1973). It is also evident that ions enter the xylem in parts of the root where all xylem maturation has been completed (Läuchli *et al.*, 1978).

Microelectrode insertions into the roots of maize showed that there was only a slight gradient of electrical potential across the cells of the root (Dunlop & Bowling, 1970) but that the xylem sap was positive in relation to the xylem parenchyma surrounding it. Using this information and some reasonable estimates of the ionic concentrations in the parenchyma, they concluded that K^+ and Cl^- (and by inference all other monovalent ions) were at a lower electrochemical potential in the xylem than in the surrounding tissue. The gradient for diffusion was, therefore, tipped towards the xylem; by implication no active transport would be necessary to bring about ion movement into the sap. Since the publication of this paper, our views about the distinction between active and passive transport have changed profoundly with the realization that most ions probably move by gradient-coupled transport as a consequence of the primary active transport of protons or (in animals) Na^+. Since the xylem sap is normally pH 5 to 6 and the cytoplasm pH 7 to 8 it is evident that protons travelling from the xylem parenchyma to the xylem move uphill thermodynamically. If the movements of other ions across this boundary are coupled in any way with this proton gradient, then we would expect to find that their release into the xylem was metabolically dependent.

7.5.2 Metabolic regulation of release

Several lines of evidence lead to the conclusion that there is a set of

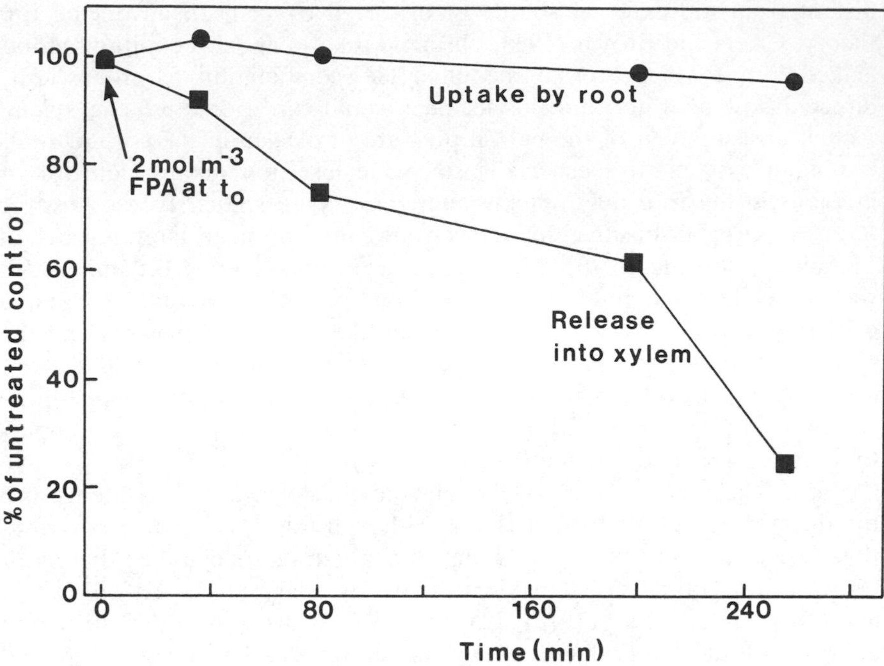

Fig. 7.16 Differential effect of FPA in the uptake of $^{36}Cl^-$ by the root and its release into the xylem of barley roots. (Based on data in Table 3, Schaeffer *et al.* 1975.)

metabolic controls over ion fluxes into the xylem which are partially independent of those regulating initial absorption on the outer side of the symplast. One source of this idea can be seen in Fig. 7.16. In the experiments of Schaeffer *et al.* (1975), barley roots were treated with the phenylalanine analogue, fluorophenylalanine (FPA), which becomes incorporated into proteins and renders them non-functional (= junk protein). The consequences of this treatment for the initial uptake of Cl^- and its release into the xylem were quite different (Table 7.4). Very shortly after the treatment began, Cl^- movement into the xylem sap declined sharply; it was not for several hours after this that Cl^- uptake by the root became inhibited. This experiment seemed to say that there was a protein involved in either symplast movement, or release from the symplast, which turned over rapidly and that was rendered ineffective by substitution of FPA for phenylalanine. The turnover of the protein(s) involved in initial uptake would have been slower. Essentially similar results were obtained using azetidine carboxylic acid (AZ) which subsitutes for proline and forms junk protein (Pitman *et al.*, 1977). These experiments, particularly the FPA one, are now being re-evaluated because of the finding by Erwee and Goodwin (1984) that aromatic amino acids seem greatly to decrease move-

Table 7.4 Effects of FPA on the distribution of labelled Cl^- in root tissues of *Zea mays* and release into the xylem 150 min after treatment had commenced†

Transport into xylem	Tracer Content	
	Cortex	Stele
	(% of control untreated by FPA)	
25	174	110

† Data from Schaeffer *et al.* (1975).

ment of substances through plasmodesmata when they are injected into cells of *Egeria densa*. Could it be that the observed effects of FPA are caused by blockage of the symplast? This question was anticipated by Schaefer *et al.* (1975) who showed that FPA treatment did not diminish the accumulation of labelled Cl^- in the stele; thus, its radial movement across the cortex had not been disturbed in the way one would expect had there been some interference with passage through plasmodesmata.

It is important to emphasize what these experiments do *not* show. They do not show that Cl^- release depends on an active transport across a membrane but merely that it depends on some protein which turns over rapidly; this protein does not necessarily have to be a membrane protein. The results are not, therefore, incompatible with the interpretation of Dunlop & Bowling (1970) as Bowling (1981) has pointed out.

The appearance of xylem parenchyma cells in the electron microscope has encouraged the view that ions are actively secreted into the xylem (Läuchli *et al.*, 1974; Pitman, 1977). Although they are not growing, the relatively large cytoplasmic volume fraction, numerous mitochondria, ER profiles and cytoplasmic vesicles all indicate intense metabolic activity; this may be related to their considerable capacity to pump protons (see below).

7.5.3 The role of stelar proton pumps

Several attempts have been made to study the movement of ions across the symplast – xylem boundary by perfusing the xylem vessels of roots (Clarkson *et al.*, 1984; De Boer & Prins, 1985) and hypocotyls (Okamoto *et al.*, 1984). By controlling directly the composition of the xylem sap it is possible to see how ionic concentrations, pH, oxygenation and metabolic inhibitors influence ion fluxes into the xylem. A relatively simple set up has been devised to do this is onion roots (Fig. 7.17) and a much more sophisticated one for *Vigna sesquipedalis* hypocotyls (Okamoto *et al.*, 1984).

In onion roots either exuding sap under root pressure or perfused with unbuffered solutions, the pH of the sap is usually in the range pH 5.5 to 6.5. The pH seems to be closely controlled so that perfusion of the xylem

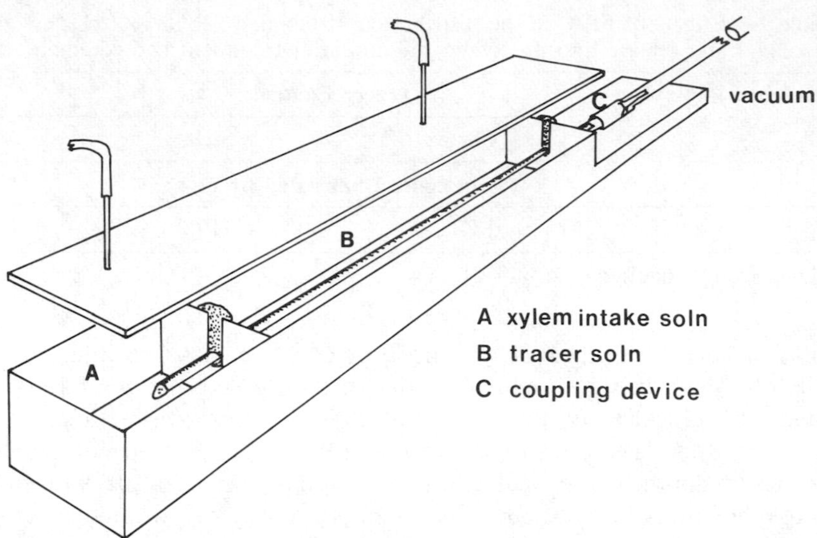

Fig. 7.17 A simple apparatus used to perfuse the xylem vessels of onion roots with defined solutions and to measure fluxes of ions from the stelar parenchyma into the xylem. Perfusion solutions are introduced into A and collected at C. Tracer solutions can be placed in B. Apoplast flow of solutions into the collection tube is prevented by the special coupling device (see Clarkson *et al.*, 1984).

with unbuffered acid or alkaline solutions scarcely affects the pH of the outflow. It is evident that protons have been either neutralized (equivalent to OH^- addition) or added to bring the pH back to its set value (Table 7.5). This rather curious response – an apparent attempt to control the pH of the apoplast at a pH not dissimilar to that of the vacuole – may be related to optimizing conditions for the net release of ions, but this is highly speculative. If the xylem is perfused with buffered solutions which hold the pH well away from its set value, the rate of proton addition or neutralization is greatly increased (Table 7.5). The output of protons across the symplast – xylem boundary can be greater than 1 pmol mm^{-2} s^{-1}; this is equivalent to a rate of about 200 μmol g^{-1} stelar parenchyma h^{-1} and is thus comparable to the H^+ output from epidermal cells in iron-deficient dicots (Römheld *et al.*, 1984). Proton pumping into the xylem was stimulated when fusicoccin (FC) was added to the perfusion stream but there was no effect on H^+ flux across the symplast–xylem boundary when FC was added to the external solution (Clarkson & Hanson, 1986).

At this moment it is not exactly clear how H^+ fluxes across the xylem – symplast boundary are coupled to the movements of other ions but there are precedents for proton symports and antiports in other plant membrane systems (see Chs. 3 and 4). Some time ago Hanson (1978) proposed a scheme for ion uptake and release by roots which depended on: (a) H^+-

Table 7.5 Effect of perfusing the xylem of onion roots with phosphate buffer of increasing strength on the release of protons from the xylem parenchyma†

Buffer‡ strength	**Perfusion solution pH**		**Net proton addition§**
(mol m^{-3})	**Inflow**	**Outflow**	**(pmol mm^{-2} s^{-1})**
1	8.0	6.56	0.40
3	8.0	7.10	0.78
6	8.0	7.33	1.00
10	8.0	7.50	1.21
Unbuffered	8.0	5.90	0.10

† Results from Fig. 4 in Clarkson & Hanson (1986).
‡ Na_2HPO_4/KH_2PO_4 buffer
§ Calculated on the measured junctional area between xylem parenchyma and xylem vessels.

pumps in a back-to-back arrangement at either end of the symplast; and (b) at least four types of carrier dependent on either the proton gradient or the membrane potential. These ideas are summarized in Fig. 7.18. It shows that the outer side (uptake) of the symplast there is H^+-anion co-transport and cation uptake via uniports dependent on $\Delta\psi$, while on the inner side (release) cation efflux is via a H^+-cation antiport and anions are driven down their free energy gradient by $\Delta\psi$. The scheme depends on the relative activity (or abundance) of the four types of carrier at the two sides. Much of this is speculative but electrophysiological measurements indicate that the trans-root electric potential (i.e. between the xylem sap and the external solution) is generated by two independent electrogenic mechanisms (Okamoto *et al.*, 1978; De Boer *et al*, 1983). The work with onion roots suggests that the electrogenic component in the stele is a proton-pumping ATPase. Thus one of the cardinal features of the Hanson hypothesis is probable; direct evidence about the carriers will be much harder to obtain. The use of xylem-perfused roots should allow observation of the relationships between the proton gradient and the fluxes of ions across the symplast – xylem boundary.

7.6 Ion movement into and out of phloem

The cells of the phloem form part of the symplast and are provided with abundant plasmodesmata (Warmbrodt, 1985). It is not clear whether there is any net input (loading) of ions into the root phloem. Tracer studies using segment-labelling techniques (see above) indicate that ^{32}P and $^{86}Rb(K^+)$ applied to mature zones of the seminal axis in barley and marrow move rapidly towards the root tip as well as to the shoot. This may be little more than tracer exchange, however. The concentration of some ions, notably

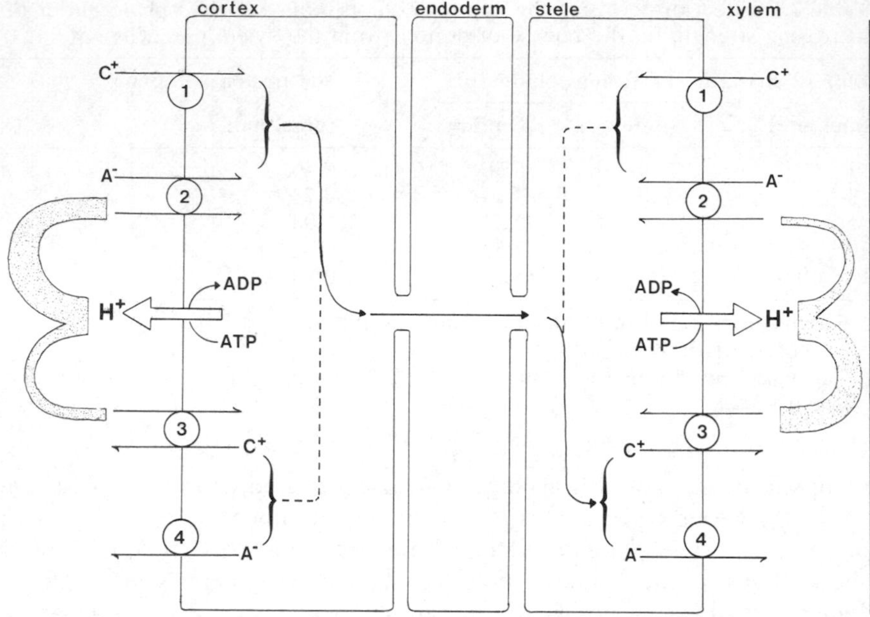

Fig. 7.18 A model, involving 'back-to-back' ATP-dependent H^+ translocation, to explain net uptake of ions on the cortical side of the endodermis and net release of ions into the xylem. Vectorial transport of ions is achieved by different activities of the four types of carriers on either side. The endodermis serves as an osmotic barrier. The uniports are dependent on the membrane potential, $\Delta\psi$; the symports on $\Delta\bar{\mu}_{H^+}$. (Adapted from Hanson, 1978.)

K^+, is large in the phloem sap (see Ch. 8); in tomato plants it was estimated that from 20 to 22 % of the K^+ in the xylem was derived from ions cycling within the plant, implying net unloading of K^+ from the phloem of roots (Armstrong & Kirkby, 1979). Ions which enter the companion cells or sieve tubes via plasmodesmata must do so against substantial flows of sucrose and amino compounds in the opposite direction. It is not known whether ions can be loaded into the phloem from the stelar apoplast in roots.

7.7 Nutrient demand and ion fluxes to the xylem

Several kinds of experiment lead to the conclusion that the rate of ion uptake and their subsequent movement into the root xylem is influenced by the demand for nutrients elsewhere in the plant. Demand, in this context, means simply the quantity of a given nutrient necessary to sustain a given rate of growth. Thus, transient changes in the growth rate of the shoot are quickly reflected in the rate of net ion uptake by a given unit of root. In other words the capture of light and CO_2 by leaves and the capture of resources from the rooting medium are coordinated, especially where none of these resources are growth limiting.

Changes in nutrient demand of a rather different kind are brought about when a plant becomes deficient in a given mineral nutrient. Deficiency in this context implies that, in some part of the plant, the supply of the given nutrient has, or will shortly become, growth limiting. A number of interesting changes occur in the physiology and development of plant roots in these circumstances (see Clarkson, 1985); for the present discussion three changes are of particular significance:

1. The absorption capacity for the deficient ion may increase, i.e. V_{max} increases.
2. The affinity of the absorption mechanism for the deficient ion may increase, i.e. K_m decreases.
3. The proportion of the uptake moving to the xylem and translocated within the plant may increase.

There is a great deal of variation in the response of species with respect to the changes and the response varies with different ions even in a given species. For instance, in barley K^+-deficiency produced changes (1) to (3) while P-deficiency did not alter the K_m of the phosphate transport system.

The control of transport mechanisms at the plasmalemma is incompletely understood, but it seems widely agreed that they must interact with some substance in the cytoplasm which varies in concentration. This substance could be some pool of the ion itself or some product of its assimilation in the case of the nitrogenous, phosphate and sulphate ions. We will now consider how radial movement of ions across the root might interact with this regulatory pool of ions or ion products in the cytoplasm.

The results discussed below show that the flux of ions through the symplast to the xylem can be changed by nutrient stress; this-change is quite independent of the nutrient status of the root which is actually doing the ion uptake. The data are from a paper by Drew & Saker (1984) in which young barley plants were arranged with a single root axis separated from its neighbours. Mild nutrient deficiency in the plants could be induced if the supply of K^+ or $H_2PO_4^-$ was restricted to this root, the remainder of the root system being placed in a culture solution lacking these nutrients. After a short time in such regimes, marked changes in the rate of ion uptake by the single root were observed in comparison with a single root on a control plant which had its roots uniformly supplied with the ions. In Table 7.6 the results for phosphate have been selected but those for K^+ are similar in most respects.The single roots were treated with ^{32}P-labelled solution for 24 h prior to harvesting. The restriction of phosphate supply to a single root for 3 d did not influence the size of that root nor its $[P_{total}]$ and $[P_i]$ relative to the control. The shoots and remaining roots had, however, significantly lower P-status, especially P_i relative to the

Table 7.6 Effect of 'demand' for phosphate on the uptake of individual roots, themselves well supplied with phosphate. In P-stressed plants no phosphate was applied to the bulk of the root system for 3 d prior to the experiment

Measurement	**Control plant (single root + P) (other roots + P)**	**P-stressed plant (single root + P) (other roots − P)**	**% increase or decrease**
Phosphate uptake (μmol g^{-1} wt root d^{-1})			
Total	50	159†	318
Distribution of uptake			
in the single root	27	42†	156
shoot	19	97†	510
other roots	4	20†	400
Tissue P concentration (mol g^{-1} dry wt)			
single root			
Total	245	215 NS	–
P_i	78	100 NS	–
shoot			
Total	246	172†	−30
P_i	98	31†	−67
other roots			
Total	236	146†	−38
P_i	72	19†	−74
Plant dry weight (mg)			
single root	27	26 NS	–
shoot	420	427 NS	–
Other roots	213	216 NS	–

† Significantly different from control P = 0.05
Adapted from Table 4 in Drew & Saker (1984).

control even though the plants had been growing at a similar rate over the 3 d period. Thus, growth had been 'diluting' the average shoot [P]. This was accompanied by a three-fold increase in P_i uptake by the single root; a compensatory absorption to offset the lack of uptake by the remainder of the root system. There are numerous examples of this behaviour: e.g. K^+ uptake in *Lolium multiflorum* (Drew & Nye, 1969); P_i uptake in potato (Cogliatti & Clarkson, 1983); NO_3^- uptake in barley (Drew & Saker, 1975); SO_4^{2-} uptake in *Macroptilium atropurpureum* (Clarkson *et al.*, 1983). The essential similarity in all of these examples was that the increased uptake was induced in roots which were not themselves demonstrably deficient in the nutrient concerned. A most important clue leading to an explanation of how this is brought about can be found for the translocation data in Table 7.6. In the P-deficient treatment there was a five-fold increase in translocation to shoots and a four-fold increase to the remaining roots but only a 1.6-fold increase in the labelled P in the single root itself. *Nutrient stress had increased both the flux into the xylem and the flux of tracer back down the phloem to the untreated roots.* In view of the much lower [P_i] in the shoots it seems highly probable that the [P_i] in the phloem would have been lower, but the specific activity of ^{32}P higher than in the controls. A rather similar conclusion might be reached from an experiment on foliar uptake and redistribution of labelled P in barley (Clarkson & Scattergood, 1982).

In Fig. 7.19 there are two schemes which could account for the results in the experiment we have just considered. In scheme 1, the flux to the xylem is governed by some undefined resistance R_2. The other principal resistances are those between the vacuole and the cytoplasmic regulatory pool, R_3 and between the pool and the outside solution, R_1. It is proposed that nutrient stress lowers R_2 to a value much below R_3 causing ions to flow rapidly to the xylem and diluting the cytoplasmic pool. This dilution would de-repress the activity of the carrier mechanism in the plasmalemma (see Glass, 1983) causing, in effect a lowering of R_1. This is all very well, but how could the value of R_2 be changed? Clearly this could be done if the activity of the carrier involved in P_i efflux from the xylem parenchyma to the xylem were to be increased (see Hanson model, Fig. 7.18). If some substance, transported in the phloem from shoot to root, changed its concentration and was released into the stelar symplast at an increased rate one might envisage how the properties of the membranes of the xylem parenchyma might be modified. The problem is one of specificity. The responses described above are all highly specific for the deficient ion. Its uptake and its translocation alone are affected by deficiency (see Lee, 1982). A general increase in the titre of some 'stress' hormone, such as abscisic acid (ABA), would not carry such a specific message to the xylem parenchyma.

A specific message from shoot to root might be the [P_i] in the phloem sap. Scheme 2 differs fundamentally from the other one because R_2 is envisaged as a fixed resistance between the xylem and two flows of P_i one from the phloem and the other from the cortical symplast. If the flux from the phloem is high, i.e. $[P_i]_{phloem}$ is large, the flux from the cortical symplast will be small. A diminution in $[P_i]_{phloem}$ such as one might expect when the $[P_i]_{shoot}$ is lowered by P-stress, would increase the flux through the cortical symplast and operate on the cytoplasmic regulatory pool as discussed in scheme 1. In practice the flux of P_i from the phloem becomes a variable resistance between the cytoplasmic pool and R_2. As drawn, scheme 2 implies short-term modulation of the fluxes through the symplast. It seems unlikely, however, that $[P_i]_{phloem}$ will vary rapidly. Experiments with barley plants which had been deprived of any source of P have shown that the enhanced capacity for uptake and translocation is maintained in P-deficient detached root systems for at least 2 h after excision (Clarkson *et al.*, 1979). During this time, of course, delivery of P_i via the phloem would cease.

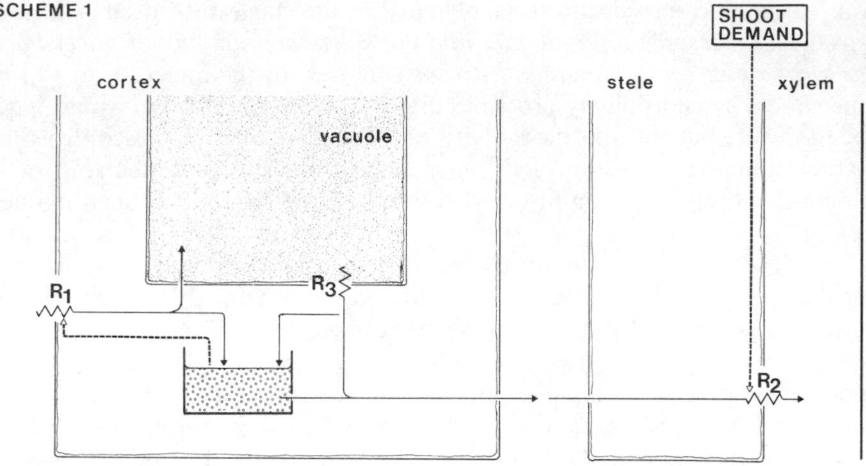

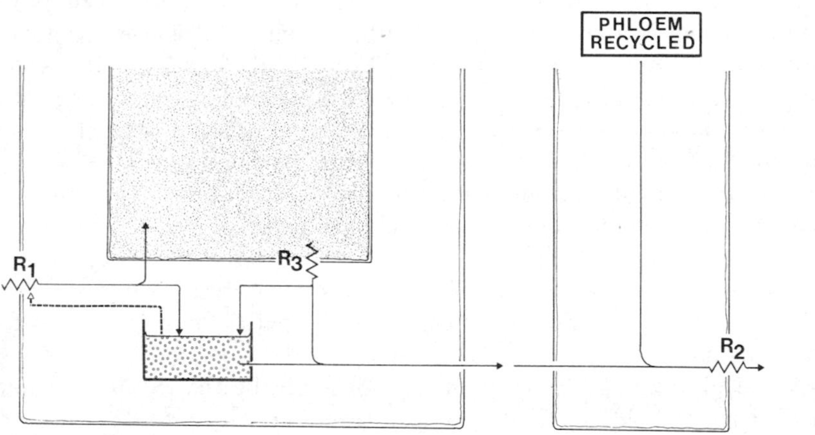

Fig. 7.19 Two schemes to explain how events affecting the stelar parenchyma may influence transport across the plasmalemma of cortical cells. In scheme 1, R_1 is the resistance (variable) across cortical plasmalemma, R_2 is the resistance (variable) across stelar parenchyma plasmalemma; R_3 is the resistance across tonoplast. When shoot demand is high, R_2 is lowered much below the value of R_3 – the level in the regulatory pool falls and the value of R_1 is lowered. Scheme 2 is similar except that R_2 is a fixed resistance and the ion flow via the root symplast is joined by variable amounts of ions recycling via the phloem. When this latter contribution is large (e.g. in a nutrient-sufficient plant) flow out of the regulatory pool is reduced and the value of R_1 rises. When few ions are recycled the level in the regulatory pool drops. It is important to realize that the regulatory pool could be either the whole cytoplasm, or a small volume of it close to the plasmalemma.

There is, then, some more permanent change in the characteristics of the symplast pathway, more like those suggested in scheme 1. Some combination of the concepts in the two schemes may turn out to be the most satisfactory explanation.

What the experiment of Drew & Saker (1984) and the above speculations reveal most clearly, is how little we know about the factors which regulate radial ion fluxes in roots.

7.8 Interaction of transpiration and ion fluxes across the root

All of the foregoing discussion implies that material fluxes are constrained to make them cross a pair of membranes at some point between the outside solution and the xylem sap. The pathway conducting water across membranes is quite distinct from those which conduct ions, thus there is probably no direct coupling of water and ion fluxes at the molecular level. Water molecules are smaller than hydrated ions and may enter spaces (known as 'kinks') between the hydrocarbon chains of the membrane lipids. The permeability of both artificial and biological membranes to water appears anomalously high if one's expectations are based on the solubility of water in the lipid. That ions tend to move much faster across the root than the mass flow of water can be verified simply; the sap exuding from detached roots is commonly ten-fold more concentrated in the major nutrients than is a dilute culture solution. Even when transpiration is high, the concentration of ions in the external solution is lowered by roots. What relationships would one expect to find between the bulk flow of water through the root and the movement of ions?

Let us consider first ions, such as Ca^{2+}, which may move radially in the cortical apoplast. We saw in Fig. 7.13 that the pattern of water- and Ca^{2+} uptake along the length of a marrow root was strikingly similar. This leads to the expectation that Ca^{2+} transport across the root should show a marked dependence on transpiration; this has been found to be the case in barley (Lazaroff & Pitman, 1966, see Fig. 9.9). In general, the influence of transpiration is most obvious if the ions in the external solution are concentrated, i.e. above the concentration which saturates high-affinity transport mechanisms (Russell & Barber, 1960). This is compatible with the view that absorption at higher concentrations involves movement into the apoplast to gain access to the cortical cells (see Sect. 7.4.1). High rates of bulk water movement would obviously add an important convective component to ion diffusion into the apoplast.

The second general influence one might expect is that increased water flow to the xylem, J_v, will reduce the concentration of ions in the sap and therefore reduce the tendency, implicit in the Hanson model (Fig. 7.18) for ions to diffuse back from the sap into the xylem parenchyma. This 'sweeping' action would tend to increase the net flux of ions from the symplast. It is tempting to study such matters in de-topped root systems in chambers where water flow can be varied by the application of hydrostatic pressure. This approach has been extensively used by Fiscus (e.g. 1977a, b) in soybean roots. Salim & Pitman (1984) compared solute fluxes to the shoots of intact plants of four species with the fluxes into the xylem of pressurized detached roots with equivalent J_v. The fluxes of Na^+ and Cl^- under applied pressure were ten-fold greater than for an equivalent J_v in transpiring plants of all four species (see Ch. 9). In roots of tomato and bean the fluxes of K^+ were also much greater. In his excellent review, Pitman (1982) suggests that the strong correlation between J_v and the solute flux, J_s found by Fiscus (1977b) might be a property of detached root systems not seen in intact plants. Application of quite small pressures (0.05 MPa) to roots of *Plantago maritima* drove air out of the intercellular spaces of the cortex and the resulting decrease in the diffusion coefficient of O_2 in the root tissue left the stelar cells anoxic and inhibited electrogenic ion pumping (De Boer & Prins, 1984).

7.9 Future developments

The two matters which are presently the most obscure in radial ion transport in roots are the mechanisms of ion release into the xylem and the regulation of symplast transport by the general demand for nutrients. These matters are probably interrelated. Their study will provide the basis for a much better integration of ion transport with whole-plant physiology than exists at the moment. It is accepted that there is interplay between the activity of the shoot in fixing CO_2 and elaborating substrates and that of the root in providing water and mineral nutrients. It is not clear whether the regulation of ion fluxes into or across the root is simply an expression of this mutual dependence on substrates or whether activities are coordinated by circulating messengers (hormones) which inform transport mechanisms how hard they should be working. A message, however small or discrete, might be intercepted by an experimenter if he puts his probes in the right place. A battery of techniques is now available for analysis of nano- and pico-gram amounts of plant growth regulators; what we lack, in all but a few well tried species, is access to phloem sap, especially in roots. If messages are sent to the root, this seems the most likely channel of communication; movement through the phloem is rapid and its exits are close to the stelar parenchyma at the unloading end of the symplast. Ingenuity is needed to find a way of studying this matter.

REFERENCES

Armstrong, M. J. & Kirkby, E. A. (1979). Estimation of potassium recirculation in tomato plants by comparision of the rates of potassium and calcium accumulation in their tops with their fluxes in the xylem stream. *Plant Physiology* **63**, 1143–8.

Attia, A. el F. & Jeanjean, R. (1983). Influence of osmotic shock and of plasmolysis on phosphate uptake by excised corn roots. *Physiologie Végétale* **21**, 39–47.

Bange, G. G. H. (1973). Diffusion and absorption of ions in plant tissue. III. The role of the root cortex in ion absorption. *Acta Botanica Neerlandica* **22**, 529–42.

Barclay, G. F., Peterson, C. A. & Tyree, M. T. (1982). Transport of fluorescein in trichomes of *Lycopersicon esculentum. Canadian Journal of Botany* **60**, 397–402.

Bonnett, H. T., Jr. (1968). The root endodermis: fine structure and function. *Journal of Cell Biology* **37**, 199–205.

Bostrom, T. E. & Walker, N. A. (1976). Intercellular transport in plants. II. Cyclosis and the rate of intercellular transport of chloride in *Chara. Journal of Experimental Botany* **27**, 347–57.

Bowen, G. D. & Rovira, A. D. (1967). Phosphate uptake along attached and excised wheat roots measured by an automatic scanning method. *Australian Journal of Biological Sciences* **20**, 369–78.

Bowling, D. J. F. (1973). Measurement of a gradient of oxygen partial pressure across the intact root. *Planta* **111**, 323–8.

Bowling, D. J. F. (1981). Release of ions to xylem in roots. *Physiologia Plantarum* **53**, 392–7.

Branton, D. & Deamer, D. W. (1972). *Membrane Structure*. Vienna, New York, Springer-Verlag.

Bryant, A. E. (1934). A demonstration of the connection of the protoplasts of the endodermal cells with the Casparian strips in the roots of barley. *New Phytologist* **33**, 231–9.

Burgess, J. (1971). Observations on structure and differentiation in plasmodesmata. *Protoplasma* **73**, 83–95.

Cailloux, M. (1972). Metabolism and the absorption of water by root hairs. *Canadian Journal of Botany* **50**, 557–73.

Capaccio, L. C. M. & Callow, J. A. (1982). The enzymes of polyphosphate metabolism in vesicular-arbuscular mycorrhizas. *New Phytologist* **91**, 81–91.

Carpita, N., Sabulase, O., Montezinos, D. & Delmer, D. P. (1979). Determination of the pore size of cells walls of living plants. *Science* **105**, 1144–7.

Clarkson, D. T. (1974). *Ion Transport and Cell Structure in Plants*, p. 313, London, McGraw-Hill.

Clarkson, D. T. (1984). Calcium transport between tissues and its distribution in the plant. *Plant, Cell and Environment* **7**, 449–56.

Clarkson, D. T. (1985). Factors affecting mineral nutrient acquisition in higher plants. *Annual Review of Plant Physiology* **36**, 77–115.

Clarkson, D. T., Gerloff, G. C. & Scattergood, C. B. (1979). Movement of endogenous and exogenous phosphate into xylem sap of detached root system of barley. Agricultural Research Council Letcombe Laboratory Report (1978), pp 54–6.

Clarkson, D. T. & Hanson, J. B. (1986). Proton fluxes and the activity of stelar proton pump in onion roots. *Journal of Experimental Botany* **37**, 1136–50.

Clarkson, D. T. & Robards, A. W. (1975). The endodermis, its structural development and physiological role. In Torrey, J G & Clarkson, D T. (eds.) *The Development and Function of Roots*. pp. 415–46. London, Academic Press.

Clarkson, D. T., Robards, A. W., Stephens, J. E. & Stark, M. (1987). Suberin lamellae in the hypodermis of maize (*Zea mays*) roots; development and factors affecting the permeability of hypodermal layers. *Plant, Cell and Environment* **10**, 83–93.

Clarkson, D. T. & Scattergood, C. B. (1982). Growth and phosphate transport in barley and tomato plants during the development of, and recovery from phosphate-stress. *Journal of Experimental Botany* **33**, 865–75.

Clarkson, D. T., Smith, F. W. & Vanden Berg, P. J. (1983). Regulation of sulphate transport in a tropical legume, *Macroptilium atropurpureum*, cv. siratro. *Journal of Experimental Botany* **34**, 1463–83.

Clarkson, D. T., Williams, L. & Hanson, J. B. (1984). Perfusion of onion root xylem vessels: a method and some evidence of control of the pH of the xylem sap. *Planta* **16**, 361–9.

Cogliatti, D. H. & Clarkson, D. T. (1983). Physiological changes in, and phosphate uptake in potato plants during development of, and recovery from phosphate deficiency. *Physiologia Plantarum* **58**, 287–94.

Cox, G. & Tinker, P. B. (1976). Translocation and transfer of nutrients in vesicular–arbuscular mycorrhizas. I. The arbuscule and phosphorus transfer: a quantitative ultrastructural study. *New Phytologist* **77**, 371–8.

Cox, G., Moran, K. J., Sanders, F., Nockolds, C. & Tinker, P. B. (1980). Translocation and transfer of nutrients in vesicular–arbuscular mycorrhizas. III. Polyphosphate granules and phosphorus translocation. *New Phytologist* **84**, 649–59.

Crafts, A. S. & Broyer, T. C. (1938). Migration of salts and water into xylem of the roots of higher plants. *American Journal of Botany* **25**, 529–35.

De Boer, A. H. & Prins, H. B. A. (1984). Transport electrical potential in roots of *Plantago media* L. as affected by hydrostatic pressure: the induction of an O_2-deficient root core. *Plant and Cell Physiology* **25**, 643–55.

De Boer, A. H. & Prins, H. B. A. (1985). Xylem perfusion of tap root segments of *Plantago maritima*: the physiological significance of electrogenic xylem pumps. *Plant Cell and Environment* **8**, 587–94.

De Boer, A. H., Prins, H. B. A. & Zanstra, P. E. (1983). Biphasic composition of trans-root potential in roots of *Plantago* species: involvement of spatially separated electrogenic pumps. *Planta* **157**, 259–66.

Drake, G. A. (1979). Electrical coupling, potentials and resistances in oat coleoptiles: effects of azide and cyanide. *Journal of Experimental Botany* **30**, 719–25.

Drake, G. A. Carr, D. J. & Anderson, W. P. (1978). Plasmolysis, plasmodesmata and the electrical coupling of oat coleptile cells. *Journal of Experimental Botany* **29**, 1205–14.

Drew, M. C., Chamel, A., Garrec, J.-P. & Fourcy, A. (1980). Cortical air spaces (Aerenchyma) in roots of corn subjected to oxygen stress. Structure and influence on uptake and translocation of ^{86}rubidium ions. *Plant Physiology* **65**, 506–11

Drew, M. C. & Nye, P. H. (1969). The supply of nutrient ions by diffusion to plant roots in soil. II. The effect of root hairs on the uptake of potassium by roots of rye grass (*Lolium multiflorum*). *Plant and Soil* **31**, 407–24.

Drew, M. C. & Saker, L. R. (1975). Nutrient supply and the growth of the seminal root system in barley. II. Localized, compensatory increases in lateral root growth and rates of nitrate uptake when nitrate supply is restricted to only part of the root system. *Journal of Experimental Botany* **26**, 79–90.

Drew, M. C. & Saker, L. R. (1984). Uptake and long distance transport of phosphate, potassium and chloride in relation to internal ion concentrations in barley: evidence of non-allosteric regulation. *Planta* **160**, 500–7.

Drew, M. C. & Saker, L. R. (1986). Transport to the xylem in aerenchymatous roots of *Zea mays* L. *Journal of Experimental Botany* **37**, 22–33.

Dumbroff, E. B. & Pierson, D. R. (1970). Probable sites for passive movement of ions across the endodermis. *Canadian Journal of Botany* **49**, 35–8.

Dunlop, J. & Bowling, D. J. F. (1970). The movement of ions to the xylem exudate of maize roots. III. The location of the electrical and electrochemical potential differences between the exudate and the medium. *Journal of Experimental Botany* **22**, 453–64.

Ehwald, U. R., Sammler, P. & Göring, H. (1973). Die Bedentung der Diffusion im Freien Raum für die Konzentrationsabhängigkeit der

Aufnahme von Zuckern unt Ionen durch pflanzliche Gewebe. *Biochemie und Physiologie der Pflanzen* **164**, 596-613.

Erwee, M. G. & Goodwin, P. B. (1983). Characterization of the *Egeria densa* Planch. leaf symplast. I. Inhibition of intercellular movement of fluorescent probes of group II ions. *Planta* **158**, 320–8.

Erwee, M. G. & Goodwin, P. B. (1984). Characterization of *Egeria densa* leaf symplast: response to plasmolysis, deplasmolysis and aromatic amino acids. *Protoplasma* **122**, 162–8.

Erwee, M. G. & Goodwin, P. B. (1985). Symplast domains in extracellular tissues of *Egeria densa* Planch. *Planta* **163**, 9–19.

Falk, H., Lüttge, U. & Weigl, J. (1965). Untersuchungen zur Physiologie plasmolysierter Zellen. II. Ionenaufnahme, O_2-Wechsel, Transport. *Zeitschrift für Pflanzenphysiologie* **54**, 446–62.

Ferguson, I. B. & Clarkson, D. T. (1976a). Ion uptake in relation to the development of root hypodermis. *New Phytologist* **77**, 11–14.

Ferguson, I. B. & Clarkson, D. T. (1976b). Simultaneous uptake and translocation of magnesium and calcium in barley (*Hordeum vulgare* L.) roots. *Planta* **128**, 267–9.

Fiscus, E. L. (1977a). Determination of hydraulic and osmotic properties of soybean root systems. *Plant Physiology* **59**, 1013–20.

Fiscus, E. L. (1977b). Effects of coupled solute and water flow in plant roots with special reference to Brouwer's experiment. *Journal of Experimental Botany* **28**, 71–7.

Foster, R. G. (1981). The ultrastructure and histochemistry of the rhizosphere. *New Phytologist* **89**, 263–73.

Glass, A. D. M. (1983). Regulation of ion transport. *Annual Review of Plant Physiology* **34**, 311–26.

Glass, A. D. M. & Perley, J. E. (1979). Cytoplasmic streaming in the root cortex and its role in the delivery of potassium to the shoot. *Planta* **145**, 399–401.

Gmaj, P., Murer, H. & Kinne, R. (1979). Calcium ion transport across plasma membranes isolated from rat kidney cortex. *Biochemical Journal* **178**, 540–57.

Goodwin, P. B. (1983). Molecular size limit for movement in the symplast of the *Elodea* leaf. *Planta* **157**, 124–30.

Gunning, B. E. S. & Hughes, J. E. (1976). Quantitative assessment of symplastic transport of pre-nectar into the trichomes of *Arbutilon* nectaries. *Australian Journal of Plant Physiology* **3**, 619–37.

Gunning, B. E. S. & Robards, A. W. (eds.). (1976) *Intercellular Communication in Plants: Studies on Plasmodesmata*. Berlin, Springer-Verlag.

Haas. D. L. & Carothers, Z. B. (1975). Some ultrastructural observations on endodermal cell development in *Zea mays* roots. *American Journal of Botany* **62**, 336–48.

Hanson, J. B. (1978). Application of the chemiosmotic hypothesis to ion transport across the root. *Plant Physiology* **62**, 402–5.

Hanson, J. B. (1984). The functions of calcium in plant nutrition, advances in plant nutrition. In Jinker, P. B. & Läuchli, A (eds.). *Advances in Plant Nutrition*, vol. 1, pp. 149–208. New York, Praeger.

Harrison-Murray, R. S. & Clarkson, D. T. (1973). Relationships between structural development and the absorption of ions by the root system of *Cucurbita pepo*. *Planta* **114**, 1–16.

Haüssling, M., Leisen, E. Marschner, H. & Römheld, V. (1985). An improved method for non-destructive measurements of pH at the root–soil interface (Rhizosphere). *Journal of Plant Physiology* **117**, 371–5.

Hepler, P. K. (1982). Endoplasmic reticulum in the formation of the cell plate and plasmodesmata. *Protoplasma* **111**, 121–33.

Hepler, P. K. & Wayne, R. O. (1985). Calcium and plant development. *Annual Review of Plant Physiology* **36**, 397–439.

Hiatt, A. J. (1967). Relationship of the cell sap pH to organic acid change during ion uptake. *Plant Physiology* **42**, 294–8.

Higinbotham, N., Davis, R. F., Mertz, S. E. & Shumway, L. K. (1973). Some evidence that radial transport in maize roots is into living vessels. In Anderson, W. P. (ed.). *Ion Transport in Plants*, pp. 493–506. London, Academic Press.

Hooymans, J. J. M. (1974). Role of cell compartments in the redistribution of K and Na ions absorbed by the roots of intact barley plants. *Zeitschrift für Pflanzenphysiologie* **73**, 234–42.

Hooymans, J. J. M. (1976). Competition between vacuolar accumulation and upward translocation of K^+ ions in barley plants. *Zeitschrift für Pflanzenphysiologie* **79**, 182–6.

Hylmö, B. (1953). Transpiration and ion absorption. *Physiologia Plantarum* **6**, 333–405.

Itoh, S. & Barber, S. A. (1983). Phosphorus uptake by 6 plant species as related to root hairs. *Agronomy Journal* **75**, 457–461.

Jaffe, L. F. & Nuccitelli, R. (1974). An ultrasensitive vibrating probe for measuring steady extracellular currents. *Journal of Cell Biology* **63**, 614–28.

Kauss, H., Köhle, H. & Jeblick, W. (1983). Proteolytic activation and stimulation by Ca^{2+} of glucan synthase from soybean cells. *FEBS Letters* **158**, 84–8.

Kochian, L. V. & Lucas, W. J. (1982). Potassium transport in corn roots. I. Resolution of kinetics into a saturable and linear components. *Plant Physiology* **70**, 1723–31.

Kochian, L. V. & Lucas, W. J. (1983). Potassium transport in corn roots. II. The significance of the root periphery. *Plant Physiology* **73**, 208–15.

Kochian, L. V. & Lucas, W. J. (1985). Relationship of spatial localisation of ion fluxes along *Zea mays* roots to root salt status. *Plant Physi-*

ology **77** (suppl.), 20 (Abstract).

Läuchli, A., Kramer, D. Pitman, M. G. & Lüttge, U. (1974). Ultrastructure of xylem parenchyma cells of barley roots in relation to ion transport to the xylem. *Planta* **11**, 85–99.

Läuchli, A., Pitman, M. G., Lüttge, U., Kramer, D. & Ball, E. (1978). Are developing xylem vessels the site of ion exudation from root to shoot? *Plant, Cell and Environment* **1**, 217–223.

Lazaroff, N. & Pitman, M. G. (1966). Calcium and magnesium uptake by barley seedlings. *Australian Journal of Biological Science* **19**, 991–1005.

Lee, R. B. (1982). Selectivity and kinetics of ion uptake by barley plants following nutrient deficiency. *Annals of Botany* **50**, 429–49.

Lefebvre, D. D. & Clarkson, D. T. (1984). Compartmental analysis of phosphate in roots of intact barley seedlings. *Canadian Journal of Botany* **62**, 1076–1080.

Lucas, W. J. & Kochian, L. V. (1985). Ion transport processes in corn roots: an approach utilizing micro-electrode techniques. In Gensler, W. (ed.). *Proceedings NATO Advanced Study Workshop*.

McCully, M. E. (1975). The development of lateral roots. In Torrey, J. G. & Clarkson, D. T. (eds.). *The Development and Function of Roots*, pp. 105–24. New York, Academic Press.

Marmé, D. (1983). Calcium transport and function. In Läuchli, A. & Bieleski, R. L. (eds.). *Inorganic Plant Nutrition, Encylopedia of Plant Physiology*, New Series, vol. 15B, pp. 599–625. Berlin, Springer-Verlag.

Marschner, H., Römheld, V. & Ossenberg-Neuhaus, H. (1982). Rapid method for measuring changes in pH and reducing processes along roots of intact plants. *Zeitschrift für Pflanzenphysiologie* **105**, 407–16.

Marx, C., Dexheimer, J., Gianinazzi-Pearson, V. & Gianinazzi, S. (1982). Enzymatic studies on the metabolism of vesicular–arbuscular mycorrhizas. IV. Ultracytoenzymological evidence (ATPase) for active transfer processes in the host-arbuscule interface. *New Phytologist* **90**, 37–43.

Miller, A. L., Raven, J. A., Sprent, J. I. & Weisenseel, M. H. (1986). Endogenous ion currents traverse growing roots and root hairs of *Trifolium repens*. *Plant, Cell and Environment* **9**, 79–83.

Nagahashi, G., Thomson, W. W. & Leonard, R. T. (1974). The Casparian strip as a barrier to the movement of lanthanum in corn roots. *Science* **183**, 670–1.

Nellans, H. N. & Popovitch, J. E. (1981). Calmodulin-regulated, ATP-driven calcium transport by basolateral membranes of rat small intestine. *Journal of Biological Chemistry* **256**, 9932–6.

Nye, P. H. & Tinker, P. B. (1977). *Solute Movement in the Soil-Root System*. Oxford, Blackwell.

Okamoto, H. Ichino, K. & Katou, K. (1978). Radial electrogenic activity in the stem of *Vigna sesquipedalis*: involvement of spatially separate pumps. *Plant, Cell and Environment* **1**, 279–84.

Okamoto, H., Mizuno, A., Katou, K, Ono, Y., Matsemura, Y. & Kojima, H. (1984). A new method in growth electrophysiology: pressurized intra-organ perfusion. *Plant, Cell and Environment* **7**, 139–47.

Olesen, P. (1980). A model of a possible sphincter associated with plasmodesmatal neck regions. *European Journal of Cell Biology* **22**, 250.

Overall, R. L., & Gunning, B. E. S. (1982). Intercellular communication with *Azolla* roots. II. Electrical coupling. *Protoplasma* **111**, 134–50.

Overall, R. L., Wolfe, J., & Gunning, B. E. S. (1982). Intercellular communication with *Azolla* roots. I. Ulstrastructure of plasmodesmata. *Protoplasma* **111**, 134–80.

Pang, P. C. & Paul, E. A. (1980). Effects of vesicular-arbuscular mycorrhiza on ^{14}C and ^{15}N distribution in nodulated faba beans. *Canadian Journal of Soil Science* **60**, 241–50.

Patterson, B. D. & Graham, D. (1977). Effect of chilling temperatures on protoplasmic streaming of plants from different climates. *Journal of Experimental Botany* **28**, 736–44.

Peterson, C. A. & Emanuel, M. E. (1983). Casparian strips occur in onion root hypodermal cells: evidence from band plasmolysis. *Annals of Botany* **51**, 135–7.

Peterson, C. A. Emanuel, M. E. & Humphreys, G. B. (1981). Pathway of movement of apoplastic fluorescent dye tracers through the endodermis at the site of secondary root formation in corn (*Zea mays*) and broad bean (*Vicia faba*). *Canadian Journal of Botany* **59**, 618–25.

Peterson, C.A., Emanuel, M.E. & Wilson, C. (1982). Identification of a Casparian band in the hypodermis of onion and corn roots. *Canadian Journal of Botany* **60**, 1529–35.

Peterson, C. A. & Perumalla, C. J. (1984). Development of the hypodermal Casparian band in corn and onion roots. *Journal of Experimental Botany* **35**, 51–7.

Peterson, C. A., Peterson, R. L. & Robards, A. W. (1978). A correlated histochemical and ultrastructural study of the epidermis and hypodermis of onion roots. *Protoplasma* **96**, 1–21.

Pitman, M. G. (1977). Ion transport into the xylem. *Annual Review of Plant Physiology* **28**, 71–88.

Pitman, M. G. (1982). Transport across plant roots. *Quarterly Reviews of Biophysics* **15**, 481–554.

Pitman, M. G., Wildes, R. A., Schaefer, N. & Wellfare, D. (1977). Effect of Azetidine 2-carboxylic acid on ion uptake and ion release to the xylem of excised barley roots. *Plant Physiology* **60**, 240–6.

Robards, A. W. (1968). A new interpretation of plasmodesmatal ultrastructure. *Planta* **82**, 200–10.

Robards, A. W. (1975). Plasmodesmata. *Annual Review of Plant Physiology* **26**, 13–29.

Robards, A. W., Clarkson, D. T. & Sanderson, J. (1979). Structure and permeability of the epidermal/hypodermal layers of the sand sedge (*Carex arenaria* L.). **101**, 331–47.

Robards, A. W., Jackson, S. M., Clarkson, D. T. and Sanderson, J. (1973). The structure of barley roots in relation to the transport of ions into the stele. *Protoplasma* **77**, 291–312.

Robards, A. W., Newman, T. M. & Clarkson, D. T. (1980). Demonstration of the distinctive nature of the plasmamembrane of the endodermis in roots using freeze-fracture electron microscopy. In Spandswick, R. M., Lucas, W. J. & Dainty, J. (eds.), *Plant Membrane Transport: Current Conceptual Issues*, pp. 395–6. Amsterdam, Elsevier/North Holland.

Römheld, V. & Kramer, D. (1983). Relationship between proton efflux and rhizodermal transfer cells induced by iron deficiency. *Zeitschrift für Pflanzenphysiologie* **113**, 73–83.

Römheld, V., Müller, C. & Marschner, H. (1984). Localization and capacity of proton pumps in roots of intact sunflower plants. *Plant Physiology* **76**, 603–6.

Rona, J-P., Van De Sype, G., Cornel, D., Grignon, C. & Heller, R. (1980). Plasmolysis effect on electrical characteristics of free cells and protoplasts of *Acer pseudoplatanus L. Bioelectrochemistry Bioenergetics* **7**, 377–91.

Rufty, T. W., Jr., Thomas, J. F., Remmler, J. L., Campbell, W. H. & Volk, R. J. (1986). Intercellular localization of nitrate reductase in roots. *Plant Physiology* **82**, 675–80.

Russell, R. S. & Barber, D. A. (1960). The relationship between salt uptake and the absorption of water by intact plants. *Annual Review of Plant Physiology* **11**, 127–40.

Russell, R. S. & Clarkson, D. T. (1975). Ion transport in root systems. In Sunderland, N. (ed.). *Perspectives in Experimental Biology*, vol. 2, pp. 401–11. Oxford, Pergamon Press.

Russell. R. S. & Sanderson, J. (1967). Nutrient uptake by different parts of the intact roots of plants. *Journal of Experimental Botany* **18**, 491–508.

Salim, M. & Pitman, M. G. (1984). Pressure-induced water and solute flow through plant roots. *Journal of Experimental Botany* **35**, 869–81.

Sanderson, J. (1983). Water uptake by different regions of the barley root. Pathways of radial flow in relation to development of the endodermis. *Journal of Experimental Botany* **34**, 240–53.

Schaefer, N., Wildes, R. A. & Pitman, M. G. (1975). Inhibition by *p*-fluorophenylalanine of protein synthesis and of ion transport across roots in barley seedlings. *Australian Journal of Plant Physiology* **2**, 61–73.

Scott, F. M. (1963). Root hair zone of soil-grown roots. *Nature* **199**, 1009–10.

Scott, M. G. & Peterson, R. L. (1979a). The root endodermis in *Ranunculus acris*. I. Structure and ontogeny. *Canadian Journal of Botany* **57**, 1040–62.

Scott, M. G. & Peterson, R. L. (1979b). The root endodermis in *Ranunculus acris*. II. Histochemistry of the endodermis and the synthesis of phenolic compounds in roots. *Canadian Journal of Botany* **57**, 1063–77.

Snellgrove, R. C., Splittstoesser, W. E., Stribley, D. P. & Tinker, P. B. (1982). The distribution of carbon and the demand of the fungal symbiont in leek plants with vesicular arbuscular mycorrhizas. *New Phytologist* **92**, 75–87.

Spanswick, R. M. (1972). Electrical coupling between cells of higher plants. A direct demonstration of intercellular communication. *Planta* **102**, 215–27.

Stephens, J. S. (1981). Effects of temperature on the hydraulic conductivity of the roots of *Zea mays*. Ph.D. thesis, 150 pp. University of Reading.

Thomson, W. W. and Liu, L. L. (1967). Ultrastructural features of the salt gland of *Tamarix aphylla* L. *Planta* **73**, 201–20.

Toth, R. & Miller, R. M. (1984). Dynamics of arbuscule development and degeneration in a *Zea mays* mycorrhiza. *American Journal of Botany* **71**, 449–60.

Vakhmistrov, D. B. & Kurkova, E. B. (1979). Symplastic connections in the rhizodermis of *Trianea bogotensis*. *Soviet Plant Physiology* **26**, 763–71.

Vakhmistrov, D. B., Kurkova, E. B. & Zlotnikova, I. F. (1981). Symplastic connections and intracellular activity of potassium in the rhizodermis of *Raphanus sativus*. *Soviet Plant Physiology* **28**, 826–33.

Van Fleet, D. S. (1961). Histochemistry and function of the endodermis. *Botanical Review* **27**, 165–221.

Van Iren, F. & Boers van der Sluijs, P. (1980). Symplasmic and apoplasmic radial ion transport in plant roots. Cortical plasmalemmas lose absorption capacity during differentiation. *Planta* **148**, 130–7.

Walker, N. A. & Pitman, M. G. (1976). Measurement of fluxes across membranes. In *Encyclopedia of Plant Physiology*, vol. 2A, pp. 93–126. Berlin, Heidelberg, New York, Springer-Verlag.

Warmbrodt, R. D. (1985). Studies on the root of *Hordeum vulgare* L. – ultrastructure of the seminal root with special reference to the phloem. *American Journal of Botany* **72**, 414–32.

Weisenseel, M. H., Dorn, A. & Jaffe, L. F. (1979). Natural H^+ currents traverse growing roots and root hairs of barley (*Hordeum vulgare* L.). *Plant Physiology* **64**, 512–18.

Williamson, R. E. & Ashley, C. C. (1982). Free Ca^{2+} and cytoplasmic streaming in the alga *Chara*. *Nature* **296**, 647–51.

8 Phloem transport – with emphasis on loading and unloading

T. E. Humphreys

8.1 Introduction

From experiments in which plants were grown under high light intensities and increased CO_2 concentrations, it is concluded that rates of photosynthesis usually limit crop yield. The various plant parts compete for a limited supply of photosynthate, and improvements in crop yield have resulted primarily from selecting plants in which the part of economic interest competes well (Gifford & Evans, 1981). What determines success in this competition for photosynthate is not well understood, but certainly, it involves the three processes required for import of photosynthate: phloem loading, long-distance transport, and phloem unloading. Movement of photosynthate within the phloem occurs in sieve tubes which extend throughout the plant. In the stem, the many sieve tubes run parallel to the main axis; for example, Schumacher (1948) counted 300–400 tubes in a cross-section of the stem of *Byronia dioica* (a cucurbitaceous vine). The *Byronia* plant was 7.4 m long and the combined length of the branches was 13.8 m. Schumacher (1948) estimated that the total internal surface area of the sieve tubes of the entire plant was 1 m^2, and she estimated that this surface enclosed a volume of only about four 4 ml! The sieve tubes of the branches, leaves and inflorescences contained a volume about equal to that of the main axis but with 75 % of the total surface area. Clearly, the phloem presents a large surface across which water and solutes can move and which encloses a relatively, very small volume of transport solution. The reader should keep these surface–volume relations in mind while perusing this chapter.

The intent of this chapter is to explore the pathways for photosynthate leading to and from the phloem, and then to examine the processes of phloem loading and unloading. But first a general account of phloem transport is given to orient the reader.

8.2 Phloem transport: a survey

8.2.1 Flows of solutions in phloem and xylem

In the sieve tubes of the phloem, sucrose and other substances move long distances in a flow of solution from places of synthesis, say, mature leaves, to parts of the plant where they are used for respiration and growth. In the vessels and tracheids of the xylem there is a companion flow of solution whose direction is usually from root to leaf. The phloem solution is concentrated ($0.3–0.9 \times 10^3$ mol m^{-3} sucrose), slow moving (usually less than 200 cm h^{-1}) and under pressure. The xylem solution is dilute (0.1–0.4 % solids), fast moving (m h^{-1}) and under tension. In a mature leaf during the day, the volume of solution that enters in the xylem may be tenfold or so greater than the volume of solution that exits in the phloem, the difference being water lost in transpiration. During the night, however, the volumes of solution entering in the xylem and leaving in the phloem are nearly the same. Because water readily moves between xylem and phloem, these flows of solution constitute a circulation of water within the plant. Water may also move from phloem to xylem. For example, during the development of many fruits solutes moving in with the sieve tube stream are converted to insoluble reserves creating an excess of water which, if not transpired, moves out of the fruit in the xylem (Pate *et al.*, 1985).

There also is movement of solutes (amino acids, for example) between xylem and phloem, but a large difference in concentration of most solutes is maintained, whereas the difference in water potential between adjacent sections of xylem and phloem is probably always near zero. Some solutes, such as K^+ and phosphorus anions, circulate in the plant by alternating between xylem and phloem, but most do not. Sucrose moves in the phloem from sites of synthesis to sites of use, with little of it being lost to the xylem. However, carbon may be said to circulate in that it moves as sucrose in the phloem from mature leaf to root where some of it is used to synthesize amino acids, amides or ureides; and these return again to the mature leaf in the xylem.

During the day, the flow of the xylem solution is driven primarily by the difference in water potential between leaf and atmosphere. At night, with the stomates closed, the flow of the xylem solution is driven by the difference in water potential between xylem and aerial tissues, especially phloem and rapidly growing tissues. The flow of the phloem solution is driven by a turgor pressure difference between two sections along the same sieve tube. This is the pressure flow hypothesis originally proposed by Münch (1930), and it appears to be the one favoured by most phloem physiologists. The difference in turgor pressure results from an active and selective accumulation of solutes, 'phloem loading', at one place and removal of solutes, 'phloem unloading', at another place along the sieve tube. For instance, sucrose may be loaded in the phloem of the minor veins of a mature leaf and be unloaded from the phloem of the root, fruit or immature leaf. The loading process requires metabolic energy to move solutes up a chemical potential gradient, and this is the energy input that drives the flow of the phloem solution according to the pressure flow hypothesis.

8.2.2 Symplast/apoplast

Before being loaded into the phloem, sucrose must move to the phloem from cells where it is synthesized or stored; after being unloaded from the phloem, sucrose must move from the phloem to cells where it is used or stored. The routes sucrose takes during these movements pass through living cells, cell walls and intercellular spaces, i.e. pass through symplast and apoplast. All the protoplasts connected to one another by plasmodesmata form a continuum called the 'symplast' (see Spanswick, 1976; and Chs. 6, 7). The cell walls and intercellular spaces form another continuum called the 'apoplast' (see Läuchli, 1976; and Chs. 6, 7) which surrounds and contains the symplast. Some protoplasts may be isolated, having no plasmodesmatal connections (guard cells, for instance, see Ch. 12), but these are few. The apoplast may be interrupted by a layer of suberized cell walls such as that of the root endodermis (see Ch. 7).

The plasmalemmas of the protoplasts form a continuous boundary between symplast and apoplast across which solutes are selectively transported and water moves readily. Vacuoles are islands within the symplast; they make up most of the symplast volume and serve as reservoirs for solutes. In some plants (e.g. barley), much of the carbon fixed in photosynthesis and not immediately exported from the leaf in the sieve tubes of the phloem is stored as sucrose in vacuoles for export during the night (Kaiser & Heber, 1984).

The apoplast contains both water- and gas-filled spaces. In leaves, gas space may be as high as 30 % of the leaf volume. Only a few measurements have been made of the water-filled space (the one of interest for this chapter), and they range from 4–20 % of the volume of the tissue (Hawker, 1965; Minchin & Thorpe, 1984). In transpiring leaves, the water of the mesophyll apoplast is mostly confined to the cell wall; in wheat leaves, the volume of the water in the wall is roughly the same as that of the cytosol (Altus & Canny, 1985). The term 'apoplast' is not synonymous with 'apparent free space' (see Ch. 1). The latter term refers to the volume into which a particular exogenous solute moves rapidly and passively. This volume may include, depending on the solute, part or all of the protoplast as well as the cell wall and intercellular space. 'Apoplast' and 'symplast', on the other hand, refer to anatomical spaces without regard to the ability of solutes to enter them.

8.2.3 Source/sink

A 'source' is a plant part from which sucrose and other substances are exported after being loaded in the phloem. A 'sink' is a plant part that imports substances from the source; a plant part where substances are unloaded from the phloem and used for growth and storage. A leaf serves as an example of both source and sink. A young leaf is a sink; it imports sucrose, and it requires more reduced carbon for respiration and growth than its limited capacity for photosynthesis can supply. As a leaf matures it goes through a transition from sink to source which starts at the tip and moves to the base. This means that during the transition there is a period when the leaf both imports and exports. Of course, this bidirectional movement occurs in separate sieve tubes. For example, sugar-beet leaves start exporting sucrose when they reach about 35 % of their final length and stop importing sucrose when they reach about 45 % of their final length (Fellows & Geiger, 1974). During the transition period the osmotic pressure of the sieve tube contents was found to be highest at the tip (sucrose-exporting region) and lowest at the base (sucrose-importing region) of the leaf. Interpreted according to the Münch pressure flow hypothesis, these data indicate that export is initiated during leaf development when phloem loading of sucrose causes the turgor pressure of a sieve tube in the leaf tip to exceed that of the same sieve tube in the stem. The transition stage of leaf development appears to be irreversible so that a mature leaf does not revert to a sink even when shaded (see Turgeon, 1984a). Indeed, when albino tobacco shoots (grown *in vitro*) were grafted to de-topped green tobacco plants, the developing albino leaves acted like green leaves in that they terminated import progressively from tip to base; but the albino leaves died soon after cessation of import (Turgeon, 1984b). Clearly, in this case the cessation of import was not a result of the initiation of export.

A plant has many sinks and they compete with one another for the limited supply of photosynthate. How well a particular sink competes is determined by position, size and a factor called 'sink strength'. Sink strength is a measure of the rate per gram fresh weight at which the sink can unload the photosynthate delivered to it in the sieve tubes. There are other factors involved in sink competition, one being the length of time a plant part remains a sink. For example, the share of total photosynthate obtained by a kernel of wheat is partly determined by the duration of kernel development, and this is an important determinant of crop yield. According to this view, phloem unloading (rate and duration) plays a dominant role in determining how well a particular sink fares in the competition for photosynthate. As a general rule, a sink is supplied by its nearest source. The lower leaves supply the roots, the upper leaves supply the stem apex and young leaves at the top and the middle leaves supply both roots and top. But along much of the route from source to distant sink phloem passes through sink tissue that it supplies.

8.2.4 Phloem structure

The sieve tube is made up of highly specialized, elongated cells called 'sieve elements' joined end-to-end. Where two sieve elements join, the walls are modified to form a sieve plate which contains pores lined with plasmalemma connecting the cytoplasmic contents of the two cells. The pores of the sieve plate are much larger in diameter than plasmodesmata, from 50–100 times larger in most angiosperms. Sieve elements are modified in other ways. During development they go through a senescence-like phase, a selective autophagy, in which the nucleus and vacuole disappear and the endoplasmic reticulum loses its ribosomes and takes a parietal position. These modifications allow the sieve tube to function as a conduit containing a flow of solution. There is much controversy about what drives the flow. If the flow is driven by a turgor pressure difference between source and sink, as most phloem physiologists believe, the pressure difference must be great enough to overcome the resistance to flow of the sieve plates and to cause a flow speed of about 100 cm h^{-1}. There is uncertainty, however, as to the size of the opening in the pores which often appear plugged with protein. Because the sieve tube contents are under pressure, there is a surge of solution when the tissue is prepared for sectioning, and this may cause the plugs seen in many electron micrographs. If steps are taken to reduce surge artifacts during tissue preparation, more open pores are obtained. Nevertheless, Cronshaw (1981) concludes that the evidence in favour of unplugged pores is still inconclusive.

But even assuming open pores in the sieve plates, Canny (1984) using the Poiseuille equation calculated that in some plants, especially gymnosperms whose sieve cells have very small pores, the pressure gradients necessary to drive the required flow are unrealistically high. Because of the seemingly high resistance to flow across sieve plates, other model mechanisms have been proposed as alternatives to the pressure flow mechanism of Münch. These have not found favour with most physiologists. To quote Weatherley (1975), 'The Münch hypothesis seems, then, to be rising to a position of unassailable glory. However, it should be remembered that it is still based on few facts, and its position still derives more from the weakness of its adversaries than from its own inherent strength.' The reader should see the book edited by Zimmermann & Milburn (1975) for detailed discussions of this controversy.

Of course, any model of phloem transport requires phloem loading and unloading regardless of what is proposed as the driving force for solution flow. The beauty of the Münch hypothesis is that phloem loading not only charges the phloem solution with photosynthate but also provides the energy that moves it.

Sieve elements are surrounded by parenchyma tissue containing a range of cell types including companion cells, phloem parenchyma and bundle sheath cells. Sometimes, as in minor veins of the leaf, a tracheary element lies along one side of the sieve element. (In most cases, even if xylem and phloem elements are not in actual contact, they are in close proximity. This is an important consideration when devising models for loading and unloading as they should not allow sucrose to have direct access to the transpiration stream.) The bundle sheath cells of C_4 plants have many chloroplasts, whereas in C_3 plants, chloroplasts are few or absent in the bundle sheath. Chloroplasts may sometimes be found in phloem parenchyma and companion cells. The sieve element and companion cell are derived from the same precursor, and there is extensive cytoplasmic continuity between the two cells through lateral sieve areas. In large veins, the companion cells are smaller in diameter than the sieve elements, but in the minor veins of the leaf where most phloem loading occurs the tiny sieve elements are dwarfed by their companion and phloem parenchyma cells (see Turgeon & Webb, 1976). Some companion and phloem parenchyma cells in the minor veins of dicots develop extensive wall ingrowths with the result that there is a great increase (about five-fold) in the plasmalemma surface area. These cells are called 'transfer cells' because they are thought to be especially active in moving substances between symplast and apoplast (Gunning *et al.*, 1974). In *Pisum*, the development of wall ingrowths in the transfer cells of the minor veins occurs about the time the leaves start to export (Gunning & Pate, 1974). This suggests that transfer cells are involved in phloem transport, possibly in loading sucrose from the apoplast into the phloem.

8.2.5 Substances transported in sieve tubes

Sieve tube exudates contain many solutes, inorganic as well as organic, but only a few of special interest for this chapter are mentioned here. Ziegler (1975) gives an extensive listing and discussion of transported substances.

Sugars or sugar alcohols make up the bulk of the organic solutes found in the sieve tube exudates of most species. Sucrose is the most common, and usually most abundant, sugar. In some species, galactose derivatives of sucrose (the raffinose family of sugars: raffinose, stachyose and verbascose) are present, and these may predominate. Zimmermann and Ziegler (1975) list the sugar contents of sieve tube exudates from over 500 woody species of dicots. Sucrose was found in all of them. Sugar alcohols are widely distributed throughout the plant kingdom, and are major transport compounds in many species. Why sugar alcohols have arisen as important metabolites and transport compounds in place of hexoses and sucrose is not known, but interesting speculations about their roles are to be found in Bieleski (1982) and Loescher *et al.* (1985). It is noteworthy that hexoses are not found in sieve tube exudates and that all the transport sugars are non-reducing. Apparently, this selectivity resides in the phloem loading process.

Nitrogenous compounds are usually second to sugars or sugar alcohols in order of abundance in sieve tube exudates. Depending on the species and on the stage of growth, asparagine, glutamine and ureides are usually most abundant; but many amino acids may also be found (Ziegler, 1975; Pate, 1980). Mature leaves, because of their rapid transpiration, receive most of the organic nitrogen (amides, amino acids and ureides) and NO_3^- of the xylem stream, and yet they maintain a nearly constant nitrogen content. The excess nitrogen is exported in the phloem from the mature leaf to young leaves, inflorescences and fruits, i.e. to sinks. The amides and amino acids may be exported without change. Ureide nitrogen (if present) and NO_3^- are converted to NH_4^+, which then is used to form glutamate, glutamine and asparagine for export in the phloem. Similarly, during leaf senescence and seed germination, nitrogen in leaf protein and seed storage protein is mobilized for export in the phloem by forming glutamine, asparagine and amino acids.

K^+ and ATP are found in sieve tube exudates. K^+ is the most abundant inorganic solute, being present at concentrations of 50–200 mol m^{-3}. K^+ is balanced by inorganic and organic phosphate, Cl^-, and organic acid anions in a sieve tube solution of from pH 7.2–8.5. The ATP concentration is usually in the 0.2–0.5 mol m^{-3} range. The 'adenylate energy charge' (defined as the ratio: [ATP] + $\frac{1}{2}$ [ADP]/[ATP] + [ADP] + [AMP]) was calculated to be 0.8–0.9 in *Robinia* and *Tilia* sieve tube exudates (see Ziegler, 1975), and this is the range of values found in normal, actively metabolizing cells (Pradet & Raymond, 1983). The 'adenylate energy charge' of the sieve tube contents may reflect the activity of the companion cells, as these are thought to provide the energy metabolism for the mitochondria-poor sieve elements. It is likely that sieve tube ATP is an energy source for phloem loading. However, it is not known if ATP is also a transport solute, and whether it enters the phloem at one place and exits at another.

Four of the five plant growth substances (auxin, gibberellins, abscisic acid, and cytokinins) move in the phloem as well as in the xylem. Some sinks are supplied entirely by the phloem (e.g. apical meristems), others are supplied by both phloem and xylem (e.g. young leaves), while mature leaves are supplied entirely by the xylem. This undoubtedly has important implications for phloem transport processes because the growth substances may affect sink strength (e.g. rate and duration of growth) and source function (e.g. leaf senescence). Discussions of this subject are to be found in papers by King (1976) and Vreugdenhil (1983) and in references therein.

8.3 Phloem loading

8.3.1 Synthesis of transport sugars and sugar alcohols

In leaves, the transport sugars and sugar alcohols are synthesized in mesophyll and bundle sheath cells not far from the minor vein phloem.

Table 8.1 Reactions for the synthesis and breakdown of transport sugars and sugar alcohols

Synthesis:
1. UDPG + fructose ⟷ sucrose + UDP (sucrose synthase)
2. UDPG + fructose-6-P ⟷ sucrose-P + UDP (sucrose phosphate synthase)
3. sucrose-P + H_2O ⟶ sucrose + P_i (sucrose-P phosphatase)
4. UDP-galactose + *myo*-inositol ⟷ galactinol + UDP (galactinol synthase)
5. galactinol + sucrose ⟷ raffinose + *myo*-inositol (raffinose synthase)
6. galactinol + raffinose ⟷ stachyose + *myo*-inositol (stachyose synthase)
7. glucose-6-P + NADPH + H^+ ⟷ sorbitol-6-P + $NADP^+$ (glucose-6-P reductase)
8. sorbitol-6-P + H_2O ⟶ sorbitol + P_i (specific phosphatase)
9. mannose-6-P + NADPH + H^+ ⟷ mannitol-1-P + $NADP^+$ (mannose-6-P reductase)
10. mannitol-1-P + H_2O ⟶ mannitol + P_i (specific phosphatase)

Breakdown:

11. sucrose + UDP ⟷ UDPG + fructose (sucrose synthase)
12. sucrose + H_2O ⟶ glucose + fructose (invertase)
13. raffinose (stachyose) + H_2O ⟶ sucrose + 1 (2) galactose (α-galactosidase)
14. galactose + ATP ⟶ galactose-1-P + ADP (galactokinase)
15. galactose-1-P + UTP ⟷ UDPgalactose + PP_i (UDPgalactose pyrophosphorylase)
16. UDPgalactose ⟷ UDPG (UDPG-4-epimerase)
17. sorbitol + NAD^+ ⟷ fructose + NADH + H^+ (sorbitol dehydrogenase)
18. mannitol + NAD^+ ⟷ fructose + NADH + H^+ (mannitol dehydrogenase)

In sugar-beet for example, mesophyll cells are situated no more than two to three cell diameters (65–100 μm) from a minor vein (Geiger & Cataldo, 1969). Sucrose, raffinose and stachyose are synthesized in the cytosol by the reactions shown in Table 8.1. The starting compound for synthesis is triose phosphate from the chloroplast where it is formed either directly during photosynthesis or from starch breakdown (see Ch. 2). Less is known about the synthesis of sugar alcohols (see Bieleski, 1982). Glucose-6-phosphate reductase (reaction 7, Table 8.1) has been localized in the chloroplasts of apple cotyledons (Yamaki, 1981) and the chloroplast would supply the necessary quantity of NADPH. In celery, mannose-6-phosphate reductase (reaction 9, Table 8.1) is a cytoplasmic enzyme (Rumpho *et al.*, 1983), non-reversible glyceraldehyde-3-P dehydrogenase supplying the NADPH. Since hexitols do not readily penetrate the chloroplast membrane (indeed, sorbitol is commonly used as an osmoticum in chloroplast isolation), the phosphatases (reactions 8, 10) are probably located in the cytosol. It is not known how sorbitol-6-P crosses the chloroplast membrane, if indeed, it is synthesized in the chloroplast.

One might expect that enzymes for synthesis would be in the source tissue and that enzymes for breakdown (Table 8.1) would be in the sink tissue, but such a separation of enzymes is not always found. During leaf development the transition from sucrose importer to sucrose exporter may be accompanied by an increase in sucrose phosphate synthase activity (reaction 2, Table 8.1) but the enzymes for sucrose breakdown, sucrose synthase (reaction 11) and invertase (reaction 12), persist even in the mature leaf. Clearly, these enzymes must be controlled to prevent a futile cycle of sucrose synthesis followed by sucrose breakdown; control may involve inhibition or sequestration in a cellular compartment away from sucrose. In apple, a sorbitol translocator, developing leaves initially contain high activities of sorbitol dehydrogenase (reaction 17), and at this stage they import sorbitol. As the leaves mature from tip to base and begin exporting sorbitol, glucose-6-phosphate reductase activity increases and sorbitol dehydrogenase activity decreases (Loescher *et al.*, 1982).

These changes in the levels of the enzymes involved in synthesis and breakdown of transport substances are part of a complex series of events leading to development of export capability in a leaf (Fellows & Geiger, 1974; Giaquinta, 1983). The initiation of phloem loading (accompanied, presumably, by the cessation of phloem unloading) is the last event in this development.

8.3.2 Pathways of photosynthate to the phloem

Phloem loading is defined as an 'active and selective accumulation' of photosynthate in the phloem. This means that there is a step-up in photosynthate concentration between the cells of the mesophyll and bundle sheath and the cells of phloem. In the minor veins of sugar-beet the osmotic potential of the sieve elements and companion cells was estimated to be −3 MPa whereas it was −0.8 MPa in the vascular parenchyma cells and −1.3 MPa in the mesophyll cells (see Geiger, 1975). Since most of the sieve tube solute in sugar-beet is sucrose, the step-up in sucrose concentration must occur at the plasmalemmas of the sieve element and companion cell (se–cc complex). Possible pathways to the phloem that include this selective step-up in concentration are shown in Fig. 8.1 which shows a file of cells from mesophyll to xylem with plasmodesmata. The diagram can serve for a monocot or dicot leaf from a C_3 or C_4 plant.

The only known way to accomplish a step-up in concentration along a path is for sucrose to be actively taken up from the apoplast. If the apoplast is in the path to the phloem, a sucrose molecule must cross a plasmalemma twice, once when entering the apoplast and once when entering the se–cc complex from the apoplast. Release into the apoplast may occur from all cell types, mesophyll, bundle sheath and vascular parenchyma (Fig. 8.1, A), or it may be limited to bundle sheath or vascular parenchyma cells near the se–cc complex (Fig. 8.1, B). When the outer tangential and radial walls of the bundle sheath cells are suberized as they are in minor veins of the maize leaf, a variation in the B-type path could occur in which sucrose released into the apoplast from bundle sheath

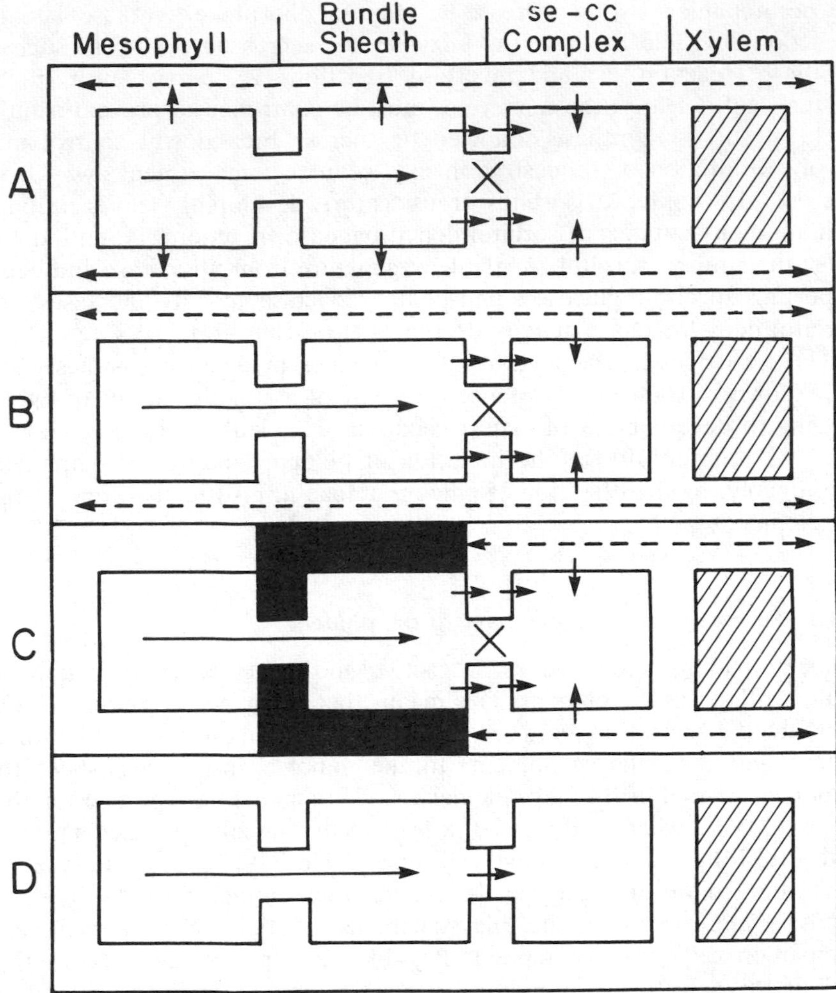

Fig. 8.1 Possible pathways of photosynthate to the phloem in the leaf. Long, solid arrows represent symplastic transport; short, solid arrows represent membrane transport; and dashed arrows represent apoplastic transport. The heavy shading in C represents a suberized cell wall. In A, B and C, the plasmodesmata between the bundle sheath cells and the se–cc complex are closed (indicated by X). In D, the plasmodesmatal continuity between the bundle sheath cells and the se–cc complex is interrupted by a hypothetical membrane (see text). In B and C, the localized release of photosynthate into the apoplast could also include, or even be limited to, phloem parenchyma cells which have been omitted for the sake of simplicity.

or vascular parenchyma cells is prevented from diffusing back into the mesophyll apoplast (Fig. 8.1, C).

There is evidence (see below) for a completely symplastic path to the phloem. Seemingly, a step-up in concentration in a symplastic path would have to occur across a plasmodesma, and in the diagram (Fig. 8.1, D), a membrane is inserted across the lumen of a plasmodesma to indicate this. A membrane in this position is not in accord with the present view of plasmodesmatal structure, although the structure is not well understood (see Gunning & Overall, 1983 and discussion in Ch. 7). A plasmodesma has the shape of a narrow cylinder (30–80 nm in diameter) with a solid rod running down the middle. The walls of the cylinder are lined with a membrane that is continuous with the plasmalemmas of the two interconnected cells. The central rod is an extension of the endoplasmic reticulum, and it is so tightly curled that it has no lumen. The space between the central rod and the inner face of the cylinder is called the 'cytoplasmic annulus'; it is presumed to be an open channel connecting the cytosol compartments of the two cells. Probably, both diffusion and mass flow of solution can occur through the cytoplasmic annulus. Experiments using fluorescent peptides of various molecular weights indicate that the molecular size cut-off for passage through plasmodesmata is in the range of 700–900 (see Gunning & Overall, 1983). It is noteworthy that verbascose, a pentasaccharide of M_r 829, is the largest sugar translocated in any quantity (ajugose, a hexasaccharide, is occasionally found in trace amounts in sieve tube exudates). It is thought that a plasmodesma can be open or closed (perhaps the central rod is a valve), and this would explain how two cells with plasmodesmata between them (e.g. bundle sheath and companion cells) can have greatly different turgor pressures. However, there is no evidence that the cytoplasmic annulus is interrupted by a selectively permeable membrane. This view of plasmodesmatal structure is in accord with the idea that phloem loading takes place from the apoplast (Fig. 8.1 A, B or C).

Evidence for phloem loading from the apoplast comes mainly from work with sugar-beet (see Geiger, 1975; Madore & Webb, 1981; Giaquinta, 1983). Briefly, the evidence is as follows:

1. There is a steep step-up in sucrose concentration between the mesophyll and the se–cc complex, and this cannot be explained by passive diffusion through the symplast.
2. Labelled sucrose (20 mol m^{-3}) introduced into the apoplast through the abraded surface of a leaf was translocated at rates similar to those found from leaves photosynthesizing in $^{14}CO_2$.
3. When the abraded portion of the leaf was bathed in unlabelled sucrose while the leaf was photosynthesizing in $^{14}CO_2$, ^{14}C-sucrose was trapped in the unlabelled sucrose of the bathing solution.
4. Application of the relatively non-permeable sulphydryl reagent, PCMBS, inhibited translocation of ^{14}C-sucrose from leaves photosynthesizing in $^{14}CO_2$ or from abraded leaves bathed in ^{14}C-sucrose. but did not inhibit photosynthesis or respiration.

Clearly, the above evidence supports the idea that the apoplast is in the path to the phloem. But if we accept the possibility that a step-up in concentration can be achieved along a symplastic path, then the evidence in (2) and (3) above can be interpreted to favour a completely symplastic path (Madore & Webb, 1981). The evidence in (2) could be interpreted as an uptake of ^{14}C-sucrose from the apoplast into the mesophyll cells followed by symplastic transport to the phloem. This interpretation is consistent with the results of Fondy & Geiger (1977) who estimated the apoplastic sucrose concentration in translocating sugar-cane leaves to be 70 mmol m^{-3}, whereas the addition of 20 mol m^{-3} sucrose was necessary to achieve high rates of translocation from the apoplast. The evidence in (3) could be interpreted as loss of ^{14}C-sucrose from the mesophyll or vascular tissue because the leaves were abraded or because of naturally leaky membranes. That mesophyll cells can take up sucrose from the apoplast is agreed upon and is interpreted as evidence for a retrieval system by proponents of an apoplastic step in the path to the phloem (where it is needed to retrieve sucrose that has diffused back into the mesophyll after being secreted into the apoplast near the phloem, Fig. 8.1, B) as well as by proponents of a completely symplastic path (where it is needed to retrieve sucrose that has leaked from mesophyll, bundle sheath, or phloem cells). The evidence in (4) above that PCMBS inhibits translocation strongly supports an apoplastic step in the path. If PCMBS acts only at the outer surface of the plasmalemma, the translocation of sucrose produced in photosynthesis should not be inhibited if the path to the phloem is completely symplastic. However, Madore & Webb (1981) suggest that PCMBS may affect the plasmodesmatal connections. Summing up: The sugar-beet results favour phloem loading from the apoplast; an interpretation favouring a completely symplastic path requires an assumption about plasmodesmatal function (ability to perform a PCMBS-sensitive, selective step-up in concentration) that, as yet, does not have an experimental basis.

The results of Madore & Webb (1981) with squash leaves can be interpreted as evidence for a completely symplastic path to the phloem. Squash translocates stachyose, sucrose and smaller amounts of verbascose, and these sugars are synthesized in the mesophyll (Hendrix, 1973; Madore & Webb, 1982). After exposure of a squash leaf to $^{14}CO_2$. label was translocated predominantly in the form of ^{14}C-stachyose. However, when the leaf was fed ^{14}C-labelled stachyose, sucrose or galactose through the apoplast, label was translocated predominantly in the form of ^{14}C-sucrose, even though ^{14}C-stachyose predominated in the leaf cells. Clearly, sugars supplied to the apoplast were not treated by the leaf in the same way as those derived from photosynthesis. Furthermore, after photosynthesis in $^{14}CO_2$, the leaf 'free space' extract contained the same ^{14}C-sugars (transport as well as non-transport sugars) as the ethanolic extract of the whole leaf, but in amounts two orders of magnitude lower. The 'free space' extract was obtained by flushing the apoplast (vacuum infiltration followed by centrifugation), and therefore, amounts of sugars in the extract may overestimate the amounts in the undisturbed apoplast. Notably, the 'free space' extract did not contain an excess of transport sugars, and it is likely that the 'free space' sugars are from leaky cells. From these results Madore & Webb conclude that a completely symplastic pathway for minor vein loading in squash remains a possibility.

Sugar-beet and squash are C_3 plants and both appear to have adequate plasmodesmatal connections between the cells in the path from mesophyll to sieve tube. A major difference, however, is that nitrogenous substances are the predominant solutes in the sieve tubes of squash, whereas sucrose makes up about 80 % of the sieve tube solutes in sugar-beet. It is possible, therefore, that the gradients of stachyose and sucrose are in the downhill direction from mesophyll to sieve tube in squash, and sugar movement is completely symplastic and passive. But if the plasmodesmatal channels are open enough to allow sugars to leak into the se-cc complex, they would also allow nitrogenous substances (present in high concentration in the se–cc complex) to leak out, and, of course, the plasmodesmatal route would not be selective (see Ho & Baker, 1982).

Maize, like sugar-beet, translocates sucrose and has a step-up in concentration at the se–cc complex. However, unlike sugar-beet, the maize bundle sheath cells contain suberin on their outer tangential and radial walls (Fig. 8.1,C) that presumably prevents lateral diffusion from the apoplastic space of the small vein, and in maize there is a paucity of plasmodesmata between the se–cc complex and bundle sheath or vascular parenchyma cells (Evert *et al.*, 1977). These structural features of maize favour the idea that phloem loading occurs from the apoplast. In addition, xylem exudates from maize leaf vascular bundles, which are presumed to reflect the contents of the sequestered apoplastic space (Fig. 8.1,C), contained sucrose and no other sugar (Heyser *et al.*, 1978). Again, the question: Is the apoplastic sucrose on the main route to the sieve tube, or is it sucrose that has strayed from the path through leaky membranes? At the present time this question cannot be answered. Phloem loading from the apoplast is favoured primarily because there is no evidence that a selective, step-up in concentration can be achieved during symplastic transport.

Recently, fluorescent dyes have been used to trace symplastic transport in leaves (Erwee *et al.*, 1985; and Ch. 7). These dyes did not cross the plasmalemma, but moved readily from cell to cell after being injected with a microelectrode into a single cell. The dyes appear to be non-toxic as they did not inhibit cytoplasmic streaming. Dye injected into a mesophyll cell of *Commelina cyanea* moved readily into vascular cells and then moved freely along the minor veins. The exact path of this movement in the minor vein, whether it included the sieve tube, is not known. These dyes are very sensitive probes of symplastic continuity (for example, they were used to show that guard cells were isolated from other cells), and they should prove invaluable for the study of plasmodesmatal function.

8.3.3 Membrane transport of sugars and amino acids

Probably most types of leaf cell can take up sugars and amino acids. In mesophyll cells, these transport mechanisms could serve for retrieval of solutes that entered the leaf apoplast in the transpiration stream or from leaky cells. In addition, if phloem loading takes place from the apoplast, the plasmalemmas of the se–cc complex must be sites of intense uptake activity, and there must be a group of cells that release sugars and amino acids into the apoplast (cf. Fig. 8.1, B C). Transports into the se–cc complex are (or can be) uphill for those sugars and amino acids that can be loaded (i.e. accumulated), and therefore, these are active processes requiring metabolic energy. Release of sugars and amino acids is presumed to be downhill from a high (symplastic) to a low (apoplastic) concentration, and therefore, release processes could be passive.

Active and passive membrane transport of sugars and amino acids are discussed below for higher plant cells generally. Although the phloem loading step is one of prime interest here, it is not possible to isolate uninjured se–cc tissue for uptake studies, and many experiments purporting to investigate phloem loading have been run using leaf pieces or other tissues that contain only a small proportion of phloem cells. Furthermore, the cells that release sugars and amino acids into the apoplast have not been identified and may differ from species to species. Detailed accounts of membrane transport of sugars and amino acids are to be found in Komor (1982) and Reinhold & Kaplan (1984).

8.3.3.1 Active transport across the plasmalemma

When sugars and amino acids are transported into a cell against a concentration gradient, each transport mechanism must be coupled to another process capable of making the transport spontaneous. In higher plants the process coupled to sugar and amino acid transport is proton transport (see Baker, 1978). Protons moving down a proton gradient can drive sugars and amino acids up their gradients because protons and the driven solute move together across the plasmalemma (see also Chs. 2, 3 and 4). This coupled transport is called 'proton co-transport' (e.g. sucrose–proton co-transport) and it is the only mechanism known in higher plants for active transport of sugars and amino acids. Co-transport is the general term; when the substrate and the proton cross the membrane in the same direction, the transport is termed 'symport', but when they cross in opposite directions it is termed 'antiport'. A model for sucrose–proton symport is shown in Fig. 8.2. This model also will serve for transport of neutral amino acids

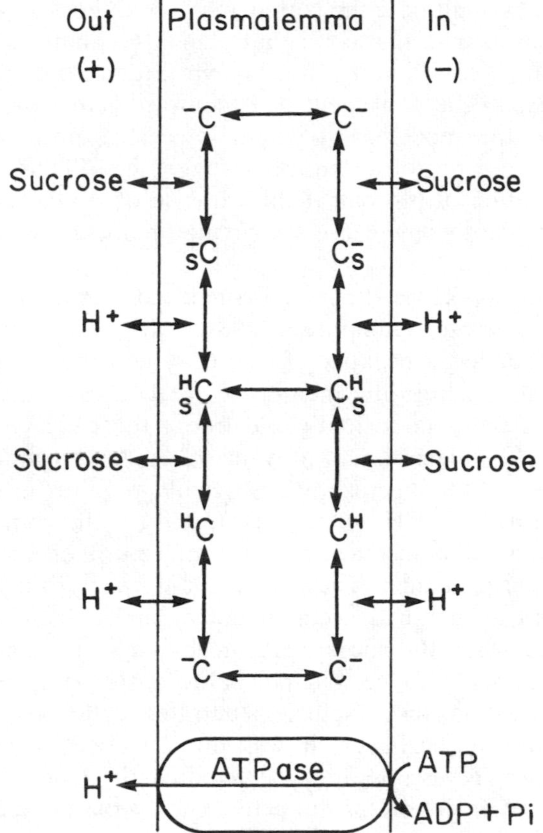

Fig. 8.2 A model for sucrose–proton symport across the plasmalemma. See text for details.

but it would need to be modified to serve for transport of acidic and basic amino acids (see Rottenberg, 1976). The model contains a carrier (C^-) and an ATPase (proton pump); both are proteins that span the thickness of the plasmalemma. The ATPase creates and maintains the proton gradient across the plasmalemma (pH high inside, membrane potential negative inside). The free carrier has a negative charge and binds specifically with sucrose and protons in one-to-one ratio, but the order of binding is random. The uncharged proton–carrier–sucrose complex (HCS) and the free carrier (C^-) can move in either direction across the plasmalemma, but the carrier cannot move when complexed only with a proton (HC) or only with a sucrose molecule (SC^-).

This model incorporates most of what is known about sugar–proton symport in plants and also a few unknowns. The charge on the free carrier is not known. With a negative charge, movement of the free carrier (C^-) to the inner face of the membrane would be retarded by the membrane potential, thereby limiting efflux of sucrose. A one-to-one sugar–proton ratio has been measured in a few higher plants (Humphreys, 1981; Komor *et al.*, 1982; Lin, 1985a), but it is not known whether the order of binding is random or fixed. The movement of HC inward across the plasmalemma is forbidden in the model as it would lower the proton gradient and increase the rate of proton extrusion through the ATPase, i.e. it would cause a recirculation of protons at the expense of ATP. Likewise, movement of SC^- outward would cause a recirculation of sucrose at the expense of ATP.

Experimental support for the model comes from experiments on release of accumulated sucrose (Humphreys, 1981, 1985). If, as proposed in the model, the carrier has a negative charge (C^-) and the protonated carrier (CH) cannot move across the membrane, then a low outside pH and a negative electric potential inside would freeze the carriers at the outside surface of the plasmalemma and prevent release of accumulated sucrose. The model predicts that sucrose release would require (a) an increase in outside pH to form C^- ($CH \rightarrow C^- + H^+$), and (b) the abolishment of the membrane potential to increase the rate of return of C^- to the inside surface of the plasmalemma. These predictions were realized when it was found that sucrose that had accumulated in maize scutellum slices was rapidly released when the slices were incubated in triphenylmethylphosphonium ion ($TPMP^+$) above, but not below, pH 7 (Fig. 8.3). $TPMP^+$ is a lipid-soluble cation that rapidly penetrates cells and abolishes the membrane potential. However, it was not effective in causing sucrose release until the pH was raised as predicted by the model. The results of Fig. 8.3 suggest that the p*K* of the active site of the carrier is above 7.5. This assumes that the pH at the active site is the same as that of the bulk solution, but it may be lower. The electrical potential across the plasmalemma (which is probably small in the presence of $TPMP^+$) and the electrical potential at the surface of the plasmalemma, which results from the fixed negative charges on the membrane, would cause a considerable decrease in the pH at the membrane surface (Barber, 1982).

The model predicts that upon initiation of sucrose uptake there would occur an increase in outside pH, a decrease in inside pH and a decrease in the magnitude (a depolarization) of the membrane potential. These would be transient changes whose magnitude and time course would depend on the time lag between discharge of the proton gradient by the proton co-transport and activation of the ATPase which would act to restore the proton gradient. (The pH optimum for plasmalemma ATPase is about 6.5 (see Ch. 3), whereas the cytoplasmic pH is in the 7 to 8 range. Therefore, a decrease in cytoplasmic pH would activate the ATPase.) One or more of these transient changes have been observed with a number of higher plant tissues upon initiation of sucrose or amino acid uptake. For example, when glutamine was added to a weakly buffered solution bathing castor bean cotyledons, the pH first increased and then decreased until a

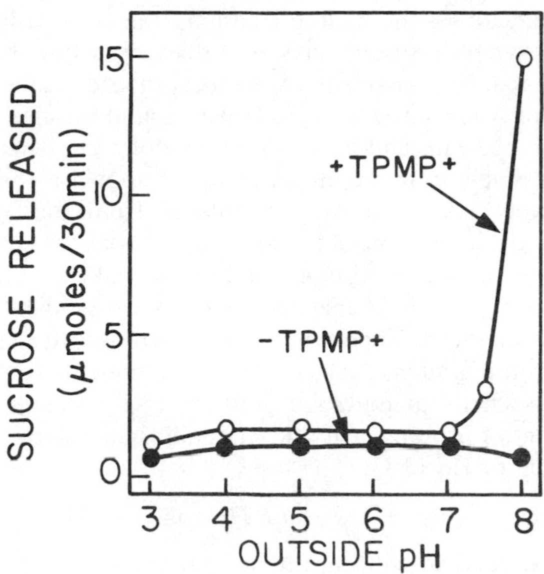

Fig. 8.3 Release of accumulated sucrose from maize scutellum cells required a high pH and a lipid-soluble cation. Triphenylmethylphosphonium bromide ($TPMP^+$, 15 mol m^{-3}) was added to buffered solutions bathing maize scutellum slices (0.5 g) which contained 60–70 μmol of sucrose; after 30 min, the sucrose content of the bathing solution was measured. See text for an explanation. (Adapted from Humphreys, 1981, 1985.)

new steady state was reached in about 10 min. If, after 10 min, sucrose was added to the glutamine solution, a second transient pH change occurred, indicating that the uptakes of both glutamine and sucrose were driven by the proton gradient (Robinson & Beevers, 1981). Recently, all three of the transient changes were observed upon initiation of sucrose uptake in protoplasts from developing soybean cotyledons (Lin, 1985a).

There also is direct evidence that sucrose–proton symport occurs across the plasmalemmas of the se–cc complex. Heyser (1980) forced a sucrose solution through the xylem of maize leaf strips and obtained a transient pH increase in the xylem perfusate. Hexoses, mannitol and raffinose did not elicit an increase in pH. It is thought that the xylem perfusate flushed only the apoplastic space of the vascular bundles, being excluded from the rest of the apoplast by the suberized walls of bundle sheath cells (cf. Fig. 8.1, C). This evidence taken together with the anatomical features of the maize minor vein strongly suggests that phloem loading in maize leaves takes place from the apoplast via a proton co-transport mechanism. Wright & Fisher (1981) measured the sieve tube membrane potential in willow bark by inserting a microelectrode into the sap exuding from a severed aphid stylet (aphid stylets penetrate into the sieve tube lumen), and obtained a rapid depolarization of about 60 mv when sucrose (but not mannitol) was added to the solution bathing the bark strips. Perhaps a sucrose–proton symport system in sieve tubes of willow bark serves to retrieve sucrose that had passively leaked from the sieve tubes. This agrees with the ideas of Minchen *et al.* (1984) who found an apoplastic sucrose pool in the stem of bean, and concluded that along the length of the stem sucrose leaks passively from the phloem into the apoplast while at the same time it is actively taken up by the phloem from the apoplast. This apoplastic pool of sucrose would buffer out sudden changes in sucrose supply at the source and sucrose demand at the sink.

In a co-transport system, the sucrose and proton gradients are coupled but opposite in direction. Therefore, if the co-transporter can use *all* the energy in the proton gradient, sucrose would continue to accumulate inside the cell until its chemical potential gradient ($\Delta\mu_s$) equals the chemical potential gradient of protons ($\Delta\mu_{H^+}$). At equilibrium the two gradients are related according to Eq (8.1):

$$RT/F \log [S]_i/[S]_o = n(\psi_o - \psi_i) - nRT/F(pH_o - pH_i) \qquad (8.1)$$

where [S] is the sucrose concentration, n is the sucrose–proton stoichiometry ($n = 1$ in the model of Fig. 8.2), ψ is the electrical potential, RT/F equals 60 mV at 30°, and the subscripts o and i refer to outside and inside the plasmalemma. Equation (8.1) states that a proton gradient of 180 mV (say, a pH gradient of 2 units and a membrane potential of 60 mV, negative inside) can support a 1000-fold accumulation of sugar when $n = 1$. Thus, with an outside concentration of 0.1 mol m^{-3}, a cytosolic concentration of 10^2 mol m^{-3} would be attained at equilibrium. However, the amount of sugar that can be accumulated is limited, and in the above example a $[S]_o$ of 10 mol m^{-3} would not result in a $[S]_i$ of 10^4 mol m^{-3}. The reasons for this are not entirely clear. But high inside sugar concentrations would result in higher rates of passive sugar leakage and also would inhibit movement of the carrier to the outside surface of the plasmalemma by increasing the formation of SC^- at the inside surface (cf. Fig. 8.2).

In plant cells sugar accumulation ratios probably reflect accumulation across the tonoplast as well as the plasmalemma (Thom *et al.*, 1982; Humphreys, 1983) so Eq. (8.1) cannot be applied directly. However, sieve tube elements have no vacuoles and Eq. (8.1) can be used to calculate sucrose and proton gradients from values reported in the literature for the six parameters. At a pH_o of 5.7, Wright & Fisher (1981) measured a membrane potential of 155 mV in the sieve tubes of willow bark. Assuming that willow bark sieve tubes have a pH_i of 8 (a value in the middle of the pH range reported for the sieve tube solutions from many plants), the proton gradient calculated from Eq. (8.1) would be 293 mV at 30°. This gradient could support a 76,000-fold accumulation of sucrose with a one-to-one co-transport stoichiometry. In the apoplast of sugar-beet leaves, $[S]_o$ has been estimated to be 0.07 mol m^{-3} (Fondy & Geiger, 1977); and in the sieve tubes of the minor veins of sugar-beet leaves, $[S]_i$

has been estimated to be 800 mol m^{-3} (Sovonick *et al.*, 1974), making $[S]_i/[S]_o$ = 11,429 (or $RT/F \log [S]_i/[S]_o$ = 242 mV). Clearly, if the proton gradient across the sieve tube plasmalemma in sugar-beet leaves is similar to that in willow bark, it is sufficient to support the observed sucrose accumulation.

Equation (8.1) states that both components of the proton gradient, the pH gradient and the membrane potential, can be used by the co-transporter to drive sucrose uptake. Thermodynamically, the two components are interchangeable. In energy-transducing systems, such as ATP synthesis, they are equally effective and a sucrose uptake in the maize scutellum, a proton symport system, takes place at similar rates when either component of the proton gradient is predominant (Humphreys, 1985). There is evidence that protonation of the co-transporter increases its affinity for the sugar (Komor & Tanner, 1974), and this may be the way co-transporters use the pH gradient to drive sugar uptake. How the membrane potential is used is not known. Mitchell (1968) proposed the idea of a 'proton well'. The well is imagined to be a channel or pore penetrating into but not through the membrane. Protons, in moving down this well, move down the electric field of the membrane. Thus, with distance down the well, the proton activity increases; the difference in electrical potential becomes less but the proton chemical potential remains the same. In this way, a proton co-transporter containing a proton well in which the proton-binding site was situated would convert the membrane potential (or part of it, depending on the depth of the well) to a pH gradient. Although the idea of a proton well is speculative (and untestable), it is presented here to illustrate how a membrane potential might be used by a co-transporter to drive sugar uptake.

The broad outlines of proton co-transport systems in higher plants have largely been worked out: ATPases are present in the membranes, the measured proton gradients are of sufficient magnitude and respond in the expected way upon initiation of uptake, and co-transporters have been identified by their transport activity. However, that is not to say that we understand the mechanism of co-transport. Perhaps the elegant studies on lactose transport being carried out with lipid vesicles that have the highly purified lactose–proton co-transporter of *Escherichia coli* inserted into their lipid membrane (Kaback, 1983) will prove enlightening as to mechanism.

8.3.3.2 Passive transport across the plasmalemma

Passive membrane transport of sugars can occur by simple diffusion through the lipid part of the membrane or through water-filled pores or by carrier-mediated (facilitated) diffusion. Diffusion through the lipid is relatively slow even when driven by a steep gradient, whereas diffusion through water-filled pores can be fast. In fungi, rates of mediated diffusion of sugars (1 mol m^{-3}) are at least 100-fold faster than rates of simple diffusion into most higher plant cells (Komor, 1982). Mediated diffusion is a saturable system; the rate reaches a plateau as the sugar concentration is increased. Also, mediated diffusion shows competition between substrates and the phenomenon of 'counterflow' (see Ch. 1).

The rate of simple diffusion of sugar into a plant tissue can be measured by the uptake of metabolically inert sugars or sugar alcohols (e.g. L-glucose). This method is based on the assumption that the uptake of a natural sugar, D-glucose or sucrose for instance, has a diffusive component of the same magnitude as the rate of uptake of L-glucose. In this way it was shown that the diffusive component makes a considerable contribution (40% at a sucrose concentration of 20 mol m^{-3}) to the sucrose uptake rate in protoplasts from soybean cotyledons (Schmitt *et al.*, 1984). Since soybean cotyledon cells accumulate sucrose, the large diffusive component of sucrose uptake indicates that accumulation occurs primarily across the tonoplast in these cells. Also, it indicates that the rate of simple diffusion across the tonoplast is much lower than that across the plasmalemma. Perhaps the difference lies in the number and size of water-filled pores in the two membranes. Diffusion into maize scutellum cells is thought to occur through water-filled pores that allow rapid penetration of hexoses and sugar alcohols but not sucrose (Garrard & Humphreys, 1965). The hexoses and sugar alcohols enter a cellular space that is 12–14 % of the water volume of the cell. Its size and other evidence suggest that this space is the cytosol. Therefore it is concluded that the scutellum cell tonoplast is relatively impermeable to hexoses, which nevertheless rapidly penetrate the plasmalemma. This rapid movement of hexoses across the scutellum plasmalemmas did not show saturation or competition, ruling out carrier-mediated diffusion.

Mediated diffusion mechanisms are found in fungi and erythrocytes, cells that live in high sugar environments. According to Komor (1982), there is no clear evidence for the occurrence of mediated diffusion of sugars in higher plants (for evidence of mediated diffusion, see the section on transport at the tonoplast). However, mediated diffusion mechanisms may be in relatively few cells and difficult to identify.

Efflux of sugars from plant tissues and isolated protoplasts is usually very slow and is near zero from isolated vacuoles of some plants (see Reinhold & Kaplan, 1984). This indicates that the tonoplast of these vacuoles lacks mechanisms for mediated diffusion, and also lacks water-filled pores of a diameter great enough to allow hexose or sucrose passage. However, the small extent of efflux from whole cells may indicate only a low sugar concentration in the cytosol. Nevertheless, efflux of sugars and amino acids from cells that provide a symplastic path for photosynthate would be expected to be slow. During a 1 h light period, isolated mesophyll cells from poppy (*Papaver somniferum*) released only 0.75 % of the total carbon fixed into the bathing solution (Kaiser *et al.*, 1979).

8.3.3.3 Sugar transport at the tonoplast

In source leaves of some species, considerable amounts of sucrose are stored in vacuoles during the day, and this sucrose is exported from the leaf at night. In spinach leaves, for instance, the vacuolar sucrose concentration was found to rise to 45 mol m^{-3} at the end of a 9 h light period and then to fall to about 1 mol m^{-3} during the subsequent 15 h dark period (Gerhardt & Heldt, 1984). On the other hand, in a sink such as sugar-beet root, sucrose is accumulated in vacuoles where it is stored for long periods. Sequestration of sucrose in vacuoles serves the same purpose as sequestration of starch in plastids: it buffers the supply of exportable sucrose at the source and removes imported carbon from the metabolic compartments of the sinks.

A mediated diffusion mechanism in the tonoplast would keep the cytosolic and vacuolar sucrose concentrations equal, and this would prevent rapid changes in cytosolic sucrose concentration because of the relatively large volume of the vacuole. Mediated diffusion appears to be the transport mechanism for sucrose in vacuoles isolated from barley mesophyll protoplasts (Kaiser & Heber, 1984). Sucrose uptake in these vacuoles was saturable (K_m = 21 mol m^{-3}) and competitively inhibited by raffinose and maltose. There was a rapid exit of sucrose from preloaded vacuoles which was complete within 15 min, and both exit and entrance of sucrose were strongly inhibited by PCMBS. In contrast, vacuoles of spinach leaves, which have been shown to be filled with and emptied of sucrose on a diurnal cycle, appear not to have a mediated diffusion mechanism for sucrose. This is concluded because vacuoles prepared from spinach leaf protoplasts by a procedure that took 1.5 h at 15 °C still contained about 50 % of the initial protoplast sucrose and because, after protoplast photosynthesis in $^{14}CO_2$, less than 10 % of the ^{14}C-sucrose was recovered in the isolated vacuoles (Asami *et al.*, 1985). (A mediated diffusion mechanism should rapidly equilibrate ^{14}C-sucrose between vacuole and cytosol.)

There is evidence for active sugar uptake in vacuoles isolated from both sink and source tissue. Vacuoles from red beetroot took up sucrose against a concentration gradient and uptake was stimulated by ATP (Doll *et al.*, 1979). Likewise, sucrose uptake into tonoplast vesicles isolated from sugar-beet tap-root required ATP, and the vesicles contained an inwardly directed, proton-translocating ATPase (Briskin *et al.*, 1985). Vacuoles from pea mesophyll protoplasts took up hexose; uptake was increased by ATP and was abolished when the proton gradient across the tonoplast was abolished (Guy *et al.*, 1979). These results are consistent with the operation of sugar-proton antiport mechanisms during active uptake across the tonoplast. Proton antiport is a proton co-transport in which the transports of protons and sugar are in opposite directions. Sugar–proton antiport is analogous to the symport model of Fig. 8.2, the difference being that the sugar-binding and proton-binding sites on the carrier face opposite sides of the membrane instead of the same side. The proton gradient across the tonoplast is high on the vacuole side: the vacuole is more acid than the cytosol and the electric potential across the tonoplast, although small, is positive on the vacuole side. The proton gradient is maintained by an ATPase pumping protons into the vacuole (Sze, 1985; and Ch. 3).

Recently, another type of transport known as 'group transport' was found in vacuoles from sugar-cane suspension cells (Thom & Maretzki, 1985). These vacuoles take up UDPG, releasing sucrose and sucrose phosphate inside the vacuole and UDP outside the vacuole. Apparently, there is an enzyme complex in the tonoplast consisting of five enzymes: UDPG pyrophosphorylase, phosphoglucomutase, phosphoglucose isomerase, sucrose phosphate synthase and sucrose phosphate phosphatase. Half of the UDPG that reacts within the enzyme complex is used to produce fructose-6-phosphate and half reacts with fructose-6-phosphate to produce sucrose phosphate (Table 8.1, reaction 2). Hexoses and hexose phosphates added to the incubation medium were not used; only UDPG was used. UDPG pyrophosphorylase (cf. reaction 15, Table 8.1) requires inorganic pyrophosphate (PP_i). Since it was not necessary to add PP_i to the sugar-cane vacuoles, perhaps UDPG phosphorylase is present in the tonoplast enzyme complex instead of UDPG pyrophosphorylase. UDPG phosphorylase catalyses the reaction: UDPG + P_i $\rightarrow$ UDP + glucose-1-P. Although group transport systems have been known in bacteria for some time, the sugar-cane system is the only one reported in higher plants. Note that this group transport system does not actively transport sucrose; sucrose is accumulated but not transported (see also Ch. 3).

Release of sucrose carbon from vacuoles is not well understood with the exception of vacuoles having a mediated diffusion mechanism, which has been reported, so far, only for barley mesophyll vacuoles. In mature red beetroots most of the sucrose is in the vacuoles and acid invertase activity is very low. When slices of the mature beet were washed in aerated water for 1–3 d, acid invertase activity rose in the vacuole and sucrose levels fell (Leigh *et al.*, 1979). Beets (*Beta vulgaris*) are biennials. The sucrose accumulated in vacuoles of the storage root during the first year is used to support the sprouting and initial growth of the reproductive shoots at the beginning of the second year. The increase in vacuolar invertase is a necessary step in the conversion of the storage root from a sink to a source. How the hexoses are released from the vacuole is not known.

8.3.3.4 Membrane transport of sucrose: Summary

The mechanisms for sucrose transport across the plasmalemma and tonoplast that were discussed in the previous sections of this chapter are listed in Fig. 8.4.

(a) Plasmalemma and tonoplast proton-translocating ATPases (Fig. 8.4, 1 and 2) are different enzymes which maintain the cytosol both alka-

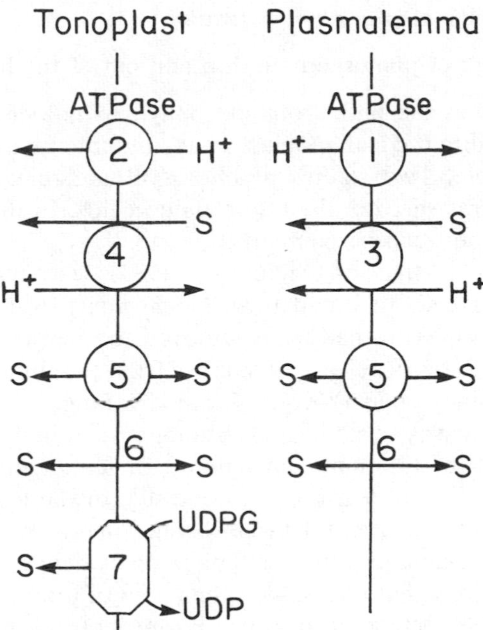

Fig. 8.4 Some transport functions of plasmalemma and tonoplast. Numbers 1 and 2 represent ATPases that pump protons from the cytosol into the apoplast and vacuole. Numbers 3 to 7 represent types of sucrose transport. See text for details.

line and electrically negative relative to apoplast and vacuole.

(b) The plasmalemma sucrose-proton symporter (3) and the tonoplast sucrose–proton antiporter (4) are examples of proton co-transporters, proton co-transport being the only mechanism known for active transport of sugars and amino acids in higher plants. There are reports suggesting the occurrence of sucrose–K^+ symport at both the plasmalemma and the tonoplast (see Reinhold & Kaplan, 1984), but the evidence is equivocal.

(c) Mediated diffusion (5) is not well documented in higher plants. There is good evidence that it occurs in barley mesophyll tonoplasts, but the evidence for its occurrence in other cells and membranes is equivocal.

(d) It is commonly observed that sucrose and hexoses leak (6) from isolated tissues and cells. This leakage is usually slow and may occur through water-filled pores as well as by diffusion through the membrane lipid.

(e) Recently, group transport of sucrose (7) was discovered in vacuoles of sugar-cane suspension cells. Undoubtedly, it will be found in other plants.

(f) Most likely, mechanisms for sucrose release other than mediated diffusion (5) exist in both tonoplast and plasmalemma of some cells.

8.3.4 The apoplastic step in phloem loading

8.3.4.1 Movement of photosynthate into and out of the leaf apoplast

If the apoplast is in the path from mesophyll to phloem, photosynthate must be released into the leaf apoplast in an amount equal to that exported through the petiole. Two studies of photosynthate release are discussed below: one concerns sucrose, the other amino acids. In these studies rates of release and export can be compared.

The first study is that of Anderson (1983) who measured sucrose released from leaf discs of broad bean (*Vicia faba*) from which both the upper and lower epidermis had been removed to open the apoplast to the bathing solution. Rates of release of sucrose from the broad bean leaf discs (1.7 nmol cm^{-2} min^{-1} of leaf) were within the range of rates of sucrose export shown by leaves of pepper (1.2 nmol cm^{-2} min^{-1} of leaf: Grange, 1985) and sugar-beet (3.8 nmol cm^{-2} min^{-1} of leaf: Geiger & Swanson, 1965). (Note that cm^{-2} of leaf refers to one side of the leaf only.) Sucrose export data are not available for broad bean. Sucrose release from broad bean leaf discs was inhibited by uncouplers of oxidative phosphorylation; therefore it is unlikely that release was the result of injury. The mechanism of release, whether active or passive, is not known, but the data are consistent with a mechanism of mediated diffusion. When the leaf discs were incubated with $^{14}CO_2$ in the light, a constant rate of ^{14}C-sucrose release was attained after 20 min. Significantly, this is about the time required for the peak level of ^{14}C-sucrose to be reached in leaf veins after giving leaves of intact broad bean plants a pulse of $^{14}CO_2$ (Outlaw *et al.*, 1975). In this pulse labelling experiment, the rapid movement of sucrose from mesophyll to vein can be interpreted to mean either that sucrose was released from a few cells near the veins into a small, rapidly turning-over, apoplastic pool whence it was taken up into the se–cc, or that the path to the se–cc complex was entirely symplastic. Madore & Webb (1981) interpret it as favouring an entirely symplastic path, but the leaf disc experiments of Anderson (1983) support the first interpretation. Sucrose release from leaf discs showed the proper timing and was of sufficient rate for it to be the release step on the path to the phloem. Of course, sucrose release could have come *after* an entirely symplastic transfer to the se–cc complex, but the high rate of release makes this unlikely.

Most plant cells have ion fluxes in the range of 1–10 pmol cm^{-2} s^{-1} of membrane, and MacRobbie (1971) points out that, until there are experimental data to the contrary, models of phloem loading should not demand sucrose fluxes much higher than this. (It is notable that the few sugar flux data available are within or not far above the range for ion fluxes.) To estimate the minimum flux required for the measured sucrose release in the experiments of Anderson (1983), it is necessary to know the surface area of the mesophyll plasmalemma. Surface areas of dicot mesophyll cells fall in the range of 25–50 cm^2 of mesophyll surface for each square centimetre (cm^2) of leaf surface (one side only) of light grown leaves depending on the species (see Nobel *et al.*, 1975, and references therein). (Leaves grown in full sunlight are thicker and have a higher mesophyll surface area per square centimetre (cm^{-2}) of leaf than shade-grown leaves.) On the assumption that the area of the mesophyll of broad bean falls within this range, and from the rate of release of 1.7 nmol cm^{-2} min^{-1} of leaf (i.e. 28 pmol cm^{-2} s^{-1}), the sucrose flux would be 0.6–1.2 pmol cm^{-2} s^{-1} of membrane, assuming that all mesophyll cells of the leaf disc are contributing. Clearly, even if only about 10 % of the leaf cells were releasing sucrose, the flux would be in the required range.

If the phloem is loaded from the apoplast in the intact leaf, the rate of sucrose uptake into the se–cc complex should equal the rate of release into the apoplast. How great a flux would be required at the plasmalemma of companion cells of broad bean if uptake equalled release? Gunning *et al.* (1974) have estimated the plasmalemma surface of the companion cells of the broad bean leaf to be about 5 cm^2 cm^{-2} of leaf, and therefore, a flux of about 6 pmol cm^{-2} s^{-1} of plasmalemma would be required. If the plasmalemma surfaces of the sieve tube elements were included in these calculations, the required sucrose flux would be lowered but not halved. The companion cells of the broad bean leaf are transfer cells, and the wall ingrowths result in about four-fold amplification of plasmalemma surface. In many plants, the companion cells of the minor veins do not have wall ingrowths, and higher fluxes may be required.

The calculated sucrose fluxes for broad bean leaves, taken together with the data showing a rapid and sharp peaking of the ^{14}C-sucrose content in the veins following a pulse of $^{14}CO_2$ (Outlaw *et al.*, 1975), are consistent with a model of phloem loading in which sucrose is released into the apoplast from cells near or in the minor veins and then is taken up into the se–cc complex (Fig. 8.1, B).

The second study of photosynthate release into the apoplast is that of Secor & Schrader (1984) who measured the release of amino acids from isolated leaf cells of soybean. During seed development (pod filling) soybean leaf proteins are degraded and the amino acids are translocated to the developing seeds. Secor & Schrader (1984) found that amino acids were released from isolated soybean leaf cells at rates that could account for the loss of protein in the intact, attached leaf. Twenty amino acids were released. The mechanism for release is not known, but the results indicate more than one carrier is involved and are consistent with a mediated diffusion mechanism. The soybean data also indicate that amino acids were released from leaf cells generally, not just from a few specialized cells near the se–cc complex. These results support the idea that the apoplast is in the path for transfer of substances from mesophyll to phloem. However, protein breakdown and amino acid transport are part of the senescence syndrome in these leaves, and it is certainly possible that export of photosynthate from a healthy leaf takes a different path.

8.3.4.2 Concentration of sucrose in the leaf apoplast

If an apoplastic step is obligatory in the path of sucrose to the phloem, the rate of sucrose release into the apoplast, the rate of sucrose uptake from the apoplast and the rate of sucrose export through the petiole should be the same or nearly so. Knowing rates of export through the petiole (pepper and sugar-beet) and the rate of release into the apoplast (broad bean), what concentrations of sucrose in the apoplast would be necessary to obtain rates of sucrose uptake equal to these? In sugar-beet leaves, 20 mol m^{-3} sucrose added to the apoplast through an abraded leaf surface was exported through the petiole at a rate similar to that for export of photosynthetically derived sucrose from unabraded leaves (see Geiger, 1975). In broad bean leaf discs (lower epidermis removed), rates of sucrose uptake in the range of rates of sucrose release found by Anderson (1983) required sucrose concentrations of 20–25 mol m^{-3} in the bathing solution (Delrot & Bonnemain, 1981). In the corn scutellum, a net sucrose influx of 6 pmol cm^{-2} s^{-1} of plasmalemma (the calculated sucrose flux across the plasmalemma of broad bean companion cells; see previous section) required a sucrose concentration of 20 mol m^{-3} (Humphreys, 1973).

Are such high sucrose concentrations found in the apoplast? Since the apoplast contains both gas and liquid spaces, estimates of sucrose concentration must be rough at best. Delrot *et al.* (1983) estimated that the apoplast of the broad bean leaf contained 1.0–4.5 mol m^{-3} sucrose depending on the time of day it was harvested. On the low side, Fondy & Geiger (1977) estimated that the apoplast of sugar-beet leaves contained only 0.07 mol m^{-3} sucrose. High concentrations of apoplastic sucrose (10–200 mol m^{-3}) have been reported in stems and developing seeds (e.g. Thorne, 1985), but these resulted from phloem unloading (see next Sect 8.3.4.3).

Proponents of phloem loading from the apoplast suggest that high concentrations of photosynthate are restricted to apoplastic regions near or within the minor veins. This could be achieved by localized release (from phloem parenchyma, for example) coupled with retrieval by mesophyll cells of any photosynthate that diffused away from the vein (Fig. 8.1, B) or by localized release into an apoplastic space enclosed by suberized cell walls (Fig. 8.1, C). Whether or not these schemes for confining photosynthate to the vein apoplast are practicable is not known. Retrieval requires expenditure of energy. If apoplastic movement of photosynthate away from the vein to the mesophyll were to be helped along by a flow of water from the xylem, the energy expenditure might be too high or retrieval might be too slow to be effective. Gunning *et al.* (1974) point out that in many leaves the path of least resistance for the transpiration stream is provided by the vein ribs, dorsiventral extensions leading from the small veins to the upper and lower epidermis; and they suggest that the apoplast of the central leaf tissues is bypassed by the main transpiration stream.

Nevertheless, it is clearly possible that the transpiration stream could interfere with phloem loading from the apoplast by preventing a local build-up in the concentration of photosynthate at the surface of the se–cc complex. In grasses, the walls of the bundle sheath are suberized and these are presumed to prevent loss of photosynthate to the mesophyll apoplast (Fig. 8.1, C). However, Peterson *et al.* (1985) found in rye and maize leaves that only the secondary walls of the bundle sheath cells were suberized, leaving an apoplastic path for water and solutes from vein to mesophyll through the radial primary walls. Therefore, in grasses as in dicots, the transpiration stream might prevent a local build-up in photosynthate concentration.

If the localized release of photosynthate takes place by mediated diffusion there would be gradients of photosynthate concentrations high in the mesophyll cytosol and low in the local apoplast space into which release takes place. For sucrose, this means that the cytosol of the mesophyll cells should have a concentration of 20 mol m^{-3} or greater. In spinach leaves at the end of a 9 h light period, the cytosolic sucrose concentration has been estimated at 75 mol m^{-3} (Gerhardt & Heldt, 1984). Although, these and other data are consistent with a mechanism of mediated diffusion for sucrose release (see Anderson, 1983), they do not rule out an active release mechanism. Indeed, Huber & Moreland (1981) postulate a sucrose–K^+ symport mechanism for sucrose efflux from mesophyll protoplasts of wheat and tobacco.

Some recent papers, however, question the idea that phloem loading takes place from the leaf apoplast.

Kaiser & Martinoia (1985) measured assimilate efflux from leaf slices, mesophyll protoplasts and a unicellular alga. Efflux from leaf slices was slow in the presence of Ca^{2+}, but was greatly increased by EDTA. In contrast, the slow effluxes from protoplasts and the alga were insensitive to EDTA. Kaiser & Martinoia suggest that assimilate efflux from leaf slices occurred through the cut ends of veins and through open plasmodesmata. (Ca^{2+} enhances the formation of plugs in severed sieve tubes, and presumably, also would enhance the closing of ruptured plasmodesmata.) They interpret their results as favouring a symplastic path for assimilates from mesophyll to phloem.

Anderson (1986) concluded that sucrose release from leaf slices of soybean and twelve other species occurred by two paths: one PCMBS-sensitive, the other PCMBS-insensitive. The PCMBS-insensitive release was stimulated by EGTA, and Anderson suggests that this sucrose originated in the phloem. But even assuming that all of the sucrose released was on its way to the phloem, Anderson estimated that the rate of sucrose release from slices of soybean leaf was only 20 % of the rate of sucrose export from the intact, attached leaf (see Sect. 8.3.4.1).

In source leaves of *Ipomea tricolor*, Madore *et al.* (1986) demonstrated symplastic continuity from mesophyll to minor vein using a fluorescent dye (see Sect. 8.3.2). PCMBS in the leaf apoplast did not inhibit symplastic movement of the dye. However, when the pH of the apoplast was raised from 5.5 to 8, movement of the dye into and along the minor veins was inhibited, although there was no effect on dye movement between mesophyll cells. Clearly, a better understanding of plasmodesmatal function is necessary if the phloem loading puzzle is to be solved.

8.3.4.3 The problem presented by cell wall-bound invertase

Acid invertases (pH optima 3.0–5.5) bound strongly to cell walls have been found in source tissues, e.g. mature leaves (Pollock & Lloyd, 1977) and maize scutella (Echeverria & Humphreys, 1984). Although often found in sink tissues, the presence of wall-bound invertase in source tissues has seldom been investigated. Where wall-bound invertase is known to occur, it throws doubt on the idea that the phloem is loaded with sucrose from the apoplast. A good example of this is the mature leaf of *Lolium temulentum* (annual rye grass). In *L. temulentum*, wall-bound invertase activity started to increase as the leaf matured and continued to do so into senescence (Pollock & Lloyd, 1977). The wall-bound invertase had a K_m of 1.0 mol m^{-3}, and the activities found in the mature leaf were high enough effectively to prevent a build-up of sucrose in the apoplast. Perhaps invertase and sucrose are confined to different parts of the leaf apoplast in *L. temulentum*, but this remains to be shown. A wall-bound invertase also has been found in maize leaves, and some of the activity was associated with walls of the bundle sheath (De Fekete & Vieweg, 1976). Nevertheless, xylem exudates from maize leaf strips that are presumed to reflect the contents of the apoplastic space enclosed by the suberized secondary walls of the bundle sheath (Fig. 8.1, C) contained sucrose but not hexose (Heyser *et al.*, 1978).

It may be, then, that source leaves are designed so as to keep apoplastic sucrose apart from wall-bound invertase. But this point is crucial and warrants thorough investigation. If they are, in fact, kept apart, what is the role of wall-bound invertase?

Of course, wall-bound invertase is not present in all source leaves. For example, discs from sugar-beet source leaves took up asymmetrically labelled sucrose (^{14}C-fructosyl sucrose) intact from the apoplast. The accumulated ^{14}C-sucrose in these discs retained the asymmetry of label, indicating that the apoplastic sucrose was not hydrolysed by invertase (Giaquinta, 1977).

8.4 Phloem unloading

8.4.1 Apoplastic versus symplastic pathway

In sinks, the path of photosynthate from phloem to other tissues may include the apoplast or be entirely symplastic. For example, in developing seeds the apoplast must necessarily intervene between the maternal tissues of the seed coat and the embryo, whereas, in young leaves, the path is entirely symplastic (Schmalstig & Geiger, 1985). The inclusion of an apoplastic step places two membrane transport mechanisms in the path: one for release of photosynthate from phloem cells (or from cells connected to the phloem by open plasmodesmata). and one for uptake of photosynthate by sink cells. Tonoplast transport also has a role in some sinks. For example, in sugar-cane internodes and sugar-beet storage roots, where sucrose in the storage cells may reach concentrations approaching those of sieve tubes, the apoplast is included in the path of phloem unloading even though sucrose is accumulated across the tonoplast, not the plasmalemma (Saftner & Wyse, 1980; Thom & Maretzki, 1985). These membrane transports are selective and can be controlled. Furthermore, when the apoplast contains a cell–wall-bound invertase, as does the sugar-cane internode (Glasziou & Gayler, 1972), a build-up of sucrose in the apoplast to concentrations that might inhibit sucrose release is prevented.

In contrast, the symplastic path seems to have few places where controls can be exerted. Rapid growth of sink tissue coupled with accumulation of sugars in vacuoles could maintain steep gradients of photosynthates between sieve tube and sink cells, but these gradients are steep whatever the growth rate. Does this mean that transport across plasmodesmata is controlled? Animal cells are connected by 'gap junctions' (the animal-cell counterpart of plasmodesmata) whose permeability is regulated by cytoplasmic Ca^{2+} concentration (see Peracchia, 1980). Erwee & Goodwin (1983) found that Ca^{2+} or Mg^{2+} injected into a single cell of a leaf of the aquatic plant *Egeria densa* inhibited movement of a fluorescent dye from the injected cell through plasmodesmata to other cells, and suggested that the concentration of these ions may directly regulate the permeability of the plant symplast. As yet, however, there is no direct evidence that plasmodesmatal permeability is regulated. Clearly, our understanding of phloem unloading is hampered by our lack of knowledge about how plasmodesmata function.

8.4.2 Examples of pathways for phloem unloading

Since all cells of higher plants are sinks during part or all of their lives, the pathways and mechanisms for delivery of photosynthate to them from the phloem are necessarily diverse. Two examples from the soybean plant illustrate this diversity.

The first example is from a developing soybean seed. In the developing seed, the maternal vascular system terminates in the seed coat, and there are no vascular connections between maternal tissue and the developing embryo. The seed coat has an extensive minor vein network, and phloem tissue is embedded in parenchyma near the inner surface of the seed coat.

It is thought that phloem unloading is symplastic from sieve element to companion cell to vascular parenchyma and that sucrose and nitrogenous compounds are released into the apoplast across the plasmalemma of each of the three cell types (see Thorne, 1985). These compounds then diffuse in the apoplast to the growing embryo where they are taken up. The pathway of unloading in other legume seeds is similar to that in soybean, although the initial symplastic part of the path may be longer and include non-vascular parenchyma (Offler & Patrick, 1984).

In soybean, sucrose traverses the path from sieve tube to embryo unchanged, but extensive transformations of nitrogenous compounds occur in the path. For example, ureides, which may constitute from 10 to 15 % of the nitrogen in the phloem of the seed pod, are not released from the seed coats. Instead, the ureides are metabolized in the pod and seed coat and the nitrogen incorporated mainly into glutamine and asparagine which are the principal nitrogenous compounds released into the apoplast from the seed coat.

The reduced carbon, water, mineral ions and hormones (those not synthesized in the embryo) required for embryo development are supplied by the phloem. The xylem is not extensive in the seed coat, and the direction of the xylem stream is out of the seed coat, not in. The excess water of the embryo (the result of soluble photosynthates being converted to insoluble compounds in the embryo) exits via the seed coat xylem, and presumably, is transpired from the pod wall. Some of the photosynthate released into the apoplast must also be captured by the xylem stream and taken to the pod wall where it could re-enter the phloem and move again to the seed coat. The extent of this recycling of photosynthate between seed apoplast and pod wall is not known; probably, it is limited by the paucity of xylem tissue in the minor veins of the seed coat (see Thorne, 1985).

The second example of phloem unloading is from a nodulated soybean root where there is symbiotic association between root cortical cells and *Rhizobium japonicum* (a nitrogen fixer). In an infected cortical cell, there are many *Rhizobium* bacteroids, each enclosed in a membrane derived from the host plasmalemma (Verma, 1982). Therefore, although the bacteroids appear to be intracellular, they are really extracellular, and the host cytosol is separated from the bacteroid cytoplasm by three membranes: host plasmalemma and the outer and inner membranes of the bacteroid envelope.

Reduced carbon for bacteroid development and nitrogen fixation comes from the host, and it must cross the three membranes separating host cytosol from bacteroid cytoplasm (analogous to photosynthate release into the seed coat apoplast followed by uptake into the embryo). First, however, the reduced carbon must move from the phloem, through the stele and endodermis, and into the cortical cells. In peas and corn roots, the path from sieve tube to cortical cell has been shown to be symplastic

(Dick & Ap Rees, 1975; Giaquinta *et al.*, 1983). (Briefly, the evidence for a symplastic route is that, when sucrose labelled with ^{14}C only in the fructosyl moeity was supplied to corn roots via the phloem, the asymmetry of label was maintained, whereas, when it was supplied via the root apoplast, a randomization of label occurred.) It is quite likely that the path in soybean roots also is symplastic; for if sucrose were released into the apoplast of the stele, it would be swept up in the xylem stream and returned to mature leaves whence it came. That this xylem route back to the leaves is available for some carbon compounds is shown by the fact that upwards to one-half of the carbon in the photosynthate entering the nodule in the phloem exits in the xylem in the form of amino acids and ureides (see Gunning *et al.*, 1974; Verma, 1982). If the path from sieve tube to stelar cell is symplastic, nitrogenous compounds should be unloaded along with sucrose. (Indeed, since the symplastic path is not selective, all compounds of the sieve tube solution should be unloaded, each compound moving at a rate commensurate with its concentration gradient.) Presumably, in the nodule these nitrogenous compounds entering the stele from the phloem mix with those synthesized in the host cytosol (from NH_3 produced by the bacteroid) and then are released into the stelar apoplast whence they enter the xylem stream.

Soybean nodules have much higher levels of sucrose than non-infected roots (Streeter, 1980). The host cytosol has alkaline invertase and sucrose synthase (Table 8.1, reactions 11 and 12), but the bacteroid does not (Morell & Copeland, 1984). Most likely, therefore, the sucrose delivered to the host cytosol from the phloem is hydrolysed, and the resulting hexoses move from host to bacteroid. The membrane transport mechanisms involved are not known.

8.4.3 Membrane transport during phloem unloading

Much use has been made of developing seeds in studies of phloem unloading (see Thorne, 1985 for a detailed account). The development of the empty seed coat technique has made the unloading step (release of photosynthate from seed coat tissues) directly accessible to experimental attack. In this technique a window is cut in the wall of an attached soybean pod (for example) exposing a seed. The distal half of the seed is cut away and the remaining half embryo is removed from the attached half seed. This leaves an empty half seed coat in the shape of a cup still attached to the pod. The cup is filled with experimental trapping solutions, and release of photosynthate into the cup is monitored. When source leaves were fed $^{14}CO_2$, labelled sucrose and amino acids were released into the cup, and in some cases rates of release were similar to rates of import into intact seeds. The mechanism of photosynthate release, whether active or passive, is not known. In some seeds the evidence favours active unloading (a sucrose–H^+ antiport mechanism has been suggested), while in others the evidence is consistent with a mediated diffusion mechanism. Active unloading might be necessary to attain sucrose concentrations of 100–200 mol m^{-3} found in the seed apoplast of some legumes. Surprisingly, in pea and broad bean seeds phloem unloading from the seed coat was enhanced by apoplastic solute concentrations up to 350 mol m^{-3}, and this may be a general phenomenon, in legume seeds at least. The mechanism of this turgor-sensitive release is not known.

Uptake of sucrose and amino acids by the developing embryos also is enhanced by high apoplastic concentrations of these substances. With developing embryos (and many other plant tissues), a plot of uptake rate versus concentration of sucrose or amino acids contains two components: one saturable and one linear. It is supposed that the saturable component represents the proton co-transport system, but the linear component is puzzling. Undoubtedly, part of the linear uptake represents simple diffusion, but recent evidence (Lin, 1985b) suggests that the non-diffusive part of the linear uptake occurs via a proton symport mechanism. The two components of uptake are incorporated in Eq. (8.2):

$$v = V_{max}\,[S]/([S] + K_m) + k[S] \qquad (8.2)$$

where v is the rate, [S] is the sugar or amino acid concentration, V_{max} and K_m are the Michaelis–Menten constants (the maximum rate and the [S] giving half maximal rate, see Ch. 1), and k is the first-order rate coefficient. A plot of v versus [S] is shown in Fig. 8.5, curve 1. Curves similar to this in shape have been obtained with many plant tissues by measuring entrance of ^{14}C-sugars or amino acids into the tissue after short incubation periods in a range of substrate concentrations. The first-order rate coefficient, k is obtained from the linear portions of curve 1. The linear component of uptake, curve 2, is constructed from k and [S], and the saturable component of uptake, curve 3, is obtained by subtracting curve 2 from curve 1. From curve 3, K_m and V_{max} are calculated (see Ch. 1).

From Fig. 8.5 it can be seen that either the saturable or the linear component can dominate uptake, depending on the substrate concentration. For example, in developing soybean seeds, phloem unloading from the seed coat resulted in a high (150–200 mol m^{-3}) apoplastic sucrose concentration which was necessary to maintain the sucrose level of about 90 mol m^{-3} in the cotyledons (Gifford & Thorne, 1985). The curve for sucrose uptake into developing cotyledons had a saturable and a linear component; the K_m for the saturable component was about 8 mol m^{-3} and the linear component dominated uptake above 50 mol m^{-3} (Thorne, 1982). It appears from these results that, at concentrations of 150–200 mol m^{-3} in the apoplast, the saturable component has virtually no role to play in sucrose uptake into the cotyledons.

Sucrose uptake into source leaf tissue (leaf discs, isolated mesophyll tissue or vascular tissue) also showed the saturable and linear uptake

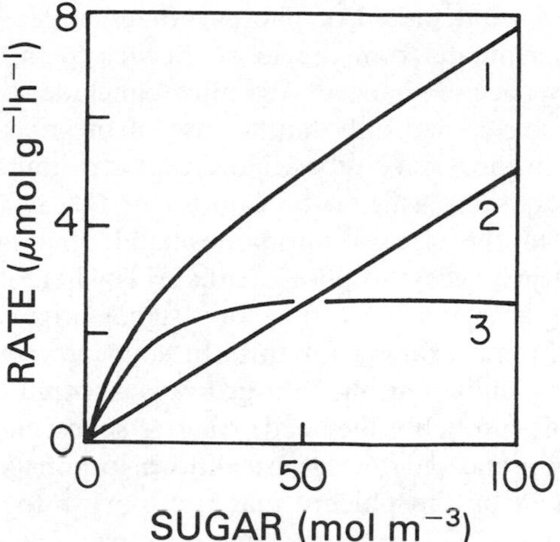

Fig. 8.5 The rate of sugar uptake into plant cells as a function of the outside (apoplastic) sugar concentration. Curve 1 shows the observed uptake rate. Curves 2 and 3 were derived from curve 1 using Eq. (2) as described in the text.

components depicted in Fig. 8.5 (Wilson *et al.*, 1985; Van Bel & Koops, 1985). However, in the discussion of phloem loading in a previous section, it was concluded that sucrose concentrations in the leaf apoplast need be no higher than 20 mol m^{-3}; in the concentration range of 0–20 mol m^{-3} the saturable component has been shown to dominate uptake (Maynard & Lucas, 1982).

8.4.4 Sink strength

Sink strength is a measure of the ability of a sink to attract photosynthate to itself. For example, a single cucumber fruit during its period of rapid growth attracted 40 % of the photosynthate produced by the entire plant (Pharr *et al.*, 1985). During this period, root growth was strongly inhibited and leaf growth less so. The question of how the fruit won the competition for photosynthate is not easily answered. Strong sinks are rapidly growing sinks or sinks that are accumulating storage compounds. It has been suggested that these sinks obtain a larger share of photosynthate because of their ability to take up sugars from the apoplast, thereby maintaining steep sugar and pressure gradients in the sieve tubes from source to sink. In the case of developing legume seeds discussed in the previous section, active transport from the seed coat symplast into the seed apoplast might be the step that maintains the steep gradients.

However, this simple explanation does not account for the inhibition of root and leaf growth during rapid growth of the cucumber fruit in the above example. There is no evidence that the sieve tubes in the young roots and leaves were depleted of photosynthate; indeed, if this were so, the flow of photosynthate from the source leaves to the young roots and leaves should have been enhanced. We must conclude, therefore, that the young roots and leaves had only limited use of the rich supply of photosynthate present in their sieve tubes. Note that this limitation was exerted along a completely symplastic path. Control of the symplastic path may include changes in the size of plasmodesmatal openings, but probably growth itself, not just access to photosynthate, is where control is exerted. Control of growth can involve a hormone signal originating in the fruit; soybean is an extreme example of this. In soybean, both fruit and root nodules are strong sinks. During fruit growth, a hormone-like signal originating in the fruit (probably the seed) triggers senescence in nodules and leaves (both young and old) where breakdown products of the senescent state are scavenged by the phloem and transported to the growing fruit (Lindoo & Noodén, 1977). Clearly, one sink can influence and control other sinks, but how this is done is poorly understood.

Sinks also can influence source leaves (see Herold, 1980). In the cucumber example above, the photosynthetic rates of leaves above and below the fruiting node were double those of the non-fruiting, control plants. In addition, the activity of stachyose synthase (reaction 6, Table 8.1) in leaves of the fruiting plant was double that in control plants, whereas fruiting had no effect on sucrose-P synthase activity (reaction 2, Table 8.1). Here too, a hormonal or perhaps turgor signal originating in the fruit might be involved in triggering the response of the source leaves, but this remains to be shown.

Although we speak of competition among organs for a limited supply of photosynthate, the partitioning of photosynthate is not analogous to hogs feeding at a trough. Rather, it is under a complex set of controls; the same set that governs development of which partitioning of photosynthate is a part. Most likely, some control is exerted along the paths leading to and from the phloem and on the loading and unloading processes themselves.

References

Altus, D. P. & Canny, M. J. (1985). Loading of assimilates in wheat leaves. II. The path from chloroplast to vein. *Plant, Cell and Environment* **8**, 275–85.

Anderson, J. M. (1983). Release of sucrose from *Vicia faba L.* leaf discs. *Plant Physiology* **71**, 333–40.

Anderson, J. M. (1986). Sucrose release from soybean leaf slices. *Physiologia Plantarum* **66**, 319–27.

Asami, S., Hara-Nishimura, I., Nishimura, M. & Akazawa, T. (1985). Translocation of photosynthates into vacuoles in spinach leaf protoplasts. *Plant Physiology* **77**, 963–8.

Baker, D. A. (1978). Proton co-transport of organic solutes by plant cells. *New Phytologist* **81**, 485–97.

Barber, J. (1982). Influence of surface charges on thylakoid structure and function. *Annual Review of Plant Physiology* **33**, 261–95.

Bieleski, R. L. (1982). Sugar alcohols. In Loewus, F. A. & Tanner, W. (eds.). *Encyclopedia of Plant Physiology*, New Series, vol. 13A. pp. 158–92. Heidelberg, Springer-Verlag.

Briskin, D. P., Thornley, W. R. & Wyse, R. E. (1985). Membrane transport in isolated vesicles from sugar beet taproot. II. Evidence for a sucrose/H^+-antiport. *Plant Physiology* **78**, 871–5.

Canny, M. J. (1984). Translocation of nutrients and hormones. In Wilkins, M. B. (ed.). *Advanced Plant Physiology*, pp. 277–96. London, Pitman.

Cronshaw, J. (1981). Phloem structure and function. *Annual Review of Plant Physiology* **32**, 465–84.

De Fekete, M. A. R. & Vieweg, G. H. (1976). Wandgebundene disaccharidhydrolasen des Maisblattes. *Berichte der Deutschen Botanischen Gesellschaft* **89**, 313–19.

Delrot, S. & Bonnemain, J.-L. (1981). Involvement of protons as a substrate for the sucrose carrier during phloem loading in *Vicia faba* leaves. *Plant Physiology* **67**, 560–4.

Delrot, S., Faucher, M., Bonnemain, J.-L. & Bonmort, J. (1983). Nycthemeral changes in intracellular and apoplastic sugars in *Vicia faba* leaves. *Physiologie Végétale* **21**, 459–67.

Dick, P. S. & Ap Rees, T. (1975). The pathway of sugar transport in roots of *Pisum sativum*. *Journal of Experimental Botany* **26**, 305–14.

Doll, S., Rodier, F. & Willenbrink (1979). Accumulation of sucrose ln vacuoles isolated from red beet tissue. *Planta* **144**, 407–11.

Echeverria, E. J. & Humphreys, T. E. (1984). Involvement of sucrose synthase in sucrose catabolism. *Phytochemistry* **23**, 2173–78.

Erwee, M. G. & Goodwin, P. B. (1983). Characterisation of the *Egeria densa* Planch. leaf symplast. *Planta* **158**, 320–8.

Erwee, M. G., Goodwin, P. B. & Van Bel, A. J. E. (1985). Cell–cell communication in the leaves of *Commelina cyanea* and other plants. *Plant, Cell and Environment* **8**, 173–8.

Evert, R. F., Eschrich, W. & Heyser, W. (1977). Distribution and structure of the plasmodesmata in mesophyll and bundle sheath cells of *Zea mays* L. *Planta* **136**, 77–89.

Fellows, R. J. & Geiger, D. R. (1974). Structural and physiological changes in sugar beet leaves during sink to source conversion. *Plant Physiology* **54**, 877–85.

Fondy, B. R. & Geiger, D. R. (1977). Sugar selectivity and other characteristics of phloem loading in *Beta vulgaris*. *Plant Physiology* **59**, 953–60.

Garrard, L. A. & Humphreys, T. E. (1965). Glucose space of the corn scutellum. *Nature* **207**, 1095–6.

Geiger, D. R. (1975). Phloem loading. In Zimmermann, M. H. & Milburn, J. A. (eds.). *Encyclopedia of Plant Physiology*, New Series, vol. 1, pp. 395–431. Heidelberg, Springer-Verlag.

Geiger, D. R. & Cataldo. D. A. (1969). Leaf structure and translocation in sugar beet. *Plant Physiology* **44**, 45–54.

Geiger, D. R. & Swanson, C. A. (1965). Sucrose translocation in the sugar beet. *Plant Physiology* **40**, 942–7.

Gerhardt, R. & Heldt, H. W. (1984). Measurement of subcellular metabolite levels in leaves by fractionation of freeze stopped material in nonaqueous media. *Plant Physiology* **75**, 542–7.

Giaquinta, R. T. (1977). Phloem loading of sucrose: pH dependence and selectivity. *Plant Physiology* **59**, 750–5.

Giaquinta, R. T. (1983). Phloem loading of sucrose. *Annual Review of Plant Physiology* **34**, 347–87.

Giaquinta, R. T., Lin, W., Sadler, N. L. & Franceschi, V. R. (1983). Pathway of phloem unloading of sucrose in corn roots. *Plant Physiology* **72**, 362–7.

Gifford, R. M. & Evans, L. T. (1981). Photosynthesis, carbon partitioning, and yield. *Annual Review of Plant Physiology* **32**, 485–509.

Gifford, R. M. & Thorne, J. H. (1985). Sucrose concentration at the apoplastic interface between seed coat and cotyledons of developing soybean seeds. *Plant Physiology* 77, 863–8.

Glasziou, K. T. & Gayler, K. R. (1972). Storage of sugars in stalks of sugarcane. *Botanical Review* **38**, 471–90.

Grange, R. I. (1985). Carbon partitioning and export in mature leaves of pepper (*Capsicum annuum*). *Journal of Experimental Botany* **36**, 734–44.

Gunning, B. E. S. & Overall, R. L. (1983). Plasmodesmata and cell-cell transport in plants. *BioScience* **33**, 260–5.

Gunning, B. E. S. & Pate, J. S. (1974). Transfer cells. In Robards, A. W. (ed.). *Dynamic Aspects of Plant Ultrastructure*, pp. 441–80. London, McGraw-Hill.

Gunning, B. E. S., Pate, J. S., Minchin, F. R. & Marks, I. (1974). Quantitative aspects of transfer cell structure in relation to vein loading in leaves and solute transport in legume nodules. *Symposium of the Society of Experimental Biology* **28**, 87–126.

Guy, M., Reinhold, L. & Michaeli, D. (1979). Direct evidence for a sugar transport mechanism in isolated vacuoles. *Plant Physiology* **64**, 61–4.

Hawker, J. S. (1965). The sugar content of cell walls and intercellular spaces in sugar cane stems and its relation to sugar transport. *Australian Journal of Biological Sciences* **18**, 959–69.

Hendrix, J. E. (1973). Translocation of sucrose by squash plants. *Plant Physiology* **52**, 688–99.

Herold, A. (1980). Regulation of photosynthesis by sink activity – The missing link. *New Phytologist* **86**, 131–44.

Heyser, W. (1980). Phloem loading in the maize leaf. *Berichte der deutschen botanischen Gesellschaft* **93**, 221–8.

Heyser, W., Evert, R. F., Fritz, E. & Eschrich, W. (1978). Sucrose in the free space of translocating maize leaf bundles. *Plant Physiology* **62**, 491–4.

Ho, L. C. & Baker, D. A. (1982). Regulation of phloem loading and unloading in long distance transport systems. *Physiologia Plantarum* **56**, 225–30.

Huber, S. C. & Moreland, D. E. (1981). Co-transport of K^+ and sugars across the plasmalemma of mesophyll protoplasts. *Plant Physiology* **67**, 163–9.

Humphreys, T. E. (1973). Sucrose transport at the tonoplast. *Phytochemistry* **12**, 1211–19.

Humphreys, T. E. (1981). Sucrose–proton efflux from maize scutellum cells. *Phytochemistry* **20**, 2319–23.

Humphreys, T. E. (1983). Proton electrochemical gradients and sucrose accumulation in the maize scutellum. *Phytochemistry* **22**, 2669–74.

Humphreys, T. E. (1985). The influence of external pH on sucrose uptake and release in the maize scutellum. In Heath, R. L. & Preiss, J. (eds.). *Regulation of Carbohydrate Partitioning in Photosynthetic Tissue*, pp. 215–30. Rockville, Maryland, American Society of Plant Physiologists.

Kaback, H. R. (1983). The *lac* carrier protein in *Escherichia coli*. *Journal of Membrane Biology* **76**, 95–112.

Kaiser, G. & Heber, U. (1984). Sucrose transport into vacuoles isolated from barley mesophyll protoplasts. *Planta* **161**, 562–8.

Kaiser, W. M. & Martinoia, E. (1985). Absence of a apoplastic step in assimilate transport to the phloem? A comparison of assimilate efflux from leaf slices, mesophyll protoplasts and a unicellular green alga. *Journal of Plant Physiology* **121**, 463–74.

Kaiser, W. M., Paul, J. S. & Bassham, J. A. (1979). Release of photosynthates from mesophyll cells *in vitro* and *in vivo*. *Zeitschrift für Pflanzenphysiologie* **94**, 377–385.

King, R. W. (1976). Implication for plant growth of the transport of regulatory compounds in phloem and xylem. In Wardlaw, I. F. & Passioura, J. B. (eds.). *Transport and Transfer Processes in Plants*, pp. 415–31. New York, Academic Press.

Komor, E. (1982). Transport of sugar. In Loewus, F. A. & Tanner, W. (eds.). *Encyclopedia of Plant Physiology*, New Series, vol. 13A, pp. 635–76. Heidelberg, Springer-Verlag.

Komor, E. & Tanner, W. (1974). The hexose–proton cotransport system of *Chlorella*. pH-dependent change in K_m-values and translocation constants of the uptake system. *Journal of General Physiology* **64**, 568–81.

Komor, E., Thom, M. & Maretzki, A. (1982). The mechanism of sugar uptake by sugar cane suspension cells. *Planta* **153**, 181–92.

Läuchli, A. (1976). Apoplastic transport in tissues. In Lüttge, U. & Pitman, M. G. (eds.). *Encyclopedia of Plant Physiology*, New Series, vol. IIB, pp. 3–34. Berlin, Springer-Verlag.

Leigh, R. A., Ap Rees, T., Fuller, W. A. & Banfield, J. (1979). The location of acid invertase activity and sucrose in the vacuoles of storage roots of beetroot (*Beta vulgaris*). *Biochemical Journal* **178**, 539–47.

Lin, W. (1985a). Energetics of sucrose transport into protoplasts from developing soybean cotyledons. *Plant Physiology* **78**, 41–45.

Lin, W. (1985b). Linear sucrose transport in protoplasts from developing soybean cotyledons. *Plant Physiology* **78**, 649–51.

Lindoo, S. J. & Noodén, L. D. (1977). Studies on the behavior of the senescence signal in Anoka soybeans. *Plant Physiology* **59**, 1136–40.

Loescher, W. H., Fellman, J. K., Fox, T. C., Davis, J. M., Redgwell, R. J. & Kennedy, R. A. (1985). Other carbohydrates as translocated carbon sources: Acyclic polyols and photosynthetic carbon metabolism. In Heath, R. L. & Preiss, J. (eds.). *Regulation of Carbon Partitioning in Photosynthetic Tissue*, pp. 309–32. Rockville, MD., American Society of Plant Physiologists.

Loescher, W. H., Marlow, G. C. & Kennedy, R. A. (1982). Sorbitol metabolism and sink–source interconversion in developing apple leaves. *Plant Physiology* **70**, 335–9.

MacRobbie, E. A. C. (1971). Phloem translocation, facts and mechanisms: A comparative survey. *Biological Reviews* **46**, 429–81.

Madore, M. & Webb, J. A. (1981). Leaf free space analysis and vein loading in *Cucurbita pepo*. *Canadian Journal of Botany* **59**, 2550–7.

Madore, M. & Webb, J. A. (1982). Stachyose synthesis in isolated mesophyll cells of *Cucurbita pepo*. *Canadian Journal of Botany* **60**, 126–30.

Madore, M. A., Oross, J. W. & Lucas, W. J. (1986). Symplastic transport in *Ipomea tricolor* source leaves. *Plant Physiology* **82**, 432–42.

Maynard, J. W. & Lucas, W. J. (1982). A reanalysis of the two-component phloem loading system in *Beta vulgaris*. *Plant Physiology* **69**, 734–9.

Minchin, P. E. H., Ryan, K. G. & Thorpe, M. R. (1984). Further evidence of apoplastic unloading into the stem of bean: identification of the phloem buffering pool. *Journal of Experimental Botany* **35**, 1744–53.

Minchin, P. E. H. & Thorpe, M. R. (1984). Apoplastic phloem unloading in the stem of bean. *Journal of Experimental Botany* **35**, 538–50.

Mitchell, P. (1968). *Chemiosmotic Coupling and Energy Transduction*. Glynn Research, Bodmin England.

Morell, M. & Copeland, L. (1984). Enzymes of sucrose breakdown in soybean nodules. Alkaline invertase. *Plant Physiology* **74**, 1030–34.

Münch, E. (1930). *Die Stoffbewegungen in der Pflanze*. Jena, Fischer.

Nobel, P. S., Zaragoza, L. J. & Smith, W. K. (1975). Relation between mesophyll surface area, photosynthetic rate, and illumination level during development for leaves of *Plectranthus parviflorus* Henckel. *Plant Physiology* **55**, 1067–70.

Offler, C. E. & Patrick, J. W. (1984). Cellular structures, plasma membrane surface areas and plasmodesmatal frequencies of seed coats of *Phaseolus vulgaris* L. in relation to photosynthate transfer. *Australian Journal of Plant Physiology* **11**, 79–99.

Outlaw, W. H., Jr., Fisher, D. B. & Christy, A. L. (1975). Compartmentation in *Vicia faba* leaves. II. Kinetics of ^{14}C-sucrose redistribution among individual tissues following pulse labeling. *Plant Physiology* **55**, 704–11.

Pate, J. S. (1980). Transport and partitioning of nitrogenous solutes. *Annual Review of Plant Physiology* **31**, 313–40.

Pate, J. S., Peoples, M. B., Van Bel, A. J. E., Kuo, J. & Atkins, C. A. (1985). Diurnal water balance of the cowpea fruit. *Plant Physiology* **77**, 148–56.

Peracchia, C. (1980). Structural correlates of gap junction permeation. *International Review of Cytology* **66**, 81–146.

Peterson, C. A., Griffith, M. & Huner, N. P. A. (1985). Permeability of the suberized mestome sheath in winter rye. *Plant Physiology* **77**, 157–61.

Pharr, D. M., Huber, S. C. & Sox, H. N. (1985). Leaf carbohydrate status and enzymes of translocate synthesis in fruiting and vegetative plants of *Cucumis sativus* L. *Plant Physiology* **77**, 104–8.

Pollock, C. J. & Lloyd, E. J. (1977). The distribution of acid invertase in developing leaves of *Lolium temulentum*. *Planta* **133**, 197–200.

Pradet, A. & Raymond, P. (1983). Adenine nucleotide ratios and adenylate energy charge in energy metabolism. *Annual Review of Plant Physiology* **34**, 199–224.

Reinhold, L. & Kaplan, A. (1984). Membrane transport of sugars and amino acids. *Annual Review of Plant Physiology* **35**, 48–83.

Robinson, S. P. & Beevers, H. (1981). Evidence for amino acid:proton cotransport in *Ricinus* cotyledons. *Planta* **152**, 527–33.

Rottenberg, H. (1976). The driving force for proton(s) metabolites cotransport in bacterial cells. *FEBS Letters* **66**, 159–63.

Rumpho, M. E., Edwards, G. E. & Loescher, W. H. (1983). Pathway for photosynthetic carbon flow to mannitol in celery leaves. *Plant Physiology* **73**, 869–73.

Saftner, R. A. & Wyse, R. E. (1980). Alkali cation/sucrose cotransport in the root sink of sugar beet. *Plant Physiology* **66**, 884–9.

Schmalstig, J. G. & Geiger, D. R. (1985). Phloem unloading in developing leaves of sugar beet. I. Evidence for pathway through the symplast. *Plant Physiology* **79**, 237–41.

Schmitt, M. R., Hitz, W. D., Lin, W. & Giaquinta, R. T. (1984). Sugar transport into protoplasts isolated from developing soybean cotyledons. II. Sucrose transport kinetics. Selectivity and modeling studies. *Plant Physiology* **75**, 941–6.

Schumacher, A. (1948). Beitrag zur kenntnis des Stofftransportes in dem Siebröhrensystem höherer Pflanzen. *Planta* **35**, 642–700.

Secor, J. & Schrader, L. E. (1984). Characterization of amino acid efflux from isolated soybean cells. *Plant Physiology* **74**, 26–31.

Sovonick, S. A., Geiger, D. R. & Fellows, R. J. (1974). Evidence for active phloem loading in the minor veins of sugar beet. *Plant Physiology* **54**, 886–91.

Spanswick, R. M. (1976). Symplastic transport in tissues. In Lüttge, U. & Pitman, M. G. (eds.). *Encyclopedia of Plant Physiology*, New Series, vol. IIB, pp. 35–53. Berlin, Springer-Verlag.

Streeter, J. G. (1980). Carbohydrates in soybean nodules. II. Distribution of compounds in seedlings during onset of nitrogen fixation. *Plant Physiology* **66**, 471–6.

Sze, H. (1985). H^+-translocating ATPases: Advances using membrane vesicles. *Annual Review of Plant Physiology* **36**, 175–208.

Thom, M., Komor, E. & Maretzki, A. (1982). Vacuoles from sugarcane suspension cultures. II. Characterization of sugar uptake. *Plant Physiology* **69**, 1320–5.

Thom, M. & Maretzki, A. (1985). Group translocation as a mechanism for sucrose transfer into vacuoles from sugarcane cells. *Proceedings of the National Academy of Sciences*, USA **82**, 4697–701.

Thorne, J. H. (1982). Characterization of the active sucrose transport system of immature soybean embryos. *Plant Physiology* **70**, 953–8.

Thorne, J. H. (1985). Phloem unloading of C and N assimilates in developing seeds. *Annual Review of Plant Physiology* **36**, 317–43.

Turgeon, R. (1984a). Efflux of sucrose from minor veins of tobacco leaves. *Planta* **161**, 120–8.

Turgeon, R. (1984b). Termination of nutrient import and development of vein loading capacity in albino tobacco leaves. *Plant Physiology* **76**, 45–8.

Turgeon, R. & Webb, J. A. (1976). Leaf development and phloem transport in *Cucurbita pepo*: maturation of the minor veins. *Planta* **129**, 265–9.

Van Bel, A. J. E. & Koops, A. J. (1985). Uptake of [^{14}C] sucrose in isolated minor-vein networks of *Commelina benghalensis* L. *Planta* **164**, 362–9.

Verma, D. P. S. (1982). Plant–rhizobium interactions in symbiotic nitrogen fixation. In Smith, H. & Grierson, D. (eds.). *The Molecular Biology of Plant Development*, pp. 437–66. Berkeley, Calif., Universitv of California Press.

Vreuadenhil, D. (1983). Abscisic acid inhibits phloem loading of sucrose. *Physiologia Plantarum* **57**, 463–7.

Weatherley, P. E. (1975). Summary of the conference. In Aranoff, S., Dainty, J., Gorham, P. R., Srivastava, L. M. & Swanson, C. A. (eds.). *Phloem Transport*, p. 623. *New York*, Plenum Press.

Wilson, C., Oross, J. W. & Lucas, W. J. (1985). Sugar uptake into *Allium cepa* leaf tissue: an integrated approach. *Planta* **164**, 227–40.

Wright, J. P. & Fisher, D. B. (1981). Measurement of the sieve tube membrane potential. *Plunt Physiology* **67**, 845–8.

Yamaki, S. (1981). Subcellular localization of sorbito1-6-phosphate dehydrogenase in protoplast from apple cotyledon. *Plant and Cell Physiology* **22**, 359–367.

Ziegler, H. (1975). Nature of transported substances. In Zimmermann, M. H. & Milburn, J. A. (eds.). *Encyclopedia of Plant Physiology*, New Series, vol. 1, pp. 59–100. Heidelberg, Springer-Verlag.

Zimmermann, M. H. & Milburn, J. A. (eds.) (1975). *Transport in Plants, Encyclopedia of Plant Physiology*, New Series, vol. 1. Heidelberg, Springer-Verlag.

Zimmermann, M. H. & Ziegler, H. (1975). List of sugars and sugar alcohols in sieve-tube exudates. In Zimmermann, M. H. & Milburn, J. A. (eds.) *Transport in Plants, Encyclopedia of Plant Physiology*, New series, vol. 1, pp. 480–503. Heidelberg, Springer-Verlag.

9 Whole plants

M. G. Pitman

9.1 Introduction

Along the coasts of the warmer parts of the world it is common to find mangroves, plants able to grow with their roots in sea water and adapted to tolerate the high levels of salinity in sea water (see Ch. 10). Many species have glands that excrete excess salt from the plant (see Ch. 13); others exclude salt from the roots. Far more of the world's vegetation grows where ions are much less abundant, especially where the soils are derived from well weathered tertiary materials, as in Australia. Experience in growing plants for agriculture shows that after adequate water, needs for plant nutrients are the next major requirement for increased plant growth. The association of 'fertiliser' and growth of plants goes back many thousands of years.

Uptake and use of ions (nutrients) is a common feature of plants growing in the very wide range of environments on land and in water. The plant may have both to scavenge for nutrients in short supply and to cope with others present at levels that could be toxic; many necessary elements are toxic at high levels, including P, B, Zn and Mn. The question is, what processes for ion uptake are there in common among plants, and what is the relationship between different parts of the plant?

Roots are the absorbing organs for the plant. For many purposes it has been convenient to follow the abstraction made by plant physiologists and study roots in isolation from the rest of the plant. Excised roots continue to respire, will take up salt and excrete ions at rates similar to those of whole plants, and will even transport ions into the xylem where a standing gradient of osmotic pressure leads to exudation of solution from the cut end of the root (see Ch. 7). However, in other ways an excised root reflects the activity of the whole to about the same degree as an arm cut from the body, in that sugars can no longer enter the organ nor can messages pass between it and the rest of the organism.

Transport of ions from the root to the shoot takes place in the water flow through the xylem vessels. Within the stem and leaves the living cells around the xylem act as sinks to reabsorb ions from the xylem sap and perhaps transport these ions to other living cells not in contact with the xylem. The vascular bundle in the leaf appears to be a system for lateral ion transport similar to the stele in the root, but removing ions from, instead of secreting ions into, the xylem. The phloem too is a pathway for the redistribution of ions in the shoot as well as the main supply of sugars to the root. Mature leaves are supplied with ions in the xylem but export ions in the phloem to young leaves and fruits. Phloem transport also returns ions from the shoot to the root though the amount moving in this way may be small. The intact plant is ingeniously organized for the recycling of impressive amounts of nutrients from redundant to juvenile sites or to storage systems.

The movement of ions in the plant is related to the supply of carbohydrates from the shoot to the root in the phloem (see Ch. 8) and probably to some hormonal 'messages' that may regulate the output from the root to the shoot.

This chapter is concerned with models for uptake of ions by whole plants and the interaction of processes involved in transport within the plant.

9.2 Amounts of ions in plants

The amounts of nutrients in plants are expressed in various ways. The simplest measure is perhaps amount per plant, either on a weight or gram molecule basis, but as this quantity increases with plant size it is more common to express the content relative to the weight of the plant. Very often the amount is quoted relative to dry weight as 'ppm' or percentage or mmol gDW^{-1}. This is a convenient measure as dry weight is often simpler to determine than fresh weight in large-scale field experiments. It has the disadvantage, however, that it is more difficult to relate measurements based on dry weight to the function of the nutrient than if a water basis is used. For example, on a dry weight basis, K^+ deficiency is noticeable in many crop plants when K^+ is less than 8–10 mg gDW^{-1} in the leaves. The K^+ content of leaves of various plant species growing on nutrient-poor sandstone near Sydney was about 2–3 mg gDW^{-1}. Apparently the sandstone plants are very efficient in using K^+, but in fact the low K^+ level was due to a very high ratio of dry material to water. When expressed relative to leaf water, the concentration of K^+ in crop plants and sandstone plants is about the same (50 mol m^{-3}) and this value is close to the minimal concentration found in healthy cells and needed for maximum activity of many of the K^+-requiring enzymes, e.g. Leigh & Wyn Jones (1984).

Table 9.1 Approximate levels of various minerals in plant shoots

Mineral	**Relative content (μmol gFW^{-1})**
N	150–300
K	50–100
Ca	10–50
Mg	10–50
Na	5–50(150)
P	25
S	10–20
Fe	0.5–1
Zn	0.1–0.2
Mo	0.001

Nutrient levels will be quoted here either as concentrations (mol m^{-3}, relative to water content of the tissue) or as 'relative content' in units of mmol gFW^{-1}, (relative to fresh weight) or mmol gDW^{-1} (relative to dry weight).

The orders of magnitude of relative content of various ions are given in Table 9.1 as a general guide. Of course, the content of ions relative to each other and in absolute terms will vary from one species to another or according to the availability in the soil or solution, or ontogenetically as the plant passes through different stages of growth.

The concentration of nutrients in plants is related to their function. The main inorganic, osmotic cations in cell vacuoles are Na^+ and K^+ and their total concentration can be related to the water potential developed in the leaves, ranging from 150–200 mol m^{-3} to 800–1000 mol m^{-3} in some halophytes (see Ch. 10). Nitrogen is mainly in amino compounds or protein in the cytoplasm, but may also be present as NO_3^- in the vacuole. Phosphorus and sulphur are usually present in the cytoplasm though occasionally phosphate or sulphate may accumulate in the vacuoles of some root cells. Trace elements such as Zn or Mo are parts of specific enzymes and minimum levels are related to the amount of that enzyme present. Frequently much of the Ca^{2+} present in plants is precipitated in the cell as oxalates or carbonate and only very low concentrations are required in the cytoplasm (Fig. 9.1).

9.3 Interactions between uptake, growth and concentration

The levels of individual nutrients needed by the plant establish a 'demand' that the roots must take up from the soil. If adequate amounts of one nutrient are not available then the growth of the plant can be reduced. The concentrations of nutrients in the plant represent a balance between the

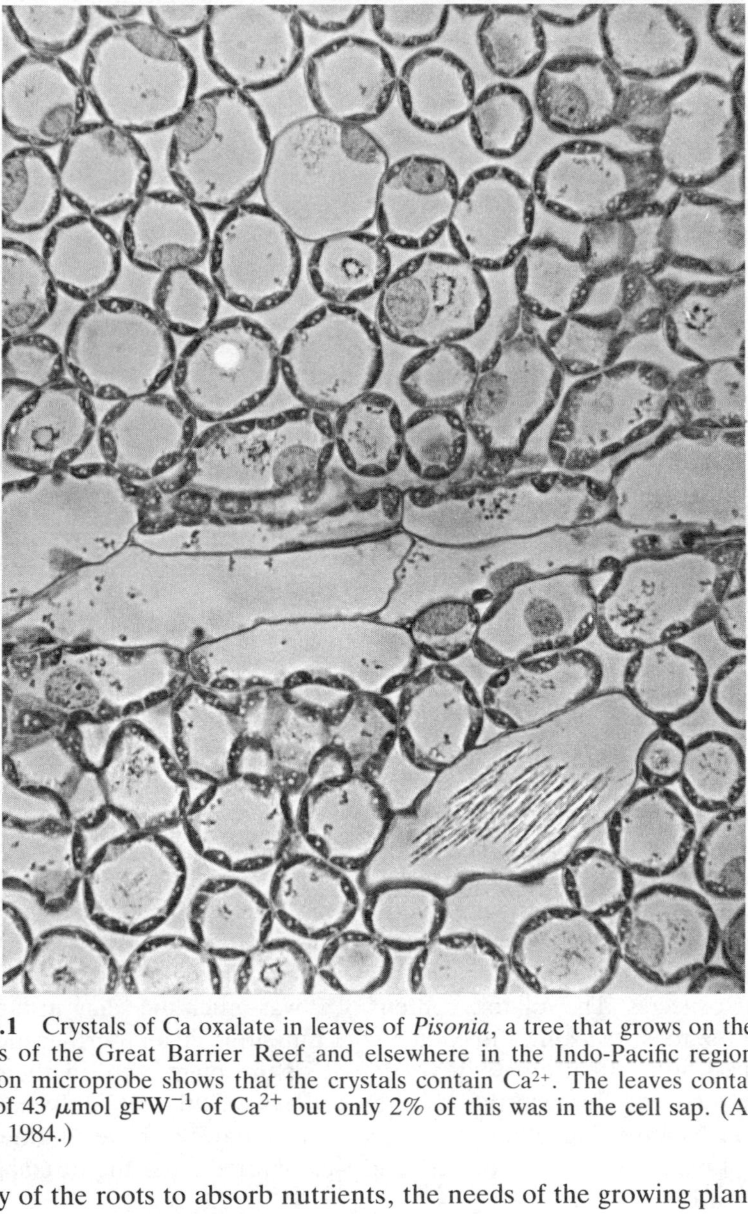

Fig. 9.1 Crystals of Ca oxalate in leaves of *Pisonia*, a tree that grows on the coral islands of the Great Barrier Reef and elsewhere in the Indo-Pacific region. The electron microprobe shows that the crystals contain Ca^{2+}. The leaves contained a total of 43 μmol gFW^{-1} of Ca^{2+} but only 2% of this was in the cell sap. (Allaway *et al.*, 1984.)

ability of the roots to absorb nutrients, the needs of the growing plant, and the rate of growth of the plant.

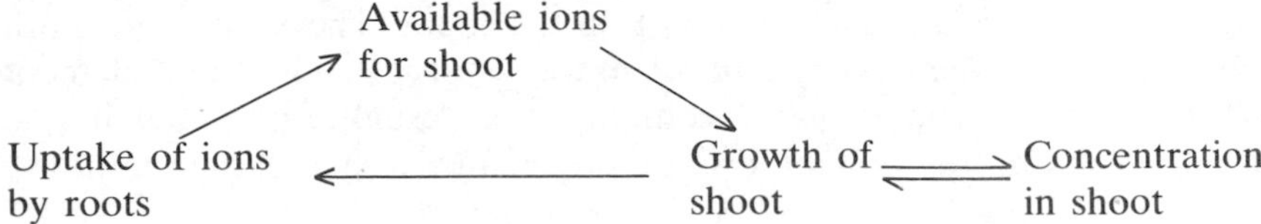

9.3.1 Effects of external concentration: P and K^+

Availability of ions in the soil is difficult to express with precision. The ion may be part of soil minerals; it may be adsorbed on to soil particles or associated with charged exchange sites, or small amounts may be free in the soil solution. Each of these sources may interact in supplying the plant root with the nutrient it requires, but in general the solid phase of the soil is a reservoir of nutrients that can be exploited by the root. (Concentrations in various soils and problems of measuring nutrient availability are discussed by Epstein, 1972; Milthorpe & Moorby, 1974; Nye & Tinker 1977; Lüttge & Higinbotham, 1979).

To avoid these interactions between components of nutrient supply many studies of uptake have been made using culture solutions. To be comparable with soil solutions, concentrations need to be as low as 10 or 100 mmol m^{-3} for some ions and then there is a problem of continuously supplying adequate nutrient to the roots since uptake depletes the solution. Asher *et al.* (1965) overcame this problem by growing plants in large volumes of solution that could be circulated past the roots. Clement *et al.* (1974) describe an adaptation to flowing culture solution systems using ion-sensitive electrodes to monitor concentrations. Use of flowing culture solutions is described in a review by Asher & Edwards (1983).

Results from a series of papers by Asher *et al.* using such culture systems are given here to show how plant growth and nutrient conent may be affected by external concentration (Asher & Loneragan, 1967; Asher & Ozanne, 1967; Loneragan, 1968; Loneragan & Snowball, 1969a, b). A large number of plant species were grown in this series of experiments but data have been averaged here to emphasize general trends. The concentrations were intended to be representative of concentrations in the soil.

Table 9.2 shows the relative content of P in tops and roots of seven species of crop and pasture plants grown in culture solutions containing 0.04 to 25 mmol m^{-3} of P. Growth is shown relative to the largest yield for each species. The relative content of P was much the same in the root and in the tops, reaching a maximum at 1 mmol m^{-3} P and further increase in P had little effect on relative content of the plant, though there was further increase in growth rate. In these species growth and uptake appears to be in balance, but other species can accumulate P above the 'needed' levels. Lupin is one such species, and was omitted from the original data in preparation of Table 9.2.

A somewhat similar pattern of response was found for K^+ (Table 9.3) except that there is more difference between shoot and root and a greater range of relative content of K^+ than P in the shoot. These differences are mainly related to the dual role of K^+ in the plant, as an essential element in the cytoplasm and a major component of osmotic potential in the vacuole (see Ch. 6), whereas P is predominantly a cytoplasmic nutrient.

Table 9.2 Relative content of P, and relative yield and rates of uptake of seven species†

		Relative content (μ mol gFW^{-1})	
Concentration of P in solution (mmol m^{-3})	**Relative yield (%)**	**Tops**	**Roots**
0.04	50	3.8	3.9
0.2	76	10.5	7.4
1	89	23	15
5	97	25	20
25	98	27	22

† Data from Asher & Loneragan (1967).

Table 9.3 Relative K^+ content and relative yield of fourteen species†

		Relative content (μ mol gFW^{-1})	
K^+ in solution (mmol m^{-3})	**Relative yield (%)**	**Tops**	**Roots**
1	7	22	12
8	64	110	42
24	86	144	66
95	91	162	83
1000	97	190	112

† Data from Asher & Ozanne (1967).

Growth and relative content of K^+ were near maximum values in 24 mmol m^{-3} K^+ solution.

The concentrations of these solutions are low compared with the range of K^+ and Na^+ solutions plants can tolerate. At higher concentration there is a tendency in many species for the K^+ (or $K^+ + Na^+$) concentration in the shoot to increase proportionately to the external concentration (Fig. 9.2), as these ions contribute to cellular osmotic potential (but see Sect. 9.7). The interaction of K^+ and Na^+ in plant shoots are considered separately in Sect. 9.5.

9.3.2 Calcium

The effect of external Ca^+ concentration on the amount in the plant is shown in Table 9.4. Two separate groups of plants are shown since the amount in legumes seems to be substantially higher than in cereals and grasses.

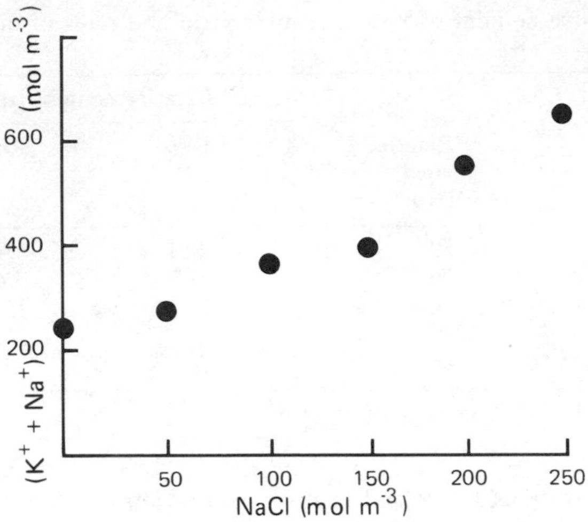

Fig. 9.2 Concentration of K^+ + Na^+ in barley plants. (Data from Storey & Wyn Jones, 1978.)

At lower external concentrations the relative content of Ca^{2+} was nearly constant, but then rose almost proportionately to concentration, and the level in the shoot was then much larger than in the root. For barley at least, this trend continues to much higher external Ca^{2+} concentrations than the 1 mol m^{-3} maximum shown in Table 9.4, and appears to be due to an effect of transpiration on uptake to the plant (Fig. 9.9). Magnesium behaves to some extent like Ca^{2+} in this respect. At the lower range, growth was found to be impaired below about 2.5 mmol m^{-3} for legumes and below 10 mmol m^{-3} Ca^{2+} for cereals and grasses.

Table 9.4 Relative content of Ca^{2+} in sixteen legume species and eleven grass and cereal species†

	Relative content of Ca^{2+} (μmol gFW^{-1})			
	Legumes		**Graminae**	
Concentration of Ca^{2+} in solution (mmol m^{-3})	**Tops**	**Roots**	**Tops**	**Roots**
0.3	4.5	1.1	1.7	0.9
0.8	4.4	1.4	1.5	0.8
2.5	4.7	1.6	1.7	0.7
10	12	1.7	3.3	0.9
100	39	2.2	10.5	1.5
1000	54	3.3	18	2.3

† Data calculated from Loneragan & Snowball (1969a) using percentage dry weights from Asher & Ozanne (1967).

More complicated responses of growth to Ca^{2+} level are found (Fig. 9.3) in which high concentrations of Ca^{2+} may be inhibitory to certain species (calcifuges) but not to others. Such differences can have marked effects on plant competition in natural communities (Kinzel, 1983).

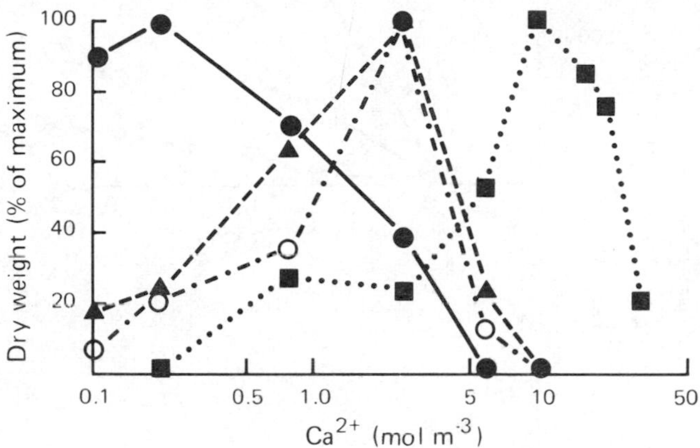

Fig. 9.3 The growth of plants on cultures containing different levels of Ca^{2+} supplied as NO_3^-. For each plant the weight of the shoot is expressed as a percentage of the maximum weight attained. Note that *Juncus squarrosus* (● – ●) grew best at very low Ca^{2+} concentrations and *Origanum vulgare* (■ – ■) required higher concentrations. *Nardus stricta* (Δ – Δ) and *Sieglingia decumbens* (○ – ○) both showed inhibition of growth at higher Ca^{2+} concentrations. (Redrawn from Jefferies & Willis, 1964.)

9.3.3 Nitrogen

Figure 9.4 (a) shows the amounts of N in the shoots of wheat plants grown on soils with different fertilizer applications during 22 weeks after sowing. The amounts extracted from the soil depend markedly on the fertilizer level. Figure 9.4 (b) shows, however, that the relative content of N in the plants was much less different. The plants grew better and contained more N per plant at the higher N-level but there was much less difference in relative content of N, and it appears that the supply of N in the soil restricted plant growth. Studies with flowing culture solution where the concentration can be changed but the *supply* of N is not restricted have shown that uptake was 70 % of maximum at 7 mmol m^{-3} NO_3^- and there was only a small effect on growth between 7 and 710 mmol m^{-3} NO_3^- (Clement *et al.*, 1974).

Figure 9.4 and Tables 9.2 to 9.4 show some of the problems of expressing nutrient data. *Plants with the same relative content of P or N or K^+ can be growing at quite different rates.* The concentration in the plant depends both on the availability of nutrients in the soil, on the rate of growth of the plants, and on the distribution of the nutrient in the plant

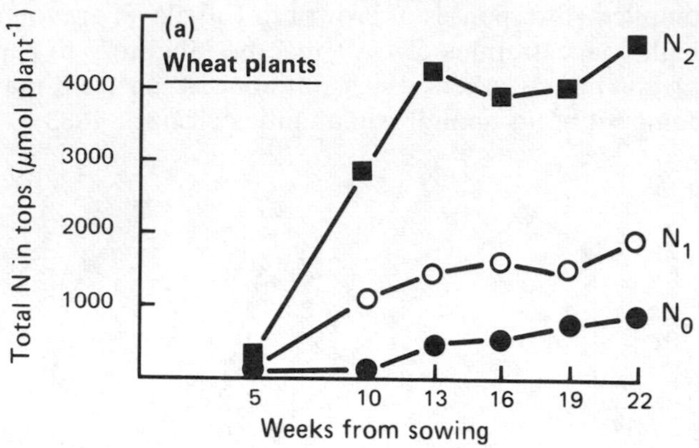

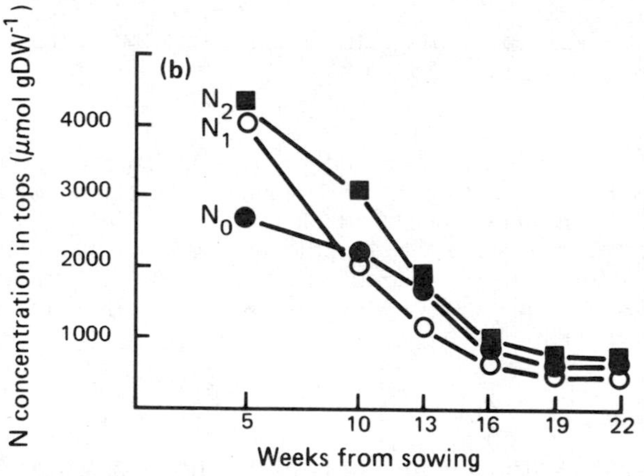

Fig. 9.4 Data for wheat plants grown at three different N-levels ($N_2 > N_1 > N_0$). There were about 180 plants m^{-2}. (a) Average concentration of N in above ground parts of the plant. (b) Uptake of N above ground plotted against time from sowing. (From Halse *et al*, 1969, with scales changed.)

tissue (especially K^+). In general, comparison of many different species of plants shows that there can be varied responses of plants to nutrient level as a result of selection and adaptation to particular environments.

9.3.4 Relating uptake, growth and plant content

The previous sections show a qualitative interaction between growth and uptake. This section discusses how plant growth, rate of uptake by the roots (J) and relative content (C) can be related.

The rate of uptake by the roots is usually expressed relative to the fresh weight of the roots (W_R). If M is the total ion content of the plant, and M_S, M_R of shoot and root, then

$$J = \frac{1}{W_R} \cdot \frac{dM}{dt} \tag{9.1}$$

and

$$J = \frac{1}{W_R} \cdot \left(\frac{dM_S}{dt} + \frac{dM_R}{dt}\right) \tag{9.2}$$

The rate of export from root to shoot

$$J_S = \left(\frac{1}{W_R} \cdot \frac{dM_S}{dt}\right) \tag{9.3}$$

is often referred to as 'specific root transport' (SRT: Shennan, 1981), and is particularly useful in experiments primarily concerned with changes of concentration in the shoot. The average 'concentration' or relative content of the shoot is $C_S = M_S/W_S$ and by differentiating and substituting for J_S it can be shown that

$$\frac{dC_S}{dt} + C_S R_S - \frac{J_S W_R}{W_S} = 0 \tag{9.4}$$

where R_S is the relative growth rate of the shoot, i.e.

$$R_S = \frac{1}{W_S} \cdot \frac{dW_S}{dt} \tag{9.5}$$

It can be seen that at one extreme, when C_S is constant during growth, then d/dt is zero and transport to the shoot is

$$J_S = W_S \cdot C_S \cdot R_S/W_R \tag{9.6}$$

(This is then a convenient way to calculate J_S if R_S is known.) Figure 9.5 gives an example for barley seedlings. Firstly, the open symbols show J_S for plants grown on solution culture (○), soils (▲), or sand (□). For those plants with the same ratio W_S/W_R, J lies on a straight line but the other examples with different W_S/W_R have values of W_S well off the line. When the effect of W_S/W_R is allowed for by plotting $J_S \cdot W_R/W_S$, the points then lie on one line (solid points) even though from different experiments and in different external concentration, since the relative content of the shoot was constant. These data also illustrate a difficulty in using an average value of J_S based on total root weight which can include young and old parts of the root. Differences in efficiency of uptake along the root are then ignored. A similar problem occurs if J_S is converted to a flux based on root area. In the example given in Fig. 9.5 the lower value of J_S for plants in sand could be due either to reduced rates of uptake along the whole of the root or to restriction of uptake to a smaller proportion of the root surface.

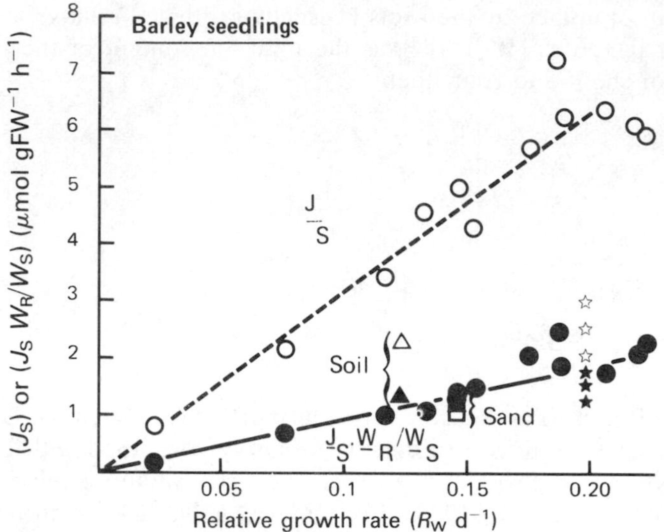

Fig. 9.5 Rate of ($K^+ + Na^+$) uptake to shoots of barley seedlings relative to root weight (J_S) (open symbols) and the product J_S time W_R/W_S plotted against relative growth rate. (○, ●), plants grown on 10 mol m^{-3} K: (Δ, Δ), plants grown in soil: (□, ■), plants grown in sand (data from Pitman, 1972); (☆, ★), plants grown on 20 200 and 2000 mmol m^{-3} K^+ (data from Johansen *et al.*, 1968.)

Reduction in W_S/W_R is commonly found as a response to low nutrient availability. For the plants studied in Table 9.2 W_S/W_R was 4 to 6 at full nutrient level falling to about 1 in lower P concentrations. Bradshaw *et al.* (1964) found a similar response to various grass species to N. For *Agrostis stolonifera* the response was:

N-level (ppm)	1	3	9	27	81	243
W_S/W_R	1.3	1.5	1.9	2.3	3.4	5.2

Figure 9.5 refers to plants growing under varied conditions to alter relative growth rate, which was constant for each condition during the experiment. Figures 9.6 and 9.7 show an example for the halophytic daisy *Aster tripolium* in which growth rates and rates of transport changed during growth over several weeks on the same culture solution (containing 100 mol m^{-3} NaCl). Relative growth rate decreased from 19 d but the rate of transport to the shoot was less affected and the plant showed an increase in content relative to dry weight (○). However, due to a change in ratio of shoot fresh weight to dry weight (●), the content relative to cell water was relatively steady at about 270 mol m^{-3} (▲). During the later part of the experiment the reverse changes in relative growth weight and content relative to dry weight occurred. This example is relevant to consideration

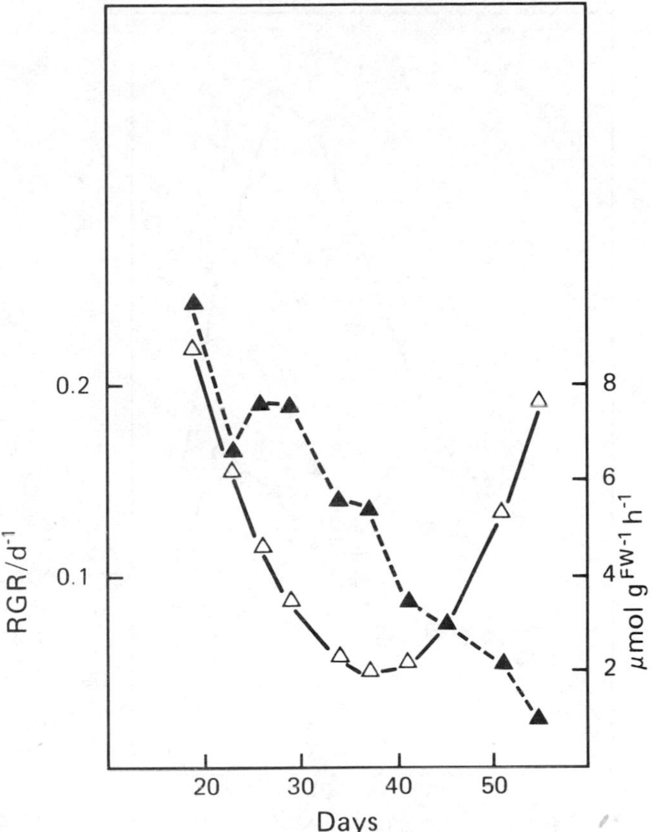

Fig. 9.6 Relative growth rate (Δ) and rate of transport of ($K^+ + Na^+$) from root to shoot (μmol gFW root^{-1} h^{-1}, ▲) for *Aster tripolium* grown on solution containing 100 mol m^{-3} NaCl (see Fig. 9.7.)

of regulation of solute levels in plant leaves as part of the plants adjustment to environmental water potentials (see Ch. 10; Sect. 9.7). Experimentally, values of J_S or R_S may need to be calculated from estimates of content or weight at a few harvests, i.e.

$$J_S = \frac{1}{W_R} \cdot \frac{(M_2 - M_1)}{T_2 - T_1} \tag{9.7}$$

where W_R is an average root weight and M_2, M_1 are shoot contents at times T_2 and T_1. An alternative approach is to fit curves to sets of data for M_S, W_R and W_S and use the analytical curves to give estimates of R_S and J_S at particular times. This approach was used for data in Figs. 9.4, 9.6 and 9.7. Hunt (1982) gives a good account of suitable approaches to calculation and application of statistics in growth experiments.

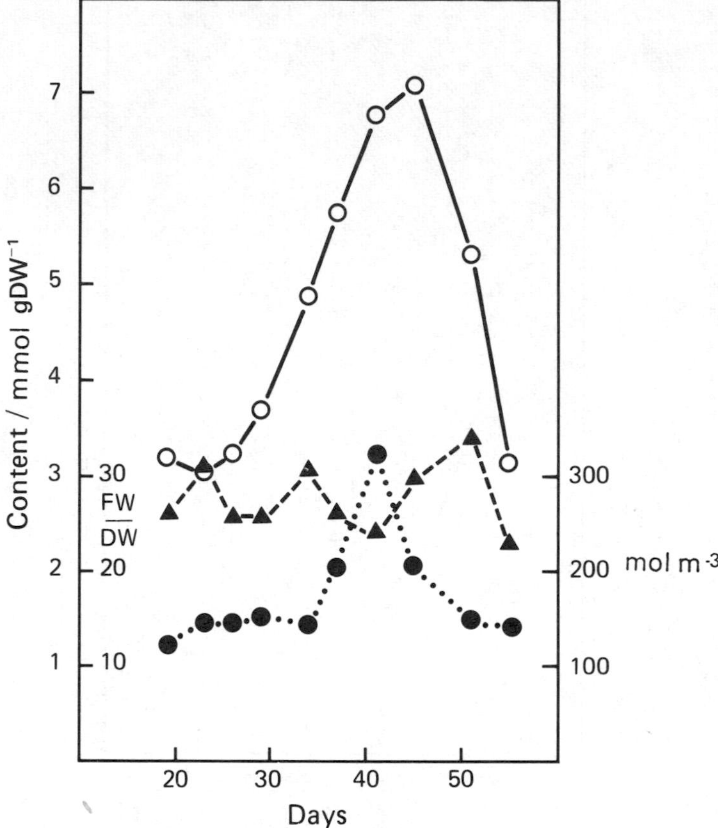

Fig. 9.7 As Fig. 9.6, but here are shown concentration of (K^+ + Na^+) in the shoot relative to tissue water (mol m^{-3}, Δ) or to shoot dry weight (μmol gDW^{-1}, ○) and ratio or shoot fresh-to-dry weights (●). (Data from Shennan, 1981, figures from Pitman, 1983.)

9.3.5 External concentration and rates of uptake

Studies of the effect of concentration on ion influxes have been important as a way of characterizing the uptake mechanism in terms of apparent 'enzyme constants' (Ch. 1). The effect of concentration on uptake to the whole plant is more complicated, since at high concentrations the demand of the shoot for growth is the major factor limiting uptake, while at low concentrations growth may be reduced to a level limited by the rate of nutrient absorption. A further complication is that the plant may have different morphology at low concentrations or may have higher concentrations of absorption sites (see Sect. 9.8). Net rates of absorption of various nutrients have been collected in Fig. 9.8 from the sources shown. The dashed lines show where growth was limited by the low concentration of the nutrient. Over the range shown by the solid lines, relative growth rates were about 0.1–0.15 d^{-1}. As discussed above rates of absorption are affected both by relative growth rate and W_R/W_S. At the lower concentrations where uptake (and growth) appears to be limited by availability of nutrient there seems greater efficiency of the roots at taking up P (and perhaps K^+) than the other ions. Measurements of P uptake by many plant species shows that P can be absorbed at a rate to meet the demand of the plant from extremely low concentrations (less than 10 mmol m^{-3}). Since those concentrations are common in soils it seems that roots are adapted for more efficient uptake of P than other nutrients. The ability of roots to absorb P at low concentrations has already been discussed (Ch. 7).

For most ions there is a range of concentration above which plant growth is depressed. In Fig. 9.8 growth was reduced by Zn^{2+} above 8 mmol m^{-3}. Loneragan and Asher (1967) suggested growth is reduced by phosphate toxicity when the level in the leaves increases above about 50 μmol gFW^{-1} and therefore if absorption by the roots was greater than about

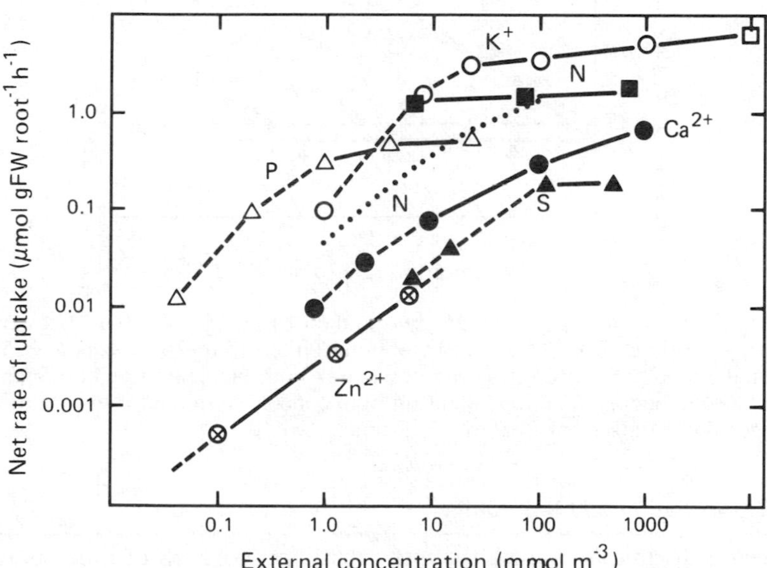

Fig. 9.8 Rates of net uptake of various elements calculated from change in external concentration or internal content. Dashed lines show where growth limitation was observed. Note this is log/log to increase the range of data given. (○) K^+, data from, Asher & Ozane (1967), assuming R_S = 0.13 d^{-1}, mean of fourteen species; (□) from Pitman (1972) for same R_S for barley; (Δ) P, from Asher & Loneragan (1967), Loneragan and Asher (1967), mean of eight species; (■) N, from Clement *et al.* (1974) and dotted line, N, from Lycklama (1963), both from *Lolium perenne*; (●) Ca^{2+}, from Loneragan & Snowball (1969b), mean of thirty species; (Δ) S, from Bouma (1967a, b, c), for *Trifolium subterraneum*; (0) Zn^{2+}, from Carroll & Loneragan (1968, 1969), mean of eight species.

0.4 μmol gFW^{-1} h^{-1}. For some species they found phosphate toxicity in concentrations as low as 24 mmol m^{-3}, though other species (and varieties) were unaffected.

Figure 9.8 showed rates of Ca^{2+} uptake increasing up to 1 mol m^{-3}, separate experiments show this trend continues at least up to 100 mol m^{-3} for barley; Mg^{2+} uptake follows a similar pattern. This increased uptake only occurs at high transpiration and appears partly due to water flow in the plant (Fig. 9.9) (see Sect. 9.4).

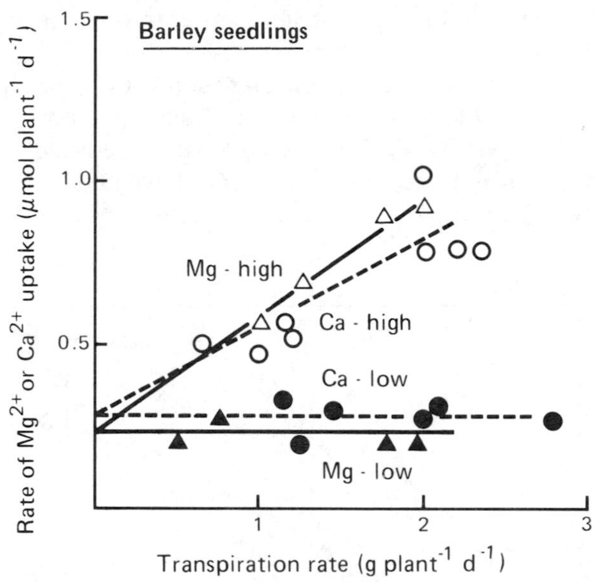

Fig. 9.9 Effect of transpiration rate on uptake of Ca^{2+} (○, ●) and Mg^{2+} (Δ, ▲) by barley seedlings from solutions containing either 15 mol m^{-3} (open symbols, high or 0.5 mol m^{-3}; closed symbols, low). For both elements uptake appears to have two components, one dependent on both concentration and transpiration and the other not. (Data from Lazaroff & Pitman, 1966.)

9.4 Transpiration and ion uptake

Earlier results in this chapter show that large proportions of the ions taken into the roots are transported to the shoot. Some of this transport takes place in the phloem (see Ch. 8), but by far the greater part is transported in the flow of water from root to shoot in the xylem. Rates of water flow in plants are commonly about 0.5–1.0 g h^{-1} gFW of $shoot^{-1}$ so using the average estimates of ion transport from root to shoot, such as in Fig. 9.8, gives estimates of 'concentrations' in the xylem sap of about 1–2 mol m^{-3} K^{+}, 0.1–0.2 mol m^{-3} P and 0.1–0.2 mol m^{-3} Ca^{2+} (taking $W_S/W_R = 3$). The concentrations would be lower, of course, if the rate of water flow were raised or higher under saline conditions (see Ch. 10).

Measurements of concentrations in the xylem sap show that it may contain high concentrations of nitrogen compounds usually in the form of amide or amino nitrogen. The concentrations normally may fluctuate depending on the amount of transpiration, time of day and particularly on the season. Table 9.5 shows concentrations in xylem sap of apple trees showing peaks of P, N and K^+ during the spring-summer growing season, when the demand by the plant is higher.

Table 9.5 Concentrations (mol m^{-3}) of nutrients in xylem sap of an apple tree†

Month	N	K†	P
August	1.1	0.85	1.14
September	1.0	0.92	0.15
October	9.8	1.8	0.7
November	6.2	3.1	0.4
December	5.4	3.6	0.5
January	2.7	2.4	0.3
February	1.3	1.8	0.4
March	1.3	1.8	0.4
April	0.4	1.6	0.4
May	0.4	0.7	0.3
June	0.4	0.7	0.3
July	0.4	0.7	0.3
August	0.4	0.5	0.3

† Data calculated from Bollard (1953) as monthly averages

Reference to Table 9.1 shows that the amounts of cations taken up by the plant can be somewhat higher than the amounts of anions such as NO_3^-, $H_2PO_4^-$ and SO_4^{2-}. Within the plant NO_3^- may also be reduced to —NH_2 forming protein. Where there are imbalances between inorganic cations and anions in the sap, it is usual to find xylem sap contains equivalent amounts of organic acid anions such as malate or citrate, which may also be important in transport of trace elements to the shoot.

9.4.1 Interactions of water and solute flows

One question that is raised by considering uptake by the whole plant is the extent to which water flow in transpiration affects ion transport across the root and into the shoot.

Chapter 7 showed that excised roots exude solution from the cut end due to secretion of ions into the xylem vessels. Concentrations of 30–50 osmol m^{-3} in the exudate are common mainly due to salts of univalent cations. The secretion of ions seems to be an active process drawing on energy metabolism in the root and is sufficiently large to account for rates of transport of K^+ and Na^+ from root to shoot measured with intact plants. The concentrations are higher than in transpiring plants, but the volumes of exudation are much lower than transpiration rates.

In some situations uptake of solutes to the shoot appears to behave as if transport across the root were regulated by active transport. In other situations uptake is evidently strongly affected by the flow of water in transpiration.

Operation of active transport into the xylem of the root has been discussed in Ch. 7. At low external concentrations the 'pump' maintains a concentration in the xylem across a boundary that is less permeable to ion diffusion. The activity of the pump appears to be geared to plant demand and can be controlled by supply of sugar to the roots (Sect. 9.9). For example, Table 9.6 shows the effect of transpiration on transport of univalent cations to shoots of mustard and barley plants showing little effect of water flow on total transport ($K^+ + Na^+$).

Table 9.6 Effect of transpiration on K^+ and Na^+ content of roots and shoot of barley and mustard seedlings. Solution contained 15 mol m^{-3} K^+ and 45 mol m^{-3} Na^+

Plant	Transpiration (%)	Ratio of K^+/Na^+		Content of shoot
		Roots	Shoots	
Barley	100	3.1 ± 0.1	2.5 ± 0.1	108 ± 2†
	50	2.9 ± 0.1	3.56 ± 0.25	106 ± 4
Mustard	100	2.5 ± 0.1	0.73 ± 0.07	3.20 ± 0.1‡
	7	2.4 ± 0.1	2.6 ± 0.1	3.25 ± 0.05

† μmol/plant: Pitman (1956b).
‡ mmol gDW^{-1}: Pitman (1966).

In other situations there are clear effects of transpiration on transport of ions to the shoot. Shone *et al.* (1973) described uptake of triazines (simazine, atrazine, hydroxyatrazine), showing that uptake to the shoot of these organic molecules depended on transpiration and the exchangeable content in the root, and suggested that they moved with transpiration flow across the root.

Silica is mainly present in soils as the uncharged SiO_2 molecule and this has been found to be taken up proportionally to water flow and external concentration by oats (Jones & Handreck, 1965) as expected for uptake due to flow of external solution into the plant. Other species of plants, though, show restriction of uptake of silica by the roots in ways analagous to nutrient ions (Barber & Shone, 1966; Handreck & Jones, 1967). Figure 9.10 shows transport of sulphate increasing with water flow in sunflower seedlings. Greenway (1965) measured Cl^- transport to shoots of barley plants at varied transpiration rates showing that transport to the shoot could be described as the sum of a constant component that could be inhibited by the uncoupler DNP (pump), plus a component unaffected by

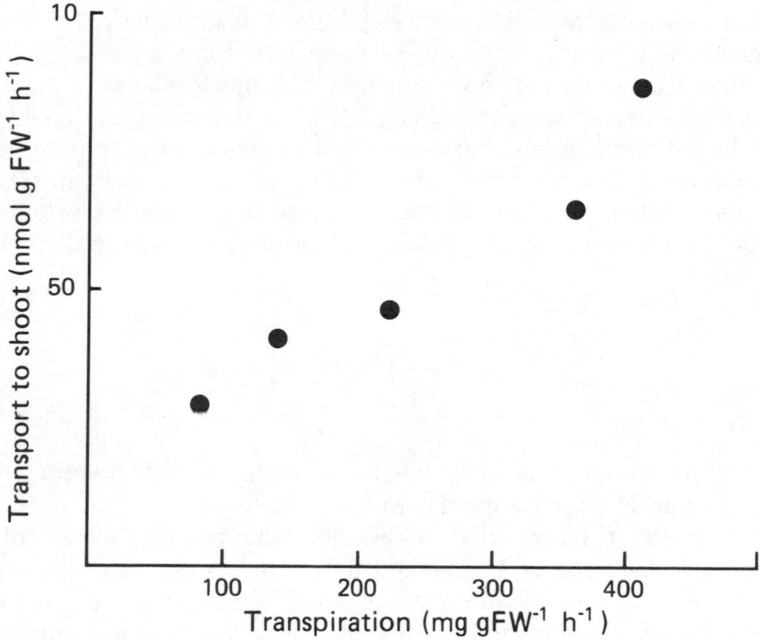

Fig. 9.10 Transport of sulphate to sunflower seedlings at varied transpiration rates. The solution contained 0.5 mol m^{-3} sulphate. (Data from Petersson, 1960.)

DNP and proportional to water flow. Similar evidence for an 'active' component and a non-active component of Cl^- uptake in barley was obtained by Lüttge and Laties (1967). Figure 9.9 showed divalent cation transport increasing with transpiration from solutions of high concentration, though independent of transpiration at lower concentrations (as if there were a small 'pump'). Other evidence for active Ca^{2+} uptake comes from Barber and Koontz (1963) who found that DNP inhibited Ca^{2+} transport to the shoot very strongly both from 0.5 and 5.0 mol m^{-3}, though over a longer period in DNP (24 h) there was increased Ca^{2+} uptake as if the barrier to diffusion in the root had been broken down. This effect of DNP makes it difficult to analyse uptake into 'pump' and 'flow-dependent' components by use of inhibitors for there may be no assurance that the inhibitor has not changed the dependence of transport on flow. Uptake of NaCl by halophytes is commonly found to increase with the rate of transpiration (Ch. 10.).

One way of explaining the flow-dependent component has been to assume that a small percentage of water entering the xylem bypasses the membranes around the symplast. Such a flow need be only 2–5 % of total water flow to explain observed results. Tests of such a pathway have been made by measuring flow using root systems under pressure to develop flow and have shown mannitol flows of 2–6 % (Perry & Greenway, 1973; Salim & Pitman 1984) but this technique is suspect as it can increase the flows of Na^+ and Cl^- across the root compared with intact plants.

For certain solutes (simazine, SiO_2) it is clear that molecules could pass across the cell membranes, but even when solutes are transported across the membrane there could still be a relation of net transport to transpiration flow if there is a flux across the membrane from the xylem proportional to xylem concentration. Net transport would then be

$$J_S = \phi_{ox} - \phi_{xo} = \phi_{ox} - kC_x \tag{9.8}$$

As $C_x = J_S/J_V$

$$J_S = \frac{\phi_{ox} \cdot J_V}{J_V + k} \tag{9.9}$$

which will be proportional to J_V when J_V is small compared with k, rising to ϕ_{ox} when J_V is large compound with k.

The location of the barrier to diffusion and possible bypass of the symplast are discussed in Chs. 7 and 8. The important issue here is to recognize that the combination of a pump and diffusion barrier in the roots can either allow accumulation of ions into the xylem of transpiring plants from the very low concentrations normally found in the soil, or the barrier can prevent excessive inflow of ions from solutions of high concentration. An extreme example of resistance to influx is shown by mangroves (Ch. 10) in which the xylem sap of non-secreting species has been shown to contain about 5–20 mol m^{-3} Cl^- when the roots were in sea water of about 400 mol m^{-3} Cl^- (Scholander *et al.*, 1962).

9.4.2 Ion reabsorption and efflux

Concentrations of ions reaching the leaves may be lower than the estimates for whole shoot averages made above since cells along the path of the xylem extract ions from the solution. As the xylem solution passes through the stem and the leaf petiole it becomes depleted (Klepper & Kaufmann, 1966). Removal of Ca^{2+} and Na^+ in the stem may also reduce the amounts of these ions transported to the leaves and reduce the 'load' of non-essential ions in the leaf (see Ch. 10).

9.5 Selective uptake of K^+ relative to Na^+

Selective uptake of K^+ in preference to Na^+ is found in many plant cells due to operation of an active K^+ influx and active Na^+ efflux (Ch. 5). Cells of plant roots show similar selectivity for K^+ (Pitman & Saddler, 1967;

Jeschke, 1970, 1973, 1984). Within the whole plant the overall proportions of K^+/Na^+ are determined by a number of other factors relating to transport from root to shoot and translocation of ions from older to younger leaves. It is generally found that the ratio of K^+/Na^+ in the shoot is higher than in the roots and higher again than in the solution. Plants of different species are found to vary in selectivity of K^+ relative to Na^+ (Collander, 1941), with grasses and certain legumes showing high selectivity for K^+ and plants such as *Atriplex* at the other extreme.

A convenient way to study selective uptake to whole plants is to grow them in culture solution. The average content, over young and old leaves, of K^+ and Na^+ in the shoot can be more or less constant in proportions over long periods of growth although total content increass exponentially (e.g. barley: Greenway *et al.*, 1965). It appears that there is a 'steady-state' transport (J_S) of K^+ and Na^+ from the root to the shoot.

The 'steady state' appears to be due to regulation of net import to meet the demands of growing leaves that in turn contain ions at levels determined by the osmotic conditions in which the plant grows. An issue that has not been resolved is the extent to which the net transport to the shoot is due to regulation of export from root to shoot in the xylem or due to recirculation of ions excessive to the needs of the leaves (and fruits) in the phloem. In barley, the proportions of K^+ to Na^+ in the xylem exudate from de-topped seedlings is very nearly the same as in the shoot as a whole (Table 9.7) (and the phloem would contain very high K^+/Na^+ ratio) so it appears that the xylem is the main pathway of transport into the shoot.

Table 9.7 Comparison of K^+/Na^+ in shoots of whole barley seedlings and exudate from de-topped seedlings (data from Pitman, 1965a)

Solution concentration (mol m^{-3})		Ratio of K^+/Na^+				Exudate concentration
K^+	Na^+	Solution	Roots	Shoots	Exudate	($K^+ + Na^+$) (mol m^{-3})
0.5	9.5	0.052	0.8	3.4	2.1	37
1.0	9.0	0.11	1.1	4.0	3.1	32
2.5	7.5	0.33	1.9	8.1	7.1	34
6.0	4.0	1.5	6.7	18	22	37
8.0	2.0	4.0	8.9	30	36	37

This view has been confirmed for K^+ transport (Pitman, 1972). However, within the shoot K^+ and Na^+ are distributed unevenly due to K^+ retranslocation in the phloem out of older to younger leaves. Jeschke (1984), however, considers that recirculation of K^+ may also contribute to regulation of shoot content overall (see also Sect. 9.8).

Figure 9.11 shows that the total cation and the total ($K^+ + Na^+$) contents of the tropical grass *Chloris gayana* were virtually independent

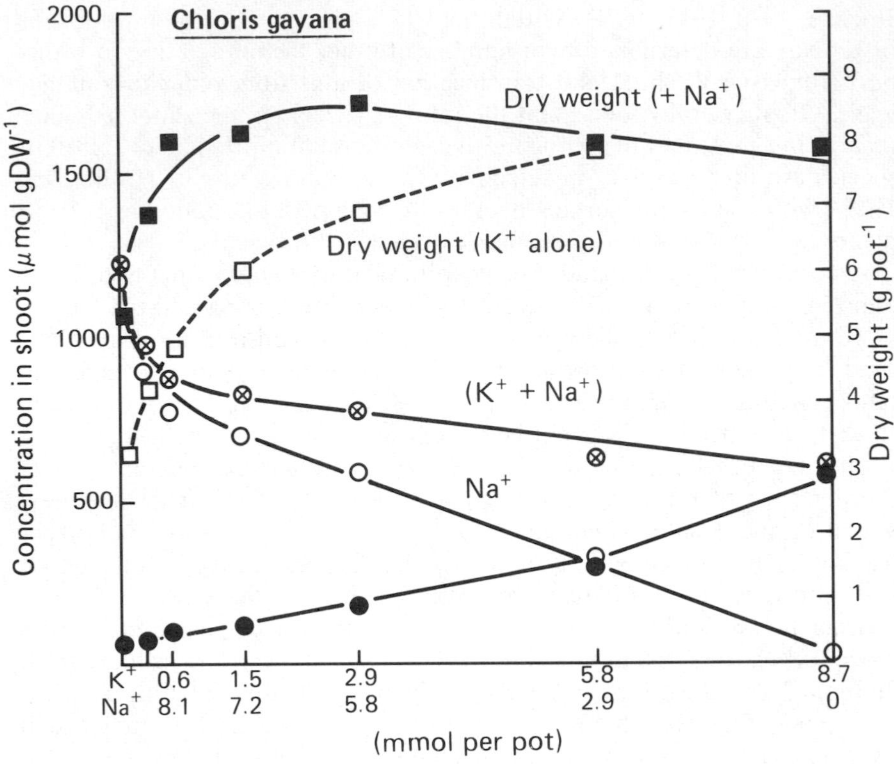

Fig. 9.11 Growth response of plants of *Chloris gayana* grown in pots with varied K^+ content (□), or (K^+ + Na^+) content (■) shows 'sparing' effect of Na^+ on K^+. The content of the plants grown on (K^+ + Na^+) shows interchange of Na^+ for K^+, but the total content is less affected: (○), Na^+; (●), K^+; (○) K^- + Na^+. (Redrawn from Smith, 1974.)

of the ratio of K^+/Na^+ in the soil (Smith, 1974). Similar results have been found for barley (Pitman, 1965a) and mustard (Pitman, 1966).

Table 9.7 (and the data of Fig. 9.11) also show that the ratio of K^+/Na^+ in the plant depends very much on the proportions of K^+/Na^+ in the soil or solution. However, the ratio of K^+/Na^+ in the plant may also be affected by the total concentration of (K^+ + Na^+) around the roots. Thus, when the external ratio of K^+/Na^+ was kept constant but total concentration of (K^+ + Na^+) increased, there was a marked decrease in K^+/Na^+ in the shoots of both barley and mustard (Fig. 9.12). The patterns of response of the two species were remarkably alike except that the reduction in selectivity occurred at lower external concentration for mustard than for barley. At low concentrations (in the range found in many soils) mustard can be as selective for K^+ as barley, but it becomes less effective in more saline conditions.

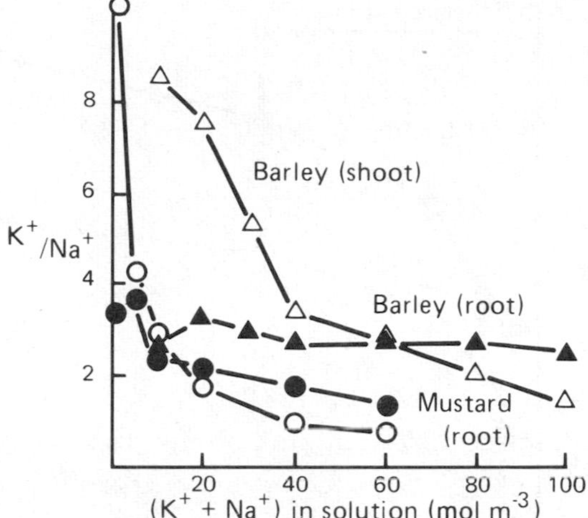

Fig. 9.12 Ratios of K^+/Na^+ in root (closed symbols) and shoot (open symbols) when external concentration of ($K^+ + Na^+$) was varied but K^+/Na^+ constant and equal to 1 : 3. Note that K^+/Na^+ in the shoot is more affected by external concentration than in the roots. (Pitman 1965b, 1966.)

The ratio of K^+/Na^+ in the roots was much less affected by external concentration than that in the shoots, particularly for barley. Presumably the content of the roots is determined by fluxes into the cells but the content of the shoot is also affected by the rate of transpiration (Table 9.6). At higher concentrations, Na^+ uptake was increased at the expense of K^+ when transpiration was occurring, but at low concentrations transpiration had no effect on K^+ and Na^+ uptakes. Jeschke (1984) analysed K^+ and Na^+ fluxes in barley roots under different conditions of water flow and also showed greater transport of Na^+ to the shoot at higher transpiration rates, but in this case there was a parallel effect on the ratio of K^+/Na^+ in the roots. As will be shown below, there are other situations where changes in selectivity in the shoot appear to be due to changed selectivity at the cellular level (Fig. 9.13).

There are many other published data for uptake of K^+ and Na^+ that show similar trends in selectivity. Often, though, comparison has been made by changing Na^+ concentration when K^+ was kept constant so that both ($K^+ + Na^+$) concentration and the ratio of K^+/Na^+ varied. In this case it is less easy to separate effects due to concentration from those due to the K^+/Na^+ ratio.

9.5.1 Divalent cations and selectivity of K^+/Na^+

The ability of plants to take up K^+ selectively appears to depend on availability of divalent cations. Figure 9.13 shows the ratio of K^+/Na^+ in roots

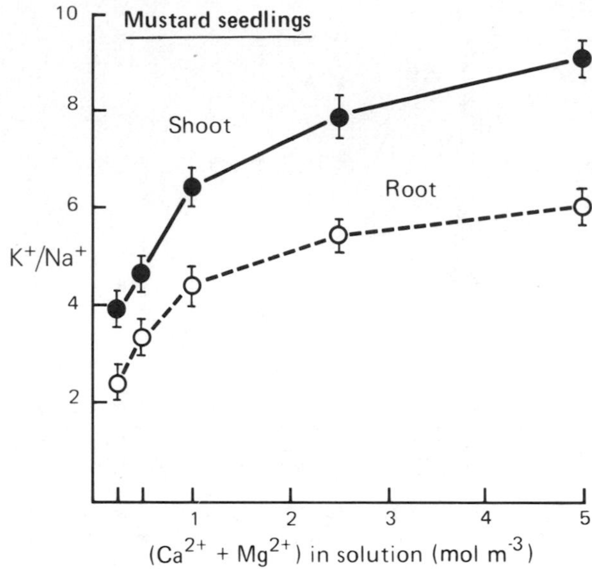

Fig. 9.13 Effect of varying divalent cation concentration on the ratio of K^+/Na^+ in root and shoot of mustard seedlings, when $K^+ = 0.5$ and $Na^+ = 1.5$ mol m^{-3} in the solution. The ratio of Ca^{2+}/Mg^{2+} was constant at 3 : 2. (Data from Pitman, 1966.)

and shoots of mustard seedlings when the concentration of Ca^{2+} and Mg^{2+} was varied. Note that selectivity of uptake to both root and shoot was affected. The concentration of divalent cations needed for maximum selectivity varied with external concentration of $(K^+ + Na^+)$ and with the kind of plant. Barley was much less sensitive to Ca^{2+} level than mustard, but in both cases maximum selectivity was only found when Ca^{2+} was greater than about 2.5 mol m^{-3}. Hyder and Greenway (1965) have suggested this role of divalent cations may be important in the resistance of plants to high salinity.

9.5.2 Conclusions

The selectivity of a plant for K^+ transport into the xylem appears to be determined at two levels. Firstly, during selective uptake of K^+ into the root cells and secondly, during transport of K^+ and Na^+ across the root to the xylem. The selectivity of the uptake process to the root cells is determined by K^+/Na^+ in the solution, availability of Ca^{2+} and to some extent by interaction with water flow. It thus has some properties like selective K^+ and Na^+ transport in algal cells (Ch. 5) and, like (H^+)-ATPase systems, requires Mg^{2+} for activation (Ch. 3). Transport across the root operates proportionally to the selectivity in the root cells (Pitman, 1966) but can be affected too by the water flow across the root to the shoot and the concentration of ions external to the root. A further component of K^+ and Na^+ uptake is translocation in the phloem and this too will affect the selectivity of K^+ in various parts of the shoot as discussed below.

9.6 Translocation in the phloem

The xylem is the primary pathway for movement of nutrients to the shoot, but the phloem is involved in redistribution of ions within the shoot, and possibly from shoot to root. Ions are retranslocated from older to younger leaves, to the roots and to developing fruits and seeds, which in many cases are supplied entirely via the phloem.

Cereal plants show redistribution of nutrients during the growth of the plant. As the stem elongates the oldest leaves become senescent and eventually, when the grain is being formed, photosynthesis is carried out almost entirely by the flag leaf and the ear itself. Eventually, the grain weighs about 45 % of the total plant and contains 70–80 % of the total N, 35 % of the total K^+ and 85 % of the total P in the plant. As the grain develops, N and other nutrients are translocated out of the leaves and stem into the grain, so that at this stage the adult plant is a reservoir for N and uptake from the soil is negligible (Fig. 9.14).

Greenway (Table 9.8) has measured translocation of K^+, Na^+ and Cl^- into developing grain of barley plants grown under different salinities, and

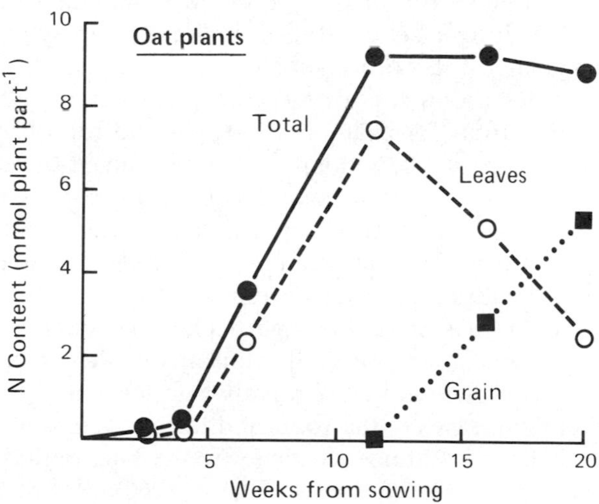

Fig. 9.14 Amounts of N in oats plants growing at high N and P levels. Note the redistribution from leaves to grain from about 12 weeks onwards. (Data from Williams, 1936.)

Table 9.8 Translocation of K^+, Na^+ and Cl^- to grain of barley plants and levels in the leaves†

Ion	Concentration in solution (mol m^{-3})	Concentration in leaves (mmol gDW^{-1})	Concentration in grain (mmol gDW^{-1})	Amount translocated (mmol)
(a) K^+	6	1.3	0.29	2.36
Na^+	1	0.1	0.02	0.16
Cl^-	1	0.2	0.05	0.40
(b) K^+	6	0.5	0.25	2.00
Na^+	50	1.0	0.04	0.32
Cl^-	50	1.0	0.09	0.72

† Data from Greenway *et al.* (1965).

shown how efficient the transport system is in discriminating against Na^+. Despite large Na^+ and Cl^- concentrations in the leaves, only small amounts of Na^+ and Cl^- were transported to the grain of the saline-treated plants, whereas K^+ translocation was nearly the same from each treatment. The amount of K^+ translocated to the grain was about 1 % of the K^+ in the plant as a whole.

Mature barley grains weigh about 44 mg and contain 0.18 mmol gDW^{-1} K^+, 0.08 mmol gDW^{-1} Mg^{2+} and 0.01 mmol gDW^{-1} Ca^{2+}, i.e. Mg^{2+} is translocated to the grain almost as fast as K^+, and very much more rapidly than Ca^{2+}. Translocation of Ca^{2+} is generally slow both into fruits and out of leaves. Use of tracers shows small amounts of ^{45}Ca move in the opposite direction to flow in the xylem (and so presumably in the phloem) but only when levels of Ca^{2+} in the leaves are high (e.g. Millikan and Hanger, 1966; Ringoet *et al.*, 1968). Often the supply of Ca^{2+} in the phloem is inadequate to meet the demands of growing fruits, and must be supplied in the xylem; a number of diseases of cultivated fruits are due to lack of Ca^{2+} in the fruit (for example, 'blossom end rot' of tomato).

The mobility of ions in the phloem has been studied extensively using the technique of applying a radioactive nuclide to the surface of a leaf and observing its subsequent redistribution either by autoradiography, or by measuring radioactivity. Basal movement from the site of application has been used as a measure of relative efficiency of translocation in the phloem. The method has limitations, particularly as there is some uncertainty about the specific activity of the ion at the site of loading to the phloem. Another problem is that a rapidly translocated nutrient, such as P, may pass from the leaves to the roots and then be returned to the leaves in the xylem so that measurements made over a long period may underestimate translocation (Greenway & Gunn, 1966). There is the basic assumption too that water always flows acropetally in the xylem.

Steucek and Koontz (1970) collected data for translocation of foliar-applied nutrients in bean plants (*Phaseolus*) (Table 9.9). It can be seen that

Table 9.9 Export of radioactive tracers from leaves of bean plants (*Phaseolus*) expressed relative to the amount absorbed in the leaf

Tracer	Tracer exported in 24 h (%)
^{22}Na	39†
^{86}Rb	34†
^{42}K	25†
^{32}P	25†
^{35}S	12†
^{36}Cl	7†
^{28}Mg	7‡
^{45}Ca	0.05§

† From Bukovac and Wittwer (1957).
‡ From Steucek and Koontz (1970).
§ From Biddulph *et al.* (1959).

Mg^{2+} is very much more labile than Ca^{2+}, as already discussed for seeds, and Na^+ translocation was very rapid, confirming other measurements showing large net export of Na^+ from *Phaseolus* leaves. *Phaseolus* seems to be different in this respect from many other plants where Na^+ translocation is usually very slow. In the leaves of barley plants, for example, the ratio of K^+/Na^+ falls as the leaves become older because there is greater export of K^+ than Na^+ from older to younger leaves. Plants grown on a solution containing 2.5 mol m^{-3} K^+ + 7.5 mol m^{-3} Na^+ maintained a steady proportion of K^+/Na^+ in the shoot as a whole over 10 days of seedling growth, but the ratios of K^+/Na^+ in first second and third leaves were 7.5, 15.4 and 23, respectively (Pitman, 1965a).

Export of K^+, Na^+, Cl^- and P from leaves has been estimated from measurements of total content and tracer uptake to the leaf from the roots. For example, the total content of K^+ in the oldest leaf of barley was found to change by only 0.34 ± 0.4 μmol d^{-1} $plant^{-1}$ and the amount of tracer supplied to the leaf from the roots increased by 1.9 μmol d^{-1} $plant^{-1}$. Apparently, there was an export of K^+ from this leaf at a rate of 1.6 μmol d^{-1}. In fact, the rate of import was higher due to dilution of tracer with K^+ from the vacuoles of root cells (Greenway & Pitman, 1965). Rates of export from the first leaf of barley expressed as an amount per day and as a percentage of content of the leaf were:

K^+	3.1 μmol d^{-1} (7%)	(Greenway & Pitman, 1965)
Na^+	0.7 μmol d^{-1} (0.8%)	(Greenway *et al.*, 1965)
Cl^-	1.6 μmol d^{-1} (2.7%)	(Greenway *et al.*, 1965)
P	0.8 μmol d^{-1} (14%)	(Greenway & Gunn, 1966)
Sugars	7 μmol d^{-1} (50–100%)	(Greenway & Pitman, 1965)

Measurements of P and K^+ retranslocation in leaves of different ages show that three stages can be distinguished, which roughly parallel patterns of sugar translocation. During the early stages of development of the leaf (to about 20 % of its final weight) nutrients are supplied predominantly in the phloem. There then follows an intermediate stage when the leaves are increasing their content of the ion, possibly now via the xylem, but are exporting the ion in the phloem. At the final stage when the leaf is mature there is concurrent import and export, but little increase in net content. (At even later stages when the leaf becomes senescent the export may be larger than the import; see Thrower, 1962; Hopkinson, 1964; Greenway & Gunn, 1966).

Leaves of pine which stay on the tree for 2–3 years can show cycles of accumulation and export of nutrients (Fig. 9.15).

In general, the mobility of ions in the phloem is related to the concentration in the phloem sap. MacRobbie (1971) quotes the following values from various sources:

K^+ = 20–85 mol m^{-3} Ca^{2+} = 0.25–0.5 mol m^{-3}
Na^+ = 0.06–0.3 mol m^{-3} Mg^{2+} = 2.3–23 mol m^{-3}
P (total) = 3–10 mol m^{-3}

NO_3^- and SO_4^{2-} were both absent and these elements are presumably translocated as organic compounds. Further information is given by Ziegler (1975). It appears that the efficiency of translocation of elements is largely due to the efficiency of loading into the phloem; once in the phloem, redistribution can take place. Differences in Na^+ mobility between *Phaseolus* and barley (for example) are then presumably due to greater efficiency of the phloem in *Phaseolus* at absorbing Na^+. The low mobility of Ca^{2+} in the plant also reflects its low rate of entry into the phloem, and its generally low concentration in plant cell cytoplasm.

One of the limitations of understanding the role of phloem translocation is the nature of phloem transport itself. It makes some difference, for example, if bulk flow in the phloem transports all solutes to a young leaf or fruit proportional to the concentration in the phloem or if transport is determined by the rate of unloading and hence by demand of the tissue (see Ch. 8).

9.7 Regulation of nutrient content in leaves

The leaf is a closed system receiving nutrients via the xylem and at early stages of development via the phloem. Ions may be exported again in the phloem. In most physiological experiments these are the only mechanisms contributing to the amount in the leaf, though other components of loss from the leaf need to be recognized. Under very humid conditions, guttation may occur. In field conditions, water will be present occasionally on the leaves due to rain or mist, and may leach ions from the leaves (see Sect. 9.7.3).

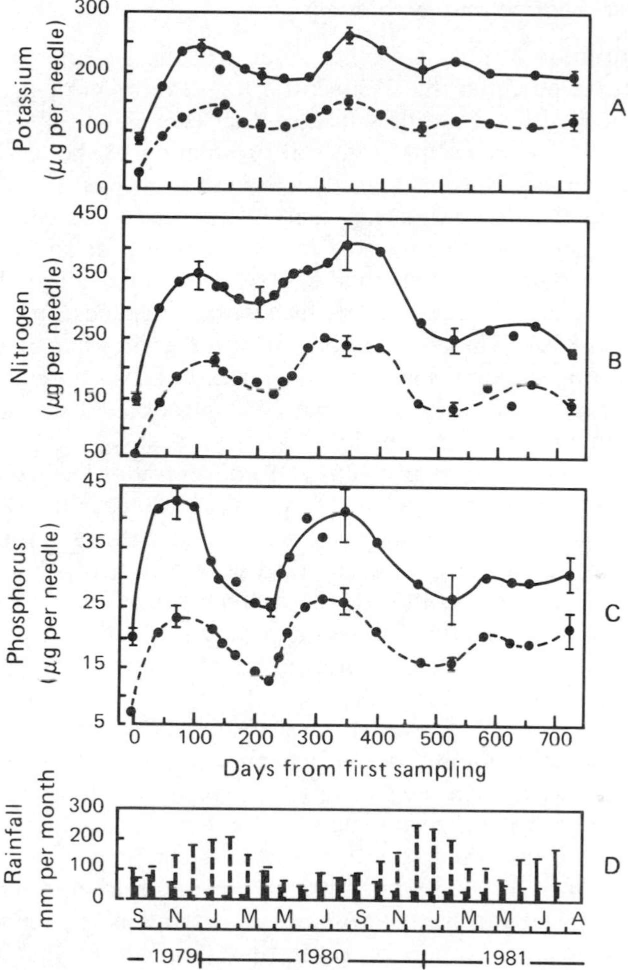

Fig. 9.15 Fluctuations in the content from September 1979 to September 1981 of A, potassium; B, nitrogen; C, phosphorus in primary branch needles (●——●) and secondary branch needles (●----●) formed in spring 1979 on radiata pine planted in 1977 (experiment 1); D, monthly rainfall (T) and evaporation (T). (From Fife & Nambiar, 1984.)

The amount of an ion in the leaf is perhaps less useful as a measure of physiological activity than its concentration (expressed for convenience in any of the ways given earlier). The question discussed here is the extent to which the concentration in the leaf is regulated and by what mechanisms.

9.7.1 Water relations and ion content

The concentration of ions and other solutes in the vacuoles of plant cells has an osmotic potential that is essential for generation of the low water potentials needed for water flow in the plant (see Baker, 1984).

When flow of water occurs from soil or solution to the leaf, the water potential in the leaf is lower than that around the roots. If the soil is dry or the solution has low osmotic potential then the water potential of the leaf must be correspondingly lower for water transport and transpiration to occur, and the associated uptake of CO_2 to take place through the stomata of the leaves. Opening of the stomata depends on maintenance of turgor in the leaf. Under conditions of water stress when external water potential is low, turgor maintenance requires lower osmotic potentials in the vacuoles and hence larger vacuole concentrations of non-metabolized ions or organic solutes (see Ch. 12).

There is increasing evidence that the concentrations of solutes in the vacuoles of leaf cells are regulated to a level related to the difference in water potential between transpiring leaves and the external soil and solution. For example, the increased concentration of NaCl in barley leaves in Fig. 9.2 can be thought of as an adaptation to the increased concentration in the external solution and its associated lower water potential. (The relationship of vacuolar and cytoplasmic concentrations is discussed in Chs. 6 and 12.)

The data of Fig. 9.7 showed the concentration of (K^+ + Na^+) maintained at about 270 mol m^{-3} despite changes in fresh to dry weight ratio, and other data from the same source (Shennan, 1981) show this controlled concentration rising proportionately to external concentration (Table 9.10). Often the increased concentration is accompanied by a decreased leaf area and net carbon assimilation. One simple interpretation of this observation is that increased vacuolar concentration is accompanied by increased cytoplasmic solute concentration (see Ch. 6) that can inhibit photosynthesis and other cytoplasmic processes. The plant's growth is a balance then between its ability to develop low enough osmotic potentials

Table 9.10 Concentration in leaves of *Aster tripolium* (mol m^{-3})

	Age (days)		
NaCl (mol m^{-3})	**26**	**37**	**45**
0	221	245	244
10	205	231	282
100	270	320	308
300	438	456	460
500	654	672	686

† Data from Shennon (1981).

in leaf cells that will allow sufficient turgor for the stomata to maintain CO_2 influx, and the damaging effect of solutes needed to produce the same low osmotic potential in the cytoplasm on enzymic reactions. Support for the view that the concentration in leaf cells responds both to external solute concentration and to water stress in the plants comes from growing plants at high humidity which reduces transpiration and so reduces the difference in water potential needed to maintain water flow between soil and leaves. Table 9.11 gives data for a number of species grown on culture solution containing 50 mol m^{-3} NaCl which shows that there is generally better growth in humid conditions and in some cases (mung beans, *Atriplex*) there is then decreased (K^+ + Na^+) concentration in the shoot. Figure 9.17 shows a striking example for a range of NaCl concentrations for *Atriplex* (Gale & Poljakoff-Mayber, 1970; Gale *et al.*, 1970). In normal air the growth was low in solutions containing little NaCl, increased to a maximum and then decreased with further increase in the NaCl concentration. In humid air the growth was higher in solutions of low NaCl concentration. This growth response was interpreted as showing a need for the plant to take up enough NaCl to maintain high concentrations in the leaves when transpiring under water stress. Measurements of water potential of the leaf sap (Fig. 9.16) showed that this changed with external water potential to about the same extent as the change in water potential of the solution, due

Table 9.11 Relative growth rates, concentration of (K^+ + Na^+) in the leaves, water flow and net (K^+ + Na^+) transport to the shoot for a number of species grown in a full nutrient solution containing 5 mol m^{-3} K^+ and 50 mol m^{-3} Na^+.† Relative humidities were 30–35 % and 95 % for high and low rates of water flow for all species

Species	Relative growth rate (d^{-1})	Concentration in leaves (mol m^{-3})	Water flow (mg g^{-1} h^{-1}) root	K^+ + Na^+ transport (μmol g^{-1} h^{-1}) root
Mung bean	0.11	300	390	2.7
	0.13	170	250	1.7
Red kidney bean	0.13	140	270	1.3
	0.15	135	140	1.5
Sunflower	0.21	165	530	1.7
	0.23	155	270	2.3
Tomato	0.28	95	790	2.6
	0.25	85	390	3.2
Triticale	0.18	255	640	1.6
	0.16	260	310	2.1
Atriplex	0.16	370	1190	19.1
	0.24	330	750	16.4

† Data from M. Salim (unpublished).

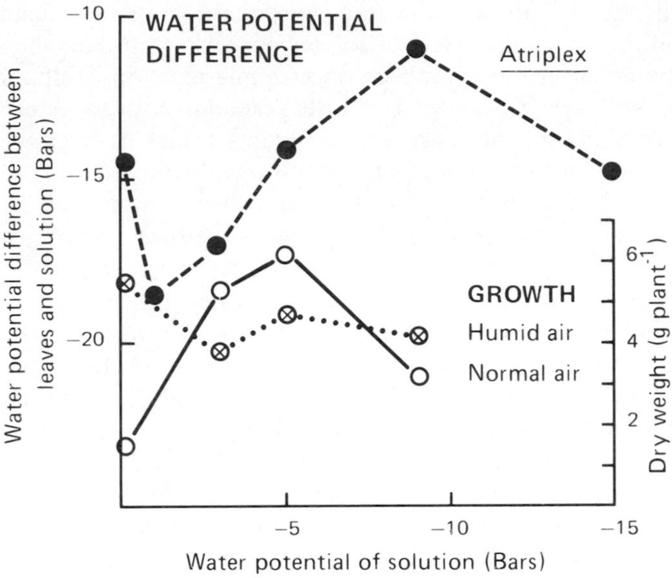

Fig. 9.16 Effect of water stress on growth and water potential difference of leaves of *Atriplex*. (●), water potential difference between leaves and solution of NaCl; (○), growth of plants in normal (dry) air; (⊗), growth of plants in humid air. (Data from Gale & Poljakoff-Mayber, 1970 & Gale *et al.*, 1970.)

to changes in solute content of the leaves. In other species than haloplytes the plant may adapt to water stress when external inorganic solutes are not available by accumulating organic solutes such as sugar in the vacuoles. This can happen, for example, in barley seedlings grown on a dilute solution of calcium sulphate with no external K^+ so that the amount of K^+ available is restricted to that in the seed. Such seedlings can accumulate 50–80 mol m^{-3} glucose and fructose.

9.7.2 Changes during leaf development

Measurements show that the concentrations of many ions in developing leaves is often low at first as growth of the leaf outstrips import but then concentrations of (K^+ + Na^+) may be steady as the leaf adapts to water potentials needed for water flow in the leaf. The steady level occurs while import of solutes continues in water flow to the leaf and there must either be a balance between growth and solute uptake to the leaf or export of ions in the phloem at the same time. It is shown in Sect. 9.6 that export of K^+ and P does occur from older to younger leaves. Changes seen in concentrations in the leaf depend on this balance between import, export and growth.

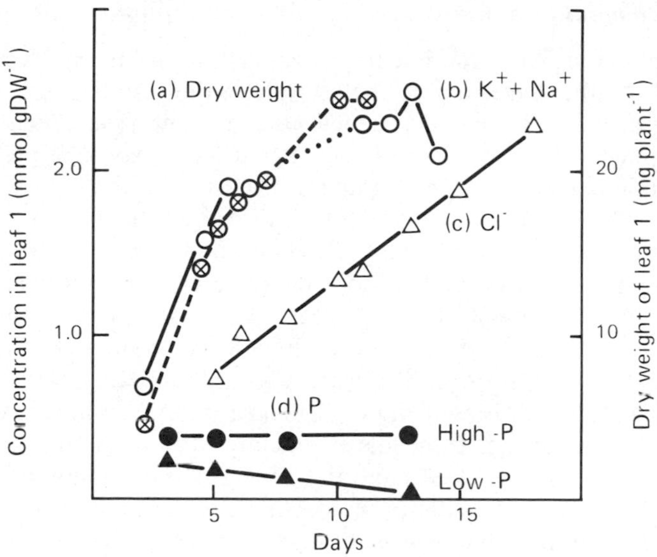

Fig. 9.17 Changes in dry weight of the first leaf of barley seedlings and in concentration of certain ions. (a) dry weight; (b) ($K^+ + Na^+$) concentration, plants grown in 10 mol m^{-3} K^+ or ($K^+ + Na^+$) (data of Pitman, 1965a and Greenway & Pitman, 1965); (c) Cl^- concentration in leaves of plants transferred to 100 mol m^{-3} NaCl (data of Greenway *et al.*, 1966); (d) changes in P concentration for plants grown on high and low P (data of Greenway & Gunn, 1966.)

Figure 9.17 gives collected data for growth of the first leaf of barley seedlings showing the rise in ($K^+ + Na^+$) to a steady level, the continuous rise of Cl^-, and the fall in P level at low external P concentrations due to export from the leaves.

While salt bush is adapted to make use of high levels of Na^+ and Cl^- in the leaves and other plants can tolerate concentrations of 300–400 mol m^{-3} (Ch. 10), at the other end of the range there are many species that are very sensitive to increased salinity. Greenway, in various publications, has stressed the importance of regulation of Cl^- and Na^+ concentration in the leaves as contributing to the ability of moderately resistant plants to tolerate saline conditions. Particularly in young developing leaves the continuing growth provides space for accumulation of Na^+ and Cl^- taken into the leaf (Greenway & Thomas, 1965). In older leaves there seemed to be less ability to regulate Na^+ and Cl^--concentrations, but the increased content of mature/senescent leaves which leads to premature death and abscission in some species, provides a means for the plant to remove NaCl (See also Ch. 10). It is interesting that the salt-sensitive *Phaseolus* has a well developed mechanism for removal of Na^+ and Cl^- via the phloem (Greenway *et al.*, 1966).

9.7.3 Leaching of ions from leaves and foliar absorption of nutrients

In most physiological experiments using culture solutions, leaching of nutrients from the leaves is not likely to affect the results given in Tables 9.2–9.4 and Fig. 9.8, nor the conclusions drawn about regulation of nutrient uptake. However, large losses of nutrients may take place under natural conditions where water from rain or mist can wash over the leaves. Table 9.12 gives examples of content of litter or foliage from forests in North American and north-west England and the relative amounts removed each year in rainfall. (These figures allow for nutrients entering the system in the rainfall above the canopy.) Large proportions of Na^+ were removed from the leaves even though the content of the leaves differed. K^+, Ca^{2+}, Mg^{2+} and S were also removed in relatively large amounts from the leaves, but there was some difference in loss of N and P. Leaching of nutrients from leaves (mainly of horticultural crops) has been reviewed by Tukey (1970) and the general pattern follows the results of Table 9.12. Cations tend to be leached preferentially to anions and the solution from the leaves is usually slightly alkaline. The rate of leaching is often lower for young than for older leaves and, of course, leaves differ from one species to another in the development of cuticle. Loss of ions was suggested to depend on the content of ions in the apoplast and so be related to supply and retranslocation of ions within the leaf. Under natural conditions it is clearly a process contributing to the overall balance of ions in the leaf, though care should be taken to put the losses into perspective by expressing them in the same units as entry of ions to the shoot. Leaching of ions from leaves by rain has its converse in supplying deficient plants with nutrients as a spray on the leaves.

Using isolated cuticles, measurements have been made of the rate of

Table 9.12 Effect of rain in leaching nutrients from leaves of deciduous forest trees

	Quercus petraea		Maple/birch/beech	
Element	**Content of leaves† (μmol gDW^{-1})**	**Percentage removed by rain during season**	**Content of leaves†(μmol gDW^{-1})**	**Percentage removed by rain during season**
N	760	−1.8	1700	13
P	19	40	19	12
K	69	240	270	91
Na	19	1200	0.8	500
Ca	155	42	160	30
Mg	38	120	46	50
S	–	–	15	370

† At leaf fall, Carlisle *et al.* (1966).
‡ Standing crop, Eaton *et al.* (1973).

penetration of ions (McFarlane & Berry, 1974). The results suggest that cuticle contains negatively charged sites and show K^+ penetrating slightly more rapidly than Na^+ and univalent cations diffusing faster than divalent cations and faster again than trivalent ions. The ratio of K^+/Ca^{2+} was 4.5 and of K^+/Fe^{3+} was 10. No information was given on anion permeability. This is an area of study that could well be integrated with studies of turnover in natural communities or crops. A separate issue is the extent to which loss could be due to guttation rather than diffusion from the apoplast through the cuticle. Guttation occurs most extensively under high humidity conditions.

9.7.4 Conclusions

Processes affecting the concentrations of the non-metabolized ions such as K^+, Na^+ or Cl^- in the leaf are shown in Fig. 9.18. The difference between import to the leaf and export in the phloem determines net changes in the amount of each ion in the leaf, but the concentration in the leaf cells is determined by active transport acting against the efflux from the vacuole. (As for example in algal cells, see Ch. 5.) Export from the leaf and net uptake to cells of the leaf compete for ions in the free space, if removal of ions from the free space is slower than delivery in the phloem, as may occur under saline conditions, the build-up of ion concentration around the cells may lead to increased concentration in the cytoplasm and so to

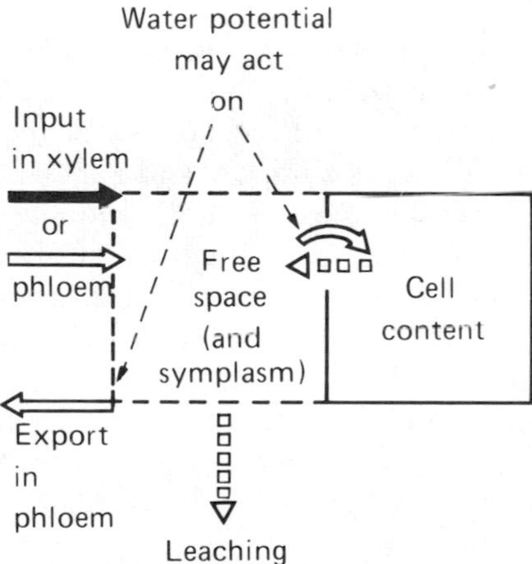

Fig. 9.18 Regulation of ion content in the leaf involves many processes. The observed concentration is that in the cells of the leaf, but this is in contact with the free space and the symplasm in the leaf and so can change in response to variation of input in xylem or phloem, export in the phloem or leaching across the cuticle. The cell content also appears to be affected by water potential in certain species.

inhibition of growth. As discussed above, the concentrations in the leaf are also dependent on the rate of growth of the leaf compared with the net accumulation in the leaf. The content of the apoplast also relates to the general question of how the regulation of cell content that is found in examples given in this chapter is achieved.

Is rate of uptake regulated by feedback from vacuolar content or does it 'float', controlled only by the availability of ions in the free space? This question is discussed in detail in Ch. 6. In general, the constant concentrations in leaf cells appear to be due to regulation of vacuolar or cytoplasmic concentration to a set level. The set level may be related to in intrinsic properties of the cell, such as the ionic levels needed in the cytoplasm, but is also determined to some extent by the water relations of the plant. The leaf is a closed system but export of ions in the phloem (or glands, see Ch. 13) provides one means of balancing import of ions in the xylem and the needs of the leaf cells. The ability of the leaf to balance input and export of ions affects the response of the plant to high or low external concentrations. This is an important issue in understanding plant responses to stress since deviations from the acceptable range of solute concentrations appear to be able to feedback to inhibit growth. The alternative to locating the regulatory process in the leaf (balance of input and output) is to locate it in the root (control of input to the leaf).

9.8 Regulation of ion content of the whole plant

Discussion of regulation of ion content in plant leaves considered maintenance of constant levels in leaves to be due to regulation at the cellular level in response to a combination of growth demands and leaf water potentials and a balance between import and export. Similar conclusions can be drawn about the shoot as a whole and the question raised in particular is the extent to which the content of the shoot is determined by activity of the export system from the root, or the interaction of xylem and phloem transport. The 'steady state' of (K^+ + Na^+) content in the shoot as a whole has already been described and discussed (Sect. 9.5). It has also been shown that the content of the shoot is connected with growth of the plants (Fig 9.3). Mechanisms of regulation must consider, then, the interactions of uptake and growth, regulation at the leaf cell level and roles of xylem and phloem transport.

A major question is the extent to which export from shoot to root in the phloem may remove surplus ions from the shoot and so contribute to regulation of nutrient concentrations in the shoot. Ben Zioni *et al.* (1971) suggested that NO_3^- uptake in tobacco could be regulated by the rate at which potassium malate was translocated from the shoot to the root. Using lupins, Jeschke *et al.* (1984) collected phloem sap from the stems of plants fed different nutrient levels and found the phloem concentrations increased with nutrient status. From estimates of phloem concentration and volume flow they calculated net transport to the root and proposed that phloem translocation was able to return substantial amounts of solutes to the roots. There are other situations though where evidence points to only a minor role for phloem translocation. In barley seedlings grown on full nutrient solution the export of K^+ from shoot to root was not greater than 0.5 μmol gFW^{-1} h^{-1} compared with net upward transport of 4.0 μmol gFW^{-1} h^{-1} (Pitman, 1972). Greenway and Thomas (1965) and Greenway *et al.* (1966) found translocation of Na^+ and Cl^- from shoots to roots of barley, *Phaseolus* or *Atriplex* was much less important in regulation of NaCl content than limitation of uptake to the root. There is doubt about the ability of plants to translocate malate in the phloem to the extent needed, however.

Transport of ions from root to shoot in the xylem is clearly dependent on the supply of metabolites to the root. Bowling (1968) cooled the stem of *Helianthus* plants to reduce translocation of sugars to the root, and found there was a 60 % reduction in the K^+ uptake by the roots, though transpiration was unaffected. Hatrick and Bowling (1973) have also shown that Rb^+ uptake (and respiration) was correlated with translocation of sugar from the shoot to the root of barley seedlings. In other experiments the uptake of K^+ by the roots and transport to the shoot increased by nearly 100 % 1 h after the start of the light period in barley seedlings grown in 2 h light/22 h dark, and this increase coincided with the appearance in the root of sugars produced in photosynthesis (Fig. 9.19). However, no difference in the rate of K^+ uptake was found between light and dark using plants grown in 16 h light/8 h dark. It seems that while translocation of nutrients to the root may regulate uptake of ions, particularly when growth is restricted, other controls may also be involved when growth rates are high.

Differences in rates of uptake can be due to the amount of 'carrier' available in the roots. Clark *et al.* (1973) showed that a maize mutant that was inefficient at using Fe, lacked the ability to transport Fe from the root to the xylem. In other ways the mutant was like normal varieties; in particular, there was little difference in uptake of Fe to the root, production of citrate (which chelates with Fe) and transport of P to the shoot. A further example of changed levels of 'carrier' is that certain plants grown under nutrient deficiency have shown exaggerated rates of uptake when transferred to normal soils (P uptake: Bowen, 1970; Bieleski & Ferguson, 1983; P and S uptake: Bouma, 1967a).

A further interesting aspect of Fe uptake is that H^+ release by the roots is related to the Fe status of the plants (see Ch. 7). Fe-deficient sunflower plants have been shown to acidify the culture solution but then, when Fe is available, make the culture solution alkaline as the plants re-green. Cycles of pH change were induced in the culture solution as the plants

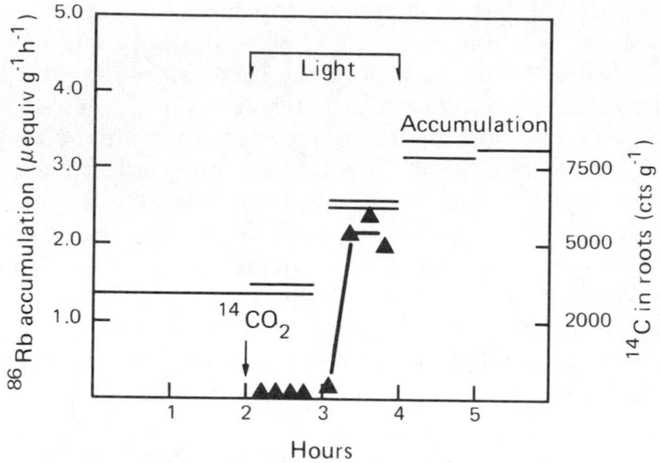

Fig. 9.19 Rates of ^{86}Rb accumulation by roots of seedlings growing in 10 mol m^{-3} K^+ culture solution in 2 h photoperiods; $t = 0$ is 2 h before the end of the 22 h dark period. Also shown is rate of translocation of ^{14}C to the roots following a pulse of $^{14}CO_2$ at the start of the light period (Pitman & Cram, 1973, data from Pitman, 1972.)

became Fe-stressed and recovered (Raju *et al*, 1972).

Much interest has been given in recent years to the possible involvement of phytohormones in regulation of ion transport from root to shoot. Plant roots produce cytokinins (and gibberellins) which have been suggested to affect growth of the plant as a whole (e.g. Vaadia & Itai, 1969). Specifically, changes in cytokinin levels in xylem exudate from roots have been shown to be associated with water stress (ltai & Vaadia, 1965) and other environmental conditions. Wagner and Michael (1971) have shown that cytokinin in exudate from sunflower seedlings was reduced when N was omitted from the culture solution but increased when N was replaced. Skene and Kerridge (1967) showed temperature affected the level of cytokinin in root exudate from *Vitis vinifera* and Atkin *et al.* (1973) have shown that cytokinin and gibberellic acid in exudate from maize plants increased strongly with temperature to a maximum at 28 °C. There are many other such examples.

The shoot too produces hormones which may be translocted to the roots in the phloem. In particular, abscisic acid (ABA) seems to be produced mainly in leaves and its level responds rapidly to water stress (e.g. Hiron & Wright, 1973). Production of ABA in wilted roots is doubtful, but it can be translocated from the shoot to the root (Hocking *et al.*, 1972). See Walton *et al.* (1976) for production of ABA in roots.

Both ABA and cytokinins have been shown to have a specific inhibitory effect on transport of ions from root to shoot. Transport into the xylem (i.e. exudate from the cut end of a root) can be inhibited by stopping energy metabolism by an inhibitor such as DNP but then uptake at the root surfaces is inhibited too (Ch. 7). Cytokinins have been shown to inhibit transport of certain ions to the xylem without inhibiting uptake to the root as a whole and so appear to be having a direct effect on movement of ions into the xylem and hence on transport from root to shoot. ABA can also inhibit transport without effect on uptake to the root, but only at specific conditions of nutrient status and temperature. ABA has also been reported to stimulate ion transport and to affect selectivity between K^+ and Na^+. (Cram & Pitman, 1972; Collins & Kerrigan, 1974; Pitman *et al.*, 1974; Van Steveninck *et al.*, 1982).

Another group of substances which act on transport to the xylem without inhibition of uptake are certain inhibitors of enzyme synthesis. Cycloheximide (CHM), for example, inhibits protein synthesis and also inhibits ion transport to the xylem without effect on uptake in short-term experiments (Läuchli *et al.*, 1973). In longer experiments CHM disturbs nucleotide levels in the cell (Cocucci & Marrè, 1973) which may lead to reduced uptake of ions. The analogue of phenylalanine, *p*-fluorophenylalanine (FPA), also inhibits transport but not uptake to the cells (Schaefer *et al.*, 1975). This compound does not stop protein production but renders the protein ineffective by incorporation of the analogue instead of phenylalanine. Similar results have been obtained with the analogue of proline, azetidine-2-carboylic acid (AZ), which is thought to affect plasmodesmata (see Ch. 7).

It seems that transport into the xylem involves a protein which turns over rapidly in the root. Both ABA and cytokinin have been implicated in regulation of RNA and protein metabolism and it is clearly an attractive hypothesis that plant hormones can act on ion transport through control of the number of active sites available.

9.9 Overview

The hormone–protein model provides one system for regulation of transport but there is also the role of carbohydrate transport from root to shoot which could provide another, a broad integration of root–shoot activity. These ideas are shown in Fig. 9.20.

The xylem is the pathway for upward movement of ions and hormones (cytokinins and gibberellins?). The phloem is the pathway for downward movement of carbohydrates, hormones and possibly ions (K^+) to the roots as well as possible upward movement of all of these to the growing points in the shoot. The balance of supply and export of nutrients and activity of transport into the cells determines the content in the leaves, and also the retranslocation of ions to younger organs. This is clearly a simplification, as messages carried into the xylem and phloem are likely to be

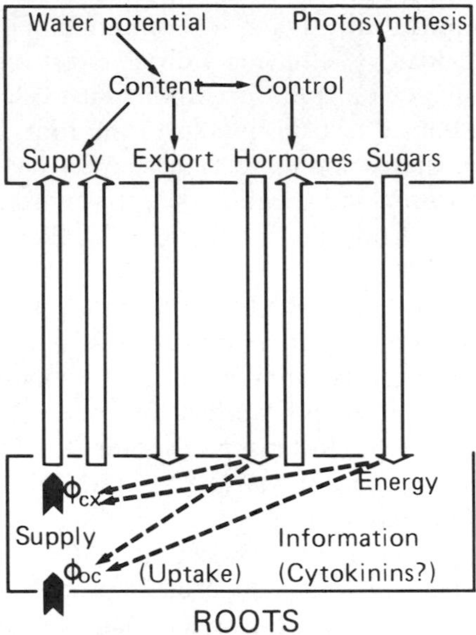

Fig. 9.20 Interaction of processes between root and shoot. In the roots ions enter by the active flux ϕ_{oc} and are transported to the xylem by ϕ_{cx}. Energy for these processes is supplied by the phloem (blank arrow, dashed lines). Control may be affected by hormones in the phloem or produced in the roots. Hormone transport in the xylem (dashed arrow) may interact with processes in the leaves. The water potential of the leaves can act on both energy and control factors exported to the root and so affect the content of leaf cells. (See also Pitman, 1972 and Lüttge, 1973.)

based on combinations of phytohormones rather than absolute levels.

There are clearly a number of questions that need to be answered about regulation of ion transport in plants. What leads to control of levels of solutes in leaf cells? Is it determined at the leaf cell level in response to leaf water relation or is it controlled by net import of carbohydrates to the root? What is the role, if any, of phytohormones? What interactions are there between water and solute flows?

Key issues are the regulatory role of phloem transport and the control of transport from root to xylem, but in particular there is a need to develop techniques to measure these flows in intact plants when there can be no artefacts of the measuring process. There have been many theories about regulation of transport in whole plants but little evidence that can resolve conflicting models.

References

Allaway, W. G., Pitman, M. G., Storey, R., Tyerman, S. & Ashford, A. E. (1984). Water relations of coral cay vegetation on the Great Barrier Reef: water potentials and osmotic content. *Australian Journal of Botany* **32**, 449–64.

Asher, C. J. & Edwards, D. G. (1983). Modern solution culture techniques. In Laüchli, A. Bieleski, R. L. (eds:). *Encyclopedia of Plant Physiology*, New Series, vol. 15A, pp. 94–119. Berlin, Heidelberg, New York, Tokyo, Spring-Verlag.

Asher, C. J. & Loneragan, J. F. (1967). Response of plants to phosphate-concentration in solution culture. 1. Growth and phosphorus content. *Soil Sciencc* **103**, 225–33.

Asher, C. J. & Ozanne, P. G. (1967). Growth and potassium content of plants in solution cultures maintained at constant potassium concentrations. *Soil Science* **103**, 155–61.

Asher, C. J., Ozanne, P. G. & Loneragan, J. F. (1965). A method for controlling the ionic environment of plant rools. *Soil Science* **100**, 149–56.

Atkin, R. K., Barton, E. G. & Robinson, D. K. (1973). Effect of root-growing temperature on growth substances in xylem exudate of *Zea mays*. *Journal of Experimental Botany* **24**, 475–87

Baker, D. A. (1984). Water relations. In Wilkins, M. B. (ed.). *Advanced Plant Physiology* pp. 297–318. London, Pitman.

Barber, D. A. & Koontz, H. V. (1963). Uptake of dinitrophenol and its effect on transpiration and calcium accumulation in barley seedlings. *Plant Physiology* **38**, 60–5.

Barber, D. A. & Shone, M. G. T. (1966). The absorption of silica from aqueous solutions by plants. *Journal of Experimental Botany* **17**. 569.

Ben Zioni, A., Vaadia., Y, Lips, S. H. (1971). Nitrate uptake by roots as regulated by nitrate reduction products of the shoot. *Plant Physiology* **24**, 288–90.

Biddulph, O., Cory, R. & Biddulph, S. (1959). Translocation of calcium in the bean plant. *Plant Physiology* **34**, 512–19.

Bieleski, R. L. & Ferguson, I. B. (1983). Physiology and metabolism of phosphate and its compounds. In Läuchli, A. & Bieleski, R. L. (eds). *Encyclopedia of Plant Physiology*, New Series, vol. 15A, pp. 422–45. Berlin, Heidelberg, New York, Tokyo, Springer-Verlag.

Bollard, E. G. (1953). The use of tracheal sap in the study of apple tree nutrition. *Journal of Experimental Botany* **4**, 363–8.

Bouma, D. (1967a). Growth changes of subterranean clover during recovery from phosphorus and sulphur stresses. *Australian Journal of Biological Sciences* **20**, 51–66.

Bouma, D. (1967b). Nutrient uptake and distribution in subterranean clover during recovery from nutritional stresses. I. Experiments with phosphorus. *Australian Journal of Biological Sciences* **20**, 601–12.

Bouma, D. (1967c). Nutrient uptake and distribution in subterranean clover during recovery from nutritional stresses. II. Experiments with sulphur. *Australian Journal of Biological Sciences* **20**, 613–21.

Bowen, G. D. (1970). Early detection of phosphate deficiency in plants. *Soil Science Plant Analysis* **1**, 293–8.

Bowling, D. J. F, (1968). Translocation at 0 °C in *Helianthus annuus*. *Journal of Experimental Botany* **19**, 381–8.

Bradshaw, A. D., Chadwick, M. J., Jowett, D. & Snaydon, R. W. (1964). Experimental investigations into the mineral nutrition of several grass species. *Journal of Ecology* **52**, 665–76.

Bukovac, M. J. & Wittwer, S. H. (1957). Absorption and mobility of foliar-applied nutrients. *Plant Physiology* (*Lancaster*) **31**, 428–35.

Carlisle, A., Brown, A. H. F. & White, E. J. (1966). The organic matter and nutrient elements in the precipitation beneath a sessile oak (*Quercus petraea*) canopy. *Journal of Ecology.*, **54**, 87–98.

Carroll, M. D. & Logeragan, J. F. (1968). Response of plant species to concentrations of zinc in solution. I. Growth and zinc content of plants. *Australian Journal of Agricultural Research* **19**, 859–68.

Carroll, M. D. & Loneragan, J. F. (1969). Response of plant species to concentrations of zinc solution. II. Rates of zinc absorption and their relation to growth. *Australian Journal of Agricultural Research* **20**, 457–63.

Clark, R. B., Tiffin, L. O. & Brown, J. C. (1973). Organic acids and iron translocation in maize genotypes. *Plant Physiology* **52**, 147–50.

Clement, C. R., Hopper, M. J., Canaway, R. J. & Jones, L. H. P. (1974). A system for measuring uptake of ions by plants from flowing solutions of controlled composition. *Journal of Experimental Botany* **25**, 81–99.

Cocucci, M. C. & Marrè, E. (1973). The effects of cycloheximide on respiration, protein synthesis and adenosine nucleotide levels in *Rhodotorula gracilis*. *Plant Science Lettes* **1**, 293–307.

Collander, R. (1941). Selective absorption of cations by higher plants. *Plant Physiology* **16**, 691–720.

Collins, J. C. & Kerrigan, A. P. (1914). The effect of kinetin and abscisic acid on water and ion transport in isolated maize roots. *New Phytologist* **73**, 309–314.

Cram, W. J. & Pitman, M. G. (1972). The action of abscisic acid on ion uptake and water flow in plant roots. *Australian Journal of Biological Sciences* **25**, 1125–31.

Eaton, J. W., Likens, G. E. & Bormann, F. H. (1973). Throughfall and stemflow chemistry in a northern hardwood forest. *Journal of Ecology* **61**, 495–508.

Epstein, E. (1972). *Mineral Nutrition of Plants: Principles and Perspective.* New York, London, Sydney, Toronto, John Wiley.

Fife, N. D. & Nambiar, E. K. S. (1984). Movement of nutrients in radiata pine needles in relation to the growth of shoots. *Annals of Botany* **54**, 303–14.

Gale, J., Naaman, R. & Poljakoff-Mayber, A. (1970). Growth of *Atriplex halimus* L. in sodium chloride salinated culture solutions as affected by the relative humidity of the air. *Australian Journal of Biological Sciences* **23**, 947–52.

Gale, J. & Poljakoff-Mayber, A. (1970). Interrelations between growth and photosynthesis of salt bush (*Atriplex halimus* L.) grown in saline media. *Australian Journal of Biological Sciences* **23**, 937–45.

Greenway, H. (1965). Plant responses to saline substrates. IV. Chloride uptake by *Hordeum vulgare* as affected by inhibitors, transpiration and nutrients in the medium. *Australian Journal of Biological Sciences* **18**, 249–68.

Greenway, H. Gunn, A. (1966). Phosphorus retranslocation in *Hordeum vulgare* during early tillering. *Planta* **71**, 43–67.

Greenway, H. & Gunn, A. Pitman, M. G. & Thomas, D. A. (1965). Plant responses to saline substrates. VI. Chloride, sodium and potassium uptake and distribution within the plant during ontogenesis of *Hordeum vulgare. Australian Journal of Biological Science* **18**, 525–40.

Greenway, H., Gunn, A. & Thomas, D. A. (1966). Plant resonse to saline substrates. VIII Regulation of ion concentrations in salt-sensitive and halophytic species. *Australian Journal of Biological Sciences* **19**, 741–56.

Greenway, H. & Pitman, M. G. (1965). Potassium retranslocation in seedlings of *Hordeum vulgare. Australian Journal of Biological Sciences* **18**, 235–47.

Greenway, H. & Thomas, D. A (1965). Plant response to saline substrates. V. Chloride regulation in the individual organs of *Hordeum vulgare* during treatment with sodium chloride. *Australian Journal of Biological Sciences* **18**, 505–24.

Halse, N. J., Greenwood, E. A. N., Lapins, R. & Boundy, C. A. P. (1969). An analysis of the effects of nitrogen deficiency on the growth and yield of a Western Australian wheat crop. *Australian Journal of Agricultural Research* **20**, 987–98.

Handrek, K. A. & Jones, L. H. P. (1967). Uptake of monosilicic acid by *Trifolium incarnatum (L.). Australian Journal of Biological Sciences* **20**, 483–85.

Hatrick, A. A. & Bowling, D. J. F. (1973). A study of the relationship between root and shoot metabolism. *Journal of Experimental Botany* **24**, 607–13.

Hiron, R. W. P. & Wright, S. T. C. (1973). The role of endogenous abscisic acid in the response of plants to stress. *Journal of Experimental Botany* **24**, 69–81.

Hocking, T. J., Hillman, J. R. & Wilkins, M. B. (1972). Movement of abscisic acid in *Phaseolus vulgaris*. *Nature New Biology* **235**, 124–25.

Hopkinson, J. M. (1964). Studies on the expansion of the leaf surface. IV. The carbon and phosphorus economy of a leaf. *Journal of Experimental Botany* **15**, 145–37.

Hunt, R. (1982). *Plant Growth Curves: The Functional Approach to Plant Growth Analysis*. London, Edward Arnold.

Hyder, S. Z. & Greenway, H. (1965). Effects of Ca^{++} on plant sensitivity to high NaCl concentrations. *Plant Soil* **23**, 258.

Itai, C. & Vaadia, Y. (1965). Kinetin-like activity in root exudate of water stressed sunflower plants. *Physiologia Plantarum* **18**, 941–4.

Jefferies, R. L. & Willis, A. J. (1964). Studies on the calcicole-calcifuge habit. II. The influence of calcium on the growth and establishment of four species in soil and sand cultures. *Journal of Ecology* **52**, 691–707.

Jeschke, W. D. (1970). Evidence for a K^+-stimulated Na^+ efflux at the plasmalemma of barley root cells. *Planta* **94**, 240–5.

Jeschke, W. D. (1973). K^+- stimulated Na^+ efflux and selective transport in barley roots. In Anderson, W. P (ed.). *Ion Transport in Plants*, pp 285–96. London, New York; Academic Press.

Jeschke, W. D. (1984). Effects of transpiration on potassium and sodium fluxes in root cells and the regulation of ion distribution between roots and shoots of barley seedlings. *Journal of Plant Physiology* **117**, 267–85.

Jeschke, W. D, Atkins, C. A. & Pate, J. S. (1984). Ion circulation via phloem and xylem between root and shoot of nodulated white lupin. *Journal of Plant Physiology* **117**, 319–30.

Johansen, C., Edwards, D. G. & Loneragan, J. F. (1968). Interactions between potassium and calcium in their absorption by intact barley plants. II. Effects of calcium and potassium concentration on potassium absorption. *Plant Physiology* **43**, 1722–26.

Jones, L. H. P. & Handrek, K. A. (1965). Studies of silica in the oat plant. III. Uptake of silica from soils by the plant. *Plant and Soil* **23**, 79–96.

Kinzel, H. (1983). Influence of limestone, silicates and soil pH on vegetation. In Lange, O. L., Nobel, P. S., Osmond, C. B. & Ziegler, H. (eds). *Encyclopedia of Plant Physiology*, New Series, vol. 12C, pp. 201–44. Berlin, Heidelberg, New York, Tokyo, Springer-Verlag.

Kleeper, B. & Kaufmann, M. R. (1966). Removal of salts from xylem sap by leaves and stems of guttating plants. *Plant Physiology* **41**, 1743–47.

Läuchli, A., Lüttge, U. & Pitman, M. G. (1973). Ion uptake and transport through barley seedlings; differential effect of cycloheximide. *Zeitschrift für Naturforschung* **28c**, 431–4.

Lazaroff, N. & Pitman, M. G. (1966). Calcium and magnesium uptake by barley seedlings. *Australian Journal of Biological Sciences* **19**, 991–1005.

Leigh, R. A. & Wyn Jones, R. G. (1984) A hypothesis relating critical potassium concentrations for growth to the distribution and functions of this ion in the plant cell. *New Phytologist* **97**, (1), 1–13.

Loneragan, J. F. (1968). Nutrient concentration, nutrient flux and plant growth. *Proceedings of the Ninth International Congress of Soil Science, Adelaide* **2**, 173–82.

Loneragan, J. R. & Asher, C. J. (1967). Response of plants to phosphate concentration in solution culture. II. Rate of phosphate absorption and its relation to growth. *Soil Science* **103**, 311–18.

Loneragan, J. F. & Snowball, K. (1969a). Calcium requirements of plants. *Australian Journal of Agricultural Research* **20**, 465–78.

Loneragan, J. F. & Snowball, K. (1969b). Rate of calcium absorption by plant roots and its relation to growth. *Australia Journal of Agricultural Research* **20**, 479–90.

Lüttge, U. (1973). *Stofftransport der Pflanzer* Berlin Springer-Verlag.

Lüttge, U. & Higinbotham, N. (1979). *Transport in Plants* New York, Heidelberg, Berlin. Springer-Verlag.

Lüttge, U. & Laties, G. G. (1967). Selective inhibition of absorption and long distance transport in relation to the dual mechanisms of ion absorption in maize seedlings. *Plant Physiology* **42**, 181–5.

Lycklama, J. C. (19G3). The absorption of ammonium and nitrate by perennial rye-grass. *Acta Botanica Neerlaandica* **12**, 361–423.

McFarlane, J. C. & Berry, W. L. (1974). Cation penetration through isolated leaf cuticles. *Plant Physiology* **53**, 723–7.

MacRobbie, E. A. C. (1971). Phloem translocation. Facts and mechanisms: A comparative survey. *Biological Reviews* **46**, 429–81.

Millikan, C. R. & Hanger, B. C. (1966). Movement of previously deposited ^{45}Ca in subterranean clover (*Trifolium subterraneum* L.) by foliar injections of certain cations. *Australian Journal of Biological Sciences* **19**, 1–14.

Milthorpe, F. L. & Moorby, J. (1974). *An Introduction to Crop Physiology*. Cambridge University Press.

Nye, P. H. & Tinker, P. B. (1977). *Solute Movement in the Soil-Root System*. Oxford, London, Edinburgh, Melbourne, Blackwell Scientific Publications.

Perry, M. W. & Greenway, H. (1973). Permeation of uncharged molecules and water through tomato roots. *Annals of Botany* **37**, 225–32.

Pettersson, S. (1960). Ion absorption in young sunflower plants. I. Uptake and transport mechanisms for sulphate. *Physiologia Plantarum* **13**, 133–47.

Pitman, M. G. (1965a). Sodium and potassium uptake by seedlings of *Hordeum vulgare*. *Australian Journal of Biological Sciences* **18**, 10–24.

Pitman, M. G. (1965b). Transpiration and the selective uptake of potassium by barley seedlings (*Hordeum vulgare cv bolivia*). *Australian Journal of Biological Sciences* **18**, 987–98.

Pitman, M. G. (1966). Uptake of potassium and sodium by seedlings of *Sinapis alba*. *Australian Journal of Biological Sciences* **19**, 257–69.

Pitman, M. G. (1972). Uptake and transport of ions in barley seedlings. III. Correlation between transport to the shoot and relative growth rate. *Australian Journal of Biological Sciences* **25**, 905–19.

Pitman, M. G. (1982). Transport across plant roots. *Quarterly Reviews of Biophysics* **15**, 481–54.

Pitman, M. G. (1984). Transport across the root and shoot/root interactions. In Staples, R. C. (ed.). *Salinity Tolerance in Plants: Strategies for Crop Improvement*, pp. 93–123. New York, John Wiley & Sons.

Pitman, M. G. & Cram, W. J. (1973). Regulation of inorganic ion transport in plants. In Anderson, W. P. (ed). *Ion Transport in Plants*, pp. 465–81. London, New York, Academic Press.

Pitman, M. G., Lüttge, U., Läuchli, A. & Ball, E. (1974). Ion uptake to slices of barley leaves and regulation of K content in cells of the leaves. *Zeitschriff für Pflanzenphysiologie* **72**, 75–88.

Pitman, M. G. & Saddler, H. D. W. (1967). Active sodium and potassium transport in cells of barley roots. *Proceedings of the National Academy of Sciences USA* **57**, 44–9.

Raju, V. K. & Marschner, H. & Römheld, V. (1972). Regulation of iron uptake from relatively insoluble iron compounds by sunflower plants. *Zeitschrift für Pflanzenernahrung und Bodenkunde* **133**, 227–39.

Ringoet, A., Sauer, G. & Gielink, A. J. (1968). Phloem transport of calcium in oat leaves. *Planta* **80**, 15–20.

Salim, M. & Pitman, M. G. (1984). Pressure-induced water and solute flow through plant roots. *Journal of Experimental Botany* **35**, 869–81.

Schaefer, N., Wildes, R. A. & Pitman, M. G. (1975). Inhibition by *p*-fluorophenylalanine of protein synthesis and of ion transport across the roots in barley seedlings. *Australian Journal of Plant Physiology* **2**, 61–74.

Scholander, P. F., Hammel, H. T., Hemmingsen, E. & Garey, W. (1962). Salt balance in mangroves. *Plant Physiology* **37**, 722–9.

Shennan, C. (1981). Salt tolerance in *Aster tripolium* L. PhD thesis,

University of Cambridge.
Shone, M. G. T, Clarkson, D. T., Sanderson, J. & Wood, A. V. (1973). A comparison of the uptake and translocation of some organic molecules and ions in higher plants. In Anderson, W. P. (ed.) *Ion Transport in Plants*, pp. 571–82. London, New York, Academic Press.
Skene, K. G. M. & Kerridge, G. H. (1967). Effect of root temperature on cyfokinin activity in root exudate of *Vitis vinifera* L. *Plant Physiology* **42**, 1131–9.
Smith, F. W. (1974). The effect of sodium on potassium nutrition and ionic relations in rhodes grass. *Australian Journal of Agricultural Research* **25**, 407–14.
Steucek, G. L. & Koontz, H. V. (1970). Phloem mobility of magncsium. *Plant Physiology* (*Lancaster*) **46**, 50–2.
Storey, R. & Wyn Jones, R. G. (1978). Salt stress and comparative physiology in the gramineae. I. Ion relations of two salt- and water-stressed barley cultivars, California Mariout and Arimar. *Australian Journal of Plant Physiology* **5**, 801–16.
Thrower, S. L. (1962). Translocation of labelled assimilates in the soybean. II. The pattern translocation in intact and defoliated plants. *Australian Journal of Biological Sciences* **15**, 629–49.
Tukey, H. B., Jr. (1970). The leaching of substances from plants. *Annual Review of Plant Physiology* **21**, 305–24.
Vaadia, Y. & Itai, C. (1969). Interrelationships of growth with reference to the distribution of growth substances. In Whittington, W. J. (ed.). *Root Growth*, pp. 65–79. London, Butterworths.
Van Steveninck, R. F. M., Van Steveninck, M. E., Stelzer, R. & Läuchli, A. (1982). Studies on the distribution of Na and Cl in two species of lupin (*Lupinus luteus* and *Lupinus angustifolius*) differing in salt tolerance. *Physiologia Plantarum* **56**, 465–73.
Wagner, H. & Michael, G. (1971). Der Einfluss unterschiedlicher Stickstoffversorgung auf die Cytokininbildung in Wurzeln von Sonnenblumenpflanzen. *Biochemie und Physiologie der Pflanzen* **162**, 147–58.
Walton, D. C., Harrison, M. A. & Cote, P. (1976). The effects of water stress on abscisic-acid levels and metabolism in roots of *Phaseolus vulgaris* L. and other plants. *Planta* **131**, 141–4.
Williams, R. F. (1936). Physiological ontogeny in plants and its relation to nutrition. 2. The effect of phosphorus supply on the growth of the plant and its parts. *Australian Journal of Experimental Biological and Medical Sciences* **14**, 165–85.
Ziegler, H. (1975). Nature of transported substances. In Zimmermann, M. H. & Milburn, J. A. (eds.). *Encyclopedia of Plant Physiology*, New Series, vol. 1 pp. 59–60. Berlin, Springer-Verlag.

10 Ion relations of salt tolerance

T. J. Flowers and A. R. Yeo

10.1 Salinity: the global background

It is difficult to judge exactly how much land is salt affected, largely due to problems of soil mapping: consequently, estimates vary widely – between 3.4 million (Ponnamperuma, 1984) and more than 9 million km^2 (Massoud, 1974; Kovda, 1980). However, even the lower figure is an enormous area – about half the size of the USA. What is more, about 13 % (Mudie, 1974) of land under cultivation (14.9×10^6 km^2: FAO, 1978) and 30 to 50 %. of irrigated land (from 2.3 to 2.4×10^6 km^2: Kovda, 1980) is salt affected (Maas & Hoffman, 1977; Kovda, 1980, respectively). This alarming link between salinity and irrigation reflects the cumulative effect over decades or centuries of adding water on to the soil in order to grow crops in regions whose rainfall is otherwise inadequate. Were this water as pure as rainwater, there would be few if any problems, but it is not. Dissolved salts are contained in it, which remain in the soil when the water evaporates. Saline soils and their effects on plants are therefore of great importance to agriculture; this is the background to our wish to understand the physiology of the effects of salinity on plants.

As anybody who has visited a maritime salt marsh will know, salinity is not inimical to all plant life: there is a large range of plants which occurs naturally on saline soils. These plants are collectively known as 'halophytes'. In some ways the existence of this extensive salt-tolerant flora is not surprising, since 71 % of the world's surface is covered by a solution containing about 480 mol m^{-3} Na^+ and 560 mol m^{-3} Cl^- (that is a chlorinity of 35 ppt). Although the absolute concentration of salts in sea waters varies between different oceans, the proportion of ions is virtually independent of the chlorinity. Consequently, Na^+ and Cl^- are the most common ions in soils which are now, or have been in the past, affected by sea water and we shall chiefly be concerned with their effects on plants. Other anions, particularly carbonates, bicarbonates and sulphates do also occur at high concentrations and are locally important.

10.2 Salinity: the plant background

Plant species, then, differ in their tolerance of salinity between those that can withstand sea-water concentrations of NaCl for prolonged periods (the halophytes) and those which are killed by exposure to lower (even one-tenth sea water) salt concentrations for a short period. The latter are known as 'glycophytes' (plants of sweet or fresh water). Since there have been few studies on the effect of salinity on growth and especially reproduction of plants, relative to the total number of species, the designation of a plant as 'halophyte' or 'glycophyte' has usually to be based on a knowledge of the natural habitat. By this means it is possible to partition plants into the broad categories of halophytes and glycophytes, but much less easy to subdivide the groups further. True halophytes are clearly very different from glycophytes in the concentration of salt in which they will grow and reproduce. However, the boundary between the relatively salt-resistant glycophyte and the marginal halophyte is a blurred one. There is no absolute division and the two categories merge into each other (see Flowers *et al.*, 1977).

Calculations made about 10 years ago showed that (ecologically) salt tolerance is widespread amongst the families of flowering plants (Flowers *et al*, 1977). More recently, Aronson (1985) has prepared a list of species of plants which includes all those with a reputation for salt tolerance. There are at present close to 1500 species on his list and about 500 genera. Of these, few are pteridophytes (four genera) or gymnosperms (four genera) and there are no bryophytes: the majority are monocotyledons (110 genera) and dicotyledons (405 genera). The angiospermous halophytes are members of 129 families, about one-third of the total (354 families: Heywood, 1978). Thus salt tolerance is extremely widespread amongst the families of higher plants, although not uniformly: about twenty families contain more than 50 % of the halophytic genera (Table 10.1). Heading the lists for the mono- and dicotyledonous plants, respectively, are the Poaceae and the Chenopodiaceae. The latter is the family of halophytes *par excellence*, with 312 species (see Table 10.1).

It is worth emphasizing that salt tolerance in the ecological sense of inhabiting a saline habitat does not necessarily go hand-in-hand with a high growth rate under saline conditions. Under natural conditions, it is survival that is all important, and this may be a function of seed production or vegetative perennation rather than a rapid growth rate. In fact, the growth of many monocotyledonous halophytes is progressively inhibited with increasing salinity and contrasts with that of the Chenopodiaceae, where growth is often optimal in the presence of about half-seawater salt concentrations (Greenway & Munns, 1980). In general, then, we have a rather scanty knowledge of the growth characteristics of most halophytes: in the majority of cases plants are known as halophytes because they occur naturally in saline habitats.

Table 10.1 Families of flowering plants with either more than ten genera, or more than twenty species, which are halophytic†

Family	Number of halophytic genera	Number halophytic species	Percentage halophytic genera‡
Monocotyledones			
Poaceae	45	109	7
Cyperaceae	13	83	14
Arecaceae	13	22	6
Zosteraceae	3	20	100
Juncaceae	1	22	11
Dicotyledones			
Chenopodiaceae	44	312	44
Asteraceae	34	53	3
Aizoaceae	21	48	15
Papilionaceae	19	35	3
Apiaceae	19	31	6
Euphorbiaceae	15	33	5
Brassicaceae	15	30	4
Scrophulariaceae	13	21	6
Caryophyllaceae	9	16	11
Rhizophoraceae	8	24	50
Plumbaginaceae	6	31	50
Solanaceae	5	21	6
Tamaricaceae	2	43	50

† Taken from a list of halophytic species prepared by Aronson (1985).
‡ The percentages were calculated using the numbers of genera per family quoted by Heywood (1978).

10.3 Characteristics of salt-tolerant plants

10.3.1 Osmotic adjustment

All plants growing under saline conditions, be they halophytes or glycophytes, face the twin problems of water deficit and ion toxicity: the Scylla-Charybdis analogy of Greenway & Munns (1980). The osmotic potential of sea water is about −2.3 MPa and so a plant must have a water potential lower than this if it is to maintain inward water movement to its symplast. The plants must adjust osmotically while restricting the osmotic potential generated by ions in the cytoplasm to about −1.0 MPa, since high ion concentrations, particularly with a high Na^{+}/K^{+} ratio, are toxic to the metabolism (Flowers *et al.*, 1977). Failure in either osmotic adjustment or restricting the excessive entry of ions, results in the death of the plant. It has been clear for many years that halophytes generally achieve their osmotic balancing act utilizing the ions available to them in the medium (see Flowers, 1975 for references). Two examples serve to illustrate this

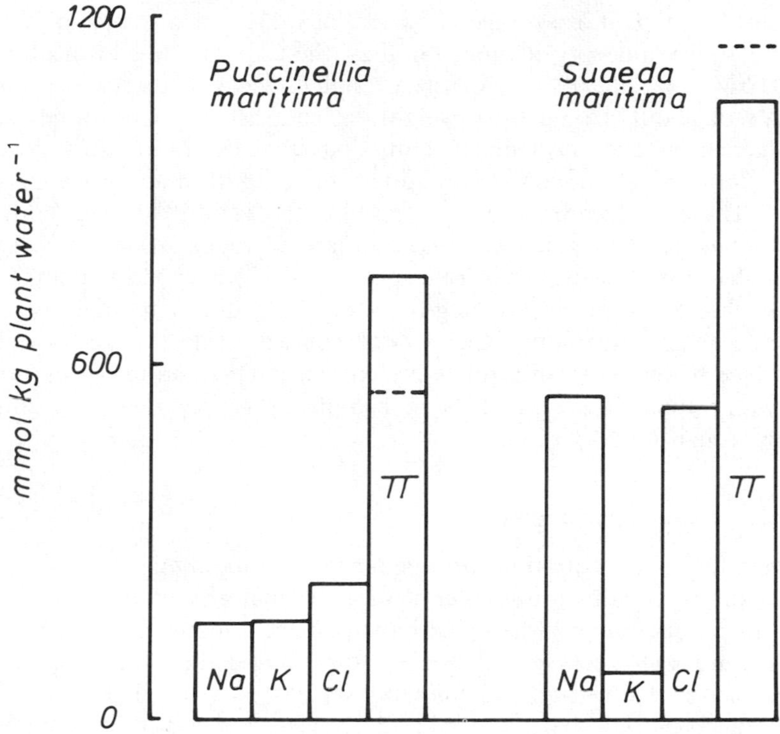

Fig. 10.1 The ion concentrations (expressed per unit of plant water) and the osmotic potential of two species of halophyte, one a member of the Poaceae (*Puccinellia maritima*) and the other a member of the Chenopodiacea (*Suaeda maritima*. The broken line represents the value of π for Na + K + Cl. [The figure is drawn from the data of Gorham *et al.* (1980) collected for plants growing on a salt marsh in North Wales.]

(Fig. 10.1); one is a member of the Poaceae and the other of the Chenopodiaceae – the largest groups of mono- and dicotyledonous halophytes, respectively (see Table 10.1). In the grass, *Puccinellia maritima* the osmotic potential of the shoots was equivalent to about −1.8 MPa and 70% of this potential was accounted for by Na^+ K^+ and Cl^- (Na^+ and K^+ are of approximately equal concentration, Fig. 10.1). In *Suaeda maritima* the osmotic potential was somewhat more negative (−2.6 MPa) and this was completely accounted for by Na^+, K^+ and Cl^-. The two species differ in the proportion of the osmotic potential accounted for by monovalent ions and in the ratio of Na^+ to K^+ (1.0 in *Puccinellia maritima* but 7.4 in *Suaeda maritima*); both, however, rely on ions for osmotic adjustment. Even in the Poaceae, only about a third of the osmotic potential is not accounted for by monovalent ions and here free sugars are particularly important (Albert & Kinzel, 1973; Gorham *et al.*, 1980).

These examples are, then, typical of their respective families: in the

Poaceae, Na^+/K^+ ratios average about 1.3, a value which is similar to that in some other monocotyledonous families, the Cyperaceae and the Juncaceae (Flowers *et al.*, 1986). All these families have relatively low water contents of about 1.5 g g^{-1} dry weight. In contrast, the Chenopodiaceae and other dicotyledonous families tend to have Na^+/K^+ ratios much greater than 1 (16 in the Chenopodiaceae) and water contents of about 7 g g^{-1} dry weight. These differences were recognised in the early 1970s and the term 'physiotypes' used to describe the two different groups (Albert & Kinzel, 1973). The term, however, has little specific value other than to emphasize the fact that different taxonomic groups with very different anatomy must meet the same constraints. High water contents reflect large vacuoles, which (see below) contain a relatively concentrated solution of Na^+ salts. Low water contents suggest a large cytoplasm/vacuole ratio and consequently a high K^+/Na^+ ratio.

10.3.1.1 Compartmentation

In general, ion concentrations are higher in the shoots than the roots and since shoot/root ratios are greater than 1 the majority of the ions in the plant are in the shoot, where halophytes must tolerate monovalent ion concentrations in excess of 300 mol m^{-3} for osmotic adjustment. Although the evolution of a salt-tolerant metabolism has occurred in the so-called 'halophilic bacteria' (Larsen, 1967), it is clear that a parallel course of events did not take place in the flowering plants: rather they have relied on compartmentation to avoid the toxic effects of high concentrations of monovalent ions (Na^+, K^+, Cl^-) on enzyme activity. There is now a substantial body of evidence from *in vitro* studies of enzyme activity that the monovalent cations, both Na^+ and K^+, are inhibitory to enzymes especially above concentrations of about 200 mol m^{-3}. Such studies made *in vitro* may be equivocal, however, as the enzyme response can often be modified by changes in the assay conditions such as substrate concentration or pH. However, Cl^- ions are especially damaging as far as protein synthesis (Gibson *et al.*, 1984) and enzymes with anionic substrates (Gimmler *et al.*, 1984) are concerned. Thus, while the shoot must contain high (in excess of 300 mol m^{-3}) ion concentrations for osmotic adjustment, these ions must be excluded to some extent from the cytoplasm.

Although studies on enzyme activity provide circumstantial evidence that cytoplasmic ion concentrations in halophytes are not the same as in the cell as a whole, they are no substitute for direct measurements of subcellular ion concentrations. There are, however, considerable technical difficulties in making such measurements, chiefly because of the high solubility of the ions concerned and because the cell wall prevents the fractionation of the cell by other than violent means. Basically three approaches have been used for plant tissues. The simplest technically is to contrast the apparent ion concentrations in cells with differing degrees of vacuolation

– that is to contrast meristematic cells with those of a mature leaf or root. More complex procedures are flux analysis (where the efflux of radioisotope can be analysed in terms of subcellular compartments: Walker & Pitman, 1976) and X-ray microanalysis (where cells prepared for transmission electron microscopy under strictly anhydrous conditions are analysed directly: Harvey, 1980). The particular strengths and weaknesses of the various methods are discussed at greater length by Wyn Jones *et al*. (1979) and Flowers & Läuchli (1983).

The compartmentation of ions is perhaps best illustrated in *Suaeda maritima*, where all three methods have contributed to our knowledge of the distribution of ions within the cells. X-ray microanalysis is the most powerful of the analytical tools, since it can give information on individual subcellular compartments. However, it only provides concentrations per unit of analysed volume in the prepared resin-embedded sample. This is not always the same as the concentration *in vivo*, since the water concentration varies from compartment to compartment. It is generally assumed that in the vacuole there is a simple solution, and no correction needs to be made to the apparent concentration obtained from the X-ray analysis. However, a unit volume of cytoplasm contains materials other than water and ions: we have assumed its water content to be 85 %. v/v and that 80 % of the water is available to dissolve solutes (from 70 to 90 %: Wyn Jones and Pollard, 1983) giving an effective water content of about 70 %, in making the calculations that follow. It has been argued (see Ling, 1979) that macromolecules, water and ions of the cytoplasm form a highly structured matrix, where ions are bound by fixed charges on macromolecules. The cytoplasm acts as an ion-exchange resin with a marked preference for K^+ over Na^+ – hence the observed cytoplasmic Na^+/K^+ ratios. We do not take such an extreme position. However, certainly some cytoplasmic water, close to the surface of macromolecules, will not behave as a solvent, although we do not know how much (Wyn Jones and Pollard, 1983): we have just to make an informed guess. Furthermore, in most cases we do not know the activity coefficient of ions in the various cell compartments and this is an important limitation especially in any attempt to evaluate the energetics of ion transport. For the cell walls, the water content is probably about 35 % (see Flowers, 1985; Flowers & Yeo, 1986). Armed with these assumptions it is then possible to compare concentrations within subcellular compartments. Within the cytoplasm, Na^+, K^+ and Cl^- concentrations are 166, 27 and 86 mol m^{-3}, respectively, allowing for the water content (Table 10.2): Na^+ and Cl^- are about 28 % of the respective vacuolar concentrations. Happily, the mean Na^+ ion concentration estimated from leaves by flux analysis (Yeo, 1981) is very similar to that obtained from X-ray microanalysis (Table 10.2).

If mature leaves of similar mean ion concentrations to those subjected to X-ray microanalysis are extracted and analysed by conventional means, the respective concentrations of Na^+, K^+ and Cl^- are 461, 57 and

Table 10.2 Ion concentrations in *Suaeda maritima* determined for various subcellular compartments† by a variety of methods

	Ion								
	Na^+			K^+			Cl^-		
Method	**Vac**	**Cyt**	**CW**	**Vac**	**Cyt**	**CW**	**Vac**	**Cyt**	**CW**
X-ray microanalysis									
mol m^{-3} vol‡	494	116	194	20	19	14	352	60	138
mol m^{-3} H_2O	494	166	554	20	27	40	352	86	394
Flux analysis§									
mol m^{-3} H_2O	–	165	–	–	–	–	–	–	–
Direct analysis									
means¶	461	–	–	57	–	–	339	–	–
derived‖	513	–	–	54	–	–	381	–	–

† Vac, vacuole, Cyt, cytoplasm, CW, cell wall.
‡ Harvey *et al.* (1981).
§ Yeo (1981).
¶ analysis of whole tissue extract: Yeo (1974); not a true vacuolar concentrations (see text).
‖ both mol m^{-3} H_2O calculated assuming volume fractions, V_v, of 0.8, 0.15 and 0.05 for vacuole, cytoplasm and cell wall, respectively.

339 mol m^{-3} (Table 10.2). These are reasonably close to the vacuolar ion concentrations (Table 10.2). However, since these concentrations are the sum of those in the cytoplasm, vacuole and the cell wall, some adjustment must be made if the actual vacuolar concentrations are to be compared. This can only be done if the proportion of ions in the various compartments and the compartmental volumes are known. X-ray microanalysis (Harvey *et al.*, 1981) reveals the former and cytometric analysis (Hajibagheri *et al.*, 1984) the latter. Using the ratio of concentrations in various compartments and the cytometric data, it is then possible to make a correction to the bulk vacuolar concentration: these corrected values agree very closely to those obtained by X-ray microanalysis (Table 10.2, except for the small discrepancy for K^+: this may be a consequence of the calibration carried out in early X-ray microanalysis).

So far we have not mentioned comparisons of meristematic with mature tissues. The data for shoots which are in the literature are difficult to interpret, since there is no indication of the degree of vacuolation in the regions which have been analysed and the growth conditions differ from those for the plants already described. Even so, the Na^+/K^+ ratio in *Suaeda maritima* is clearly much less in the young than the old tissues (Table 10.3). It appears that K^+ is concentrated, relative to Na^+, in shoot tissues which are actively dividing. This is confirmed for roots, where the Na^+/K^+ ratios are also greater in the older regions than in the root tips (Table 10.3). There remains a discrepancy in the absolute K^+ concentrations seen in the plants raised by different groups, but this is probably related to the age at which salt was applied [the plants used by Gorham & Wyn Jones (1983) were grown for 6 months in the absence of NaCl, while the others were salinized after about 1 month and harvested a month later]. Although there are fewer data for other halophytic species, the conclusions hold: Na^+/K^+ ratios in densely cytoplasmic tissue is 2 or less (Table 10.3), any variation probably reflecting the sampling precision of the various techniques used and the continuous rather than abrupt change from 'meristematic' to 'vacuolate' cells. Overall, then, the evidence available suggests that the cytoplasm of meristematic cells contains rather higher concentrations of K^+

Table 10.3 Sodium/potassium ratios in a some halophytic species and families

Species	Comments	$(Na^+)_0$ (mol m^{-3})	Na^+/K^+ ratio	Reference†
Suaeda maritima	Shoot apex	50	0.9	1
	Old leaf	50	3.7	1
	Leaf primordium‡	300	0.84	1
	Mesophyll, old leaf vacuole‡	300	17	1
	Mesophyll, mature leaf cytoplasm‡	340	6.1	2
	Mesophyll, mature leaf vacuole‡	340	25	2
	Root, apical mm	340	1.9	3
	root, 7–8 mm behind tip	340	4,7	3
Triglochin maritima	Root, cytoplasm	500	2.1	4
	Root, vacuole	500	3.6	4
Atriplex spongiosa	Root meristem‡	400	0.38	5
	Shoot apical meristem‡	400	0.35	6
	Mature mesophyll vacuole†	400	14.0	6
A. lentiformis	Whole shoot	360	6.2	7
A. patula	Whole shoot	360	29	7
Chenopodiaceae	Shoots	sea water	7.3	8
Poaceae	Shoots	sea water	1.6	8

†References:
1. Gorham & Wyn Jones (1983).
2. See Table 10.2 calculated from Harvey *et al.* (1981).
3. Hajibagheri *et al.* (1985).
4. Jefferies (1973).
5. Storey *et al.* (1983a).
6. Storey *et al.* (1983b).
7. Glenn & O'Leary (1984).
8. Flowers *et al.* (1986).

‡ Results from X-ray microanalysis.

than in the mature cells both in shoots (Gorham & Wyn Jones, 1983) and roots (Hajibagheri *et al.*, 1985). This may be important in promoting the rate of protein synthesis since translational efficiency increases with increasing K^+/Na^+ ratio (Wyn Jones *et al.*, 1979). Thus, although we only have a detailed account of the ion concentrations within *Suaeda maritima* cells, we conclude that the cytoplasmic Na^+ and Cl^- concentrations are regulated well below those found in the vacuoles and necessary for osmotic adjustment. As cells mature, the K^+ concentration falls and some of the roles of this element may be taken over by Na^+ (see also Flowers & Läuchli, 1983). Where K^+ is a major osmoticum (monocotyledons), this ion will also be maintained at less than about 200 mol m^{-3} in the cytoplasm.

10.3.1.2 Compatible solutes

The imbalance between ion concentrations in cytoplasm and vacuole must be made up if it is not to lead to osmotic imbalance. Furthermore, whatever osmoticum is used in the cytoplasmic phase, it must not itself be metabolically inhibitory. A variety of compounds have now been isolated from halophytes, the most common of which are proline, methylated quaternary ammonium compounds (e.g. glycinebetaine), sugars and polyols (both acyclic and alicyclic). They are known as 'compatible' solutes – compatible with the metabolism. That proline and glycinebetaine do not interfere with the activity of at least some enzymes is well established (see Flowers *et al.*, 1986), but the position with some of the other putative compatible solutes is less certain. It is clear, however, that sugars such as sucrose are not compatible, being inhibitory to a number of enzymes. This being so, they are probably contained within the vacuoles of the monocotyledonous species where they are used as part of the overall osmotic adjustment. Since the putative compatible solutes must maintain concentrations of several hundreds of mol m^{-3} in the cytoplasm to be effective, they are unlikely to be intermediates in normal metabolism. Unfortunately, it is more difficult to check the localization of these compatible solutes than it is for the ions we have discussed already. Carbon cannot be detected by X-ray microanalysis, although a complex between glycinebetaine and iodoplatinic acid has been shown to be largely cytoplasmic in *Suaeda maritima* (Hall *et al.*, 1978). Other evidence suggests that proline and glycine betaine may be present in both cytoplasm and vacuolar phases of the protoplast (Leigh *et al.*, 1981: Pahlich *et al.*, 1983). This being so, it is possible for osmotic adjustment of the cytoplasm to be achieved under conditions of increasing salinity without net synthesis of the compatible solute: it is simply necessary to move the solute between vacuole and cytoplasm during adjustment.

10.3.1.3 Protoplast versus apoplast

Although we have dwelt at some length on the maintenance of an osmotic balance between cytoplasm and vacuole, there is also a potential osmotic imbalance between the apoplast and the protoplast – that is between the cell well and the cytoplasm across the plasmalemma. The ion concentrations recorded by X-ray microanalysis for the cell walls are similar to those in the vacuoles if allowance is made for the low water content of the apoplastic phase. At a water potential of −1.5 MPa, the water content of cellulose is 35 % (Wiebe, 1966): this is consistent with the view that only about a third of the cell wall volume is available to water (Nobel, 1983). Consequently, a relatively small number of ions in the apoplast can generate a high concentration, with concomitant effects on the cellular water relations. The situation is exacerbated as far as the leaf cells are concerned by the relatively small volume of the apoplast – it is only about 3% of the total leaf volume (for references, see Flowers & Yeo, 1986). So, if ions which reach the shoot are not accumulated into the protoplast (and then into the cell vacuoles, the only major compartment for ion storage), they will remain in the cell walls where their concentration will rise rapidly to generate low external solute potentials (Fig. 10.2). This will in turn lead to the continuous need for osmotic adjustment within the protoplast or, if this fails, cellular death through dehydration. Much of the remainder of this chapter will be concerned with the importance of the apoplast in the overall response of the plant to salinity.

10.3.2 Ion relations

10.3.2.1 Shoot ion concentrations

We have already established that halophytes adjust osmotically using the ions available within their saline environment, although the monocotyledonous species also utilize sugars to generate the necessary solute potentials. Cytoplasmic ion concentrations are probably from 25 to 30 % of those in the vacuoles and the balance of the cytoplasmic solute potential is generated by some form of organic 'compatible' solute. In general, the ion concentrations expressed per unit of plant water, in the leaves of the succulent dicotyledonous halophytes (e.g. Chenopodiaceae) are maintained at a constant ratio with the external concentration above an external concentration of about 100 mol m^{-3}. Below this concentration, these halophytes are rather variable with respect to the concentrations of ions that they accumulate (see Storey & Wyn Jones, 1979; Flowers & Yeo, 1986). However, it is unlikely that the ion concentrations in the soil ever fall very low. Measurements made on salt marshes suggests a minimum concentration of about 100–200 mol m^{-3} with a maximum value of about

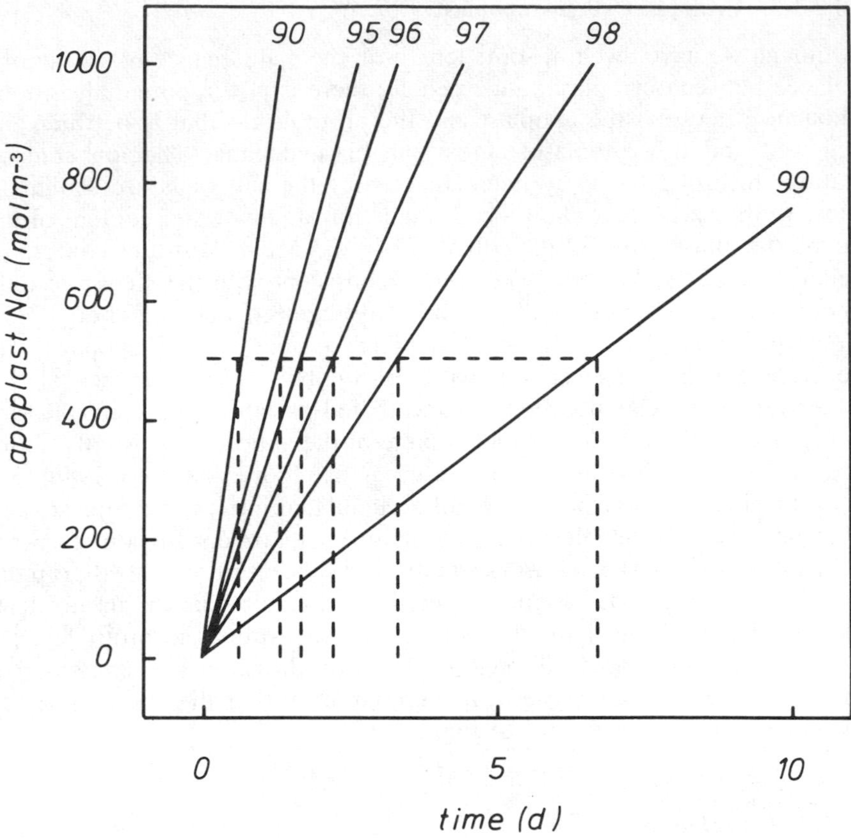

Fig. 10.2 The effect of the proportion of ions accumulated into the protoplast on the change in apoplastic ion concentration with time. The figures 90 to 99 represent the percentage of ions delivered which are accumulated into the protoplast and the broken lines indicate the time taken for the apoplastic ion concentration to reach 500 mol m^{-3}.

1 kmol m^{-3} (Flowers, 1985). Consequently, salt-marsh halophytes probably do not need to be able to withstand external osmotic potentials higher than about −1 MPa and their response to low salt concentrations for which they are quite unsuited might be expected to be erratic. They may have to withstand potentials as low as −5 MPa, however. Although the data in the literature are very limited, it appears that halophytes such as *Suaeda maritima* adjust to the external salinity very rapidly – adjustment to a 300 mol m^{-3} change in the NaCl concentration is complete within about 24 h (Yeo & Flowers, 1986). Thereafter, at a steady-state salinity, the leaf ion concentrations are rather constant with time. This suggests that the ion concentrations are regulated either because shoot growth controls net transport through the root or because ion supply to the shoot controls

growth. For less tolerant species, the principles of osmotic adjustment to salinity remain as for halophytes, although regulation of ion supply to the shoot presumably fails at a lower external concentration than it does in halophytes. For the salt-sensitive plants, it is often not possible to tell exactly when the homeostatic mechanisms fail, and so to separate adaptation from necrosis.

10.3.2.2 Salt uptake by the roots

For plants growing under normal field conditions, K^+, the only monovalent cation of significance as a major nutrient may be present at between 1 and 2 mol m^{-3}: in sea water K^+ is some 10 mol m^{-3} and Na^+ around 500 mol m^{-3} (Flowers & Läuchli, 1983). So whereas under non-saline conditions the function of the roots is to accumulate sufficient K^+ to supply the nutritional requirements (metabolic and osmotic for K^+: see Flowers & Läuchli, 1983) of the plant, under saline conditions K^+ must be selected from the mixture of Na^+ and K^+ presented to the root plasmalemma. The selectivity of the root system must be sufficient to supply the K^+ necessary for the metabolic requirements of the plant (this also applies to other ions than K^+) and the transport of saline ions (Na^+ and Cl^-) regulated in such a way that osmotic adjustment is achieved in the cells of the shoot without excessive delivery resulting in lethal ion concentrations. The selectivity by a root system for K^+ over Na^+ can be estimated from the plant ion contents by the so-called 'selectivity ratio'. This is the ratio of K^+/Na^+ in the plant divided by K^+/Na^+ in the medium and given the symbol $S_{K,Na}$. The ratio is, however, influenced by a considerable number of factors and when calculated from the whole shoot contents is the sum of the various component selectivities. For example, within the roots, selectivity may be imparted at the root epidermal plasmalemma, at the cortical cell tonoplasts and at the boundary between root symplast and xylem apoplast (Jeschke, 1983). Furthermore, retranslocation of K^+ in the phloem can bring about changes in the ratio of K^+ to Na^+ within different parts of the plant (see also Table 10.3). However, the similarity of $S_{K,Na}$ between xylem exudate and shoots of barley seedlings indicates that the ratio of shoot ion contents is often a good indication of root selectivity (see Pitman, 1975; 1982 and Ch. 9). For plants growing in non-saline conditions, the $S_{K,Na}$ ratio can range from less than 1 to more than 20 [see examples calculated from Collander (1941) in Table 10.4]. For barley growing in 10 mol m^{-3} ($Na^+ + K^+$), $S_{K,Na}$ values ranged from 7.5 to 65 as the Na^+/K^+ ratio in the medium was increased from 0.25 to 19 (Pitman, 1965). As the external concentration of Na^+ rises, however, selectivity may fall due to the non-selective uptake of Na^+ related to water flow (Pitman, 1975). For barley at a Na^+ plus K^+ concentration of 100 mol m^{-3} ($Na^+/K^+ = 3/1$) shoot $S_{K,Na}$ fell to 4.4 (Table 10.4). In *Suaeda maritima* $S_{K,Na}$ is maintained above 4 even with an external Na^+/K^+ ratio of about 50 (Table 10.4). Selectivity

Table 10.4 Selectivities† for a range of plant species

Species	Culture solution $Na^{+}+K^{+}$	Na^{+}/K^{+}	Shoot $S_{K.Na}$	References‡
Atriplex hortensis	8	1	2.0	1
	2.05	0.025	0.51	1
Plantago maritima	8	1	1.4	1
	2.05	0.025	1.3	1
Spinacia oleracea	8	1	12	1
	2.05	0.025	9.8	1
Avena sativa	8	1	20	1
	2.05	0.025	5.9	1
Hordeum vulgare	10	3	24	2
	100	3	4.4	2
Suaeda maritima	347	48.5	4.4	3
Puccinellia maritima	406	66.7	66	4

† $S_{K,Na} = (K^{+}/Na^{+})_{plant}/(K^{+}/Na^{+})_{medium}$.
‡ References:
1. Calculated from Collander (1941).
2. Calculated from Pitman (1975).
3. Yeo & Flowers (1986).
4. Thirty-five day old plants: Ansari (1982).

for K^{+} is even greater in the grass *Puccinellia maritima*, as might be predicted from the greater K^{+}/Na^{+} ratio in the halophytic grasses (see above) than in the dicotyledonous halophytes. Clearly, halophytes are well able to accumulate the K^{+} they require, for whatever purpose, in the presence of a considerable excess of Na^{+}.

As far as the supply of osmoticum to the shoots is concerned, this is measured by the net transport of Na^{+} ad Cl^{-} and (K^{+} in the case of halophytic monocotyledons) to the shoot, symbolized as J_{Na} and J_{Cl} (J_{K} for K^{+}), which is a minimal estimate of the unidirectional flux to the shoot (which it will equal if the return flux is zero). Values of J_{Na} for fast-growing halophytes such as *Suaeda maritima* are very much greater than for relatively slow growing monocotyledonous species and for K^{+} transport in glycophytes (Table 10.5). For example, *Suaeda maritima* transports some 9.5 mmol Na^{+} g^{-1} dry weight d^{-1} under optimal growth conditions. Transport to the shoot is closely allied to the growth rate (Yeo & Flowers, 1986), as it must be if overall shoot concentrations are to remain relatively constant in time, which indeed they do (Fig. 10.3; Flowers & Yeo, 1986). Net transport in *Puccinellia* is much slower, but then its growth rate is also lower (Table 10.5); Na^{+} and K^{+} are transported at similar rates.

It is pertinent to ask how this quantity of ions crosses the root without damaging concentrations arising in the symplasm. Ion transport across roots has received much attention in the literature, but only as far as plants growing in what might loosely be called 'normal' conditions; little or no

Table 10.5 Relative growth rates (RGR) and Na^+ and K^+ fluxes (expressed as mol ion transported to the shoot g^{-1} dry weight root d^{-1}) for a number of plant species

Species	**$(Na^+)_0$ (mol m^{-3})**	**RGR (d^{-1})**	**J_{Na} (mmol g^{-1}**	**J_K d^{-1})**	**Reference†**
Suaeda maritima	340	0.18	7.2	0.57	1
Puccinellia maritima	200	0.10	0.3	0.34	2
Hordeum vulgare	10	0.24	0.16	1.3	3
Sinapsis alba	10	0.20	3.1‡		4

† References:
1. Yeo (1974); mean figures for 35–56 d.
2. Ansari (1982).
3. Calculated from Pitman (1965).
4. Pitman & Cram (1977).

‡ K + Na.

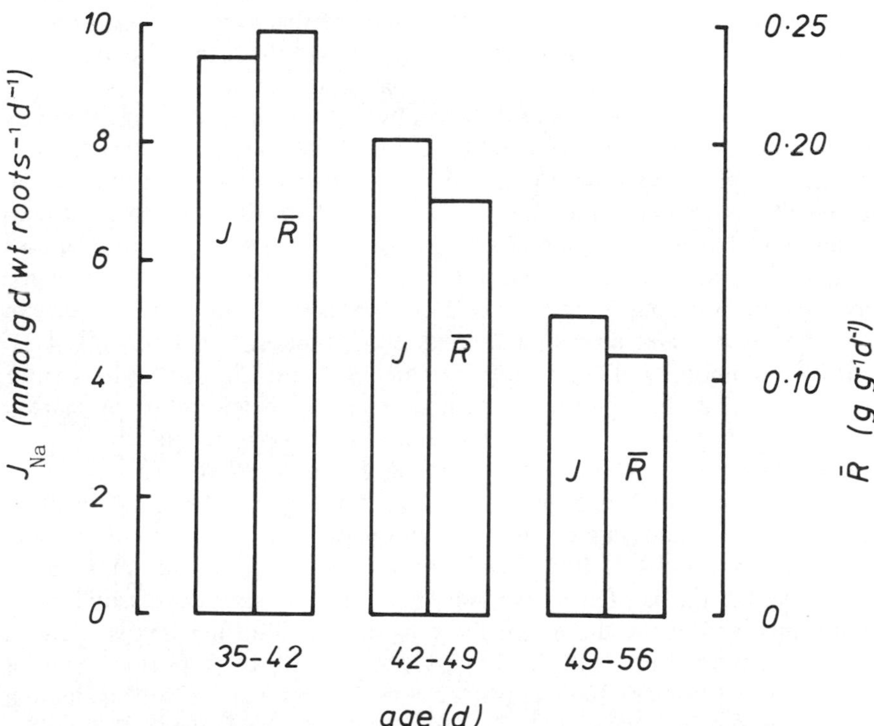

Fig. 10.3 The relationship between the flux of Na^+ *to the shoot* (J_{Na}) and the relative growth rate (R) for plants of *Suaeda maritima* of varying age. (Drawn from the data of Yeo and Flowers, 1986.)

attention has been paid to halophytes. In principle, ions can cross to the endodermis either from cell to cell (the so-called 'transcellular path' involving multiple crossings of plasmalemma), in the symplast, which the ions may enter at any point between epidermis and endodermis, or in the apoplast (either by diffusion or mass transfer in solution associated with the water movement of transpiration: see Ch. 7 for a detailed discussion). Some interesting facts come to light in considering the transport of ions across the roots of *Suaeda maritima*, which are of general importance if halophytes and glycophytes differ only quantitatively and not qualitatively.

Two lines of argument suggest that few ions move in the apoplast and hence that the great majority of water movement is not apoplastic. Firstly, the rate of water uptake through the roots of *Suaeda maritima* is about three orders of magnitude greater (9 mol g^{-1} dry weight root d^{-1}) than for Na^+ uptake (9 mmol g^{-1} dry weight root d^{-1}). This means that if solution moved in the apoplast to the endodermis (which we assume to be a relatively tight barrier between the outside and the stele) ions would be left in the apoplast as the water moved across the stele: this would amount to about 25 mmol g^{-1} dry weight root d^{-1} or 174 mol m^{-3} min^{-1}. This clearly cannot happen as the salt would crystallize at the endodermis within an hour. Secondly, knowing the dimensions of the roots of *Suaeda maritima*, it is possible to calculate the net flux of Na^+ through the endodermis should this be the site of ion uptake: for a J_{Na} of 8.5 mmol g^{-1} dry weight root d^{-1} (which is conveniently 100 nmol g^{-1} dry weight roots s^{-1}) the flux at the endodermal surface would be some 6000 nmol m^{-2} s^{-1}. This is implausibly high given that the ion fluxes across plant membranes reported in the literature fall within the range from 10 to 100 nmol m^{-2} plasmalemma s^{-1} (Lüttge & Higinbotham, 1979). These two simple calculations suggest that the ions are transferred across the roots in a pathway which involves the protoplasts: this conclusion is consistent with recent views of the pathway of water movement expressed by Steudle and Jeschke (1983).

It is a commonly held view (see Pitman, 1982 and Ch. 7) that ions enter the symplast at the epidermal plasmalemma. If this were so in *Suaeda maritima*, the requisite net flux of Na^+ ions would be about 750 nmol m^{-2} plasmalemma s^{-1} (Yeo & Flowers, 1986). This is still a rather high figure in relation the the ion fluxes generally found in plant cells (loc. cit.) and higher than the probable rate of proton pumping (100–200 nmol m^{-2} s^{-1}: Raven & Rubery, 1982) reported for plant cells. It may be that this is the true net Na^+ flux, but if the cortical plasmalemma is also involved in the uptake of the ions the resultant net flux is reduced to about 250 nmol m^{-2} s^{-1}: this begins to close with the reported range of ion fluxes. However, the supply of ions to the cortical plasmalemma depends upon their rate of diffusion through the free spaces of the root (the apoplast). Calculations suggest that an adequate supply of ions could be maintained by diffusion, especially since the diffusion path is short (as the roots are very fine: see below), but only if the diffusion coefficent of

Na^+ is as high as that of Cl^- (cf. Pitman, 1984): cations can be expected to be impeded by charge interaction in the cell wall pathway. We can conclude that loading is primarily at the epidermis with a possible contribution from the cortical cells. It then remains to explain the transport through the symplast.

Symplastic movement of ions relies on diffusion enhanced by cytoplasmic streaming: important parameters are the concentration of ions in the symplast and their gradient across the root as well as the rate of cytoplasmic streaming. Calculations made by Yeo & Flowers (1986) suggest that a concentration gradient of 40 mol m^{-3} between epidermal cytoplasm and xylem would be the maximum necessary to drive adequate transplasmodesmatal diffusion. The smallness of this gradient is very much a consequence of the fact that in cross-section the cortex of the roots of *Suaeda maritima* is only three cells wide, so the ions need not pass more than four plasmodesmata in series. From literature values of cytoplasmic streaming we (Yeo & Flowers, 1986) have calculated the cytoplasmic ion concentration in the innermost cortical layer necessary to sustain the net flux as 80 mol m^{-3} This would mean a gradient, at most, from 120 to 80 mol m^{-3} from outside to inside of the cortex and quite within the range which is feasible without causing problems to the metabolism (Flowers *et al.*, 1986). However, the rate of cyclosis could easily be rate limiting to the net flux of Na^+ as well as crucial to minimizing the mean concentration necessary to sustain the observed flux. To conclude, it appears that transport across the root is unlikely to occur in the apoplast and likely to involve the protoplasts of the cortical cells. The flux may be limited by the rate of cytoplasmic streaming within the cortical cells.

The final stage of the passage of ions across the root is their release to the xylem (see Ch. 7). If the root is in a steady state, then this release must proceed at the same rate as the symplasm is loaded. Were the whole of the pericycle and the parenchyma of the stele available for ion transport into the xylem vessels, then the flux would be some 1200 nmol m^{-2} plasmalemma s^{-1} (Yeo & Flowers, 1986). We cannot conceive that this is an active flux, but equally, cannot say whether this is simply a passive leakage or involves an ion channel in 'downhill' carrier-mediated transport.

We have not yet touched upon the nature of the uptake process itself – is it active or passive? We can only speculate at the moment. Since the likely flux we have calculated (750 nmol m^{-2} plasmalemma s^{-1}) is about seventy-five times the value for most active-transport processes in plant cells (10 nmol m^{-2} s^{-1}: Penning de Vries, 1975), we can guess that the flux is passive. This does not present any particular problems for the movement of Na^+ into cells with internally negative membrane potentials and cytoplasmic concentrations of less than half those in the external solution. The situation is less clear for Cl^-, although even for the negative ion providing the membranes are relatively hypopolarized (−30 mV) downhill transport could still occur, given the relatively low cytoplasmic Cl^- concentrations *Suaeda maritima* maintains – at least in its mesophyll cells. Advection to the surface could increase the chemical concentration of NaCl external to the epidermal plasmalemma still further. However, by whatever means the uptake of ions occurs, it is quite clear that the influx must be regulable. An example will serve to illustrate this. At the steady state, the rate of influx to the xylem must equal the rate at which ions are transferred to the stelar apoplast. If the transpiration rate is decreased, whether by submergence of the plant by tidal water or the onset of darkness, the removal of ions from the endodermal symplast will decrease. If it ceases altogether the ion concentration therein would rise by 0.3 mol m^{-3} s^{-1} and exceed 1 kmol m^{-3} in an hour. This would kill the cells. Consequently, when the transpiration rate decreases, net ion uptake at the plasmalemma must decrease. This could be achieved either through decreased influx or increased efflux (an increased flux to the vacuoles could also act as a temporary buffer). Increased efflux is likely to be a very expensive option, energetically, as it would have to involve the use of metabolic energy to remove Na^+ ions (and possibly Cl^-: see above). We presume, therefore, that influx is decreased as the cytoplasmic ion concentration rises (perhaps above a set point). It is implicit in this that the cytoplasmic ion concentration is used as a signal. The manner in which the putative ion channel is gated remains entirely unknown (but see Edmonds, 1982).

10.3.2.3 Ion supply to the shoots

We have already posed the question of whether shoot growth controls net transport through the root or whether ion supply from the roots limits growth of the shoot. The analysis of root fluxes suggests the latter (see also Yeo and Flowers, 1986) and so ion fluxes in the leaves might be expected to be much lower than those in the roots. Leaves receive ions through the xylem: for individual leaf cells the supply may be apoplastic or symplastic depending upon their position and the nature of ion transport through the leaf. The quantity of ions supplied is the product of the xylem ion concentration and the transpiration rate. Once in the shoot, ions must be accumulated therein (that is within the leaf vacuoles: see above) or be re-exported. Export from the shoot can take place though the operation of salt glands (see Ch. 13) or, in theory, in the phloem (see Ch. 8). Phloem transport would return ions to the root medium, but since phloem transport is presumably supplying carbohydrate and K^+ to the meristems any exported Na^+ and Cl^- would also find their way to these regions. What little evidence there is suggests that phloem-fed tissues in halophytes growing in NaCl receive relatively little Na^+ in relation to K^+: this applies to seeds (O'Leary *et al.*, 1985), ray florets (Gorham *et al.*, 1980), developing leaves (Gorham & Wyn Jones, 1983) and root apices (Hajibagheri *et al*, 1985). We do not believe the phloem to play a significant role in the re-export of Na^+ ions from leaves (see also Yeo, 1981): rather, under saline conditions salt-sensitive plants may be unable to exclude Na^+ ions from their phloem. There is evidence for this view in that Lessani and Marschner (1978) reported that less tolerant species transported more Na^+ from the leaf apoplast to the root medium than did more tolerant species. Such a response should be separated from that which may occur in the presence of low external Na^+ concentrations (a few mol m^{-3}), where the recycling in the phloem may be of significance (e.g. Winter, 1982), although even this is yet to be established unequivocally. Ions, then, must be accommodated within the leaves or excreted by glands. Since many halophytes do not have glands, they cannot be essential to the control of leaf ion concentrations. The possession of glands can perhaps be seen as a final safety valve. We will not consider them further here, as they are not essential: to function without them is more demanding of the balancing of supply and demand in the leaves than to function with them. Within the leaf three compartments are important to the ion relations, the cell walls, the cytoplasm and the vacuoles. The precise ratio between vacuole and cytoplasm will depend upon the degree of expansion of the leaf. Unfortunately, there is little quantitative information for halophytes and none for rapidly expanding leaves that we are aware of. For *Suaeda maritima*, the relative proportions of cell wall, air spaces, cytoplasm and vacuole in the mesophyll are 3 %, 33 %, 15 % and 49 %, respectively: if the volume of the air spaces is ignored, the proportional volumes of cell wall, cytoplasm and vacuole are 4 %, 23 % and 73 %, respectively (Hajibagheri *et al*., 1984). In the large central parenchymatous cells, the vacuole is some 97 % of the cell volume. The vacuoles, then, are the only repository of any significant volume for ions. Injury to the cell can result either from a failure of compartmentation (ion toxicity) or an excess of ions in the apoplast bringing about cellular dehydration. The osmotic consequences of high ion concentrations remaining in the leaf apoplast were first outlined in detail by Oertli (1968). Whether ions remain in the apoplast or are accumulated within the protoplast is a consequence of the balance between delivery of ions and the rates of uptake at the plasmalemma of the leaf cells. If more than 1 % of those delivered to the leaves remain in the apoplast, the apoplastic ion concentration will rise faster than that in the protoplast (that is in the vacuole assuming that ion compartmentation between vacuole and cytoplasm is maintained (see Fig. 10.2).

The delivery of ions to the leaves is the product of the rate of transpiration and the xylem ion concentration. Although there are data on xylem ion concentrations in the literature for a range of halophytes (see Flowers, 1985), we can only analyse the fluxes in detail for *Suaeda maritima*. With a transpiration rate of 0.1 g water g^{-1} fresh weight h^{-1} (equivalent to 0.1 m^3 m^{-3} h^{-1}) and a xylem ion concentration of about 50 mol m^{-3} (data of Clipson: see Flowers, 1985), the delivery is some 5 mol m^{-3} shoot h^{-1}. Although we do not know the average surface-to-volume ratio for a shoot of *Suaeda maritima* we do know that it is 0.08×10^{-6} m^2 m^{-3} for the mesophyll cells of a mature leaf. With this surface-to-volume ratio the mean flux to the leaf cells would be about 17 nmol m^{-2} s^{-1}. In fact over a fourfold range of surface-to-volume ratios (0.05 to 0.2×10^{-6} m^2 m^{-3}), which covers a range of cell radii from 62 to 13 μm, and so most of the cells likely to be found in *Suaeda maritima* leaves, the flux varies from 7 to 28 nmol m^{-2} s^{-1}. These values fall well within the range of ion fluxes reported for plant cells (loc. cit.). Overall then, the fluxes required to remove the ions supplied to the shoots are not great, especially when compared with the fluxes which are apparent in the roots. The capacity of the shoot system for ion transport need not be excessive in order to remove the ions supplied by the transpiration stream.

The shoot is clearly a mixture of cell types and cell sizes, of which the variation in cell size with leaf age is of particular interest in the present context. Operationally we can separate three classes of cells, the meristematic which are small and highly cytoplasmic, the cells of young rapidly expanding leaves and the cells of what we can call 'mature leaves'. In *Suaeda maritima* we believe the cells of these 'mature' leaves are still expanding, albeit at a reduced rate, as the leaf fresh weight and water content continue to increase (Fig. 10.4; Flowers & Yeo, 1986). This being so we can designate the three classes of cells as meristematic, rapidly expanding and slowly expanding. By following the change in leaf ion contents with time it is possible to calculate the net fluxes into the slowly and rapidly expanding leaves. Again making assumptions about the surface-to-volume ratios, the fluxes into the two categories of leaf are some 6 and 0.8 nmol m^{-2} s^{-1} (Flowers & Yeo, 1986). Clearly, the majority of the 17 nmol m^{-2} s^{1} must be going to the 'meristematic' class. As the leaf ages the net flux falls. There is presumably a decreasing supply to the leaves as they age, associated with a declining transpiration (although the transpiration rates of individual leaves of *Suaeda maritima* have not been determined). A decrease in uptake with increasing age could result from decreased energy supply or simply from a fixed number of ion channels in the surface: in the latter case, as the cell expands, the surface density of ion channels will decrease. If in the later stages of the life of the leaf, supply exceeds uptake, then the apoplastic ion concentration will rise and the cell turgor fall as outlined in the Oertli (1968) hypothesis and noted by Clipson *et al.* (1985).

Although the Oertli hypothesis was outlined some 20 years ago and has received considerable attention in recent years (Greenway & Munns, 1980; Leigh & Tomos, 1983; Munns & Passioura, 1984; Yeo *et al.*, 1985; Flowers & Yeo, 1986), there is little experimental evidence either for or against it. The predictions of the hypothesis to be tested are that apoplastic ion concentrations will be much higher than in the xylem and consequently that turgor pressures will be lower than otherwise expected. The lack of evidence is largely the result of difficulties in determining ion concen-

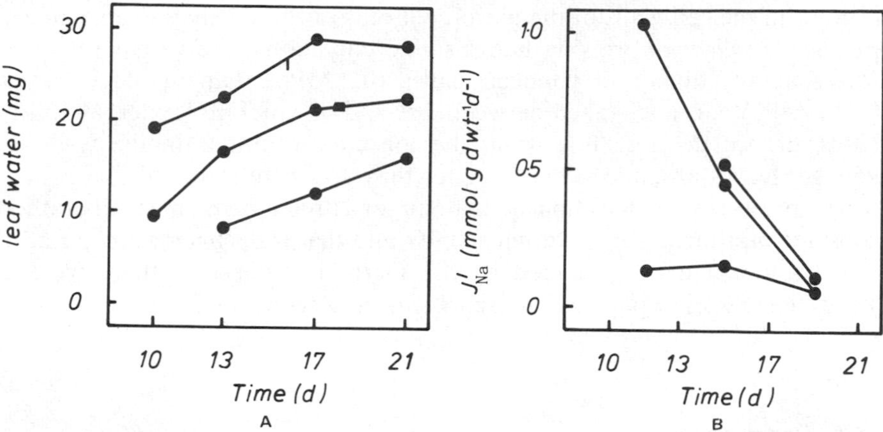

Fig. 10.4 Changes in water content, A, and Na^{+} flux, B, for leaves of decreasing age (increasing number) of seedlings of the halophyte *Suaeda maritima*. (Redrawn from Flowers and Yeo, 1986.)

trations within the apoplast and in measuring the turgor pressure in cells. Unless leaves are particularly large, something which is not characteristic of halophytes, it is not possible to measure the ion contents of the apoplast by elution (Bernstein, 1971) since efflux from cut cells with high ion concentrations obscures that from the cell walls. Generally it is not possible to elute ions from leaves of halophytes without damaging the cells in some way, due to the impermeability of the cuticle to aqueous solutions – an important characteristic for plants which undergo periodical submergence with sea water. The properties of the cuticle also interfere with measurements of the plant water potential using thermocouple psychrometers since leaf tissues do not equilibrate even over 24 h (Yeo & Flowers, 1986). Again, if the tissue is cut to facilitate equilibration, the potential that is measured is not the water potential, but is dominated by the osmotic component from damaged cells. This being so, it is never possible to deduce the turgor pressure of halophytes from the difference between osmotic and water potentials; only direct measurement will suffice. We are only aware of two such investigations into the turgor of a halophyte. For *Suaeda maritima* turgor pressures were measured by Tomos & Wyn Jones (1982) and Clipson *et al.* (1985) using the miniaturized pressure probe (Husken *et al.*, 1978). For mature leaves, the turgor was very low (less than 0.1 MPa) in relation to the mean osmotic potential (from −1.0 to −2.5 MPa) and relatively low in comparison with many other species, where turgor pressures range from 0.1 to 0.7 MPa (see Clipson *et al.*, 1985, and Ch. 6 for references). Were the apoplastic ion concentrations similar to those in the xylem sap (about 50 mol m^{-3} or −0.23 MPa), higher turgor pressures might have been anticipated. The relatively low turgor pressure measured in the mature leaves is entirely consistent with high ion concentrations in the cell walls of the mesophyll cells of these leaves (Table 10.2). In *Suaeda maritima* growing under saline conditions, the turgor pressure was generally higher in younger leaves (0.2 MPa) than in older leaves (<0.1 MPa) and unchanged between 200 and 400 mol m^{-3} external NaCl. However, we do not know what the ion concentrations in the walls of younger leaves are, and hence whether they are correlated with the higher turgor pressures. It does appear that turgor is regulated, since it is rather constant with increasing external salinity and that it declines with leaf age. The latter would be predicted by the Oertli hypothesis if there were a steady increase in apoplastic ion concentration with time.

10.4 Conclusions

The ability of a plant to grow in high external salinity depends upon the maintenance of a fine balance between the supply of ions to the shoots and ion uptake therein. Sufficient ions for osmotic adjustment must be transferred to the leaves without exceeding the capacity of the leaf cells to accumulate them. Failure in either respect has lethal consequences.

References

Aronson, J. A. (1985). *World Halophyte Database*. Applied Research Institutes, Beer Sheva, Israel, Ben-Gurion University of the Negev. In preparation.

Albert, R. & Kinzel, H. (1973). Unterscheidung von physiotypen bei Halophyten des Neusiedlerseegebietes (Osterreich). *Zeitschrift für Pflanzenphysiologie* **70**, 138–57.

Ansari, R. (1982). *Salt Tolerance Studies in some Halophytes*. D. Phil. Thesis, Brighton, University of Sussex.

Bernstein, L. (1971). Method for determining solutes in the cell walls of leaves. *Plant Physiology* **47**, 361–5.

Clipson, N. J. W., Tomos, A. D., Flowers, T. J. & Wyn Jones, R. G. (1985). Salt tolerance in the halophyte *Suaeda maritima* (L.) Dum. The maintenance of turgor pressure and water potential gradients in plants growing at different salinities. *Planta* **165**, 392–6.

Collander, R. (1941). Selective absorption of cations by higher plants. *Plant Physiology* **16**, 691–720.

Edmonds, D. T. (1982). The ordered water channel. In Franks, F. & Mathias, S. F. (eds.). *Biophysics of Water*, pp. 148–50. New York, John Wiley.

FAO (1978). How much good land is left? *CERES* **13**, 12–16.

Flowers, T. J. (1975). Halophytes. In Baker, D. A. & Hall, J. L. (eds.). *Ion Transport in Plant Cells and Tissues*, pp. 309–34. Amsterdam and New York, North Holland/Elsevier.

Flowers, T. J. (1985). Physiology of halophytes. *Plant and Soil* **89**, 41–56.

Flowers, T. J. & Läuchli, A. (1983). Sodium versus potassium: substitution and compartmentation. In Läuchli, A. & Pirson, A. (eds.). *Inorganic Plant Nutrition*, pp. 651–81. *Encyclopedia of Plant Physiology*, New Series, vol. 15B, Berlin, Springer-Verlag.

Flowers, T. J., Hajibagheri, M. A. & Clipson, N. J. W. (1986). Halophytes. *Quarterly Review of Biology*, **61**, 313–17.

Flowers, T. J., Troke, P. F. & Yeo, A. R. (1977). The mechanism of salt tolerance in halophytes. *Annual Review of Plant Physiology* **28**, 89–121.

Flowers, T. J. & Yeo, A. R. (1986). Ion relations of plants under drought and salinity. *Australian Journal of Plant Physiology* **13**, 77–91.

Gibson, T. S., Speirs, J. & Brady, C. J. (1984). Salt tolerance in plants. II. *In vitro* translation of m-RNAs from salt-tolerant and salt-sensitive plants on wheat germ ribosomes. Responses to ions and compatible organic solutes. *Plant Cell and Environment* **7**, 579–83.

Gimmler, H., Kaadan, R., Kirchner, U. & Weyand, A. (1984). The chloride sensitivity of *Dunaliella parva* enzymes. *Zeitschrift für Pflanzenphysiologie* **114**, 131–50.

Glenn, E. P. & O'Leary, J. W. (1984). Relationship between salt accumulation and water content of dicotyledonous halophytes. *Plant Cell Physiology* **7**, 253–61.

Gorham, J., Hughes, L. L. & Wyn Jones, R. G. (1980). Chemical composition of salt marsh plants from Ynys Môn (Angelsey): the concept of physiotypes. *Plant Cell and Environment* **3**, 309–18.

Gorham, J. & Wyn Jones, R. G. (1983). Solute distribution in *Suaeda maritima. Planta* **157**, 344–9.

Greenway, H. & Munns, R. (1980). Mechanisms of salt tolerance in non-halophytes. *Annual Review of Plant Physiology* **31**, 149–90.

Hajibagheri, M. A., Hall, J. L. & Flowers, T. J. (1984). Stereological analysis of leaf cells of the halophyte *Suaeda maritima* (L.) Dum. *Journal of Experimental Botany* **35**, 1547–57.

Hajibagheri, M. A., Yeo, A. R. & Flowers, T. J. (1985). Salt tolerance in *Suaeda maritima* L. (Dum): fine structure and ion concentrations in the apical region of roots. *New Phytologist* **99**, 331–43.

Hall, J. L., Harvey, D. M. R. & Flowers, T. J. (1978). Evidence for the cytoplasmic localization of betaine in leaf cells of *Suaeda maritima. Planta* **140**, 59–62.

Harvey, D. M. R. (1980). The preparation of botanical samples for ion localisation studies at the subcellular level. In *Scanning Electron Microscopy*, vol. II, pp. 409–19. O'Hare, Ill., Scanning Electron Microscopy Inc., AMI.

Harvey, D. M. R., Hall, J. L., Flowers, T. J. & Kent, B. (1981). Quantitative ion localisation within *Suaeda maritima* leaf mesophyll cells. *Planta* **151**, 555–60.

Heywood, V. H. (ed.) (1978). *Flowering Plants of the World*. Oxford University Press.

Hüsken, D., Steudle, E. & Zimmermann, U. (1978). Pressure probe techniques for measuring water relations of cells in higher plants. *Plant Physiology* **61**, 158–63.

Jefferies, R. L. (1973). The ionic relations of seedlings of the halophyte *Triglochin maritima* L. In Anderson, W. P. (ed.). *Ion Transport in Plants*, pp. 297–321. London, Academic Press.

Jeschke, W. D. (1983). Cation fluxes in excised and intact roots in relation to specific and varietal differences. *Plant and Soil* **72**, 197–212.

Kovda, V. A. (1980). *Land Aridization and Drought Control*. Boulder, Colo., Westview Press.

Larsen, H. (1967). Biochemical aspects of extreme halophilism. *Advances in Microbial Physiology* **1**, 97–132.

Leigh, R. A, Ahmad, N. and Wyn Jones, R. G. (1981). Assessment of glycinebetaine and proline compartmentation by analysis of isolated beet vacuoles. *Planta* **153**, 35–41.

Leigh, R. A. & Tomos, D. (1983). An attempt to use isolated vacuoles to determine the distribution of sodium and potassium in cells of storage roots of red beet (*Beta vulgaris* L.). *Planta* **159**, 469–75.

Lessani, H. & Marschner, H. (1978). Relation between salt tolerance and long-distance transport of sodium and chloride in various crop species. *Australian Journal of Plant Physiology* **5**, 27–37.

Ling, G. N. (1979). The polarized multilayer theory of cell water and other facets of the association–induction hypothesis concerning the distribution of ions and other solutes in living cells. In Keith, A. D. (ed.). *The Aqueous Cytoplasm*, pp. 23–60. New York, Dekker.

Lüttge, U. & Higinbotham, N. (1979). *Transport in Plants*. Amsterdam, North Holland.

Maas, E. V. & Hoffman, G. J. (1977). Crop salt tolerance: evaluation of existing data. In Dregne, H. E. (ed.). *Managing Saline Water for Irrigation*, pp. 187–98. Lubbock, Tex., Texas Technical University.

Massoud, F. I. (1974). Salinity and alkalinity. In *A World Assessment of Soil Degradation. An International Programme of Soil Conservation*. Report of an Expert Consultation on Soil Degradation, pp. 16–17. Rome, FAO, UNEP.

Mudie, P. J. (1974). The potential economic uses of halophytes. In Reimold, P. J. & Queen, W. H. (eds.). Ecology of Halophytes, pp. 565–97. New York, London, Academic Press.

Munns, R. & Passioura, J. (1984). Effect of prolonged exposure to NaCl on the osmotic pressure of leaf xylem sap from intact, transpiring barley plants. *Australian Journal of Plant Physiology* **11**, 497–507.

Nobel, P. S. (1983). *Biophysical Plant Physiology and Ecology*. San Francisco, Calif., W. H. Freeman.

Oertli, J. J. (1968). Extracellular salt accumulation, a possible mechanism of salt injury in plants. *Agrochimica* **12**, 461–9.

O'Leary, J. W., Glenn, E. P. & Watson, M. C. (1985). Agricultural production of halophytes irrigated with seawater. *Plant and Soil* **89**, 311–12.

Pahlich, E., Kerres, R. & Jager, H.-J. (1983) Influence of water stress on the vacuole/extravacuole distribution of proline in protoplasts of *Nicotiana rustica*. *Plant Physiology* **72**, 590–1.

Penning de Vries, F. W. T. (1975). The cost of maintenance processes in plant cells. *Annals of Botany* **39**, 73–92.

Pitman, M. G. (1965). Sodium and potassium uptake by seedlings of *Hordeum vulgare*. *Australian Journal of Biological Sciences* **18**, 10–24.

Pitman, M. G. (1975). Whole plants. In Baker, D. A. & Hall, J. L. (eds.). *Ion Transport in Plant Cells and Tissues*, pp. 267–308. Amsterdam, New York, North Holland/Elsevier.

Pitman, M. G. (1982). Transport across plant roots. *Quarterly Reviews of Biophysics* **15**, 481–514.

Pitman, M. G. (1984). Transport across the root and root/shoot interactions. In Staples, R. C. & Toennissen, G. H. (eds.). *Salinity Tolerance in Plants: Strategies for Crop Improvement*, pp. 93–123. New York, John Wiley.

Pitman, M. G. & Cram, J. (1977). Regulation of ion contents in whole plants. In Jennings, D. H. (ed.). *Integration of Activity in the Higher Plant*, pp. 391–424. Cambridge University Press.

Ponnamperuma, F. N. (1984). Role of cultivar tolerance in increasing rice production on saline lands. In Staples, R. C. & Toennissen, G. H. (eds.). *Salinity Tolerance in Plants: Strategies for Crop Improvement*, pp. 255–71. New York, John Wiley.

Raven, J. A. and Rubery, P. H. (1982). Coordination of development: hormone receptors, hormone action and hormone transport. In Smith, H. and Grierson, D. (eds.). *Molecular Biology of Plant Development*, pp. 28–48. Oxford, Blackwell.

Steudle, E. & Jeschke, W. D. (1983). Water transport in barley roots: measurement of root pressure and hydraulic conductivity of roots in parallel with turgor and hydraulic conductivity of root cells. *Planta* **158**, 237–48.

Storey, R., Pitman M. G. & Stelzer, R. (1983a). X-ray microanalysis of cells and cell compartments of *Atriplex spongiosa*. II. Roots. *Journal of Experimental Botany* **34**, 1196–206.

Storey, R., Pitman, M. G., Stelzer, R. & Carter, C. (1983b). X-ray microanalysis of cells and cell compartments of *Atriplex spongiosa*. I. Leaves. *Journal of Experimental Botany* **34**, 778–94.

Storey, R. & Wyn Jones, R. G. (1979). Responses of *Atriplex spongiosa* and *Suaeda monoica* to salinity. *Plant Physiology* **63**, 156–62.

Tomos, A. D. & Wyn Jones, R. G. (1982). Water relations in the epidermal cells of the halophyte *Suaeda maritima*. In Franks, F. & Mathias, S. F. (eds.). *Biophysics of Water*, pp. 327–31. New York, John Wiley.

Walker, N. A. & Pitman, M. G. (1976). Measurement of fluxes across membranes. In Lüttge, U. & Pitman, M. G. (eds.). *Transport in Plants*, II, *Encyclopedia of Plant Physiology*, New Series, vol. 2A, pp. 93–126. Berlin, Heidelberg, New York, Springer-Verlag.

Wiebe, H. H. (1966). Matric potential of several plant tissues and biocolloids. *Plant Physiology* **41**, 1439–42.

Winter, E. (1982). Salt tolerance in *Trifolium alexandrinum* L. II. Ion balance in relation to its salt tolerance. *Australian Journal of Plant Physiology* **9**, 239–50.

Wyn Jones, R. G., Brady, C. J. & Speirs, J. (1979). Ionic and osmotic relations in plant cells. In Laidman, D. L. & Wyn Jones, R. G. (eds.). *Recent Advances in the Biochemistry of Cereals*, pp. 63–103. New York, Academic Press.

Wyn Jones, R. G. & Pollard, A. (1983). Proteins, enzymes and inorganic ions. In Läuchli, A. & Pirson, A. (eds.). *Inorganic Plant Nutrition. Encyclopedia of Plant Physiology*, New Series, vol. 15B, pp. 528–62. Berlin, Springer-Verlag.

Yeo, A. R. (1974). Salt tolerance in the halophyte *Suaeda maritima* (L.) Dum. D. Phil. Thesis, Brighton, University of Sussex.

Yeo, A. R. (1981). Salt tolerance in the halophyte *Suaeda maritima* (L.) Dum.: Intracellular compartmentation of ions. *Journal of Experimental Botany* **32**, 487–97.

Yeo, A. R., Caporn, S. J. M. & Flowers, T. J. (1985). The effect of salinity upon photosynthesis in rice (*Oryza sativa* L.): gas exchange by individual leaves in relation to their salt content. *Journal of Experimental Botany* **36**, 1240–8.

Yeo, A. R. & Flowers, T. J. (1986). Ion transport in *Suaeda maritima*: its relation to growth, and implications for the pathways of radial transport of ions and water across the root. *Journal of Experimental Botany* **37**, 143–59.

11 CAM plants

U. Lüttge and J. A. C. Smith

11.1 Introduction: The central role of malic acid in plant cell physiology

In an important review, Lance & Rustin (1984) have drawn attention to the central role of malic acid in plant cell physiology. They list the reasons as follows:

1. Malate is a mobile storage form for CO_2 and reducing equivalents.
2. Malic acid is a source of anions and protons.
3. Malate contributes to charge separation, i.e. splitting of water in energy-transfer systems.
4. Malate can be oxidized by plant mitochondria without control by the cell energy charge.

For the first of these reasons malate is the basis of the metabolic function of CAM. For the second, malic acid is integrated into a large variety of cell physiological functions where processes of transport across membranes are involved. There are various expressions used to denote different forms of malic acid, viz.:

Malic acid	Loosely used for the anion plus its two protons.
H_2mal	Refers specifically to the non-dissociated acid, H_2mal^0, emphasizing that it is electrically neutral.
malate	Loosely used for the anion, 'total malate (Σmalate)' referring to what is measured in an enzymatic test.
Δmal	Day–night changes of total malate levels during the CAM rhythm.
malate^{2-}	Refers specifically to the divalent anion, abbreviated mal^{2-}.
Hmal$^-$	Refers specifically to the univalent anion.

11.1.1 Crassulacean acid metabolism (CAM)

The essence of the metabolic function in CAM is that during the dark period CO_2 is fixed via PEP carboxylase and stored in the form of malic acid. During the light period malic acid is remobilized and decarboxylated, and the CO_2 is assimilated via RuBP carboxylase. This is associated with the well known inverted stomatal rhythm of CAM.

11.1.2 Malate accumulation in the vacuoles of higher plant cells

Malic acid synthesis plays a role in 'plant nutrition' by balancing surplus cation uptake. It provides protons that exchange for cations with the medium and anions for accumulation in the vacuole together with the cations taken up (Tori & Laties, 1966; Osmond & Laties, 1969). Another rather similar example is NO_3^- and SO_4^{2-} reduction. These metabolic processes leave behind OH^- ions and the cations taken up together with anions. H^+ from malic acid neutralizes the OH^-, and malate accumulates together with the cations in the vacuole.

In an equivalent way malic acid serves 'extension growth'. Protons are extruded to provide the cell wall acidification required in the terms of the 'acid-growth hypothesis'. K^+ and malate are transported into the vacuoles and provide for the adjustment of osmotic pressure and turgor pressure in the extending cells (Haschke & Lüttge, 1977).

Extremely similar as well is the mechanism by which most 'stomatal guard cells' can increase stomatal aperture. Protons from malic acid exchange for K^+, and K_2-malate is accumulated in the vacuoles as an osmoticum (see Ch. 12; and Raschke, 1979).

In general, all these processes involve maintenance of charge balance, maintenance of cytoplasmic pH (Smith & Raven, 1979) and osmotic adjustments (Cram, 1976). With the exception of the stomata, all these examples are slow processes related to growth and maintenance. Accumulation of malate in the vacuoles together with inorganic cations as the predominating counterions proceeds continuously. Such processes are certainly also involved in growth of leaves of CAM plants. *Kalanchoë* leaves, for example, contain large amounts of Ca^{2+} (Lüttge & Smith, 1984a) and organic-acid anions such as citrate and isocitrate, whose concentrations do not oscillate much during the day–night rhythm of CAM (Vickery, 1952, 1957; Phillips & Jennings, 1976; Kenyon *et al.*, 1985). There is also a small fraction of the total malate pool that does not participate metabolically in the day–night rhythm (e.g. Phillips, 1980; see also Fig. 11.2). But on top of this malic acid oscillations with very high amplitudes are observed.

As with CAM, movements of stomatal guard cells are rhythmic phenomena, and the two have often been compared (Willmer *et al.*, 1973;

Cockburn, 1981). However, in guard cells, malate accumulation is accompanied by K^+, whereas in the CAM rhythm changes in malate levels are always associated with stoichiometric changes in titratable protons. A stoichiometry of 2 H^+ : 1 malate is invariably observed. Attempts to disturb this stoichiometry by supplying a surplus of K^+ or lipophilic cations have proved unsuccessful (Lüttge *et al.*, 1975; Lüttge & Ball, 1980). This difference need not be a qualitative one. It probably has quantitative reasons. Although for the guard cells themselves concentration changes are large, they are small for the plant tissue as a whole, and H^+/K^+ exchanges between guard cells and subsidiary cells or surrounding epidermis cells can be readily accommodated. In contrast, the malic acid changes in CAM plants can involve entire leaves or large parts of the above-ground biomass. Day–night changes of malic acid levels in *K. daigremontiana*, one of the most widely used CAM species in laboratory studies, can amount to 200 mol m^{-3} (Lüttge *et al.*, 1982). The highest Δmalic acid values ever observed for a CAM plant were found during field studies with the bromeliad *Aechmea nudicaulis* in Trinidad, where Δmalic acid estimated by measuring titratable protons on one occasion was 237 mol m^{-3}; one individual at dawn contained 320 mol m^{-3} (Smith *et al.*, 1986). This even may be an underestimate for the chlorenchyma tissue because *Aechmea* leaves possess a central non-green water-storage tissue, and it is known that such tissues do not participate in the CAM rhythm (e.g. Kluge *et al.*, 1979). Diurnal oscillations of K^+ levels of this size would have to involve a rhythm of import via the xylem and export via the phloem. This not only appears unfeasible energetically but often would also be impossible anatomically, as large groups of epiphytic CAM plants, e.g. in the Bromeliaceae and Orchidaceae, do not have functional roots. The solution is acid accumulation, since metabolism provides 2 H^+ and 1 malate^{2-} in stoichiometric amounts.

The driving forces for solute movements in plant cells are now generally thought to be generated by membrane ATPases at the plasmalemma and tonoplast, which extrude protons from the cytoplasm by primary active transport (see Ch. 3; and Spanswick, 1981; Sze, 1984, 1985; Reinhold and Kaplan, 1984; Marrè and Ballarin-Denti, 1985). As summarized in Fig. 11.1, growth, plant nutrition, stomatal guard cells and CAM may use the same basic transport mechanisms.

11.2 Thermodynamics and energetics of malic acid transport at the tonoplast of CAM plants

11.2.1 The Nernst criterion

The Nernst equation describes passive ion distribution at membranes (see Ch. 1). If it is not fulfilled by experimental observations, one can assume

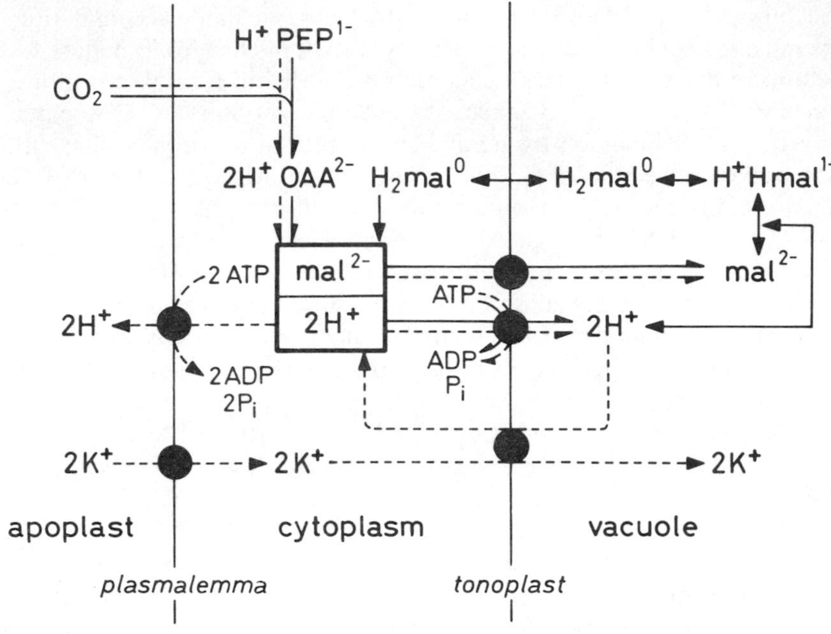

Fig. 11.1 Transport scheme involving fluxes of H^+, K^+, divalent malate anions (mal^{2-}) and non-dissociated malic acid (H_2mal^o). Circles with arrows at the membranes indicate primary active H^+ transport (ATP consumption) or secondarily coupled transport mechanisms (uniport or exchange mechanisms). Simple arrows across membranes denote passive lipid diffusion. Dashed arrows indicate events predominating in malate accumulation during various physiological processes as discussed in Sect. 11.1.2. Solid arrows show events important in CAM.

that metabolic energy is involved in distribution of the ionic species in question (Nernst criterion). Written for the tonoplast:

$$\Delta\psi = -\frac{RT}{zF}\ \ln\ \frac{c^{\text{cytoplasm}}}{c^{\text{vacuole}}} \qquad (11.1)$$

it can be used to assess the distribution of the ionic species arising from dissociation of malic acid, viz. H^+, $Hmal^{1-}$, mal^{2-}. The membrane potential at the tonoplast, $\Delta\psi$, of *Kalanchoë* was found from transcellular electrical profiles to be + 25 mV (inside of vacuoles positive) (Rona *et al.*, 1980).

Vacuolar pH and malate concentrations can be readily obtained from analyses of sap expressed from the tissue. Estimations using the Nernst criterion have been made with *K. daigremontiana* and *K. tubiflora*, where the vacuoles occupy ~ 97 % of the cell volume (Steudle *et al.*, 1980; Lüttge *et al.*, 1982). Malate concentrations are typically $\leqslant$ 40 mol m^{-3} at the end of the light period and up to 220 mol m^{-3} at the end of the dark period. Vacuolar pH changes between ~ 6.0 at the end of the light period and ~ 3.3 at the end of the dark period (Fig. 11.2).

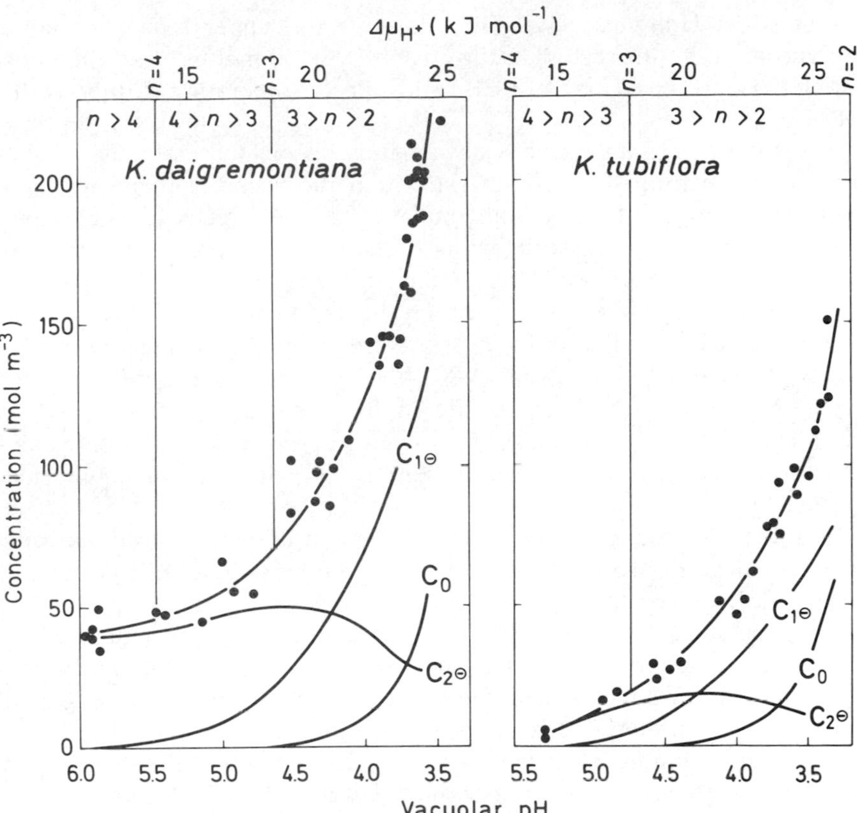

Fig. 11.2 Relations between total malate concentrations (points representing enzymatic determinations of malate) and pH in the vacuoles or $\Delta\bar{\mu}_{H^+}$ at the tonoplasts of *K. daigremontiana* and *K. tubiflora*. Using the p*K* values of malic acid dissociation determined in the vacuolar sap of *K. daigremontiana* ($pK_1 = 3.18$, $pK_2 = 4.25$), the relative contributions of non-dissociated malic acid ($H_2mal^o = c_0$), univalent malate anion ($Hmal^1 = c_1\ominus$) and divalent malate anion ($mal^{2-} = Kc_2\ominus$) were also plotted versus vacuolar pH and $\Delta\bar{\mu}_{H^+}$. It should be noted that low malate levels and high pH values (left-hand end of the diagrams) correspond to the beginning and high malate levels and low pH values (right-hand end of the diagrams) to the end of the dark period. The ranges of total malate accumulation, where thermodynamically a pumping of $4H^+$ ($n \geqslant 4$), $3H^-$ ($4 > n > 3$) or $2H^+$ ($3 > n > 2$) per ATP hydrolysed represents the highest possible stoichiometry are also indicated. (See Lüttge *et al.*, 1981; Lüttge and Smith, 1984b; Smith *et al.*, 1982.)

Cytoplasmic pH is assumed to be about 7.5 to 8.0 (Lüttge *et al.*, 1982; Marigo *et al.*, 1982). Cytoplasmic malate concentration has never been measured directly. An upper limit can be assumed on the basis of the feed-back inhibition of PEP carboxylase by malate. K_i^{malate} for PEP carboxylase is in the range between 0.005 and 1.0 mol m^{-3} (Winter, 1974; Greenway et al., 1978; Bollig and Wilkins, 1979; Buchanan-Bollig and Kluge, 1981;

Osmond and Holtum, 1981). On this basis the upper limit for malate concentration in the cytoplasm is from 1 to 10 mol m^{-3} at the most during the dark period when PEP carboxylase is operating (Lüttge *et al.*, 1982).

At cytosolic pH, malic acid is always fully dissociated and malate exists entirely in the form of mal^{2-}. However, in the vacuoles there are large day–night changes of pH in the range of the pK values of malic acid. Therefore, malic acid dissociation also changes. The pK values given in the literature (as estimates at zero ionic strength) are $pK_1 = 3.54$ and $pK_2 = 5.12$. However, pK values depend on the exact concentrations and ionic environment. Determinations with the expressed sap of *K. daigremontiana* gave values of $pK_1 = 3.18$ and $pK_2 = 4.25$ (Lüttge and Smith, 1984b). Using these values the total malate in the vacuoles, as determined by enzymatic tests, can be separated for the corresponding vacuolar pH values into the different ionic species present in the dissociation equilibrium (Fig. 11.2).

With these data and assumptions, the Nernst criterion shows that only mal^{2-} could be passively distributed across the tonoplast. As can be seen in Fig. 11.2, the concentration of mal^{2-} changes little while total malate shows large oscillations in the vacuole. The remaining two species of the dissociation equilibrium, i.e. H^+ and $Hmal^{1-}$, would have to be transported into the vacuoles against their electrochemical gradient (Lüttge and Ball, 1979). A primary active transport of $Hmal^{1-}$ is considered unlikely. One of the reasons is the very strict $2\ H^+ : 1$ malate stoichiometry pertaining to the day–night changes in the CAM rhythm. If $Hmal^{1-}$ were transported actively with a cation following passively, attempts to design experimental conditions where protons at least were partially replaced by other cations should not have failed (Lüttge *et al.*, 1975; Lüttge and Ball, 1980).

Hence, we assume that the primary active tranport involved in the vacuolar malic acid accumulation of CAM is H^+ pumping and that divalent malate anions (mal^{2-}) follow the electrical gradient established in this way.

11.2.2 Active H^+ transport into the vacuole

Assuming a constant cytoplasmic pH of 7.5 and a constant electrical potential at the tonoplast of *Kalanchoë* cells during the CAM rhythm, the pH scale, i.e. the lower *x*-axis in Fig. 11.2, where malate levels were originally plotted versus vacuolar pH, can be replaced by a scale giving the proton–electrochemical difference, $\Delta\bar{\mu}_{H^+}$, obtained for the corresponding vacuolar pH values as follows (upper *x*-axis in Fig. 11.2):

$$\Delta\bar{\mu}_{H^+} = RT\cdot\ln\frac{10^{-\mathrm{pH\ vacuole}}}{10^{-\mathrm{pH\ cytoplasm}}} + F\Delta\psi \tag{11.2}$$

The energy content of $\Delta\bar{\mu}_{H^+}$ (kJ mol^{-1}) can be compared with the free energy of ATP hydrolysis, ΔG_{ATP}, in the cytosol:

$$\Delta G_{ATP} = \Delta G'_{0_{ATP}} + RT \cdot \ln \frac{[ADP]\,[P_i]}{[ATP]} \qquad (11.3)$$

where $\Delta G_{0_{ATP}}$ is the Gibbs free energy ($\simeq -33.5$ kJ mol^{-1}: Guynn and Veech, 1973) and [ADP] [P_i] / [ATP] is the mass action ratio for ATP hydrolysis. The apparent number of protons pumped per ATP hydrolysed, n, near equilibrium is given by:

$$\Delta\bar{\mu}_{H^+} = \frac{\Delta G_{ATP}}{n} \qquad (11.4)$$

This assumes that coupling is 100 % (see Smith *et al.*, 1982). As pointed out by Rottenberg (1979), the degree of coupling in energy-transducing membranes may be somewhat lower but is at least 85%. Table 11.1 gives data for *K. daigremontiana* and *K. tubiflora*. The ΔG_{ATP} in the cytoplasm of from ~ 50 to ~ 54 kJ mol^{-1} agrees with values from the literature on other systems (Wilson *et al.*, 1974; Veech *et al.*, 1979; Raven, 1984).

Table 11.1 also gives values for n. It is thermodynamically possible that at the beginning of the dark period 3 (*K. tubiflora*) or even 5 (*K. daigremontiana*) protons are transported per ATP hydrolysed. However, the more important information is contained in Fig. 11.2. The lines for the

Table 11.1 Comparison of the free energy available from ATP hydrolysis (ΔG_{ATP}) with the H$^+$ electrochemical potential difference across the tonoplast ($\Delta\bar{\mu}_{H^+}$) at different times of the CAM rhythm in *K. daigremontiana* and *K. tubiflora*†

	K. daigremontiana			***K. tubiflora***	
	beginning	**middle**	**end**	**beginning**	**end**
	of dark period			**of dark period**	
$\Delta G_{0_{ATP}}$ (kJ mol^{-1})	−33.5	−33.5	−33.5	−33.5	−33.5
‡$RT \cdot \ln \frac{[ADP][P_i]}{[ATP]}$ (kJ mol^{-1})	−20.0	−22.4	−22.8	−19.6	−20.0
ΔG_{ATP} (kJ mol^{-1})	−53.5	−55.9	−56.3	−53.1	−53.5
$\Delta\bar{\mu}_{H^+}$ (kJ mol^{-1})	10.8	22.4	25.0	16.2	26.0
$n = -\Delta G_{ATP}/\Delta\bar{\mu}_{H^+}$	5.0	2.5	2.3	3.3	2.1

† Smith *et al.* (1982); Lüttge and Smith (1984b); Pistelli and Lüttge (1985).
‡ To calculate the mass action ratio, [ADP] [P_i] / [ATP], concentration of ATP, ADP and P_i in the tissue were determined, and it was assumed that no adenine nucleotides were contained in the vacuoles (Stitt *et al.*, 1980, 1982). It was also assumed that P_i was equally distributed between cytoplasm and vacuole. In contrast to adenine nucleotide levels (see Sect. 2.2), P_i concentration did not change during the day–night rhythm of CAM and was ~0.5 mmol kg^{-1} fresh weight in *K. daigremontiana* and ~1 mmol kg^{-1} fresh weight in *K. tubiflora*.

transitions of whole numbers of n are drawn in this figure using average values of ΔG_{ATP} (55 and 53 kJ mol^{-1} for *K. daigremontiana* and *K. tubiflora*, respectively). It is clear that a stoichiometry of > 2 H^+ per ATP ($n \geqslant 3$) in both species would only be thermodynamically possible for a malic acid accumulation of about 20 mol m^{-3}, i.e. only in the very early stages of the dark period, and certainly plays no important role in the malic acid rhythms observed. Direct measurements with isolated vacuoles of *Beta vulgaris* have shown that the tonoplast ATPase in this species functions as a 2 H^+-ATPase (Bennett and Spanswick, 1984).

Fig. 11.3 allows estimation of the energy budget for malic acid accumulation of CAM during the dark period. The PEP required as the acceptor for nocturnal CO_2 fixation is generally thought to arise from glycolysis (Pucher *et al.*, 1949; Sutton, 1975). It becomes obvious from Fig. 11.3 that, starting with free hexose, the ATP requirement for malic acid accumulation would be the same if PEP were exclusively synthesized via glycolysis and the oxidative pentose–phosphate pathway, respectively, although the latter would need 20 % more hexose to form the same amount of PEP. Assuming a $2H^+$-ATPase mechanism for the transport step a the tonoplast, the overall ATP requirement would be 1 ATP : 1 malic acid accumulated in the vacuole in either case.

For *K. tubiflora*, manometric measurements of respiratory O_2 uptake during the dark period gave rates of $\sim$ 1 μmol O_2 h^{-1} g^{-1} fresh weight (Lüttge *et al.*, 1981), corresponding to an approximate rate of oxidative phosphorylation producing 6 μmol ATP h^{-1} g^{-1} fresh weight. This is only about half of the ATP requirement of the observed accumulation rate of malic acid in this plant if free hexose were the source of PEP and the pump is a $2H^+$-ATPase (Lüttge *et al.*, 1981). It should be noted that it would be only one-quarter of the ATP requirement if the pump were a 1 H^+-ATPase and that a transport of more than 2 H^+ per ATP is ruled out thermodynamically (see above and Fig. 11.2).

The discrepancy can be resolved if PEP is derived from phosphorolysis of starch or soluble glucans, circumventing the ATP requirement of hexokinase. The remaining ATP requirement of the whole pathway including the transport step then would be 0.4 ATP : malic acid with the oxidative pentose-phosphate cycle and 0.5 ATP : malic acid with glycolysis. This could just be supplied by the observed rates of respiration. Further, during nocturnal malate synthesis in *Kalanchoë* and also in *Sedum*, PEP indeed is not derived from free hexose but exclusively from starch or soluble glucans (Vickery, 1952, 1957; Sutton, 1975; Kenyon *et al.*, 1985).

One consequence of these observations and calculations is that during the dark period, energy turnover almost exclusively serves malic acid accumulation. This should not elicit undue concern, because quantitatively malic acid accumulation in CAM is indeed a truly dominating process. The plants can use photosynthetically active radiation (PAR) during the light

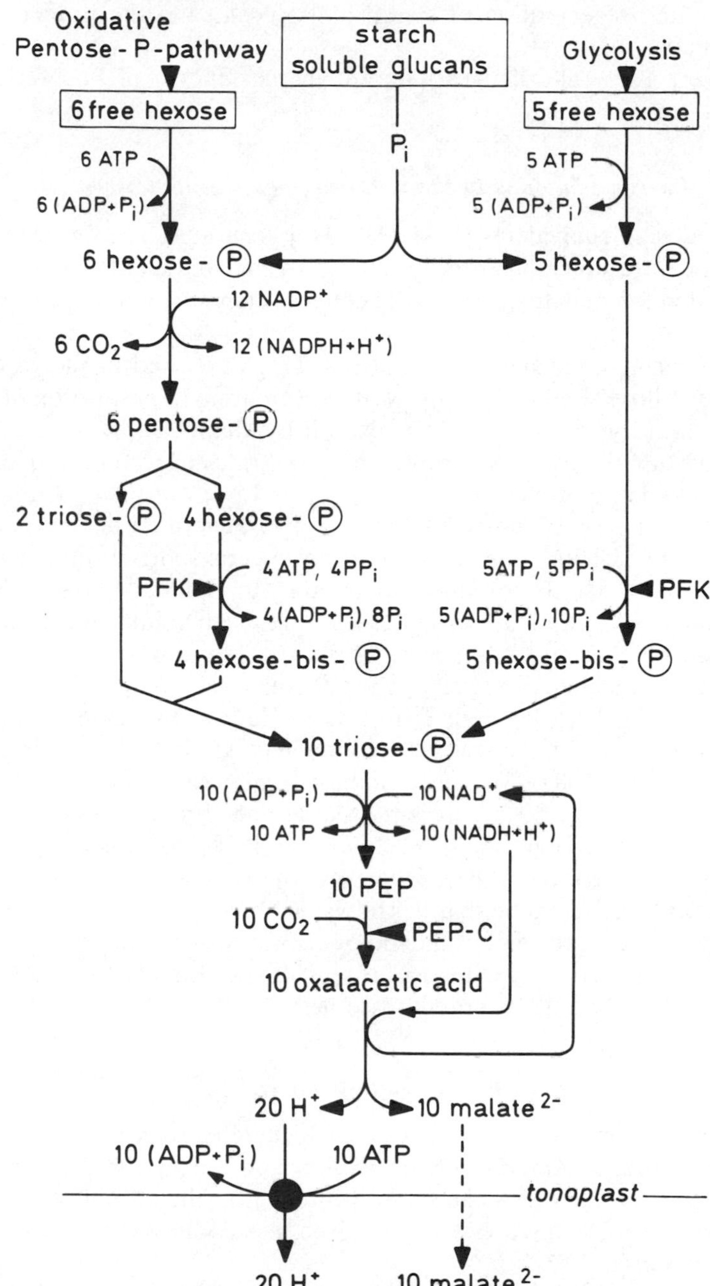

Fig. 11.3 Reaction sequences generating PEP and malic acid. The scheme was drawn for the purpose of assessing the energy budget of nocturnal malic acid accumulation of CAM, i.e. turnover of adenylate phosphates, redox states of pyridine nucleotides and involvement of PP_i. Other details were omitted.

period for the regeneration of starch and glucans and the production of new biomass.

However, a few alternatives inherent in the scheme of Fig. 11.3 need comment.

11.2.1.1 Energy contained in the reduction equivalents formed

1. The reducing equivalents [NADH + H^+] generated by triosephosphate reductase are stoichiometrically utilized by malate dehydrogenase in the formation of malate from oxaloacetate. They do not appear in the balance.
2. The reducing equivalents [NADPH + H^+] generated in the oxidative pentose-phosphate cycle, if this were contributing to production of PEP, could give rise to ATP via the respiratory chain. But this would lead to an O_2 uptake contained in the measured rates of respiration, and the balance would not change.
3. The reducing equivalents [NADPH + H^+] generated in the oxidative pentose–phosphate cycle could drive proton transport into the vacuoles directly via an oxidoreductase in the tonoplast. Indeed, proton extruding NAD(P)H-oxidoreductases have been found in the plasmalemma of higher plants (see Chs. 1 and 3). They appear to be primarily involved in iron acquisition. Rates are low (from 1 to 5 μmol h^{-1} g^{-1} fresh weight) unless induced by iron deficiency (28 μmol h^{-1} g^{-1} fresh weight) (Römheld *et al.*, 1984; review in Lüttge and Clarkson, 1985). At the tonoplast, oxidoreductase activities have also been assumed to be present (Matile and Wiemken, 1967; Matile, 1968; Leigh and Branton, 1976; Saunders, 1979; Poole *et al.*, 1984; see Lüttge and Clarkson, 1985). However, the only activity so far characterized in some detail is that of the vacuole-like lutoids of *Hevea* latex, and in this case protons are moved out of the vesicles and not into them (Chrestin, 1984). It remains to be seen if the tonoplast of CAM cells possesses NAD(P)H oxidoreductase activity.

11.2.1.2 Energy of inorganic pyrophosphate (PP_i)

1. Proton pumping. In a few species of higher plants a vacuolar PP_i-ase has been found (Karlsson, 1975; Walker and Leigh, 1981; Wagner and Mulready, 1983), which can pump protons into the vacuoles (Churchill and Sze, 1983; Bennett *et al.*, 1984; Rea and Poole, 1985; see Ch. 3 for a full description).
2. Phosphofructokinase (PFK). PP_i-dependent PFK activity has been detected in several higher plant species, including CAM plants (Carnal and Black, 1983; Kruger *et al.*, 1983; Kombrink *et al.*, 1984; Kowalczyk *et al.*, 1984; Smyth *et al.*, 1984; Wu *et al.*, 1984). It is evident that with an H^+ pumping tonoplast PP_i-ase and with PP_i–PFK reactions, much

ATP would be saved during the dark period if PAR were used to accumulate PP_i or polyphosphate during the light period. This needs to be analyzed in CAM plants. The evidence so far available for other plant species shows that PP_i levels are very low, i.e. in the range of cofactor concentrations (from 5 to 40 mmol m^{-3}) (Edwards *et al.*, 1984; Smyth and Black, 1984). Thus it appears unlikely that this possibility of energy storage during the light period is realized.

Two additional observations require interpretation. First, in *K. daigremontiana* and *K. tubiflora*, ATP levels increase during the dark period (Smith *et al.*, 1982; Pistelli and Lüttge, 1985). In *K. daigremontiana*, ADP and AMP were shown to decrease and the 'adenylate energy charge', AEC, as given by

$$AEC = \frac{1}{2} \cdot \frac{[ADP] + 2\,[ATP]}{[AMP] + [ADP] + [ATP]} \qquad (11.5)$$

increases from 0.61 ± 0.08 ($n = 7$) at the end of the light period to 0.86 $\pm$ 0.11 ($n = 7$) (errors are SD) at the end of the dark period (Pistelli and Lüttge, 1985). This intriguing observation has been partially explained by a decrease of PEP levels during the dark period (Cockburn and McAulay, 1977; Pierre and Queiroz, 1979), i.e. a shift of the 'general-energy charge' towards adenylate. In view of the tight energy budget during the dark period it remains unexplained, however, why the adenylate energy charge should be so high at the end of the dark period. It needs to be shown whether this is typical of CAM and how other energy-rich compounds, particularly PEP and PP_i, behave quantitatively.

The second observation which requires interpretation is that, in contrast to *Kalanchoë* and *Sedum*, in the bromeliad *Ananas* free hexose is the predominant precursor for PEP during the dark period (Kenyon *et al.*, 1985). Hexose is stored in the vacuoles. The energy for this transport process as for hexose synthesis can be provided by PAR. A nocturnal energy budget is not available for *Ananas*. If the thermodynamic limits are similar to those in *Kalanchoë*, respiration in *Ananas* must provide a higher proportion of the energy needed for nocturnal malic acid accumulation. It would be extraordinarily interesting to know if this were also the case in other CAM bromeliads. The bromeliads are a large neotropical family with many terrestrial and particularly epiphytic CAM species. They show the highest rates of nocturnal malic acid accumulation observed so far (see Sect. 11.1.2). Observations in the field show that much respiratory CO_2 is recycled via PEP carboxylase into malate at mean night temperatures between 19.2 and 26.2°C (Griffiths *et al.*, 1986). Owing to the high night temperatures, respiration presumably is also rather high.

In conclusion, it emerges from this section that CAM provides a unique example where an active transport process every night dominates the entire energy metabolism of the cells. Thermodynamic considerations, however, do not allow conclusions about mechanisms, and different approaches are needed for the elucidation of the molecular basis of active malic acid transport into the vacuoles (Sects. 11.3 and 11.4).

11.2.3 Passive malic acid efflux from the vacuole

Malic acid efflux from the vacuoles of CAM plants is a passive process (Lüttge *et al.*, 1975; Lüttge and Ball, 1977, 1979; Marigo *et al.*, 1983). The rates of total malate efflux from the vacuoles of *K. daigremontiana* cells were determined with tissue slices suspended in an aqueous medium and related to the total malate concentration in the vacuoles (Fig. 11.4). Similar maximum rates of total malate efflux from the vacuoles were obtained from measurements of malate utilization during the light period *in vivo* (Lüttge and Smith, 1984b). With the relations between vacuolar pH, total malate concentration and ionic forms given in Fig. 11.2, several possibilities can be assessed. Considering the possible ionic forms, viz. mal^{2-}, $Hmal^{1-}$ and H_2mal^0, and assuming in turn that malate efflux occurs exclusively as one of these species, the putative fluxes can be plotted versus the vacuolar concentration of the respective species. A straight line, which is expected for a process of passive diffusion across the lipid phase of the

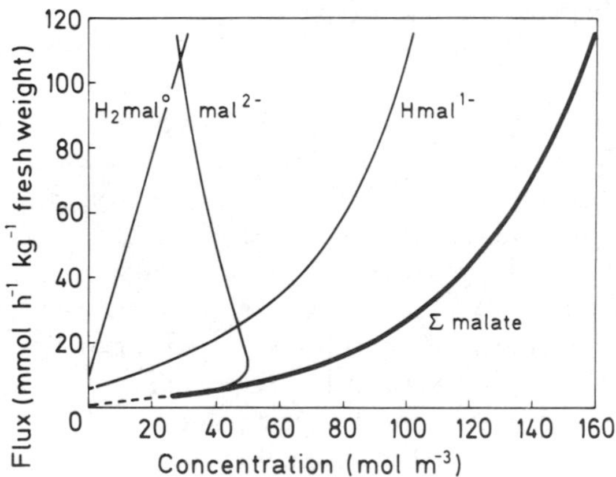

Fig. 11.4 Malate efflux from 2 mm wide leaf slices of *K. daigremontiana* corrected for extracellular diffusion limitation plotted versus total malate concentration (Σmalate). The curve of Σmalate was obtained by calculating the linear regression line between log of rates of efflux and concentrations (Lüttge and Ball, 1977) and replotting the line on a linear scale. The correction factor for extracellular diffusion limitation was 1.8 (Lüttge and Smith, 1984b). The lines for mal^{2-}, $Hmal^{1-}$ and H_2mal^o give the rates of efflux for each species plotted versus the concentrations of the divalent malate anion ($c_{2\theta}$), the univalent malate anion ($c_{1\theta}$) and the non-dissociated acid (c_o) corresponding to each total malate concentration, i.e. they reflect the concentration dependence of efflux if efflux were exclusively in the form of mal^{2-}, $Hmal^{1-}$ or H_2mal^o, respectively (Lüttge and Smith, 1984b).

membrane, is only obtained for the undissociated acid, H_2mal^0. Moreover, the slope of the log–log plot relating efflux to the tissue concentration of H_2mal^0 was close to 1.0, which is required of passive diffusion (Lüttge and Smith, 1984b).

Fick's first law:

$$P = J \cdot \Delta c \qquad (11.6)$$

and the Goldman–Hodgkin–Katz constant field equation (see Ch. 1):

$$P = \frac{J}{c} \cdot \frac{RT}{zF\Delta\psi} \cdot \frac{1 - e^{zF\Delta\psi/RT}}{e^{zF\Delta\psi/RT}} \qquad (11.7)$$

can be used to calculate permeability coefficients for H_2mal^0 and the electrically charged species $Hmal^{1}$ and mal^{2-}, respectively. (J is the flux rate, P the permeability coefficient, c the concentration, Δc the concentration difference.)

The permeability coefficient of H_2mal^0 obtained was $0.94 \cdot 10^{-8}$ m s^{-1} (submerged leaf slices) to $2.22 \cdot 10^{-8}$ m s^{-1} (malate utilization *in vivo*). A similar value of $4 \cdot 10^{-8}$ m s^{-1} was estimated from Collander plots on the basis of the partition coefficient of malic acid between ether and water (Lüttge and Smith, 1984b). Conversely, the calculated permeability coefficients required for $Hmal^{1-}$ and mal^{2-} were too large by orders of magnitude as compared with literature values. In other words, the assumption that efflux is exclusively by lipid permeation of $Hmal^{1-}$ or mal^{2-} gives unreasonably large permeability coefficients, whereas a reasonable value is obtained for H_2mal^0.

This analysis suggests that, at high tissue malic acid concentrations, efflux of malate from the vacuole occurs predominantly by passive, non-catalyzed diffusion of the undissociated acid, H_2mal^0. However, it can be seen from Fig. 11.4 that at low tissue malic acid concentrations (≤ 65 mol m^{-3}), H_2mal^0 is no longer present in significant amounts. Malate efflux in this range must therefore be in the form of $Hmal^{1-}$ and/or mal^{2-}. These anionic species also cross the tonoplast passively, but their transport is presumably carrier-mediated (cf. Buser-Suter *et al.*, 1982).

Seen in relation to the day–night rhythm of CAM, the efflux of H_2mal^0 from the vacuole will only be important at the beginning of the light period while the tissue malic acid concentrations are still high. It may even be possible to interpret the concept of 'gates' controlling malate transport during circadian oscillations (Wilkins, 1983, 1984) in terms of the concentrations of H_2mal^0 at different stages of the cycle. At any rate, the back-flux of H_2mal^0 into the cytoplasm constitutes a potential 'leak' against which the proton pump must work during the dark period to bring about malic acid accumulation. In addition to thermodynamic restrictions on the activity of the proton pump (Sect. 11.2.2), the amounts of malic acid accumulated during the dark period are thus also likely to be limited by the passive permeability of the tonoplast to H_2mal^0.

11.3 Vacuolar ATPase of CAM plants

11.3.1 Isolation of vacuoles, demonstration of vacuolar ATPase activity and its characterization

A prerequisite for more direct studies on malic acid transport at the tonoplast has been the development of methods for isolating intact vacuoles. The first methods to be applied to CAM plants were based on lysis of isolated protoplasts induced either by diethylaminoethyl (DEAE)–dextran (Buser and Matile, 1977) or EGTA (Kringstad *et al.*, 1980). In our laboratory we have developed a method for *K. daigremontiana* based on that of Boudet *et al.* (1981), full details of which are given in Smith *et al.* (1984b). Briefly, small sections of leaf chlorenchyma tissue are incubated for 45 min in a medium containing 0.450 % (w/v) cellulase 'Onozuka' RS and 0.025 % (w/v) pectolyase Y-23. The protoplasts thereby released are washed and then loaded on to a discontinuous density gradient containing four layers of Ficoll 400. The first layer (starting from the top) contains 4.0 kg m^{-3} DEAE–dextran and 5.0 mol m^{-3} EDTA to lyse the protoplasts, and the second contains 4.0 kg m^{-3} dextran sulphate to neutralize the polybase. Centrifugation at 1200 *g* for 10 min releases vacuoles from the protoplasts; these can be collected at the final interface in the gradient. The advantages of this method are that it is relatively quick and that lysis of the intact protoplasts to yield vacuoles is both rapid and controlled.

One problem that has not helped progress in this field is the limitation on the number of vacuoles that can be prepared from CAM plants. Based on the number of leaf cells in the starting material, the highest yields of isolated vacuoles attained with the present method are about 5 % (Smith *et al.*, 1984b). This is not so much lower than the yields obtained with other species, but the starting material is generally not available in commercial quantities. Further, the isolated vacuoles from CAM plants are very fragile, presumably on account of their large size (~100 μm diameter). Nevertheless, the yields have proved sufficient for enzymic studies and a vacuolar ATPase of high catalytic activity has been identified. Measurements of cytoplasmic marker enzymes have shown that the ATPase activity cannot be attributed to contamination of the vacuole fraction (Smith *et al.*, 1984b).

The most important properties of the vacuolar ATPase activity from *K. daigremontiana* are summarized in the left-hand part of Table 11.2. All assays are carried out in the presence of 100 mmol m^{-3} ammonium molybdate to suppress unspecific phosphatase activity (Smith *et al.*, 1984a). Using the method of Kringstad *et al.* (1980), Aoki and Nishida (1984) have also isolated vacuoles from *K. daigremontiana* and characterized an ATPase activity with essentially identical properties. The most diagnostic features of the vacuolar ATPase activity compared with other ATPases are its relatively high pH optimum (pH 8.0), its insensitivity to cations and its

Table 11.2 Properties of the vacuolar ATPase activity and H^+-transport activity in membrane vesicles from *Kalanchoë daigremontiana*

Property	ATPase activity	H^+-transport activity
Nucleotide specificity	ATP > > ADP, GTP, ITP > UTP, AMP, PP_i	ATP only; other nucleotides ineffective
Divalent cation requirement	Mg ≃ Mn >> Ca; Co; Zn no effect, Cu inhibitory	Mg (others not tested)
pH optimum	8.0	–
Apparent K_m (Mg-ATP^{2-}), pH 8.0	0.31 mol m^{-3}	–
Ion sensitivity: cations	No effect	No effect
anions	*stimulation*: Cl^- > HCO_3^-, malate^{2-}	Malate^{2-} > Cl^- > Mes^- iminodiacetate^{2-} > SO_4^{2-}
	no effect: Mes, iminodiacetate^{2-}, SO_4^-	
	inhibition: NO_3^- > citrate^{3-}	NO_3^-
Inhibitor sensitivity	Diethylstilbestrol (DES)	DES
	N, *N'*-dicyclohexylcarbodiimide (DCCD)	DCCD
	Tributyltin	Tributyltin
		4,4'-Diisothiocyano-2,2'-stilbenesulphonic acid (DIDS)

ATPase activity was measured by following the release of P_i from ATP colorimetrically (Jochem *et al.*, 1984; Smith *et al.*, 1984a,b), whereas H^+-transport activity was measured using the technique of quinacrine–fluorescence quenching (J. A. C. Smith, P. Jochem & U. Lüttge, unpublished observations; cf. Bennett & Spanswick, 1983).

marked sensitivity to anions (Smith *et al.*, 1984a). In desalted extracts, both Cl^- and malate^{2-} at 50 mol m^{-3} stimulated the vacuolar ATPasc activity about 40 %, whereas NO_3^- at 50 mol m^{-3} inhibited the activity about 40 % (Jochem *et al.*, 1984). Vanadate, azide and oligomycin are without effect on the vacuolar ATPase activity, which provides a useful means of distinguishing this activity from that of the plasmalemma and mitochondrial ATPases (Smith *et al.*, 1984a, b). The marked sensitivity of the vacuolar ATPase activity to DCCD and tributyltin in particular provides strong circumstantial evidence that the enzyme might function as an H^+-translocating ATPase (see Smith *et al.*, 1984a).

Overall, the properties of the vacuolar ATPase activity from the leaf chlorenchyma cells of *K. daigremontiana* show no major qualitative differences from those that have now been described for vacuolar ATPases from a variety of tissues and organs from other plants (see Ch. 3 and Leigh, 1983; Sze, 1984, 1985; Marin, 1985). We can now turn to its physiological importance in CAM plants, which is directly concerned with its proposed function in energizing the vacuolar accumulation of malic acid during the day–night rhythm.

11.3.2 Physiological relevance of the vacuolar ATPase activity for the function of CAM in vivo

Table 11.3 shows that the mean ATPase activity detected in preparations of isolated vacuoles of *K. daigremontiana* rather closely corresponds to the highest rates of malate accumulation observed in the intact plants. This means that, with a stoichiometry of 1 ATP hydrolysed : 2 H^+ pumped into the vacuole : 1 malate^{2-} accumulated, the ATPase activity obtained with vacuoles *in vitro* is of the right order for explaining the activity required for the physiological function *in vivo*. There are a number of plant species that are intermediate between C_3 photosynthesis and CAM, and in which CAM can be induced, e.g. by drought in the Bromeliaceae *Guzmania monostachia* (Medina *et al.*, 1977; Griffiths and Smith, 1983; Smith *et al.*, 1985), by photoperiod in the Crassulaceae *Kalanchoë blossfeldiana cv*. Tom Thumb (Queiroz, 1965; Brulfert *et al.*, 1972, 1975) and by drought or salinity in the Aizoaceae *Mesembryanthemum crystallinum* (Winter and von Willert, 1972; Winter, 1979; Winter and Lüttge, 1979). Studies with the two latter species showed that vacuolar ATPase activity increased

Table 11.3 Rates of nocturnal malate accumulation and ATPase activity of isolated vacuoles in *K. daigremontiana*† and in *M. crystallinum* in the C_3 and the CAM state‡

Plants	**Malate accumulation**	**Mean vacuolar ATPase activity**
	(nmol m^{-2} s^{-1})	
K. daigremontiana, CAM	62	46
M. crystallinum, C_3	0	9
M. crystallinum, CAM	47	47

† Smith *et al.* (1984a).
‡ Struve *et al.* (1985).
To obtain the numbers in this table, nocturnal malic acid accumulation (per unit of fresh weight) was assumed to be at constant rate and exclusively in the vacuoles. Rates were related to tonoplast area on the basis of measurements of cell sizes and cell numbers per unit fresh weight. Size and density of vacuoles were determined in the suspensions used for tests of ATPase activities at 25, 30 or 37 °C. A Q_{10} of 2.7 (D'Auzac, 1977) was assumed, and the measured activities were reduced by this factor to the activities pertaining at 16 °C (the temperature during the dark period).

considerably during the transition from C_3 photosynthesis to CAM (Struve *et al.*, 1985).

More detailed analyses were performed with *M. crystallinum*. Vacuolar ATPase activity was determined by differential inhibition in homogenates of isolated protoplasts. Aliquots were used to inhibit plasmalemma ATPase with vanadate, mitochondrial ATPase with azide and chloroplast ATPase with phlorizin. All aliquots contained molybdate to inhibit unspecific acid phosphatases. Total ATPase activity was obtained with molybdate only and the vacuolar ATPase activity calculated as the difference between total activity and the sum of the activities of non-vacuolar ATPases. It was shown that in the CAM state vacuolar ATPase activity was 5.1 ± 1.7 times (SD, n = 5) that observed in the C_3 state (Struve *et al.*, 1985).

The approach of differential inhibition was also used in a number of experiments of the kind illustrated in Fig. 11.5. *M. crystallinum* plants in soil culture were subjected to salinity stress by increasing the NaCl concentration in the irrigation solution in steps of 50 mol m^{-3} per day until 400 mol m^{-3} was reached, and the plants then continued to be irrigated with this concentration. It can be seen that on the day on which 400 mol m^{-3} was reached (day 0 on the abscissa of Fig. 11.5) Na^+ levels in the tissue were already considerably increased above those of the controls. But vacuolar ATPase activity was still as low as in the controls; CAM was not yet induced, as indicated by the absence of day–night

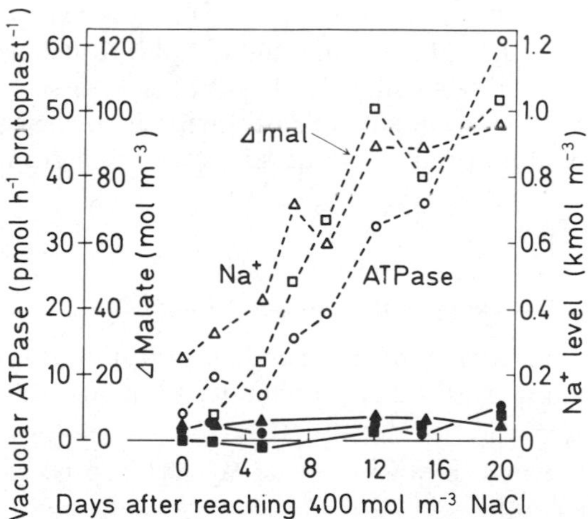

Fig. 11.5 Increase of Na^+ levels (triangles) and induction of CAM (day–night changes of malate levels, Δmal, squares) with concomitant increase in vacuolar ATPase activity (circles) in *M. crystallinum*. Open symbols: plants were treated with daily increments of 50 mol m^{-3} NaCl in the irrigation solution; the abscissa gives days after 400 mol m^{-3} NaCl was reached and subsequently kept at this level. Closed symbols: no NaCl in the irrigation water.

changes of malate levels. ΔMalate began to become discernible after 4 d in 400 mol m^{-3} NaCl and increased up to day 12 accompanied by an increase in vacuolar ATPase activity. This sequence of events, with salt accumulation commencing much earlier than expression of CAM and increase of vacuolar ATPase activity, shows that the latter is physiologically associated with CAM and not with NaCl accumulation (Struve *et al.*, 1985).

Vacuoles were also prepared from *M. crystallinum* in the C_3 and the CAM state. As shown in Table 11.3, vacuolar ATPase activity in the C_3 state is not sufficient to explain the rates of malic acid accumulation observed in the CAM state. The observed increase is necessary for the operation of CAM. Indeed, in the CAM state malic acid accumulation *in vivo* and vacuolar ATPase activity *in vitro* closely match the stoichiometry of 1 malic acid : 1 ATP. This is also an independent confirmation that the transport process at the tonoplast should operate with such a stoichiometry.

11.4 H^+-transport into isolated vacuoles and tonoplast vesicles of CAM plants

11.4.1 Measurements with microelectrodes

Glass microsalt bridges and micro-pH-electrodes were inserted into isolated vacuoles of *K. tubiflora* to measure $\Delta\psi$ between the suspension medium and the vacuole interior and internal pH, respectively. The results are given in Table 11.4. Addition of Mg-ATP to the suspension medium as a substrate for the vacuolar ATPase increases the electrical potential difference by about 9 mV and acidifies the vacuole interior by about 1 pH unit.

11.4.2 Fluorescence quenching in tonoplast vesicles

Another means of studying ATP-dependent proton transport is by the use of fluorescent amine dyes such as 9-aminoacridine and quinacrine. As the

Table 11.4 Microelectrode measurements of Mg-ATP dependent H^+-transport into isolated vacuoles of *K. tubiflora*

Measurement	No Mg-ATP	7 mol m^{-3} Mg-ATP
$\Delta\psi$ (mV)	+9.2 ± 2.8 (13)	+ 17.8 ± 3.7 (12)
Vacuolar pH	5.4 ± 0.2 (11)	4.3 ± 0.4 (12)
$\Delta\bar{\mu}_{H^+}$ (kJ mol^{-1})	10.4	15.7

The medium was buffered at pH 7.4. Errors are SD (Jochem *et al.*, 1984).

uncharged but not the charged form of the amines can pass rapidly through membranes, the dye will equilibrate in such a way that its concentration inside a membrane-bound compartment is determined by the pH difference between the inside and outside of the compartment. The essence of the technique is that the fluorescence of the dye is completely quenched when the dye is located within the membrane-bound compartment (Deamer *et al.*, 1972; Schuldiner *et al.*, 1972; see also Ch. 3).

Using the quinacrine technique as described by Bennett and Spanswick (1983), we have carried out studies using a partially purified preparation of tonoplast vesicles (J. A. C. Smith, P. Jochem and U. Lüttge, unpublished observations). After isolation in the normal way, protoplasts of *K. daigremontiana* are ruptured by passage through a steel canula. Centrifugation of this protoplast homogenate for 5 min at 1000 g pellets 99.7 % of the chlorophyll; about 20 % of the total ATPase activity in the protoplast remains in the supernatant, but of this about 40 % is vacuolar ATPase (measured by its nitrate sensitivity). This supernatant contains large quantities of sealed vesicles that are very amenable to fluorescence–quenching studies.

Figure 11.6 shows an example of the spectrofluorometric traces recorded in these studies. After the initial equilibration of the dye with the vesicle preparation, addition of 3.0 mol m^{-3} Mg/ATP causes a rapid fluorescence quenching; the highest quench rates observed are around 50 % in 10 min. This quenching indicates an acidification of the vesicle interior, which causes more of the basic dye to be trapped within the membrane compartment. Addition of the cation ionophore gramicidin immediately restores the fluorescence to its previous level in the absence of Mg/ATP, showing that acidification is dependent on the membranes having a relatively low permeability to protons. High concentrations of NH_4Cl, which would also

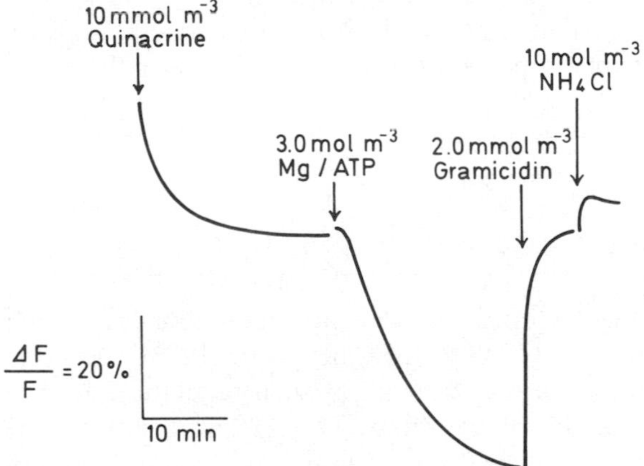

Fig. 11.6 Time course of quinacrine–fluorescence quenching in a membrane–vesicle fraction from chlorenchyma cells of *Kalanchoë daigremontiana*.

collapse any pH difference across the membrane, have almost no further effect.

Some of the characteristics of the H^+-transport activity observed in this vesicle preparation are summarized in the right-hand part of Table 11.2. (Transport activity is measured as the initial rate of fluorescence quenching on addition of substrate.) H^+ transport is extremely substrate-specific. No other nucleotide supports transport at more than 5 % of the rate observed with Mg/ATP. The slight sensitivity of H^+ transport to vanadate, azide and oligomycin indicates that about 25 % of the activity is of plasmalemma and mitochondrial origin. But H^+ transport is extremely sensitive to NO_3^-, suggesting that most of the activity is associated with tonoplast vesicles; it is also almost completely eliminated by low (< 30 mmol m^{-3}) concentrations of DES, DCCD and tributyltin. Furthermore, assayed in the presence of vanadate and azide, the transport activity shows a strong anion sensitivity. To a large extent the anion effects parallel those on the ATPase activity (Table 11.2), with the important exception that malate^{2-} is three times more effective than Cl^- in stimulating H^+ transport. Malate itself may therefore be directly involved in modulating the activity of the proton pump during the CAM rhythm.

Membranes in the supernatant of the 1000 *g* centrifugation described above have been further fractionated by centrifugation for 120 min at 10^5 *g* on a continuous sucrose gradient (15 % to 60 %). This allowed isolation of a purified tonoplast fraction of *K. daigremontiana*. Studies of ATP hydrolysis and quinacrine quenching with this preparation confirmed that the ATPase and H^+-transport properties described in Sects. 11.3.1 and 11.4.2 (Table 11.2) are attributable to the tonoplast (P. Jochem, personal communication). Similar gradients were prepared for *M. crystallinum*. The ATPase and the H^+-transport activity of tonoplast vesicles were correlated, and both increased about four to six times during the transition from C_3 photosynthesis to CAM (I. Struve, personal communication). This corroborates the results obtained with isolated vacuoles. It also underlines the importance of increased H^+-transport activity after the induction of CAM (see Sect. 11.3.2)

11.5 Regulation

It is evident that cytoplasmic pH and metabolite levels must play a role in the regulation of CAM (Kluge and Ting, 1978; Osmond, 1978). Some of the more conspicuous elements of such regulation are the inhibition of PEP-carboxylase by malate (Sect. 11.2.1) and the stimulation of malate synthesis by increased pH and of malate decarboxylation by lowered pH (Davies, 1973a,b; Smith and Raven, 1979). But certainly the regulation of the entire metabolic system of CAM is more subtle.

The question arising here is to what extent the proton-pumping ATPase at the tonoplast is involved. Its activity is increased during CAM induction in intermediate species (Sect. 11.3.2). It has a rather pronounced pH optimum near pH 8.0 (Table 11.2). It is also stimulated by malate (Table 11.2).

Experiments with equilibration of tissue slices with low concentrations of the weak acid DMO (labelled with ^{14}C) have shown qualitatively that within certain limits cytoplasmic pH changes during the CAM rhythm. Supply of weak acids and bases at larger concentrations to isolated phyllodia of *K. tubiflora* tended to reduce and increase, respectively, the pH of the cytoplasm. This led to an inhibition of daytime malic acid remobilization and nocturnal malic acid accumulation, respectively (Fig. 11.7; Marigo *et al.*,1982, 1983).

Quantitatively, the amplitudes of cytoplasmic pH changes and possible

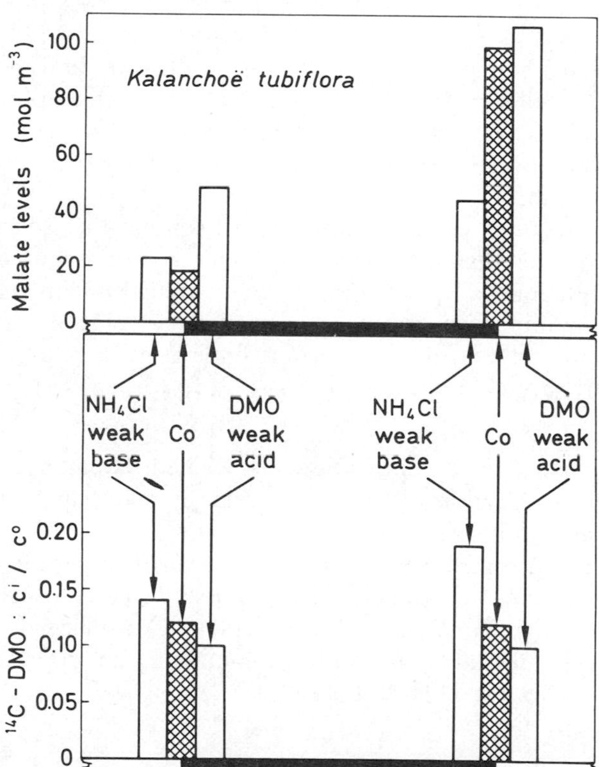

Fig. 11.7 Effects of a weak base (NH_4Cl) and a weak acid (DMO), both at 50 mol m^{-3}, fed to phyllodia of *Kalanchoë tubiflora*, on the day–night rhythm of malate levels. An increase of the ratio of ^{14}C-DMO in the cells to ^{14}C-DMO in the medium (c^i/c^o) indicates more alkaline and a decrease less alkaline or more acid cytoplasmic pH.

changes of malate levels in the cytoplasm during the CAM rhythm are not known. Before these parameters can be measured it will be impossible to evaluate the regulatory role of the tonoplast ATPase.

11.6 Osmotic consequences of malic acid oscillations during CAM

11.6.1 Succulence

CAM plants are always succulent. If not by physiognomic appearance, they are succulent at least at the cellular level with large relative volumes of vacuoles (e.g. *Tillandsia usneoides*, Kluge *et al.*, 1973; Kluge and Ting, 1978).

Attempts to characterize the succulence of CAM plants biophysically have used the basic water relations equation

$$\Psi = P - \pi \qquad (11.8)$$

where Ψ is the water potential, P the turgor pressure and π the osmotic pressure of the cell sap (Passioura, 1982; Nobel, 1983). Experiments have shown that in CAM plants Ψ is generally high, P low and the cell sap diluted, i.e. π is also low. Relative water contents are correspondingly high (Lüttge and Smith, 1984a).

The large day–night changes in malic acid content that characterize the CAM rhythm will obviously have important osmotic consequences for plant water relations. These have now been studied under both field and laboratory conditions for several species of CAM plants. Here we shall summarize some of the most important findings with reference to three species contrasting markedly in their life-forms and morphology (Fig. 11.8):

1. *Aechmea nudicaulis*, an epiphytic leaf-succulent Bromeliaceae in which a considerable fraction of the leaf volume is made up by a parenchymatous water-storage tissue.
2. *Cereus validus*, a stem-succulent Cactaceae with a peripheral chlorenchyma tissue and a central, relatively massive water-storage tissue.
3. *Kalanchoë daigremontiana*, a leaf-succulent Crassulaceae, where cells of the leaves are rather uniform, all containing chloroplasts and large central vacuoles (Steudle *et al.*, 1980).

11.6.2 Day–night changes in plant water relations

Typical day–night changes in gas exchange and plant water relations are shown for the three contrasting CAM species in Fig. 11.9. The epiphytic bromeliad *A. nudicaulis* was studied in its natural habitat in Trinidad,

Fig. 11.8 Three CAM species of markedly contrasting life-form and morphology. Left: *Aechmea nudicaulis* L. Grisebach var. *nudicaulis*, an epiphytic member of the Bromeliaceae growing in its natural habitat in Trinidad. Centre: *Cereus validus* Haworth, a member of the Cactaceae from northern Argentina and Bolivia. Right: *Kalanchoë daigremontiana* Hamet et Perrier de la Bâthie, a member of the Crassulaceae from Madagascar. The latter two were grown in the glasshouse at Darmstadt Botanic Garden.

whereas the other two species were investigated under controlled environmental conditions. Nevertheless, basic similarities can be seen in all three. Gas exchange is predominantly nocturnal, but late afternoon ('phase IV') fixation also occurs. Malate concentration and cell sap π increase in parallel and at an almost constant rate during the night. The highest nocturnal increase in malate concentration (Δmal) yet observed for a CAM plant was in fact recorded for *A. nudicaulis* in Trinidad (see Sect. 11.1.2).

By plotting the nocturnal increases in osmotic pressure against the corresponding changes in malate concentration, we can also consider quantitatively the osmotic effectiveness of the malate synthesized in the CAM rhythm. This has been done in Fig. 11.10 with the data collated from the three different species (osmotic pressure is here given as the measured cell sap osmolality). Figure 11.10 reveals two important features. First, the relationship is not discernibly different for the three contrasting species. And second, the data show a reasonable fit to a line of slope 1.00. In other words, the malate participating in the CAM rhythm is apparently behaving as an ideal solute (in the sense of having an osmotic coefficient of unity: see Tyree and Richter, 1981). The associated protons are of no osmotic consequence, because even at the lowest pH values observed (pH 3.3) the

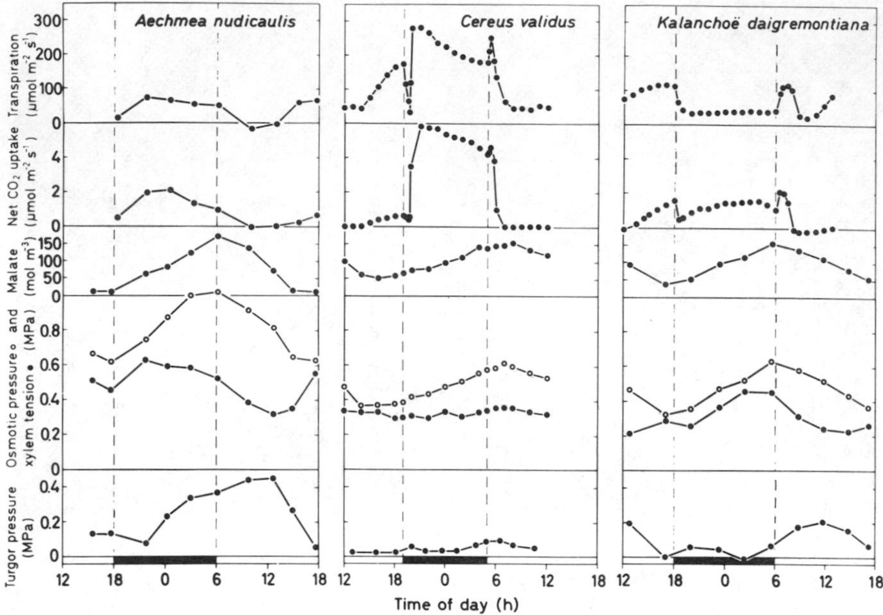

Fig. 11.9 Day–night changes in gas exchange and plant water relations observed in three contrasting species of CAM plants. *Aechmea nudicaulis* studied *in situ* in a forest clearing in Trinidad (data from Fig. 1 of Smith *et al.*, 1986). Malate concentration was calculated from the measured titratable acidity values on the assumption of a stoichiometry of 2 H^+: 1 malate. Values for turgor pressure were derived using the relationship between xylem sap π and leaf malate content found for *K. daigremontiana* (Smith & Lüttge, 1985). *Cereus validus* studied under laboratory conditions (data from Nobel *et al.*, 1984). The gas exchange measurements were made on plants under growth cabinet conditions, whereas the other values refer to plants sampled under glasshouse conditions. Turgor pressure was measured on individual cells in the chlorenchyma using the pressure probe. *Kalanchoë daigremontiana* studied under growth cabinet conditions (data from Smith & Lüttge, 1985). Turgor pressure was calculated after correcting xylem tension for xylem sap π to obtain leaf Ψ.

concentration of free protons is only 0.50 mol m^{-3}. The great majority of the protons are buffered – to a large extent by the accumulating malate itself – and thus make no additional contribution to cell sap π.

The three species do show somewhat different behaviour with respect to the day–night changes in xylem tension as measured with the pressure chamber (Fig. 11.9). In both *A. nudicaulis* and *K. daigremontiana*, xylem tension reaches its maximum value at night and its minimum around midday, reflecting concurrent transpiration rates. The magnitude of this day–night difference is also relatively large, reaching 0.32 MPa in *A. nudicaulis* (Smith *et al.*, 1986) and 0.28 MPa in *K. daigremontiana* (Smith and Lüttge, 1985). In contrast, day–night changes are barely detectable

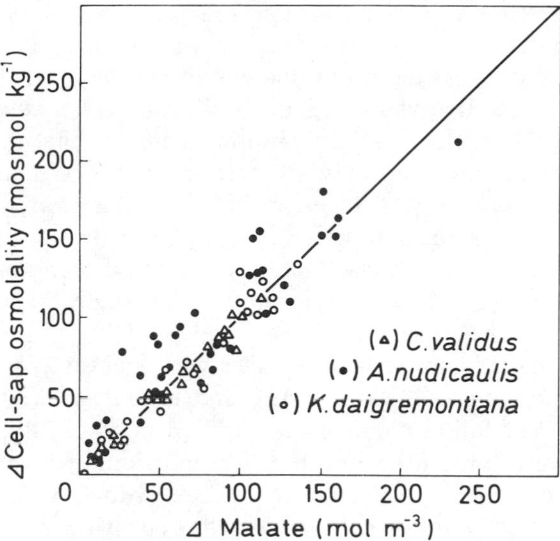

Fig. 11.10 Relationship between changes in cell-sap osmolality and changes in malate concentration during the CAM rhythm relative to values at the start of the dark period. The line indicated has a slope of 1.00. Addition of pure L-malic acid to the sap yields exactly the same relationship (Lüttge & Nobel, 1984; Smith & Lüttge, 1985), showing that the malic acid is solely responsible for the observed changes in osmolality.

in *Cereus validus*, even though this species shows the highest transpiration rates on a unit area basis. The reason is that the pressure chamber technique only provides bulk-averaged values of xylem tension for the whole organ. As the stem-succulent cacti have a very low surface area : volume ratio, this means that the transpiration rates expressed on a unit mass basis are also low.

Finally, we can examine what happens to turgor pressure (P) during the day–night cycle. Two approaches are possible: one is to estimate P from the algebraic difference between Ψ and π (Eq. 11.8), the other is to measure it directly using the pressure probe.

The first requires a correction for xylem sap π, because the pressure chamber technique measures xylem tension and not Ψ. Moreover, xylem sap π itself changes during the CAM rhythm (Smith and Lüttge,1985), so that this relationship must first be worked out for any given set of conditions. This approach has as yet only been used for *K. daigremontiana*, in which P is low during the late afternoon and for much of the night and then increases to a maximum of 0.20 MPa around midday (Fig. 11.9). In effect, the day–night changes in P reflect the balance between changes in cell sap π (as influenced by the cycle of malic acid accumulation and degradation) and Ψ (as influenced by transpiration rate). At the start of the light period the decrease in xylem tension (and increase in Ψ) exceeds in magnitude the decrease in cell sap π, the net result being that *P* increases.

We should emphasize that the terms defined in this way represent bulk-averaged values for the whole organ. In *K. daigremontiana* about 95 % of the leaf mass is made up by the chlorenchyma, so that the values here relate closely to the tissue actually involved in the CAM rhythm. In leaves with more heterogeneous cell types, such as *A. nudicaulis*, the bulk-averaged values are less meaningful. Values of *P* have nevertheless been calculated in Fig. 11.9 for *A. nudicaulis* (using the xylem sap data from *K. daigremontiana*): they show that a similar day–night trend exists, with maximum values of *P* around midday.

However, a better approach for species with markedly different tissues is to use the pressure probe (see Ch. 6 and Hüsken *et al.*, 1978; Steudle *et al.*, 1980). With this technique an oil-filled micropipette connected to a pressure transducer can be inserted into individual cells of a particular tissue. Such experiments have been carried out with *C. validus*, where the *P* values (Fig. 11.9) refer to cells in the chlorenchyma tissue. In contrast to the other two species, the *P* values here more closely follow the day–night changes in cell sap π. This can be attributed to the near constancy of xylem tension (and Ψ) in the relatively massive stems characteristic of cacti.

11.6.3 Water uptake and water storage

The pattern of day–night changes in plant water relations described above shows that the malic acid involved in the CAM rhythm is important not only as the transient storage form for the nocturnally assimilated CO_2. It is also central to the plant's water economy. This is seen particularly in the changes in turgor pressure, which ultimately correspond to alterations in leaf or stem water content. In increasing cell sap π, the nocturnally accumulated malic acid provides a major part of the driving force for the osmotic charging of the tissue 'storage capacitor' with water at the start of the day. This is the time at which stomatal closure causes a marked decrease in xylem tension (and increase in Ψ) in the leaf-succulent CAM species (cf. Fig. 11.9).

Quantitatively, the magnitude of the storage capacitance will be important in determining how water is partitioned between storage in the cells and evaporative loss in transpiration. The cell water storage capacitance, C_c, is given by the equation

$$C_c = V/(\varepsilon + \pi) \qquad (11.9)$$

where *V* is the cell volume and ε the volumetric elastic modulus of the cell wall, which is inversely proportional to elasticity (Dainty, 1976). The value for the chlorenchyma cells of *K. daigremontiana* is around 0.2×10^{-9} m^3

Pa^{-1} kg^{-1} fresh weight (Steudle *et al.*, 1980). (The actual value of C_c varies with turgor pressure, because ε itself is strongly pressure dependent.) For *K. daigremontiana*, this means that the amount of water stored by the cells for the 0.1 to 0.2 MPa increase in turgor pressue (cf. Fig. 11.9) is of the same order as that lost in transpiration during the entire night (Steudle *et al.*, 1980; Lüttge and Smith, 1984a).

One feature of the time course of changes in turgor pressure during the day–night cycle may be of great ecological importance for CAM plants in their natural habitats (Smith and Lüttge, 1985). The main increase in turgor pressure occurs in the latter part of the night and the first part of the day. Some of the most characteristic habitats of leaf-succulent CAM plants are found in Namibia and Madagascar, where coastal fogs and dew formation represent a potentially important source of water for the vegetation. (This is true also for many of the coastal regions of Chile, Peru and Ecuador.) But this water, in so far as it moistens the upper layers of the soil at all, will tend only to be available to the plants in the early hours before the land mass heats up. This is the time at which the CAM plants, which have superficial root systems, also have their lowest water potentials. There may therefore be a close correspondence between the time at which CAM plants can most effectively take up available water and the time at which it tends to be available during the day–night cycle. However, the relative importance of this source of water compared with that from rain-events has not yet been assessed for CAM plants.

11.7 Concepts

The search for the mechanism of malic acid transport at the tonoplast of CAM cells, i.e. the investigation of passive malic acid efflux and particularly the study of vacuolar ATPase is not only motivated by the aim of understanding a transport mechanism *per se*. CAM is considered as a biochemical adaptation to environmental stress due to limited water supply. It is the role of malic acid transport in the physiological function of CAM *in vivo* and under real ecological situations in the field that is at issue.

The present survey shows that thermodynamic approaches, membrane biophysics and biochemistry are required for elucidation of the role of malic acid transport at the tonoplast as an integral part of the biochemical reaction sequence of CAM. This establishes the biochemical basis for nocturnal CO_2 fixation with stomata opened and daytime CO_2 assimilation using PAR with stomata closed, and the secondary beneficial consequences for the water relations of the plants. But at the same time it reveals direct effects on water relations due to the osmotic consequences of malic acid oscillations. So it is quite clear that biochemical and biophysical processes are inseparably integrated in the 'physiological function' of CAM *in vivo*.

However, analyses of the behaviour of plants under real environmental conditions in the field are essential for understanding the 'ecophysiological function' *in situ*. Modulations of the schematic physiological performance observed under laboratory conditions can arise from intraspecific and interspecific variations among CAM species, from differences between habitats and sites, and even from short-term changes of weather conditions. This requires a considerable range of comparative quantitative ecophysiological measurements in the field.

In the studies of the present survey these criteria have been applied to the problem of malic acid accumulation and water relations of CAM plants. In general terms, however, the concept is more a programme than an achievement. It produces the challenge and chance to reintegrate the diverse individual disciplines developing in biology.

References

Aoki, K. & Nishida, K. (1984). ATPase activity associated with vacuoles and tonoplast vesicles isolated from the CAM plant, *Kalanchoë daigremontiana. Physiologia Plantarum* **60**, 21–5.

Bennett, A. B., O'Neill, S. D. & Spanswick, R. M. (1984). H^+-ATPase activity from storage tissue of *Beta vulgaris*. I. Identification and characterization of an anion-sensitive H^+-ATPase. *Plant Physiology* **74**, 538–44.

Bennett, A. B. & Spanswick, R. M. (1983). Optical measurements of $\triangle$pH and $\triangle\psi$ in corn root membrane vesicles. Kinetic analysis of Cl^- effects on a proton-translocating ATPase. *Journal of Membrane Biology* **71**, 95–107.

Bennett, A. B. & Spanswick, R. M. (1984). H^+-ATPase activity from storage tissue of *Beta vulgaris*. II. H^+/ATP stoichiometry of an anion-sensitive H^+-ATPase. *Plant Physiology* **74** 545–8.

Bollig, I. C. & Wilkins, M. B. (1979). Inhibition of the circadian rhythm of CO_2 metabolism in *Bryophyllum* leaves by cycloheximide and dinitrophenol. *Planta* **145**, 105–12.

Boudet, A. M., Canut, H. & Alibert, G. (1981). Isolation and characterization of vacuoles from *Melilotus alba* mesophyll. *Plant Physiology* **68**, 1354–8.

Brulfert, J., Guerrier, D. & Queiroz, O. (1972). Photopériodisme et activité enzymatique: présence en jours long d'un système inhibiteur d'une voie métabolique active en jours courts. *Comptes Rendues de l'Académie Scientifique Paris* **274 D**, 2671–4.

Brulfert, J., Guerrier, D. & Queiroz, O. (1975). Photoperiodism and enzyme rhythms: kinetic characteristics of the photoperiodic induction of crassulacean acid metabolism. *Planta* **125**, 33–44.

Buchanan-Bollig, I. C. & Kluge, M. (1981). Crassulacean acid metabolism (CAM) in *Kalanchoë daigremontiana*: temperature response of phosphoenol-pyruvate (PEP)-carboxylase in relation to allosteric effectors. *Planta* **152**, 181–8.

Buser, C. & Matile, P. (1977). Malic acid in vacuoles isolated from *Bryophyllum* leaf cells. *Zeitschrift für Pflanzenphysiologie* **82**, 462–6.

Buser-Suter, C., Wiemken, A. & Matile, P. (1982). A malic acid permease in isolated vacuoles of a crassulacean acid metabolism (CAM) plant. *Plant Physiology* **69**, 456–9.

Carnal, N. W. & Black, C. C. (1983). Phosphofructokinase activities in photosynthetic organisms. The occurrence of pyrophosphate-dependent 6-phosphofructokinase in plants and algae. *Plant Physiology* **71**, 150–5.

Chrestin, H. (1984). Le compartiment vacuo-lysosomal (les lutoïdes) du latex d'*Hevea brasiliensis*. Son rôle dans le maintien de l'homeostasie et dans les processus de senescence des cellules laticifères. Thèse de Docteur ès Sciences Naturelles, Université des Sciences et Techniques du Languedoc, Montpellier.

Churchill, K. & Sze, H. (1983). Anion-sensitive, H^+-pumping ATPase in membrane vesicles from oat roots. *Plant Physiology* **71**, 610–17.

Cockburn, W. (1981). The evolutionary relationship between stomatal mechanism, crassulacean acid metabolism and C_4 photosynthesis. *Plant, Cell and Environment* **4**, 417–18.

Cockburn, W. & McAulay, A. (1977). Changes in metabolite levels in *Kalanchoë daigremontiana* and regulation of malic acid accumulation in crassulacean acid metabolism. *Plant Physiology* **59**, 455–8.

Cram, W. J. (1976). Negative feedback regulation of transport in cells. The maintenance of turgor, volume and nutrient supply. In Lüttge, U. & Pitman, M. G. (eds.). *Transport in Plants* II, *Cells, Encyclopedia of Plant Physiology*, New Series, Vol. 2A, pp. 284–316. Berlin, Heidelberg, New York, Springer-Verlag.

D'Auzac, J. (1977). ATPase membranaire de vacuoles lysosomales: les lutoïdes du latex d'*Hevea brasiliensis*. *Phytochemistry* **16**, 1881–5.

Dainty, J. (1976). Water relations of plant cells. In Lüttge, U. & Pitman, M. G. (eds.). *Transport in Plants* II, *Cells, Encyclopedia of Plant Physiology*, New Series, Vol. 2A, pp. 12–35. Berlin, Heidelberg, New York, Springer-Verlag.

Davies, D. D. (1973a). Control of and by pH. *Symposium of the Society for Experimental Biology* **27**, 513–29.

Davies, D. D. (1973b). Metabolic control in higher plants. In Milborrow, B. V. (ed.). *Biosynthesis and its Control in Plants*, pp. 1–20. London, New York, Academic Press.

Deamer, D. W., Prince, R. C. & Crofts, A. R. (1972). The response of fluorescent amines to pH gradients across liposome membranes. *Biochimica et Biophysica Acta* **274**, 323–35.

Edwards, J., ap Rees, T., Wilson, P. M. & Morrell, S. (1984). Measurement of the inorganic pyrophyosphate in tissues of *Pisum sativum* L. *Planta* **162**, 188–91.

Greenway, H., Winter, K. & Lüttge, U. (1978). Phosphoenolpyruvate carboxylase during development of crassulacean acid metabolism and during a diurnal cycle in *Mesembryanthemum crystallinum. Journal of Experimental Botany* **29**, 547–59.

Griffiths, H., Lüttge, U., Stimmel, K.-H., Crook, C. E., Griffiths, N. M. & Smith, J. A. C. (1986). Comparative ecophysiology of CAM and C_3 bromeliads. III. Environmental influences on CO_2 assimilation and transpiration. *Plant, Cell and Environment.* **9**, 385–93.

Griffiths, H. & Smith, J. A. C. (1983). Photosynthetic pathways in the Bromeliaceae of Trinidad: relations between life-forms, habitat preference and the occurrence of CAM. *Oecologia* **60**, 176–84.

Guynn, R. W. & Veech, R. L. (1973) The equilibrium constants of adenosine tri-phosphate hydrolysis and the adenosine triphosphate-citrate lyase reactions. *Journal of Biological Chemistry* **248**, 6966–72.

Hampp, R., Goller, M. & Ziegler, H. (1982). Adenylate levels, energy charge, and phosphorylation potential during dark-light and light-dark transition in chloroplasts, mitochondria, and cytosol of mesophyll protoplasts from *Avena sativa* L. *Plant Physiology* **69**, 448–55.

Haschke, H.-P. & Lüttge, U. (1977). Auxin action on K^+–H^+-exchange and growth, $^{14}CO_2$ fixation and malate accumulation in *Avena* coleoptile segments. In Marrè, E. & Ciferri, O. (eds.). *Regulation of Cell Membrane Activities in Plants*, pp. 243–8. Amsterdam, Elsevier/North Holland Biomedical Press.

Hüsken, D., Steudle, E. & Zimmermann, U. (1978). Pressure probe technique for measuring water relations of cells in higher plants. *Plant Physiology* **61**, 158–63.

Jochem, P., Rona, J.-P., Smith, J. A. C. & Lüttge, U. (1984). Anion-sensitive ATPase activity and proton transport in isolated vacuoles of species of the CAM genus *Kalanchoë. Physiologia Plantarum* **62**, 410–15.

Karlsson, J. (1975). Membrane-bound potassium and magnesium ion stimulated inorganic pyrophyosphatase from roots and cotyledons of sugar beet (*Beta vulgaris* L.). *Biochimica et Biophysica Acta* **399**, 356–63.

Kenyon, W. H., Severson, R. F. & Black, C. C. (1985). Maintenance carbon cycle in crassulacean acid metabolism plant leaves. Source and compartmentation of carbon for nocturnal malate synthesis. *Plant Physiology* **77**, 183–9.

Kluge, M., Knapp, I., Kramer, D., Schwerdtner, I. & Ritter, H. (1979). Crassulacean acid metabolism (CAM) in leaves of *Aloë arborescens* Mill. Comparative studies of the carbon metabolism of chlorenchym and central hydrenchym. *Planta* **145**, 357–63.

Kluge, M., Lange, O. L., von Eichmann, M. & Schmid, M. (1973). Diurnaler Säurerhythmus bei *Tillandsia usneoides*: Untersuchungen über den Weg des Kohlenstoffs sowie die Abhängigkeit des CO_2-Gaswechsels von Lichtintensität, Temperatur und Wassergehalt der Pflanze. *Planta* **112**, 357–72.

Kluge, M. & Ting, I. P. (1978). *Crassulacean Acid Metabolism. Analysis of an Ecological Adaptation*. Berlin, Heidelberg, New York, Springer-Verlag.

Kombrink, E., Kruger, N. J. & Beevers, H. (1984). Kinetic properties of pyrophosphate: fructose-6-phosphate phosphotransferase from germinating castor bean endosperm. *Plant Physiology* **74**, 395–401.

Kowalczyk, S., Januszewska, B., Cymerska, E. & Mastowski, P. (1984). The occurrence of inorganic pyrophosphate: D-fructose-6-phosphate 1-phosphotransferase in higher plants. I. Initial characterization of partially purified enzyme from *Sansevieria trifasciata* leaves. *Physiologia Plantarum* **60**, 31–7.

Kringstad, R., Kenyon, W. H. & Black, C. C. (1980). The rapid isolation of vacuoles from leaves of crassulacean acid metabolism plants. *Plant Physiology* **66**, 379–82.

Kruger, N. J., Kombrink, E. & Beevers, H. (1983). Pyrophosphate: fructose-6-phosphate phosphotransferase in germinating castor bean seedlings. *FEBS Letters* **153**, 409–12.

Lance, C. & Rustin, P. (1984). The central role of malate in plant metabolism. *Physiologie Végétale* **22**, 625–41.

Leigh, R. A. (1983). Methods, progress and potential for the use of isolated vacuoles in studies of solute transport in higher plant cells. *Physiologia Plantarum* **57**, 390–6.

Leigh, R. A. & Branton, D. (1976). Isolation of vacuoles from root storage tissue of *Beta vulgaris* L. *Plant Physiology* **58**, 656–62.

Lüttge, U. & Ball, E. (1977). Concentration and pH dependence of malate efflux and influx in leaf slices of CAM plants. *Zeitschrift für Pflanzenphysiologie* **83**, 43–54.

Lüttge, U. & Ball, E. (1979). Electrochemical investigation of active malic acid transport at the tonoplast into the vacuoles of the CAM plant *Kalanchoë daigremontiana*. *Journal of Membrane Biology* **47**, 401–2.

Lüttge, U. & Ball, E. (1980). $2H^+$: 1malate^{2-} stoichiometry during crassulacean acid metabolism is unaffected by lipophilic cations. *Plant, Cell and Environment* **3**, 195–200.

Lüttge, U., Ball, E. & Tromballa, H.-W. (1975). Potassium independence of osmoregulated oscillations of malate^{2-} levels in the cells of CAM-leaves. *Biochemie und Physiologie der Pflanzen* **167**, 267–83.

Lüttge, U. & Clarkson, D. T. (1985). II. Mineral nutrition: Plasmalemma and tonoplast redox activities. *Progress in Botany*. **47**, 73–86.

Lüttge, U. & Nobel, P. S. (1984). Day-night variations in malate concentration, osmotic pressure, and hydrostatic pressure in *Cereus validus*. *Plant Physiology* **75**, 804–7.

Lüttge, U. & Smith, J. A. C. (1984a). Structural, biophysical, and biochemical aspects of the role of leaves in plant adaptation to salinity and water stress. In Staples, R. C. & Toenniessen, G. H. (eds.). *Salinity Tolerance in Plants: Strategies for Crop Improvement*, pp. 125–150. New York, John Wiley.

Lüttge, U. & Smith, J. A. C. (1984b). Mechanism of passive malic-acid efflux from vacuoles of the CAM plant *Kalanchoë daigremontiana*. *Journal of Membrane Biology* **81**, 149–58.

Lüttge, U., Smith, J. A. C. & Marigo, G. (1982). Membrane transport, osmoregulation, and the control of CAM. In Ting, I. P. & Gibbs, M. (eds.). *Crassulacean Acid Metabolism*, pp. 69–91. *Proceedings of the Fifth Annual Symposium in Botany*, University of California, Riverside, Rockville, Md., American Society of Plant Physiologists.

Lüttge, U., Smith, J. A. C., Marigo, G. & Osmond, C. B. (1981). Energetics of malate accumulation in the vacuoles of *Kalanchoë tubiflora* cells. *FEBS Letters* **126**, 81–4.

Marigo, G., Ball, E., Lüttge, U. & Smith, J. A. C. (1982). Use of the DMO technique for the study of relative changes of cytoplasmic pH in leaf cells in relation to CAM. *Zeitschrift für Pflanzenphysiologie* **108**, 223–33.

Marigo, G., Lüttge, U. & Smith, J. A. C. (1983). Cytoplasmic pH and the control of crassulacean acid metabolism. *Zeitschrift für Pflanzenphysiologie* **109**, 405–413.

Marin, B. P. (1985). *Biochemistry and Function of Vacuolar Adenosinetriphosphatase in Fungi and Plants*. Berlin, Heidelberg, New York, Tokyo, Springer-Verlag.

Marrè, E. & Ballarin-Denti, A. (1985). The proton pumps of the plasmalemma and the tonoplast of higher plants. *Journal of Bioenergetics and Biomembranes* **17**, 1–21.

Matile, P. (1968). Lysosomes of root tip cells in corn seedlings. *Planta* **79**, 181–96.

Matile, P. & Wiemken, A. (1967). The vacuoles as the lysosome of the yeast cell. *Archiv für Mikrobiologie* **56**, 148–55.

Medina, E., Delgado, M., Troughton, J. H. & Medina, J. D. (1977). Physiological ecology of CO_2 fixation in *Bromeliaceae*. *Flora* **166**, 137–52.

Nobel, P. S. (1983). *Biophysical Plant Physiology and Ecology*. San Francisco, Calif., W. H. Freeman.

Nobel, P. S., Lüttge, U., Heuer, S. & Ball, E. (1984). Influence of applied NaCl on Crassulacean acid metabolism and ionic levels in a cactus, *Cereus validus*. *Plant Physiology* **75**, 799–803.

Osmond, C. B. (1978). Crassulacean acid metabolism: a curiosity in context. *Annual Review of Plant Physiology* **29**, 379–414.

Osmond, C. B. & Holtum, J. A. M. (1981). Crassulacean acid metabolism. In Hatch, M. D. & Boardman, N. K. (eds.). *The Biochemistry of Plants*, Vol. 8, pp. 283–328. New York, Academic Press.

Osmond, C. B. & Laties, G. G. (1969). Compartmentation of malate in relation to ion absorption in beet. *Plant Physiology* **44**, 7–14.

Passioura, J. B. (1982). Water in the soil-plant-atmosphere continuum. In Lange, O. L., Nobel, P. S., Osmond, C. B. & Ziegler, H. (eds.). *Physiological Plant Ecology*, II, *Water Relations and Carbon Assimilation, Encyclopedia of Plant Physiology*, New Series, Vol. 12B, pp. 5–33. Berlin, Heidelberg, New York, Springer Verlag.

Phillips, R. D. (1980). Deacidification in a plant with crassulacean acid metabolism associated with anion-cation balance. *Nature* **287**, 727–8.

Phillips, R. D. & Jennings, D. H. (1976). Succulence, cations and organic acids in leaves of *Kalanchoë daigremontiana* grown in long and short days in soil and water culture. *New Phytologist* **77**, 599–611.

Pierre, J. N. & Queiroz, O. (1979). Regulation of glycolysis and level of the crassulacean acid metabolism. *Planta* **144**, 143–51.

Pistelli, L. & Lüttge, U. (1985). Nocturnal changes in adenine nucleotide levels in the CAM plant *Kalanchoë daigremontiana. Giornale Botanico Italiano* **119S**, 86–7.

Poole, R. J., Briskin, D. P., Krátký, Z. & Johnstone, R. M. (1984). Density gradient localization of plasma membrane and tonoplast from storage tissue of growing and dormant red beet. Characterization of proton-transport and ATPase in tonoplast vesicles. *Plant Physiology* **74**, 549–56.

Pucher, G. W., Vickery, H. B., Abrahams, M. D. & Levenworth, C. S. (1949). Studies in the metabolism of Crassulacean plants: diurnal variation of organic acids and starch in excised leaves of *Bryophyllum calycinum. Plant Physiology* **24**, 610–20.

Queiroz, O. (1965). Sur le métabolisme acide des Crassulacées. I. Action à long terme de la température de nuit sur la synthèse d'acide malique par *Kalanchoë blossfeldiana* "Tom Thumb" placée en jours courts. *Physiologie Végétale* **3**, 203–13.

Raschke, K. (1979). Movements of stomata. In Haupt, W. & Feinleib M. E. (eds.) *Physiology of Movements, Encyclopedia of Plant Physiology*, New Series, Vol. 7, pp. 383–441. Berlin, New York, Springer-Verlag.

Raven, J. A. (1984). *Energetics and Transport in Aquatic Plants*. New York, Alan R. Liss.

Rea, P. A. & Poole, R. J. (1985). Proton translocating inorganic pyrophosphatase in red beet (*Beta vulgaris* L.) tonoplast vesicles. *Plant Physiology* **77**, 46–52.

Reinhold, L. & Kaplan, A. (1984). Membrane transport of sugars and amino acids. *Annual Review of Plant Physiology* **35**, 45–83

Römheld, V., Müller, C. & Marschner, H. (1984). Localization and capacity of proton pumps in roots of intact sunflower plants. *Plant Physiology* **76**, 603–606.

Rona, J.-P., Pitman, M. G., Lüttge, U. & Ball, E. (1980). Electrochemical data on compartmentation into cell wall, cytoplasm, and vacuole of leaf cells in the CAM genus *Kalanchoë*. *Journal of Membrane Biology* **57**, 25–35.

Rottenberg, H. (1979). Non-equilibrium thermodynamics of energy conversion in bioenergetics. *Biochimica et Biophysica Acta* **549** 225–53.

Saunders, J. A. (1979). Investigations of vacuoles isolated from tobacco. I. Quantitation of nicotine. *Plant Physiology* **64**, 74–8.

Schuldiner, S., Rottenberg, H. & Avron, M. (1972). Determination of $\triangle$pH in chloroplasts. 2. Fluorescent amines as a probe for the determination of $\triangle$pH in chloroplasts. *European Journal of Biochemistry* **25**, 64–70.

Smith, F. A. & Raven, J. A. (1979). Intracellular pH and its regulation. *Annual Review of Plant Physiology* **30**, 289–311.

Smith, J. A. C., Griffiths, H., Bassett, M. & Griffiths, N. M. (1985). Day-night changes in the leaf water relations of epiphytic bromeliads in the rain forests of Trinidad. *Oecologia*. **67**, 475–85.

Smith, J. A. C., Griffiths, H., Lüttge, U., Crook, C. E., Griffiths, N. M. & Stimmel, K.-H. (1986). Comparative ecophysiology of CAM and C_3 bromeliads. IV. Plant water relations. *Plant, Cell and Environment*. **9**, 395–410.

Smith, J. A. C. & Lüttge, U. (1985). Day-night changes in leaf water relations associated with the rhythm of crassulacean acid metabolism in *Kalanchoë daigremontiana*. *Planta* **163**, 272–82.

Smith, J. A. C., Marigo, G., Lüttge, U. & Ball, E. (1982). Adenine-nucleotide levels during crassulacean acid metabolism and the energetics of malate accumulation in *Kalanchoë tubiflora*. *Plant Science Letters* **26**, 13–21.

Smith, J. A. C., Uribe, E. G., Ball, E., Heuer, S. & Lüttge, U. (1984a). Characterization of the vacuolar ATPase activity of the crassulacean acid metabolism plant *Kalanchoë daigremontiana*. *European Journal of Biochemistry* **141**, 415–20.

Smith, J. A. C., Uribe, E. G., Ball, E. & Lüttge, U. (1984b). ATPase activity associated with isolated vacuoles of the crassulacean-acid-metabolism, plant *Kalanchoë daigremontiana*. *Planta* **162**, 299–304.

Smyth, D. A. & Black, C. C. (1984). Measurement of the pyrophosphate content of plant tissues. *Plant Physiology* **75**, 862–4.

Smyth, D. A., Wu, M.-X. & Black, C. C. (1984). Pyrophosphate and fructose 2,6-bisphosphate effects on glycolysis in pea seed extracts. *Plant Physiology* **76**, 316–20.

Spanswick, R. M. (1981). Electrogenic ion pumps. *Annual Review of Plant*

Physiology **32**, 267–89.

Steudle, E., Smith, J. A. C. & Lüttge, U. (1980). Water-relation parameters of individual mesophyll cells of the crassulacean acid metabolism plant *Kalanchoë daigremontiana*. *Plant Physiology* **66**, 1155–63.

Stitt, M., Lilley, R. McC. & Heldt, H. W. (1982). Adenine nucleotide levels in the cytosol, chloroplasts, and mitochondria of wheat leaf protoplasts. *Plant Physiology* **70**, 971–7.

Stitt, M., Wirtz, W. & Heldt, H. W. (1980). Metabolite levels during induction in the chloroplast and extrachloroplast compartments of spinach protoplasts. *Biochimica et Biophysica Acta* **593**, 85–102.

Struve, I., Weber, A., Lüttge, U., Ball, E. & Smith, J. A. C. (1985). Increased vacuolar ATPase activity correlated with CAM induction in *Mesembryanthemum crystallinum* and *Kalanchoë blossfeldiana* cv. Tom Thumb. *Journal of Plant Physiology* **117**, 451–68.

Sutton, B. G. (1975). Glycolysis in CAM plants. *Australian Journal of Plant Physiology* **2**, 389–402.

Sze, H. (1984). H^+-translocating ATPases of the plasma membrane and tonoplast of plant cells. *Physiologia Plantarum* **61**, 683–91.

Sze, H. (1985). H^+-translocating ATPases: advances using membrane vesicles. *Annual Review of Plant Physiology* **36**, 175–208.

Torii, K. & Laties, G. G. (1966). Organic acid synthesis in response to excess cation absorption in vacuolate and nonvacuolate sections of corn and barley roots. *Plant Cell Physiology* **7**, 395–403.

Tyree, M. T. & Richter, H. (1981). Alternative methods of analysing water potential isotherms: some cautions and clarifications. I. The impact of non-ideality and of some experimental errors. *Journal of Experimental Botany* **32**, 643–53.

Veech, R. L., Lawson, J. W. R., Cornell, N. W. & Krebs, H. A. (1979). Cytosolic phosphorylation potential. *Journal of Biological Chemistry* **254**, 6538–47.

Vickery, H. B. (1952). The behavior of isocitric acid in excised leaves of *Bryophyllum calycinum* during culture in alternating light and darkness. *Plant Physiology* **27**, 9–17.

Vickery, H. B. (1957). The formation of starch and the behavior of isocitric acid in excised leaves of *Bryophyllum calycinum* cultured in darkness. *Plant Physiology* **32**, 220–4.

Wagner, G. J. & Mulready, P. (1983). Characterization and solubilization of nucleotide-specific, Mg^{2+}-ATPase and Mg^{2+}-pyrophosphatase of tonoplast. *Biochimica et Biophysica Acta* **728**, 267–80.

Walker, R. R. & Leigh, R. A. (1981). Mg-dependent, cation-stimulated inorganic pyrophosphatase associated with vacuoles isolated from storage roots of red beet (*Beta vulgaris* L.). *Planta* **153** 150–5.

Wilkins, M. B. (1983). The circadian rhythm of carbon-dioxide metabolism in *Bryophyllum*: the mechanism of phase-shift induction by thermal stimuli. *Planta* **157**, 471–80.

Wilkins, M. B. (1984). A rapid circadian rhythm of carbon-dioxide metabolism in *Bryophyllum fedtschenkoi*. *Planta* **161**, 381–4.

Willmer, C. M., Pallas, J. E., Jr. & Black, C. C., Jr. (1973). Carbon dioxide metabolism in leaf epidermal tissue. *Plant Physiology* **52**, 448–52.

Wilson, D. F., Erecińska, M. & Dutton, P. L. (1974). Thermodynamic relationships in mitochondrial oxidative phosphorylation. *Annual Review of Biophysics and Bioengineering* **3**, 203–30.

Winter, K. (1974). Der Einfluβ von Wasserstreβ auf die Aktivität der Phosphoenolpyruvat-Carboxylase bei *Mesembryanthemum crystallinum*. *Planta* **121**, 147–53.

Winter, K. (1979). Effect of different CO_2 regimes on the induction of crassulacean acid metabolism in *Mesembryanthemum crystallinum* L. *Australian Journal of Plant Physiology* **6**, 589–94.

Winter, K. & Lüttge, U. (1979). C_3-Photosynthese und Crassulaceen-Säurestoffwechsel bei *Mesembryanthemum crystallinum* L. *Berichte der Deutschen Botanischen Gesellschaft* **92**, 117–32.

Winter, K. & von Willert, D. J. (1972). NaCl-induzierter Crassulaceen-Säurestoffwechsel bei *Mesembryanthemum crystallinum*. *Zeitschrift für Pflanzenphysiologie* **67**, 166–70.

Wu, M.-X., Smyth, D. A. & Black, C. C. (1984). Regulation of pea seed pyrophosphate-dependent phosphofructokinase: Evidence for interconversion of two molecular forms as a glycolytic regulatory mechanism. *Proceedings of the National Academy of Sciences USA* **81**, 5051–5.

Note added in proof: Techniques for purification of tonoplast vesicles of CAM plants and the characterization of the ATPase have advanced (Jochem & Lüttge, 1987). A pyrophosphatase activity at the tonoplast has been detected and characterized (Marquardt & Lüttge, 1987). Both enzymes have been solubilized and separated by liquid chromatography (Bremberger *et al.*, 1987). Properties of vacuolar ATPases of *Mesembryanthemum crystallinum* in the C_3 and the CAM state have been compared (Struve & Lüttge, 1987).

In *Kalanchoe*, PP_i levels were determined (Pistelli *et al.*, 1987). Respiration measurements extended energy budgets also to bromeliads (Lüttge & Ball, 1987). Mg/ATP-dependent malate transport at the tonoplast of isolated vacuoles of *K. daigremontiana* has been demonstrated (Nishida & Tominaga, 1987). The reader may also wish to refer to a more recent review (Lüttge, 1987).

Bremberger, C., Haschke, H.-P., Grötsch, S. & Lüttge, U. (1987). XIV International Botanical Congress, Abstracts, p 45. Berlin.

Jochem, P. & Lüttge, U. (1987). *Journal of Plant Physiology* **129**, 251–68.

Lüttge, U. (1987) *New Phytologist* **106**, 593–629.

Lüttge, U. & Ball, E. (1987). *Plant Physiology and Biochemistry* **25**, 3–10.

Marquardt, G. & Lüttge, U. (1987). *Journal of Plant Physiology* **129**, 269–86.

Nishida, K. & Tominaga, O. (1987). *Journal of Plant Physiology* **127**, 385–93.

Pistelli, L., Marigo, G., Ball, E. & Lüttge, U. (1987). *Planta*, in press.

Struve, I. & Lüttge, U. (1987). *Planta* **170**, 111–20.

12 Stomatal guard cells

E. A. C. MacRobbie

12.1 Introduction

Stomatal guard cells are exceptional in that they show large fluctuations of salt accumulation, and hence of volume and turgor, during their normal diurnal cycle or in response to a range of environmental signals. In most plant cells a large central vacuole is formed early in development, and thereafter, in the absence of drastic changes in the external salt load, the level of solute accumulation remains relatively stable, thereby maintaining relatively constant cell volume and turgor; in general a high proportion of the vacuolar solute is inorganic salt, with K^+ in excess of Na^+, often considerably, balanced by a mixture of Cl^- and organic acid anions, in proportions dependent on the plant, the type of cell and the growth conditions. In general, the process of vacuolation, the result of sequestering solute in internal compartments within the bulk cytoplasm, is not reversible, and the capacity for vacuolar solute accumulation, once initiated, is maintained throughout the life of the cell. Vacuolar salt accumulation in stomatal guard cells is not strikingly different from that in plant cells in general, but is peculiar in its instability and in its sensitivity to various external conditions; the capacity of guard cells to maintain vacuolar solutes shows striking fluctuations throughout the normal diurnal cycle, and vacuolar solute accumulation is reversed and then re-established regularly in the normal cell function. The consequent changes in cell volume and turgor are responsible for the associated changes in stomatal aperture. The phenomenon is illustrated in Fig. 12.1, showing the appearance of guard cells of *Commelina communis* having different apertures.

It is clearly established that large changes in the accumulation of potassium salts are involved, and a partial description of the associated processes, or at least their end results, can be provided, but we do not have any real understanding of the nature of the controls on salt accumulation, or of the sequence of events involved in the initiation of such drastic

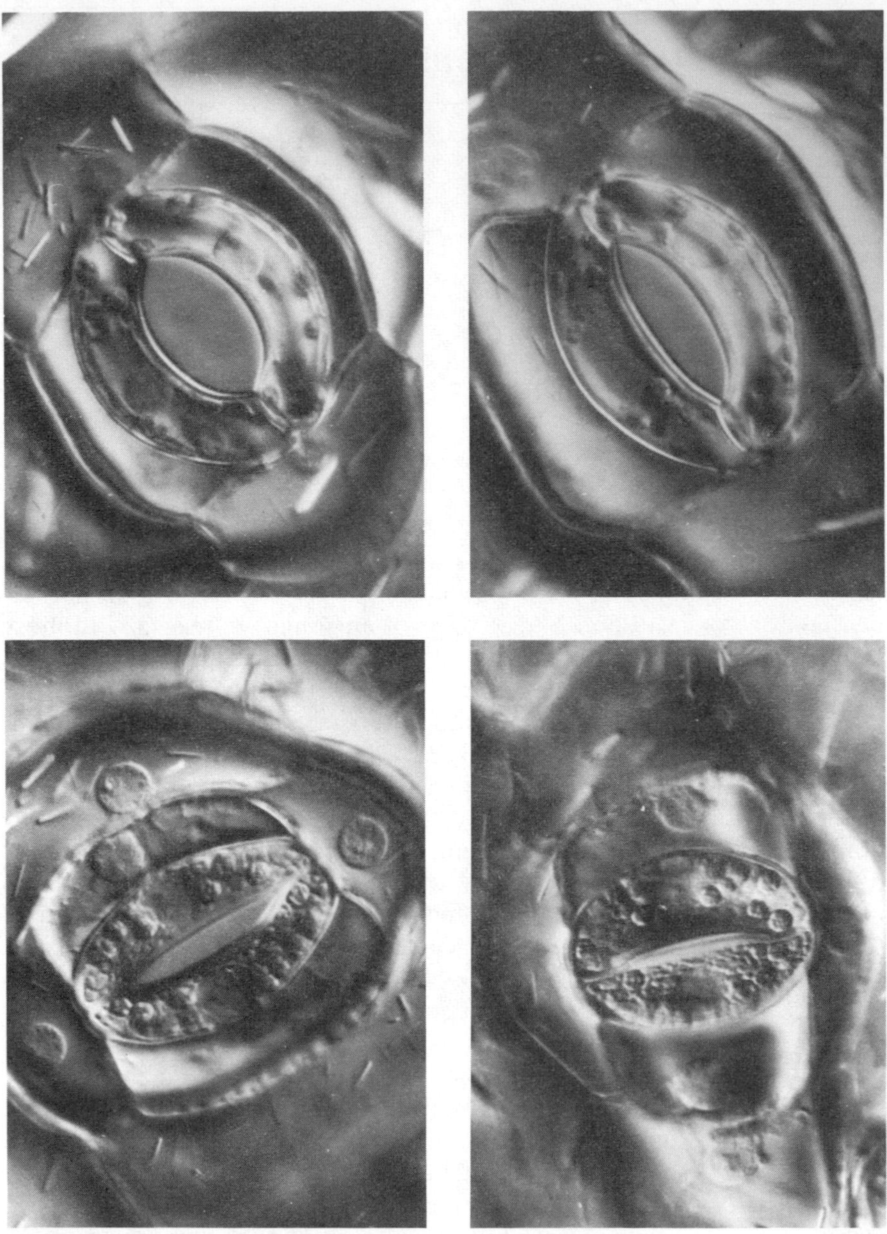

Fig. 12.1 Intact guard cells of *Commelina communis* L. of different apertures, showing swelling of guard cells and compression of inner lateral subsidiary cells associated with stomatal opening. Nomarski optics.

changes in the state of the guard cells concerned. We are still trying to define the nature of the problem, to describe the processes requiring explanation; only after this detailed description is available can we hope to understand the phenomenon.

There are two distinct requirements. The first is to provide a quantitative description of the two extreme conditions, the high-salt, swollen, turgid cell and the low-salt, low-volume, low-turgor cell. The second, perhaps more important and more informative, is to identify the transients involved in the initiation of the transitions between these two states, in response to specific environmental signals. The aim of this chapter is to review the information available on guard cells, in the light of these dual requirements, and to compare what we know with what we need to know if we are to gain any understanding of the processes involved.

Before discussing work on the ionic relations of guard cells it is worth considering, briefly, their metabolic properties. Guard cells are unique among cells of the epidermis in containing chloroplasts. There are exceptions, in that guard cells of *Paphiopedilum* lack chloroplasts, but in general both chloroplasts and abundant mitochondria are a feature of the ultrastructure of guard cells. It is now established that both photosystems are present (Outlaw *et al.*, 1981; Zeiger *et al.*, 1981a; Melis & Zeiger, 1982; Ogawa *et al.*, 1982; Hipkins *et al.*, 1983), and that the capacity for either cyclic or pseudocyclic photophosphorylation exists (Shimazaki & Zeiger, 1985). High rates of both oxygen uptake in the dark, and oxygen evolution in the light, have been demonstrated, and the activities of the key enzymes of heterotrophic carbon metabolism are high, when expressed on the basis of chlorophyll or protein content (Shimazaki *et al.*, 1982, 1983; Hampp *et al.*, 1982; Outlaw *et al.*, 1985). After earlier negative reports, the presence of high activities of NADP-triose phosphate dehydrogenase in guard cells has been demonstrated (Outlaw *et al.*, 1985), and the question of ribulose bisphosphate carboxylase remains one of debate. Attempts to demonstrate its activity enzymatically have always failed, but there is now immunological evidence for its presence, and for its inactivation before enzyme assay (Zemel & Gepstein, 1985). It may be therefore that guard cells have full autotrophic and heterotrophic carbon metabolism, at very high rates. Certainly the capacity for ATP production, by both mitochondria and chloroplasts, is very high, as might be expected to be necessary to support high rates of ion transport.

12.2 Potassium accumulation in open and closed guard cells

The hypothesis that the extent of K-salt accumulation in stomatal guard cells determines their turgor, and hence stomatal aperture, was developed in the late 1960s, with the first publication in English of Fujino's work

(1967), and from independent work by Fischer (1968a, b; Fischer & Hsiao, 1968). But evidence of the importance of K-salts in stomatal opening was available, much earlier, from extended work by Imamura (1943) who showed that guard cells could retain their internal solutes, and maintain very high osmotic pressures, only in the presence of high concentrations of K^+ in the external solution bathing epidermal strips. Above a threshold concentration of external K^+ (of about 80 mol m^{-3}) osmotic pressures at incipient plasmolysis were 4.7 MPa in *Vicia*, 6 MPa in *Commelina* and 7–9 MPa in *Zebrina*, compared with values of only about 1 MPa or less, when treated with K^+-free solutions. Failure to take account of this observation casts doubt on the results of many subsequent plasmolytic determinations with extended exposure to K^+-free plasmolysing solutions; Raschke (1979) shows how rapid is the loss of solute in such conditions. Fischer (1973) also showed the importance of external K^+ in the plasmolyticum, and its effect in the retention of guard cell solute contents.

The first histochemical demonstration of very high levels of potassium salts in open, but not in closed, guard cells was by Macallum (1905), and Yamashita (1952) showed correlation between K-content of guard cells and stomatal aperture. The evidence for the role of K-salts in guard cells has been extensively considered in earlier reviews (Rashke, 1975, 1979; Thomas, 1975; Hsiao, 1976; MacRobbie, 1977; Outlaw, 1983), and may therefore be given only briefly. The marked accumulation of K-salts in guard cells during opening has been demonstrated histochemically in some fifty species, including examples of pteridophytes and gymnosperms, as well as a wide range of angiosperms (Willmer & Pallas, 1973; Dayanandan & Kaufman, 1975). Quantitative information is available for only very few species, as shown in Table 12.1, giving estimates of K^+ concentrations in open and closed guard cells. In Table 12.1 the results for *Vicia faba* and *Commelina communis* have been separated, these being the only species for which multiple measurements have been made, and for which related information is available. The concentrations calculated in Table 12.1, C suffer from considerable uncertainty in volume estimates. However, from these figures the general conclusion can be drawn, that the K-content of guard cells increases markedly with opening, by amounts which are significant osmotically. The observed correlation between K-content and aperture holds for opening/closing on the intact leaf or in epidermal strips, in response to different stimuli or treatments, and using a range of methods for K-determination. The K-hypothesis is also strengthened by observations that opening of guard cells in epidermal strips is in many species (but not in maize), dependent on the presence of K-salts in the bathing medium.

It is in fact only for the two species, *Vicia faba* and *Commelina communis*, that a wider range of different measurements is available, of ion contents, fluxes, membrane potentials, enzyme activities and metabolic

Table 12.1 Potassium concentrations in open and closed guard cells. K^+ concentration (mol m^{-3}) and (aperture)

	Open	Closed	Reference†
A. Intact guard cells: *Vicia* and *Commelina*			
Vicia faba			
	880 (12 μm)	77 (2 μm)	1
	552 (16.5 μm)	112 (6.5 μm)	2
	460–760 (10 μm)	80 (0)	3
Commelina communis			
	448	90	4‡*
	385–600 (14 μm)	75–90 (0–1 μm)	5‡*
B. Isolated guard cells: *Vicia* and *Commelina*			
Vicia faba			
	276 (13 μm)	52 (7 μm)	6
	357 (11 μm)	0 (4 μm)	7
	322 (9 μm)	80 (6 μm)	8
Commelina communis			
	157 (15 μm)	51 (5 μm)	9‡*
	167–74 (15 μm)	20–56 (5 μm)	9
C. Other intact guard cells			
Nicotiana tabacum	500 (8 μm)	210 (1 μm)	10
Zea mays	400	176	11
Allium cepa	70–160 (9 μm)	38–84 (2 μm)	12
Tradescantia albiflora	633	152	13‡

† References:
1. Humble & Raschke (1971). Electron microprobe analysis.
2. Allaway & Hsiao (1973). Aperture set on intact leaf then epidermal cells killed by rolling epidermal strips; flame photometry.
3. Outlaw & Lowry (1977). Dissection of guard cell pairs from frozen–dried tissue: enzymic assay for K^+.
4. Penny & Bowling (1974). K^+-sensitive electrode.
5. MacRobbie & Lettau (1980b). K^+-sensitive electrode.
6. Fischer (1972). $^{42}K^+$ uptake.
7. Pallaghy & Fischer (1974). $^{42}K^+$ uptake.
8. Fischer (1968a); Fischer & Hsiao (1968). $^{86}Rb^+$ uptake.
9. MacRobbie & Lettau (1980a). K^+-sensitive electrode (*), or $^{86}Rb^+$ uptake.
10. Sawney & Zelitch (1969). Microprobe analysis.
11. Raschke & Fellows (1971). Microprobe analysis. Guard cell assumed 2 pl open and 1.7 pl closed.
12. Schnabl & Raschke (1980). Concentrations calculated for volumes of guard cell pair of 12 and 5.4 pl open, 8 and 3.6 pl closed. Volumes from dimensions quoted by Meidner & Mansfield (1968) and Willmer (1983).
13. Zlotnikova *et al.* (1977).

‡ *Measurement of potassium activity rather than K^+ content.
Where K^+ content was measured the K^+ concentrations are calculated from the guard cell volumes measured by the original authors, where possible; allowance is made for volume change during opening.

patterns, from which we might hope to achieve some integrated picture of events and processes associated with stomatal opening. For this reason, this review will concentrate on these two species. The basic problem to be considered is preliminary defined by the figures in Table 12.1, A, showing potassium contents in open and closed 'intact' guard cells of *Commelina* and of *Vicia*, from different authors, by different methods. 'Intact' guard cells have opened or closed on the intact leaf, when opening requires turgor high enough to overcome the resistance to cell swelling provided not only by the guard cell wall itself, but also by the very considerable back-pressure from subsidiary or epidermal cells. In these conditions very high turgor is required for opening, reflected in the very high K^+ concentrations observed in open guard cells, much higher than the maximum value reached in more typical plant cells. If guard cells have feedback controls to regulate either turgor or internal ionic concentrations, of the kind found in a number of other plant cells and tissues, then their set points much be at much higher levels (higher, for example, than the maximum KCl accumulation of 160 μmol gFW^{-1} in barley roots (Pitman, 1969), or 80 μmol gFW^{-1} in carrot (Cram, 1980), or maximum turgor of 0.64 MPa in beet (Cram, 1980). The question of what regulatory properties feature in the ionic relations of guard cells is an important problem for future work.

In spite of the long history of the now well-established 'potassium hypothesis' for guard cell opening/closing, our understanding of the process remains rudimentary, largely because our ability to describe the system is still so limited. For this reason, before discussing the current state of knowledge of the ionic relations of guard cells, it is worth setting out the necessary aims of such study, the information that is required for proper consideration of the mechanisms – even if some of these requirements are not easily met.

It is first necessary to describe the various steady states, in guard cells of different apertures, including intermediate states between open and closed; such descriptions should include figures for steady-state ion fluxes at different apertures, in different conditions, and not simply for ion concentrations in guard cells. More critically, it is then necessary to identify the changes in ionic state in response to signals for opening/closing, the flux changes responsible for the transition to a new steady state with different ion content, cell volume and stomatal aperture. In opening, the end result is an increased accumulation of salt (commonly K^+ balanced by a variable proportion of Cl^- and malate, depending on species and conditions); in closing, the end result is the reverse, a marked fall in the level of salt accumulation, not necessarily achieved simply by switching off the processes of accumulation. We need to describe the two different transients, which are likely to be achieved by controls on different ion fluxes. The key to understanding guard cells must lie in understanding what happens in these two transients, and the nature of the associated control processes. While it is clearly important to understand the control of *opening*, the initiation of processes of ion accumulation in guard cells, I believe that it is probably more important to understand how the ability to *stay open* is controlled, and the reason for the failure to maintain high levels of vacuolar accumulation in closing conditions, in response to quite ordinary signals, such as darkening, or increased CO_2 levels. This is the most striking feature distinguishing the behaviour of guard cells from that of other plant cells, and we are likely to find the particular responses and properties which are specific to guard cells associated with the *closing* process, rather than with opening.

Given the need to measure ion fluxes, and not simply their end results, the ion contents of the cell, the difficulties of proceeding further by the study of guard cells in the intact leaf are at once apparent. First, in the intact leaf the extracellular ion concentrations are ill defined and cannot be controlled experimentally; Blatt (1985) has measured very low K^+ concentrations in the bulk leaf apoplast (of < 50 mmol m^{-3}), but this may differ markedly from that in the guard cell wall itself, with diffusion restricted by the underlying air space to the narrow longitudinal path along the subsidiary or neighbouring epidermal cell walls. Figure 12.2 shows the spatial relations of guard cells to the substomatal air space in a maize leaf, and emphasizes the position of guard cells at the end of the line as far as diffusive access is concerned. In this context the lack of protoplasmic connections between mature guard cells and their neighbours is of critical importance, requiring any ion exchange with guard cells to involve transport across the guard cell plasmalemma, to and from the apoplast. The lack of plasmodesmata in mature guard cells was convincingly shown by the electron microscopical study of Wille & Lucas (1984), and is also confirmed in the study of the spread (or lack of spread) from cell to cell of injected dyes by Palevitz & Hepler (1985).

Secondly, for the measurement of fluxes in guard cells it is necessary that the two-way fluxes into and out of guard cells can be distinguished from the contributions to tracer exchange of the much larger subsidiary/epidermal cells. For these reasons progress demands the use of 'isolated' guard cells, in which all cells other than guard cells have been removed, or killed. So far the system most widely used has been 'isolated' epidermal strips, in which cells other than guard cells have been killed, either mechanically or by treatment at low pH, but suspensions of isolated guard cells can be prepared by partial enzymic digestion (of thinner epidermal/subsidiary cell walls but not of thicker guard cell walls), and may prove useful for future work.

Table 12.1, B shows figures for K^+ concentrations in 'isolated' guard cells of *Commelina communis* and *Vicia faba*, opened or closed in experimentally defined conditions, in the absence of any back-pressure from subsidiary/epidermal cells. It is clear that the effect of such back-pressure is very considerable. Opening against turgid surrounding cells requires

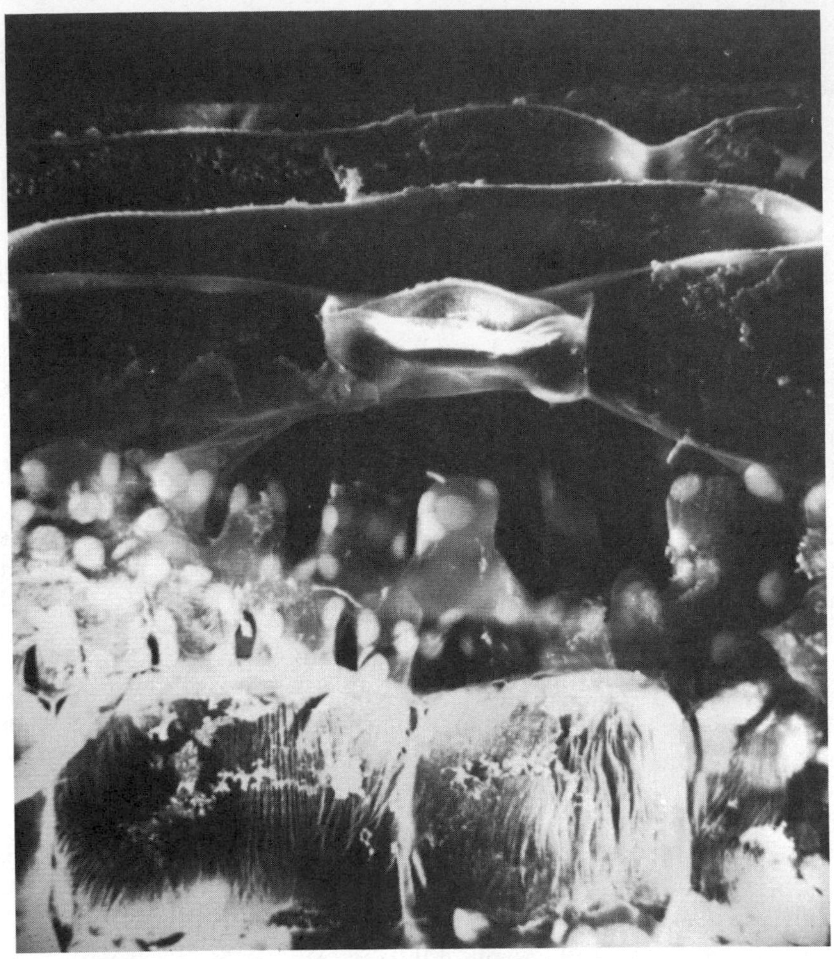

Fig. 12.2 Scanning electron micrograph of frozen fractured maize leaf, showing guard cell with air space beneath.

guard cell K^+ concentrations in *Vicia* of 460–880 mol m^{-3}, compared with only 276–357 mol m^{-3} in isolated guard cells; in *Commelina* corresponding concentrations are 385–600 mol m^{-3} in intact guard cells, compared with only 157–174 mol m^{-3} in isolated guard cells. But the role of such back-pressure is most critical for closing, and isolated guard cells tend not to close completely. It is important to recognize this fact, and such failure to

Table 12.2 Comparison of isolated and intact guard cells: *Commelina*

Aperture (μm)	**K^+ concentrations** (mol m^{-3})		$\Delta\pi_{KCl}$	P_{gs}	**Estimated A_K^{intact}** (μm)
	Intact	**Isolated**	(mosmol kg^{-1} μm^{-1})		
6	110	50	110	89	closed
10	205	90	210	222	4
14	385	150	420	355	8
18	–	208	–	–	10

K^+ concentrations from MacRobbie & Lettau (1980a, b).
π_{KCl} is the osmotic value of a KCl solution of the same concentration of K^+ as measured.
$\Delta\pi_{KCl}$ is the difference in π_{KCl} in intact and isolated guard cells.
P_{gs} is the measure of the contribution of subsidiary cell turgor to the water relations of the guard cell (MacRobbie, 1980).
A_K^{intact} is the estimated aperture of intact guard cells having the same K^+ content as that measured in the isolated guard cells.

close completely should not be interpreted as a failure of the normal inherent closing response. For *Commelina communis*, figures are available to allow comparison of isolated and intact guard cells, as shown in two ways in Table 12.2. First, it can be seen that the estimate of the osmotic difference associated with the difference in K-salt accumulation in the two conditions is, over the range 6–14 μm aperture, close to the estimate of P_{gs}, the contribution of subsidiary cell turgor to the water relations of the guard cell (MacRobbie, 1980); thus, the extra K-salt measured in intact guard cells is close to the amount calculated to be necessary to overcome the back-pressure. Secondly, it is possible to calculate the range of apertures in intact guard cells which correspond to the measured K^+ contents in isolated guard cells in the experimental conditions used. Table 12.2 shows that the observed aperture range of 6–18 μm in isolated guard cells relates to guard cells with K-contents for the aperture range 0–10 μm in intact guard cells. Again, it is important to recognize that in our studies of isolated guard cells we are likely to be looking at the events and metabolic/ionic changes associated with incomplete opening in intact guard cells, in *Commelina* up to about 10 μm only.

12.3 Contribution of potassium salts to the osmotic changes during opening

It is important to look at the relation between ion content and aperture over the full range of apertures, and not simply in open and closed stomata, and to consider the extent to which measured changes in K-salt concentrations can account for the osmotic changes required to change the aperture, over every stage of the opening or closing process. It is only for *Commelina* that figures allowing this comparison are available over a range of apertures, for both intact and isolated guard cells (MacRobbie & Lettau, 1980a, b; MacRobbie, 1981a).

With intact guard cells K^+ changes were estimated from measurements with an intracellular K^+-sensitive electrode. In isolated epidermal strips the aperture reached in the steady state can be varied by the use of different external concentrations of K(Rb) Cl (Br), by incubation in light or dark, by the addition of abscisic acid (ABA), or fusicoccin (FC). In addition to measurements of K^+ activity using the K^+-sensitive electrode, the ion content of isolated guard cells may be measured by the tracer accumulation in the steady state, using $^{86}Rb^+$ as an analogue of K^+, in solutions of RbCl, and $^{82}Br^-$ as an analogue of Cl^-, in solutions of KBr. Under these conditions, with isolated guard cells floating on K(Rb) Cl(Br), there appears to be uptake of anion and cation in equivalent amounts, with no synthesis of malate to balance a fraction of the K^+ uptake. The tracer content (Q^*; $^{86}Rb^+$ or $^{82}Br^-$) increased linearly with aperture (A), and the regressions of Q^* on A for cation and anion could be compared. In experiments with external concentrations setting aperture the estimates of ion content at 10 μm aperture were 58–91 pmol mm^{-2} for $^{86}Rb^+$ (five experiments), compared with 72 and 73 pmol mm^{-2} for $^{82}Br^-$ (two experiments); at 14 μm aperture the values were 84–150 pmol mm^{-2} for $^{86}Rb^+$ and 127 and 137 pmol mm^{-2} for $^{82}Br^-$. Comparison of contents of $^{86}Rb^+$ and $^{82}Br^-$ when light/dark was the variable setting aperture gave very similar figures. Thus under these conditions, with halide anion supplied in the medium, its uptake appears to balance the cation uptake.

Estimates of the osmotic effects of the measured changes in K^+ concentration with aperture in intact guard cells are shown in Table 12.3, together with the estimate of the osmotic requirement for the aperture change. It is clear that in the early stages of opening, up to about 10 μm aperture, the osmotic effects of the observed changes in K-salt are much less than those required to change the aperture; some other solute must also be involved. Only in the later stages of wide opening is the osmotic contribution of the increases in K-salt accumulation comparable with osmotic contribution required.

With isolated guard cells there was reasonable agreement between the estimates of changes in potassium with aperture from tracer measurements and from the K^+-sensitive electrode. Estimates of the osmotic changes associated with the measured K^+ changes in the isolated guard cells are shown in Table 12.4. The results show that in isolated guard cells, whether opening is controlled by the external concentration or by light dark, the salt changes are much too small to account for the osmotic changes, over the whole range of aperture observed. However, it should be remembered that the K-contents by which an aperture range of 6–18 μm is achieved in isolated guard cells would be capable of supporting openings of only

Table 12.3 Comparison of osmotic effects of accumulated K-salt and osmotic changes required: intact guard cells of *Commelina*

	Osmotic changes (mosmol kg^{-1} μm^{-1})		
Aperture range (μm)	$\frac{d\pi_m}{dA}$†	$\frac{d\pi}{dA}$†	$\frac{d\pi_{KCl}}{dA}$‡
7–9	91 ± 6 (33)	116	34–58, 38–50
11–13	137 ± 7 (20)	162	110–225, 73–85
above 15	162 ± 17 (4)	225	

† MacRobbie (1980). $d\pi_m/dA$ is the measured osmotic slope, the osmotic change required to change the aperture, against the resistance to expansion of the guard cell wall, when subsidiary cell turgor is removed. $d\pi/dA$ allows also for the effect of P_{gs}, and its change with aperture; $d\pi/dA$ is therefore the osmotic change required to open against the resistance of both the guard cell and the subsidiary cell turgor.
‡ MacRobbie and Lettau (1980b). Two sets of measurements of K^+ activity with K^+-sensitive electrode. $d\pi_{KCl}/dA$ is the estimate of the osmotic effect of the changes in K-salt accumulation, assuming Cl^- is the balancing anion.
Results shown as mean ± SEM (number of values on which mean is based).

0–10 μm in intact guard cells, opening against turgid subsidiary cells (Table 12.2); the discrepancy between $d\pi_{KCl}/dA$ an $d\pi/dA$ might therefore be expected to extend also in the higher aperture range in isolated guard cells.

The only condition in Table 12.4 in which the increase in K-salt can account for the osmotic change is in the immediate effects of adding FC; this would be consistent with a primary effect of FC on a proton pump, with other ion fluxes dependent on the consequent gradients of pH and membrane potential (as suggested by Marrè, 1979).

Thus in *Commelina*, although K-salt accumulation plays a major role in generating guard cell turgor, particularly for wide opening, other changes in the guard cell are also involved in the early stages of opening. An important goal for the future must be the identification of the nature of these changes.

In *Vicia* there are considerable uncertainties in the figures on which the view has been accepted that K-salt accumulation can account, quantitatively, for the osmotic changes in guard cell opening. K^+ changes have been measured by a number of authors, by a number of methods, with values in the range 0.1–0.25 pmol cell^{-1} μm^{-1} (see review by Outlaw, 1983) for detailed references. Humble and Raschke's argument (1971) that K-salt changes are adequate is based on the lowest estimates of guard cell volumes in *Vicia* (1.3 pl opening to 2.4 pl), and the lowest quoted osmotic requirement (0.16 MPa μm^{-1}, based on 30 min exposure to K^+-free plasmolysing solutions). Fischer (1972) used tracer methods to measure the increase in K^+-content with aperture in isolated guard cells of *Vicia*; his

Table 12.4 Comparison of osmotic effects of accumulated salt and osmotic changes required: isolated guard cells of *Commelina*

Method	Variable	Aperture range (μm)	$d\pi_{KCl}/dA$ (mosmol kg^{-1} μm^{-1})	$d\pi/dA$ (mosmol kg^{-1} μm^{-1})
Steady-state tracer				
^{86}Rb	C_0†	6–14	13–30	
	Light/dark‡	6–14	11	
	Light/dark‡	10–14	31	
	ABA§	7–9	28	
	ABA§	8.5–10.5	42.5	
	FC 0–20 minπ	6–7	62, 124	
	FC0–135 minπ	6–12	32, 35	
$^{82}Br^-$	C_0†	6–15	28, 34	
	Light/dark‖	6–17	16.5	
	Light/dark‖	11–14	27	
K^+-electrode	††	7–9	10	
		11–13	20–40	
Osmotic requirement	External sucrose‡‡	7–9		74 ± 6
		11–13		121 ± 8
		above 15		188 ± 21

† MacRobbie & Lettau (1980a); MacRobbie (1981a). External concentrations in the range 10–90 mol m^{-3} in different experiments; ^{86}RbCl or K^{82}Br, pH 3.9. Regressions of steady-state tracer content on aperture in six experiments using ^{86}Rb and two using K^{82}Br; slope in pmol mm^{-2} μm^{-1} converted to concentration changes in guard cells using values of 120 guard cells per mm^2 of epidermis and the estimated relation between single guard cell volume (V, pl) aperture (A, μm) of $V = 4.0 + 0.2A$.

‡ MacRobbie (1983). Incubation in 20 mol m^{-3} ^{86}RbCl, pH 3.9 in light or dark. Value is calculated from $\Delta Q^*/\Delta A$, the difference in means of aperture and content in light or dark.

§ MacRobbie (1981b). Calculated from changes in aperture and tracer content after 3–4 h in 20 mmol m^{-3} ABA, added to steady-state tissue in light, in 20 or 60 mol m^{-3} ^{86}RbCl.

π Clint & MacRobbie (1984). Calculated from changes in aperture and content after adding 30 mmol m^{-3} FC to steady-state tissue, in 10 mol m^{-3} ^{86}RbCl in light; changes measured over 20 or 135 min.

‖ MacRobbie (1984). Incubation in 30 mol m^{-3} K^{82}Br in light or dark. Calculated from $\Delta Q^*/\Delta A$.

†† MacRobbie & Lettau (1980a).

‡‡ MacRobbie (1980). Osmotic measurements on isolated guard cells.

figure of 2.6 nmol K^+ cm^{-2} μm^{-1} (based on epidermal area) was estimated to give an osmotic contribution of up to 0.19 MPa μm^{-1}, depending on the anion. However, Raschke (1979) quotes a revised estimate of 0.48 MPa μm^{-1} for the osmotic requirement, in view of considerable solute leakage during extended exposure to the plasmolyticum. Thus it remains to be clearly established how much of the osmotic change required to open the pore in *Vicia* can be accounted for in terms of K-salts. In particular it is necessary to provide an accurate balance sheet for the early stages of opening. In this connection the suggestion by Outlaw & Kennedy (1978), that little malate is synthesized in the early stages of opening in *Vicia*, is relevant, and there is also evidence from Outlaw's group that both hexose and sucrose increase in the guard cells during opening (Outlaw & Lowry, 1977; Outlaw & Manchester, 1979). Work on protoplasts also suggests that other changes are important during cell swelling. Gotow *et al.* (1982) measured changes in guard cell protoplasts of *Vicia* during swelling induced by exposure to KCl; although KCl in the medium was necessary for swelling to occur, only 13 % of the increase in protoplast volume could be attributed to the increase in K-salt. However, in contrast to these results, Fitzsimons & Weyers (1986) found that the increase in K-salt *was* adequate to account for the solute increase during KCl-induced swelling of *Commelina* protoplasts, and suggested that either the presence of damaged protoplasts, or incomplete sedimentation of protoplasts through pure silicone oil, could account for the discrepancy in the earlier work.

12.4 Mechanism of ion accumulation in guard cells

What we know of ion uptake mechanisms in guard cells is consistent with the general pattern of salt accumulation in plants, with a primary, ATP-powered, proton extrusion at the plasmalemma, generating gradients of pH and membrane potential for secondary active transport of other ions. The differences seem likely to be quantitative rather than qualitative, and there is no evidence of anything very special about the uptake mechanisms in guard cells, at least on present knowledge.

There is some direct evidence for net proton extrusion from guard cells. This was first shown by Raschke & Humble (1973), in the continued excretion of protons from guard cells of *Vicia faba* into potassium iminodiacetate solutions during the opening process, of an amount consistent with 1 H^+ : 1 K^+, whereas in non-opening conditions, in calcium iminodiacetate, only initial exchange of free space protons was observed. Later, Gepstein *et al.* (1982) measured rates of proton release from isolated guard cells in epidermal strips of *Vicia faba*, floating on 10 mol m^{-3} KCl, 0.1 mol m^{-3} $CaCl_2$, although again, unfortunately only in relative terms; using median figures for the area of strips per sample, frequencies of 60 complexes per mm^2 of epidermis, and single guard cell areas of 17×10^{-10} m^2 $cell^{-1}$, their rates convert to fluxes of 30–100 nmol m^{-2} s^{-1} on the basis of guard cell area, comparable with the K^+ fluxes measured in *Vicia*. Proton extrusion, together with stomatal opening, was inhibited in the dark, or by 1 mol m^{-3} vanadate in the light, and both were stimulated by FC. Proton extrusion was only partially inhibited by ABA, although K^+ uptake was abolished. Among the diagnostic features of the proton pump in the plasmalemma of many other plant cells are its stimulation by FC (see Marrè, 1979), and its sensitivity to vanadate (see Ch. 3; and Sze, 1985); the results therefore suggest mechanisms of ion transport in common between guard cells and other plant cells. Zeiger (1983) has argued for the primary role of the proton pump in salt accumulation in guard cells, and has suggested that most of the factors controlling stomatal aperture, including light-stimulated opening, act through modulation of the proton pump.

The clearest evidence for the existence of active proton extrusion at the plasmalemma of guard cells comes from recent very elegant studies on the blue light response in guard cells. It has been recognized for some years that there is a specific effect of blue light on stomatal aperture and net ion accumulation, which saturates at low light intensities, and is quite distinct from the photosynthetic light response with its peaks in red and blue (Hsiao *et al.*, 1973; Ogawa *et al.*, 1978; Zeiger & Hepler, 1977; Karlsson *et al.*, 1983; Zeiger, 1983; Karlsson, 1985). The kinetic properties of the blue light response in *Commelina communis* have recently been characterized by measurements of the changes in stomatal conductance which follow a brief pulse (1–100 s) of blue light (Iino *et al.*, 1985; Zeiger *et al.*, 1985). Such a pulse induces a transient peak in stomatal conductance, which rises within about 2 min to a peak at about 15 min, and returns to the base value within 50–60 min. These observations are not consistent with blue light as the energy source for the processes involved in stomatal opening, but instead suggest that a blue light-dependent photoconversion of some molecular species is involved in the activation of the opening processes. In this connection the demonstration of blue light-reducible cytochromes in the plasma membrane of *Neurospora* is particularly interesting (Borgeson & Bowman, 1985). The involvement of the proton pump in the blue light response is established in further work using the pulse technique. Assmann *et al.* (1985) used patch-clamping of *Vicia* guard cell protoplast membranes to characterise the electrical responses to 30 s blue light pulses. After a 25–35 s delay there was activation of an electrogenic ion pump, seen as a hyperpolarization of the membrane potential, or as a blue light-stimulated outward current in voltage–clamp conditions, dependent on the presence of ATP in the solution bathing the inside of the plasmalemma. Hyperpolarizations of up to 45 mV, and currents up to 0.88 μA cm^{-2} (equivalent to 9 pmol cm^{-2} s^{-1} of monovalent ion) were observed. The much greater noise in the voltage response in the presence of K^+ than in the presence of *n*-methylglucamine, a non-permeant cation, was attributed to the passive flux of K^+ through discrete channels in the membrane, of the kind identified in earlier patch–clamp studies by Schroeder *et al.* (1984) (see further description later in this section).

In a parallel study (Shimazaki *et al.*, 1986) net extrusion of H^+ from *Vicia* guard cell protoplasts in response to such blue light pulses was directly observed. H^+ extrusion continued for about 10 min after the pulse, and had a maximum rate of about 2–6 pmol cm^{-2} s^{-1}, similar to the pump current. The blue light-induced H^+ extrusion was abolished by diethylstilbesterol (DES) and by carbonyl cyanide-*m*-chlorophenylhydrazone, but was only partially inhibited (by 40 %) by ABA; this is similar to the partial ABA-inhibition of H^+ extrusion observed by Gepstein *et al.* (1982).

Thus the role of blue light is clearly identified as the activation of an electrogenic proton pump, but the link between the blue light photoreceptor and the activation of the membrane ATPase remains a challenging problem. The detailed kinetic study of the blue light effects on conductance suggested that in continuous light a photostationary state of activating molecular species was established, by which the blue light photosystem would drive stomatal opening at dawn (Zeiger *et al.*, 1981b), but would then act as a 'light-on' signal during the day with photophosphorylation in guard cell chloroplasts as the energy source for ion transport.

At this stage it is worth noting that the role of phytochrome in modulation of stomatal movement is quite different from the blue light response. Holmes & Klein (1985) showed that phytochrome had no effect in determining steady-state stomatal aperture, but that removal of the far-red absorbing form of phytochrome, P_{fr}, increased the rate of both opening and closing; they suggest that the effects should be interpreted as an increase in K^+ permeability on removal of P_{fr}, similar to that seen in mungbean subhypocotyl hook sections (Brownlee & Kendrick, 1977). Thus both the primary active transport, the proton pump and the secondary ion fluxes such as K^+, may be modulated by environmental signals.

The mechanism of K^+ influx is not established. It is difficult to estimate the equilibrium concentration for K^+ in guard cells, in the intact leaf. In such conditions the placing of the reference electrode for measurements of the membrane potential poses problems, and a reference electrode in the solution bathing the petiole may not register the potential in the guard cell apoplast, nor is the extracellular K^+ concentration well defined. Earlier measurements of membrane potentials in guard cells in epidermal strips, bathed in defined solutions, give surprisingly low potentials, which respond to external monovalent cations with little specificity for K^+, and which appear to have surprisingly small contributions from an electrogenic pump (Zeiger *et al*, 1977; Moody & Zeiger, 1978; Saftner & Raschke, 1981). Leakage of KCl into such small cells may affect the potential registered with electrodes containing 3000 mol m^{-3} KCl (Blatt & Slayman, 1983). Recent measurements suggest that the electrogenic component is higher than has previously been recorded. Blatt (1987) found that membrane potentials and resistance were low when measured using electrodes containing the usual 1000–3000 mol m^{-3} KCl as filling solution, but that much more negative membrane potentials, and much higher membrane resistances, were observed with electrodes containing 50 or 200 mol m^{-3} K^+-acetate. With 0.1 mol m^{-3} KCl and 1 mol m^{-3} Ca^{2+} outside, the membrane potential and resistance were -182 ± 7 mV and 1.6 ± 0.2 Ωm^2 respectively (n = 54). Potentials as large as -282 mV were recorded, and impalements held stable potentials and resistances for up to 2 h. Cells were highly selective for K^+ over Na^+, and were pH-sensitive in the range 4.5–7.4, with a 53 mV/pH unit slope; the potential was not affected by CO_2 in the range below the air-saturated level, but the response to high CO_2 levels was not studied. The membrane potential is also very sensitive to cyanide inhibition, as would be expected for a state in which the electrogenic pump makes a major contribution to the membrane electrical properties (Blatt, unpublished).

Edwards & Bowling (1985) observed a depolarization of the guard cell membrane potential on exposure to high CO_2, followed by a very marked hyperpolarization on its removal, to an extent which depended on the time of CO_2 treatment; they interpret the results in terms of inhibition of the proton pump by high CO_2, acidification of the cytoplasm during CO_2 treatment, and enhanced proton extrusion as the pump restarts on removal of CO_2. However, it is important to exclude changes in pH as the cause of the observed changes. Inhibition of the proton pump in *Nitella* by high CO_2 (1 mol m^{-3}) was shown by Spanswick & Miller (1977), and it may be that such sensitivity is a general feature in all plant cells. In guard cells it would explain the closing in response to high CO_2.

A recent development, which promises well for the future, is the introduction of patch–clamp techniques into guard cell electrophysiology (Schroeder *et al.*, 1984). Measurements on protoplasts of *Vicia faba* identified a K^+-selective channel, with a single channel conductance in symmetrical 225 mol m^{-3} KCl of 37 pS; the gating mechanism was voltage-insensitive, with a mean open time of 7–10 ms and a mean closed time of about 13 ms. The authors calculate that such channels would support a net K^+ flux of about 20 pmol cm^{-2} s^{-1}, given a membrane potential 30 mV displaced from the K^+-equilibrium potential, and a density of 1 channel per 15 μm^2, open one-third of the time. This is comparable with K^+ fluxes during stomatal opening, but considerably less than the net K^+ fluxes required to effect stomatal closure at the rates which can be observed (see Sect. 12.7, below). Thus it may well be that only passive K^+ influx through specific K^+ channels is required for stomatal opening, driven by the membrane potential set up by the electrogenic proton pump. However, a direct demonstration that K^+ levels never exceed their equilibrium values is required, before ruling out some active mechanism for K^+ influx, as for example the H^+–K^+ co-transport system in *Neurospora* (Rodriguez-Navarro *et al.*, 1986).

The influx of Cl^- into guard cells must be active, but there is no evidence identifying its mechanism. *Chara* remains the only cell in which the mechanism of Cl^- influx has been established, as a 2 H^+ : 1 Cl^- co-transport system (Sanders, 1980; Sanders & Hansen, 1981; see Ch. 5). If the mechanisms of salt accumulation in guard cells are those established in *Chara* then the requirements for ATP may be compared with the measured capacities for ATP synthesis by photophosphorylation in guard cell chloroplasts. The proton pump in *Chara* is now shown to be 1 H^+ : 1 ATP (Lucas, 1982; Beilby, 1984) and hence the (indirect) ATP requirement for Cl^-influx in *Chara* is 2 ATP : $1Cl^-$. If K^+ entry is passive, driven by the membrane potential, then this is also the stoichiometry for KCl entry. In guard cells in the leaf the anion balancing K^+ is a mixture of malate (or other organic acid anion) and Cl^-, depending on species and conditions. Potassium malate accumulation will demand the extrusion of only 1 H^+ per 1 K^+ accumulated, and also has the further yield of ATP from the glycolytic production of phosphoenolpyruvate (of 0.5 ATP per K^+), though at the expense of potential respiratory substrate. Thus the ATP requirement for a salt influx of 100 nmol m^{-2} s^{-1} as observed in opening wi11 be between 0.3 and 0.6 pmol $cell^{-1}$ h^{-1}, for the plasmalemma transport alone. Mechanisms of ion transport at the tonoplast are even less well characterized, but will presumably add (directly or indirectly) a further ATP requirement. The ATP synthetic capacitics in *Vicia* protoplasts measured by Shimazaki & Zeiger (1985) were about 0.2 pmol $cell^{-2}$ h^{-1} in non-cyclic photophosphorylation, and 0.4 pmol $cell^{-1}$ h^{-1} in cyclic photophosphorylation. Thus both processes are likely to be necessary to supply ATP at the rate required, and there seems to be little leeway for supporting any further ATP requirement (direct or indirect) for K^+ entry.

12.5 Opening of guard cells in epidermal strips

It is necessary to use isolated guard cells for flux measurements, but it is then important to consider the extent to which their behaviour differs from that of intact guard cells. In this connection the comparison simply of apertures is misleading. Isolated guard cells, in the absence of subsidiary/epidermal cell turgor, are mechanically in a different environment, particularly in 'closing' conditions, and this will be reflected in their apertures. This difference is secondary, and unimportant for our studies, provided we do not attempt to make invalid comparisons. What is, however, important is the question of whether isolated and intact guard cells differ significantly in their metabolic and ionic properties and capabilities. It is my view that the evidence available suggests that they do not.

In epidermal strips of *Vicia faba* opening requires 1–10 mol m^{-3} KCl in the bathing medium, and is still sensitive to light/dark, to ABA and to CO_2; much higher concentrations (100 mol m^{-3}) of NaCl, LiCl or CsCl are required to support opening, although the cation specificity is lost in the absence of Ca^{2+} (Humble & Hsiao, 1969). The opening in 10 mol m^{-3} KCl, 0.1 mol m^{-3} $CaCl_2$, is insensitive to pH in the range 4–7 (Fischer, 1972). Opening is reduced with increasing Ca^{2+} above 0.1 mol m^{-3}, particularly at low concentrations of KCl (Willmer & Mansfield, 1969: Fischer, 1972). Extreme sensitivity to external Ca^{2+} seems to be a feature of guard cell behaviour, and one of the more interesting questions for future work.

With epidermal strips of *Commelina* surprisingly high concentrations of external K^+ are required to achieve opening, particularly for intact guard cells. At lower concentrations of external KCl, intact guard cells will open only in CO_2-free conditions in light; Zeiger (1983) suggests 20–30 mol m^{-3} is adequate, while Travis & Mansfield (1979) suggested 50 mol m^{-3}. In air, higher concentrations are required, with intact guard cells requiring 60–100 mol m^{-3} KCl outside, although isolated guard cells can achieve the same aperture with only 20–30 mol m^{-3} KCl (Willmer & Mansfield, 1969; MacRobbie & Lettau, 1980a). Opening is abolished by the addition of 1 mol m^{-3} Ca^{2+}, and severely inhibited even at 0.1 mol m^{-3} Ca^{2+} (Fujino, 1967; Willmer & Mansfield, 1969). The degree of Ca^{2+} sensitivity is surprisingly high, as are the K^+ concentrations required, particularly in view of the very low K^+ concentrations measured in the leaf apoplast by Blatt (1985). In the intact leaf, low CO_2 in the air space, and possibly reduced turgor in subsidiary cells exposed to the air space on their inner walls, may promote opening, but the possibility remains of some important and unrecognized difference in conditions between guard cells in the intact leaf, and those in intact epidermal strips.

Comparisons between the behaviour of guard cells in intact and isolated epidermal strips of *Commelina* suggest that no drastic alterations in the metabolic/ionic properties are associated with the isolation procedure. Isolated guard cells are still sensitive to the proper signals for opening and closing, such as light, CO_2, presence of ABA. The osmotic measurements (MacRobbie, 1980) showed that the osmotic requirement was not significantly different in isolated and intact epidermal strips, suggesting that the properties of the guard cell wall are not altered by the isolation treatment at low pH. Incubation of intact epidermal strips on varying concentrations of KCl produces smaller apertures than would be achieved in isolated guard cells in corresponding conditions, but if we assume that, after incubation in a given external concentration, the ion concentrations in guard cells are equal in the two sets, then the intact guard cells have apertures close to those predicted for their assumed ion contents. (This rests on combinations of figures from MacRobbie & Lettau 1980a, b.)

In *Commelina* guard cells, as in *Vicia*, there seems to be little effect of external pH; pH 3.9 is used to isolate guard cells, and they appear able to survive for long periods at this pH. There is little difference between the degree of opening achieved in isolated guard cells on given concentrations of KCl at pH 3.9 and at pH 6.7 (MacRobbie & Lettau, 1980a; Clint & MacRobbie, 1984). Indeed such tolerance of low external pH may be a requirement for guard cell function; the extrusion of significant quantities of protons into the restricted volume of the guard cell wall, with poor diffusion access into the rest of the apoplast, may give rise to transient conditions of low external pH; the greater sensitivity of subsidiary cells to low pH may lead to leakage and mobilization of some of their solutes for uptake by the guard cell.

The comparisons which can be made suggest that isolated guard cells retain their ion transport properties, and can reasonably be used as a model for the behaviour of guard cells in the intact leaf. They will, however, provide information on their responses in given experimental conditions, conditions which are not necessarily close to those to which they are exposed in the intact leaf. While we can expect the behaviour of isolated guard cells to throw light on processes in the intact leaf, we must not expect them to reproduce exactly the patterns of their behaviour *in vivo*. As has already been argued, isolated guard cells are likely to serve as models for events in guard cells whose apertures in the leaf are up to about 10 μm; the experimental conditions so far used do not reproduce the internal state of the intact guard cell associated with wide opening.

12.6 Regulation of ion levels in guard cells

Another question of interest is whether there are any sort of homeostatic controls on salt levels in guard cells, superimposed on the more dramatic responses of ion fluxes to environmental signals. In both *Vicia* and *Commelina* intact guard cells accumulate salt to much higher levels, and achieve much higher turgor, than the regulated levels, the set points, seen in many more typical cells (Cram, 1976, 1980). Hence if there are feedback controls to regulate either turgor or internal ion concentration these can only be activated at very high levels. However, comparison of the behaviour of intact and isolated guard cells suggests that neither turgor nor internal concentration is such a regulatory factor. For example, isolated guard cells of *Commelina* will open up to 14–18 μm when bathed with 30 mol m^{-3} K(Rb)Cl(Br) in light, with somewhat greater openings in KCl, than in KBr, than in RbCl (MacRobbie, 1983); the intracellular salt concentration is then of the order of 160–200 mol m^{-3}. Accumulation to much higher levels (400 mol m^{-3} and more) is possible in intact guard cells, with consequently higher turgor. The same difference is seen between ion levels in intact and isolated *Vicia* guard cells (Table 12.1), and the question therefore arises why the isolated guard cells do not continue to accumulate ions, and open to the wider apertures which are mechanically possible. One possibility is that ion accumulation in guard cells is regulated by volume, rather than by turgor or internal concentration, as for example in the wall-less alga *Poterio-ochromonas* (Kauss, 1978).

Some preliminary experiments suggest that homeostatic regulatory controls should be included among the factors determining stomatal aperture. Table 12.5 shows that isolated guard cells of *Commelina*, incubated

Table 12.5 Effect of external sucrose on steady-state opening†

^{86}RbCl (mol m^{-3})	Sucrose	Aperture (μm)	Q^* (pmol mm^{-2})	C^* (mol m^{-3})	ΔC (mol m^{-3})
30	0	10.3 ± 0.2 (116)	29 ± 5(7)	40 ± 7	
30	200	10.4 ± 0.2 (121)	54.7 ± 5.5(8)	75 ± 7.5	35 ± 10
30	0	12.2 ± 0.3 (87)	25 ± 1 (6)	33 ± 2	
30	200	12.1 ± 0.3 (68)	34 ± 3 (5)	45 ± 3.5	12 ± 4
60	0	11.5 ± 0.2 (75)	56 ± 3 (5)	74 ± 4	
60	200	9.4 ± 0.2 (85)	77 ± 9 (6)	109 ± 13	35 ± 14

† MacRobbie, unpublished.
Isolated guard cells incubated overnight, in light, pH 3.9, in ^{86}RbCl.
Q^*, steady-state content.
C^*, Internal concentration Q^*/V; volume of single guard cell V(pl) = 4.0 + 0.2A (μm), 120 guard cells per mm^2.
ΔC, increase in internal ^{86}Rb concentration with 200 mol m^{-3} sucrose outside.
200 mol m^{-3} sucrose, 214 mosmol kg^{-1}, is equivalent osmotically to 115 mol m^{-3} NaCl.

overnight in ^{86}RbCl with 0 or 200 mol m^{-3} (214 mosmol kg^{-1}) sucrose in the medium, open to the same extent; in the presence of external sucrose the internal concentration of RbCl is increased, though the increase is much smaller than the osmotic increase externally, and therefore again some other solute must also be involved. It remains to be established which flux is subject to control, and the mechanism by which it is achieved.

Also consistent with the suggestion that control may be exerted on cell volume are observations by Clint (1985b) that guard cell protoplasts, incubated in 30 mol m^{-3} KCl with varying sucrose concentrations, showed tendencies towards volume regulation. At high sucrose, with initial volumes less than about 1.8 pl per protoplast, there was an increase in volume with time, whereas in lower sucrose, with initial volumes greater than about 1.8 pl per protoplast, the volume tended to decrease with time. Thus although in the short term protoplasts act as osmometers, in the long term there is evidence for volume regulation, provided KCl is present in the medium. (It may be added at this point that for isolation of protoplasts in best condition, Clint (1985a) found that it was essential to have KCl in the medium at all stages; this is not unexpected in the light of the behaviour of guard cells as established by Imamura (1943), but KCl has not been included in most media for the isolation of guard cell protoplasts. It would seem important that this should become standard practice in future work.)

12.7 Flux measurements in guard cells of *Vicia*

The first measurements of ion flux in isolated guard cells were made by Fischer, using epidermal strips of *Vicia faba* in which cells other than guard cells had been killed (Fischer, 1972; Pallaghy & Fischer, 1974), but no

Table 12.6 Flux measurements in guard cells of *Vicia faba*†

Pre-incubation	Aperture (μm)	Net tracer influx (nmol m^{-2} s^{-1})		Efflux (nmol m^{-2} s^{-1})	
		‡	§	‡	§
–	about 7	165	110		
300 min‡ or 200 min§, light, CO_2- free air	about 13	20	27	37	26
300 min‡ or 200 min§, dark	about 7–8	122	85		

† Fischer (1972).
‡§ Fluxes were measured using ^{42}K over 90 min(‡), or ^{86}Rb over 60 min (§), in light plus CO_2-free air, on 10 mol m^{-3} KCl, 0.1 mol m^{-3} $CaCl_2$. Figures are based on guard cell areas, assuming 65 complexes per mm^2 of epidermal strip, and single guard cell areas of 14×10^{-10} m^2.

systematic study of ion fluxes in *Vicia* guard cells in different conditions has been undertaken.

Fischer's results, shown in Table 12.6, show that K^+-influx during the initial opening phase is much higher than the steady-state fluxes in open guard cells, when steady content and aperture have been reached. They argue therefore for some form of regulation of ion fluxes, of both influx and efflux. After opening, such isolated guard cells will have reached an internal K^+ concentration of only about 280 mol m^{-3}, much less than the level reached in open guard cells in the intact leaf, with turgid epidermal cells. Neither turgor nor ion content seems likely, therefore, as the regulatory factor controlling ion influx; instead the results may suggest regulation based on cell volume.

The measured fluxes in these conditions may be compared with the net ion movements required to open or close the pore over the observed opening/closing times. In both *Vicia* and *Commelina* a net movement of about 1.2 mmol m^{-2} is required (based on guard cell area). Thus opening over 1–2 h will require an average net influx of 300–150 nmol m^{-2} s^{-1}, of the same order as the influx during initial opening of isolated guard cells on 10 mol m^{-3} KCl. However, the flux required to close guard cells over 20 min, of 1000 nmol m^{-2} s^{-}, is much greater than the steady-state fluxes in open guard cells. These flux measurements suggest that initiation of closing must involve a massive stimulation of ion efflux, and cannot be achieved by simply deactivating the processes of accumulation.

12.8 Flux measurements in guard cells of *Commelina*

Although they remain very incomplete, measurements of ion fluxes in isolated guard cells of *Commelina*, in a range of conditions, both in the steady states and in transitions between different states, throw some light on the processes involved in the control of salt accumulation, and hence aperture, in these cells (MacRobbie, 1981a, b, 1983, 1984; Clint & MacRobbie, 1984).

Before discussing the results, a brief consideration should be given to the conditions necessary for flux measurements, and the information available from such experiments. For measurement of plasmalemma fluxes it is necessary to be able to distinguish tracer exchange in the free space from that of the cells, requiring a wash period long enough to remove free space tracer but short enough to involve negligible loss of intracellular tracer. Unfortunately, because of the very slow cation exchange in the cell wall, this requirement is best met by the use of low pH outside, with reduced free space cation. At pH 6.7 it is more difficult to separate free space and cell tracer. For this reason most flux measurements have been done at pH 3.9, the pH used for isolation of guard cells, but one in which they are stable for many hours. To allow direct comparison, this pH has been used for both cation measurements ($^{86}Rb^+$ in RbCl, as an analogue of K^+), and anion measurements [$^{82}Br^-$ in KBr, as an analogue of Cl^- or $^{36}Cl^-$ in (K + Na) Cl]. However, it seems likely that the pH in the guard cell wall must often reach very low values, in the initial transients of opening, and this low pH may not be as 'unphysiological' as plant physiologists conditioned to more commonly used tissues may fear. Provided we are cautious in extrapolation to the intact leaf (where ionic conditions are in any case unknown), we may hope that the measurements on isolated guard cells will throw light on the patterns of behaviour *in vivo* on the type of process involved.

Isolated guard cells, incubated overnight on K(Rb) Cl(Br), reach steady states of aperture and ion content, dependent on the external concentration and conditions. Measurements of ion influx and efflux were made over the subsequent 4–8 h, influx over a 30 min period of exposure to tracer, and efflux by collection of tracer lost to non-radioactive bathing solution in successive time periods. In constant conditions it may be possible to use analysis of efflux kinetics to distinguish vacuolar and cytoplasmic tracer, to allow separate calculation of the ion contents of cytoplasm and vacuole, and of the plasmalemma and tonoplast fluxes. The necessary condition is that, relative to the plasmalemma fluxes, the tonoplast fluxes should not be too high; there is then a sizeable component of faster exchanging cytoplasmic tracer in the efflux, and fast and slow components can be distinguished in the period after free space exchange. In most experiments with $^{82}Br^-$, this condition was met, as the plasmalemma fluxes were high relative to the tonoplast fluxes, and cytoplasmic and vacuolar contributions could be distinguished. With $^{86}Rb^+$ the condition was not met, and only a single component appeared in the tracer exchange, providing measurements only of the plasmalemma flux.

It is important to recognize the assumptions on which the calculations of the efflux analysis are based. These are that cytoplasm and vacuole are in series, that the contents of cytoplasm and vacuole, and the fluxes, are constant during the efflux period (but not during the whole of the loading period), and that the internal specific activity at the start of the efflux is equal to that outside; the last is justified because the loading period was long enough relative to the half-time for exchange, or because there was little or no Br^- in the tissue at the start of the loading.

Two types of flux experiment have been done. The first concerns measurements in the steady states in various conditions, and the second concerns the flux transients, the responses to changes in conditions leading to changes in ion content and aperture. The measurements of transients, in particular, help throw light on the mechanisms of control of ion accumulation in guard cells.

12.8.1 *Steady-state fluxes at the plasmalemma*

Plasmalemma fluxes of ^{86}Rb and ^{82}Br have been measured in steady-state tissue of different apertures, after incubation in light at different external concentrations of RbCl or KBr. For both ions the steady-state flux at the plasmalemma estimated from the efflux curves was approximately equal to the influx measured over short time periods in tissue given the same treatment as the efflux tissue, thus confirming the steady-state condition.

Plasmalemma fluxes of up to 50 pmol mm^{-2} h^{-1} for $^{86}Rb^+$ were measured (on the basis of epidermal strip area). Plasmalemma fluxes of $^{82}Br^-$ were of the same order, but increased more steeply with concentration; in different conditions fluxes of 7–112 pmol mm^{-2} h^{-1} were measured. On the basis of guard cell area these figures correspond to up to 50 nmol m^{-2} s^{-1} for Rb^+ and 10–120 nmol m^{-2} s^{-1} for Br^-. Thus although the higher values are large in comparison with fluxes in other cells, they are still of the same order. Again it might be noted that the steady-state ion effluxes, even those in open guard cells, are much lower than the net ion efflux required to produce closing in anything like the observed time in the intact leaf; thus a simple deactivation of the uptake mechanisms would not be adequate to produce the rate of closure observed, and a massive stimulation of efflux seems to be required.

12.8.2 *$^{82}Br^-$ efflux analysis*

In general the time course of tracer efflux from $^{82}Br^-$ loaded tissue could be split into two exponential components, and cytoplasmic and vacuolar components could be separately estimated, from the intercepts and rate constants of these components. The faster exchanging compartment has a half-time of about 38 min, independent of aperture, while the rate constant for the slower component increased with aperture, having half-times of 3–35 h in tissue of different apertures.

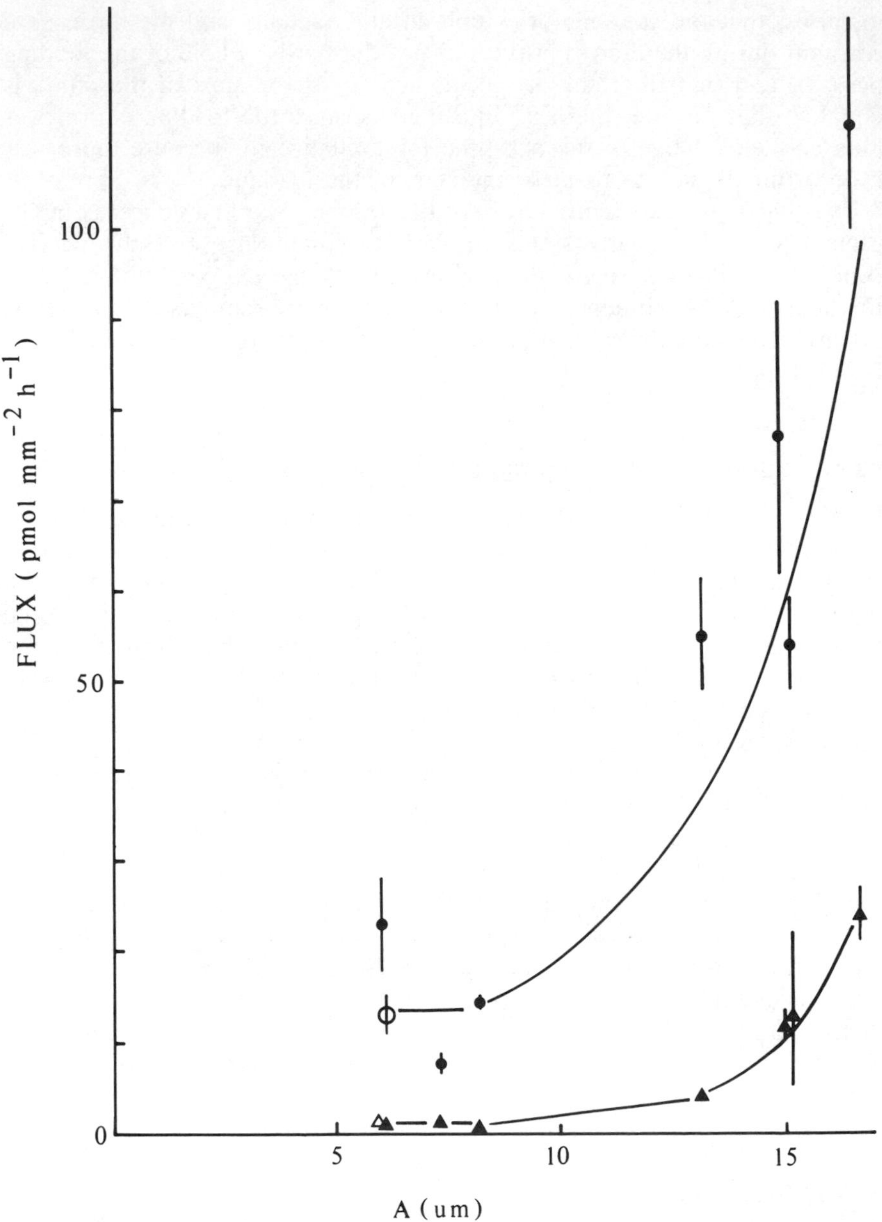

Fig. 12.3 Steady-state fluxes of $^{82}Br^-$ at plasmalemma (● and ○) and tonoplast (▲ and Δ) in guard cells of different apertures. Different degrees of opening produced by different external concentrations of KBr in light (●, ▲) or by incubation in the dark (○, Δ). (MacRobbie, 1981a, 1983, 1984.)

Figure 12.3 shows the values of plasmalemma and tonoplast fluxes, ϕ_p and ϕ_t, in different experiments, with aperture, A, varied by different external concentration, C_o, or by incubation in dark rather than light, or by the inherent variability in different batches of tissue. It shows that the plasmalemma flux is considerably higher than that at the tonoplast, by a factor of 5–30, and that both increase with aperture. The tonoplast fluxes are surprisingly low in comparison with the plasmalemma fluxes, or with those in other cells. To assess differences between different treatments it is worth trying to separate out the direct effects of changes in internal and external concentrations from changes in the state of the transport systems. Figure 12.4 shows the relations between aperture and the rate constants for the various fluxes, the quantities ϕ_p/C_o, ϕ_p/Q_c, ϕ_t/Q_c and ϕ_t/Q_v. (The last three of these quantities are in units of h^{-1}, and can be compared directly, but ϕ_p/C_o is in units of (pmol mm^{-2} h^{-1})/(mol m^{-3}), and cannot be compared with the others). The results show that two of the rate constants, ϕ_p/C_o for influx at the plasmalemma, and ϕ_t/Q_v for flux from vacuole to cytoplasm, increase markedly with increasing aperture. The relation between ϕ_t/Q_c and aperture is less clear; the overall regression is just significant at $P = 0.05$, with a less steep slope in the regression, but in the experiment with replicate tissue at different C_o this rate constant did not change with aperture. Thus open guard cells have a more ready transfer of Br^- from vacuole to cytoplasm, and perhaps also from cytoplasm to vacuole. The increasing tonoplast fluxes with aperture are not simply the result of an increase in tonoplast area with opening. In opening the guard cell volume increases from about 4 pl to about 7 pl, and the increase in cell area will be by a factor of about 1.32. Because open guard cells have fewer, larger vacuoles than do closed guard cells the area increase for the tonoplast will be by much less than this factor, and indeed tonoplast area may not change. Thus the increase in the tonoplast rate constants, ϕ_t/Q_c and ϕ_t/Q_v, by factors of about 10 must represent real changes in the transport systems involved, and activation of the tonoplast transport processes.

In the experiment in which aperture was varied in replicate tissue by varying C_o, the ratio of fluxes at plasmalemma and tonoplast was constant, but when replicate tissue was compared in light and dark the tonoplast flux was reduced much more drastically in the dark than was the plasmalemma flux.

The variation of ϕ_p/C_o with aperture is very marked. Figure 12.4 shows that this rate constant, for influx at the plasmalemma, increases markedly with aperture in the general relation for all experiments; this was also true in the two experiments with replicate tissue, whether aperture was reduced by lowering C_o, or by overnight incubation in dark rather than light. The higher plasmalemma influx in light than in dark is not unexpected, but the tenfold increase in plasmalemma influx produced by a threefold increase in C_o (from 20 to 60 mol m^{-3}) is more surprising.

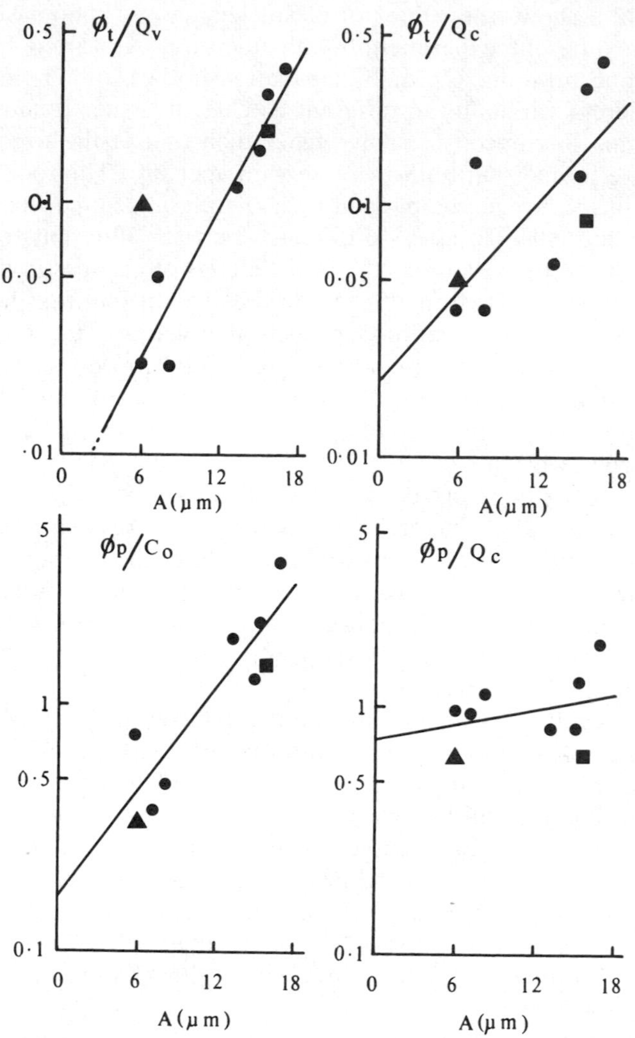

Fig. 12.4 Relation between rate constants for plasmalemma and tonoplast fluxes of $^{82}Br^-$ and aperture, data as in Fig. 3.
ϕ_p, ϕ_t: steady-state fluxes at plasmalemma and tonoplast (pmol mm^{-2} h^{-1})
Q_c, Q_v: estimated Br^- content of cytoplasm and vacuole (pmol mm^{-2})
C_o: external concentration of KBr (mol m^{-3})
Different apertures produced by different concentrations of KBr (●), or by incubation of replicate tissue in light, (■) or dark (▲). Significance of regressions of log (rate constant) on A:
ϕ_p/C_o: $r_7 = 0.894$ $P < 0.01$
ϕ_p/Q_c: $r_7 = 0.316$ not significant
ϕ_t/Q_c: $r_7 = 0.691$ $P < 0.05$
ϕ_t/Q_v $r_7 = 0.854$ $P < 0.01$
$r_6 = 0.925$ $P < 0.001$ (light points, line shown)
(Calculated from data in MacRobbie, 1981a, 1983, 1984.)

It is very unlikely that the increase in plasmalemma influx can be attributed to an effect of changing membrane potential. If the mechanism of Br^- influx at the plasmalemma is, as in *Chara*, a 2 H^+ : 1 Cl^- co-transport system (Sanders, 1980), then the depolarization as C_o increases would, if anything, cause a reduction in this rate constant, rather than an increase. The results suggest that opening is the result of activation of the plasmalemma transport processes, that this has occurred in the light-treated but not in the dark-treated tissue, but also that the activation is sensitive to external ion concentration. The effects of C_o may well be mediated through changes in internal conditions, and may reflect the same sensitivity of the state of guard cells to external K^+ as was demonstrated by Imamura (1943). The result suggests that transport systems may be controlled by the internal state of the cells, by mechanisms which need to be established, and that it is likely that cytoplasmic ion concentrations are important in such regulation.

12.8.3 Cytoplasmic and vacuolar concentrations of Br^-

The intracellular distribution of the tracer Br^- was assessed from the amounts in cytoplasm and vacuole estimated from the efflux kinetics, with rather surprising results. The efflux analysis gives values for Q_c and Q_v, the cytoplasmic and vacuolar tracer contents in the various steady states reached at the start of the efflux period. The surprising result was that, while both Q_c and Q_v increased with aperture, the slope of the regression of Q_c on A was greater than that of Q_v on A. The conclusion seems inescapable that the slower exchanging phase, with content Q_v, must indeed represent the vacuole, yet the vacuolar volume, at least in open guard cells, must be greater than the cytoplasmic volume, and the result therefore implies relatively smaller changes in vacuolar ion concentrations with aperture.

There are various uncertainties in trying to convert these contents to concentrations, the most serious being the estimation of cytoplasmic and vacuolar volumes, but none of them can alter the basic conclusions. Figure 12.5 shows the estimated values of cytoplasmic and vacuolar concentrations, Br_c^-, and Br_v^-, in the various experiments, including the two with replicate tissue in different conditions. Calculations were done on the basis of constant cytoplasmic volume during opening, with the increase in volume going to the vacuole. Campbell & Garber (1980) showed that cytoplasmic volume was constant during swelling of *Albizzia* pulvinar motor cells, and the same may be true in guard cells: it is certain that the fraction of cell volume occupied by the guard cell vacuole increases during opening. The figure of 70 % cytoplasmic volume in closed guard cells, used for calculation of the results shown in Fig. 12.5, is likely to be an overestimate; its use will give an upper limit for the vacuolar ion concen-

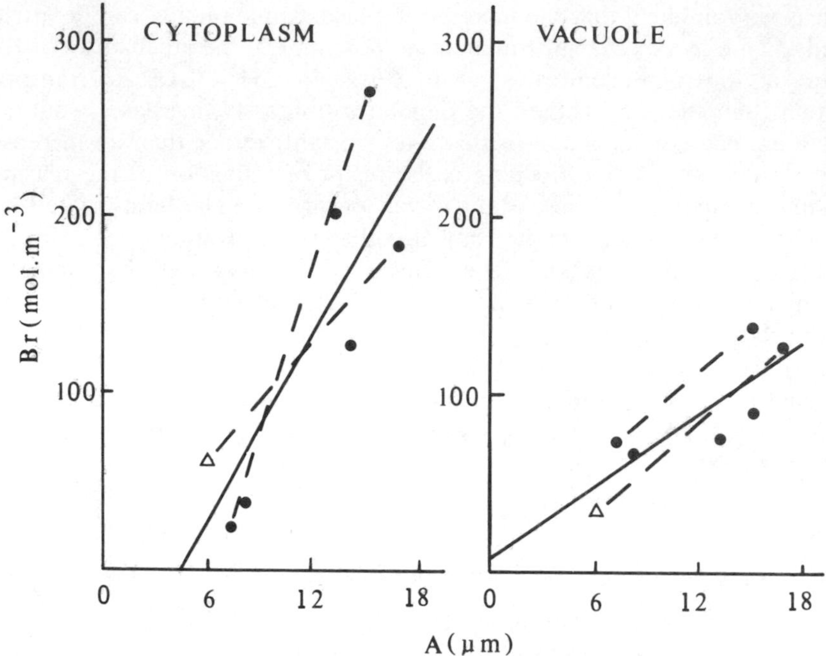

Fig. 12.5 Estimated concentrations of Br^- in cytoplasm and vacuole in steady state guard cells of different aperture; (●) incubation in light, various external concentrations; (Δ) incubation in dark. Dotted lines join values for replicate tissue in different conditions. Full lines are calculated regressions for all points.
Br_c: $r_5 = 0.822$ $P < 0.05$; slope 17.6 ± 5.5 mol m^{-3} μm^{-1}
Br_v: $r_5 = 0.839$ $P < 0.02$; slope 6.9 ± 2.0 mol m^{-3} μm^{-1}
(Calculated from data in MacRobbie, 1981a, 1983, 1984.)

trations, which will therefore provide an underestimate of the vacuolar osmotic discrepancy revealed by these results. The cytoplasmic Br^- concentrations are relatively high, particularly in open guard cells. This may be a consequence of the conditions of incubation, with high external KBr, and in the unknown conditions of guard cells in the intact leaf the cytoplasmic ion levels may well not reach the highest of these levels. Nevertheless, it is worth examining the osmotic consequences of the observed changes in the isolated guard cells.

The most important conclusion from such examination concerns the rate of increase of cytoplasmic and vacuolar concentration with aperture, the estimate of the consequent osmotic effects, and their comparison with the osmotic requirement to change the aperture. Table 12.7 shows the estimates of rates of increase in ion concentrations in cytoplasm and vacuole. Neither of the slopes reached the osmotic requirement, although the higher estimates of cytoplasmic changes are close, and with slightly different estimates of cytoplasmic volume might be stretched to do so; thus the

Table 12.7 Changes in cytoplasmic and vacuolar Br^- concentrations with aperture

	$\frac{\Delta Br_c}{\Delta A}$ (mol m^{-3} µm^{-1})	$\frac{\Delta Br_V}{\Delta A}$	$\Delta\pi/\Delta A$ (mosmol kg^{-1} µm^{-1})
Collected results†	25, 18	6.3, 6.9	
L/D	16, 11	7.5, 8.6	
C_o	45, 32	8.0, 8.4	
Average over 7–15 µm‡			112

Br_c^-, Br_V^-: estimated cytoplasmic and vacuolar concentrations; the two figures shown are based on 50 % and 70 % cytoplasmic volume in closed guard cells, respectively; cytoplasmic volume is assumed constant with aperture.
A: aperture.
Collected results: slopes of the calculated regressions for all experiments.
L/D: experiment based on comparison of replicate tissue in light and dark.
C_o: comparison of replicate tissue in 20 and 60 mol m^{-3} KBr, in light.
† MacRobbie (1981a, 1984).
‡ MacRobbie (1980).

increase in cytoplasmic KBr accounts for a large fraction of the required osmotic increase, and the cytoplasmic discrepancy is relatively small. In contrast, the vacuolar salt increase is nowhere near that required osmotically; the contribution of increasing vacuolar KBr is only 12–17 mosmol kg^{-1} µm^{-1}, certainly less than 20 mosmol kg^{-1} µm^{-1}, compared with more than 100 mosmol kg^{-1} µm^{-} required to change the aperture, over the range concerned. This is an important, albeit unexpected, result. The discrepancy between measured salt accumulation and osmotic solute required is primarily a vacuolar, and not a cytoplasmic, discrepancy. It does not appear to be a situation in which osmotic balance is achieved by salt accumulation in the vacuole, and accumulation of a non-salt, 'compatible' solute in the cytoplasm, as Outlaw (1983) has suggested. Instead, the results suggest that as the plasmalemma transport is activated (seen as high ϕ_p/C_o) and the cytoplasmic salt level increases, the vacuolar transfer of both salt and some other solute is increased. Thus both plasmalemma transport, and the tonoplast transport processes for both salt and some other solute, seem to be sensitive to the state of the tissue, as affected both by light/dark and more surprisingly, by the level of external salt.

12.8.4 Flux responses to closing signals

One of the most striking peculiarities of guard cells, and therefore one of the key sequences to understand, is their loss of the ability to retain solutes, to maintain high turgor, in response to a closing signal, such as high CO_2, a light-off signal, or exposure to ABA. It is first necessary to identify the sensitive process(es), to describe the change in terms of flux transients, by recording the response of both influx and efflux to such signals. As yet only two such responses have been characterized, that seen on the transfer of open guard cells from light to dark, and that seen on adding ABA to open guard cells. Interestingly, both turn out to be very similar.

12.8.4.1 Effect of transfer from light to dark

Isolated guard cells do reduce their aperture on transfer to the dark, although in the absence of subsidiary cell turgor they do not close completely; the reduction in aperture is associated with a reduction in ion content. This reduction is not the result of reduced ion influx, but instead, of a 'transient' stimulation of efflux of both cation and anion (MacRobbie, 1983, 1984).

To assess the effect of this transfer on the ion influxes, guard cells were opened by overnight incubation on non-radioactive solution in light, to reach a steady state of content and aperture, and influx was then measured over a period of about 30 min in light or dark. In 30 mol m^{-3} $^{86}RbCl$ the ratio of the influx in the dark to that in the light was 0.97 ± 0.06 (4), while in 25 mol m^{-3} $K^{82}Br$ the corresponding ratios were 1.09 and 0.96 in two experiments. Thus neither influx is significantly reduced by transfer to the dark.

The effects on efflux are most clearly seen in the changes in the apparent rate constant for efflux k the quantity tracer efflux/tracer content. For $^{86}Rb^+$, for which only one component appears in the efflux kinetics, k is constant during steady-state efflux; for $^{82}Br^-$, when two exponential components are distinguishable in the efflux curve, k will fall with time in the steady-state tissue, but an increase in efflux will be seen as an increase in k. Figure 12.6 shows the effect on k of the transfer of open guard cells from light to dark, and a clear transient stimulation of tracer efflux of both cation and anion is seen.

It is important to recognize that we are not dealing with a simple difference in efflux between light and dark conditions, with a simple permeability change, but with a specific transient high efflux from open guard cells, induced by transfer to the dark. Thus although transfer to the dark induces a transient increase in Rb^+ efflux, by more than twofold, the rate constant falls again to a value very similar to that before the change. By contrast the transfer from dark to light produces a reduction in k only if done during the short period of stimulated efflux; there is no change if the transfer is made after a longer dark period, or on transfer of steady-state dark tissue to light.

The amount of excess loss of $^{86}Rb^+$ during the transient is difficult to estimate accurately, since the specific activity is uncertain after the change, but it amounted to about 11 ± 1 (12) pmol mm^{-2}, a relatively small fraction of the total ion content in the initial steady-state tissue.

The effects of transfer to the dark on the efflux of $^{82}Br^-$ and $^{36}Cl^-$ are

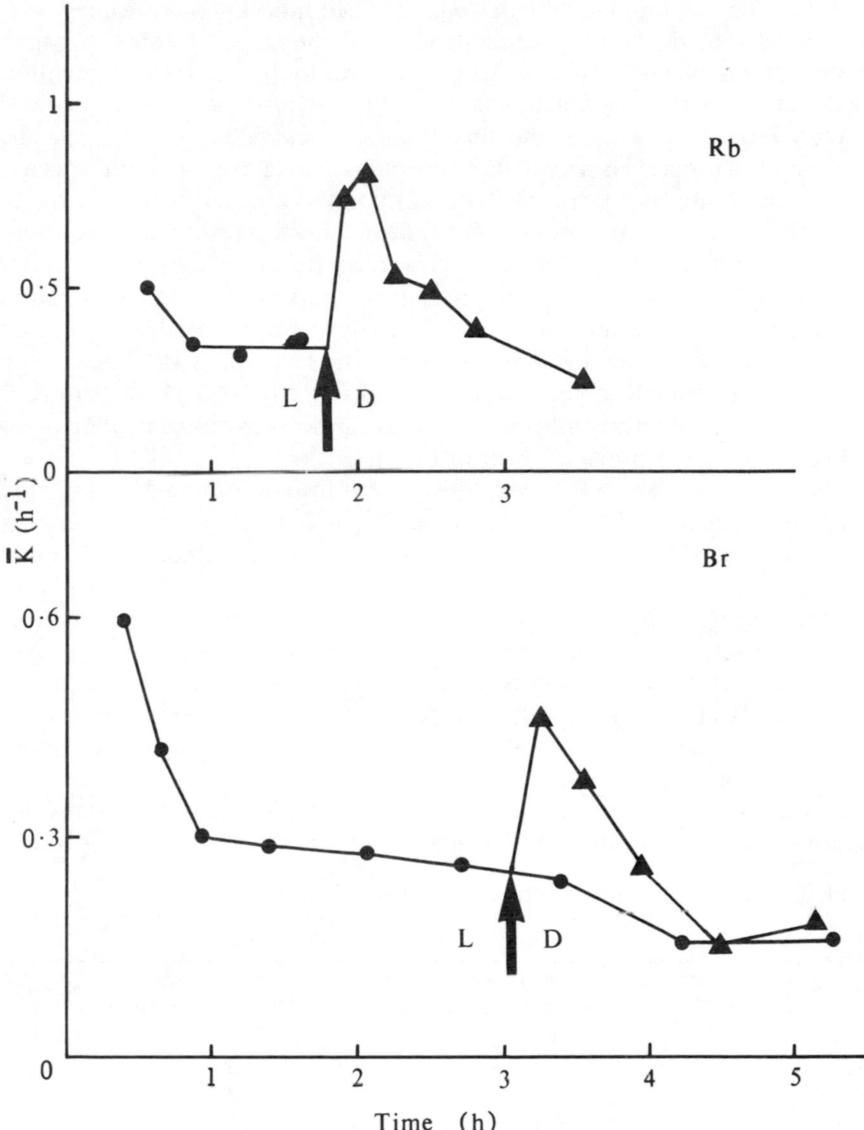

Fig. 12.6 Effect of transfer from light to dark on efflux of $^{86}Rb^+$ and $^{82}Br^-$ from steady-state tissue, loaded overnight in light. Apparent rate constant for exchange, $\bar{k}$ = rate/content. Rb: mean of four strips. Br: initial wash-out of five strips in light; transfer of three strips to dark (▲), remaining two strips left in light (●). (MacRobbie, 1983, 1984.)

similar to those seen with $^{86}Rb^+$, as shown in Fig. 12.6. The interpretation of the anion transients is more complex than with $^{86}Rb^+$, but nevertheless the marked increase in k after the transfer from light to dark shows clearly that the efflux is stimulated by this change. The ratio $\bar{k}_D/\bar{k}_L$ associated with first transfer to the dark was independent of the stage of efflux at which the transfer was made; this means that the response reflects a stimulation of the efflux at the plasmalemma.

With $^{82}Br^-$ the subsequent time course, dependent on the specific activity in the cytoplasm, will be determined by fluxes at both plasmalemma and tonoplast. Borle *et al.* (1982) pointed out that a peak in tracer efflux in such an experiment may, in some circumstances, reflect a transient in specific activity in the phase from which tracer efflux occurs, under conditions of a constant stimulation of true efflux. However, examination of the detailed time course of $\bar{k}$, and its computer simulation, showed that the transient in $\bar{k}$ does indeed represent a true transient in efflux at the plasmalemma (MacRobbie, 1984). On transfer to the dark the efflux is increased more than twofold, and then declines over the next hour or so (though not to the previous level in light).

The amount of excess Br^- loss during the transient is difficult to estimate accurately, because of the uncertain specific activity after the transfer, but it was similar to the amount of excess Rb^+ loss, of the order of 10–25 pmol mm^{-2}.

After the transfer to the dark the tissue is no longer in a steady state, and ion fluxes and contents are no longer constant. As a result the calculation of cytoplasmic and vacuolar contents, and tonoplast fluxes, from the efflux kinetics is not valid, and the information which can be derived from the time course of tracer loss is limited. However, something of the nature of the flux changes can be deduced by comparison of the efflux kinetics for tissue transferred to the dark at the start of the efflux with those for steady-state tissue washed out in light. The results of fitting such efflux

Table 12.8 Br efflux kinetics from open guard cells in light and dark

	Light	**Dark**
k_1	2.2 ± 0.1	1.73 ± 0.07
k_2	0.27 ± 0.01	0.096 ± 0.019
A_1	39 ± 5	67 ± 13
A_2	91 ± 7	42 ± 4
Q_T	130 ± 7	109 ± 16
Q_c	62 ± 4	
Q_v	68 ± 5	
Number of strips	4	3

A_1, A_2, Q_T, Q_c, Q_v: pmol mm^{-2}.
k_1, k_2: h^{-1}.
Q_T, the total tracer in the tissue is fitted to the sum of two exponentials:

$$Q_T = A_1\,e^{-k_1 t} + A_2\,e^{-k_2 t}$$

Q_c, Q_v: cytoplasmic and vacuolar contents, calculated from tissue washed out in light (steady state).
Tissue was loaded in 30 mol m^{-3} $K^{82}Br$ for 15–16 h in light, then washed out in light or dark (MacRobbie, 1984).

curves to the sum of two exponentials are shown in Table 12.8, together with the calculated fluxes and contents for the steady-state tissue. The figures show that in the dark much more of the total tracer appears in the fast phase of the wash-out, and that the slow rate constant is much reduced. This means that the fluxes at the tonoplast are strongly inhibited in the dark, and in fact after transfer to the dark the whole of the cytoplasmic tracer seems to be lost to the solution, rather than transferred in part to the solution and in part to the vacuole during the wash-out. Thus the transport of ions from cytoplasm to vacuole is largely suppressed in the guard cells transferred to the dark.

The immediate flux responses to the transfer of open guard cells from light to dark may be compared with the long-term differences in fluxes in steady-state light-adapted and dark-adapted tissue. Figures 12.3 and 12.4 showed that both the steady-state fluxes of Br^-, and the corresponding rate constants, were greatly reduced in dark-adapted tissue, of low aperture. Thus, although on transfer of open guard cells to the dark there is no immediate reduction in influx, and only a transient rise in the rate constant for efflux, nevertheless the steady-state fluxes in the dark, when low levels of content and aperture have been reached, are low. Thus influxes differ between open and closed guard cells, rather than between light and dark. The results therefore suggest that these low fluxes are not a direct effect of darkness, but instead reflect the subsequent changes in the internal state of the tissue, and in the activation of the transport systems. The results are consistent with the view that the activation of the uptake systems, reflected in the rate constant ϕ_p/C_o, is very sensitive to the cytoplasmic salt concentration; low internal concentrations, whether the result of low external concentration or of prolonged treatment in the dark, seems to be associated with low capacity for ion uptake. The results suggest that this question deserves more detailed study, which may throw light on the nature of the control.

12.8.4.2 Effect of addition of ABA to open guard cells

The responses of guard cell fluxes to the second closing signal so far studied, the addition of ABA, are in fact very similar to those seen on transfer to the dark, suggesting a common mechanism (MacRobbie, 1981b).

Again there was no significant inhibition of the influx of either $^{86}Rb^+$ or $^{82}Br^-$, on adding 20 mmol m^{-3} ABA to open guard cells. There was, however, a very marked transient increase in efflux of both ions, with a time course very similar to that seen on transfer to the dark, and involving a very similar amount of excess ion loss (Fig. 12.7).

The effect is best seen in the ratio of the maximum rate constant for efflux after the addition of ABA to that before the addition ($\bar{k}_{max}/\bar{k}_c$). In different experiments this ratio was in the range 1.9–7.4 for Rb^+, and

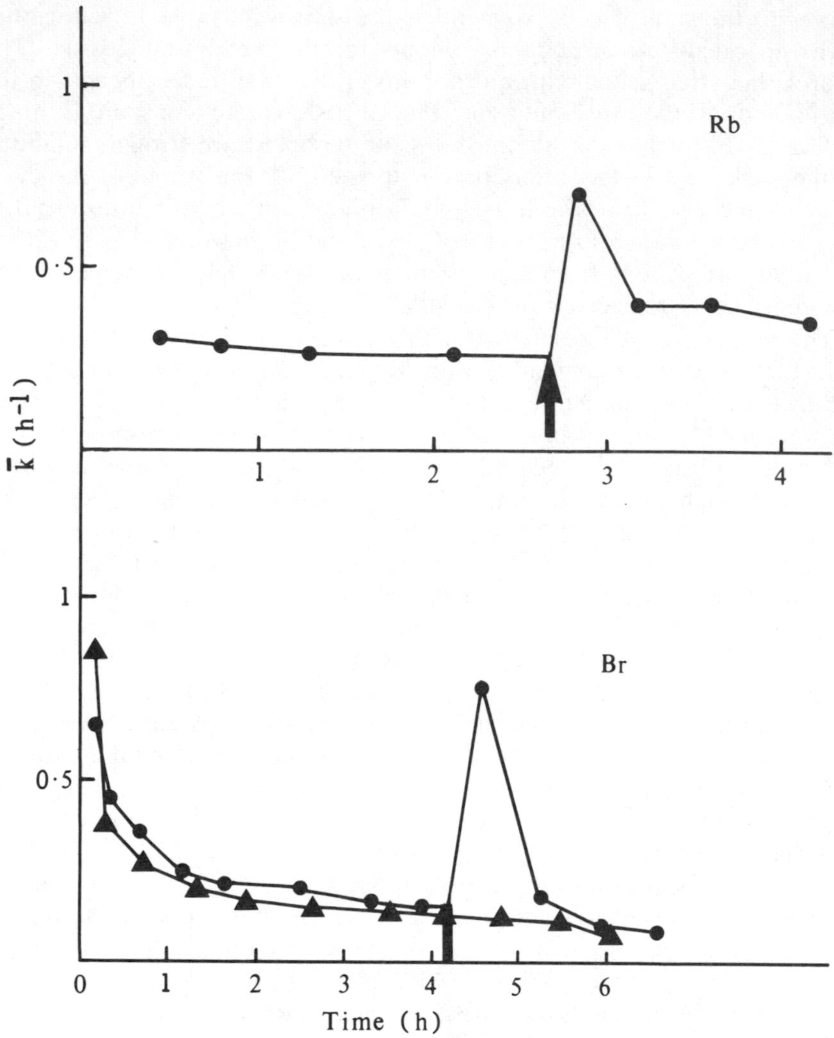

Fig. 12.7 Effect of adding 20 mmol m^{-3} ABA during the efflux of $^{86}Rb^{+}$ and $^{82}Br^{-}$ from steady-state guard cells in light. Rb: mean of three strips, ABA added at arrow. Br: (●) single strip, ABA added at arrow, (▲) single strip, in absence of ABA throughout. (MacRobbie, 1981b).

4.1–16.3 for Br^{-}. Again it should be stressed that the effect is transient, and that the efflux falls again within about 20 min.

After the addition of ABA the steady-state condition no longer applies, and the specific activity becomes uncertain, preventing the calculation of tonoplast fluxes from the efflux kinetics. Nevertheless, the time course of the $^{82}Br^{-}$ efflux transient makes it clear that the flux ϕ_{vc}, from vacuole to cytoplasm, must also be greatly increased after the addition of ABA, and not simply the efflux at the plasmalemma, ϕ_{co}.

An enhanced leakage of $^{86}Rb^+$ from intact epidermal strips of *Commelina* was also seen by Weyers & Hillman (1980), on adding ABA, for about 10 min. While autoradiography did not allow them to distinguish guard cell tracer from that in the subsidiary cells, it seems likely that enhanced leakage from guard cells was involved, as in isolated strips, and that both reflect processes responsible for ABA-induced closure in the intact leaf.

12.8.4.3 The closing response

Taken together the flux results suggest that ABA-induced closure and dark-induced closure are achieved by similar mechanisms. They suggest the activation of a specific process of salt excretion, followed by its deactivation, but the nature of the controls remains to be elucidated. In connection with these results it is interesting that both ABA-induced closure and dark-induced closure are sensitive to a number of metabolic inhibitors (Fujino, 1967; Pemadasa & Jeyaseelen, 1976; Pemadasa & Koralege, 1977; Weyers *et al.*, 1982), suggesting that metabolism is necessary for the processes by which the reduction in ion content is achieved.

The flux results identify the process involved in the closing response, as the transient stimulation in salt efflux. The most pressing need is now to identify the control factors, the nature of the cytoplasmic regulator; it is tempting to speculate that transient changes in cytoplasmic Ca^{2+} concentration may be involved, in view of the extreme sensitivity of guard cells to external Ca^{2+}, but this remains for future work.

12.8.5 Flux responses to opening signals

Very few measurements have been made of ion fluxes during the reverse response, the initiation of opening by appropriate signals. Hence we have very little understanding of the mechanisms involved in the activation of the processes of net salt accumulation by which the opening is achieved. Nevertheless, it is clear from the results so far that control of ion influx is critical in the initiation of opening. This contrasts with the results discussed in the previous sections, showing that control of ion efflux, rather than ion influx, is the critical factor in determining the ability to stay open.

Flux responses during fusicoccin-induced opening have been determined, and some preliminary measurements have been made of $^{86}Rb^+$ influx during opening, and of the effects of light/dark and of Ca^{2+} on this flux.

12.8.5.1 Effects of fusicoccin (FC) on ion fluxes

When FC (30 mmol m^{-3}) is added to isolated epidermal strips guard cells open more widely. Changes in $^{86}Rb^+$ influx and efflux have been measured during the FC-induced opening response (Clint & MacRobbie, 1984); changes in flux and aperture are measurable within the first 15 min period, but continue to develop over much longer, at least 30–45 min, and some-

Table 12.9 Effect of FC on ^{86}Rb fluxes in *Commelina*[†]

Rb (mol m^{-3})	pH	FC	Efflux (pmol mm^{-2} h^{-1})	Influx	Net flux
30	3.9	−	22.8	25.5	+2.7
		+	17.1	77.0	+59.9
10	3.9	−	5.4	6.0	+ 0.6
		+	2.3	36.3	+34.0
30	6.7	−	43.4	44.0	+ 0.6
		+	3.0	59.8	+56.8
10	6.7	−	9.8	13.5	+ 3.7
		+	0.9	61.0	+60.1

† Clint & MacRobbie (1984).
−: fluxes measured before the addition of 30 mmol m^{-3} FC.
+: fluxes measured in the period 30–45 min after the addition of FC.

times up to 135 min or more. Table 12.9 shows the $^{86}Rb^+$ fluxes measured before the addition of FC and in the period 30–45 min after its addition. The large net influx in the presence of FC is the result of both increased influx and reduced efflux, and in three of the four conditions tested the change in influx makes the major contribution. The time course of the changes in Rb-content of the guard cells was estimated in two ways, from the changing influx and efflux in successive periods after FC-addition, and from measurements of total $^{86}Rb^+$ in fully loaded tissue, by adding FC to the radioactive solution bathing tissue already labelled in the absence of FC; there was good agreement between the two estimates.

In the first 20 min after the addition of FC the estimated osmotic contribution of the increase in Rb-salt is comparable with the osmotic requirement for the observed change in aperture, but thereafter the Rb-salt contribution falls short of that required. The results suggest that the primary effect of FC is in stimulating net salt accumulation, but that there follow other osmotic changes within the tissue, perhaps as a result of the increased salt levels.

The effects of FC on $^{82}Br^-$ fluxes have also been measured (Clint, unpublished) and appear to be more complex. At pH 3.9 the addition of FC inhibits both influx and efflux of $^{82}Br^-$ at the plasmalemma, whereas at pH 6.7 the influx is unaffected but the efflux increases. In either case, in the presence of FC the net Br^- influx is much less than the net Rb^+ influx, suggesting that significant malate synthesis must be induced by FC, although not quantitatively important in control strips floating on K(Rb) Cl(Br). Thus although the overall effects of FC are consistent with a primary activation of the proton pump, leading to enhanced influx of $K^+(Rb^+)$, the subsequent changes seem to be complex, and remain to be established in detail, together with their causal relationships.

12.8.5.2 $^{86}Rb^+$ influxes during opening

Measurements of $^{86}Rb^+$ influx into guard cells opening at different rates in different conditions make it clear that control of ion influx is important in determining the rate of opening. Opening is inhibited in the dark, or by the presence of increased Ca^{2+} in the medium (Fujino, 1967; Willmer and Mansfield, 1969). Guard cells were isolated, starting from the closed state in dark-treated leaves, and $^{86}Rb^+$ influx was then measured during the early stages of opening of the newly isolated guard cells (MacRobbie, 1986); by incubation in light or dark, and in the presence of either 5 mol m^{-3} Mg^{2+} or 5 mol m^{-3} Ca^{2+}, different rates of opening were induced. The results are shown in Table 12.10, and two conclusions emerge. First, the $^{86}Rb^+$ influx is inhibited in the presence of Ca^{2+} rather than Mg^{2+}, but secondly, there is an inhibition of influx in the dark only in the presence of Ca^{2+}. Thus the effect of dark in inhibiting $^{86}Rb^+$ influx is a Ca^{2+}-dependent process. It was further shown that the dark-induced inhibition in the presence of Ca^{2+} is abolished by the addition also of 2 mol m^{-3} $LaCl_3$ to the incubation medium; this is consistent with the hypothesis that the effect involves Ca^{2+} influx, and that this is blocked by La^{3+}. The effects of Ca^{2+} on fluxes during the initiation of closing remains to be investigated, and may reveal further Ca^{2+}-dependent regulation of ion fluxes. The elucidation of the mechanism by which Ca^{2+} inhibits the Rb^+-influx during opening is an important aim for future work.

In connection with these flux measurements, showing Ca^{2+}-dependence of the dark effect on stomatal opening, it is interesting that De Silva *et al.* (1985b) have recently shown that inhibition of stomatal opening by ABA is also a Ca^{2+} dependent process. They have also tested the effects

Table 12.10 $^{86}Rb^+$ influxes during opening: effects of Ca^{2+}/Mg^{2+} and light/dark†

	$^{86}Rb^+$ influx in MgL (pmol mm^{-2} h^{-1})	**Ratio of $^{86}Rb^+$ influxes (number of strips)** CaL/MgL	CaD/CaL	MgD/MgL
I	19.3 ± 1.4 (6)	0.71 ± 0.09 (7,6)	0.66 ± 0.10 (7,7)	1.15 ± 0.10 (5,6)
II	44.3 ± 2.4 (8)	0.71 ± 0.09 (8,8)	0.70 ± 0.09 (8,8)	1.16 ± 0.07 (6,8)
III	43.0 ± 3.4 (12)	0.83 ± 0.10 (10,12)	0.58 ± 0.06 (12,10)	

† MacRobbie (1986).
Leaves dark-treated to close stomata before the isolation of guard cells by low pH treatment.
Influxes then measured in the isolated guard cells over 30 min in the early stages of opening, in light (L) or dark (D).
All solutions contained 30 mol m^{-3} RbCl, 10 mol m^{-3} MES, pH 3.9.
Mg: 5 mol m^{-3} Mg^{2+}, 10 mmol m^{-3} Ca^{2+}.
Ca: 5 mol m^{-3} Ca^{2+}.

of Ca^{2+}-channel blockers (La^{3+}, verapamil and nifedipine), and of the calmodulin antagonists (trifluoperazine, W7 and compound 48/80), on the stomatal responses to ABA; both sets of agents reduced the inhibition of stomatal opening by ABA (De Silva *et al.*, 1985a). They argue that the effect of ABA involves influx of Ca^{2+} into the guard cell, and that the subsequent responses involve Ca^{2+}-calmodulin-dependent enzymes of some kind. Effects of calmodulin antagonists on stomatal aperture were also observed by Donovan *et al.* (1985), but were obviously complex, involving more than one process.

Overall, the evidence suggests that control of ion influx into guard cells is important in inducing stomatal opening, that Ca^{2+} can act as a second messenger in certain stomatal responses, and that at least some of the ion fluxes may be Ca^{2+}-calmodulin regulated. Future work must establish the detailed chain of events involved in such regulation.

12.9 Control of stomatal aperture

Understanding of the control of guard cell ion fluxes is essential for an understanding of the control of stomatal aperture. The flux work has not yet given a detailed picture of the mechanisms of control, but it has done something to define the nature of the problems requiring solution. The results suggest that two separate sets of processes are important in the control of the level of ion accumulation and aperture in guard cells.

The ability to open must be determined by regulation of the influx processes, probably primarily by activation of the proton pump. The very elegant blue light-pulse work from Zeiger's group identifies such activation as the mechanism by which blue light stimulates stomatal opening. The flux work shows how control of ion influx is important in determining the effects of Ca^{2+} and of darkness on the rate of opening, providing evidence for the regulation of ion influx by Ca^{2+}; other work suggests Ca^{2+}-calmodulin mediated effects on the opening process.

In contrast to this, the ability to stay open seems to be controlled by regulation of the efflux process. Closing is initiated by activation of a process of salt excretion, as a transient response to a closing signal, and low efflux is then re-established at a lower salt content. Ca^{2+} effects on the flux responses to closing signals remain to be investigated.

In addition to the specific flux responses to opening or closing signals there is also evidence for regulation of the fluxes by internal conditions. Here volume regulation may be more likely than turgor regulation, with the degree of stretching of the plasmalemma as a possible control signal. There are also indications (but not yet proof) that cytoplasmic ion levels may be important in the regulation of fluxes, both of influx processes at the plasmalemma and of transfer across the tonoplast. In this connection the recent observations by Kottmeier & Schnabl (1986) are interesting, namely that K^+-induced swelling of guard cell protoplasts of *Vicia* was accompanied by a decrease in the K_m of phosphoenolpyruvate (PEP) carboxylase from 0.2 mol m^{-3} to 0.01 mol m^{-3}. Thus both ion fluxes and metabolic activities in guard cells may be sensitive to the internal ionic state.

The problems of understanding the mechanisms of control, and of identifying the detailed chain of events involved in any of the stomatal responses, have not yet been touched. We have a somewhat better description of the gross events than 10 years ago, but the molecular interactions remain unknown.

References

Allaway, W. G. & Hsiao, T. C. (1973). Preparation of rolled epidermis of *Vicia faba* L. so that stomata are the only viable cells: analysis of guard cell potassium by flame photometry. *Australian Journal of Biological Sciences* **26**, 309–18.

Assmann, S. M., Simoncini, L. & Schroeder, J. I. (1985). Blue light activates electrogenic ion pumping in guard cell protoplasts of *Vicia faba*. *Nature* **318**, 285–7.

Beilby, M. J. (1984). Current–voltage characteristics of the proton pump at *Chara* plasmalemma. I. pH-dependence. *Journal of Membrane Biology* **81**, 113–25.

Blatt, M. R. (1985). Extracellular potassium activity in attached leaves and its relation to stomatal function. *Journal of Experimental Botany* **36**, 240–51.

Blatt, M. R. (1987). Electrical characteristics of stomatal guard cells: The ionic basis of the membrane potential and the consequence of KCl leakage from microelectrodes. *Planta* **170**, 272–87.

Blatt, M. R. & Slayman, C. L. (1983). KCl leakage from micro-electrodes and its impact on the membrane parameters of a non-excitable cell. *Journal of Membrane Biology* **72**, 223–34.

Borgeson, C. E. & Bowman, B. J. (1985). Blue light-reducible cytochromes in membrane fractions from *Neurospora crassa*. *Plant Physiology* **78**, 433–7.

Borle, A. B., Uchikawa, T. & Anderson, J. H. (1982). Computer simulation and interpretation of ^{45}Ca efflux profile patterns. *Journal of Membrane Biology* **68**, 37–46.

Brownlee, C. & Kendrick, R. E. (1977). Phytochrome and potassium uptake by mung bean hypocotyl sections. *Planta* **137**, 61–4.

Campbell, N. A. & Garber, R. C. (1980). Vacuolar reorganisation in the motor cells of *Albizzia* during leaf movement. *Planta* **148**, 251–5.

Clint, G. M. (1985a). The investigation of stomatal ionic relations using guard cell protoplasts. I. Methodology. *Journal of Experimental Botany* **36**, 1726–38.

Clint, G. M. (1985b). The investigation of stomatal ionic relations using guard cell protoplasts. II. Osmotic relations of guard cell protoplasts in short and long-term incubation. *Journal of Experimental Botany* **36**, 1739–48.

Clint, G. M. & MacRobbie, E. A. C. (1984). Effects of fusicoccin in 'isolated' guard cells of *Commelina communis* L. *Journal of Experimental Botany* **35**, 180–92.

Cram, W. J. (1976). Negative feedback regulation of transport in cells. In Lüttge, U. & Pitman, M. G. (eds.). *Transport in Plants*, II, *Encyclopedia of Plant Physiology*, New Series, vol. 2A, pp. 284–316. Berlin, Springer-Verlag.

Cram, W. J. (1980). Chloride accumulation as a homeostatic system: Negative feedback signals for concentration and turgor maintenance differ in a glycophyte and a halophyte. *Australian Journal of Plant Physiology* **7**, 237–49.

Dayanandan, P. & Kaufman, P. B. (1975). Stomatal movements associated with potassium fluxes. *American Journal of Botany* **62**, 221–31.

De Silva, D. L. R., Cox, R. C., Hetherington, A. M. & Mansfield, T. A. (1985a). Suggested involvement of calcium and calmodulin in the responses of stomata to abscisic acid. *New Phytologist* **101**, 555–63.

De Silva, D. L. R., Hetherington A. M. & Mansfield, T. A. (1985b). Synergism between calcium ions and abscisic acid in preventing stomatal opening. *New Phytologist* **100**, 473–82.

Donovan, N., Martin, S. & Donkin, M. E. (1985). Calmodulin binding drugs trifluoperazine and compound 48/80 modify stomatal responses of *Commelina communis* L. *Journal of Plant Physiology* **118**, 177–87.

Edwards, A. & Bowling, D. J. F. (1985). Evidence for a CO_2 inhibited proton extrusion pump in the stomatal cells of *Tradescantia virginiana*. *Journal of Experimental Botany* **36**, 91–8.

Fischer, R. A. (1968a). Stomatal opening: role of potassium uptake by guard cells. *Science* **160**, 784–5.

Fischer, R. A. (1968b). Stomatal opening in isolated epidermal strips of *Vicia faba*. I. Response to light and CO_2-free air. *Plant Physiology* **43**, 1947–52.

Fischer, R. A. (1972). Aspects of potassium accumulation by stomata of *Vicia faba*. *Australian Journal of Biological Sciences* **25**, 1107–23.

Fischer, R. A. (1973). The relationship of stomatal aperture and guard cell turgor pressure in *Vicia faba*. *Journal of Experimental Botany* **24**, 387–99.

Fischer, R. A. & Hsiao, T. C. (1968). Stomatal opening in isolated epidermal strips of *Vicia faba*. II. Responses to KCl concentration and role of potassium absorption. *Plant Physiology* **43**, 1953–8.

Fitzsimmons, P. J. & Weyers, J. D. B. (1986). Potassium uptake by swelling *Commelina communis* guard cell protoplasts. *Physiologia Plantarum* **66**, 469–75.

Fujino, M. (1967). Role of adenosinetriphosphate and adenosinetriphosphatase in stomatal movement. *Scientific Bulletin of the Faculty of Education, Nagasaki University* **18**, 1–47.

Gepstein, S., Jacobs, M. & Taiz, L. (1982). Inhibition of stomatal opening in *Vicia faba* epidermal tissue by vanadate and abscisic acid. *Plant Science Letters* **28**, 63–72.

Gotow, K., Kondo, N. & Syono, K. (1982). Effects of CO_2 on volume change of guard cell protoplast from *Vicia faba* L. *Plant and Cell Physiology*, **23**, 1063–70.

Hampp, R., Outlaw, W. H. & Tarczynski, M. C. (1982). Profile of basic carbon pathways in guard cells and other leaf cells of *Vicia faba* L. *Plant Physiology* **70**, 1582–5.

Hipkins, M. F., Fitzsimons, P. J. & Weyers, J. D. B. (1983). The primary processes of photosystem II in purified guard-cell protoplasts and mesophyll-cell protoplasts from *Commelina communis* L. *Planta* **159**, 554–60.

Holmes, M. G. & Klein, W. H. (1985). Evidence for phytochrome involvement in light-mediated stomatal movement in *Phaseolus vulgaris* L. *Planta* **166**, 348–53.

Hsiao, T. C. (1976). Stomatal ion transport. In Lüttge, U. & Pitman, M. G. (eds.). *Transport in Plants*, II, *Encyclopedia of Plant Physiology*, New Series, vol. 2B, pp. 195–221. Berlin, Springer-Verlag.

Hsiao, T. C., Allaway, W. G. & Evans, L. T. (1973). Action spectrum for guard cell Rb^+ uptake and stomatal opening in *Vicia faba*. *Plant Physiology* **51**, 82–8.

Humble, G. D. & Hsiao, T. C. (1969). Specific requirement for potassium for light-activated opening of stomata in epidermal strips. *Plant Physiology* **44**, 230–4.

Humble, G. D. & Raschke, K. (1971). Stomatal opening quantitatively related to potassium transport. Evidence from microprobe analysis. *Plant Physiology* **48**, 447–53.

Iino, M., Ogawa, T. & Zeiger, E. (1985). Kinetic properties of the blue light response of stomata. *Proceedings of the National Academy of Sciences, USA* **82**, 8019–23.

Imamura, S. (1943). Untersuchungen über den Mechanisms der Turgorschwankung der Spaltoffnungsschliesszellen. *Japanese Journal of Botany* **12**, 251–346.

Karlsson, P. E. (1985). Blue light regulation of stomata in wheat. II. Action spectrum and search for action dichroism. *Physiologia Plantarum* **66**, 207–10.

Karlsson, P. E., Höglund, H. O. & Klockare, R. (1983). Blue light induces stomatal transpiration in wheat seedlings with chlorophyll deficiency caused by SAN 9789. *Physiologia plantarum* **57**, 417–21.

Kauss, H. (1978). Osmotic regulation in algae. *Progress in Phytochemistry* **5**, 1–27.

Kottmeier, C. & Schnabl, H. (1986). The K_m-value of phosphoenolpyruvate carboxylase as an indicator of the swelling state of guard cell protoplasts. *Plant Science* **43**, 213–17.

Lucas, W. J. (1982). Mechanism of acquisition of exogenous bicarbonate by internodal cells of *Chara corallina*. *Planta* **156**, 181–92.

Macallum, A. B. (1905). On the distribution of potassium in animal and vegetable cells. *Journal of Physiology* **32**, 95–118.

MacRobbie, E. A. C. (1977). Functions of ion transport in plant cells and tissues. In Northcote, D. H. (ed.). *International Review of Biochemistry, Plant Biochemistry*, II, vol. 13, pp. 211–47. Baltimore, Md., University Park Press.

MacRobbie, E. A. C. (1980). Osmotic measurements on stomatal cells of *Commelina communis* L. *Journal of Membrane Biology* **53**, 189–98.

MacRobbie, E. A. C. (1981a). Ion fluxes in isolated guard cells of *Commelina communis* L. *Journal of Experimental Botany* **32**, 545–62.

MacRobbie, E. A. C. (1981b). Effects of ABA in 'isolated' guard cells of *Commelina communis* L. *Journal of Experimental Botany* **32**, 563–72.

MacRobbie, E. A. C. (1983). Effects of light/dark on cation fluxes in guard cells of *Commelina communis* L. *Journal of Experimental Botany* **34**, 1695–710.

MacRobbie, E. A. C. (1984). Effects of light/dark on anion fluxes in isolated guard cells of *Commelina communis* L. *Journal of Experimental Botany* **35**, 707–26.

MacRobbie, E. A. C. (1986). Calcium effects on stomatal guard cells. In Trewavas, A. & Marmé, D. (eds.). *Molecular and Cellular Aspects of Calcium in Plant Development*, pp. 383–4. New York, Plenum Press.

MacRobbie, E. A. C. & Lettau, J. (1980a). Ion content and aperture in 'isolated' guard cells of *Commelina communis* L. *Journal of Membrane Biology* **53**, 199–205.

MacRobbie, E. A. C. & Lettau, J. (1980b). Potassium content and aperture of 'intact' stomatal and epidermal cells of *Commelina communis* L. *Journal of Membrane Biology* **56**, 249–56.

Marrè, E. (1979). Fusicoccin: A tool in plant physiology. *Annual Review of Plant Physiology* **30**, 273–88.

Meidner, H. & Mansfield, T. A. (1968). *Physiology of Stomata*. New York, McGraw-Hill.

Melis, A. & Zeiger, E. (1982). Chlorophyll *a* fluorescence transients in mesophyll and guard cells. Modulation of guard cell photophosphorylation by CO_2. *Plant Physiology* **69**, 642–7.

Moody, W. & Zeiger, E. (1978). Electrophysiological properties of onion guard cells. *Planta*, **139**, 159–65.

Ogawa, T., Grantz, D., Boyer, J. & Govindjee (1982). Effects of cations and abscisic acid on chlorophyll *a* fluorescence in guard cells of *Vicia faba*. *Plant Physiology* **69**, 1140–4.

Ogawa, T., Ishikawa, H., Shimida, K. & Shibata, K. (1978). Synergistic action of red and blue light and action spectra for malate formation in guard cells of *Vicia faba* L. *Planta* **142**, 61–5.

Outlaw, W. H. (1983). Current concepts on the role of potassium in stomatal movements. *Physiologia Plantarum* **59**, 302–11.

Outlaw, W. H. & Kennedy, J. (1978). Enzymic and substrate basis for the anaplerotic step in guard cells. *Plant Physiology* **62**, 648–52.

Outlaw, W. H. & Lowry, O. H. (1977). Organic acid and potassium accumulation in guard cells during stomatal opening. *Proceedings of the National Academy of Sciences, USA* **74**, 4434–8.

Outlaw, W. H. & Manchester, J. (1979). Guard cell starch concentration quantitatively related to stomatal aperture *Plant Physiology* **64**, 79–82.

Outlaw, W. H., Mayne B. C., Zenger, V. E. & Manchester, J. (1981). Presence of both photosystems in guard cells of *Vicia faba* L. Implications for environmental signal processing. *Plant Physiology* **67**, 12–16.

Outlaw, W. H., Springer, S. A., & Tarczynski, M. C. (1985). Histochemical technique. A general method for quantitative enzyme assays of single cell 'extracts' with a time resolution of seconds and a reading precision of femtomoles. *Plant Physiology* **77**, 659–66.

Palevitz, B. A. & Hepler, P. K. (1985). Changes in dye coupling of stomatal cells of *Allium* and *Commelina* demonstrated by microinjection of Lucifer yellow. *Planta* **164**, 473–9.

Pallaghy, C. K. & Fischer, R. A. (1974). Metabolic aspects of stomatal opening and ion accumulation by guard cells in *Vicia faba*. *Zeitschrift für Pflanzenphysiologie* **71**, 332–44.

Pemadasa, M. A. & Jeyaseelen, K. (1976). Some effects of sodium azide on stomatal closure in *Stachyarpheta indica*. *Annals of Botany* **40**, 655–8.

Pemadasa, M. A. & Koralege, S. (1977). Stomatal responses to 2,4-dinitrophenol. *New Phytologist* **78**, 573–8.

Penny, M. G. & Bowling, D. J. F. (1974). A study of potassium gradients in the epidermis of intact leaves of *Commelina communis* L. in relation to stomatal opening. *Planta* **119**, 17–25.

Pitman, M. G. (1969). Simulation of Cl^- uptake by low-salt barley roots as a test of models of salt uptake. *Plant Physiology* **44**, 1417–27.

Raschke, K. (1975). Stomatal action. *Annual Review of Plant Physiology* **26**, 309–40.

Raschke, K. (1979). Movements of stomata. In Haupt, W. and Feinleib, M. E. (eds.) *Physiology of Movements, Encyclopedia of Plant Physiology*, New Series, vol. 7, pp. 383–441. Berlin, Springer-Verlag.

Raschke, K. & Fellows, M. P. (1971). Stomatal movements in *Zea mays*. Shuttle of potassium between guard cells and subsidiary cells. *Planta* **101**, 296–316.

Raschke, K. & Humble, G. D. (1973). No uptake of anions required by opening stomata of *Vicia faba*: guard cells release hydrogen ions. *Planta* **115**, 47–57.

Rodriguez-Navarro, A., Blatt, M. R. & Slayman, C. L. (1986). A potassium–proton symport in *Neurospora. Journal of General Physiology* **87**, 649–74.

Saftner, R. A. & Raschke, K. (1981). Electric potentials in stomatal complexes. *Plant Physiology* **67**, 1124–32.

Sanders, D. (1980). The mechanism of Cl^- transport at the plasma membrane of *Chara corallina*. I. Co-transport with H^+. *Journal of Membrane Biology 53*, 129–41.

Sanders, D. & Hansen, U. P. (1981). Mechanism of Cl^--transport at the plasma membrane of *Chara corallina*. II. Transinhibition and the determination of H^+/Cl^- binding order from a reaction kinetic model. *Journal of Membrane Biology* **58**, 139–53.

Sawney, B. L. & Zelitch, L. (1969). Direct determination of potassium ion accumulation in guard cells in relation to stomatal opening in light. *Plant Physiology* **44**, 1350–54.

Schnabl, H. & Raschke, K. (1980). Potassium chloride as stomatal osmoticum in *Allium cepa* L., a species devoid of starch in guard cells. *Plant Physiology* **65**, 88–93.

Schroeder, J. I., Hedrich, R. & Fernandez, J. M. (1984). Potassium-selective single channels in guard cell protoplasts of *Vicia faba. Nature* **312**, 361–2.

Shimazaki, K., Gotow, K. & Kondo, N. (1982). Photosynthetic properties of guard cell protoplasts from *Vicia faba* L. *Plant Cell Physiology* **23**, 871–9.

Shimazaki, K., Gotow, K., Sakaki, T. & Kondo, N. (1983). High respiratory activity of guard cell protoplasts from *Vicia faba* L. *Plant Cell Physiology* **24**, 1049–56.

Shimazaki, K., Iino, M. & Zeiger, E. (1986). Blue-light-dependent proton extrusion by guard cell protoplasts of *Vicia faba. Nature* **319**, 324–6.

Shimazaki, K. & Zeiger, E. (1985). Cyclic and noncyclic photophosphorylation in isolated guard cell chloroplasts from *Vicia faba* L. *Plant Physiology* **78**, 211–14.

Spanswick, R. M. & Miller, A. G. (1977). The effect of CO_2 on the Cl^- influx and electrogenic pump in *Nitella translucens*. In Thellier, H., Monnier, A., Demarty, M. & Dainty, J. (eds.). *Transmembrane Ionic Exchanges in Plants*, pp. 239–45. Paris, CNRS.

Sze, H. (1985). H^+-translocating ATPases: advances using membrane vesicles. *Annual Review of Plant Physiology* **36**, 175–208.

Thomas, D. A. (1975). Stomata. In Baker, D. A. & Hall, J. L. (eds.). *Ion Transport in Plant Cells and Tissues*, pp. 377–412. Amsterdam; Elsevier.

Travis, A. & Mansfield, T. (1979). Stomatal responses to light and CO_2 are dependent on KCl concentration. *Plant, Cell and Environment* **2**, 319–23.

Weyers, J. D. B. & Hillman, J. R. (1980). Effects of abscisic acid on $^{86}Rb^+$ fluxes in *Commelina communis* L. leaf epidermis. *Journal of Experimental Botany* **31**, 711–20.

Weyers, J. D. B., Paterson, N. W., Fitzsimons, P. J. & Dudley, J. M. (1982). Metabolic inhibitors block ABA-induced stomatal closure. *Journal of Experimental Botany* **33**, 1270–8.

Wille, A. C. & Lucas, W. J. (1984). Ultrastructural and histochemical studies on guard cells. *Planta* **160**, 129–42.

Willmer, C. M. (1983). *Stomata*. London, Longman.

Willmer, C. M. & Mansfield, T. A. (1969). A critical examination of the use of detached epidermis in studies of stomatal physiology. *New Phytologist* **68**, 363–75.

Willmer, C. M. & Pallas, V. E. (1973). A survey of stomatal movements and associated potassium fluxes in the plant kingdom. *Canadian Journal of Botany* **51**, 37–42.

Yamashita, T. (1952). Influence of potassium supply upon properties and movement of the guard cell. *Sieboldia Acta Biologia* **1**, 51–70.

Zeiger, E. (1983). The biology of stomatal guard cells. *Annual Review of Plant Physiology* **34**, 441–75.

Zeiger, E., Armond, P. & Melis, A. (1981a). Fluorescence properties of guard cell chloroplasts. *Plant Physiology* **67**, 17–20.

Zeiger, E., Field, C. & Mooney, H. A. (1981b). Stomatal opening at dawn: possible roles of the blue light response in nature. In Smith, H. (ed.). *Plants and the Daylight Spectrum*. pp. 391–407. London, Academic Press.

Zeiger, E. & Hepler, P. K. (1977). Light and stomatal function: blue light stimulates swelling of guard cell protoplasts. *Science* **196**, 887–9.

Zeiger, E., Iino, M. & Ogawa, T. (1985). The blue light response of stomata: pulse kinetics and some mechanistic implications. *Photochemistry and Photobiology* **42**, 759–63.

Zeiger, E., Moody, W., Hepler, P. & Varela, F. (1977). Light-sensitive membrane potentials in onion guard cells. *Nature* **270**, 270–1.

Zemel, E. & Gepstein, S. (1985). Immunological evidence for the presence of ribulose bisphosphate carboxylase in guard cell chloroplasts. *Plant Physiology* **78**, 586–90.

Zlotnikova, I. F., Gunar, I. I. & Panichkin, L. A. (1977). Measurements of intracellular potassium activity in *Tradescantia* leaf epidermal cells. *Izvestia Timiryazevskoi selsko-khozyaistvennoi Akademii* **2**, 10–16.

13 Salt glands

W. W. Thomson, C. D. Faraday and J. W. Oross

13.1 Introduction

Salt glands are epidermal structures that occur on aerial surfaces, particularly the leaves, in a number of dicotyledonous plant families and in the Chloridoideae and Panicoideae subfamilies of the grass family (Lüttge, 1971, 1975; Waisel, 1972; Hill & Hill, 1976; Fahn, 1979; Liphschitz & Waisel, 1982). Although salt glands drew the interest of several investigators in the latter half of the nineteenth century and the early years of the present one (Volkens, 1884; Wilson, 1890; De Fraine, 1910; Schtscherback, 1910; Ruhland, 1915; Sutherland & Eastwood, 1916), interest flagged until the appearance of a major study by Arisz *et al.* (1955) on the salt glands of *Limonium* and a cogent review on the subject by Helder (1956). During the 1960s and early 1970s, several studies were published on salt glands and the subject was again reviewed in the mid-1970s by Lüttge (1971, 1975), Hill & Hill (1973a, 1976), and Thomson (1975). Most investigators hold the view, first proposed by Schtscherback in 1910, that the adaptive role of salt glands is to remove excess mineral ions delivered to leaves by the transpiration stream. However, salt glands are also thought to regulate the internal ionic composition of the leaf and they are known to secrete a wide variety of ions. Measurements indicate that they are capable of high rates of secretion, and they appear to be highly specialized for ion transport. Thus these glands have attracted the interest of a broad spectrum of investigators, from ecologists interested in the role of the glands in the adaptation of plants to saline environments to physiologists and cytologists who view the glands as a means to determine and analyse the pathway of solute movement in plant tissues. They also provide a system for the study of the correlations between cell structure as it relates to processes of ion transport and secretion. Much research has been done on these interesting complexes during the past 10 years or so; thus, it is appropriate to review and assess their structural and functional properties at this time.

13.2 General structure

Based on their structural organization, there are three types of glands (Thomson, 1975): the two-celled glands of the grasses, the bladder cells of the Chenopodiaceae, and the multicellular glands which occur in other dicotyledonous families.

13.2.1 Salt glands of grasses

The salt glands of the grasses (Fig. 13.1) consist of an outer cap cell and a subtending basal cell (Sutherland & Eastwood, 1916; Skelding & Winterbothom, 1939; Levering & Thomson, 1971; Liphschitz *et al.*, 1974; Liphschitz & Waisel, 1974, 1982; Oross & Thomson, 1982a). Early studies indicate that these glands develop from a single protoderm cell (Sutherland & Eastwood, 1916; Skelding & Winterbothom, 1939). The mature glands vary somewhat in different genera in that they may be sunken, semi-sunken, or extend outward from the epidermis (Liphschitz *et al.*, 1974; Liphschitz & Waisel, 1974, 1982). Also, certain general variations in the two cells occur. The basal cell in *Spartina* and *Cynodon*, for example (Levering & Thomson, 1971; Oross & Thomson, 1982a, b), extends inward into the leaf mesophyll and is quite large and flask-shaped (Fig. 13.1), while in *Bouteloua*, the basal cell is apparently in lateral register with the epidermal cells, but extends somewhat outward from the leaf surface (Liphschitz & Waisel, 1974, 1982). There is also considerable variation in the morphology of the cap cell, e.g. with *Spartina* it is somewhat hemispherical in shape, atop a neck-like extension of the basal cell (Fig. 13.1) (Skelding & Winterbothom, 1939; Levering & Thomson, 1971), while in *Bouteloua* it protrudes as a finger-like structure from the basal cell (Liphschitz & Waisel, 1974, 1982). The leaf cuticle extends over the outer surface of the glands but the inner walls of the glands are not cutinized (Fig. 13.1; and see Levering & Thomson, 1971, Oross & Thomson, 1982a, b, 1984; Oross *et al.*, 1985).

The walls of the basal cells are in apoplastic continuum with those of mesophyll cells (Fig. 13.1). However, the lateral cell walls in the neck region of the basal cells are lignified (Liphschitz & Waisel, 1974; Oross & Thomson, 1982b) and the apoplastic pathway between the two gland cells is restricted in this region (Oross & Thomson, 1982a, b; Oross *et al.*, 1985).

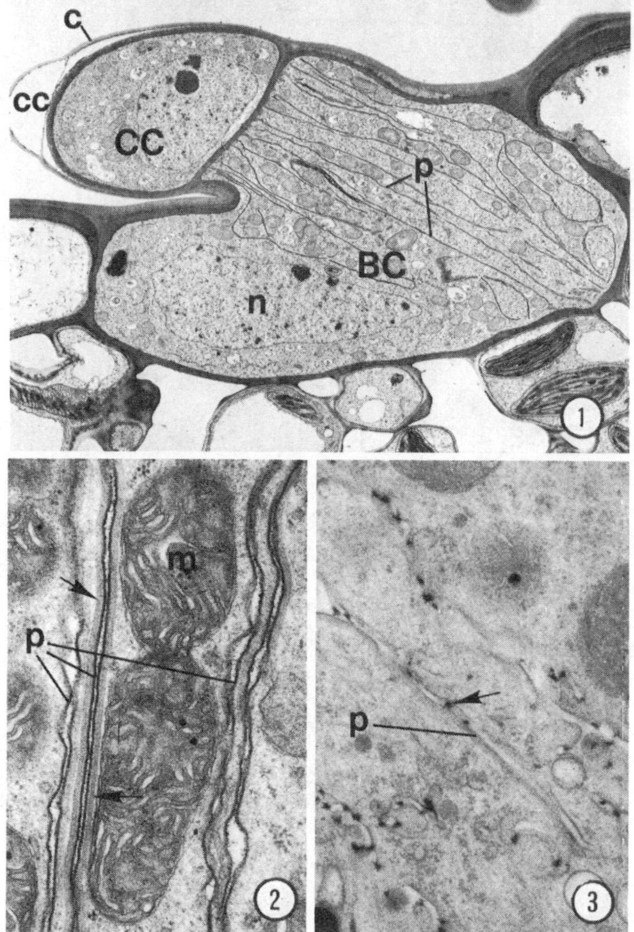

Fig. 13.1 A near median section of the two-celled salt gland of the grass, *Cynodon*. Note the partitioning membranes, p, in the basal cell, BC. CC, cap cell; cc, collecting compartment; c, cuticle, n, nucleus.

Fig. 13.2 High magnification view of the partitioning membranes, p, within the basal cell of the gland of *Cynodon*. Note the microtubles, arrows, adjacent to the partitioning membranes; m, mitochondrion.

Fig. 13.3 The arrow indicates a lanthanum deposit within the channel of the partitioning membranes, p, within the basal cell of a salt gland of the grass, *Cynodon*.

Ultrastructural studies of salt glands of the grasses, *Spartina*, *Cynodon*, and *Distichlis*, have shown that at this level they are remarkably similar in organization (Levering & Thomson, 1971; Oross & Thomson, 1982a, b, 1984). The cap cells show little structural specialization and contain a large nucleus, a few rudimentary plastids, short segments of ER, dictyo-

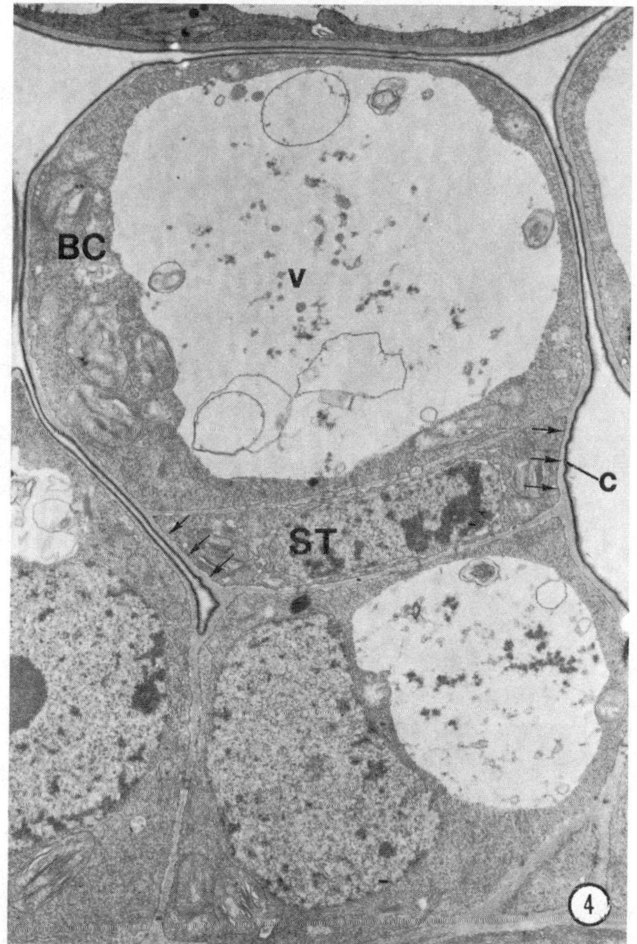

Fig. 13.4 A micrograph of the expanding bladder cell, BC, and stalk cell, ST, of *Atriplex*. Note the internal primary wall (arrows) of the stalk cell is not impregnated by the cuticle, c; v, vacuole.

somes, mitochondria, and small vacuoles of varying size (Fig. 13.1; and see Oross & Thomson, 1984). A separation of the cuticle from the outer cap cell wall forms a 'collecting chamber' (Fig. 13.1), and openings or pores that penetrate the raised cuticle are common (Oross & Thomson, 1982a).

The general features of basal cells are a large nucleus, numerous mitochondria, rudimentary plastids, small vacuoles, some dictyosomes and short segments of ER. The most striking and unique feature are the numerous and extensive invaginations of the plasmalemma, 'partitioning membranes' (Figs. 13.1, 13.2), which extend into the basal cell from the wall between the basal cell and the cap cells (Oross & Thomson, 1982a,

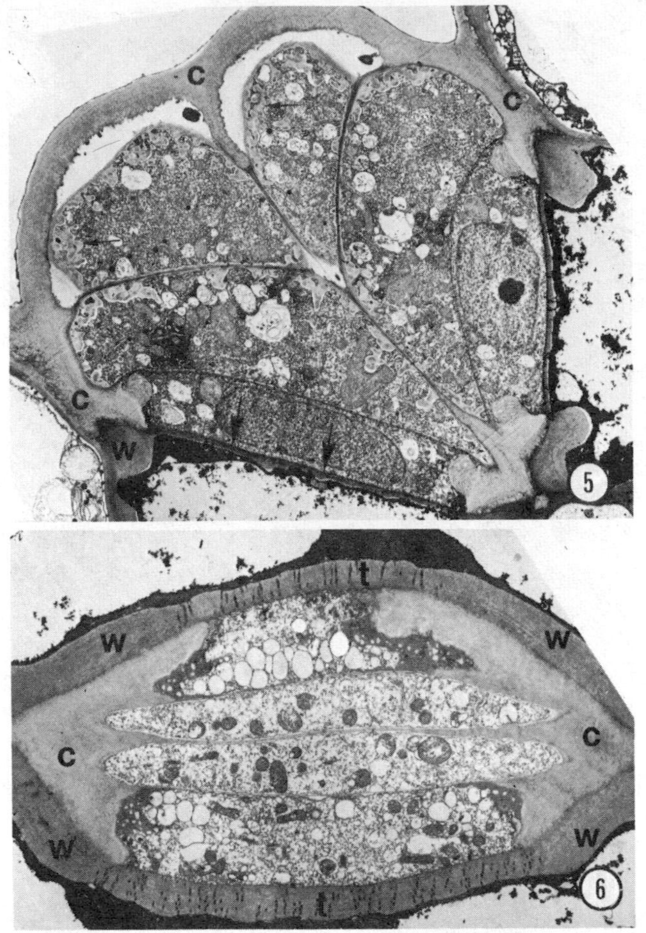

Fig. 13.5 A near median section of the salt gland of *Frankenia*. Note the wall of the transfusion zone (large arrows) is continuous with the lateral wall, w, of the subbasal cell and is not impregnated by the cuticle, c. The small arrows identify wall protuberances in the secretory cells.

Fig. 13.6 A peridermal section near the base of a salt gland illustrating both transfusion zones, t. Note that transfusion zone walls are continuous laterally with walls, w that are not impregnated by the cuticle, c.

1984). Of particular interest, although their function is not clear, is the association of many microtubules with the partitioning membranes (Fig. 13.2).

Symplastic continuity, as defined by the presence of plasmodesmata, occurs between the basal and cap cells, and between the basal and adjacent mesophyll cells (Oross *et al.*, 1985). In *Cynodon* and *Distichlis*, plasmodesmata also interconnect the basal cells and epidermal cells, but were not

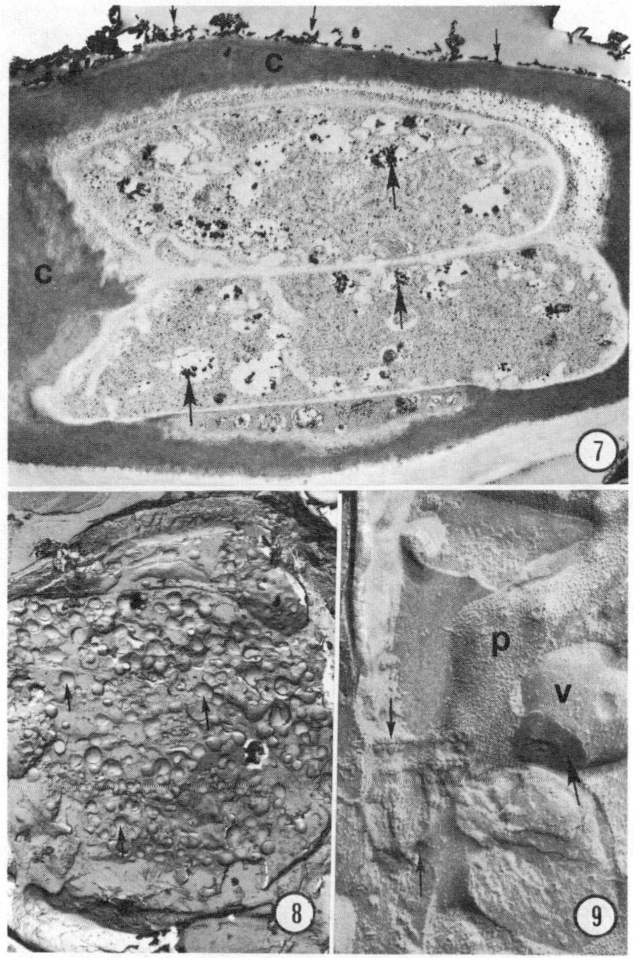

Fig. 13.7 A slightly tangential section of a *Tamarix* salt gland from material that was secreting calcium and treated with antimonate for the localization of Ca^{2+}. The large arrows indicate calcium antimonate deposits in small vacuoles. The small arrows indicate calcium antimonate deposits on the surface of the gland and represent Ca^{2+} that had been secreted. Identification of the deposits was made with an X-ray microprobe.

Fig. 13.8 A freeze–fracture replica of a secreting gland of *Tamarix*. Note the many microvacuoles (arrows) in the secretory cells.

Fig. 13.9 A freeze–fracture replica showing the p-face, p, of the plasmalemma of a secretory cell of an active gland of *Tamarix*. The apparent fusion site of a microvacuole, v, is indicated by the large arrow and is recognized by a smooth, particle-free fracture face. The small arrows indicate plasmodesmata.

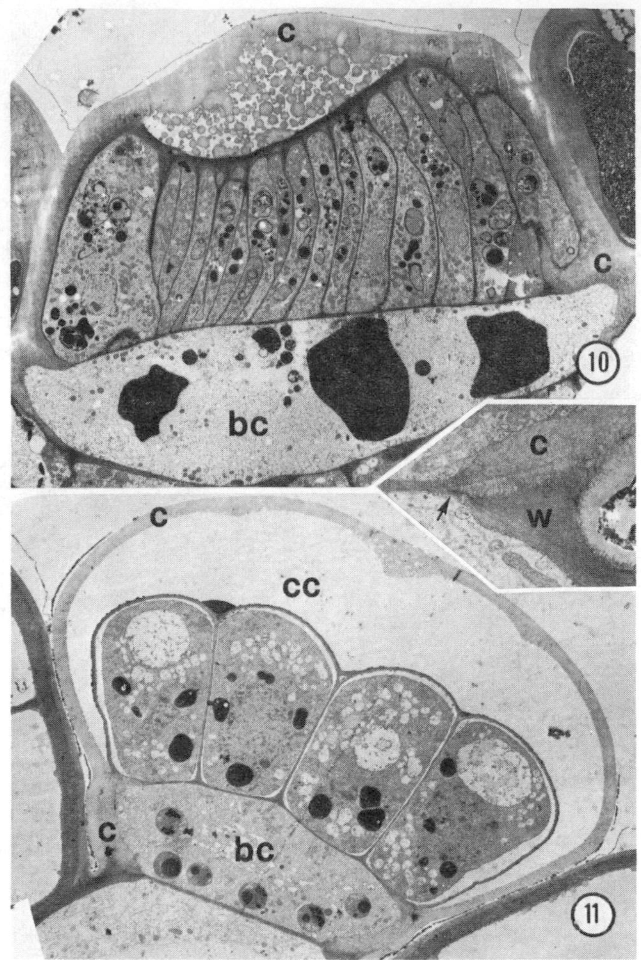

Fig. 13.10 A micrograph of the multicellular salt gland of the mangrove, *Aegiceras*; c, cuticle; bc, basal cell.

Fig. 13.11 A micrograph of the salt gland of the mangrove, *Avicennia*; c, cuticle; cc, collecting compartment; bc, basal cell.

reported in this region for the glands of *Spartina* (Levering & Thomson, 1971).

13.2.2 Bladder Cells – Chenopodiaceae

The trichome-type salt gland of the Chenopodiaceae develops from a single protoderm cell and consists of one or more stalk cells and a large terminal bladder cell (Fig. 13.4; and see Black, 1954; Osmond *et al.*, 1969; Samoui,

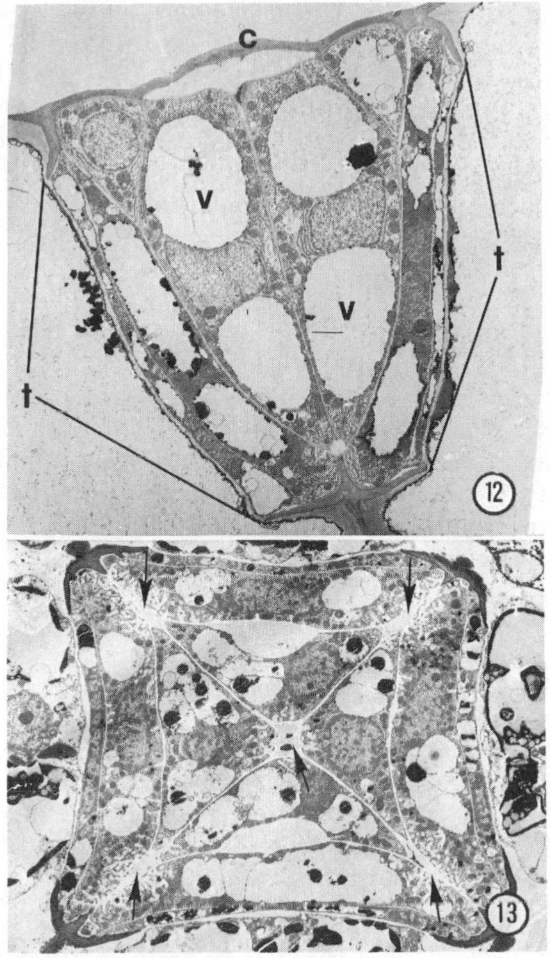

Fig. 13.12 A micrograph of the salt gland of *Ceratostigma*. The length of the transfusion, t, is illustrated; c, cuticle; v, vacuole.

Fig. 13.13 A paradermal section of a salt gland of *Limonium*. The five intercellular spaces are identified by arrows and note the wall protuberances associated with these regions.

1971; Thomson & Platt-Aloia, 1979). The bladder cell is highly vacuolate with a thin peripheral cytoplasm that contains typical cytoplasmic structures when intact. These include partially developed chloroplasts, mitochondria, dictyosomes, ribosomes, a flattened nucleus, and segments of ER (Fig. 13.4; and see Osmond *et al.*, 1969; Samoui, 1971; Thomson & Platt-Aloia, 1979). The stalk cells are richly cytoplasmic with one or more small vacuoles, several mitochondria, ER, undeveloped plastids and a large nucleus (Fig. 13.4; and see Osmond *et al.*, 1969; Samoui, 1971;

Thomson & Platt-Aloia, 1979). A symplastic continuum exists between the bladder cells and the mesophyll (Fig. 13.4; and see Osmond *et al.*, 1969).

The bladder and stalk cells are covered externally by a cuticle (Fig. 13.4), but all micrographs of sufficient resolution indicate that the inner primary wall of the stalk cells is not cutinized (Fig. 13.4; and see Samoui, 1971; Campbell *et al.*, 1974; Thomson, 1975; Thomson & Platt-Aloia, 1979) as determined by the difference in staining characteristics of the cuticle and the wall.

13.2.3 Multicellular glands

The multicellular glands found on other dicotyledonous plants are more varied in structure than those of the grasses or the trichomes of the Chenopodiaceae, although there are aspects of commonality. The most obvious gland variation is in cell number, varying from six to forty or more cells depending on the genus (e.g. Cardale & Field, 1971; Thomson, 1975; Faraday & Thomson, 1986a). Considerable variation also occurs in the position of these glands. For example, in *Frankenia* (Fig. 13.5; and see Thomson, 1975; Campbell & Thomson, 1976a), *Limonium* (Ruhland, 1915), *Aegialitis* (Thomson, 1975), and *Aegiceras* (Fig. 13.10; and see Cardale & Field, 1971), the glands appear to be more or less in lateral register with the adjacent epidermal cells. In *Tamarix*, the glands are often found in lateral register with the adjacent epidermal cells as well as sunken in epidermal crypts (Liphschitz & Waisel, 1982; Bosabalidis & Thomson, 1984). With *Glaux*, the glands occur in epidermal depressions (Rozema *et al.*, 1977). In the mangrove, *Avicennia*, the glands on the abaxial surface project from the leaf and appear as trichomes (Fig. 13.11; and see Fahn & Shimony, 1977; Liphschitz & Waisel, 1982). Drennan & Berjak (1982) reported that the adaxial glands become sunken into crypts with leaf expansion.

The cellular organization of these multicellular glands shows considerable variation. In *Tamarix* and *Frankenia* (Fig. 13.5), the glands consist of six outer, highly cytoplasmic, secretory cells arranged in pairs with the lower secretory cells associated inwardly with two large, vacuolated, subbasal cells (Thomson & Liu, 1967). Except for the gland of *Aegialitis*, which has forty secretory cells, a careful examination of the salt glands in the family Plumbaginaceae [including *Limonium* (Fig. 13.13), *Plumbago*, *Ceratostigma* (Fig. 13.12), *Goniolimon*, and *Armeria*] showed all of these glands to have sixteen secretory cells (Faraday & Thomson, 1986a). In paradermal sections, these glands are subdivided into four sectors each consisting of four secretory cells (Fig. 13.13; and see Faraday & Thomson, 1986a). In longitudinal, median sections, the general gland form is somewhat hemispherical with the secretory cells in layers angling upward from the base of the gland (Campbell *et al.*, 1974; Faraday & Thomson, 1986a).

In all these glands, vacuolate, subbasal cells of unknown function subtend the basal secretory cells (Fig. 13.12).

The glands of *Avicennia*, (Shimony *et al.*, 1973; Thomson, 1975; Fahn & Shimony, 1977), *Glaux* (Rozema *et al.*, 1977), and *Aegiceras* (Thomson, 1975), consist of several secretory cells positioned above a single, disc-shaped stalk or basal cell (Figs. 13.10 and 13.11). The stalk cells are subtended by one or more subbasal cells.

All the multicellular glands are covered by a cuticle which is continuous with that of adjacent epidermal cells (Figs. 13.5, 13.10, and 13.11; and see Thomson, 1975). In mature glands the cuticle is expanded above the secretory cells, as it is with the glands of the grasses, and forms a 'collecting chamber', between the secretory cells and the expanded cuticle, (Fig. 13.11; and see Cardale & Field, 1971; Thomson, 1975; Rozema *et al.*, 1977). Small pores occur in the expanded portion of the cuticle in all glands examined except for *Aegiceras corniculatum*, where Cardale & Field (1971) and Field *et al.*, 1984), were unable to detect cuticular pores (also see Fig. 13.10).

Of importance relative to structural and functional considerations, is that the cuticle extends inward along the lateral walls of these glands (Ruhland, 1915; and see Figs. 13.5, 13.10, 13.11 and 13.12), but contrary to most interpretations (Lüttge, 1971, 1975; Hill & Hill, 1973a, 1976), cuticular material does not extend into the walls between the innermost secretory or stalk cells and the subtending, subbasal cells (Figs. 13.5, 13.10, 13.11 and 13.12). The so-called 'encapsulating' cuticle extends to these walls, but a careful survey indicates that the cuticle does not impregnate them entirely or extend across them as determined by the marked difference in staining properties of the cuticle as compared to the wall in these regions (see Fig. 13.6; and see e.g. Thomson & Liu, 1967; Cardale & Field, 1971; Campbell *et al.*, 1974; Campbell & Thomson, 1975, 1976a,b; Fahn & Shimony, 1977; Rozema *et al.*, 1977). These walls are contiguous with those of the mesophyll and thus, structurally, there exists an apoplast continuum from the vascular tissue, to the base of the glands. In contrast, the apoplast within the glands is isolated from that of the leaf mesophyll by a complete cutinization of the side walls of the innermost secretory cells or stalk cells depending on the gland. The cutinization is of the lateral walls of the innermost secretory or stalk cells and extends to the border of the transfusion zone walls (Figs. 13.6 and 13.11; and see Thomson & Liu, 1967; Thomson *et al.*, 1969; Thomson, 1975; Fahn & Shimony, 1977; Rozema *et al.*, 1977; Bosabalidis & Thomson, 1984). Symplastic continuity between the mesophyll and all cells of these glands apparently exists since all common walls between these cells have plasmodesmata transversing them (Ziegler & Lüttge, 1966; Thomson & Liu, 1967; Cardale & Field, 1971; Campbell & Thomson, 1975; Thomson, 1975; Rozema *et al*, 1977). Since the apoplast within the glands is isolated from that of the mesophyll, the uncutinized walls and the plasmodesmata that traverse them at the base or lateral boundaries of the glands (see Fig. 13.6) represents the only apoplastic as well as symplastic interface between the glands and the mesophyll. Thus, this interface has been termed the transfusion zone (Thomson, 1975). In *Limonium*, however, the frequency of plasmodesmata in these walls is not particularly high. Faraday *et al*. (1986) determined the plasmodesmatal frequencies per unit area of the transfusion zone and found it (2–10 μm^{-2}) was well within the range normally found in less specialized tissues (Robards, 1976).

As with the basal cell of the salt glands of the grasses, the secretory cells of these multicellular glands have rudimentary plastids, ER, ribosomes, some microbodies, generally large but elongated and flattened nuclei, and numerous mitochondria (Lüttge, 1971, 1975; Hill & Hill, 1973a, 1976; Thomson, 1975). The large number of mitochondria is particularly significant considering that most evidence indicates that secretion by the glands is an active process. Using serial sections and stereological methods, Faraday & Thomson (1986b) have determined that the mitochondria occupy approximately 11 % of the total protoplasm volume of the secretory cells in *Limonium*. This compares to 1–3 % found in meristematic cells for example (Mauseth, 1982).

Labyrinthine arrays of wall protuberances occur along the walls of the secretory cells of some glands, *Tamarix*, *Frankenia* (Fig. 13.5), *Limonium* (Fig. 13.13; Ziegler & Lüttge, 1966; Thomson & Liu, 1967) and *Glaux* (Rozema *et al*., 1977) but not all. They are absent in *Avicennia* (Fig. 13.11), *Acanthus* (Thomson, 1975), *Aegiceras* (Fig. 13.10; and see Cardale & Field, 1971; Thomson, 1975), and *Aegialitis* (Thomson, 1975). In *Tamarix* and *Frankenia*, the wall protuberances are extensively developed in the outer secretory cells, less so in the middle cells, and very few occur in the innermost secretory cells (Thomson, 1975). With *Limonium*, the wall protuberances are most prominent adjacent to five intercellular spaces or vertical channels that occur within these glands (Fig. 3.13). With *Glaux*, the wall protuberances appear to be predominantly localized along the perimeter walls of the outer secretory cells (Rozema *et al*., 1977). No wall protuberances have been observed along the walls of the transfusion zones.

Considering that wall protuberances are definitive features of transfer cells and that they, along with the associated plasmalemma, are considered to be specializations for transport (Gunning, 1977), it is reasonable to speculate that membrane transport is maximal where the protuberances are most abundant in these glands. It may well be that the amplification of the surface area of the plasmalemma, rather than specialization within the membrane, could account for increased transport in these regions, but wall protuberances *per se* are not essential for salt secretion since they do not occur or are minimally developed in the salt glands of *Aegialitis*, *Aegiceras* and *Avicennia*. The absence of a wall labyrinth in the glands of *Aegialitis* is particularly germane since the flux estimates by Atkinson *et al*. (1967)

indicate that these glands have one of the highest transport rates known in plants (Lüttge, 1975).

The vacuolar complex in the secretory cells of the multicellular glands shows considerable variation between the different glands and often between different secretory cells within each gland.[1] Cardale & Field (1971) reported that the secretory cells of *Aegiceras* contained many small vacuoles rather evenly distributed throughout the cells (see also Thomson, 1975). A similar situation is also apparent with *Acanthus* (Thomson, 1975) and to some extent with *Avicennia*, although an occasional large vacuole is observed (Shimony *et al.*, 1973; Thomson, 1975). In *Glaux*, the secretory cells contain many small vacuoles, some of which contain electron dense material (Rozema *et al.*, 1977). In the Plumbaginaceae, such as *Plumbago*, *Limonium* and *Ceratostigma* (Fig. 13.12), the secretory cells contain from two to several modest-sized vacuoles which frequently contain electron dense material. Also, the size and morphology of the vacuoles varies in the different cells, being small and elongate in the thin, peripheral cells (Fig. 13.12) and more expanded and rounded in the larger central cells (Faraday & Thomson, 1986b). Of particular interest in these latter glands is that the vacuoles, although more prominent than in secretory cells of most other glands, only occupy approximately 44 % of the cell volume (Faraday & Thomson, 1986b). In the glands of *Tamarix* (Fig. 13.8) and *Frankenia* (Fig. 13.5), the vacuolar complex differs between non-secreting and secreting glands (Thomson *et al.*, 1969; Thomson, 1975; Campbell & Thomson, 1976b).

The cytoplasm of the stalk or basal cells of the glands of *Avicennia* (Fig. 13.10), (Shimony *et al.*, 1973), and *Glaux* (Thomson, 1975; Rozema *et al.*, 1977) is much less dense than the outer secretory cells. There are fewer mitochondria, and the plastids, although rudimentary in nature, also stain less densely.

The subbasal cells associated with these multicellular glands are generally highly vacuolate and often the vacuoles are observed to be enriched in electron-dense material, particularly with *Tamarix*, *Frankenia*, *Aegiceras* and, to some extent, with *Glaux*. The chloroplasts often tend to be poorly developed in these cells as compared to the adjacent mesophyll cells

[1] In investigations of salt glands, they are considered to occur in one of two functional states – secreting or non-secreting. Secretion occurs, and can be observed and measured, when the leaves have a high salt status such as during the daytime with whole plants growing under saline conditions or when leaf discs are floated on a salt solution in either the light or dark, or when the leaf petiole or small branches are inserted into a salt solution. Secretion is not observed, or is minimal, when the tissue has a low salt status such as occurs with whole plant material grown in the absence of salt. Also, secretion falls to a minimum or does not occur with whole plants after a few hours in the dark or when high salt status material, discs, leaves and small branches, are challenged with distilled water for a few hours.

(Ziegler & Lüttge, 1966; Thomson & Liu, 1967; Shimony & Fahn, 1968; Thomson, 1975; Rozema *et al.*, 1977; Drennan & Berjak, 1982), but a normal complement of other cytoplasmic organelles appears to be present (Thomson, 1975). No specific functional role for the subbasal cells has been elucidated relative to secretion.

13.3 Salt glands in relationship to the leaf

Since it is generally concluded that salt glands remove, by secretion, excess ions delivered to the leaves by the transpiration stream (see reviews by Lüttge, 1971, 1975; Thomson, 1975; Hill & Hill, 1976; Fahn, 1979; and more recent studies by Pollak & Waisel, 1979; Drennan & Pammenter, 1982), the pathway(s) of solute movement from the vascular tissue to the epidermal salt glands has drawn considerable attention. There are three possibilities: (a) direct apoplastic flow of the solutes to the glands; (b) entrance of ions into the cytoplasm of vascular and/or mesophyll parenchyma cells and symplastic movement via the plasmodesmata to the glands; (c) or some combination of both of these.

13.3.1 The apoplastic pathway

As pointed out previously, the inner primary walls of the stalk cells of the *Atriplex* glands are not cutinized and neither is any area or region of the inner basal walls of the multicellular glands. A similar lack of cutinization was found for the bicellular salt glands of some grasses (Oross & Thomson, 1982a,b). Thus, structurally, an apoplastic continuum from the mesophyll to the glands exists for all three gland types. Experimental evidence also indicates that an apoplastic continuum exists to the glands. The lanthanum ion has proven useful to demarcate apoplastic pathways because of its permeability in the apoplast, its inability to penetrate the plasmalemma, and its visibility as a precipitate in the electron microscope (Thomson *et al.*, 1973). When an ionic lanthanum solution is fed to cut branches, leaf disc, leaf petioles or leaf bases, lanthanum deposits were found throughout the apoplast of the leaves of *Atriplex*, *Tamarix* and *Limonium*, (Campbell *et al.*, 1974), the grass *Cynodon* (Oross & Thomson, 1982b) and the mangrove *Aegiceras* (Field *et al.*, 1984). Lanthanum was also found in the walls of the transfusion zones at the base of the glands of *Limonium*, *Tamarix* and *Aegiceras*, in the walls of the stalk and bladder cells of *Atriplex* and in the walls of the basal cell of *Cynodon*. Thus, the apoplastic route is available for the movement of solutes from the vascular tissue to all types of glands. Lanthanum deposits were not found in the apoplast of the outer secretory cells of the multicellular glands, which indicates that there is no apoplastic continuity between this region and the apoplast of the mesophyll.

Results from cytochemical studies provide further evidence of an apoplastic continuum to the glands. Campbell & Thomson (1975, 1976b) examined the leaves and glands of *Tamarix* and *Frankenia* for Cl^- localization using the silver-precipitation techniques of Komnick & Bierther (1969). In the high-salt status, secreting material, they observed a heavy enrichment of Cl^- in the mesophyll cell walls and in the walls at the transfusion zone as compared to low-salt status, non-secreting control material. These and the results of the lanthanum experiments clearly indicate that the apoplast is open to the base of the glands. There is a second implication to these studies. The distinctively heavy accumulation of Cl^- in the apoplast, including the walls of the transfusion zones in comparison to the cytoplasm and vacuoles of the mesophyll cells, indicates that the apoplastic pathway may be the primary route of ion movement to the glands particularly in high-salt status, secreting material. Also, Ramati *et al.* (1976) using electron microprobe procedures reported that in salt-challenged leaves of the grass *Sporobolus*, the mesophyll cell walls were enriched in Na^+, K^+ and Cl^- as compared to the cytoplasm.

Physiological evidence also suggests that the primary route of salt flow to the glands is the apoplast. Atkinson *et al.* (1967) fed radioactive Cl^- to high-salt status leaves of the mangrove, *Aegialitis*, and reported that the specific activity of the secreted Cl^- was higher than that of the leaf, indicating that the added Cl^- passed directly to the glands without coming to equilibrium with the major Cl^- pool within the cells of the leaf.

13.3.2 The symplastic pathway

The presence of plasmodesmata between all neighbouring cells of the mesophyll and those of the glands and the presumed function of plasmodesmata as pathways of transport between cells suggests that symplastic flow and physiological coupling probably occurs. In regard to ions, Ziegler & Lüttge (1967) using a silver-precipitation technique found Cl^- to be associated with plasmodesmata of the transfusion zone on *Limonium* glands. They interpreted their findings to mean that Cl^- was symplastically transported to the glands. Also, Van Steveninck *et al.* (1976) argued for symplastic transport to the glands of *Aegiceras* based on cytochemical localization of Cl^- associated with the plasmodesmata of the leaf mesophyll cells.

In later studies with *Tamarix*, Campbell & Thomson (1975) found silver chloride deposits associated with the plasmodesmata of the transfusion zone as well as in the walls of salt-loaded material. However, they did not observe any deposits associated with the plasmodesmata in the low-salt status controls. They suggested that salt probably moved both apoplastically and symplastically to the glands and concluded that the predominant route to the glands under salt-challenged conditions was probably the apoplast, based on the heavy enrichment of Cl^- in the walls, as compared to the lack of a concomitant increase in the cytoplasm of the mesophyll cells under these conditions.

Shimony *et al.* (1973) examined frozen–dried sections of *Avicennia* leaves using an X-ray microprobe. They reported a downhill gradient of Na^+ and K^+ from the vascular tissue to the glands using a line scan across the section. They assumed the elements were in the cytoplasm and suggested therefore that transport to the glands was symplastic. This assumption is questionable even though in correlative experiments they report a similar pattern of elemental location using the antimonate precipitation technique for Na^+ and presumably other cations such as Ca^{2+} and Mg^{2+}. Unfortunately, the quality of the micrographs provided does not permit a clear visualization of the described pattern and the technique is questionable due to penetration and dislocation problems.

Indirectly, the results of several other physiological studies are suggestive of some form of symplastic transport. When plants are presented with a salt challenge, either in the greenhouse or in the field, most reports indicate that salt secretion tends to follow a diurnal pattern, with secretion being highest near midday and low or absent at night (Scholander *et al.*, 1962; Atkinson *et al.*, 1967). This pattern is apparently related to daily transpiration cycles and not directly to light or an endogenous rhythm since secretion will occur in the dark (Scholander *et al.*, 1962) or when petioles of leaves or leaf discs are challenged with salt solutions in the dark (Arisz *et al.*, 1955; Atkinson *et al.*, 1967). However, a deviation from this daily pattern has been recently reported by Drennan and Pammenter (1982) for the mangrove, *Avicennia marina*. Of importance here is that several studies indicate that when a salt challenge to the leaves is removed, such as when transpiration ceases at night or when preloaded discs or excised leaves are transferred to distilled water, secretion continues for some time, but usually at a decreasing rate resulting in the concomitant lowering of the salt content of the leaves (Arisz *et al.*, 1955; Scholander *et al.*, 1962; Pollak & Waisel, 1979; Drennan & Pammenter, 1982). This suggests two things: first, the continued secretion is from compartments in the leaves, and second, with the depletion of these compartments the leaves are actually in a low-salt status. The probable compartments within the leaf are the cell walls, the cytoplasm and constituent organelles, and the vacuoles of the mesophyll cells. At present, there is little direct evidence indicating which one, or if all these compartments, are emptied and if it occurs for all compartments, whether it happens in series or in parallel. Nevertheless, if unloading from the vacuoles into the mesophyll cytoplasm occurs, then symplastic flow to the glands is a likely possibility.

Using the efflux analysis data of leaf discs of *Limonium*, Hill (1970) has argued for symplastic transport to the glands. From exponential derivations of the ion efflux curve obtained from washout experiments, he concluded that there were three common tissue compartments for Na^+: the apoplast, cytoplasm and vacuole, with the chloroplasts being possibly a fourth for Cl^-. They also reported that the half-time of transit for the labelled ions to appear in the glandular secretion corresponded to that for loading of the cytoplasm. They interpreted the similarity in time transits for cytoplasmic loading, as determined by the efflux studies, and appearance of label in the secreted fluid as indicative of symplastic flow and coupling to the glands. However, the results are equally supportive of apoplastic flow, if the cytoplasmic compartment is that of the secretory cells of the glands, rather than the mesophyll cells (Thomson, 1975).

The possibility of symplastic transport to the glands was also suggested by Larkum & Hill (1970), based on the accumulation of Na^+, K^+ and Cl^- in the chloroplasts of *Limonium* when leaf material was shifted from a low- to a high-salt status. They assayed non-aqueously isolated chloroplasts, and reported that there was a linear increase of these ions in the chloroplasts over a 3 h period after which there was a slight decrease. The appearance of glandular secretory activity was apparently coincident with the plateauing of ionic accumulation within the chloroplasts. They concluded that the rise in Na^+, K^+ and Cl^- in the chloroplasts during the first 3 h and the subsequent decrease as secretion began indicated that glandular activity directly influenced mesophyll cytoplasmic ion levels. In support of this, they reported that with the inhibition of glandular secretion, using choline benzenesulphonate, the ionic content of the chloroplasts increased. Similarly, if the leaf material was challenged with 100 mol m^{-3} NaCl and then shifted to 50 mol $m3^-$ NaCl + 50 mol m^{-3} KCl, the Na^+ content of the chloroplasts fell and KCl increased. Assuming the chloroplasts and the cytoplasm are more or less the same cellular compartment, they concluded that coupling to the glands was symplastic. Their observations provide no direct evidence for symplastic coupling, only that the salt status of the cytoplasm of the mesophyll cell may influence glandular activity, or glandular activity may influence the salt level of the cytoplasm, possibly symplastically. Further, if movement is apoplastic, it might be expected that the mesophyll cells become loaded first since they are nearest to the vascular system, and that glandular activity would be delayed until the solutes diffused across the mesophyll apoplast to the glands.

Pollak and Waisel (1970) found, with salt-challenged *Aeluropus* plants, that the amount of Na^+ secreted daily was approximately that present in the leaf, and accumulation in the leaf occurred at a low rate. In a subsequent study, they reported (Pollack & Waisel, 1979), that with the transfer of preloaded, excised leaves to salt-free medium, 40 % of the preloaded Na^+ remained in the leaves after 8 h of continued secretion. Similar results were also found using ^{22}NaCl followed by a cold chase. As Pollak and Waisel point out, these results suggest that Na^+ content of the leaves occurs in two fractions: a rapidly secreted fraction which may be localized in the apoplast and/or cytoplasm of the mesophyll cells, and a second component which is secreted slowly and possibly localized in cytoplasmic organelles or the vacuoles.

Symplastic continuity to the glands and possibly transport also has been suggested from results of electrophysiological data. Hill (1967a) abraded the cuticle from the lower surface of pretreated leaf discs of *Limonium* and placed the lower portion against a salt solution and examined the disc for conductance between the solution and an electrode at the outer surface of the leaf. Using voltage pulses of constant amplitude, it was claimed that electrical coupling occurred predominantly through the gland, not the cuticle of adjacent epidermal cells or the nearby stomata. Assuming the apoplastic pathway to the glands was unavailable, he concluded that the high-conductance pathway was symplastic to and through the glands (see also Hill & Hill, 1973a, 1976; Lüttge, 1975). However, the results of these experiments provide no direct evidence that this electrical coupling and presumed pathway of solutes is symplastic as opposed to apoplastic or both. It must be borne in mind that the current must pass across at least two membranes. If flow is symplastic, these would be the plasmalemma of the mesophyll cells and the outer membranes of gland secretory cells or, if flow is apoplastic, the plasmalemma of the basal cells of the glands and then the outer membranes of the secretory cell. Another complication is that in similar experiments with *Aegiceras*, Billard & Field (1974) were unable to detect a difference in voltage response above the glands as compared to the cuticle above epidermal cells. Also, in repeat experiments using *Limonium*, we have been unable to detect a specific voltage response over the glands as compared to that above epidermal cells (Faraday and Thomson, unpublished observation).

With *Atriplex*, electrical coupling between the leaf mesophyll cells and the vacuole of the bladder cells of the glands has been shown (Osmond *et al.*, 1969). Lüttge & Pallaghy (1969) have reported that light-induced changes in membrane potentials in the mesophyll cells were detectable in bladder cells. The changes were induced by wavelengths of light associated with photosynthesis, and they concluded this was dependent on the green mesophyll since no light-dependent transient changes in membrane potential were observed in photosynthetically inactive epidermal strips and bladder cells of *Chenopodium*. Assuming that *Chenopodium* and *Atriplex* were reasonably comparable in regard to these functions, they concluded that the signal moved symplastically to the bladder cells.

As will be discussed later, all evidence indicates that secretion is an active process, and if the gland cells are the primary sites of this process, it is important to consider the possibility of symplastic transport of metabolic substrate to the glands. This is of particular importance since all glands, except possibly *Atriplex*, lack chloroplasts; and with *Atriplex* the chloroplasts are minimally developed (Osmond *et al.*, 1969; Samoui, 1971; Thomson & Platt-Aloia, 1979).

In summary, with regard to the pathway of ion transport to the gland, evidence has been presented for both apoplastic as well as symplastic flow and both routes may be utilized in most plants.

13.4 Secretion and selectivity

In the previous section, we primarily considered coupling of the mesophyll to the glands in regard to transport pathways. Here we discuss the composition of the secreted fluid and it will become apparent that the composition may relate to both physiological coupling of the mesophyll to the glands, as well as to glandular function.

One of the most common assumptions many investigators have made is that the secreted salt consisted primarily of NaCl (Hill & Hill, 1973a, 1976). Nevertheless, there was early evidence (Volkens, 1884; Ruhland, 1915) that the secretion contained other components and possibly even some organic material. Results from studies where rather complete analyses have been done indicate that some plant salt glands secrete a variety of anions and cations. For example, with *Tamarix*, analyses of secreted salts have revealed the presence of Ca^{2+}, Mg^{2+}, Na^{+}, K^{+}, Cl^{-}, SO_4^{2-} NO_3^{-}, PO_4^{3-}, HCO_3^{-}, and even Br^{-} (Waisel, 1961; Berry & Thomson, 1967; Thomson *et al.*, 1969; Berry, 1970; Pollak & Waisel, 1970). Similarly, analysis with *Limonium* revealed the presence of Ca^{2+}, Mg^{2+}, Na^{+}, K^{+}, and Cl^{-} (Faraday & Thomson, 1986c), with *Armeria*, Ca^{2+}, Mg^{2+}, Na^{+} K^{+}, Cl^{-}, and HCO_3^{-} (Baumeister & Ziffus, 1981), and with the estuarine grass, *Spartina*, Ca^{2+}, Mg^{2+}, Na^{+}, K^{+}, Cl^{-}, SO_4^{2-} and PO_4^{3-} (Sutherland & Eastwood, 1916; McGovern *et al.*, 1979; Rozema *et al.*, 1981). Again, with *Tamarix*, Kleinkopf & Wallace (1974) implied that 25 % of the leaf K, Ca, Cu, Mn, B, Al, Si, Ti, Mo, Sr and Ba ions were secreted by the salt glands. However, their procedure involved extensive washing of the leaves with distilled, deionized water to collect the salt and their data could well represent mineral elements leached from the tissue (Tukey, 1971), as well as those secreted (Boon & Allaway, 1982). Considering the fact that many plants with salt glands are adapted to saline environments other than a maritime one, it is not surprising to find that a variety of different ions are secreted.

A useful technique for studying salt glands has been the floating of a leaf disc on a defined salt solution, or the placing of a leaf petiole or the insertion of a small branch into such a solution. Under these conditions, salt secretion by the glands occurs and ions not normally associated with plant tissue, e.g., Rb^{+}, Cs^{+}, Br^{-} and I^{-} have been recorded as secreted (Hill, 1967b; Thomson *et al.*, 1969).

In some instances, there is a general correlation between the composition of the secreted fluid and that of the root medium. For example, Scholander *et al.* (1962) found that the ions secreted from the salt glands of the mangrove, *Aegiceras*, occurred in approximately the same ratio as they did in seawater. With *Tamarix aphylla* growing in culture solution to which salt amendments were made, Thomson *et al.* (1969) found that the composition of the salt secreted varied in relationship to the cation concentration of the culture solution, e.g. increasing the Na^+ concentration of the culture medium resulted in an increased Na^+ secretion and a concomitant decrease in K^+ secretion. In subsequent studies, Berry (1970) did a more thorough analysis and found that the composition of the secreted salts was similar to that of the culture medium. However, there were some notable exceptions and differences; PO_4^{3-} was not secreted, and the percentage of NO_3^- equivalents was much less in the secreted fluid than in the culture solutions. Also, there were large amounts of carbonate. Similarly, Waisel (1961) reported large amounts of calcium carbonate in the secreted salt of *Tamarix* growing in particular soil types.

For the tidal marsh grass *Spartina*, McGovern *et al.* (1979) reported that the proportions of Na^+ and K^+ in the secreted salt were similar to that of seawater, but SO_4^{2-}, Ca^{2+} and Mg^{2+} were much less. PO_3^{4-}, however, was higher in the secreted salt than in sea water.

Of interest, particularly in regard to the mechanism(s) of secretion, is the balance between cations and anions of the secreted salt and the pH of the secreted fluid. To put this in perspective, a digression is necessary. Hill & Hill have consistently maintained that the secreted fluid has a neutral pH from which the argument has been presented that secretion is not a function of a proton pump (Hill & Hill, 1973a, b). Second, Hill (1967a) reported from electrophysiological studies that the electropotential signal from active glands was negative, suggesting the presence of an anion pump. Third, he maintained that secretion depended on the presence of Cl^- in the basal medium. The other halide anions Br^- and I^- could replace Cl^-, yet were somewhat less effective; but secretion did not occur if Cl^- was replaced by SO_4^{2-}, benzenesulphonate, gluconate or borate. However, the reported secretion was not dependent on a particular cation since Na^+ would be replaced by other alkali metal ions. Finally, he generally assumed that Cl^- is the balancing anion based on the earlier studies of Arisz *et al.* (1955). In none of the studies by Hill and his associates was there a direct analysis of the composition of salt, and Hill & Hill in their review (1976) have ignored the studies of Berry (1970), arguing that '. . . his presentation is rather complex. . .'.

In Berry's analyses with *Tamarix*, he found that approximately 49 % of the anion composition of the secreted salt consisted of bicarbonate and 18 % of sulphate. On the basis of equivalents, it was 63 % carbonate and 30 % sulphate. The presence of carbonate and large amounts of Ca^{2+} in the salts probably accounts for the early identification of salt glands as 'chalk' glands (Volkens, 1884). In contrast to Berry's studies, Boon &

Allaway (1982) reported Cl^- to closely balance the cation equivalent in the secreted salts of both natural populations and greenhouse grown *Avicennia* plants treated with sea water.

More recently, Faraday & Thomson (1986c) analysed the secreted salt from *Limonium perezii*, *L. latifolium*, *Plumbago auriculata* and *Ceratostigma plumbaginoides*, using the electron microprobe. These plants were grown in the greenhouse and regularly fed with half-strength Hoagland and Arnon's (1950) solution. They found a range of elements to be secreted, including Na, Mg, S, P, Cl, K and Ca. Of particular interest is that the mineral anions contributed only a fraction of the charge equivalents to balance that of the cations and that Cl^- never contributed more than 2 % of the charge equivalents. However, carbonate was present in large quantities. Using excised leaves of *Limonium perezii* in which the petiole was inserted into 100 mol m^{-3} solutions of either KI, KCl, NaCl, Na_2SO_4, $MgCl_2$, and $MgSO_4$, they found that the glands secreted mainly the ions of the challenging solution. Chloride contributed 7 % or less of the ions in the secretion when absent from the challenging solution. Equally important, measurements of the pH of the secreted fluids of leaf discs taken from these leaves and floated on comparable solutions was always 8.5 or higher. Taken together, these data contradict the hypothesis of Hill and associates relative to the presence and dominance of a chloride pump in salt glands. Further, the pH of the secreted solutions suggests that the establishment of a pH gradient across a membrane, probably within the gland complex, is directly related to the mechanism of secretion. Thus, the underlying mechanism of secretion would be more in accord with the ion transport processes that occur in other plant cells (see Ch. 3; and Leonard, 1984; Sze, 1984).

The situation appears more complicated than the glands simply secreting those ions present, or present in excess in soil or a culture solution, since there is a developing body of evidence that at some point or points in the root–shoot–leaf–gland integrated flow system selectivity occurs (Waisel, 1961, 1972; Pollak & Waisel, 1970, 1979; Rozema, 1978; McGovern *et al.*, 1979; Rozema *et al.*, 1981). Selectivity is the secretion of a specific ion in larger quantities when it is present in the medium in equal or in lower amounts than other ions (Liphschitz & Waisel, 1982).

Earlier, Waisel (1961) examined the secreted salts from *Tamarix aphylla*, reporting an apparent selective series of Na >Ca>K, and Pollak Waisel (1970, 1979) found the series to be Na>K>Ca for *Aeluropus*. In subsequent studies, Rozema *et al.* (1981) reported the following sequence for four other halophytes:

Glaux maritima	Na>K>Ca
Armeria maritima	K>Ca $\geqslant$ Na
Limonium vulgare	Na $\simeq$ K>>Ca
Spartina anglica	Na>>K>>Ca

They determined the differential secretion rates between these plants to be:

Na: *Spartina anglica* ⩾ *Limonium vulgare*>*Glaux maritima*>*Armeria maritima*

K: *Limonium vulgare* > *Armeria maritima*>*Spartina anglica*>*Glaux maritima*

Ca: *Limonium vulgare* > *Armeria maritima*>*Spartina anglica*>*Glaux maritima*

However, selectivity can occur at several levels in the plant such as at the root, stem or the leaf mesophyll, and therefore cannot be assigned as a glandular function if it occurs at one or more of these sites.

In regards to glandular activity, *per se*, the important consideration is the 'excess' ions delivered to leaves via the transpiration stream and the differential accumulation of ions in the mesophyll as compared to those secreted. The evidence now indicates that mesophyll selection probably does occur to some extent. Osmond *et al.* (1969), in their studies with *Atriplex* growing in a root medium containing 5, 50 or 250 mol m^{-3} NaCl, found that the ion concentration in the leaf lamina and bladder cells was always higher than the external solution and these increased with leaf age. The ratio of K^+ to Na^+ in the bladder cells was more similar to that of the leaf lamina than that of the external solution; however, the bladder cells had a higher proportion of Na^+. These types of studies with *Atriplex* have been recently extended by Jeschke & Stetler (1983) and Storey *et al.* (1983). The electron microprobe analyses of Storey *et al.* indicated that with Na^+-treated material there was an increasing Na^+/K^+ ratio outward across the mesophyll from the vascular bundles to the bladder cells. Also, the bladder and epidermal cells had the lowest estimated amounts of K^+ and the highest ratios of Na^+/K^+, and the $Cl^-/K^+ + Na^+$ ratio appeared to be higher in these cells. They suggested that one possible explanation for this increasing ratio across the leaf mesophyll is the selective removal by the mesophyll of K^+ as the salt solution passes through the apoplast to the epidermis and bladder cells. This implies that ion selection is a mesophyll function rather than a glandular one. Of particular interest was that in young leaves, a large proportion of the leaf NaCl was localized in the vacuoles of the bladder cells. Similarly, Jeschke & Stelter (1983) found that the bladder cells on young leaves contained most of the leaf Na^+. Both groups have suggested that the preferential accumulation of Na^+ in the bladder cells on the young leaves permits the maintenance of a high K^+ status in the lamina. The implications are that a high K^+ status is essential for growth and development of the leaves. This implication is consistent with the hypothesis put forth by Leigh & Wyn-Jones (1984) that Na^+ can replace K^+ as an osmoticum when K^+ becomes reduced in plant cells, but that K^+ is maintained at an adequate level in the cytoplasm for constancy of metabolic function, if K^+ does not become limiting (see Ch. 6).

In both these studies however, Na^+ was found to increase in the lamina of the mature leaves, from which Jeschke & Stelter (1983) noted it was not apparently compensated for by uptake in the bladders due to their decreasing contribution to leaf weight (here probably numbers) as compared to that of the expanding lamina. Storey *et al.* (1983) suggested that one aspect might be that capacity for ion accumulation by the bladder cells becomes saturated. Another possibility is that the apparent decreasing capacity of the bladder cells, as well as their diminishing contribution to leaf weight with age, is related to the fact that some species of *Atriplex*, particularly annuals (Black, 1954; Osmond *et al.*, 1969) only produce one flush of bladder cells and this is consistent with the observation of Storey *et al.* (1983) that there were many fewer bladder cells on the older leaves than on the young, expanding leaves. It would be interesting to examine the situation where several series of bladder cells form and collapse during the life of the leaf (Black, 1954; Osmond *et al.*, 1969).

There is some evidence indicating an interrelationship or interaction between ions in regard to secretion. For example, with *Aeluropus* and *Tamarix*, K^+, Mg^{2+} and Cl^- in the treatment solution lowered the secretion of Na^+ (Waisel, 1961; Pollak & Waisel, 1970) but the amount of Ca^{2+} secreted was not affected by the composition or concentration of the treatment solution. However, excess $MgCl_2$ or $CaCl_2$ lowered the amount of K^+ secreted by *Tamarix*. Similarly, in the studies by Rozema *et al.* (1981), it was reported that Ca^{2+} secretion was depressed by NaCl, but that the addition of KCl inhibited Na^+ secretion for *Limonium* and *Spartina* but this was not so with *Glaux* and *Armeria*. Again, it is difficult to determine whether the interaction between ions occurs with the tissue and organs of the plants, or is a glandular function. For example, Pollak & Waisel (1979), using detached leaves of *Aeluropus*, found that the secretion of K^+ was equal to that of Na^+ only when the concentration of applied K^+ was three times that of Na^+. It is not known whether this means the glands are one-third less efficient in transporting K^+ than Na^+, or that this reflects, in part, an aspect of mesophyll selection and accumulation.

13.5 Mechanisms

In considering the mechanisms of salt secretion, membrane transport processes must be fundamentally involved. Wherever the site or sites of membrane transport occur in the mesophyll–gland complex, relative to secretion, most evidence indicates that it is an active process. Ruhland (in 1915) found that secretion occurred against a fairly high pressure imposed from the outside of the leaf, suggesting that secretion required a work function. Further, studies with *Limonium* (Arisz *et al.*, 1955), *Tamarix* (Berry, 1970), and the grass *Aeluropus* (Pollak & Waisel, 1970) revealed that the ionic concentration or osmotic concentration of the secreted fluid is higher than that of the root medium. Also, Mozafar & Goodin (1970) and Osmond *et al.* (1969) found the NaCl concentration in the bladder cells of *Atriplex* to be higher than the external solution. These observations suggest that somewhere in the root–stem–leaf–gland continuum, accumulation against a concentration gradient occurs.

More directly related to glandular function, or at least mesophyll–gland coupling, are the observations that the osmotic or ionic concentration of the secretion was higher than that of the xylem, leaf sap or challenging solutions when leaf discs or excised leaves were used (Arisz *et al.*, 1955; Scholander *et al.*, 1962; Pollak & Waisel, 1970, 1979). Further, Osmond *et al.* (1969) recorded that Cl^- transfer into the bladder cells from the mesophyll in *Atriplex* was against a concentration gradient.

These results suggest two things: (a) there is an active, energy-requiring process involved in secretion; and (b) the active site or sites are located at the point of coupling between the glands and the mesophyll, i.e. the transfusion zone and/or where the salts are emitted from the glands, i.e. the plasma membranes of the secretory cells. It is of interest to note here, however, that Atkinson *et al.* (1967) reported that the osmotic concentration of the secretion from the glands of the mangrove, *Aegialitis*, was more or less equal to that of the xylem sap. This point will be discussed later.

Further evidence that secretion is an active process is derived from physiological studies. Arisz *et al.* (1955), using leaf discs of *Limonium*, found that the rate of secretion was temperature-dependent, that metabolic inhibitors (particularly at rather high concentration) stopped or reduced secretion, and that anoxia eliminated secretion. In general, all subsequent experiments with *Limonium* and other salt glands have given similar results (Atkinson *et al.*, 1967; Osmond *et al.*, 1969; Lüttge & Osmond, 1970; Hill & Hill, 1973a,b, 1976) indicating that a metabolically derived energy source is required for secretion.

In a more direct sense, electrophysiological measurements suggest that active transport is involved in secretion. Osmond *et al.* (1969) found with *Atriplex* that Cl^- transport from the mesophyll into the bladder cells, in the light, was against an electrochemical gradient. Hill and his associates, in a series of electrochemical studies with leaf discs of *Limonium*, have provided compelling evidence that secretion is active and some form of a 'pump' is involved (Hill, 1967a, b; Hill & Hill, 1973a, b, 1976). The main feature of this evidence is that the distribution of Na^+, K^+ and Cl^- between the secreted fluid and the challenging medium were not in electrochemical equilibrium as predicted by the Nernst equation (see Ch. 1), and all three ions had a higher electrochemical potential in the secreted fluid than in the challenging medium. Second, when short circuit techniques were used, Hill (1967b) recorded current flow between the challenging medium and the secreted fluid. Under conditions where the secretion is short-circuited against the bathing medium, any current flow is considered to be due to the active transport of charged species. Further, when he clamped the secretory potential at zero, he found that radioactive Na^+, K^+, Rb^+, Cs^+, Cl^-, Br^- and I^- was transferred from the medium to the secreted fluid.

If it is assumed that secretion is mediated by active membrane-transport processes, the source of energy and its chemical nature is of particular interest. With *Atriplex*, an apparent light-stimulated accumulation of Cl^- in the bladder cells has been described (Osmond *et al.*, 1969; Lüttge & Pallaghy, 1969). Lüttge & Osmond (1970) reported that this uptake was partially inhibited in the light by the photosystem II inhibitor, DCMU. Addition of the uncouplers FCCP and CCCP eliminated the rest of the uptake, and in the dark uncouplers effectively eliminated uptake. Further, based on CO_2-fixation studies which showed that approximately 90 % of the leaf photosynthesis occurred in the mesophyll as opposed to an estimated 10 % in the bladder cells, they assumed that the photosynthetic capacity of the bladder cells was insufficient to affect uptake. From these observations, they suggested that ATP was the energy source in the dark, but uptake into bladder cells in the light was driven by photosynthetic electron transport or photosynthetically produced reducing equivalents. They also concluded that these processes occurred in mesophyll cells and not in the bladder cells. Secondly, electrophysiological studies (Osmond *et al.*, 1969; Lüttge & Pallaghy, 1969; Lüttge & Osmond, 1970) indicated that a light-induced change in membrane potentials occurred in the mesophyll cells coincident with a similar change in the bladder cells. A change was not detectable between a reference medium and apparently undamaged bladder cells in epidermal strips removed from leaves of *Chenopodium*, from which it was suggested that the light-induced transients in membrane potential within the bladder cells must result from events within the mesophyll. Assuming that electrical coupling is only symplastic, a biochemical link by this pathway from the photosynthetic mesophyll tissue to the bladder cells was suggested (see Lüttge, 1971). These assumptions have been questioned previously (see Thomson, 1975; Hill & Hill, 1976), but there are concerns as to the degree of similarity between the bladder cells of *Atriplex* and *Chenopodium*, and the extent of damage induced in the bladder cells when the epidermis is stripped from the leaves. Further, the moderately developed chloroplasts of the bladder cells could be a major parameter in these light-dependent transients.

Earlier, Arisz *et al.* (1955) reported that light stimulated the rate of secretion from discs of *Limonium* floated on a salt solution; however, it is not known whether this effect is on the mechanism of secretion or relates to the production of metabolic substrates. More recently, Pollak and Waisel (1979) found with *Aeluropus*, that the absolute amount of salt secreted was higher in the light than in the dark, but this correlated with the increased rate of salt movement to the leaf via transpiration. However, a possible relation of photosynthesis to secretion is suggested from the work of B. S. Hill & Hill (1973). Using electrical coupling between the reference solution upon which the discs were challenged and the secretion droplets on the outer disc surface as a measure of secretory activity, they found that cyanide, anaerobic conditions, and the uncoupler CCCP inhibited secretion in the dark; however, in the light, anaerobiosis was only partially inhibitory. Also DCMU, a photosystem II inhibitor, was unable to prevent the rise in secretion under anaerobic conditions when the discs were shifted from the dark to light. From this, they suggested that ATP derived from respiration and possibly cyclic photophosphorylation in the light is utilized in the secretion process. Since the glands do not have chloroplasts, they suggestted that in the light the ATP would be derived from the mesophyll and diffuse symplastically to the glands.

13.5.1 Site(s) and nature of the pump(s)

Since the glands are the exit points of the secreted fluid, it seems logical to suggest that they could be the location of the active process or processes underlying secretion. From a structural point of view, the high mitochondrial content of the secretory cells of the multicellular glands, and of the basal cell of the glands of the grasses, implicates these cells as active in the secretory process. Similarly, the amplification of the plasmalemma by wall protuberances in the secretory cells of some glands, as well as by the partitioning membranes in the basal cell of the grass glands, is consistent with this suggestion if, indeed, these elaborations are indicative of a specialization for transport (Gunning, 1977). If we assume that secretion involves membrane transport in the gland cells, three possibilities exist for the transport site:

1. If ion movement to the glands is solely symplastic, we must consider uptake across the plasmalemma of the mesophyll cells and secretion across the plasmalemma of the secretory cells to the wall space of the glands, with *Atriplex* this second site would be the tonoplast of the vacuole within the bladder cell.
2. If movement to the glands is solely apoplastic, ions must cross the plasmalemma of the innermost secretory, basal, or stalk cell, depending on the gland, and exit across the plasmalemma of the secretory cells.
3. Some combination of the above.

Since salt glands are often considered to be highly specialized for transport (Hill & Hill, 1973a, 1976; Lüttge, 1975), the rates of transport across the gland membranes are of particular importance. These rates can be obtained by collecting and analysing secreted fluid and measuring the pertinent glandular interfaces. Ruhland (1915) found as much as 0.86 mg of fluid was secreted per hour per square centimetre (h^{-1} cm^{-2}) for leaves of *Limonium* and Arisz *et al.* (1955) reported that 70 mg of fluid was secreted over a 24 h period from 150 mg leaf discs of *Limonium* floated on a salt solution.

More recently, Rozema *et al.* (1981) reported the secretion rates of Na^+ reached values of 2.3 μmol Na^+ cm^{-2} per 6 days for *Spartina anglica*, while that for *Limonium vulgare* was 1.0 μmol Na^+cm^{-2} per 6 days when both plants were grown on a medium containing similar amounts of NaCl. With *Atriplex*, Osmond *et al.* (1969) calculated the light-stimulated flux of $^{36}Cl^-$ to the bladder cells to be about 1 μeq g^{-1} h^{-1} when leaf slices were challenged with 5 mol m^{-3} $K^{36}Cl$. Atkinson *et al.* (1967) estimated the rate of Cl^- secretion from *Aegialitis* glands to be 2.5×10^{-4} mol m^{-2} s^{-1}.

Recently, Faraday *et al.* (1986), using serial sections and morphometric methods, determined the total surface area of the plasmalemma and the cross-sectional area of the plasmodesmata of the transfusion zone of *Limonium*. They found the highest total flux across the transfusion zone to be 1.7×10^{-4} mol m^{-2} s^{-1} and the highest fluxes for Cl^-, Na^+ and K^+ were 8.5, 8.4, 1.7×10^{-5} mol m^{-2} s^{-1}, respectively, and that for Ca^{2+} was 5.8×10^{-7} mol m^{-2} s^{-1}. Again, if these are transmembrane fluxes, they are very high compared to rates reported for most other plant cells (see Raven, 1976). If, however, the movement of solutes to the glands was only symplastic, then the fluxes for Cl^-, Na^+, K^+ and Ca^{2+} were calculated to be 1.4×10^{-17}, 2.9×0^{-18} and 9.7×10^{-20} mol $plasmodesma^{-1}$ s^{-1}, respectively. These can be compared to the upper value of from 2×10^{-19} to 10^{-18} $plasmodesma^{-1}$ s^{-1} estimated for plasmodesmatal fluxes in other plant cells (Gunning & Robards, 1976).

In a subsequent study, Faraday & Thomson (1986b), using similar procedures, determined the total surface density of the plasmalemma of the secretory cells including that bounding the wall protuberances. From this they were able to calculate fluxes across these membranes relative to secretion. They found the total efflux rate of ions to be 1.0×10^{-5} mol m^{-2} s^{-1}. If, however, the flux was limited to membranes bounding the wall protuberance, it was 3.2×10^{-5} mol^{-2} s^{-1}. These values are high, but considerably lower than those calculated for transmembrane flux at the transfusion zone. One hypothesis that could be proposed is that the plasmalemma of the secretory cells is the primary site of active transport relative to secretion. If this is the case, either symplastic or apoplastic flow to the transfusion zone and across the plasmalemma at these zones would be diffusional down a concentration or electrochemical gradient.

The apparent lack of salt accumulations within the secretory cells indirectly supports the view that flow to glands is down a diffusion gradient with the gland cells being at the bottom of the gradient. For example, Ziegler & Lüttge (1967) using autoradiography, reported that Cl^- was not concentrated in the gland cells of *Limonium* over that in the mesophyll. Also, Campbell & Thomson (1975, 1976b) using a silver-precipitation technique, did not observe accumulations of Cl^- in the glands of *Tamarix* and *Frankenia* in salt-challenged, secreting material (accumulations within microvacuoles within the secretory cells will be discussed later).

It should be noted, however, that Shimony *et al.* (1973), using electron-microprobe and antimonate-precipitation techniques, found an apparent downhill gradient of cations from the vascular tissue to the glands in leaves of salt-treated *Avicennia*. They concluded that this was indicative of an active step located in the xylem parenchyma and possibly a second site in the glandular cells where secretion occurs. However, this gradient would be expected if the major active site is the plasmalemma of the secretory cells and the flux is relatively high.

The above hypothesis is complicated, however, by the electrochemical studies of Bostrom & Field (1973) and Billard & Field (1974) with the mangrove *Aegiceras*. By advancing a microelectrode into the gland they found a major potential change at the apparent membrane boundary between secretory cells and the basal cell. These studies have been criticized on technical grounds by Hill & Hill (1976), but the results at least imply that active membrane transport may well occur at this site. Another implication from these results is that apoplastic flow of solute to the base of the glands does occur (Field *et al.*, 1984) since a low-resistance pathway and good electrical coupling would be expected at the transfusion zone if flow was symplastic, that is, if the plasmodesmata are not physiologically and electrically occluded.

Fluxes for the two-celled glands of *Cynodon* have also been calculated. Oross *et al.* (1986), using serial sections and morphometric procedures, determined the surface area of the plasmalemma of the cap cells and the total length of the partitioning membranes of the basal cell. They calculated the total surface area of the partitioning membranes to be 5.25×10^{-9} m^2 gland^{-1}. Knowing the number of glands per leaf segment and the amount of salt secreted over time, they calculated the flux across the partitioning membrane to be 408 nmol m^{-2} s^{-1}. Of particular importance is that the calculated rate across the partitioning membranes is well within the range reported for the plasmalemma in other plant systems (Raven, 1976; Gronwald & Leonard, 1982) but considerably less than that determined for the secretory cells of *Limonium*. Thus, for *Cynodon*, surface amplification of the plasmalemma as reflected by the partitioning membranes can account for secretion rates and the view that these membranes may be highly specialized for transport does not have to be the case. Also, Levering & Thomson (1971, 1972) had earlier hypothesized that the partitioning membranes formed channels in which a standing gradient involving solute-linked water transport could occur. Oross & Thomson (1984), using lanthanum as an apoplastic marker (Fig. 13.3), found that the extracellular lumina of the partitioning membrane was apoplastically continuous with the walls of the mesophyll. Thus, they proposed that solutes move to and into the lumina of the partitioning membrane via the apoplast and that these membranes actively transport solute into the basal cells, probably utilizing energy provided by the numerous and closely associated mitochondria.

On the other hand, arguments that the active sites, or pumps, are located in the mesophyll are derived from electrophysiological studies in which electrical coupling has been reported between the mesophyll and bladder cells of the *Atriplex*-type of glands and with *Limonium* (see Lüttge, 1971, 1975; Hill & Hill, 1973a, 1976 for reviews). However, as Lüttge (1975) has pointed out, it is not possible from these studies to localize the membrane(s) involved since these experiments were done on multicellular systems and not between individual cells.

It is of interest that Osmond *et al.* (1969), using autoradiographic techniques for the localization of $^{36}Cl^-$, noted an apparent increase in concentration in the stalk cell and in the cytoplasm of the bladder cells. From these results they suggested the possibility of these regions being involved in the transfer of Cl^- to the vacuole of the bladder cells.

As pointed out previously, in the multicellular glands the cell walls of the secretory cells are isolated from the apoplast of the mesophyll by the lateral cutinization of the walls of the innermost gland cells. Also characteristic of all multicellular glands, is the presence of a large collecting chamber between the glands and the covering cuticle. An hypothesis that has been developed to explain the ultimate emission of salt solutions to the surface of the leaf is based on these structural features (Hill & Hill, 1973b; Thomson, 1975; Campbell & Thomson, 1975, 1976b). The essence of this hypothesis is that solutes are transported into the apoplast of the glands by the secretory cells and accumulated in the collecting chamber. Solute-linked movement of water into the chamber results in its expansion. With sufficient expansion, the saline solution is released to the surface of the leaves with the opening of pores in the cuticle. The chamber then collapses, the pores close, and a new cycle of loading occurs. Backflow of the saline solution from the collecting chamber to the mesophyll apoplast would be blocked by the cuticularized band around the inner secretory cells.

If the collecting chamber is such a space, then certain physiological observations can be explained. As alluded to previously, several investigators have noted that the secreted fluid may be isotonic to or slightly higher than the osmotic concentration of the leaf sap or challenging solution (Ruhland, 1915; Arisz *et al.*, 1955; Atkinson *et at.*, 1967; Pollak & Waisel, 1970). This seems to bc an anomaly, if secretion is an active process and therefore against a concentration or electrochemical gradient. However, if a solute-linked water transport occurs into the collecting chamber, approximate osmotic equilibrium could be obtained before sufficient hydrostatic pressure develops for full expansion of the chamber and emission of the fluid through the pores. Such a mechanism could also explain Ruhland's (1915) observation that secretion occurs under pressure. In comparison, Findlay & Mercer (1971a,b) have shown for the nectaries of *Abutilon*, that as secretion occurs, the cuticle expands above the terminal cell forming a chamber which fills with the nectar solution. Small droplets are released from this expanded chamber through the cuticle and then the chamber collapses. Findlay *et al.* (1971) concluded that as the fluid collected in the chamber, a hydrostatic pressure develops, the chamber expands and the fluid is ultimately forced through pores, probably pressure-sensitive, in the cuticle (see Ch. 14).

Evidence in support of this hypothesis for salt glands also comes from cytochemical localization of substantial amounts of Cl^- in walls and collecting chambers of the glands of *Tamarix*, *Frankenia* (Campbell & Thomson, 1975, 1976b) and *Cynodon* (Oross & Thomson, 1982b) of high-salt status secreting material. Similarly, the wall and collecting chamber show heavy accumulation of Ca^{2+} in *Tamarix* glands where Ca^{2+} is also identified as being a component of the secreted salt (Fig. 13.7, and legend for explanation).

As described in the section on structure, the vacuolar complexes vary considerably between the different glands. In *Tamarix* and *Frankenia*, plants from two closely related families, numerous small vacuoles are present in the secretory cells of high-salt status, secreting glands, but these vacuoles are virtually absent in low-salt status, non-secreting glands (Thomson & Liu, 1967; Thomson *et al.*, 1969; Campbell & Thomson, 1975, 1976a,b; Thomson, 1975). Both transmission and freeze–fracture electron microscopic observations indicate that these vacuoles fuse with the plasmalemma of the secretory cells (Fig. 13.9; and see Thomson *et al.*, 1969; Platt-Aloia *et al.*, 1983). Thomson and associates have suggested that the solutes to be secreted accumulate within these vacuoles and, when fusion occurs, they are released to the apoplast of the glands. Support for this suggestion comes from the observation that apparent accumulations of Rb^+ were observed in these vacuoles in glands secreting this ion (Thomson *et al.*, 1969). Further support is also indicated by the presence in the vacuoles of Ca^{2+} accumulation localized by the antimonate precipitation technique, in glands where this ion was being secreted (see Fig. 13.7).

Although small vacuoles and vesicles have been described in other salt glands and similarly suggested to have a role in secretion (see Sect. 13.2), a clear distinction quantitatively or qualitatively between their increase or decrease between secreting and non-secreting is not as apparent as with *Tamarix* and *Frankenia*. Further, evidence of their fusion with the plasmalemma has not been clearly presented. It may well be that this form of vesicular exocytosis of solutes is limited to the glands of *Tamarix* and *Frankenia*. In any case, it does not rule out the possibility that transport across the plasmalemma may also be occurring as suggested by Lüttge (1971). Also, the origin of the vacuoles and their method of solute accumulation is not known.

Hill & Hill (1973a) have criticized the possibility of vacuoles being involved in the secretory process and suggested from their micrographs of *Limonium* that what appears to be vesicles are balloon-like invaginations of the plasmalemma. The inflated pockets observed by Hill & Hill (1973a) do appear to be plasmalemma invaginations (see Fig. 5.6 in their review) but are probably artifacts since no such invaginations (or for that matter, numerous small vacuoles) were observed by Faraday & Thomson, 1986a, b, c), in the secretory cells of well fixed secreting *Limonium* glands. Also, Hill & Hill (1973a) suggested that the vesicular fusion hypothesis would require impossible membrane recycling rates. However, considering recent numbers and rates given for membrane recycling in other systems (Christensen, 1982), this may not be a serious criticism.

In regards to 'partitioning membranes' in the basal cells of the grasses, Oross *et al.* (1986) have suggested that if they represent the site of active transport into the glands, then solute movement out of the glands would be symplastic via a downhill diffusion gradient from the basal cells, through the plasmodesmata in the wall between the basal and cap cell, through the cap cell and into the collecting chamber between the outer cap cell wall and the cuticle. Emission from the collecting chamber to the leaf surface would be due to the development of hydrostatic pressure within this compartment, cuticle expansion, and the subsequent opening of pores as was described above for the multicellular glands. Apoplastic backflow into the mesophyll would be constricted due to the lignification of the cell walls near the base of the cap cell.

Direct evidence for the presence of an osmotic gradient between basal and cap cells is limited. However, Ramati *et al.* (1976), using plasmolysis experiments, reported the osmotic potential of the cap cell of *Sporobolus* to be 0.53 MPa lower than that of the mesophyll, and no plasmolysis of the basal cell was detected. Further, with salt-treated material and using microprobe analyses of frozen, hydrated sections, their data indicate that the concentration of Na^+ K^+ and Cl^- was higher in the basal cell than in the cap cell.

Because secretion probably involves membrane-transport processes, attention to ATPases, presumably glandular, has developed. Kylin & Gee (1970) reported the presence of a cation-stimulated ATPase in homogenates of salt-secreting leaves of *Avicennia*. Hill and associates (Hill & Hill, 1973a,b, 1976) have reported on anion-stimulated ATPase in microsomal fractions isolated from salt-challenged secreting *Limonium* material. The ATPase had low activity in non-secreting material and increased in activity concomitantly with the appearance of secretory activity of the glands when low-salt status material was challenged with salt. Thus, the activity of this ATPase correlated with the activity of the glands, suggesting the enzyme might have an important role in secretion.

For a variety of reasons, they have suggested that this enzyme activity represents a Cl^--stimulated ATPase, supposedly the electrogenic Cl^- pump of the glands. It is doubtful that this is true based on current evidence. They assayed for Cl^--dependence and reported the activity was four times higher with Cl^- than with SO_4^{2-}, concluding, therefore, that they had a Cl^--dependent ATPase. However, they did the comparative assays simply based on balancing ionic charge (100 mol m^{-3} for Cl^- and 0.03 for SO_4^{2-}. There were no attempts to determine initial rates (the assays were run for 1 h) or concentration-dependent kinetics, nor, for that matter, comparisons with other anions. The best one can conclude is they may have an anionic-dependent ATPase.

To illustrate this latter point, these studies were further extended by Hill & Hanke (1979), Auffret & Hanke (1981), and more recently by Parr & Hanke (1982). The recent work is the most pertinent. Parr & Hanke further characterized these preparations using detergents and acetate and sulphate as comparative monovalent anions to Cl^-. They found that acetate and sulphate gave virtually the same stimulation of activity over a basal level as did Cl^- at the same concentration. The basal level was determined by measuring the activity against 120 mol m^{-3} benzene sulphonate, an inhibitor. Thus, it appears it is not a Cl^--stimulated ATPase, but a benzene sulphonate-inhibited ATPase. They suggested that their results do not negate the possibility of a Cl^--stimulated ATPase being present in *Limonium*, but these results and the questionable enzymology used in the studies by Hill & Hill (1973b) cast serious doubt at this time. Further, the variable composition of the secreted salts, the fact that Cl^- is often much lower in concentration and in balancing equivalents to that of the cations, argues against a Cl^- pump being the primary transport mechanism in salt glands.

13.6 Induction

A number of investigators have reported that when non-pretreated cuttings, leaf discs or plants are challenged with salt, a lag period occurs before secretory activity is seen (Arisz *et al.*, 1955; Osmond *et al.*, 1969; Hill, 1970; Lüttge & Osmond, 1970; Pollak & Waisel, 1970; Shachar-Hill & Hill, 1970). To a certain extent, the studies of Hill and his associates on this phenomena using the *Limonium* leaf disc system have drawn the most attention (Hill, 1970; Shachar-Hill & Hill, 1970; Hill & Hill, 1973a, 1976). The major aspects and interpretations derived from these studies are as follows: first, when low-salt status discs are floated on a challenging solution, secretion droplets begin to appear in about 1 h and the secretory activity rises to a steady state in about 3–4 h. Correlated electrochemical studies indicate that the electrical activity follows the same pattern, as do ion fluxes using radioactive isotopes. When the low-salt status discs are pretreated with inhibitors to transcription and translation (e.g. puromycin and actinomyocin D) and then challenged with a salt solution, no secretion occurs. If the discs were secreting on a salt solution and then these inhibitors are added, they have no effect.

Shachar-Hill & Hill (1970) reported that the sigmoid rise in secretory activity and subsequent secretion was highly sensitive to temperature, anoxia and the usual inhibitors of ATP production (DNP, CCCP, CN) but the apparent synthetic events happening during the lag phase were less sensitive. They have concluded that the lag period and the sigmoid rise represent first a developmental process consisting of the synthesis of a required ion 'pump' followed by the rise in its activity. Once synthesized and active, the pumps would be insensitive to inhibitors of translation and transcription, but sensitive to metabolic inhibitors.

In this regard, it was also noted that when the low-salt status discs were challenged stepwise over time with increasing concentrations of salt (i.e., 100 mol m^{-3} and then 200 mol m^{-3} NaCl) secretion remained constant for about 2 h and then rose sigmoidally to a higher rate. In other words, a similar lag phase occurred. Of interest also is that when the discs were challenged with different NaCl concentrations, (50, 100, 200 mol m^{-3}) the lag phase was the same for all concentrations, but the final steady-state secretory activity reflected the concentration of the challenging solution; higher for 200 mol m^{-3} and correspondingly lower for 100 and 50 $molm^{-3}$ as measured by short-circuit current techniques. Similarly, if salt-loaded discs were transferred to a solution of lower concentration, the current flow decreased. The conclusions advanced were that when low-salt status tissues are challenged with a saline condition, secretion is induced in a genetic and synthetic sense, including a regulated inductive control mechanism in response to increasing salt challenge, and that what is synthesized are ion pumps in the form of membrane complexes within the glands.

Structurally, evidence for this inductive process is enigmatic. For example, Shachar-Hill & Hill (1970) reported no obvious ultrastructural differences between secreting and non-secreting glands of *Limonium*, although numerous ribosomes and polyribosomes were present in the secretory cells of glands in both states of activity. These observations, however, were qualitative assessments and not based on any form of quantitative determination. Similarly, Faraday & Thomson (1986a) noted no obvious differences in the ultrastructure of secreting and non-secreting glands of several Plumbaginaceae, including *Limonium*. In contrast, Campbell & Thomson (1976b) found an increase in rough ER and microvacuoles in the secetory cells of *Frankenia* when shifted from a non-secreting to a secreting state of activity.

If secretion does indeed prove to be an inductive process, there are obvious implications for further research.

13.7 Summary

Gradually our understanding of the structure and function of salt glands is developing, but much is still not known. During the past decade, a clearer picture of the structure of some of the different glands has emerged, but others have only been cursorily examined and some not at all. Also, structural studies of the glands in different functional states or during the transition between states are limited. More information is now available as to what glands secrete and, to some extent, the fluxes they are capable of, but this is only for a few glands and no generalization can be developed at this time. An understanding of the structural and functional interaction between the mesophyll and the glands is emerging, but many questions are unresolved, primarily because direct information as to how glands function is indeed limited.

Finally and unfortunately, we did not consider in this chapter the role of salt glands in an adaptive sense relative to saline environments. We suggest that those interested consult Waisel (1972), Rozema (1978). Liphschitz & Waisel (1982), and Clough (1984), for example.

Acknowledgements

Most of the original research reported herein was supported in part by grants from the National Science Foundation and, in particular, PCM. We also thank Dr K. A. Platt-Aloia for her many and continuing contributions.

References

Arisz, W. H., Camphus, I. J., Heikens, S. & van Tooren (1955). The secretion of the salt glands of *Limonium latifolium* Ktze. *Acta Botanica Neerlandia* **4**, 322–38.

Atkinson, M. R., Findlay, G. P., Hope, A. B., Pitman, M. G., Saddler, H. D. W. West, K. .R. (1967). Salt regulation in the mangroves *Rhizophora mucronata* Lam. and *Aegialitis annulata* R. Br. *Australian Journal of Biological Sciences* **20**, 589–99.

Auffret, C. A. & Hanke, D. .E. (1981). Improved preparation and assay and some characteristics of Cl-ATPase activity from *Limonium vulgare*. *Biochimica et Biophysica Acta* **648**, 186–91.

Baumeister, W. & Ziffus, G. (1981). Salt secretion by the salt glands of *Armeria maritima* L. *Zeitschrift für Pflanzenphysiologie* **102**, 273–8.

Berry, W. L. (1970). Characteristics of salts secreted by *Tamarix aphylla*. *American Journal of Botany* **57**, 1226–30.

Berry, W. L. & Thomson, W. W. (1967). Composition of salt secreted by salt glands of *Tamarix aphylla*. *Canadian Journal of Botany* **45**, 1774–5.

Billard, B. & Field, C. D. (1974). Electrical properties of the salt gland of *Aegeceras*. *Planta* **115**, 285–96.

Black, R. F. (1954). Leaf anatomy of Australian members of the genus *Atriplex*. I. *Atriplex vesicaria* Heward and *A. nummularia* Lindl. *Australian Journal of Botany* **2**, 259–86.

Boon, P. I. & Allaway, W. G. (1982). Assessment of leaf-washing techniques for measuring salt secretion in *Avicennia marina* (Forsk.) Viehr. *Australian Journal of Plant Physiology* **9**, 725–34.

Bosabalidis, A. M. & Thomson, W. W. (1984). Light microscopical studies on *Tamarix aphylla* L. *Annals of Botany* **54**, 169–74.

Bostrom, T. E. & Field, C. D. (1973). Electrical potentials in the salt glands of *Aegiceras*. In Anderson, W. P. (ed.). *Ion Transport in Plants*, pp. 385–92. London, New York, Academic Press.

Campbell, N. & Thomson, W. W. (1975). Chloride localization in the leaf of *Tamarix*. *Protoplasma* **83**, 1–14.

Campbell, N. & Thomson, W. W. (1976a). The ultrastructure of *Frankenia* salt glands. *Annals of Botany* **40**, 681–6.

Campbell, N. & Thomson, W. W. (1976b). The ultrastructural basis of chloride tolerance in the leaf of *Frankenia*. *Annals of Botany* **40**, 687–93.

Campbell, N., Thomson, W. W. & Platt, K. (1974). The apoplastic pathway of transport to salt glands. *Journal of Experimental Botany* **25**, 61–9.

Cardale, S. & Field, C. D. (1971). Thc structure of the salt gland of *Aegiceras corniculatum*. *Planta* **99**, 183–91.

Chapman, V. J. (1944). The morphology of *Avicennia nitida* Jacq. and the function of the pneumatophores. *Botanical Journal of the Linnean Society* **52**, 487–533.

Christensen, E. I. (1982). Rapid membrane recycling in renal proximal tubule cells. *European Journal of Cell Biology* **29**, 43–9.

Clough, B. F. (1984). Growth and salt balance of the mangroves *Avicennia marina* (Forsk.) Vierh. and *Rhizophosa stylosa* Griff. in relation to salinity. *Australian Journal of Plant Physiology* **11**, 419–30.

De Fraine, R. (1910). The morphology and anatomy of the genus *Statice* as represented at Blakeney point. I. *Statice binervosa* G. E. Smith & S. bellidifolia D. C. (=s. reticulata). *Annals of Botany* **30**, 239–82.

Diamond, J. M. & Bossert, W. H. (1967). Standing-gradient osmotic flow: A mechanism for coupling of water and solute transport in epithelia. *Journal of General Physiology* **50**, 2061–83.

Drennan, P. M. & Berjak, P. (1982). Degeneration of the salt glands accompanying foliar maturation in *Avicennia marina* (Forsskol) Vierh. *New Phytologist* **90**, 165–76.

Drennan, P. M. & Pammenter, N. W. (1982). Physiology of salt excretion in the mangrove *Avicennia marina* (Forsk.) vierh. *New Phytologist* **91**, 597–606.

Fahn, A. (1979). *Secretory Tissues in Plants*. London, Academic Press.

Fahn, A. & Shimony, C. (1977). Development of the glandular and nonglandular leaf hairs of *Avicennia marina* (Forsskal) Vierh. *Journal of the Linnean Society of London (Botany)* **74**, 37–46.

Faraday, C. D., Quinton, P. M. & Thomson, W .W. (1986). Ion fluxes across the transfusion zone of secreting *Limonium* salt glands determined from secretion rates, transfusion zone areas, and plasmodesmatal frequencies. *Journal of Experimental Botany*. **37**, 482–94.

Faraday, C. D. & Thomson, W. W. (1986a). Structural aspects of the salt glands of the Plumbaginaceae. *Journal of Experimental Botany*. **37**, 461–70.

Faraday, C. D. & Thomson, W. W. (1986b). Morphometric analysis of *Limonium* salt glands in relation to ion efflux. *Journal of Experimental Botany*. **37**, 471–81.

Faraday, C. D. & Thomson, W. W. (1986c). Functional aspects of the salt glands of the Plumbaginaceae. *Journal of Experimental Botany*. **37**, 1129–35.

Field, C. D., Hinwood, B. G., & Stevenson, I. (1984). Structural features of the salt gland of *Aegiceras*. In Teas, H. J. (ed.) *Physiology and Management of Mangroves*, pp 37–42. The Hague, Dr W. Junk Publishers.

Findlay, N. & Mercer, F. W. (1971a). Nectar production in *Albutilon*. I. Movement of nectar through the cuticle. *Australian Journal of Biological Science* **24**, 647–56.

Findlay, N. & Mercer, F. W. (1971b) Nectar production in *Albutilon*. II Submicroscopic structures of the nectary. *Australian Journal of Biological Science* **24**, 647–56.

Findlay, N., Reed, M. L. & Mercer, F. W. (1971). Nectar production in *Albutilon*. III. Sugar production. *Australian Journal of Biological Science* **24**, 665–75.

Gronwald, J. W. & Leonard, R. T. (1982). Isolation and transport properties of protoplasts from cortical cells of corn roots. *Plant Physiology* **70**, 1391–5.

Gunning, B. E. S. (1977). Transfer cells and their roles in transport of solutes in plants. *Science Progress, Oxford* **64**, 539–68.

Gunning, B. E. S. & Robards, A. W. (1976). Plasmodesmata: current knowledge and outstanding problems. In Gunning, B. E. S. & Robards, A. W. (eds.). *Intercellular Communication in Plants: Studies on Plasmodesmata*, pp. 297–311. Berlin, Heidelberg, New York, Springer-Verlag.

Helder, R. J. (1956). The loss of substances by cells and tissues (salt glands). In Ruhland, W. (eds.). *Encyclopedia of Plant Physiology*, vol. II, pp. 468–88; Berlin, Gottingen, Heidelberg, Springer-Verlag.

Hill, A. E. (1967a). Ion and water transport in *Limonium*. I. Active transport by the leaf gland cells. *Biochimica et Biophysica Acta* **135**, 454–60.

Hill, A. E. (1967b). Ion and water transport in *Limonium* II. Short-circuit analysis. *Biochimica et Biophysica Acta* **135**, 461–5.

Hill, A. E. (1970). Ion and water transport in *Limonium*. III. Time constants of the transport system. *Biochimica et Biophysica Acta* **196**, 66–72.

Hill, A. E. & Hill, B. S. (1973a). The *Limonium* salt gland: A biophysical and structural study. *International Review of Cytology* **35**, 299–319.

Hill, A. E. & Hill, B. S. (1973b). The electrogenic chloride pump of the *Limonium* salt gland. *Journal of Membrane Biology*, **12**, 129–44.

Hill, A. E. & Hill, B. S. (1976). Mineral ions. In Lüttge, U. & Pitman, M. G. (eds.). *Transport in Plants*, II. *Encyclopedia of Plant Physiology, New Series*, vol. 2B, pp. 225–43. Berlin, Heidelberg, New York, Springer-Verlag.

Hill, B. S. & Hanke, D. E. (1979). Properties of the chloride-ATPase from *Limonium vulgare* salt glands: Activation by, and binding to specific sugars. *Journal of Membrane Biology* **51**, 185–94.

Hill, B. S. & Hill, A. E. (1973). ATP-driven chloride pumping and ATPase activity in the *Limonium* salt gland. *Journal of Membrane Biology* **12**, 145–58.

Hoagland, D. R. & Arnon, D. (1950). The water culture method for growing plants without soil. University of California Berkeley College of Agriculture Circular, 347.

Jeschke, W. D. & Stelter, W. (1983). Ionic relations of garden orache, *Atriplex hortensis* L.: Growth and ion distribution at moderate salinity and the function of bladder hairs. *Journal of Experimental Botany* **34**, 795–810.

Kleinkopf, G. E. & Wallace, A. (1974). Physiological basis for salt tolerance in *Tamarix ramosissima*. *Plant Science Letters* **3**, 157–63.

Komnick, H. & Bierther, M. (1969). On the histochemical localization of ions by electron microscopy, with special reference to the chloride reaction. *Histochemie* **18**, 337–62.

Kylin, A. & Gee, R. (1970). Adenosine triphosphate activities in leaves of the mangrove *Avicennia nitida* Jacq. *Plant Physiology* **45**, 169–72.

Larkum, A. W. D. & Hill, A. E. (1970). Ion and water transport in *Limonium*. V. The ionic status of chloroplasts in the leaf of *Limonium vulgare* in relation to the activity of the salt glands. *Biochimica et Biophysica Acta* **203**, 133–8.

Leigh, R. A. & Wyn Jones, R. G. (1984). A hypothesis relating critical potassium concentrations for growth to the distribution and functions of this ion in the plant cell. *New Phytologist* **97**, 1–13.

Leonard, R. T. (1984). Membrane-associated ATPases and nutrient absorption by roots. In Tinker, P. B. & Lüchli, A. (eds.) *Advances in Plant Nutrition* vol. 1. New York, Praeger Publishers.

Levering, C. A. & Thomson, W. W. (1971). The ultrastructure of the salt gland of *Spartina foliosa*. *Planta* **97**, 183–96.

Levering, C. A. & Thomson, W. W. (1972). Studies on the ultrastructure and mechanism of secretion of the salt gland of the grass *Spartina*. Proceedings of the *Thirtieth Electron Microscope Society of America*, pp. 222–3.

Liphschitz, N., Shomer-Ilan, A., Eshel, A. & Waisel, Y. (1974). Salt glands on leaves of Rhodes grass (*Chloris gayana* Kth.). *Annals of Botany* **38**, 459–62.

Liphschitz, N., & Waisel, Y. (1974). Existence of salt glands in various genera of the Gramineae. *New Phytologist* **73**, 507–13.

Liphschitz, N, & Waisel, Y. (1982). Adaptation of plants to saline environments: salt excretion and glandular structure. In Sen, D. N. & Rajpurohit, K. S. (eds.). *Tasks for Vegetation Science, Contributions to the Ecology of Halophytes*, vol. 2, pp. 197–214. The Hague, Dr W. Junk Publishers.

Lüttge, U. (1971). Structure and function of plant glands. *Annual Review of Plant Physiology* **72**, 23–44.

Lüttge, U. (1975). Salt glands. In Baker, D. A. & Hall, J. L. (eds). *Ion Transport in Plant Cells and Tissues*, pp. 335–76. Amsterdam, London, North-Holland.

Lüttge, U. & Osmond, C. B. (1970). Ion absorption in *Atriplex* leaf tissue. III. Site of metabolic control of light-dependent chloride secretion to epidermal bladders. *Australian Journal of Biological Science* **23**, 17–25.

Lüttge, U. & Pallaghy, C. K. (1969). Light triggered transient changes of membrane potentials in green cells to photosynthetic electron transport. *Zeitschrift für Pflanzenphysiologie* **61**, 58–67.

McGovern, T. A., Laver, L. J. & Gram, B. C. (1979). Characteristics of the salt secreted by *Spartina alterniflora* and their relation to estuarine production. *Estuarine Coastal Marine Science* **9**, 352–76.

Mauseth, J. D. (1982). A morphometric study of the ultrastructure of *Echinocereus engelmannii* (Cactaceae). V. Comparison with the shoot apical meristems of *Trichocereus pachoni* (Cactaceae). *American Journal of Botany*. **69**, 551–5.

Mozafar, A. & Goodin, J. R. (1970). Vesiculated hairs: A mechanism for salt tolerance in *Atriplex halimus* L. *Plant Physiology* **45**, 62–5.

Oross, J. W., Leonard, R. T. & Thomson, W. W. (1985). Flux rate and a model for secretion for salt glands of grasses. *Israel Journal of Botany*. **34**, 69–77.

Oross, J. W. & Thomson, W. W. (1982a). The ultrastructure of the salt glands of *Cynodon* and *Distichlis* (Poaceae). *American Journal of Botany* **69**, 939–49.

Oross, J. W. & Thomson, W. W. (1982b). The ultrastructure of *Cynodon* salt glands: The apoplast. *European Journal of Cell Biology* **28**, 257–63.

Oross, J. W. & Thomson, W. W. (1984). The ultrastructure of *Cynodon* salt glands: secreting and nonsecreting. *European Journal of Cell Biology* **34**, 287.

Osmond, C. G., Lüttge, U., West, K. R., Pallaghy, C. K. & Shacher-Hill, B. (1969). Ion absorption in *Atriplex* leaf tissue. III. Secretion of ions to epidermal bladders. *Australian Journal of Biological Science* **22**, 797–814.

Parr, A. J. & Hanke, D. E. (1982). 'Chloride-stimulated ATPase' activity in *Limonium vulgare* Mill. *Philosophical Transactions of the Royal Society, London* series B **299**, 459–68.

Platt-Aloia, K. A., Bliss, R. D. & Thomson, W. W. (1983). Lipid–lipid interactions and membrane fusion in plant salt glands. In Thomson, W. W., Mudd, J. B. & Gibbs, M. (eds.). *Biosynthesis and Function of Plant Lipids*, pp. 160–72. Rockville, Md., American Society of Plant Physiologists.

Pollak, G. & Waisel, Y. (1970). Salt secretion in *Aeluropus litoralis* (Willd.). Parl. *Annals of Botany* **34**, 879–88.

Pollak, G. & Waisel, Y. (1979). Ecophysiological aspects salt excretion in *Aeluropus litoralis*. *Physiologia Plantarium* **47**, 177–84.

Ramati, A., Liphschitz, N. & Waisel, Y. (1976). Ion localization and salt secretion in *Sporobolus arenarius* (Gou.) Duv.-Jouv. *New Phytologist* **76**, 289–95.

Raven, J. A. (1976). Transport in algal cells. In Lüttge, U. & Pitman, M. G. (eds.). *Encyclopedia of Plant Physiology, New Series*, vol. 2A, pp. 129–88. Berlin, Heidelberg, New York, Springer-Verlag.

Robards, A. W. (1976). Plasmodesmata in higher plants. In Gunning, B. E. S. & Robards, A. W. (eds.). *Intercellular Communication in Plants: Studies on Plasmodesmata*, pp. 15–57. Berlin, Heidelberg, New York, Springer-Verlag.

Rozema, J. (1978). *On the Ecology of some Halophytes from a Beach Plain in the Netherlands*. The University of Amsterdam.

Rozema, J., Gude, H. & Pollak, G. (1981). An ecophysiological study of the salt secretion of four halophytes. *New Phytologist* **89**, 201–17.

Rozema, J., Riphagen, I. & Sminia, T. (1977). A light and electron microscopical study on the structure and function of the salt gland of *Glaux maritima* L. *New Phytologist* **79**, 665–71.

Ruhland, W. (1915). Untersuchungen über die Hautdrüsen der Plumbaginaceen. Ein Beitrag zur Biologie der Halophyten. *Jahrbuecher fur Wissenschaftliche Botanik* **55**, 409–98.

Samoui, M. A. (1971). Differenciation des trichomes chez *Atriplex halimus* L. *Comptes rendus des Séances de l'Académie des Sciences*, **273**, 1268–71.

Scholander, P. F., Hammel, H. T., Hemmingsen, E. A. & Garey, W. (1962). Salt balance in mangroves. *Plant Physiology* **37**, 722–9.

Schtscherback, J. (1910). Über die Salzausscheidung durch die Blätter von

Statice gmelini. *Berichte der Deutschen Botanischen Gesellschaft* **28**, 30–4.

Shachar-Hill, B. & Hill, A. E. (1970). Ion and water transport in *Limonium*. VI. The induction of chloride pumping. *Biochimica et Biophysica Acta* **211**, 313–17.

Shimony, C. & Fahn, A. (1968). Light and electron microscopical studies on the structure of salt glands in *Tamarix aphylla* L. *Journal of the Linnean Society of London* **60**, 283–8.

Shimony, C., Fahn, A. & Reinhold, L. (1973). Ultrastructure and ion gradients in the salt glands of *Avicennia marina* (Forssk) vierh. *New Phytologist* **72**, 27–36.

Skelding, A. D. & Winterbotham, J. (1939). The structure and development of the hydathodes of *Spartina townsendii* Groves. *New Phytologist* **38**, 69–79.

Storey, R., Pitman, M. G., Stelzer, R. & Carter, C. (1983). X-Ray microanalysis of cells and cell compartments of *Atriplex spongiosa*. *Journal of Experimental Botany* **34**, 778–94.

Sutherland, G. K. & Eastwood, A. (1916). The physiological anatomy of *Spartina townsendii*. *Annals of Botany* **30**, 333–51.

Sze, E. (1984). H^+-translocating ATPases of the plasmamembrane and tonoplast of plant cells. *Physiologia Plantarium* **61**, 683–91.

Thomson, W. W. (1975). The structure and function of salt glands. In Poljakoff-Mayber, A. & Gale, J. (eds). *Plants in Saline Environments* pp. 118–146. Berlin, Heidelberg, New York, Springer-Verlag.

Thomson, W. W., Berry, W. L. & Liu, L. L. (1969). Localization and secretion of salt by the salt glands of *Tamarix aphylla*. *Proceedings of the National Academy of Sciences, USA* **63**, 310–17.

Thomson, W. W. & Liu, L. L. (1967). Ultrastructural features of the salt gland of *Tamarix aphylla* L. *Planta* **73**, 201–20.

Thomson, W. W. & Platt-Aloia, K. (1979) Ultrastructural transitions associated with the development of the bladder cells of the trichomes of *Atriplex*. *Cytobios* **25**, 105–14.

Thomson, W. W., Platt, K. & Campbell, N. (1973). The use of lanthanum to delineate the apoplastic continuum in plants. *Cytobios* **8**, 57–62.

Turkey, H. B., Jr. (1971). Leaching of substances from plants. In Preece, T. F. & Dickinson, C. H. (eds.). *Ecology of Leaf Surface Microorganisms*. Academic Press, London.

Tyree, M. T. (1970). The symplast concept. A general theory of symplastic transport according to the thermodynamics of irreversible processes. *Journal of Theoretical Biology* **26**, 181–214.

Van Stevenink, R. F. M., Armstrong, W. D., Peters, P. D. & Hall, T. A. (1976). Ultrastructural localization of ions. III. Distribution of chloride in mesophyll cells of mangrove (*Aegiceras corniculatum* Blanco). *Australian Journal of Plant Physiology* **3**, 367–76.

Volkens, G. (1884). Die Kalkdrüsen der Plumbagineen. *Berichte der Deutschen Botanischen Gesellschaft* **2**, 334–42.

Waisel, Y. (1961). Ecological studies on *Tamarix aphylla* (L.) Karst. III. The salt economy. *Plant and Soil* **13**, 356–64.

Waisel, Y. (1972). *Biology of Halophytes*. New York, Academic Press.

Wilson, J. (1890). The mucilage and other glands of the Plumbagineae. *Annals of Botany* **4**, 231–58.

Ziegler, H. & Lüttge, U. (1966) Die salzdrüsen von *Limonium vulgare*, I. Die Feinstruktur. *Planta* **70**, 193–206.

Ziegler, H. & Lüttge, U. (1967). Die saltzdrüsen von *Limonium vulgare*. II. Mitteilung Die Lokalisierung des Chlorids. *Planta* **74**, 1–17.

14 Nectaries and other glands

N. Findlay

14.1 Introduction

Recent substantial advances in the understanding of the mechanisms of solute transport in plants have, as yet, had little influence on our understanding of secretion of solutes by plant glands. The preceding chapters have shown the advances which have been achieved in recent years in our basic understanding of the processes of transport of solutes, both inorganic and organic, into and out of cells and the various cell compartments and between cells and tissues. We know a good deal now about how active transport is coupled to an energy source by means of the proton-pumping activity of ATP-powered coupling factors at various cell membranes (see Chs. 2, 3); there is a beginning in our understanding of the importance of specific ion channels in the passive movement of ions, and, as shown by a recent review (Thorne, 1985), there is also increasing interest in how organic molecules are transported out of tissues. Maybe the time is soon to come when this knowledge will be applied and extended to plant glands.

One of the reasons for this neglect of the mechanism of secretion from plant glands is that it is regarded as a curiosity; this is misguided since secretion is a process of basic importance, with nectaries and stigmas playing an essential role in the reproduction of many flowering plants. Another reason perhaps is the experimental intractability of glands for physiological studies as compared with some other systems, but this is a problem that can be overcome.

There is much information on the structure of plant glands. Studies of ultrastructure, in particular, are of relevance to the study of solute transport. Even in this field, however, there is room for a more critical approach and the use of a wider range of techniques (Robards, 1984).

Apart from the salt glands, which are discussed in Ch. 13, nectaries are the only plant glands which have been studied from a number of different aspects and for which a considerable body of information is available.

These glands will make up the major subject matter of this chapter. Other solute-secreting glands are less common.

The secretions of many other glands contain lipophilic substances or macromolecules such as polysaccharides or proteins, that are not strictly solutes, the subject of this book. There is a considerable amount of information about the ultrastructure of these glands which provides insight into the mechanisms of secretion and this will be considered where it is of relevance to our understanding of solute transport in glands.

14.2 Different types of glands

14.2.1 Nectaries

These glands secrete solutions containing a high concentration of sugars. There are two main groups of nectaries, floral and extrafloral.

14.2.1.1 Floral nectaries

These provide a reward to insects, birds, bats or even (Main, 1981) small marsupials, which are attracted to flowers to fulfil a role in pollination. The nectaries may be situated on any part of the flower, a common location being on the disc or torus as in *Citrus*. They can also occur on the calyx as in *Abutilon*, on the corolla as for instance on a spur in *Tropaeolum*, or in the septa of the ovary as in the Liliaceae.

The secretion of nectar usually begins at the time the flower opens and continues until it wilts.

Vogel (1971, 1976) has described glands in some taxa that fulfil the same role as nectaries but secrete oils instead of sugar solution.

14.2.1.2 Extrafloral nectaries

These occur on the vegetative or reproductive parts of the plant but are not usually associated with pollination. They attract insects, particularly ants, and there is considerable evidence that their role is to protect the plant from herbivorous animals, particularly other insects (Bentley, 1977).

Thus both floral and extrafloral nectaries are plant adaptations involved in different aspects of the co-evolution of animals and plants.

14.2.2 Mucilage-secreting glands

The mucilages produced by plants are polysaccharides and may be contained in cells within their tissues or secreted by glands on the surface. They are secreted by the insect-trapping glandular trichomes of insectivorous plants such as *Drosera*, *Drosophyllum* and *Pinguicula*, and also by trichomes on the surfaces in some buds such as those of *Rumex*.

14.2.3 Protein-secreting glands

Some of the glands on insectivorous plants secrete digestive enzymes usually only after stimulation. In *Pinguicula* (Heslop-Harrison & Knox, 1971) the digestive enzymes are stored, ready for secretion on stimulation, in wall protrusions in the glands. The digestive glands of the Venus flytrap (*Dionaea muscipula*) also secrete hydrochloric acid (Rea, 1982).

14.2.4 Glands secreting lipophilic substances

Terpenes (such as essential oils and resins), fats, oils and other lipophilic substances are secreted by a large number of plants.

There is much variety of structure of these glands. Many are internal, as for example the resin canals of *Pinus* or the essential oil glands of the Myrtaceae. In some of the internal glands the release of secretion is accomplished by the breakdown of cells which have previously synthesized or stored the secreted substances.

Other glands secreting lipophilic substances are external. Some examples are the essential oil-secreting hairs of the Labiatae (Bosabalidis & Tsekos, 1982) and the flavonoid-secreting farina glands of the Primulaceae.

14.2.5 Stigmas

The glandular function of stigmas is important in their role in the reproduction of flowering plants. The secretion provides a suitable environment for the trapping of pollen grains, their germination and the growth of the pollen tube. It consists of a complex mixture usually containing lipids, polysaccharides, proteins and smaller quantities of sugars, amino acids and other substances such as phenolics (Knox, 1984).

This list of different types of glands is by no means complete. Readers should refer to Fahn (1979a) for a detailed account of the structure of plant glands.

14.3 Structural aspects of transport in glands

14.3.1 Structural organization of nectaries

In general, nectaries, like other plant glands, consist of a group of small, highly cytoplasmic cells. They are usually close to phloem or supplied by phloem endings, which supply sugar to the nectary. However, there is considerable variety in the arrangement of the nectary cells and the path by which nectar is released:

1. The cells of the nectary tissue break down to release the nectar. Nectaries of this type occur in the fern, *Pteridium*. Another example is found in the extrafloral inflorescence nectaries of *Vigna* where whole cells and cell debris are released with the nectar (Kuo & Pate, 1985).
2. The nectar is secreted into the intercellular spaces and released to the outside through stomate-like structures. Examples are found in *Citrus*, *Vinca*, *Tropaeolum* and *Passiflora* floral nectaries.
3. The nectar is secreted through an epithelium covered by a cuticle. The shape of these nectaries varies from sunken nectaries secreting into an enclosed space or slit, to flat nectaries or to raised or stalk-like structures. This complete range of shapes can be found amongst the extrafloral nectaries of species of the genus *Acacia* (Boughton, 1985).
4. The nectar is secreted by multicellular trichomes inserted on the nectary tissue. Examples are found in *Abutilon*, *Gossypium*, *Lonicera* and *Vicia*.

14.3.2 Pathways through the nectary

There has been considerable discussion of the relative importance of the various possible symplastic and apoplastic pathways through the nectary (Lüttge & Schnepf, 1976). Different species probably differ greatly in this respect. Evidence of a biochemical nature has been important in this discussion (see Sect. 14.4.2). In particular, the inversion by the extracellular enzyme, acid invertase, of the sucrose supplied through the phloem, has been used to support the view that at least part of the pathway is apoplastic. Other evidence of a more structural nature will be considered here.

There is little doubt that most of the sugar secreted by nectaries is supplied by phloem. Nectaries are usually close to phloem or supplied by separate phloem endings. The phloem endings may or may not be accompanied by xylem. Those nectaries not close to xylem tend to secrete more highly concentrated nectar. The phloem endings, as in leaves, contain companion cells which are of a similar size to the sieve tubes (Findlay & Mercer 1971b; Durkee 1982) suggesting they are of particular importance in the unloading process.

The distribution of plasmodesmata in the nectaries of cotton has been investigated by Wergin *et al.* (1975). The nectary parenchyma cells are connected by numerous plasmodesmata arranged in pit-fields in thinner areas of the cell wall. The connections within the phloem are separate, more widely spaced plasmodesmata. Gunning and Hughes (1976) have found that, in *Abutilon* nectaries, there are plasmodesmatal connections between the cells from the phloem to the apical cells of the trichomes.

Hence there is a symplastic route all the way through the nectary.

In these *Abutilon* nectaries there is a barrier in the apoplast. A single cell, the stalk cell, at the base of the trichomes, has lateral walls which are impregnated throughout their thickness with a substance having similar appearance and staining properties to cutin (Fig. 14.1). A similar cell or layer of cells occurs also in other glandular trichomes. Gunning and Hughes (1976) have shown that in *Abutilon* nectaries the apoplastic tracer,

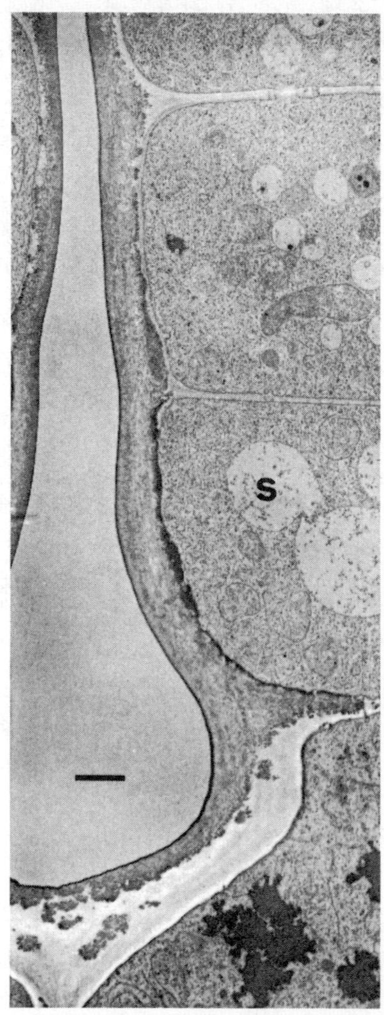

Fig. 14.1 Electron micrograph of an unstained section of the base of a nectary trichome of *Abutilon*. The outer wall of the stalk cell, S, is impregnated throughout its thickness with an electron-dense material, whereas in the epidermal cell (below) and the trichome cells (above) the electron-dense material is restricted to the outer part of the wall. Scale = 1 μm. (Reproduced from Gunning & Hughes, 1976.)

Calcofluor white M2R new, is stopped at the site of the wall impregnation. It was concluded that the transport of sugar and water is restricted to the symplast at the stalk cell.

14.3.3 Rates of transport by nectaries

Gunning and Hughes (1976) have made some calculations, based on the structure of the *Abutilon* trichome described above, to show that plasmodesmata are a feasible route for symplastic transport. The calculations are based on the dimensions of the distal wall of the stalk cells, the number and size of the plasmodesmata in it, and the rate of secretion in these nectaries. The rate of nectar secretion in these nectaries can be easily determined by observation of individual hairs in sections mounted in paraffin oil (Fig. 14.2) or alternatively, in intact nectaries. Gunning and Hughes conclude that a pressure drop of 2 kPa is sufficient to produce the water flow observed if it occurs through the cytoplasmic annulus. This is about one order of magnitude less than the pressure needed to produce a similar flux of water through two successive plasmalemmas if the hydraulic conductivity is 5×10^{-3} m s^{-1} Pa^{-1}.

For the solute, the permeability of two successive plasmalemmas would

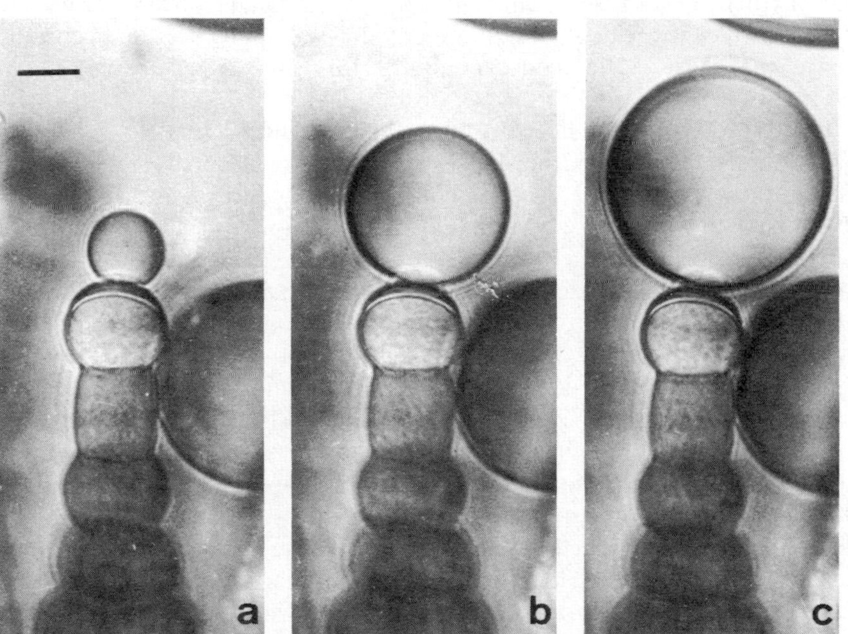

Fig. 14.2 Nectar exudation from the apical cell of an *Abutilon* trichome mounted in paraffin oil. Photos (b) and (c) are 5 and 20 min, respectively, after (a). Scale, 10 μm. (Reproduced from Findlay & Mercer, 1971a.)

need to be 1.6×10^{-4} mol m^{-2} s^{-1} if the concentration of the sugar in the prenectar were 600 mol m^{-3}. This is from three to four orders of magnitude higher than the permeability measured for efflux from storage tissues.

Thus a bulk flow of solution through plasmodesmata seems to be a more efficient process compared with the movement across two successive plasmalemmas. This assumes that there is no restriction to the movement of sugar through the annulus of the plasmodesmata. Gunning and Hughes also made calculations for movement through the desmotubule, but evidence obtained since then supports the view that the desmotubule may not be open (Gunning & Overall, 1983).

The rate of the secretion of nectar from the *Abutilon* nectary has also been examined. If secretion is limited to the apical cell of the trichome the permeability to sugar would be about 100 μmol m^{-2} s^{-1}, from two to three orders of magnitude greater than for other plant cell membranes (Lüttge, 1977; Robards, 1984). Lüttge has pointed out that if secretion is considered to occur over the total surface area of all the trichome cells above the stalk cell, the permeability would have a more reasonable value of about 0.6 μmol m^{-2} s^{-1}.

14.3.4 Ultrastructure and secretion from nectaries and other glands

Ultrastructural studies of many different types of glands from many species of plants have provided considerable evidence on how secreted substances leave the cells. A detailed account of this subject has been given by Schnepf (1974).

Except in those glands where the release of substances is accomplished by cell lysis, the secreted substances need to be transported out of the cells across the plasmalemma, often in considerable quantities and at high rates. In some glands there is evidence that this secretion is granulocrine, i.e. the secreted substances are transported across the cell membrane in membrane-bound vesicles which fuse with the plasmalemma. In others there is no evidence of granulocrine secretion; the secretion is then presumed to be eccrine, i.e. the molecules are transported across the plasmalemma directly and not packaged in membranes.

The granulocrine type of secretion is obviously suited to the transport of large or insoluble molecules across membranes, after their synthesis in a membrane-bound compartment. Such molecules are often visible in electron micrographs prepared and stained by the standard methods or may be made visible by specific stains. Also, as these substances are relatively insoluble, their synthesis and movement through the cell can be followed by high-resolution autoradiography combined with pulse labelling. These techniques have played a large part in the elucidation of polysaccharide secretion by root cap cells (Northcote & Pickett-Heaps, 1966). The polysaccharide is synthesized in the cisternae of dictyosomes and segregated into vesicles at the edges of the cisternae. These vesicles are released into the cytoplasm and they move to and fuse with the plasmalemma, thus releasing the secretion from the cells.

The dictyosomes are asymmetric bodies. There is an assembly of new cisternae at the *cis* or forming face of the dictyosome. These cisternae are then moved through the stack by the successive formation of more new cisternae. At the maturing or *trans* face the vesicles which have formed at the edges of the cisternae are released and the membrane of the central part of the cisterna disintegrates or in some instances the whole cisterna is released as a single vesicle. The membranes at the forming face of the dictyosomes are at least partly assembled from membranes derived from ER and they resemble those of the ER in thickness and staining. These characteristics change as the cisternae move through the stack of membranes by increasing in thickness and staining in material fixed with glutaraldehyde-osmium tetroxide fixative (Morré & Mollenhauer, 1974; Robinson & Kristen, 1982). The spacing of the cisternae and the staining of the intercisternal substance also changes. The vesicles formed from the dictyosomes fuse with the plasmalemma to release the secretory material to the cell exterior. Since the plasmalemma is thus derived from the dictyosomes, it is not surprising that the plasmalemma resembles the structure of the membranes of the cisternae at the maturing face.

Granulocrine secretion via dictyosomes has now been shown to occur in many cells including those of glands. It is well established that polysaccharides are secreted by this route, and it is probable that some other substances, particularly proteins, are also. However, dictyosomes seem to play no part in the transport of quite a number of other substances. In some cases there is evidence that the ER may be the source of vesicles which fuse with the plasmalemma, but this route is not as well supported by sound evidence as the route through the dictyosomes. The evidence for the different types of granulocrine and eccrine secretion in the different sorts of glands will now be considered.

14.3.4.1 Mucilage glands

These polysaccharide secreting glands have been found to form their secretion in a similar way to the root cap cells. The process has been studied in considerable detail in the stalked glands of *Drosophyllum* (Schnepf, 1974). These glands secrete the mucilage serving to trap the insects in these insectivorous plants.

14.3.4.2 Glands secreting lipophilic substances

Studies of the ultrastructure of these glands show that the secretion of these substances does not occur through the Golgi system. In those glands, visible material, where present and which appears to be secretion product or its precursor, is often located in plastids or within or closely associated with a proliferated, often tubular, ER (Schnepf, 1974). The plastids have little internal structure. Some authors have claimed that where the secretion accumulates within the ER its discharge is by the fusion of the ER with the plasmalemma, but this is poorly substantiated. A more general mode of secretion involves release into the cytoplasm before passage through the plasmalemma by an unknown mechanism, perhaps as micelles or droplets.

14.3.4.3 Nectaries

Studies of the ultrastructure of nectaries have resulted in a wide range of differing views on the processes involved in the elimination of nectar from the nectary cells. A large part of this is due to a wide variation between species and different types of nectaries. However, technical difficulties and problems in the interpretation of electron micrographs probably contribute to some extent to the perceived variety of interpretations. Two problems merit particular mention. One is the problem of localization of a small soluble molecule such as sugar. Another is the fixation of cells in which rapid membrane flow is occurring. Robards (1984) has considered some of the problems involved in studies of the nectaries of *Abutilon*.

(a) Nectaries with eccrine secretion. Nectaries are considered to be of the eccrine type if they show no marked vesicle formation by the endoplasmic reticulum or the dictyosomes. In the absence of any indication of secretion in the form of vesicles, the nectar is presumed to pass through the plasmalemma directly. Examples of this have been described in the floral and extrafloral nectaries of *Passiflora* (Durkee *et al.*, 1981; Durkee, 1982).

In many nectaries the surface area of the plasmalemma is greatly increased by the occurrence of wall protrusions, similar to those in transfer cells, at the secreting surface of the nectary. This characteristic is found in the septal nectaries of the Liliaceae. Schnepf & Pross (1976) have found that in several species these wall proliferations reach their maximum extent at the beginning of nectar secretion and begin to disappear when nectar secretion ceases. Other examples of wall proliferation are found in the floral nectaries of *Lonicera* (Fahn, 1979b) and the extrafloral stipel nectaries of *Vigna* (Kuo & Pate, 1985). An extreme case of infolding of the plasmalemma is found in *Asclepias* where the outer 3μm of the nectary epithelial cells is occupied by infoldings of the plasmalemma with only a few small wall protrusions near the wall (Schnepf & Christ, 1980). These infoldings of the plasmalemma increase the surface area available for eccrine secretion.

(b) Granulocrine secretion involving the dictyosomes In the nectaries of some species, dictyosomes which are active in the production of vesicles may be found often in addition to an abundant ER. In some instances this high activity may be associated with the secretion of polysaccharides rather than sugar secretion. One example is found in banana flowers (Fahn & Benouaiche, 1979). Banana nectar contains both protein and polysaccharide in unusually high concentration. These substances are likely to be secreted by dictyosome vesicles. In some other species the dictyosomes seem to be associated with the formation of wall protrusions. This occurs in *Gasteria* (Schnepf & Pross, 1976), the dictyosomes producing small vesicles with lightly staining contents while the wall protrusions are developing. However, the dictyosomes remain active during nectar secretion and they now produce large clear vesicles. In several species of the Bromeliaceae (Benner & Schnepf, 1975) dictyosomes are active only during secretion and have no obvious relation to polysaccharide secretion. In *Bilbergia nutans* (Schnepf & Benner, 1978) various inhibitors that reduce nectar secretion also affect dictyosome activity. Are the dictyosomes then responsible for secreting the nectar sugars? This is more difficult to establish unambiguously than the well documented role of dictyosomes in polysaccharide secretion in plants, largely because it is more difficult to localize sugars at an ultrastructural level.

Some attempts at the localization of sugars in nectaries have been made using radioactively labelled sugars (Fahn & Rachmilevitz, 1975; Heinrich, 1975b).

(c) Granulocrine secretion involving the ER In most, though not all, nectary epithelial and trichome cells there is an abundance of ER which takes various forms. During the period of secretion it is often swollen either in entire lamellae or at the edges of the lamellae. The edges of the lamellae may appear to give rise to vesicles and, in some instances, e.g. *Lonicera* (Fahn, 1979b), profiles are seen that suggest a fusion of these vesicles or the ER itself with the plasmalemma. Other authors, although they record an abundant ER and the presence of small vesicles (Baker *et al.*, 1978; Eleftheriou & Hall, 1983a), find no good evidence for the formation of vesicles from the ER or their fusion with the plasmalemma.

The general view of membrane flow through cells is that the membranes of the ER and the vesicles derived from it normally pass through the dictyosomes before they fuse with the plasmalemma. The observation in some nectaries of the ER or vesicles derived from it apparently fusing with the plasmalemma suggests that there might be an alternative pathway of membrane flow. There is some other evidence for such a pathway but it is not yet well established. In nectaries, the published evidence is limited to ultrastructural studies of material fixed by conventional chemical methods. Studies using freeze–fracture techniques should provide some useful information in a system such as nectaries where the membranes may be highly mobile. There are as yet no published studies of this sort on nectaries.

One recent study of the nectaries of *Asclepias* (Schnepf & Christ, 1980) has, unlike earlier work, taken into account the differences in membrane thickness in identifying intracellular membranes. In these nectaries, vesicles with thin membranes, i.e. membranes of similar dimensions to those of the ER, were restricted to the vicinity of the forming or *cis* face of the dictyosomes. Small vesicles with membrane thickness similar to that of the plasmalemma occurred close to the dictyosomes and were also common in the outer cytoplasm. These nectaries have a highly convoluted plasmalemma, a modification suggesting eccrine secretion. More similar careful studies of other species are obviously needed.

14.3.4.4 Stigmas

Stigmas secrete a mixture of different types of substances and this is reflected in their ultrastructure. In the secretory cells of the stigma of *Trifolium pratense* (Heslop-Harrison & Heslop-Harrison, 1983), the dictyosomes are active and there is abundant ER, both rough and smooth, during the period of protein and polysaccharide secretion. The lipids secreted by these stigmas appear as globules in the cytoplasm but it is not known how they leave the cells and enter the intercellular spaces. The stigma papilla cells of avocado contain a proliferated tubular ER and plastids, that occur in clusters, typical of lipid-secreting cells (Sedgley & Blesing, 1983). Polysaccharide secretion has been studied in the stigmas of water-melon (Sedgley & Blesing, 1985). The location of the polysaccharide was demonstrated by specific staining and its path followed by pulse labelling with tritium-labelled galactose followed by autoradiography. The polysaccharide appears in the vesicles of the dictyosomes, the secretory vesicles, cell walls and wall thickenings, and later also in the extracellular secretion.

14.3.5 Movement through the cuticle

The final barrier to the secretion of nectar is the impermeable covering on exposed plant parts, the cuticle. Some nectaries, as for instance *Tropaeolum* (Rachmilevitz and Fahn, 1975), avoid this problem as the nectar is secreted into intercellular spaces within the nectary and then moves to the exterior through stomate-like structures. Even here the larger intercellular spaces near to the stomates may be lined with thin cuticle.

In other nectaries the cuticle is ruptured by the accumulation of nectar beneath it. One example is the septal nectary of *Gasteria*. This nectary secretes into confined narrow slits within the septa of the carpels. The cuticle is very thin and can be seen to have become discontinuous in sections prepared from secreting nectaries. Rupture has also been recorded in more exposed nectaries where the cuticle is more substantial such as in the extrafloral nectaries of *Passiflora* (Durkee, 1982). In stigmas of the

Fabaceae, as for example *Trifolium* (Heslop-Harrison & Heslop-Harrison, 1983), the cuticle is impermeable until the flower is tripped and the stigma collides with the insect or the standard. The now ruptured cuticle allows adhering pollen grains to come into contact with secretion that had accumulated in the intercellular spaces in the stigma.

In many nectaries the cuticle becomes separated from the inner uncutinized layer of the cell wall by the collection of nectar between the two layers. This happens in the glandular hairs of *Lonicera* nectaries (Fahn, 1979b). In this species the cuticle is of uneven thickness. There are thick areas separated by a network of thinner cuticle. The nectar is thought to pass through the thin areas when stretched. In other nectaries, such as *Ricinus*, there is no apparent modification of the cuticle. Are the cuticles of these nectaries then more permeable than might be expected from their structure?

Some subtle structural modifications resulting in increased permeability have been found in some other glands such as those of the insectivorous plants. In the immature sessile glands of *Drosophyllum* (Joel & Juniper, 1982), the cuticle is impermeable to neutral red applied externally but becomes permeable as the glands reach maturity. Correlated with this change is a change in the structure of the cuticle. In electron micrographs of the young gland the distribution of cutin in the outer layer of the wall appears uniform and continuous but, in the mature gland, cutin particles separated by narrow channels filled with pectocellulosic wall material become visible. These channels would provide the pathway for the passage of the hydrophilic secretion. Studies of the cuticle that covers stigmas (Heslop-Harrison & Heslop-Harrison, 1982) have shown it to be modified in various ways to allow pollen grains on the surface to be hydrated. In the Caryophyllaceae, separate but closely packed rodlets of cutin are present in the outer part of the epidermal wall. The wall is covered by a thin external lamellated layer which becomes dispersed in the mature stigma.

In a few nectaries there are pores in the cuticle. In *Abutilon* (Findlay & Mercer, 1971a; Gunning & Hughes, 1976), the presence of pores is apparent from the way that the nectar escapes. The nectar collects between the inner cellulose wall and the cuticle and the latter is increasingly stretched until a small droplet of nectar escapes through it. The cuticle then returns to a position closer to the cellulose wall and the process is then repeated. A somewhat similar mode of secretion is found in the trichome hydathodes of *Cicer* (Schnepf, 1965). Nectar passes through the cutinized layer of the wall via a pore. However, it remains confined by an outer layer of the cuticle until it finds its way to another pore in this outer layer which is not aligned to the first pore. This complex pathway results in the release of discrete droplets.

14.4. Physiology of nectar secretion

14.4.1. Composition of nectar

Nectar is an aqueous solution that contains a high concentration of sugars. The sugars are predominantly glucose, fructose and sucrose that can occur in varying proportions depending on the species (van Handel *et al.*, 1972; Gottsberger *et al.*, 1984). Other sugars occur in smaller proportions. Trisaccharides that are the result of invertase action on a concentrated sucrose solution have been shown to occur in nectar (Zimmermann, 1953). The concentration of sugar varies widely between species and will be affected by environmental conditions and by evaporation after secretion, this latter being affected by the degree of exposure in the flower. Nectar collected in the field from a number of species varied in concentration from about 100–600 mg total sugar ml^{-1} nectar, the majority being in the range 200–400 mg ml^{-1} (Gottsberger *et al.*, 1984).

Small quantities of substances such as amino acids, organic acids, mineral salts, proteins, alkaloids and phenolics are also present in nectar. The amino acid content of nectar has received considerable attention because of its possible importance in the nutrition of nectar-feeding animals (Baker & Baker, 1983; Gottsberger *et al.*, 1984). Generally the amino acid content is less in the nectar than in the phloem in relation to the sugar content. However, the nectar can have a high content of amino acids in nectaries where cell lysis occurs (Pate *et al.*, 1985, Table 14.1).

14.4.2 Sugar and water transport

Nectar formation involves the transport of sugar and water from the phloem to the nectary surface. This transport will require energy from some source for its accomplishment.

In some nectaries it is possible that the driving force for nectar transport is either wholly or partly the result of active processes at the loading end of the phloem. The nectary cells or the phloem endings at the nectary would then need to be unusually permeable to sugars. The passive permeability of cell membranes to sugars is usually considered to be very low.

If nectaries such as those of *Abutilon* or *Euphorbia* are floated on water, they will secrete nectar using sugar and starch stored within the nectary. They are able to continue secretion of nectar for longer if they are supplied with sugar, and the rate of secretion may approach that on the plant. Hence, in these nectaries, there is some local driving force acting within the nectaries.

The energy requirement is borne out by the decrease of nectar secretion in isolated nectaries on the addition of respiratory inhibitors or uncouplers, or by anaerobic conditions (Matile, 1956; Findlay *et al.*, 1971; Schnepf & Benner, 1978). The inhibition is, however, only partial. The tempera-

Table 14.1 The concentration of amino acids and sugars (glucose G, fructose F, sucrose S and total sugar) in the phloem sap and nectar of the extrafloral nectaries of *Ricinus* and the two types of extrafloral nectaries of *Vigna*†.

Nectary	Amino acids (mg/ml)		Sugar (mg/ml)							
	Phloem	Nectar	Phloem				Nectar			
			G	F	S	Total	G	F	S	Total
Ricinus	5.2	1.7	–	–	80–106	80–106	297	295	280	872
Vigna stipel	16.6–33.2	0.05–0.10	6	4	180–210	190–220	170–354	170–354	127–244	550–660
Vigna Inflorescence	16.6–33.2	12.5–17.0	6	4	180–210	190–220	145–252	145–252	100–271	500–630

† The stipel nectaries of *Vigna* consist of glandular trichomes; the inflorescence nectaries consist of a tissue the cells of which separate during secretion leading to their release into the nectar. Data for *Ricinus* from Baker *et al*. (1978), for *Vigna* from Pate *et al*. (1985).

ture coefficient of 2.3 between 10 and 20 °C also supports this. In *Bilbergia nutans*, nectar secretion ceases at 0 °C (Schnepf & Benner, 1978).

The structure of the nectary cells suggests that they have a high respiratory rate. Not only do the cells contain many mitochondria, but they often have unusually well developed cristae. Findlay *et al.* (1971) have compared the rate of respiration with nectar secretion in *Abutilon*. From rates of secretion for nectaries on the plant, they calculated that only 1 ATP is available per sugar molecule secreted. The corresponding value for isolated nectaries is 5 ATP per sugar molecule. This makes no allowance for other energy needs of the tissue.

Within the nectary the sucrose supplied by the phloem is converted to a mixture of sucrose, glucose and fructose. This appears to be the result of the action of invertases. High levels of invertase activity have been found in the nectary tissues of several species (de Fekete *et al.*, 1967; Zauralov & Pavlinova, 1975). The extracellular location of acid invertase has been used as evidence that nectar moves through the nectary by an apoplastic pathway. However, not all invertase may be extracellular. Alkaline invertase, which is present in nectary tissue (de Fekete *et al.*, 1967) is of uncertain location.

In some species, carbohydrates other than sucrose are translocated in the phloem. Zauralov & Pavlinova (1975) found that ^{14}C sucrose fed to the leaves of pumpkin was translocated partly as verbascose and stachyose but neither of these labelled oligosaccharides appeared in the nectar which contained sucrose, glucose and fructose. In the Rosaceae (Bieleski & Redgwell, 1980), the carbohydrate is translocated as the sugar alcohol, sorbitol, but the nectaries secrete sucrose, glucose and fructose.

Isolated nectaries of *Euphorbia* (Frey-Wyssling *et al.*, 1954) have been found to be able to continue secretion on either glucose or fructose as well as sucrose. The ratio of the sugars in the nectar is unaffected by which of these sugars is supplied. Thus it appears that the nectary tissue is capable not only of sucrose hydrolysis but also its synthesis from glucose or fructose. This has been confirmed by feeding ^{14}C labelled glucose; the label appears in glucose, fructose and sucrose in approximately equal quantities in the nectar.

If sucrose labelled only in the glucose moiety is fed to *Abutilon* nectaries, 72 % of the activity remains in the glucose moiety of sucrose appearing in the nectar (Ziegler, 1965), showing that inversion of sucrose is not an essential step in its secretion.

Glucose, fructose and sucrose are not the only sugars that can be secreted by nectaries. Semi-quantitative experiments in which a variety of sugars, including pentoses, hexoses and oligosaccharides, were fed to nectaries, have been performed with several species (Matile, 1956; Shuel, 1956; Schnepf & Benner, 1978). The fed sugars often appeared in the nectar without affecting the rate of endogenous sugar secretion. In other instances the nectar secretion was reduced. The sugars were supplied at 5 % concentration (i.e. 280 mol m^{-3} for hexoses). In most instances it is possible that the sugars are swept along by water flow in the apoplast, but in *Abutilon*, a nectary with an apoplastic barrier, ribose, the only foriegn sugar tested, is readily secreted (Matile, 1956).

The hydrolysis of sucrose in the nectary probably contributes to the maintenance of a water potential gradient within the nectary. It may also have some effect in maintaining diffusion gradients for sugars.

While it is clear that sugars are secreted at a high concentration by nectaries, quantitative studies of the concentration relations across a nectary are more difficult. Reliable measurements of the concentration of sugar in phloem supplying the nectaries of *Ricinus* (Baker *et al.*, 1978) and *Vigna* (Pate *et al.*, 1985) (Table 14.1) have been achieved, but the *in situ* nectar concentration in these experiments may have been altered by post-secretion changes. The total sugar concentration of the nectar was at least twice that in the phloem in all cases, but the sucrose concentration in the nectar of both the nectary types of *Vigna* was little higher than that in the phloem.

Shuel (1956) has investigated nectar secretion by snapdragon flowers cultured on sucrose solution for several days. He found that the nectar and medium concentrations were nearly identical, at close to 100 % humidity. At lower relative humidities the nectar concentration was higher; at 60 % relative humidity or below the nectar concentration was three times that at 100 % relative humidity. The nectar contained mostly sucrose. Reed *et al.* (1971) found that, in covered isolated *Abutilon* nectaries, the total sugar concentration of the nectar was always higher than the sucrose concentration of the medium. However, the sucrose concentration of the nectar was lower than that of the medium, except at the lowest medium concentrations. Findlay *et al.* (1971) studied the rate of sugar secretion in these nectaries (Fig. 14.3). During an initial period of over 7 h the rate of sugar secretion was independent of the medium concentration, as a result of the large pool of sugars and starch in the nectary tissue. After 10 h the rate of sugar secretion was proportional to the sucrose concentration in the medium up to 400 mol m^{-3}. Higher medium sucrose concentrations decreased both sugar and water secretion. Sugar uptake which has a linear relation to concentration has been found in a number of other systems. The mechanism is not understood but the evidence suggests that it need not be a purely diffusional phenomenon as it is at least partly sensitive to inhibitors (Reinhold & Kaplan, 1984).

De Fekete *et al.* (1967) have investigated the presence of enzymes of sugar metabolism in the nectaries of two different species. As already mentioned the activity of invertase was high. The other enzymes necessary for the interconversion of glucose, fructose and sucrose were present though sucrose–P–synthetase was found to have a low activity in one species and appeared to be absent in the other. Possibly improved extrac-

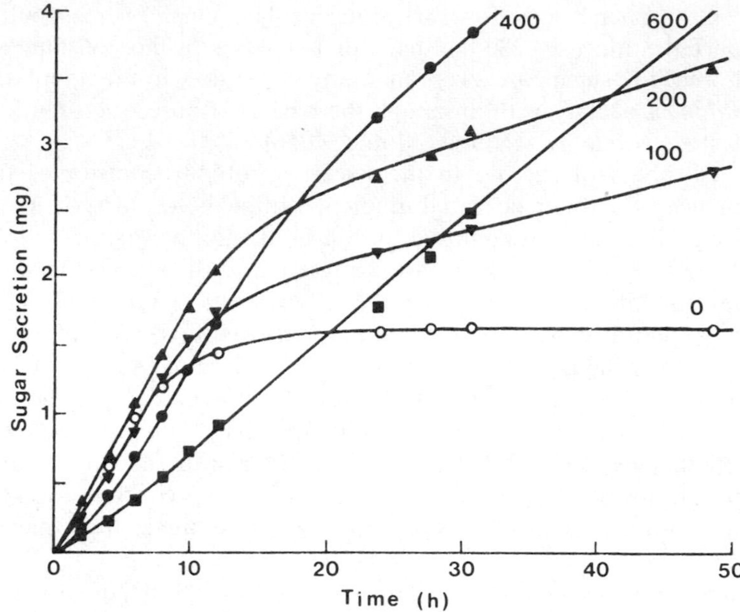

Fig. 14.3 The amount of total sugar secreted per nectary with time by isolated *Abutilon* nectaries floating on sucrose solutions of the concentration (in mol m^{-3}) shown. (Reproduced from Findlay *et al.*, 1971.)

tion methods would give a higher activity of this enzyme.

Several histochemical studies using β-glycerophosphate as substrate have shown that acid phosphatases are abundant in nectaries. The relationship of this histochemical demonstration to the biochemical or transport events in the cells is not clear. The reaction is probably produced by several different enzymes including the non-specific phosphatases of the lysosomal compartment. At an ultrastructural level, the acid phosphatase in nectaries was found to be associated with many cell components but particularly the plasmalemma at cell wall protruberances (Figier, 1968; Heinrich, 1975a).

The localization of ATPases has attracted some attention. Heinrich (1975a) found a particularly high activity on the endoplasmic reticulum of *Aloe* nectaries. Eleftheriou & Hall (1983b) found ATPase activity on the plasmalemma of cotton nectary trichomes during secretion but not in young non-secreting nectaries. The neighbouring phloem showed ATPase activity on the plasmalemma of the companion cells but not in the sieve tubes. Since plasmalemmal ATPase is thought to be associated with proton-linked sugar uptake, the presence of the ATPase in the cotton nectary cells is not easy to explain. A similar problem has arisen with phloem, where ATPase is often, though not always, found on the sieve tubes and companion cells at both loading and unloading ends of phloem.

One suggestion was that in nectaries it is involved with reabsorption of substances, such as amino acids, which occur in the nectar at a lower concentration in relation to sugar than in the supplying phloem.

Except in those nectaries where nectar is released by cell breakdown, the sugar must leave cells through a membrane. At least some of the sugar may be unloaded into the apoplast at the phloem in those nectaries without an apoplastic barrier. However, some or all of the sugar is usually secreted from nectary cells. If the secretion is by means of vesicles derived from the ER or dictyosomes, then the sugar must first be transported from the cytoplasm into the organelles through a membrane.

Little is known about transport into the ER or dictyosomes. There is now considerable understanding of the uptake of sugars into plant cells through the plasmalemma and a lesser though increasing understanding of transport into vacuoles (see Ch. 8). However, little is known about the mechanism of efflux of sugars from cells. The rate of efflux from plant storage tissues is normally extremely low. Reinhold and Kaplan (1984) point out that this may be due to the retention of sugars in the vacuoles and that the permeability of the plasmalemma may be higher than that of the tonoplast as it has not been satisfactorily measured. In nectaries the vacuoles are small and the cytoplasmic compartment would be the major pathway of sugars through the cells, and hence it is the efflux through the plasmalemma that needs to be considered.

Various possibilities for eccrine secretion would include:

1. Passive permeability of the plasmalemma high enough to allow sugar and water to flow through it – this is unlikely.
2. Facilitated diffusion.
3. Active transport. None of the mechanisms described in other cells appear directly applicable.
   (a) Proton–sugar symport is in the wrong direction.
   (b) Proton–sugar antiport as has been found for glucose transport into vacuoles (Thom & Komor, 1984), but this would need to be located on the plasmalemma.
   (c) Group translocation for which there is recent evidence in sucrose uptake by sugar cane vacuoles (see Chs. 3 and 8; and Thom & Maretzki, 1985). Again such a system would need to be located on the plasmalemma instead of the tonoplast.

Acknowledgements

I thank Drs B. G. Coombe, G. P. Findlay and M. Sedgley for reading the manuscript and making helpful criticisms and comments.

References

Baker, H. G. & Baker, I. (1983). A brief historical review of the chemistry of floral nectar. In Bentley, B. & Elias, T (eds.). *The Biology of Nectaries*, pp. 126–52. New York, Columbia University Press.

Baker, D. A., Hall, J. L. & Thorpe, J. R. (1978). A study of the extrafloral nectaries of *Ricinus communis*. *New Phytologist* **81**, 129–37.

Benner, U. & Schnepf, E. (1975). Die Morphologie der Nektarausscheidung bei Bromeliaceen: Beteiligung des Golgi-Apparates. *Protoplasma* **85**, 337–49.

Bentley, B. L. (1977) Extrafloral nectaries and protection by pugnacious bodyguards. *Annual Review of Ecology and Systematics* **8**, 407–27.

Bieleski, R. L. & Redgwell, R. J. (1980). Sorbitol metabolism in nectaries from flowers of Rosaceae. *Australian Journal of Plant Physiology* **7**, 15–25.

Bosabalidis, A. & Tsekos, I. (1982). Glandular scale development and essential oil secretion in *Origanum dictamnus* L. *Planta* **156**, 496–504.

Boughton, V. H. (1985). Extrafloral nectaries of some Australian bipinnate Acacias. *Australian Journal of Botany* **33**, 175–84.

de Fekete, M. A. R., Ziegler, H. & Wolf, R. (1967). Enzyme des Kohlenhydratstoffwechsels in Nektarien. *Planta* **75**, 125–38.

Durkee, L. T. (1982). The floral and extrafloral nectaries of *Passiflora*. II. The extrafloral nectary. *American Journal of Botany* **69**, 1420–28.

Durkee, L. T., Gaal, D. J. & Reisner, W. H. (1981). The floral and extrafloral nectaries of *Passiflora*. I. The floral nectary. *American Journal of Botany* **68**, 453–62.

Eleftheriou, E. P. & Hall, J. L. (1983a). The extrafloral nectaries of cotton. I. Fine structure of the secretory papillae. *Journal of Experimental Botany* **34**, 103–19.

Eleftheriou, E. P. & Hall, J. L. (1983b). The extrafloral nectaries of cotton. II. Cytochemical localization of ATPase activity and Ca^{2+}-binding sites, and selective osmium impregnation. *Journal of Experimental Botany* **34**, 1066–79.

Fahn, A. (1979a). *Secretory Tissues in Plants*. London, New York, San Francisco, Calif., Academic Press.

Fahn, A. (1979b). Ultrastructure of nectaries in relation to nectar secretion. *American Journal of Botany* **66**, 977–85.

Fahn, A. & Benouaiche, P. (1979). Ultrastructure, development and secretion in the nectary of banana flowers. *Annals of Botany* **44**, 85–93.

Fahn, A. & Rachmilevitz, T. (1975). An autoradiographical study of nectar secretion in *Lonicera japonica* Thunb. *Annals of Botany* **39**, 975–6.

Figier, J. (1968). Localisation infrastructurale de la phosphomonoésterase acide dans la stipule de *Vicia faba* L. au niveau du nectaire. *Planta* **83**, 60–79.

Findlay, N. & Mercer, F. V. (1971a). Nectar production in *Abutilon* . I. Movement of nectar through the cuticle. *Australian Journal of Biological Sciences* **24**, 647–56.

Findlay, N. & Mercer, F. V. (1971b) Nectar production in *Abutilon*. II. Submicroscopic structure of the nectary. *Australian Journal of Biological Sciences*. **24**, 657–64.

Findlay, N., Reed, M. L. & Mercer, F. V. (1971). Nectar production in *Abutilon*. III. Sugar secretion. *Australian Journal of Biological Sciences* **24**, 665–75.

Frey-Wyssling, A., Zimmermann, M. & Maurizio, A. (1954). Über den enzymatischen Zuckerumbau in Nektarien. *Experientia* **10**, 491–7.

Gottsberger, G., Schrauwen, J. & Linskens, H. F. (1984). Amino acids and sugars in nectar, and their putative evolutionary significance. *Plant Systematics and Evolution* **145**, 55–77.

Gunning, B. E. S. & Hughes, J. E. (1976). Quantitative assessment of symplastic transport of pre-nectar into the trichomes of *Abutilon* nectaries. *Australian Journal of Plant Physiology* **3**, 619–37.

Gunning, B. E. S. & Overall, R. L. (1983). Plasmodesmata and cell-to-cell transport in plants. *BioScience* **33**, 260–5.

Heinrich, G. (1975a). Über die Lokalisation verschiedener Phosphatasen im Nektarien von *Aloe*. *Cytobiologie* **11**, 247–63.

Heinrich, G. (1975b). Über den Glucose-Metabolismus in Nektarien zweier *Aloe*-arten und über den Mechanismus der Pronektar-Sekretion. *Protoplasma* **85**, 351–71.

Heslop-Harrison, J. & Heslop-Harrison, Y. (1982). The specialized cuticles of the receptive surfaces of angiosperm stigmas. In Cutler, D. F., Alvin, K. L. & Price, C. E. (eds.). *The Plant Cuticle*, pp. 99–119. Linnean Society Symposium Series no. 10. London, Academic Press.

Heslop-Harrison, J. & Heslop-Harrison, Y. (1983). Pollen-stigma interaction in the Leguminosae: the organization of the stigma in *Trifolium pratense* L. *Annals of Botany* **51**, 571–83.

Heslop-Harrison, Y. & Knox, R. B. (1971). A cytochemical study of the leaf-gland enzymes of insectivorous plants of the genus *Pinguicula*. *Planta* **96**, 183–211.

Joel, D. M. & Juniper, B. E. (1982). Cuticular gaps in carnivorous plant glands. In Cutler, D. F., Alvin, K. L. & Price, C. E. (eds.). *The Plant Cuticle*, pp. 121–30. Linnean Society Symposium Series no. 10. London, Academic Press.

Knox, R. B. (1984). Pollen-pistil interactions. In Linskens, H. F. & Heslop-Harrison, J. (eds.). *Cellular Interactions, Encyclopedia of plant Physiology*, New Series, 17, pp. 508–608. Berlin, Heidelberg, Springer-Verlag.

Kuo, J. Pate, J. S. (1985). The extrafloral nectaries of cowpea (*Vigna unguiculata* (L.) Walp): I. Morphology, anatomy and fine structure. *Planta*. **166**, 15–27.

Lüttge, U. (1977). Nectar composition and membrane transport of sugars and amino-acids: a review on the present state of nectar research. *Apidologie* **8**, 305–19.

Lüttge, U. & Schnepf, E. (1976). Elimination processes by glands: organic substances. In Lüttge, U. & Pitman, M. G. (eds.) *Transport in Plants*, II, Part B, *Tissues and Organs, Encyclopedia of Plant Physiology*, New Series, vol. 2B, pp. 244–77. Berlin, Heidelberg, Springer-Verlag.

Main, A. R. (1981). Plants as animal food. In Pate, J. S. & McComb, A. J. (eds.). *The Biology of Australian plants*, pp. 340–60. Nedlands, Western Australia. University of Western Australia Press.

Matile, P. (1956) Uber den Stoffwechsel und die Auxinabhangigkeit der Nektarsekretion. *Berichte der Schweizerischen Botanischen Gesellschaft* **66**, 237–66.

Morré, D. J. & Mollenhauer, H. H. (1974). The endomembrane concept: a functional integration of endoplasmic reticulum and Golgi apparatus. In Robards, A. W. (ed.). *Dynamic Aspects of Plant Ultrastructure*, pp. 84–137. London, McGraw-Hill.

Northcote, D. H. & Pickett-Heaps, J. D. (1966). A function of the Golgi apparatus in polysaccharide synthesis and transport in the root-cap cells of wheat. *Biochemical Journal* **98**, 159–67.

Pate, J. S., Peoples, M. B., Storer, P. J. & Atkins, C. A. (1985). The extrafloral nectaries of cowpea (*Vigna unguiculata* (L.) Walp): II. Nectar composition, origin of nectar solutes, and nectary functioning. *Planta*. **166**, 28–38.

Rachmilevitz, T & Fahn, A. (1975). The floral nectary of *Tropaeolum majus* L. – The nature of the secretory cells and the manner of nectar secretion. *Annals of Botany* **39**, 721–8.

Rea, P. A. (1982). Fluid composition and factors that elicit secretion by the trap lobes of *Dionaea muscipula* Ellis. *Zeitschrift für Pflanzenphysiologie* **108**, 255–72.

Reed, M. L., Findlay, N. & Mercer, F. V. (1971). Nectar production in *Abutilon*. IV. Water and solute relations. *Australian Journal of Biological Sciences* **24**, 677–88.

Reinhold, L. & Kaplan, A. (1984). Membrane transport of sugars and amino acids. *Annual Review of Plant Physiology* **35**, 45–83.

Robards, A. W. (1984). Fact or artefact – a cool look at biological electron microscopy. *Proceedings of the Royal Microscopical Society* **19**, 195–208.

Robinson, D. G. & Kristen, U. (1982). Membrane flow via the Golgi apparatus of higher plant cells. *International Review of Cytology* **77**, 89–127.

Schnepf, E. (1965). Licht- und elektronenmikroskopische Beobachtungen an den Trichom-hydathoden von *Cicer arietinum*. *Zeitschrift für Pflanzenphysiologie* **53**, 245–54.

Schnepf, E. (1974). Gland cells. In Robards, A. W. (ed.). *Dynamic Aspects of Plant Ultrastructure*, pp. 331–57. London, McGraw-Hill.

Schnepf, E. & Benner, U. (1978). Die Morphologie der Nektarausscheidung bei Bromeliaceen. II. Experimentelle und quantitative Untersuchungen bei *Billbergia nutans*. *Biochemie und Physiologie der Pflanzen* **173**, 23–36.

Schnepf, E. & Christ, P. (1980). Unusual transfer cells in the epithelium of the nectaries of *Asclepias curassavica* L. *Protoplasma* **105**, 135–48.

Schnepf, E. & Pross, E. (1976). Differentiation and redifferentiation of a transfer cell: development of septal nectaries of *Aloe* and *Gasteria*. *Protoplasma* **89**, 105–15.

Sedgley, M. & Blesing, M. A. (1983). Developmental anatomy of the avocado stigma papilla cells and their secretion. *Botanical Gazette* **144**, 185–90.

Sedgley, M. & Blesing, M. A. (1985). Polysaccharide secretion by the watermelon stigma. *Annals of Botany* **55**, 269–73.

Shuel, R. W. (1956). Studies of nectar secretion in excised flowers. I. The influence of cultural conditions on the quantity and composition of nectar. *Canadian Journal of Botany* **34** 142–53.

Thom, M. & Komor, E. (1984). H^+-sugar antiport as the mechanism of sugar uptake by sugarcane vacuoles. *FEBS Letters* **173**, 1–4.

Thom, M. & Maretzki, A. (1985). Group translocation as a mechanism for sucrose transfer into vacuoles from sugarcane cells. *Proceedings of the National Academy of Sciences, USA* **82**, 4697–701.

Thorne, J. H. (1985). Phloem unloading of C and N assimilates in developing seeds. *Annual Review of Plant Physiology* **36**, 317–43.

Van Handel, E., Haeger, J. S. & Hansen, C. W. (1972). The sugars of some Florida nectars. *American Journal of Botany* **59**, 1030–32.

Vogel, S. (1971). Ölproduzierende Blumen, die durch ölsammelnde Bienen bestäubt werden. *Naturwissenschaften* **58**, 58.

Vogel, S. (1976). *Lysimachia: Ölblumen der Holarktis. Naturwissenschaften* **63**, 44–5.

Wergin, W. P., Elmore, C. D., Hanny, B. W. Ingber, B. F. (1975). Ultrastructure of the subglandular cells from the foliar nectaries of cotton in relation to the distribution of plasmodesmata and the symplastic transport of nectar. *American Journal of Botany* **62**, 842–9.

Zauralov, O. A. & Pavlinova, O. A. (1975). Transport and conversion of sugars in nectaries in connection with the secretory function (Russian). *Fiziologiya Rastenii* **22**, 500–7.

Ziegler, H. (1965). Die Physiologie pflanzlicher Drüsen. *Berichte der Deutsche Botanische Gesellschaft* **78**, 466–77.

Zimmermann, M. (1953). Papierchromatographische Untersuchungen über die pflanzliche Zuckersekretion. *Berichte der Schweizerischen Botanischen Gesellschaft* **63**, 402–29.

Author Index

Numbers in italics refer to authors listed in the References at ends of chapters.

Subject Index